# Enzyme-Catalyzed Electron and Radical Transfer

Subcellular Biochemistry
Volume 35

# SUBCELLULAR BIOCHEMISTRY

## SERIES EDITOR

J. ROBIN HARRIS, Institute of Zoology, University of Mainz, Mainz, Germany

## ASSISTANT EDITORS

H. J. HILDERSON, University of Antwerp, Antwerp, Belgium
B. B. BISWAS, University of Calcutta, Calcutta, India

---

*Recent Volumes in This Series*

Volume 26    ***myo*-Inositol Phosphates, Phosphoinositides, and Signal Transduction**
Edited by B. B. Biswas and Susweta Biswas

Volume 27    **Biology of the Lysosome**
Edited by John B. Lloyd and Robert W. Mason

Volume 28    **Cholesterol: Its Functions and Metabolism in Biology and Medicine**
Edited by Robert Bittman

Volume 29    **Plant–Microbe Interactions**
Edited by B. B. Biswas and H. K. Das

Volume 30    **Fat-Soluble Vitamins**
Edited by Peter J. Quinn and Valerian E. Kagan

Volume 31    **Intermediate Filaments**
Edited by Harald Herrmann and J. Robin Harris

Volume 32    **α-Gal and Anti-Gal: α1,3-Galactosyltransferase, α-Gal Epitopes, and the Natural Anti-Gal Antibody**
Edited by Uri Galili and José Luis Avila

Volume 33    **Bacterial Invasion into Eukaryotic Cells**
Edited by Tobias A. Oelschlaeger and Jörg Hacker

Volume 34    **Fusion of Biological Membranes and Related Problems**
Edited by Herwig Hilderson and Stephen Fuller

Volume 35    **Enzyme-Catalyzed Electron and Radical Transfer**
Edited by Andreas Holzenburg and Nigel S. Scrutton

---

A Continuation Order Plan is available for this series. A continuation order will bring delivery of each new volume immediately upon publication. Volumes are billed only upon actual shipment. For further information please contact the publisher.

# Enzyme-Catalyzed Electron and Radical Transfer

## Subcellular Biochemistry
## Volume 35

Edited by

## Andreas Holzenburg
*Texas A&M University*
*College Station, Texas*
*(formerly of University of Leeds*
*Leeds, United Kingdom)*

and

## Nigel S. Scrutton
*University of Leicester*
*Leicester, United Kingdom*

Kluwer Academic / Plenum Publishers
New York, Boston, Dordrecht, London, Moscow

The Library of Congress cataloged the first volume of this title as follows:

Sub-cellular biochemistry.
London, New York, Plenum Press.
v. illus. 23 cm. quarterly.
Began with Sept. 1971 issue. Cf. New serial titles.
1. Cytochemistry—Periodicals.  2. Cell organelles—Periodicals.
QH611.S84    574.8'76                                          73-643479

ISSN 0306-0225

ISBN 0-306-46399-7

This series is a continuation of the journal *Sub-Cellular Biochemistry*,
Volumes 1 to 4 of which were published quarterly from 1972 to 1975

©2000 Kluwer Academic / Plenum Publishers, New York
233 Spring Street, New York, New York 10013

http://www.wkap.nl

10  9  8  7  6  5  4  3  2  1

A C.I.P. record for this book is available from the Library of Congress

To Our Parents

# Contributors

**Christopher Anthony**    Division of Biochemistry and Molecular Biology, School of Biological Sciences, University of Southampton, Southampton, SO16 7PX, United Kingdom

**Edward A. Berry**    E.O. Lawrence Berkeley National Laboratory, University of California, Berkeley, California 94720

**Brian J. Brazeau**    Department of Biochemistry, Molecular Biology, and Biophysics, University of Minnesota, Minneapolis, Minnesota 55455

**Marion E. van Brederode**    Faculty of Sciences, Division of Physics and Astronomy, Department of Biophysics and Physics of Complex Systems, Free University of Amsterdam, 1081 HV Amsterdam, The Netherlands

**Stephen K. Chapman**    Department of Chemistry, University of Edinburgh, Edinburgh EH9 3JJ, Scotland, United Kingdom

**Xiaoxi Chen**    Johnson Research Foundation, Department of Biochemistry and Biophysics, University of Pennsylvania, Philadelphia, Pennsylvania 19104

**Louise Cunane**    Department of Biochemistry and Molecular Biophysics, Washington University School of Medicine, St. Louis, Missouri 63110

**Victor L. Davidson**    Department of Biochemistry, University of Mississippi Medical Center, Jackson, Mississippi 39214-4505

**Rosemary C. E. Durley**    Department of Biochemistry and Molecular Biophysics, Washington University School of Medicine, St. Louis, Missouri 63110

**P. Leslie Dutton**     Johnson Research Foundation, Department of Biochemistry and Biophysics, University of Pennsylvania, Philadelphia, Pennsylvania 19104

**Stuart J. Ferguson**     Department of Biochemistry and Oxford Centre for Molecular Sciences, University of Oxford, Oxford OX1 3QU, United Kingdom

**Vilmos Fülöp**     Department of Biological Sciences, University of Warwick, Coventry CV4 7AL, United Kingdom

**Amanda J. Green**     Department of Chemistry, University of Edinburgh, Edinburgh EH9 3JJ, Scotland, United Kingdom

**Malcolm Halcrow**     Department of Chemistry, University of Leeds, Leeds, LS2 9JT, United Kingdom

**Russ Hille**     Department of Medical Biochemistry, Ohio State University, Columbus, Ohio 43210-1218

**Daniel E. Holloway**     Department of Biology and Biochemistry, University of Bath, Bath BA2 7AY, United Kingdom

**Li-Shar Huang**     E.O. Lawrence Berkeley National Laboratory and Department of Chemistry, University of California, Berkeley, California 94720

**Michael R. Jones**     Department of Biochemistry, School of Medical Sciences, University of Bristol, Bristol BS8 1TD, United Kingdom

**Sung-Hou Kim**     E.O. Lawrence Berkeley National Laboratory and Department of Chemistry, University of California, Berkeley, California 94720

**Peter Knowles**     School of Biochemistry and Molecular Biology, University of Leeds, Leeds, LS2 9JT, United Kingdom

**John D. Lipscomb**     Department of Biochemistry, Molecular Biology, and Biophysics, University of Minnesota, Minneapolis, Minnesota 55455

**Kirsty McLean**     Department of Pure and Applied Chemistry, University of Strathclyde, Glasgow G1 1XL, Scotland, United Kingdom

**E. Neil G. Marsh**     Department of Chemistry, University of Michigan, Ann Arbor, Michigan 48109

**F. Scott Mathews**     Department of Biochemistry and Molecular Biophysics, Washington University School of Medicine, St. Louis, Missouri 63110

**Caroline S. Miles**     Department of Chemistry, University of Edinburgh, Edinburgh EH9 3JJ, Scotland, United Kingdom

**Christopher C. Moser**     Johnson Research Foundation, Department of Biochemistry and Biophysics, University of Pennsylvania, Philadelphia, Pennsylvania 19104

**Christopher G. Mowat**     Department of Chemistry, University of Edinburgh, Edinburgh EH9 3JJ, Scotland, United Kingdom

**Andrew W. Munro**     Department of Pure and Applied Chemistry, University of Strathclyde, Glasgow G1 1XL, Scotland, United Kingdom

**Jane Murdoch**     Department of Chemistry, University of Edinburgh, Edinburgh EH9 3JJ, Scotland, United Kingdom

**Michael A. Noble**     Department of Chemistry, University of Edinburgh, Edinburgh EH9 3JJ, Scotland, United Kingdom

**Tobias W. B. Ost**     Department of Chemistry, University of Edinburgh, Edinburgh EH9 3JJ, Scotland, United Kingdom

**Christopher C. Page**     Johnson Research Foundation, Department of Biochemistry and Biophysics, University of Pennsylvania, Philadelphia, Pennsylvania 19104

**Simon Phillips**     School of Biochemistry and Molecular Biology, University of Leeds, Leeds, LS2 9JT, United Kingdom

**Stephen Ragsdale**     Department of Biochemistry, Beadle Center, University of Nebraska, Lincoln, Nebraska 68588-0664

**Emma Lloyd Raven**     Department of Chemistry, University of Leicester, Leicester LE1 7RH, United Kingdom

**Laura Robledo**     Department of Chemistry, University of Edinburgh, Edinburgh EH9 3JJ, Scotland, United Kingdom

**Margareta Sahlin**     Department of Molecular Biology, Stockholm University, SE-10691 Stockholm, Sweden

**Nigel S. Scrutton**     Department of Biochemistry, University of Leicester, Leicester, LE1 7RH, United Kingdom

**Britt-Marie Sjöberg**     Department of Molecular Biology, Stockholm University, SE-10691 Stockholm, Sweden

**Michael J. Sutcliffe**     Department of Chemistry, University of Leicester, Leicester, LE1 7RH, United Kingdom

**Shinya Yoshikawa**      Department of Life Science, Himeji Institute of Technology and CREST, Japan Science and Technology Corporation Kamigohri, Akoh, Hyogo 678-1297, Japan

**ZhaoLei Zhang**      E.O. Lawrence Berkeley National Laboratory, University of California, Berkeley, California 94720

# Preface

Major advances have been made in recent years in the field of redox enzymology. In part this has been attributable to the wealth of structural information acquired for redox systems, principally by X-ray methods. The successful application of electron transfer theory to redox proteins has also put the study of biological electron transfer onto a sound theoretical platform. Coupled with the ability to interrogate mechanism by site-directed mutagenesis, spectroscopic and transient kinetic methods these developments have contributed to the major expansion seen in recent years of research activity in the field of biological electron and radical transfer.

The purpose of this book is to bring into a single volume milestone developments in the field. The approach is one of systems: all contributions have been selected carefully to illustrate important aspects of the mechanisms of electron and radical transfer in proteins. Our aim is to unite theory and experiment to demonstrate the key principles controlling biological redox reactions. Detailed understanding of these reactions requires knowledge of the role of protein dynamics, chemistries of the redox-active prosthetic groups, the mechanisms governing the diffusional encounter and assembly of electron transfer complexes and the role of protein structure in controlling the physical parameters (reorganisational energy and electronic coupling matrix element) that govern the transfer rate. Additionally, the role of gating mechanisms, such as rate-limiting conformational change, protonation/deprotonation and ligand binding are addressed. Contributions were selected such that each system provides its own unique "spin" on some of these important issues. We have included those systems for which there is high resolution structural information, since detailed interpretation of kinetic data and the results of spectroscopic and mutagenesis experiments is less informative in its absence. The recent development of computational algorithms to calculate electron transfer rates in structurally determined protein molecules has supported, and will continue to support, experimental investigations. This aspect alone illustrates the synergy resulting from a melding of different and complementary techniques in this interdisciplinary field.

The volume begins with a theoretical treatment of biological electron transfer and descriptions of new key structures of soluble electron transfer proteins. Experimental studies of "simple" soluble systems based primarily on the quinoprotein and flavoprotein enzyme families are then introduced to illustrate key concepts in the control of electron transfer reactions. Radical chemistry is also discussed in detail in contributions on ribonucleotide reductase, vitamin $B_{12}$-dependent reactions, soluble methane monooxygenase and peroxidase-catalysed reactions. Chapters focused on nickel and molybdenum-containing enzymes discuss the chemistries of complex redox centres and their role in biological redox reactions. Finally, the remarkable advances made recently in our understanding of membrane electron transfer through structural descriptions of key membrane-embedded proteins are discussed in chapters on bovine heart cytochrome $bc_1$, cytochrome $c$ oxidase and the reaction centre complex from purple bacteria. All contributions are from leading scientists in the field, drawn from a truly international community of redox enzymologists. We hope this volume will inspire those who are less familiar with the field, but also serve as a valuable resource to experienced practitioners. The Editors found the compilation of this volume to be a pleasurable and informative experience. We hope this enthusiasm translates faithfully to the reader.

Andreas Holzenburg,
Leeds, UK

Nigel S. Scrutton,
Leicester, UK

# Contents

*Chapter 1*

**Electron Transfer in Natural Proteins: Theory and Design**
Christopher C. Moser, Christopher C. Page, Xiaoxi Chen, and
P. Leslie Dutton

| | | |
|---|---|---|
| 1. | Introduction .............................................. | 1 |
| 2. | Basic Electron Tunneling Theory......................... | 2 |
| | 2.1.   Classical Free Energy and Temperature Dependence of Tunneling ...................................... | 3 |
| | 2.2.   Quantum Free Energy and Temperature Dependence ....................................... | 7 |
| | 2.3.   A Generic Protein Tunneling Rate Expression ....... | 8 |
| 3. | Protein Structural Heterogeneity ......................... | 9 |
| | 3.1.   Monitoring Heterogeneity via Packing Density ....... | 10 |
| | 3.2.   Natural Tunneling Distances ...................... | 13 |
| 4. | Chains and Robust Electron Transfer Protein Design ...... | 14 |
| 5. | Electron Transfer Clusters ............................... | 17 |
| | 5.1.   Cluster Examples ............................... | 19 |
| | 5.2.   Control of Electron Transfer Through Escapement Mechanisms ...................................... | 21 |
| 6. | Caution and Hope ....................................... | 23 |
| 7. | References ............................................. | 24 |

*Chapter 2*

**Flavin Electron Transfer Proteins**
F. Scott Mathews, Louise Cunane, and Rosemary C. E. Durley

| | | |
|---|---|---|
| 1. | Introduction .............................................. | 29 |
| 2. | Background and Structural Properties ................... | 32 |

    2.1.  NADPH-Cytochrome P450 Reductase (CPR) ........     32
    2.2.  Flavocytochrome P450BM3 .......................     38
    2.3.  Flavocytochrome $b_2$ (FCB2) ..................     42
    2.4.  *p*-Cresol Methylhydroxylase (PCMH) ...............     45
    2.5.  Flavocytochrome c Sulfide Dehydrogenase (FCSD) ...     47
    2.6.  Trimethylamine Dehydrogenase (TMADH) ..........     48
    2.7.  Phthalate Dioxygenase Reductase (PDR) ...........     50
    2.8.  Fumarate Reductase (FUM) ......................     52
3.  Electron Transfer .......................................     55
    3.1.  General Aspects ................................     55
    3.2.  Electronic Coupling ...........................     59
          3.2.1.  BMP/FMN ..............................     59
          3.2.2.  FCB2 .................................     61
          3.2.3.  PCMH .................................     61
          3.2.4.  FCSD .................................     63
          3.2.5.  FUM ..................................     65
4.  Conclusions .........................................     65
    4.1.  Domain Interactions ...........................     65
    4.2.  Features of Electron Transfer ....................     67
5.  References ...........................................     68

*Chapter 3*

**Methanol Dehydrogenase, a PQQ-Containing Quinoprotein
Dehydrogenase**
Christopher Anthony

1.  Introduction ..........................................     73
2.  General Enzymology ...................................     75
    2.1.  The Determination of MDH Activity ..............     75
    2.2.  The Substrate Specificity of MDH ..................     76
3.  The Kinetics of Methanol Dehydrogenase ................     77
    3.1.  The Reaction Cycle of MDH ......................     77
    3.2.  The Activation of MDH by Ammonia and Amines ...     78
    3.3.  The Low *in vitro* Rate of Cytochrome $c_L$ Reduction
          by MDH .........................................     80
4.  The Absorption Spectra of Methanol Dehydrogenase ......     80
5.  Pyrrolo-Quinoline Quinone (PQQ): The Prosthetic Group of
    Methanol Dehydrogenase ..............................     84
6.  The Reaction Mechanism of Methanol Dehydrogenase .....     88
    6.1.  The Reductive Half-Reaction of MDH ..............     88
    6.2.  The Role of Ammonia in the Reductive Half-Reaction
          Mechanism of MDH ..............................     94

6.3. The Oxidative Half-Reaction of MDH .............. 94
    6.3.1. The "Methanol Oxidase" Electron
         Transport Chain .......................... 94
    6.3.2. The Interaction of MDH with Cytochrome $c_L$ .. 95
    6.3.3. The Oxidation of Reduced PQQ in MDH ..... 97
7. The Structure of Methanol Dehydrogenase .............. 97
  7.1. The Structure of the α-Subunit .................... 100
  7.2. The Tryptophan-Docking Interactions in the
     α-Subunit ...................................... 102
  7.3. Structure of the β-Subunit ....................... 104
  7.4. The Active Site of Methanol Dehydrogenase ........ 105
    7.4.1. The Novel Disulphide Ring Structure in the
         Active Site ............................... 105
    7.4.2. The Bonding of PQQ in the Active Site ...... 107
    7.4.3. The Location of Substrate in the Active Site ... 109
8. Processing and Assembly of Methanol Dehydrogenase ..... 110
9. References .......................................... 112

*Chapter 4*

**Methylamine Dehydrogenase: Structure and Function of
Electron Transfer Complexes**
Victor L. Davidson

1. Methylamine Dehydrogenase .......................... 119
  1.1. Structure and Function ........................... 120
  1.2. The TTQ Prosthetic Group ....................... 121
  1.3. Catalytic Reaction Mechanism .................... 121
  1.4. Spectral and Redox Properties ................... 124
2. Amicyanin ......................................... 125
  2.1. Physical Properties ............................. 125
  2.2. Interactions with Methylamine Dehydrogenase ....... 126
3. Methylamine Dehydrogenase-Amicyanin-Cytochrome
  $c$-551i Complex ....................................... 128
  3.1. Structure of the MADH-Amicyanin-Cytochrome
     $c$-551i Complex ................................. 128
  3.2. Interactions between Amicyanin and Cytochrome
     $c$-551i ......................................... 128
  3.3. Pathways Analysis of the Electron Transfer Protein
     Complex ....................................... 129
4. Electron Transfer Reactions in Methylamine Dehydrogenase
  Complexes .......................................... 131

    4.1.   Electron Transfer Theory ......................... 131
    4.2.   Kinetic Complexity of Protein Electron Transfer
           Reactions ...................................... 132
           4.2.1.   True Electron Transfer ..................... 132
           4.2.2.   Coupled Electron Transfer ................. 133
           4.2.3.   Gated Electron Transfer ................... 133
    4.3.   Electron Transfer from TTQ to Copper ............ 134
           4.3.1.   Non-Adiabatic Electron Transfer Reactions.... 134
           4.3.2.   Mutation of Amicyanin Alters the $H_{AB}$ for
                    Electron Transfer from MADH .............. 134
           4.3.3.   Gated Electron Transfer Reactions ........... 137
    4.4.   Electron Transfer from Copper to Heme ............ 138
  5.  Application of Marcus Theory to other Protein Electron
      Transfer Reactions .................................. 138
  6.  Conclusions ......................................... 139
  7.  References .......................................... 140

*Chapter 5*

**Trimethylamine Dehydrogenase and Electron Transferring
Flavoprotein**
Nigel S. Scrutton and Michael J. Sutcliffe

  1.  Electron Transfer Proteins in Methylotrophic Bacteria .... 145
  2.  Prosthetic Groups and Structure of TMADH ............ 148
    2.1.   Identification of the Prosthetic Groups ............. 148
    2.2.   Structure of TMADH ........................... 149
           2.2.1.   Evolution of the Crystal Structure of
                    TMADH ............................... 149
           2.2.2.   Domain and Quaternary Structure .......... 150
           2.2.3.   Large Domain and Active Site Structure ..... 150
           2.2.4.   The Medium and Small Domains ........... 153
  3.  Electron Flow in TMADH—Static Titrations, Reduction
      Potentials and Inactivation Studies ..................... 154
    3.1.   Static Titrations ................................. 154
    3.2.   Reduction Potentials and Selective Inactivation ...... 155
  4.  Single Turnover Stopped-Flow Studies of Electron Transfer 156
    4.1.   Early Stopped-Flow Investigations ................. 156
    4.2.   pH-Dependence of the Reductive Half-Reaction with
           Diethylmethylamine ............................. 158
    4.3.   pH-Dependence of the Reductive Half-Reaction
           with Trimethylamine: Native and Mutant Enzyme
           Studies ........................................ 159
    4.4.   Quantum Tunneling of Hydrogen ................. 163

5.  Control of Intramolecular Electron Transfer: pH-Jump
    Stopped-Flow Studies .............................. 164
6.  The Oxidative Half-Reaction and TMADH-ETF Complex
    Assembly.......................................... 167
    6.1.  Stopped-Flow Studies .......................... 167
    6.2.  A Model for the Electron Transfer Complex ........ 168
7.  Enzyme Over-Reduction and Substrate Inhibition:
    Multiple Turnover Studies .......................... 170
8.  Cofactor Assembly and the Role of the 6-$S$-Cysteinyl FMN    173
9.  Summary and Future Prospects ....................... 176
10. References .......................................... 177

*Chapter 6*

**Amine Oxidases and Galactose Oxidase**
Malcolm Halcrow, Simon Phillips, and Peter Knowles

1.  Introduction ......................................... 183
2.  Galactose Oxidase .................................... 184
    2.1.  Structure ...................................... 185
          2.1.1.  General Structural Properties .............. 185
          2.1.2.  Primary Structure ......................... 185
          2.1.3.  Secondary and Tertiary Structure ........... 186
          2.1.4.  Structure of the Active Site ............... 186
    2.2.  Catalytic Mechanism ............................ 189
          2.2.1.  Substrate Binding ......................... 191
          2.2.2.  Substrate Activation ...................... 191
          2.2.3.  Hydrogen Abstraction from Substrate ........ 191
          2.2.4.  Formation of $E_{reduced}$ ................. 192
          2.2.5.  Reoxidation of $E_{reduced}$ .............. 192
    2.3.  Biogenesis of the Thio-Ether Bond and other
          Processing Events .............................. 192
    2.4.  Model Studies ................................. 193
    2.5.  Biological Role of Galactose Oxidase .............. 196
3.  Amine Oxidases....................................... 197
    3.1.  Structure ...................................... 199
          3.1.1.  General Structural Properties .............. 199
          3.1.2.  Protein Structure ......................... 199
          3.1.3.  Active Site Structure ...................... 205
                  3.1.3.1.  The Copper Site ................. 205
                  3.1.3.2.  The TPQ Site.................... 205
                  3.1.3.3.  Substrate Access Channels to the
                            Active Site of Amine Oxidases ...... 207

                    3.1.3.4.   Other Structural Features . . . . . . . . . . .   207
        3.2.   Catalytic Mechanism  . . . . . . . . . . . . . . . . . . . . . . . . . . . .   208
               3.2.1.   Reductive Half Cycle . . . . . . . . . . . . . . . . . . . . .   208
                        3.2.1.1.   Substrate Binding/Substrate
                                   Schiff-Base  . . . . . . . . . . . . . . . . . . . . . .   209
                        3.2.1.2.   Product Schiff-Base . . . . . . . . . . . . . . .   211
                        3.2.1.3.   Reduced Forms of the Enzyme . . . . . .   213
               3.2.2.   Oxidative Half Cycle . . . . . . . . . . . . . . . . . . . . .   214
        3.3.   Biogenesis of TPQ and Related Cofactors . . . . . . . . . . .   217
        3.4.   Model Studies . . . . . . . . . . . . . . . . . . . . . . . . . . . . . . . . .   219
        3.5.   Biological Roles of Amine Oxidases. . . . . . . . . . . . . . .   219
               3.5.1.   Microorganisms  . . . . . . . . . . . . . . . . . . . . . . . . . .   219
               3.5.2.   Plants . . . . . . . . . . . . . . . . . . . . . . . . . . . . . . . . . . .   219
               3.5.3.   Mammals . . . . . . . . . . . . . . . . . . . . . . . . . . . . . . . .   220
    4.   Comparisons between Galactose Oxidase and Amine
         Oxidases . . . . . . . . . . . . . . . . . . . . . . . . . . . . . . . . . . . . . . . . . . . .   221
    5.   Future Directions  . . . . . . . . . . . . . . . . . . . . . . . . . . . . . . . . . . . . .   222
    6.   References . . . . . . . . . . . . . . . . . . . . . . . . . . . . . . . . . . . . . . . . . . .   223

*Chapter 7*

**Electron Transfer and Radical Forming Reactions in
Methane Monooxygenase**
Brian J. Brazeau and John D. Lipscomb

    1.   Introduction . . . . . . . . . . . . . . . . . . . . . . . . . . . . . . . . . . . . . . . . . .   233
    2.   Components . . . . . . . . . . . . . . . . . . . . . . . . . . . . . . . . . . . . . . . . . . .   237
         2.1.   MMOH . . . . . . . . . . . . . . . . . . . . . . . . . . . . . . . . . . . . . . . . . .   237
                2.1.1.   X-Ray Crystallography . . . . . . . . . . . . . . . . . . . . .   237
                2.1.2.   Spectroscopy . . . . . . . . . . . . . . . . . . . . . . . . . . . . . . .   241
                         2.1.2.1.   Diferric MMOH . . . . . . . . . . . . . . . . . .   241
                         2.1.2.2.   Mixed Valence MMOH . . . . . . . . . . . .   242
                         2.1.2.3.   Diferrous MMOH  . . . . . . . . . . . . . . . .   243
         2.2.   MMOB . . . . . . . . . . . . . . . . . . . . . . . . . . . . . . . . . . . . . . . . . .   243
                2.2.1.   NMR Solution Structure . . . . . . . . . . . . . . . . . . . .   244
         2.3.   MMOR . . . . . . . . . . . . . . . . . . . . . . . . . . . . . . . . . . . . . . . . . .   244
    3.   Component Complexes  . . . . . . . . . . . . . . . . . . . . . . . . . . . . . . . . .   245
    4.   Oxidation-Reduction Potentials . . . . . . . . . . . . . . . . . . . . . . . . .   246
    5.   Electron Transfer Kinetics . . . . . . . . . . . . . . . . . . . . . . . . . . . . . .   248
    6.   Turnover Systems . . . . . . . . . . . . . . . . . . . . . . . . . . . . . . . . . . . . .   250
    7.   Reaction Cycle Intermediates . . . . . . . . . . . . . . . . . . . . . . . . . . .   252
         7.1.   Transient Intermediates of the Reaction Cycle of
                MMOH . . . . . . . . . . . . . . . . . . . . . . . . . . . . . . . . . . . . . . . . . .   252

    7.2.  Structures of the Intermediates . . . . . . . . . . . . . . . . . . .   256
    7.3.  The Mechanism of Oxygen Cleavage . . . . . . . . . . . . . .   259
  8.  Mechanism of C—H Bond Cleavage and Oxygen Insertion     261
    8.1.  Radical Rebound Mechanism . . . . . . . . . . . . . . . . . . . .   262
    8.2.  Isotope Effects . . . . . . . . . . . . . . . . . . . . . . . . . . . . . . .   262
    8.3.  Radical Clock Substrates . . . . . . . . . . . . . . . . . . . . . . .   264
    8.4.  Modified Radical Rebound Mechanism . . . . . . . . . . . .   266
    8.5.  Concerted Oxygen Insertion Mechanisms . . . . . . . . . .   267
    8.6.  Mechanistic Theory based on Calculations . . . . . . . . . .   268
  9.  Conclusions . . . . . . . . . . . . . . . . . . . . . . . . . . . . . . . . . . . .   270
 10.  References . . . . . . . . . . . . . . . . . . . . . . . . . . . . . . . . . . . . .   270

*Chapter 8*

**Flavocytochrome $b_2$**
Christopher G. Mowat and Stephen K. Chapman

  1.  Introduction . . . . . . . . . . . . . . . . . . . . . . . . . . . . . . . . . . . . .   279
  2.  Flavin Reduction and Substrate Oxidation . . . . . . . . . . . . . .   282
  3.  Flavin to Heme Electron Transfer . . . . . . . . . . . . . . . . . . . . .   285
  4.  The Flavocytochrome $b_2$: Cytochrome $c$ Interaction . . . . . . . .   286
  5.  Engineering Substrate Specificity in Flavocytochrome $b_2$ . . . .   290
  6.  Conclusions . . . . . . . . . . . . . . . . . . . . . . . . . . . . . . . . . . . . .   292
  7.  References . . . . . . . . . . . . . . . . . . . . . . . . . . . . . . . . . . . . . .   292

*Chapter 9*

**Flavocytochrome P450 BM3—Substrate Selectivity and Electron
Transfer in a Model Cytochrome P450**
Andrew W. Munro, Michael A. Noble, Tobias W. B. Ost,
Amanda J. Green, Kirsty J. MacLean, Laura Robledo,
Carolyn S. Miles, Jane Murdoch, and Stephen K. Chapman

  1.  Introduction . . . . . . . . . . . . . . . . . . . . . . . . . . . . . . . . . . . . .   297
  2.  Bacterial Model P450 Systems . . . . . . . . . . . . . . . . . . . . . . .   302
  3.  P450 BM3 Structure and Mechanism . . . . . . . . . . . . . . . . . . .   304
  4.  Electron Transfer and its Control . . . . . . . . . . . . . . . . . . . . . .   307
  5.  Site-Directed Mutagenesis in the Study of Substrate
      Selectivity and Electron Transfer . . . . . . . . . . . . . . . . . . . . . .   309
  6.  Conclusions . . . . . . . . . . . . . . . . . . . . . . . . . . . . . . . . . . . . .   312
  7.  References . . . . . . . . . . . . . . . . . . . . . . . . . . . . . . . . . . . . . .   312

*Chapter 10*

**Peroxidase-Catalysed Oxidation of Ascorbate: Structural,
Spectroscopic and Mechanistic Correlations in
Ascorbate Peroxidase**
Emma Lloyd Raven

1.  Introduction ........................................  318
2.  cDNA Sequences and Bacterial Expression of
    Recombinant APXs....................................  319
3.  Isolation and Characterisation of APXs .................  320
4.  General Properties ....................................  321
5.  Structural Studies .....................................  321
6.  Spectroscopic and Spin-State Considerations ............  328
7.  Catalytic Mechanisms ................................  329
    7.1.  Steady-State Kinetics ...........................  329
    7.2.  Pre-Steady-State Kinetics........................  331
8.  Radical Chemistry .....................................  332
    8.1.  Fate of the Monodehydroascorbate Radical .........  332
    8.2.  Nature of the Intermediates ......................  333
          8.2.1.  Compound I .............................  333
          8.2.2.  Compound II ............................  333
          8.2.3.  Role of the Proximal Metal Ion .............  334
          8.2.4.  Role of Trp 179 ..........................  335
9.  Substrate Recognition ................................  335
10. Redox Properties .....................................  337
11. Inactivation ..........................................  339
    11.1.  Inactivation by Cyanide ........................  339
    11.2.  Inactivation by Sulfhydryl Reagents .............  339
    11.3.  Inactivation in Ascorbate-Depleted Media .........  340
    11.4.  Inactivation by Hydrogen Peroxide ..............  341
    11.5.  Inactivation by Salicylic Acid ...................  341
12. References ...........................................  342

*Chapter 11*

**Adenosylcobalamin-Dependent Enzymes**
E. Neil G. Marsh and Daniel E. Holloway

1.  Introduction ..........................................  351
    1.1.  Structure and Reactivity of Cobalamins .............  352
    1.2.  Methylcobalamin-Dependent Enzymes ..............  354

1.3.  Adenosylcobalamin-Dependent Enzymes ............    355
1.4.  Outline Mechanism of Adenosylcobalamin-Dependent
      Rearrangements ..................................    357
1.5.  Ribonucleotide Reductase ........................    358
1.6.  Carbon-Centered Radicals in Enzyme Reactions .....    359
1.7.  Longstanding Questions ...........................    360

2.  Strucural Features of Adenosylcobalamin-Dependent
    Enzymes .............................................    361
    2.1.  Cobalamin Binding by Enzymes Containing the
          D-x-H-x-x-G Motif .............................    362
    2.2.  Substrate Binding and Initiation of Catalysis .........    369
    2.3.  Structure of Diol Dehydrase ......................    371

3.  Mechanistic Aspects of Adenosylcobalamin-Mediated
    Catalysis ............................................    375
    3.1.  Homolysis of Adenosylcobalamin and the Formation
          of Substrate Radicals ..........................    375
          3.1.1.  EPR Studies on Adenosylcobalamin Enzymes .    375
          3.1.2.  Stopped-Flow Studies of Adenosylcobalamin
                  Homolysis ..............................    377
          3.1.3.  Magnetic Field Effects on Adenosylcobalamin-
                  Dependent Reactions ....................    381
          3.1.4.  Resonance Raman Experiments .............    382
          3.1.5.  Role of the Axial Ligand in Catalysis .........    384
    3.2.  Rearrangement of Substrate Radicals ..............    386
          3.2.1.  Mechanistic Studies on Methylmalonyl-
                  CoA Mutase .............................    390
          3.2.2.  Mechanistic Studies on Glutamate Mutase ....    391

4.  Perspective .........................................    394
5.  References ..........................................    397

*Chapter 12*

**Ribonucleotide Reductase—a Virtual Playground for
Electron Transfer Reactions**
Margareta Sahlin and Britt-Marie Sjöberg

1.  Introduction .........................................    405
2.  Three Different Ribonucleotide Reductase Classes ........    406
3.  Reaction Mechanism ..................................    410
    3.1.  Radical Chemistry at Work .......................    410
    3.2.  Substrate Analogues ............................    412

3.3.    Studies in Active Site Mutant Enzymes . . . . . . . . . . . . . .    413
4.  Class I ribonucleotide Reductase—The Radical Transfer
    Pathway . . . . . . . . . . . . . . . . . . . . . . . . . . . . . . . . . . . . . . . . . . . . . . .    415
    4.1.    Protein R1  . . . . . . . . . . . . . . . . . . . . . . . . . . . . . . . . . . . . .    418
    4.2.    Protein R2  . . . . . . . . . . . . . . . . . . . . . . . . . . . . . . . . . . . . .    418
            4.2.1.   The Tyrosyl Radical and the Iron Centre . . . . . .    418
            4.2.2.   A Hydrogen-Bonded Triad . . . . . . . . . . . . . . . . . .    420
            4.2.3.   The Flexible C-Terminal Domain  . . . . . . . . . . . .    421
            4.2.4.   Is the Tyrosyl Radical H-Bonded to the
                     Radical Transfer Pathway? . . . . . . . . . . . . . . . . .    422
    4.3.    Theoretical Considerations on Radical Transfer and
            Protein Dynamics  . . . . . . . . . . . . . . . . . . . . . . . . . . . . . . . . .    422
5.  Generation of the Stable Tyrosyl Radical in Protein R2 . . . . .    424
    5.1.    Radical Generation Involves the Radical Transfer
            Pathway . . . . . . . . . . . . . . . . . . . . . . . . . . . . . . . . . . . . . . . . .    424
    5.2.    Interactions between Metal Sites  . . . . . . . . . . . . . . . . . . .    428
    5.3.    Non-Native Radicals and Secondary Radical Transfer
            Pathways Observed in Mutant R2 Proteins . . . . . . . . . . .    429
    5.4.    Unexpected Hydroxylation Reactions . . . . . . . . . . . . . . .    431
6.  Stability of the Tyrosyl Radical . . . . . . . . . . . . . . . . . . . . . . . . . .    431
    6.1.    Different Tyrosyl Radical Conformers . . . . . . . . . . . . . . .    431
    6.2.    The Tyrosyl Radical Environment  . . . . . . . . . . . . . . . . . .    433
    6.3.    Radical Stability during Catalysis  . . . . . . . . . . . . . . . . . .    433
7.  Radical Transfer Reactions in Class II and III
    Ribonucleotide Reductases . . . . . . . . . . . . . . . . . . . . . . . . . . . . . .    434
    7.1.    Class II Ribonucleotide Reductases . . . . . . . . . . . . . . . . .    434
    7.2.    Class III Ribonucleotide Reductases . . . . . . . . . . . . . . . .    434
8.  References  . . . . . . . . . . . . . . . . . . . . . . . . . . . . . . . . . . . . . . . . . . .    436

Chapter 13

**Molybdenum Enzymes**
Russ Hille

1.  Introduction . . . . . . . . . . . . . . . . . . . . . . . . . . . . . . . . . . . . . . . . . . .    445
    1.1.    Molybdenum Enzymes and the Reactions
            they Catalyse  . . . . . . . . . . . . . . . . . . . . . . . . . . . . . . . . . . . .    445
    1.2.    Sequence Homologies and Classification of the
            Molybdenum Enzymes . . . . . . . . . . . . . . . . . . . . . . . . . . . . .    446
    1.3.    Structural and Catalytic Variations within the Three
            Families of Molybdenum Enzymes . . . . . . . . . . . . . . . . . .    451
2.  The Molybdenum Hydroxylases . . . . . . . . . . . . . . . . . . . . . . . . . .    453
    2.1.    Structural Studies  . . . . . . . . . . . . . . . . . . . . . . . . . . . . . . . .    453

    2.2.  Mechanistic Aspects .............................. 458
3.  The Eukaryotic Oxotransferases ........................ 465
    3.1.  Structural Studies ................................. 465
    3.2.  Mechanistic Aspects .............................. 468
4.  The Prokaryotic Oxotransferases ...................... 472
    4.1.  Structural Studies ................................. 472
    4.2.  Mechanistic Aspects .............................. 475
5.  Concluding Remarks ................................. 477
6.  References ......................................... 478

*Chapter 14*

**Nickel Containing CO Dehydrogenases and Hydrogenases**
Stephen W. Ragsdale

1.  Background ......................................... 487
    1.1.  Importance of CO and $H_2$ Metabolism for
        Anaerobic Microbes ............................ 487
        1.1.1.  Why CO? ................................. 487
        1.1.2.  Where Does the CO Come from and Where
               Does It Go? ............................. 489
        1.1.3.  Why $H_2$? ................................. 489
        1.1.4.  Where Does the $H_2$ Come from and Where
               Does It Go? ............................. 490
        1.1.5.  Supercellular Biochemistry: Interspecies $H_2$
               Transfer ................................. 490
    1.2.  Introduction to the Enzymes ...................... 491
        1.2.1.  CO Dehydrogenase and Acetyl-CoA Synthase 491
        1.2.2.  Hydrogenase ............................. 491
2.  CO Oxidation and $CO_2$ Reduction by CO Dehydrogenase .. 493
    2.1.  The Catalytic Redox Machine: a Ni—Fe—S Cluster ... 493
    2.2.  The Intramolecular Wire ......................... 495
    2.3.  The Intermolecular Wires: How Electrons Enter and
        Exit CO Dehydrogenase ......................... 495
    2.4.  By Channel or by Sea? How Carbon Enters and Exits
        CO Dehydrogenase ............................. 496
    2.5.  Coupling $CO_2$ Reduction and Acetyl-CoA Synthesis... 496
    2.6.  Acetyl-CoA Synthase: Another Catalytic Ni—Fe—S
        Cluster ........................................ 497
3.  $H_2$ Oxidation and Proton Reduction by Hydrogenase ...... 499
    3.1.  The Catalytic Redox Machine: Ni—Fe—S and
        Fe—FeS Clusters .............................. 499
        3.1.1.  The Ni—Fe Hydrogenase ................... 499

|       | 3.1.2. | The Ni—Fe Regulatory Hydrogenase | 504 |
|       | 3.1.3. | The Fe-Only Hydrogenase | 504 |
|       | 3.1.4. | Models of Hydrogenase | 505 |
|       | 3.1.5. | The Metal-Free Hydrogenase | 505 |
|  3.2. | | Proton Transfer Pathway | 506 |
|  3.3. | | The Intramolecular Wire | 508 |
|  3.4. | | The Intermolecular Wires: How Electrons Enter and Exit Hydrogenase | 510 |
|  3.5. | | Tunnel-Diodes and Catalytic Bias | 510 |
| 4. | Summary | | 511 |
| 5. | References | | 512 |

*Chapter 15*

**Cytochrome $cd_1$ Nitrite Reductase Structure Raises Interesting Mechanistic Questions**
Stuart Ferguson and Vilmos Fülöp

| 1. | Introduction | 519 |
| 2. | Structure of *Paracoccus pantotrophus* Cytochrome $cd_1$ | 522 |
| 3. | Kinetic Studies on *P. pantotrophus* Cytochrome $cd_1$ | 531 |
| 4. | Solution Spectroscopy of *P. pantotrophus* Cytochrome $cd_1$ | 532 |
| 5. | The Cytochrome $cd_1$ from *Pseudomonas aeruginosa* | 533 |
| 6. | Copper Nitrite Reductase | 536 |
| 7. | Conclusions and Outstanding Issues | 537 |
| 8. | References | 538 |

*Chapter 16*

**Mitochondrial Cytochrome $bc_1$ Complex**
ZhaoLei Zhang, Edward A. Berry, Li-Shar Huang, and Sung-Hou Kim

| 1. | Introduction | | 541 |
| 2. | Crystal Structure of a $bc_1$ Complex | | 546 |
|    | 2.1. | $bc_1$ Complex is a Homodimer as a Functional Unit | 547 |
|    | 2.2. | Cytochrome $b$ and Two Haems | 549 |
|    | 2.3. | Cytochrome $c_1$ | 550 |
|    | 2.4. | Alternative Conformations of Rieske Protein and Cross-Transfer of Electrons | 553 |
|    | 2.5. | The Two Core Proteins and Subunit 9 | 557 |
|    | 2.6. | Other Subunits without Prosthetic Groups | 559 |
| 3. | Quinone Reactions and Electron Transfer by $bc_1$ Complex | | 561 |

3.1.  Quinone Reduction at $Q_i$ Site ..................... 561
3.2.  Quinone Oxidation at $Q_o$ Site .................... 566
3.3.  The "Domain Shuttle" Mechanism ................. 569
3.4.  Bifurcated Reaction at $Q_o$ Site .................... 571
4.  Summary and Perspective.............................. 572
5.  References  ......................................... 573

Chapter 17

## Bovine Heart Cytochrome *c* Oxidase
Shinya Yoshikawa

1.  Introduction  ........................................ 581
2.  Composition of Bovine Heart Cytochrome *c* Oxidase ...... 582
    2.1.  Purification ..................................... 582
    2.2.  Metal Content  .................................. 584
    2.3.  Structures and Spectral Properties of the
          Redox-Active Metal Sites......................... 585
    2.4.  Subunit Composition and Amino Acid Sequences..... 589
3.  Function of Bovine Heart Cytochrome *c* Oxidase .......... 590
    3.1.  Enzymic Activity................................. 590
    3.2.  Reduction of $O_2$ ................................. 592
4.  Crystallisation of Bovine Heart Cytochrome *c* Oxidase ..... 596
    4.1.  Crystallisation of Membrane Proteins .............. 597
    4.2.  Crystallisation of Cytochrome *c* Oxidase............ 597
5.  X-ray Structure of Bovine Heart Cytochrome *c* Oxidase .... 599
    5.1.  Three-Dimensional Structure of the Protein Portion... 599
    5.2.  Structures and Locations of the Metal Sites ......... 603
6.  A Possible $O_2$ Reduction Mechanism .................... 607
7.  Proton Transfer in Bovine Heart Cytochrome *c* Oxidase .... 608
    7.1.  Possible Proton Transfer Pathways in Membrane
          Proteins ....................................... 608
    7.2.  A Redox Coupled Conformational Change in
          Cytochrome *c* Oxidase ......................... 611
8.  References  ......................................... 616

Chapter 18

## Reaction Centres of Purple Bacteria
Marion E. van Brederode and Michael R. Jones

1.  Introduction  ....................................... 621
2.  Structure of the Bacterial Reaction Centre ............... 622

2.1.  The Structures of the *Rhodopseudomonas viridis* and
      *Rhodobacter sphaeroides* Reaction Centres . . . . . . . . . . .   622
2.2.  Recent Structural Information Relating to Function . . .   625
2.3.  Structures of Mutant Complexes . . . . . . . . . . . . . . . . . . .   626
2.4.  Electronic Structure: The Reaction Centre Absorbance
      Spectrum . . . . . . . . . . . . . . . . . . . . . . . . . . . . . . . . . . . . .   627
3.  The Mechanism of Energy Storage by the Bacterial
    Reaction Centre . . . . . . . . . . . . . . . . . . . . . . . . . . . . . . . . . . .   628
3.1.  Primary Photochemistry . . . . . . . . . . . . . . . . . . . . . . . . . .   631
3.2.  Ubiquinol Formation and Re-Reduction of $P^+$ . . . . . . . .   633
3.3.  Light-Driven Cyclic Electron Transfer Coupled to
      Proton Translocation  . . . . . . . . . . . . . . . . . . . . . . . . . . . .   634
4.  Biological Electron Transfer . . . . . . . . . . . . . . . . . . . . . . . . . . .   634
4.1.  Non-Adiabatic Electron Transfer  . . . . . . . . . . . . . . . . . . .   635
4.2.  Adiabatic Electron Transfer . . . . . . . . . . . . . . . . . . . . . . . .   639
5.  Studies of Ultrafast Electron Transfer in a Light-Activated
    Protein . . . . . . . . . . . . . . . . . . . . . . . . . . . . . . . . . . . . . . . . . .   640
5.1.  The Role of the $B_A$ Monomeric BChl  . . . . . . . . . . . . . . .   641
5.2.  The Asymmetry of Primary Electron Transfer  . . . . . . . .   643
      5.2.1.  Evidence for the Asymmetry of Primary
              Electron Transfer  . . . . . . . . . . . . . . . . . . . . . . . . .   643
      5.2.2.  Origins of the Asymmetry of Primary Electron
              Transfer . . . . . . . . . . . . . . . . . . . . . . . . . . . . . . . . .   644
      5.2.3.  Re-Routing Primary Electron Transfer . . . . . . . .   646
5.3.  Temperature Dependence—Activationless Reactions . .   650
5.4.  Dispersive Kinetics: Heterogeneity or Protein
      Dynamics? . . . . . . . . . . . . . . . . . . . . . . . . . . . . . . . . . . . . .   651
5.5.  Femtosecond Biology: Coherent Nuclear Dynamics
      Studied in Populations of Proteins . . . . . . . . . . . . . . . . . .   654
5.6.  Modulation of the Time Constant for Primary
      Electron Transfer between 200 fs and 500 ps through
      Site-Directed Mutagenesis . . . . . . . . . . . . . . . . . . . . . . . .   656
      5.6.1.  Tyrosine M210 . . . . . . . . . . . . . . . . . . . . . . . . . . .   657
      5.6.2.  Hydrogen Bond Mutants . . . . . . . . . . . . . . . . . . .   658
5.7.  Parallel Pathways for Primary Electron Transfer . . . . . .   661
6.  Summary . . . . . . . . . . . . . . . . . . . . . . . . . . . . . . . . . . . . . . . . . .   665
7.  References  . . . . . . . . . . . . . . . . . . . . . . . . . . . . . . . . . . . . . . . .   665

Index . . . . . . . . . . . . . . . . . . . . . . . . . . . . . . . . . . . . . . . . . . . . . .   677

*Chapter 1*

# Electron Transfer in Natural Proteins
# Theory and Design

Christopher C. Moser, Christopher C. Page,
Xiaoxi Chen, and P. Leslie Dutton

## 1. INTRODUCTION

Biochemical catalysis and redox energy conversion requires the engineering of electron transfer from site to site within proteins. Yet, the protein interior is a good electrical insulator. More than 30 years ago, Devault & Chance (Devault and Chance, 1966) made it clear that Nature relies on electron tunneling to move electrons over tens of Ångstroms. In flash activated photosynthetic membranes, visible spectroscopy can monitor heme and chlorophyll oxidation and reduction electron transfer kinetics down to liquid helium temperatures. At temperatures below 100 K, electron transfer kinetics became temperature independent, the hallmark of quantum tunneling reactions. While thermal energy may be insufficient to classically carry the electron over the insulating barrier, quantum tunneling through the energy barrier is still possible. Electron tunneling, undeniable in

**CHRISTOPHER C. MOSER, CHRISTOPHER C. PAGE, XIAOXI CHEN and P. LESLIE DUTTON**    Johnson Research Foundation, Department of Biochemistry and Biophysics, University of Pennsylvania, Philadelphia, PA 19104.
*Subcellular Biochemistry, Volume 35: Enzyme-Catalyzed Electron and Radical Transfer*, edited by Holzenburg and Scrutton. Kluwer Academic / Plenum Publishers, New York, 2000.

photosynthetic membranes at low temperatures, was soon seen to be active in other systems at physiological temperatures as well.

An appreciation of the basic parameters of electron tunneling theory and a survey of the values of these parameters in natural systems allows us to grasp the natural engineering of electron transfer proteins, what elements of their design are important for function and which are not, and how they fail under the influence of disease and mutation. Furthermore, this understanding also provides us with blueprint for the design of novel electron transfer proteins to exploit natural redox chemistry in desirable, simplified *de novo* synthetic proteins (Robertson *et al.*, 1994).

## 2.  BASIC ELECTRON TUNNELING THEORY

A strict quantum mechanical calculation of a tunneling system the size of a protein quickly becomes intractably complex. Fortunately, relatively simple theory has been successful at organizing and predicting electron tunneling rates in proteins. When the donor and acceptor redox centers are well separated, non-adiabatic electron transfer theory applies Fermi's Golden Rule, in which the rate of electron transfer is proportional to two terms, one electronic, $H_{ab}^2$, and the other nuclear, *FC* (Devault, 1980).

$$k_{et} = \frac{2\pi}{\hbar} H_{ab}^2 FC \qquad (1)$$

The electronic coupling $H_{ab}^2$ between the reactant with the donor reduced, and the product with the acceptor reduced, depends on the ability of the electron wavefunction to penetrate the classical forbidden insulating barrier between the donor and acceptor.

For any given barrier height V, the rate of tunneling of an electron with mass m will fall off exponentially with distance R through the barrier.

$$H_{ab}^2 \propto \exp[-2R\sqrt{2mV}/\hbar] = \exp[-\beta R] \qquad (2)$$

The higher the barrier, the larger the exponential coefficient $\beta$ and the more dramatically the electron transfer rate decays with distance. By a fortunate coincidence of units, the $\beta$ in $\text{Å}^{-1}$ is approximated by the square root of the barrier height in eV. Thus for typical biological redox centers that must overcome a barrier of about 8 eV to be ionized in a vacuum, we can estimate the $\beta$ for exponential decay of electron transfer in vacuum to be about 2.8 $\text{Å}^{-1}$. Much less of a barrier is presented by a surrounding organic medium, where the positively charged nuclei can interact favorably with the electron. Studies of the distance dependence of electron transfer between donors and acceptors bridged by rigid covalent linkers and dissolved in an organic solvent or in monolayers on electrodes (Moser *et al.*, 1992; Smalley *et al.*, 1995) shows $\beta$ of 0.7 to 0.9 Å$^{-1}$, the larger $\beta$ corresponding to a barrier of about 0.8 eV. The protein medium presents a barrier between these extremes. In the absence of an intervening structure that is unusually well bonded or unusually loosely packed, the protein displays an average barrier around 2 eV, with a $\beta$ usually quite close to 1.4 Å$^{-1}$ (Moser *et al.*, 1992). Note that for redox centers with extremely high midpoint potentials (above 1 eV relative to hydrogen electrode, including many radical centers such as those found in ribonucleotide reductase or photosystem II) the barrier might be expected to be smaller. However, with the small set of experiments available, there is no clear indication so far that $\beta$ is different for high potential radical electron transfer mechanisms.

No matter what the value of $\beta$, the rate of electron tunneling will be fastest at short distances. There appears to be a practical upper limit on the rate of electron tunneling at distances that approach van der Waals contact of around $10^{13}$ s$^{-1}$, both in chemical systems and in many proteins (Moser *et al.*, 1992). This limiting rate may be a reflection of the characteristic time of vibration and hence nuclear rearrangement that takes place upon electron transfer. With the seemingly reasonable assumption that the $10^{13}$ limit is appropriate for proteins in general, Eq. (2) and the $\beta$ values described above permit an estimate of the maximal rate of electron tunneling in any system at a given distance, or conversely, the maximum distance at which an observed tunneling rate can take place (Figure 1).

## 2.1. Classical Free Energy and Temperature Dependence of Tunneling

The second term of Fermi's Golden Rule depends on the changing position of the atomic nuclei upon electron transfer and is abbreviated as *FC*, for Franck-Condon. The nuclei assume different lowest energy geometries when the electron is on the donor as compared to the acceptor. For example, polar solvent molecules and protein groups will point in different directions, and the lengths of chemical bonds within the redox centers themselves will change as the centers are oxidized and reduced. However, in nonadiabatic electron transfer theory, the electron tunnels from the donor to acceptor only if the nuclear geometry is such that the energies immediately before and after electron transfer are momentarily the same. Marcus provided a simple and successful description of this process by introducing the term reorganization energy $\lambda$: the energy required to distort the

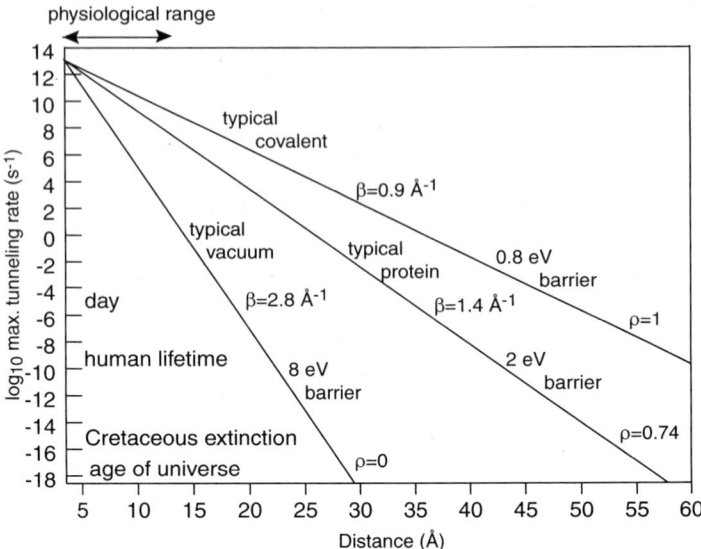

**FIGURE 1.** The maximum electron tunneling rate depends on the distance between donor and acceptor and the effective height of the tunneling barrier, decaying rapidly with distance (exponential slope $\beta$).

equilibrium geometry of the reactant into the equilibrium geometry of the product, but without allowing the electron transfer to take place (Marcus, 1956; Marcus and Sutin, 1985).

In Figure 2, the reactant and product describe simple harmonic potentials as the nuclei are displaced from their equilibrium positions at the bottom of the parabolas. If the reaction is exothermic, then the product parabola will be $\Delta G$ lower in energy. To displace the nuclei of the reactant to the geometry of the product requires an input of energy $\lambda$, the reorganization energy. In this symmetric description, the same reorganization energy $\lambda$ is required to distort product nuclei to resemble the reactant geometry. The crossing point of these surfaces where electron tunneling can take place, requires the following activation energy:

$$\Delta E^{\ddagger} = (\Delta G + \lambda)^2 / 4\lambda \tag{3}$$

Thus, classical Marcus theory predicts an electron transfer rate that has a Gaussian dependence on the free energy of the reaction (Marcus, 1956; Marcus and Sutin, 1985).

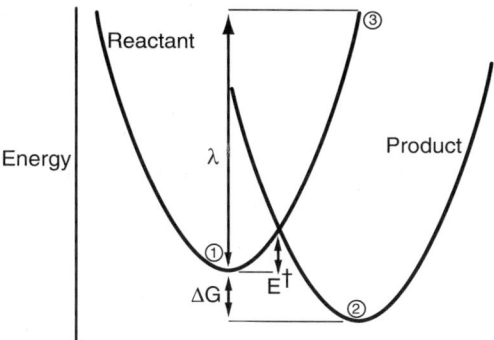

Nuclear Geometry/Reaction Coordinate

**FIGURE 2.** The Marcus expression can be understood in terms of reactant and product potential surfaces approximated as simple harmonic oscillations of the nuclei. The equilibrium position of the nuclei at the bottom of the reactant potential well (1) changes to a new equilibrium position in the product potential well (2) releasing free energy ΔG. The energy required to distort the reactant nuclei (1) so that they resemble the product equilibrium geometry, but without allowing electron transfer (3) defines the reorganization energy λ. The potential surfaces intersect and define an activation energy $E^{\dagger}$.

$$k_{et} \propto FC = \frac{1}{\sqrt{4\pi\lambda k_B T}} \exp\left[-(\Delta G + \lambda)^2 / 4\lambda k_B T\right] \qquad (4)$$

When the driving force of the reaction matches the reorganization energy, $\Delta G = -\lambda$, the surfaces intersect at the bottom of the reactant well and the reaction is activationless and free energy optimized. Because of the classical $k_B T$ term of the Marcus description, the Gaussian dependence of the rate on ΔG becomes increasingly narrow as the temperature falls, so that at cryogenic temperatures if the electron transfer is observed at all, it has an extremely steep free energy dependence.

A counter-intuitive prediction of the Marcus theory is that overdriving the reaction, so that $\Delta G < -\lambda$, causes the activation energy to rise again and the reaction to slow (Figure 3). The Marcus inverted region has been observed clearly in a number of synthetic chemical systems in which it is possible to vary the driving force of the reaction over a range greater than 1 eV (Closs *et al.*, 1986; Miller *et al.*, 1984).

The free energy dependence of a number of electron tunneling reactions in natural proteins can be approximated by a Gaussian free energy dependence (Moser *et al.*, 1992). The free energy can be varied either by extracting protein redox centers and replacing them with analogous exotic

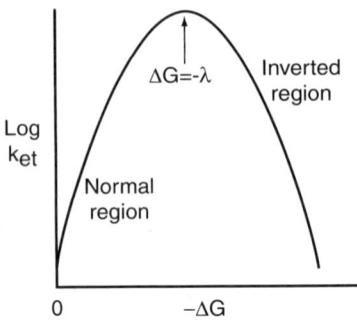

**FIGURE 3.** Marcus theory predicts a Gaussian dependence of the electron transfer rate ket on free energy ΔG, which appears as a parabolic dependence on a log plot. The maximum rate is found when the driving force matches the reorganization energy, ΔG = −λ.

centers that have different redox midpoint potentials, or by making mutational changes in the medium around the proteins to effect changes in the redox properties (Gunner *et al.*, 1986; Gunner and Dutton, 1989; Giangiacomo and Dutton, 1989; Okamura and Feher, 1992; Lin *et al.*, 1994; Labahn *et al.*, 1994). The photosynthetic reaction center (RC) provides several examples of both types of changes. It is generally difficult to vary the free energy over so great a range that the inverted region can be clearly explored. Although there appear to be some examples of the inverted region in action in natural redox proteins (Franzen *et al.*, 1992; Dutton and Moser, 1994; Moser *et al.*, 1995), more experiments are needed before the effect is no longer controversial. Somewhat greater freedom to explore the free energy dependence of tunneling in proteins can be achieved by modifying proteins to introduce extra, potentially light-activatable redox centers, usually on the outside of the proteins (Pan *et al.*, 1988; Onuchic *et al.*, 1992).

The reorganization energies experimentally defined for natural electron transfer reactions cover a range depending on the cofactors themselves and polarity of the cofactor environment. Typical reorganization energies for redox cofactors buried in a protein are 0.6 to 1 eV. Reorganization energies will tend to be larger for redox centers that are small (a few atoms) and concentrate the change in charge upon oxidation/reduction and thus tend to have large local changes in the electric field. Reorganization energies may be as large as 1.4 eV for redox centers that are near an aqueous surface where nearby polar molecules can reorient. Reorganization energies will be smaller for redox centers that are large (like chlorophylls) and distribute charge changes, are buried far from the aqueous phase, or participate in rapid electron transfers in which there is little time for dramatic reorganization. All of these conditions appear to conspire to make the reorganization of the initial charge separation in photosynthetic reaction centers as small as 0.2 eV (Parson *et al.*, 1990) or even smaller (Jia *et al.*, 1993; Parson *et al.*, 1998).

## 2.2. Quantum Free Energy and Temperature Dependence

Intraprotein electron tunneling studies in which both the free energy and the temperature have been varied over large ranges are rare. Excellent examples are provided by photosynthetic reaction centers in which quinone substitution has varied $\Delta G$ by more than 0.5 eV and temperature has ranged from 300 to 10 K (Gunner et al., 1986; Gunner and Dutton, 1989). While the free energy dependence of these reactions is roughly Gaussian, it is clear, especially at lower temperatures, that the Gaussian is considerably broader than the classical Marcus expression of Eq. (4). Apparently, these electron-tunneling reactions are coupled to nuclear rearrangements or vibrations that are larger than the room temperature thermal energy $k_BT$, about 25 meV. Thus, the classical Marcus expression should be replaced by a quantized version, such as the exact but somewhat complicated quantum mechanical harmonic oscillator expression (Levich and Dogonadze, 1959; Jortner, 1976) or the simpler semi-classical expression of Hopfield (Hopfield, 1974), which retains the Gaussian form,

$$k_{et} \propto FC = \frac{1}{\sqrt{2\pi\sigma^2}} \exp\left[-(\Delta G + \lambda)^2 / 2\sigma^2\right] \tag{5}$$

where

$$\sigma^2 = \lambda\hbar\omega \coth\left[\hbar\omega/2k_BT\right]. \tag{6}$$

$\hbar\omega$ is the characteristic frequency of the vibrations coupled to electron transfer. In the protein systems there are likely to be a great many different vibrations coupled to electron transfer, each contributing some piece to the total reorganization energy. This characteristic frequency then is the $\lambda$ weighted average of these frequencies. For the reactions in the reaction center, this value appears to be around 70 meV (Moser et al., 1992). Because of the difficulties of performing these extensive experiments, there are not enough comparable studies to determine if this value is essentially universal for all intraprotein electron transfers. Nevertheless, rate calculations using this value are usually quite successful at predicting electron transfer rates.

In principle it should be possible to determine the reorganization energy simply by examining the temperature dependence of the reaction to determine the activation energy, without the need to replace natural redox centers with exotic cofactors, or mutational changes to modulate midpoint potentials. However, this method is fraught with traps for the unwary. First, it may be more appropriate to use the semi-classical version of the

FC in Eq. (6) rather than the commonly used Marcus expression of Eq. (3). Second, the activation energy measured may instead be of a reaction coupled to electron transfer, such as proton binding or release or diffusional motions of redox centers, rather than the electron tunneling itself. This should immediately be suspected if calculated reorganization energies are larger than 1.5 eV (Davidson, 1996) or if the rates seem to be significantly slower than that predicted by Eq. (7). Thus the variation of the free energy of the reaction by chemical or even applied electric field methods provides the most secure way to determine if the observed rate is limited by the rate of electron tunneling. It is also the most secure way to find the reorganization energy and the free energy optimized tunneling rate.

Reaction center experiments show that the electron transfer rate is not very sensitive to the detailed chemical nature of the donor and acceptor. Thus ubiquinone can be replace by naphthoquinones, anthraquinones, fluorenones and even dinitrobenzenes (Warncke and Dutton, 1993; Moser *et al.*, 1992), yet all these species can still be embraced by a single broad Gaussian free energy dependence.

### 2.3. A Generic Protein Tunneling Rate Expression

Equations 1, 2 and 5 can be combined into an extremely useful empirical expression that estimates room temperature electron tunneling rates for all proteins to within about an order of magnitude (Moser and Dutton, 1992):

$$\log_{10} k_{et} = 15 - 0.6R - 3.1(\Delta G + \lambda)^2 / \lambda \qquad (7)$$

where $k_{et}$ is in units of $s^{-1}$, R in Å and $\Delta G$ and $\lambda$ in eV. In a classical Marcus expression, the coefficient 3.1 is replaced with 4.2.

The variable R, the edge-to-edge distance between cofactors, is clearly the most important variable in predicting the electron transfer rate. A practical definition of R identifies the atoms which make up a biological redox cofactors to include all atoms making up the aromatic/conjugated systems of porphyrins, chlorophylls, flavins, pterins, amino acid radicals and quinones, including oxygens attached to the quinone rings, as well as the metal atoms in redox centers and the immediate atoms liganded to these metals. Thus, a heme will have 27 atoms, a 2Fe2S center 8 atoms and a Cu center only 5.

The explosion in recent years of newly resolved crystal structures of electron transfer proteins makes it possible to estimate R and hence electron tunneling rates for many systems by using Eq. (7) together with free energies estimated from redox midpoint potentials of donor and acceptor, and estimates of $\lambda$ based on the likely polarity of the redox center environments. If $\lambda$ is moderately well defined, Eq. (7) predicts electron-tunneling rates with an uncertainty of about an order of magnitude or less. Once an appropriate crystal structure is known, an unknown $\lambda$ may contribute the largest uncertainty in the tunneling rate calculation, since many biological free energies are close to zero.

## 3.  PROTEIN STRUCTURAL HETEROGENEITY

The free energy optimized rate can be used to compare different electron tunneling reactions in different proteins and re-examine the electronic term and the tunneling barrier of Eq. (2). Because the heterogeneity of the protein medium can change the height of the barrier, it is possible in principle for natural selection to favor a well bonded tightly packed medium to accelerate productive electron transfers and to favor a poorly bonded, loosely packed medium between centers which can engage in unproductive, energy-wasting electron transfers. Indeed, one might imagine that billions of years of evolution would optimize protein structures for electron tunneling.

A general method is needed to examine the structure between redox centers to determine if the barrier for any given productive or unproductive reaction is larger or smaller than average. A full quantum mechanical calculation being impossibly complex, various methods of different degrees of complexity using different sets of approximations have been used.

One popular general method searches for the shortest set of connected bonds and short through-space gaps to define a path between redox centers, and then applies a rate penalty for each bond and a greater, distance dependent penalty for each through space gap (Beratan *et al.*, 1991; Onuchic *et al.*, 1992). Applying this method to a series of analogous reactions allows the parameters to be adjusted to fit existing experimental rates. This method successfully identifies protein regions that are more or less well bonded and are generally correlated with faster or slower tunneling.

Users of this method often give the impression that the pathway so identified, rather than the whole medium between the centers, is responsible for guiding the tunneling electron rather like a wire. The original authors have tried to guard against this interpretation by introducing the concept of "tubes" rather than a single pathway (Curry *et al.*, 1995). Nevertheless, there is often a belief that the identified pathway has been naturally selected to guide the electron and that changes to the pathway, for example through mutagenesis, will spoil the electron tunneling rate. In fact, a careful mutagenic crystallographic, electrochemical and kinetic series in the reaction centers shows this is not the case (Dohse *et al.*, 1995). Mutations often have minor effects on tunneling rates that can be understood in terms of small changes in R, $\Delta G$ and $\lambda$ (Page *et al.*, 1999).

Other methods assay the protein structure without emphasizing a best path. Kuki describes a conceptually attractive method using Monte Carlo sampling to simulate the scattering of the tunneling electron from multiple centers in the protein on its way from donor to acceptor (Kuki and Wolynes, 1987). Not surprisingly, this method shows that a cylindrical to roughly ellipsoidal region of the protein containing the redox centers as foci is most significant in determining the efficiency of electron tunneling.

## 3.1. Monitoring Heterogeneity via Packing Density

The simplest method to assay protein heterogeneity uses a basic, easily appreciated parameter, the density of the protein packing between redox centers. The packing method distinguishes regions that are relatively densely packed, well-bridged and have experimentally small $\beta$, from more loosely packed regions that are experimentally associated with larger $\beta$ (Page *et al.*, 1999).

There are a number of ways to estimate the packing density if a protein structure is available. We sample the region between donor and acceptor by drawing lines from every donor atom to every acceptor atom and determine the fraction of the lines that falls within the united van der Waals radius of a medium atom, ignoring any part of the lines that may pass through cofactor atoms. Because published structures usually do not include hydrogen atoms, we use the united-atom approximation which compensates for the missing hydrogens by multiplying the van der Waals radius of the heavier atoms by 1.4. Structures often do not include atoms such as water molecules that do not have a well defined position, but are likely to be present in any real structure. Thus we solvate structures using a molecular graphics program, such as Sybyl, to make sure we are not working with unrealistically large voids.

Packing densities of proteins estimated by various means average about 75% (Levitt *et al.*, 1997). When we examine the presumably arbitrary protein interior packing between redox centers and aromatic residues Tyr, and Phe in 6 large proteins, including cytochrome oxidase and nitrogenase, we find the packing follows a roughly Gaussian distribution with mean $\rho = 0.77$ and standard deviation $\sigma = 0.09$ (gray background of Figure 4). The packing between synthetic donor and acceptor systems linked by a rigid covalent bridges and associated with a $\beta = 0.9 \text{Å}^{-1}$, is much higher, usually between 0.9 and 1.0.

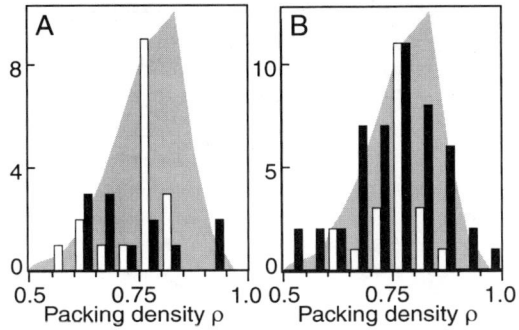

**FIGURE 4.** Histograms of protein packing for productive (dark) electron transfers, unproductive (light) electron transfers and arbitrary (gray) chemical groups. A. Packing in photosynthetic reaction centers of *Rb. sphaeroides* and *Rp. viridis*. B. Packing in 63 redox reactions in multi-center proteins available in the PDB, including only one representative species per type of oxidoreductase.

We can turn this packing density into a $\beta$ estimate by taking the packing density weighted average of $\beta = 0.9\,\text{Å}^{-1}$ for the fraction of the intervening medium within atoms and $\beta = 2.8\,\text{Å}^{-1}$ for the "vacuum" region outside atoms.

$$\beta = 0.9\,\text{Å}^{-1}(\rho) + 2.8\,\text{Å}^{-1}(1-\rho) \qquad (8)$$

This permits an enhanced tunneling rate estimate based on Eq. (7)

$$\log k_{et} = 13 - (1.2 - 0.8\rho)(R - 3.6) - 3.1(\Delta G + \lambda)^2 / \lambda \qquad (9)$$

where R is the edge-to-edge distance in Å, and $\Delta G$ and $\lambda$ are in eV as before.

A comparison of the calculated vs. experimental $\Delta G$ optimized tunneling rates for the productive charge separating electron transfers in two bacterial photosynthetic reaction centers shows that rate estimates have a standard deviation of 0.5 log units, or about a factor of 3 (Figure 5). Considering the experimental errors of determining $\Delta G$ and especially $\lambda$, or even the uncertainties in R of a dynamic protein, it is not clear that a calculation any more involved than the one we have just described is usually justified.

A histogram of the $\rho$ for these reaction centers shown in figure 4A shows that $\rho$ is distributed much as in the presumably arbitrary packing between cofactors and aromatic amino acids. There is no tendency for productive charge separation reactions to be accelerated by good packing

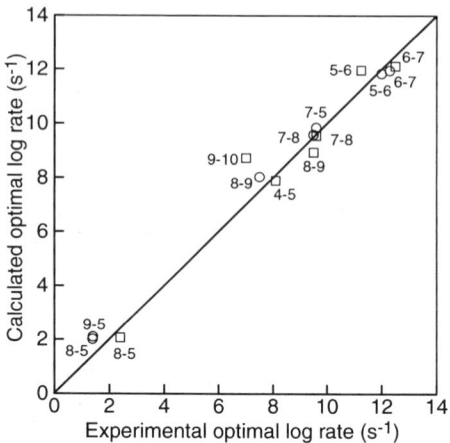

**FIGURE 5.** A packing density based estimate of the maximal electron tunneling rate in *Rp. viridis* (squares) and *Rb. sphaeroides* (circles) using Eq. (9) has a standard deviation of about 0.5 log units. Key to individual reactions found in Figure 7.

nor for unproductive, energy wasting charge recombinations to be deceler-
ated by poor packing. Apparently, natural selection has not chosen to adjust
the protein medium heterogeneity to tune reaction centers. The reaction
with the best packing is $Q_A$ to $Q_B$. Indeed, these quinones are relatively
neatly bridged by hydrogen-bonding to histidines, which in turn are linked
to a common Fe atom on the symmetry axis between the L and M subunits of the reaction centers. Accordingly, the fast phase of electron
transfer between $Q_A$ and $Q_B$ is faster than average for proteins for this distance. However, the argument that this relatively dense packing has been
naturally selected is immediately undermined by the observation that this
fast tunneling between $Q_A$ and $Q_B$ is a measurable but usually minor part
of the electron transfer kinetics (Li *et al.*, 1998). Apparently, the majority of
the electron transfer between $Q_A$ and $Q_B$ is rate limited by a non-tunneling
step, perhaps quinone or proton movement that accompanies electron
transfer.

Figure 4B shows the set of 31 productive and 21 unproductive electron
transfer reactions in multi-redox center oxidoreductases with structures
available in the Protein Data Bank (PDB). Both productive and unproductive reactions have statistically indistinguishable distributions, which are
in turn indistinguishable from the arbitrary protein packing distribution.
Thus Nature has not generally selected protein heterogeneity to assist productive and hinder counterproductive electron tunneling.

Figure 6 provides one conspicuous reason why this may be the case.
The distances between physiologically productive redox partners in these

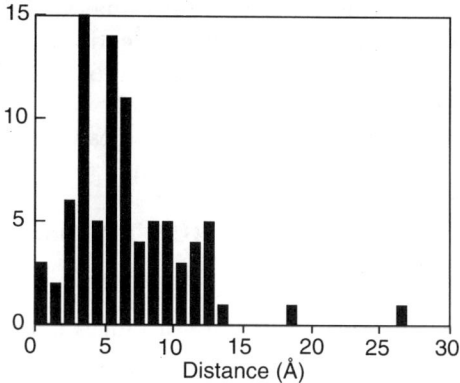

**FIGURE 6.** Distribution of edge-to-edge distances in the 42 productive reactions between redox centers within oxido-reductases deposited in the PDB. Nearly all the distances are 14 Å or less.

proteins have clearly been selected to be 14 Å or less. The two outliers may not be physiologically meaningful reactions; interdomain flexibility in sulfite oxidase may shorten the ~27 Å between heme $b_5$ and the Mo center, (Kisker *et al.*, 1997) while physiological electron transfer between non-heme iron centers separated by ~18 Å in desulfoferrodoxin has not been demonstrated (Coelho *et al.*, 1997). By maintaining tunneling distances at 14 Å or less, the tunneling rates for these reactions given by Eqs. (7) or (9) will be safely faster than the $10^4$ to $10^0$ $k_{cat}$ for physiological turnover of these proteins. Thus mutational evolutionary changes in the protein heterogeneity would have essentially no effect on overall protein electron transfer rates and would tend to remain unselected.

## 3.2. Natural Tunneling Distances

However, this observation begs the question of why cofactors are placed within 14 Å, since distances of 20 Å or more are clearly feasible. It seems likely that 14 Å is a limit that reflects natural selection for robust electron transfer protein designs which will continue to function with tunneling rates faster than typical enzyme overall catalytic rates ($k_{cat}$), even in the face of random mutational change in evolution. At less than 14 Å, tunneling rates faster than the typical $k_{cat}$ range of microseconds to seconds is virtually assured. At longer distances, random changes in the tunneling parameters can make tunneling relatively slow and rate limiting for enzyme turnover. Relatively fragile designs such as these will tend to disappear.

An even more fundamental reason why protein structural heterogeneity is not selected to assist or retard electron tunneling arises from the multiple roles that the protein medium must play in any natural protein, such as dictating folding, assembly, stability, targeting and recognition. Optimization of structure for one single function cannot be expected. As Darwin indicated in his Origin of Species (Darwin, 1872), natural selection is inhibited when an inheritable trait plays multiple roles.

## 4.  CHAINS AND ROBUST ELECTRON TRANSFER PROTEIN DESIGN

To transfer electrons over extended distances between catalytic sites of substrate oxidation and reduction and sites of energy conversion, Nature relies on redox chains. The use of chains allows biological electron transfer to escape the exponential decrease of rate with distance, and to recover an essentially linear dependence of rate over very long distances, keeping tunneling rates faster than the $k_{cat}$ of the enzymes.

Chains also contribute to the robustness of natural electron transfer protein design because the close spacing between successive redox centers means that the driving force of the reaction can usually vary widely with relatively little effect on the overall electron transfer rate through the protein. Indeed, many naturally occurring chains have uphill electron transfer steps of hundreds of meV.

Consider the c heme chain in *Rp.viridis*, illustrated in Figure 7 (Michel *et al.*, 1986; Osyczka *et al.*, 1998). Electrons are transferred from a soluble cytochrome $c_2$ (redox midpoint, Em ~0.28 V) to the low potential heme 4

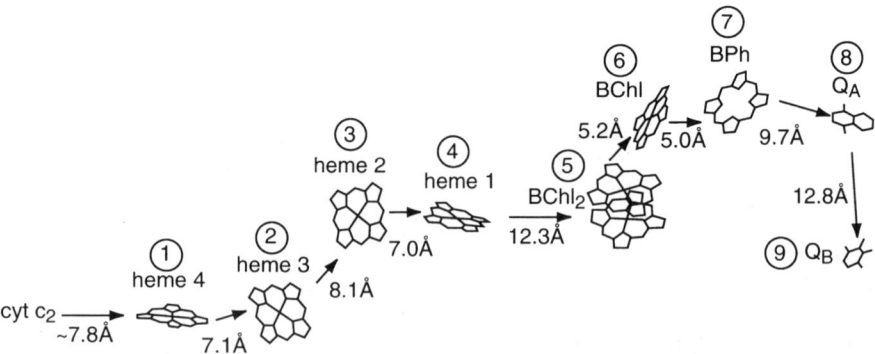

**FIGURE 7.** Two redox cofactor chains meet at the bacteriochlorophyl dimer in the photosynthetic reaction center of Rp. viridis. Electron transfer takes place by tunneling between cofactors that are spaced by no more than 14 Å, assuring overall electron transfer rates in the msec or faster range, even though a total distance of 70 Å is crossed by the c heme chain.

(Em ~–0.050 V), an endergonic step of 0.33 eV. Electron transfer to heme 3 (Em ~0.30 V) is ~0.35 eV exergonic. Transfer to heme 2 (Em ~0.05 V) is an endergonic ~0.25 eV, transfer to the final heme 1 (Em ~0.38 V) is exergonic, as is electron transfer to the photo-oxidized $BChl_2$ (Em ~0.48 V) (Alegria and Dutton, 1991). The electron transfer from $c_2$ to $BChl_2$ takes place over a distance of about 70 Å. A single step tunneling over this distance would take longer than the age of the universe, yet by using a chain, the tunneling is accomplished on the sub-millisecond time scale. No single distance is larger than 13 Å.

To analyze these uphill reactions and the electron transfer through chains, we need to extend Eqs. (7) or (9) into the endergonic region. There are a number of ways in which this may be done. The Gaussian expressions of Eqs. (7) and (9) could be extended directly into the endergonic region. In this case the forward and reverse tunneling rates would display non-Boltzman constant dependent equilibrium constants, akin to the temperature independent equilibrium constants observed at low temperatures in the nuclear tunneling reactions of Tutton salts (Trapani and Strauss, 1989). However, in the absence of experimental protein electron tunneling evidence to the contrary, it is safer to assume that the forward and backward electron tunneling rates will be in a traditional Boltzman dependent equilibrium. Thus we can simply calculate the tunneling rate for the exergonic reaction using the above equation, and then subtract a simple free energy and Boltzman thermal energy dependent term to get a generic endergonic tunneling rate expression (Page *et al.*, 1999):

$$\log_{10} k_{et}^{en} = 15 - 0.6R - 3.1(-\Delta G + \lambda)^2 / \lambda - \Delta G / 0.06 \qquad (10)$$

or an endergonic rate expression when the packing is known:

$$\log k_{et}^{en} = 13 - (1.2 - 0.8\rho)(R - 3.6) - 3.1(-\Delta G + \lambda)^2 / \lambda - \Delta G / 0.06 \qquad (11)$$

Using Eq. (10), we immediately notice that electron tunneling can often go many hundreds of meV uphill while keeping the overall chain rate within typical $k_{cat}$, provided cofactors are placed closer than 14 Å. Figure 8 gives some examples of rates for typical reaction parameters. Despite common practice, it is clearly important not to reject an active role for a redox center within an oxidoreductase simple because its midpoint potential is some hundreds of mV out of line. Similarly, small changes of a few tens of mV in the redox midpoint potential are not likely to have a significant effect on any of the rates, while larger changes of more than a hundred mV may have little effect on the overall chain activity.

Equations 10 and 11 perform quite well in estimating rates of electron transfer chains, for example, in the experimentally reported rates for several

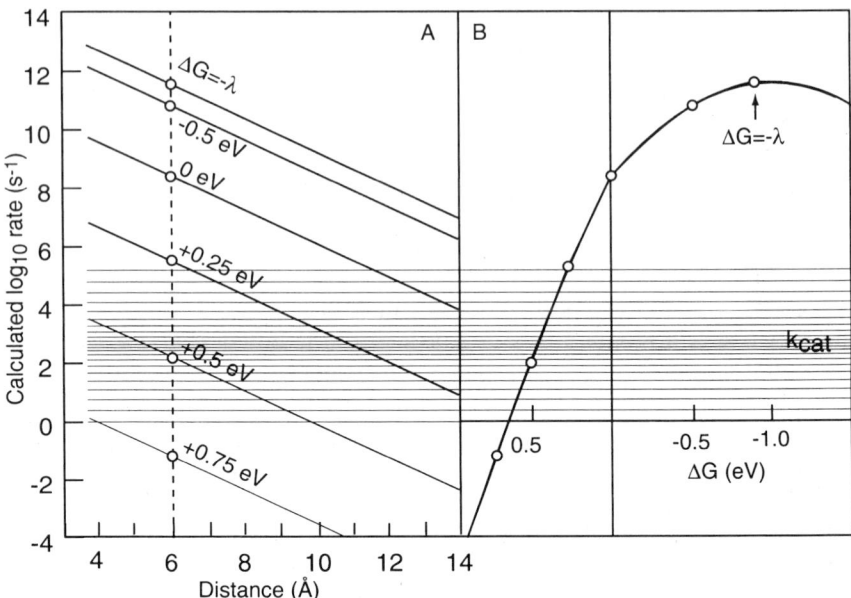

**FIGURE 8.** A. Electron tunneling rates given by Eqs. (7) and (10) for a typical reorganization energy of 1 eV. The fastest tunneling occurs when the free energy matches the reorganization energy $\Delta G = -\lambda$. When redox centers are closely spaced, even endergonic tunneling reactions of +0.6 eV will be faster than a typical enzymatic catalytic rate, $k_{cat}$. B. The tunneling rate for a reaction with a typical distance of 9 Å and a typical reorganization energy of about 1 eV can be described by Eq. (7) for the endergonic region and Eq. (10) for the exergonic region, which assumes a Boltzman dependent equilibrium constant between uphill and downhill reactions. Even chains that include a 0.5 eV uphill step can easily have rates faster than the millisecond to second catalytic turnover rates typical of oxidoreductases.

of the c heme chain electron transfers in *Rp. viridis* reaction centers. Using crystallographic distances, reported redox midpoint potentials and assuming a typical reaction center reorganization energy of 0.7 eV for heme to heme electron transfer, the estimates for the individual electron transfer rates from heme 1 to $BChl_2^+$ is $4.1 \times 10^6 s^{-1}$ compares with $5.4 \times 10^6 s^{-1}$ measured (Shopes *et al.*, 1987; Dohse *et al.*, 1995). The estimated rate for the multistep, mixed endergonic transfers from the heme 3 to heme 1 rate of $1.8 \times 10^5 s^{-1}$ compares with $2.8 \times 10^5 s^{-1}$ (Shopes *et al.*, 1987), while the overall chain rate from heme $c_2$ to $BChl_2$ of $1.1 \times 10^4 s^{-1}$ compares with $1.4 \times 10^4 s^{-1}$ (Ortega *et al.*, 1999, Meyer *et al.*, 1993).

Other examples of chains abound. In Figure 7, it can be seen that a second chain extends from the $BChl_2$ through bacteriochlorophylls and pheophytins to $Q_A$ and $Q_B$. Electron tunneling from $BChl_2$ to $Q_A$ extends over about 20 Å in 200 nsec. Hydrogenase has a chain of FeS clusters extending from the nickel-iron center towards the surface (Volbeda *et al.*, 1995; Rousset *et al.*, 1998). Calculations show that electron tunneling though this ~40 Å chain will be on the order of $2 \times 10^7 s^{-1}$, far faster than the catalytic rate (Albracht, 1994) of ~$10^4 s^{-1}$ despite a significant endergonic step.

In radical electron transfer, electron transfer chains recruit amino acids as redox centers. For example, in ribonucleotide reductase, oxidization of a di-iron center creates a tyrosine radical (Nordlund and Eklund, 1993). Because a tryptophan is in close proximity to this tyrosine, endergonic electron transfer to create an oxidized tryptophan radical should be possible, even if this tryptophan radical is not specially stabilized by charge compensation from the protein medium. Indeed, the ribonucleotide reductase structure shows a string of tryptophans and tyrosines extending about 35 Å to a Cys radical adjacent to the ribose binding site where the catalytic deoxygenation takes place (Nordlund and Eklund, 1993). Because of the relative proximity of each aromatic residue in this radical chain, the tunneling rate through the chain should be much faster than the rather slow 1 $s^{-1} k_{cat}$ of this enzyme (Rova *et al.*, 1995), even if each of these residues has no special hydrogen bonding or charge compensation to bring the redox midpoint values below about 1.5 V, some 0.5 V above the estimated starting tyrosine midpoint potential (Stubbe and van der Donk, 1998). Indeed, it appears that a relatively crude form of charge compensation in the form of local $H^+$ release and uptake may be all that is needed to make this radical chain work, without the requirement for a relatively fragile extended H atom transfer network (Siegbahn, 1998). Because the radical will only transiently occupy the intervening residues, most of the time the radical will remain on the original tyrosine, relatively safely insulated from adventitious reactions by the bulk of the ribonucleotide reductase protein.

## 5. ELECTRON TRANSFER CLUSTERS

We have described single electron tunneling reactions, yet most biochemically relevant substrates undergo two electron oxidation and reduction, reflecting the making and breaking of bonds with pairs of shared electrons. Can the single electron analysis we've described be extended to these catalytic sites? This question echoes a controversy in the organic chemistry field. Almost everyone exposed to organic chemistry becomes familiar with describing reactions in terms of a series of hopping arrows,

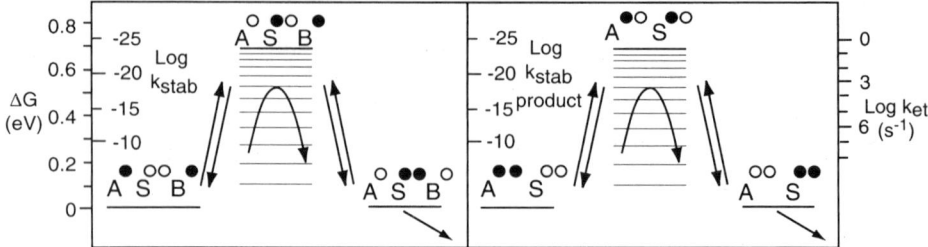

**FIGURE 9.** A general picture of redox clusters in which a substrate (S) capable of net two electron oxidation or reduction is in proximity to either two single electron redox centers (A and B in left panel) or a single center with two redox couples (A in right panel). The figure illustrates a typical biological case in which the net free energy released upon reduction of substrate (symbolized as the filling of open circles) is relatively small. Endergonic electron transfer to the unstable radical intermediate state on the substrate is followed by a second electron transfer to product. The endergonic tunneling reaction will often be rate limiting leading to the relation between $\Delta G$ for the endergonic step (left hand scale), and the overall rate of reduction of the substrate (right hand scale).

which shows the two-electron bond rearrangements that occur during a reaction. This two-electron viewpoint is a powerful way of organizing the variety of chemical reactions. Yet others, including Eberson (Eberson, 1987) and Pross (Pross, 1985), have pointed out the utility of viewing these reactions in terms of two successive single electron transfers though a radical intermediate state. Indeed, 50 years ago, Michaelis pointed out that biological oxidation and reduction of substrates takes place by means of single electron transfers, and that the role of protein is to manage the properties of the intermediate radical state (Michaelis, 1951).

The observation that significantly endergonic electron transfers are reasonable in oxidoreductase chains suggests that the single electron tunneling viewpoint described here can readily be applied to two electron substrate oxidation and reduction through an endergonic intermediate radical state. Indeed, it appears that the engineering requirements for protein modification of substrate radical states often may be minimal. Figure 9 illustrates a general example, in which two single electron redox centers, or a single redox center with two redox couples are adjacent to a substrate that undergoes net two electron oxidation and reduction. For redox centers separated by about 4 Å in a cluster, endergonic electron transfer to a radical intermediate as much as 0.6 eV above the donor level will still be fast enough to maintain $k_{cat}$ in the typical millisecond to second range.

## 5.1. Cluster Examples

A specific example of the first type of cluster is provided by the reduction of oxygen in the heme $a_3$-$O_2$-Cu cluster of cytochrome oxidase (see chapter by S. Yoshikawa). Because the heme $a_3$ and Cu centers are less than 5 Å away from $O_2$ substrate (Yoshikawa et al., 1998; Ostermeier et al., 1997), the inherent tunneling rates are fast compared to $k_{cat}$, even for an endergonic tunneling of +0.6 eV or 14 kcal/mole. Thus the unmodified oxygen radical redox levels found in solution (−0.2 V, pH 7) (Stubbe and van der Donk, 1998) should be able to participate in a pair of electron transfers in less than the $5 \times 10^3$ s$^{-1}$ turnover time of the enzyme (Proshlyakov et al., 1998).

Instead of designing a cluster with two n = 1 redox centers in proximity to a substrate binding site, a single redox center with two different n = 1 redox couples can participate in an analogous reaction scheme (Figure 9 right panel). A specific example is provided by aldehyde dehydrogenase (see chapter by R. Hille), in which the Mo metal center is capable of successive II/III and III/IV transitions on the oxidation of aldehyde to ketone (Romao et al., 1995; Huber et al., 1996).

A variation of this design involves a center with two redox transitions that may not have a completely stable intermediate, for example in the oxidation of NADH oxidation by flavin, as in xanthine oxidase (see chapter by R. Hille). Although a single step hydride transfer is often assumed in these reactions, sequential single electron transfer steps through a radical intermediate are entirely reasonable. Indeed, the semiquinone flavin intermediate state need not even be entirely stable, as long as stability product (product of the NADH and flavin stability constants) is no smaller than $10^{-20}$, to achieve the millisecond rates often observed for these proteins. Using the reported aqueous NAD$^·$/NAD$^+$ redox couple of −0.92 V (Farrington et al., 1980), we estimate an aqueous stability constant of NAD$^+$/NADH of about $10^{-20}$. Thus, even a small amount of stabilization of the NAD$^·$ radical intermediate by the protein will permit unstable flavin radical intermediates to be used. As it turns out, flavin semiquinone stability constants in proteins are commonly in the $10^3$ to $10^{-3}$ range.

Protium/deuterium/tritium kinetic isotope effects are often used to support hydride transfer mechanisms over single electron transfer mechanisms. However, sequential electron/proton/electron transfer mechanisms can easily show isotope effects as well. Even though the rate limiting step in the overall two electron reduction of flavin or NADH may be the isotope independent endergonic electron tunneling to form a radical intermediate state, once formed, this radical state can return the electron to recreate the starting state, or engage in proton transfer before the subsequent second electron transfer to form the product. This intermediate proton transfer step, while not rate limiting, nevertheless provides the opportunity for isotope sensitivity of the overall two electron transfer reaction (Powell *et al.*, 1984).

If the instability of the free substrate radical state is inherently high, or if the reorganization energies are extremely high, then a direct hydride transfer must be considered (Eberson, 1987; Kohen *et al.*, 1999; Bahnson *et al.*, 1997). However, the design of a hydride transfer site is rather stringent, requiring not only redox center proximity, but also a special geometry that allows direct atom transfer. Blind evolution may more readily uncover cluster designs in which the redox level of the donor is moved closer to the radical intermediate, or the radical intermediate is moved closer to the donor to bring the energy gap into a thermally accessible range for sequential single electron transfer.

An intriguing example of just how much can be accomplished with the strategy of raising the donor level towards the radical transition state is provided by nitrogenase (Figure 10) for which structures are now available (Schindelin *et al.*, 1997). The simple tunneling theory we have described provides a straightforward explanation of why ATP must be hydrolyzed for each pair of electrons delivered to the nitrogen center, and highlights the requirement for three redox states of the FeMoco center which forms the cluster with $N_2$ (Figure 10). The overall six-electron reduction of $N_2$ to $2NH_3$ (and the obligatory two-electron reduction of $2H^+$ to $H_2$) occurs under

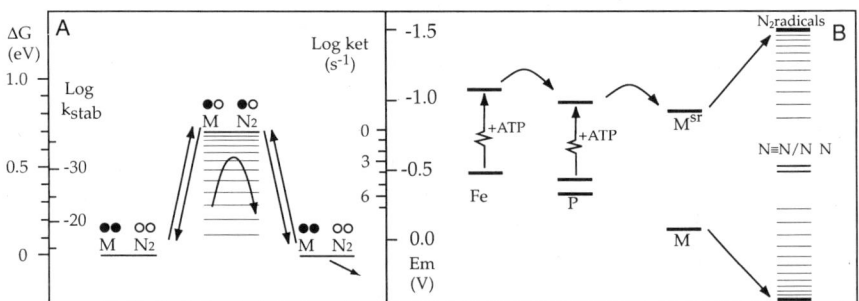

**FIGURE 10.** Electron transfer chains and catalytic cluster of nitrogenase. Binding of ATP has the effect of making the chain electron donors more negative so that a super reduced M center state is within thermal tunneling range of the $N_2$dot radical. Without a doubly reduced M center, there is no multielectron redox cluster including $N_2$, and $N_2$ reduction would not be possible.

reducing conditions (potential about −0.5 V) and is energetically favorable by −0.28 V. Nevertheless, ~4 ATP must be hydrolyzed per electron pair delivered. Two ATP binding sites have been identified on the protein containing a single 4Fe4S cluster (called Fe), but none on its partner MoFe protein, containing an 7Fe8S cluster (called P) and the FeMoco center (called M). ATP hydrolysis alters the conformation of the two proteins and lowers the redox midpoint potentials of the Fe cluster and one of the two P cluster levels from −0.30 V and −0.31 V to below −0.6 eV (Lanzilotta et al., 1998; Lanzilotta and Seefeldt, 1997). Surprisingly there is no measurable effect on the catalytic M level which remains at a relatively positive potential (−0.1 V), seemingly useless for $N_2$ reduction.

We suggest that 4 ATPs are hydrolyzed for each two electron reduction to establish the Fe and P potentials to low enough to doubly reduce the M center to the diamagnetic super-reduced state $M^{sr}$ (EPR signal disappears during turnover (Spee et al., 1998)). This redox state can then thermally access an unstable $N_2$ radical with a potential as low as −1.5 eV. Rapid subsequent electron transfer from the high potential M redox state creates a relatively stable nitrogen species. Re-reduction of the high potential M by P readies the system for protein dissociation and diffusive flavodoxin re-reduction of the Fe and P centers for the remaining cycles of nitrogen reduction.

### 5.2. Control of Electron Transfer through Escapement Mechanisms

When a redox center in a cluster has three or more redox states and is capable of transferring more than one electron, there is sometimes a great advantage in transferring the electrons to different redox partners. This may be important to assure efficient respiratory energy conversion, or to exploit special chemistry at the cluster site. However, the electrons are promiscuous, capable of tunneling in all directions in a protein and participating in physiologically productive as well as counter productive electron transfer reactions. There needs to be a mechanism to assure that the electrons don't transfer in the same direction, even if this act might be most favorable energetically. A simple redox chain may not confer directionality under these conditions, because chains generally catalyze both rapid forward and reverse electron transfer. Instead, Nature sometimes exploits the strong distance dependence of the electron tunneling rate by employing a restricted diffusive motion to move redox centers out of the way, often by 10 Å or more, after the first electron transfers. Some time is then available to complete the final electron transfers in the physiologically desirable direction, before the first redox center changes its redox state and returns for another electron transfer. We call this electron transfer direction controlling motion an escapement mechanism in analogy to the device in a pendulum clock (Figure 11).

The cytochrome $bc_1$ complex (see chapter by S. H. Kim) provides a clear example. The transfer of two electrons from reduced ubiquinone at the Qo site must follow two separate chains or conversion of redox energy into a transmembrane proton gradient will be defeated. One electron follows the chain of high potential redox centers FeS, cyt $c_1$ and cyt $c_2$ or $c_y$, while the other follows the low potential chain of cyt $b_L$ and $b_H$. In principle, FeS can be reduced by the first electron from Qo and then rapidly reoxidized by the next center in the chain, cyt $c_1$. FeS would then be available to accept the second electron from Qo, without reducing cyt $b_L$. Just as unproductive would be the return of the electron from cyt $b_L$, through the Qo radical state to the reoxidized FeS. However, it appears from crystal structures (Zhang *et al.*, 1998; Xia *et al.*, 1997) that the FeS subunit may move more than 10 Å away from the Qo site, which would slow electron tunneling dramatically and effectively insulate the high potential chain from Qo. Indeed, recent experiments have succeeded in controlling this motion and providing a direct measure of the movement time (Darrouzet *et al.*, 1999). This escapement action allows the second electron to continue down the low potential chain, charging the transmembrane electric field and reducing the second site quinone Qi leading to energy conserving proton uptake on the other side of the membrane.

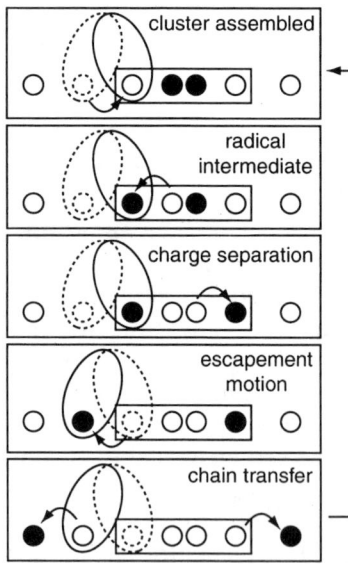

**FIGURE 11.** An escapement mechanism is sometimes used to control the direction of electron transfer within a redox cluster (small box). Here electron transfers from the substrate (close pair of circles filled with electrons) to a redox center on the left which is effectively insulated by distance from other members of a redox chain (further left) so only one electron can be transferred. The radical intermediate can transfer electrons to the chain on the right. The thermally activated escapement motion of the redox center then carries an electron to the chain at the left, and finally reassembles the cluster in preparation for the next catalysis.

Electron transfer escapement action need not be confined to separating two single electron transfers. In P450-bm3, flavin FMN is reduced by NADPH via FAD and delivers two electrons to the P450 heme which binds oxygen, leading to partial oxygen reduction (see chapter by A. Munro). In principle, two more electrons can travel down the NADPH/FAD/FMN chain and reduce $O_2$ completely to water. However, this would defeat the catalytic purpose of the enzyme: extraction of two electrons from substrate by the partially reduced $O_2$ to hydroxylate the substrate. Our analysis of a chimerical blend of two structures of cytochrome P-450-bm3 and cytochrome P-450 reductase (Wang *et al.*, 1997; Sevrioukova *et al.*, 1999) suggests an escapement motion of the FMN center returning to FAD to pick up the second set of electrons allows time for the extraction of electrons from the P450 substrate. Indeed, it seems that the FMN subunit is unusually flexible and a hinge region has been identified (Wang *et al.*, 1997).

Escapement mechanisms are real possibilities for any protein mechanism that uses oxidant induced reduction or reductant induced oxidation to control the direction of multiple electron transfer at cluster sites. Thus escapements may be active in mitochondrial Complex I (Dutton *et al.*, 1999; Dutton *et al.*, 1998). They should also be considered for anaerobic ribonucleotide reductase (see chapter by M. Sahlin and B-M. Sjöberg) and in methionine synthase (see chapter by N. Marsh).

## 6. CAUTION AND HOPE

There is a natural tendency for researchers to interpret whatever they learn about the protein system they are intensely studying as an important, perhaps evolutionarily refined purpose or function. We caution that electron transfer proteins appear to be generally designed as robust machines that are remarkably tolerant to changes in structural and energetic details. Natural electron tunneling is by design generally much faster than is necessary for the characteristic times of turnover of these enzymes. Control of the direction of electron transfer is primarily by control of distance to the nearest redox centers, hence the use of redox chains to guide electron transfer. Details of the intervening protein medium structure rarely if ever have a dramatic effect on electron tunneling. When distances to two possible redox centers are comparable, as they are in the initial charge separation in the reaction center, the direction of electron transfer can be effectively controlled by the secondary parameters $\Delta G$ and $\lambda$. This general robustness can simplify the task of understanding the essential design of large natural electron transfer proteins and allow the real possibility of redesigning biologically inspired electron transfer chemistry in simplified synthetic protein packages.

ACKNOWLEDGEMENTS. Financial support for work described in this chapter was provided by the National Institute of Health GM 41048.

## 7. REFERENCES

Albracht, S. P., 1994, Nickel Hydrogenases—In Search Of the Active-Site *Biochim. Biophys. Acta* **1188**:167–204.

Alegria, G., and Dutton, P. L., 1991, Langmuir-Blodgett monolayer films of bacterial photosynthetic membranes and isolated reaction centers: preparation, spectrophotometric and electrochemical characterization. I *Biochim. Biophys. Acta* **1057**:239–257.

Bahnson, B. J., Colby, T. D., Chin, J. K., Goldstein, B. M., and Klinman, J. P., 1997, A link between protein structure and enzyme catalyzed hydrogen tunneling *Proc. Natl. Acad. Sci. U. S. A.* **94**:12797–12802.

Beratan, D. N., Betts, J. N., and Onuchic, J. N., 1991, Protein Electron Transfer Rates Set by the Bridging Secondary and Tertiary Structure *Science* **252**:1285–1288.

Closs, G. L., Calcaterra, L. T., Green, N. J., Miller, J. R., and Penfield, K. W., 1986, Distance, Stereoelectronic Effects, and the Marcus Inverted Region in Intramolecular Electron-Transfer in Organic Radical-Anions *J. Phys. Chem.* **90**:3673–3683.

Coelho, A. V., Matias, P., Fulop, V., Thompson, A., Gonzalez, A., and Carrondo, M. A., 1997, Desulfoferrodoxin structure determined by MAD phasing and refinement to 1.9-angstrom resolution reveals a unique combination of a tetrahedral FeS4 centre with a square pyramidal FeSN4 centre *J. Biol. Inorg. Chem.* **2**:680–689.

Curry, W. B., Grabe, M. D., Kurnikov, I. V., Skourtis, S. S., Beratan, D. N., Regan, J. J., Aquino, A. J. A., Beroza, P., and Onuchic, J. N., 1995, Pathways, Pathway Tubes, Pathway Docking, and Propagators In Electron-Transfer Proteins *J. Bioenerg. Biomembr.* **27**:285–293.

Darrouzet, E., Valkova-Valchanova, M., Moser, C. C., Dutton, P. L., and Daldal, F., 1999, Uncovering the [2Fe2S] domain movement in cytochrome $bc_1$: implications for energy conversion and a target for antibiotics submitted.

Darwin, C., 1872, Origin of species by means of natural selection, Earlton House, New York.

Davidson, V. L., 1996, Unraveling the kinetic complexity of interprotein electron transfer reactions *Biochemistry* **35**:14035–14039.

Devault, D., 1980, Quantum mechanical tunnelling in biological systems *Quart. Rev. Biophys.* **13**:387–564.

Devault, D., and Chance, B., 1966, Studies of photosynthesis using a pulsed laser I. Temperature dependence of cytochrome oxidation rate in chromatium. Evidence for Tunneling *Biophysical J.* **6**:825–847.

Dohse, B., Mathis, P., Wachtveitl, J., Laussermair, E., Iwata, S., Michel, H., and Oesterhelt, D., 1995, Electron-transfer from the tetraheme cytochrome to the special pair In the *Rhodopseudomonas viridis* reaction-center—effect of mutations of tyrosine L162 *Biochemistry* **34**:11335–11343.

Dutton, P. L., and Moser, C. C., 1994, Quantum Biomechanics Of Long-Range Electron-Transfer In Protein—Hydrogen-Bonds and Reorganization Energies *Proc. Natl. Acad. Sci. U. S. A.* **91**:10247–10250.

Dutton, P. L., Moser, C. C., Sled, V. D., Daldal, F., and Ohnishi, T., 1998, A reductant-induced oxidation mechanism for Complex I *Biochim. Biophys. Acta* **1364**:245–257.

Dutton, P. L., Ohnishi, T., Darrouzet, E., Leonard, M., Sharp, R. E., Gibney, B. R., Daldal, F., and Moser, C. C., 1999, In Coenzyme Q: from molecular mechanisms to nutrition and health, Vol. in press (V. E. Kagan and P. J. Qinn, eds.) CRC Press, Boca Raton, Florida.

Eberson, L., 1987, Electron transfer reactions in organic chemistry, Springer-Verlag, New York.

Farrington, J. A., Land, E. J., and Swallow, A. J., 1980, The one-electron reduction potentials of NAD *Biochim. Biophys. Acta* **590**:273–276.

Franzen, S., Lao, K. Q., and Boxer, S. G., 1992, Electric-Field Effects On Kinetics Of Electron-Transfer Reactions: Connection Between Experiment and Theory *Chem. Phys. Lett.* **197**:380–388.

Giangiacomo, K. M., and Dutton, P. L., 1989, In Photosynthetic Reaction Centers, the Free-Energy Difference for Electron-Transfer Between Quinones Bound at the Primary and Secondary Quinone-Binding Sites Governs the Observed Secondary Site Specificity *Proc. Natl. Acad. Sci. U. S. A.* **86**:2658–2662.

Gunner, M. R., and Dutton, P. L., 1989, Temperature and -Delta-G-Degrees Dependence of the Electron-Transfer from Bph.- to Qa in Reaction Center Protein from Rhodobacter-Sphaeroides with Different Quinones As Qa *J. Amer. Chem. Soc.* **111**:3400–3412.

Gunner, M. R., Robertson, D. E., and Dutton, P. L., 1986, Kinetic Studies on the Reaction Center Protein form *Rps. Sphaeroides*: Temperature and Free Energy Dependence of Electron Transfer between various Quinones in the QA site and Oxidized Bacteriochlorophyll Dimer *J. Phys. Chem.* **90**:3783–3795.

Hopfield, J. J., 1974, Electron transfer between biological molecules by thermally activated tunneling *Proc. Natl. Acad. Sci. USA* **71**:3640–3644.

Huber, R., Hof, P., Duarte, R. O., Moura, J. J., Moura, I., Liu, M. Y., LeGall, J., Hille, R., Archer, M., and Romao, M. J., 1996, A structure-based catalytic mechanism for the xanthine oxidase family of molybdenum enzymes *Proc. Natl. Acad. Sci. U. S. A.* **93**:8846–8851.

Jia, Y. W., Dimagno, T. J., Chan, C. K., Wang, Z. Y., Du, M., Hanson, D. K., Schiffe, R. M., Norris, J. R., Fleming, G. R., and Popov, M. S., 1993, Primary charge separation in mutant reaction centers of *Rhodobacter capsulatus J. Phys. Chem.* **97**:13180–13191.

Jortner, J., 1976, Temperature dependent activation energy for electron transfer between biological molecules *J. Chem. Phys.* **64**:4860–4867.

Kisker, C., Schindelin, H., Pacheco, A., Wehbi, W. A., Garrett, R. M., Rajagopalan, K. V., Enemark, J. H., and Rees, D. C., 1997 Molecular basis of sulfite oxidase deficiency from the structure of sulfite oxidase *Cell* **91**:973–983.

Kohen, A., Cannio, R., Bartolucci, S., and Klinman, J. P., 1999, Enzyme dynamics and hydrogen tunnelling in a thermophilic alcohol dehydrogenase *Nature* **399**:496–499.

Kuki, A., and Wolynes, P. G., 1987, Electron tunneling paths in proteins *Science* **236**:1647–1652.

Labahn, A., Paddock, M. L., McPherson, P. H., Okamura, M. Y., and Feher, G., 1994, Direct Charge Recombination from D+QAQB- to DQAQB in Bacterial Reaction Centers from *Rhodobacter sphaeroides J. Chem. Phys.* **98**:3417–3423.

Lanzilotta, W. N., Parker, V. D., and Seefeldt, L. C., 1998, Electron transfer in nitrogenase analyzed by Marcus theory: Evidence for gating by MgATP *Biochemistry* **37**, 399–407.

Lanzilotta, W. N., and Seefeldt, L. C., 1997, Changes in the midpoint potentials of the nitrogenase metal centers as a result of iron protein-molybdenum-iron protein complex formation *Biochemistry* **36**, 12976–12983.

Levich, V. G., and Dogonadze, R. R., 1959, Teiriya bezizluchatelnikh electronnikh perekhodov mezhdu ionami v rastvorakh *Doklady Akad. Nauk. SSSR.* **124**:123–126.

Levitt, M., Gerstein, M., Huang, E., Subbiah, S., and Tsai, J., 1997, Protein folding: The endgame *Ann. Rev. Biochem.* 66:549–579.

Li, J. L., Gilroy, D., Tiede, D. M., and Gunner, M. R., 1998, Kinetic phases in the electron transfer from P+QA-QB to P+QAQB- and the associated processes in *Rhodobacter sphaeroides* R-26 reaction centers *Biochemistry* **37**:2818–2829.

Lin, X., Williams, J. C., Allen, J. P., and Mathis, P., 1994, Relationship between rate and free-energy difference for electron-transfer from cytochrome c(2) to the reaction center in *Rhodobacter sphaeroides Biochemistry* **33**:13517–13523.

Marcus, R. A., 1956, On the theory of oxidation-reduction reactions involving electron transfer: *I. J. Chem. Phys.* **24**:966–978.

Marcus, R. A., and Sutin, N., 1985, Electron transfers in chemistry and biology *Biochim. Biophys. Acta* **811**:265–322.

Meyer, T. E., Bartsch, R. G., Cusanovich, M. A., and Tollin, G., 1993, Kinetics of photooxidation of soluble cytochromes, hipip, and azurin by the photosynthetic reaction center of the purple phototrophic bacterium *Rhodopseudomonas viridis Biochemistry* **32**:4719–4726.

Michaelis, L., 1951, In The Enzymes, Chemistry and Mechanism of Action, Vol. 2 (J. B. Sumner and K. Myrback, eds.) Academic Press, Inc., New York, pp. 1–54.

Michel, H., Deisenhofer, J., and Epp, O., 1986, Pigment protein interactions in the photosynthetic reaction center from *Rhodopseudomonas viridis Embo J.* **5**:2445–2451.

Miller, J. R., Beitz, J. V., and Huddleston, R. K., 1984, Effect of free energy on rates of electron transfer between molecules *J. Amer. Chem. Soc.* **106**:5057–5068.

Moser, C. C., and Dutton, P. L., 1992 Engineering protein structure for electron transfer function in photosynthetic reaction centers *Biochim. Biophys. Acta* **1101**:171–176.

Moser, C. C., Keske, J. M., Warncke, K., Farid, R. S., and Dutton, P. L., 1992, Nature of biological electron-transfer *Nature* **355**:796–802.

Moser, C. C., Sension, R. J., Szarka, A. Z., Repinec, S. T., Hochstrasser, R. M., and Dutton, P. L., 1995, Initial charge separation kinetic of bacterial photosynthetic reaction centers in oriented Langmuir-Blodgett films in an applied electric field. *J. Chem. Phys.* **197**:343–354.

Nordlund, P., and Eklund, H., 1993, Structure and function of the *Escherichia coli* ribonucleotide reductase protein R2 *J. Mol. Biol.* **232**:123–164.

Okamura, M. Y., and Feher, G., 1992, Proton transfer in reaction centers from photosynthetic bacteria *Ann. Rev. Biochem.* **61**:861–896.

Onuchic, J. N., Beratan, D. N., Winkler, J. R. and Gray, H. B., 1992, Pathway analysis of protein electron-transfer reactions *Ann. Rev. Biophys. Biomol. Struct.* **21**:349–377.

Ortega, J. M., Drepper, F., and Mathis, P., 1999, Electron transfer between cytochrome c2 and the tetraheme cytochrome c in *Rhodopseudomonas viridis Photosyn. Res.* **59**:147–157.

Ostermeier, C., Harrenga, A., Ermler, U., and Michel, H., 1997, Structure at 2.7 A resolution of the *Paracoccus denitrificans* two-subunit cytochrome c oxidase complexed with an antibody FV fragment *Proc. Natl. Acad. Sci. U. S. A.* **94**:10547–10553.

Osyczka, A., Nagashima, K. V. P., Sogabe, S., Miki, K., Yoshida, M., Shimada, K., and Matsuura, K., 1998, Interaction site for soluble cytochromes on the tetraheme cytochrome subunit bound to the bacterial photosynthetic reaction center mapped by site-directed mutagenesis *Biochemistry* **37**:11732–11744.

Page, C. C., Moser, C. C., Chen, X., and Dutton, P. L., 1999, Natural engineering principles of electron tunnelling in biological oxidation-reduction *Nature* **402**:47–52.

Pan, L. P., Durham, B., Wolinska, J., and Millett, F., 1988, Preparation and characterization of singly labeled ruthenium polypyridine cytochrome c derivatives *Biochemistry* **27**:7180–7184.

Parson, W. W., Chu, Z. T., and Warshel, A., 1990, Electrostatic control of charge separation in bacterial photosynthesis *Biochim. Biophys. Acta* **1017**:251–272.

Parson, W. W., Chu, Z. T., and Warshel, A., 1998, Reorganization energy of the initial electron-transfer step in photosynthetic bacterial reaction centers *Biophysical Journal* **74**:182–191.

Powell, M. F., Wu, J. C., and Bruice, T. C., 1984, Ferricyanide oxidation of dihydropyridines and analogues *J. Am. Chem. Soc.* **106**:3850–3856.

Proshlyakov, D. A., Pressler, M. A., and Babcock, G. T., 1998, Dioxygen activation and bond cleavage by mixed-valence cytochrome *c* oxidase *Proc. Natl. Acad. Sci. U. S. A.* **95**:8020–8025.

Pross, A., 1985, The single electron shift as a fundamental process in organic chemistry: the relationship between polar and electron-transfer pathways *Acc. Chem. Res.* **18**:212–219.

Robertson, D. E., Farid, R. S., Moser, C. C., Urbauer, J. L., Mulholland, S. E., Pidikiti, R., Lear, J. D., Wand, A. J., DeGrado, W. F., and Dutton, P. L., 1994, Design and synthesis of multiheme proteins *Nature* **368**:425–432.

Romao, M. J., Archer, M., Moura, I., Moura, J. J. G., Legall, J., Engh, R., Schneider, M., Hof, P., and Huber, R., 1995, Crystal-structure of the xanthine oxidase-related aldehyde oxidoreductase from *D. gigas Science* **270**:1170–1176.

Rousset, M., Montet, Y., Guigliarelli, B., Forget, N., Asso, M., Bertrand, P., Fontecilla-Camps, J. C., and Hatchikian, E. C., 1998, [3Fe-4S] to [4Fe-4S] cluster conversion in *Desulfovibrio fructosovorans* [NiFe] hydrogenase by site-directed mutagenesis *Proc. Natl. Acad. Sci. U. S. A.* **95**:11625–11630.

Rova, U., Goodtzova, K., Ingemarson, R., Behravan, G., Graslund, A., and Thelander, L., 1995, Evidence by site-directed mutagenesis supports long-range electron-transfer in mouse ribonucleotide reductase *Biochemistry* **34**:4267–4275.

Schindelin, H., Kisker, C., Schlessman, J. L., Howard, J. B., and Rees, D. C., 1997, Structure of ADP x AIF$_4$(-)-stabilized nitrogenase complex and its implications for signal transduction *Nature* **387**:370–376.

Sevrioukova, I. F., Li, H., Zhang, H., Peterson, J. A., and Poulos, T. L., 1999, Structure of a cytochrome P450-redox partner electron-transfer complex *Proc. Natl. Acad. Sci. U. S. A.* **96**:1863–1868.

Shopes, R. J., Holten, D., Levine, L. M. A., and Wraight, C. A., 1987, Kinetics of oxidation of the bound cytochromes in reaction centers from *Rhodopseudomonas viridis Photosynth. Res.* **12**:165–180.

Siegbahn, P. E. M., 1998, Theoretical study of the substrate mechanism of ribonucleotide reductase *J. Amer. Chem. Soc.* **120**:8417–8439.

Smalley, J. F., Feldberg, S. W., Chidsey, C. E. D., Linford, M. R., Newton, M. D., and Liu, Y. P., 1995, The kinetics of electron-transfer through ferrocene-terminated alkanethiol monolayers on gold *J. Phys. Chem.* **99**:13141–13149.

Spee, J. H., Arendsen, A. F., Wassink, H., Marritt, S. J., Hagen, W. R., and Haaker, H., 1998, Redox properties and electron paramagnetic resonance spectroscopy of the transition state complex of *Azotobacter vinelandii* nitrogenase *FEBS Letters* **432**:55–58.

Stubbe, J., and van der Donk, W. A., 1998, Protein radicals in enzyme catalysis *Chem. Rev.* **98**:705–762.

Trapani, A. P., and Strauss, H. L., 1989, Orientation of NH3D+ in Tutton salt: equilibrium orientation and tunneling kinetics *J. Am. Chem. Soc.* **111**:910–917.

Volbeda, A., Charon, M. H., and Piras, C., Hatchikian, E. C., Frey, M. and Fontecillacamps, J. C., 1995, Crystal-structure of the nickel-iron hydrogenase from *Desulfovibrio gigas Nature* **373**:580–587.

Wang, M., Roberts, D. L., Paschke, R., Shea, T. M., Masters, B. S., and Kim, J. J., 1997, Three-dimensional structure of NADPH-cytochrome P450 reductase: prototype for FMN- and FAD-containing enzymes *Proc. Natl. Acad. Sci. U. S. A.* **94**:8411–8416.

Warncke, K., and Dutton, P. L., 1993, Influence of QA site redox cofactor structure on equilibrium binding, in situ electrochemistry, and electron transfer performance in the photosynthetic reaction center protein *Biochemistry* **32**:4769–4779.

Xia, D., Yu, C. A., Kim, H., Xia, J. Z., Kachurin, A. M., Zhang, L., Yu, L., and Deisenhofer, J., 1997, Crystal structure of the cytochrome $bc_1$ complex from bovine heart mitochondria [published erratum appears in Science 1997 Dec 19;278(5346):2037] *Science* **277**:60–66.

Yoshikawa, S., Shinzawa-Itoh, K., Nakashima, R., Yaono, R., Yamashita, E., Inoue, N., Yao, M., Fei, M. J., Libeu, C. P., Mizushima, T., Yamaguchi, H., Tomizaki, T., and Tsukihara, T., 1998, Redox-coupled crystal structural changes in bovine heart cytochrome *c* oxidase *Science* 280:1723–1729.

Zhang, Z. L., Huang, L. S., Shulmeister, V. M., Chi, Y. I., Kim, K. K., Hung, L. W., Crofts, A. R., Berry, E. A., and Kim, S. H., 1998, Electron transfer by domain movement in cytochrome $bc_1$ *Nature* **392**:677–684.

*Chapter 2*

# Flavin Electron Transfer Proteins

F. Scott Mathews, Louise Cunane, and
Rosemary C. E. Durley

## 1. INTRODUCTION

The vast majority of flavoenzymes catalyze oxidation-reduction reactions in which one substrate becomes oxidized and a second substrate becomes reduced and the isoalloxazine ring of the flavin prosthetic group (Figure 1) serves as a temporary repository for the substrate-derived electrons. The catalytic reaction can be broken conveniently into two steps, a reductive half reaction (from the viewpoint of the flavin) and an oxidative half reaction. The flavin ring has great utility as a redox cofactor since it has the ability to exist as a stable semiquinone radical. Thus, a flavoenzyme can oxidize an organic substrate such as lactate by removal of two electrons and transfer them as a pair to a 2-electron acceptor such as molecular oxygen, or individually to a 1-electron acceptor such as a cytochrome.

In the above example, transfer of a pair of electrons to molecular oxygen to form hydrogen peroxide during the oxidative half reaction defines an oxidase while transfer of single electrons to a hemoprotein

**F. SCOTT MATHEWS, LOUISE CUNANE, and ROSEMARY C. E. DURLEY**    Dept. of Biochemistry and Molecular Biophysics, Washington University School of Medicine, St. Louis, MO 63110, USA.
*Subcellular Biochemistry, Volume 35: Enzyme-Catalyzed Electron and Radical Transfer*, edited by Holzenburg and Scrutton. Kluwer Academic / Plenum Publishers, New York, 2000.

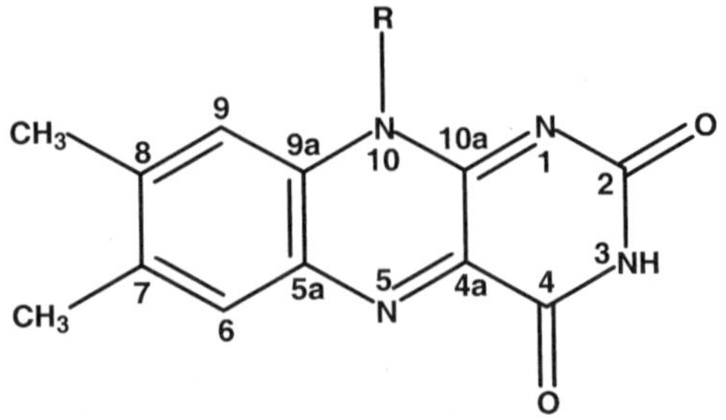

**FIGURE 1.** Isoalloxazine ring of FAD and FMN. The R represents either a ribityl phosphate group (FMN) or an adenosine ribityl phosphate group (FAD).

defines an electron transferase. There are many other 2-electron acceptors for flavoproteins. One example is glutathione or other disulfide-containing compounds, which defines the glutathione reductase family of flavoenzymes. A common substrate for the reductive half reaction of many flavoenzymes is NADH or NADPH which serves as a cellular source of electrons which can then be targeted to specific electron acceptor compounds or proteins by the flavoenzyme.

This chapter focuses on flavoenzyme electron transferases and, in particular, on those for which a crystal structure of the flavoprotein in complex with the electron acceptor partner is known. In most examples the electron acceptor is a 1-electron redox cofactor contained within another domain or subunit, but in one case the electron acceptor cofactor resides on a helix/loop motif adjacent to the domain containing the flavin. In addition to elucidating the specific chemical reactions catalyzed by them, the structures of these flavin electron-transfer complexes can provide valuable information about both the energetics of complex stabilization and the electron transfer process itself. This is important because long-range electron transfer in biological systems is a fundamental process. In the first case, the structure can define the interaction geometry of the protein components, indicating the features of the protein most pertinent for complex formation. In the second, it establishes the relative geometry of the redox centers and provides a detailed description of the intervening medium, both of which may bear on the electronic coupling between the two centers.

The structures of eight flavoprotein electron transfer complexes will be examined (Table 1). Four of these involve flavin to heme electron transfer, three involve electron transfer between flavin and an iron-sulfur center and one involves flavin to flavin electron transfer. These complexes provide a variety of domain types and arrangements, cofactor types and interdomain interactions that can help define the factors important for the electron

**Table 1**

**Electron donors and acceptors for the eight flavin-containing electron transfer complexes discussed. In the case of fumarate reductase, the electron donor and acceptor indicated in the table are for the reverse of the physiological reaction catalyzed**

| Complex | Electron Donor | Electron Acceptor | | Flavin Domain | | Primary Acceptor Domain | |
|---|---|---|---|---|---|---|---|
| | | Primary | Secondary | Fold | Cofactor | Fold | Cofactor |
| CPR | NADPH | FDX | P450 | FNR | FAD | FDX | FMN |
| BMP/FMN | NADPH | P450BMP | $O_2$ | FDX | FAD | BMP | Heme |
| FCB2 | Lactate | $CYTB_2$ | CYTC | TIM | FMN | CYTB5 | Heme |
| PCMH | p-Cresol | CYTC | Azurin | PCMH | FAD | CYTC | Heme |
| FCSD | $H_2S$ | $CYTC_4$ | $CYTC_{552}$ | GR | FAD | CYTC | Heme |
| TMADH | Trimethylamine | $Fe_4S_4$ | ETF | TIM | FMN | Loop-helix | $Fe_4S_4$ |
| PDR | NADPH | Fe2-S2 | PDO | FNR | FMN | FDN | $Fe_2$-$S_2$ |
| FUM | Fumarate | Fe2-S2 | Menaquinone | GR | FAD | FDN | $Fe_2$-$S_2$ |

transfer events. In Table 1 are summarized the electron donor and the primary and secondary electron acceptor of the eight flavin complexes discussed in this chapter. Also included are the domain topologies and the redox cofactors of the flavoprotein component and the primary electron acceptor for the eight complexes. In all but one case (fumarate reductase) the electron donor for the physiological reaction (*i.e.*, in the thermodynamically favored direction) is a small organic substrate. For fumarate reductase (FUM[1]), the organic substrate, fumarate, is actually the electron acceptor and the physiological direction of electron flow is opposite to that shown in Table 1. In all cases the primary electron acceptor (or donor) is a protein domain or structural motif carrying a redox cofactor. In all but two cases, there is a secondary electron acceptor (or donor) that is also a redox protein. The two exceptions are the expressed P450BM3 heme-binding and FMN-binding domain (BMP/FMN), where the secondary acceptor is molecular oxygen, and fumarate reductase, where the secondary electron donor (in the physiological direction) is a small electron carrier molecule, menaquinone.

There are five classes of flavin-binding structural folds presented in Table 1 that are identified by the prototype protein in which they were first discovered. These are flavodoxin (FDX), ferredoxin reductase (FNR), triosephosphate isomerase (TIM), glutathione reductase (GR) and *p*-cresol methylhydroxylase (PCMH). The topologies of four of these five domains are shown in Figure 2.[2] There are also four classes of primary acceptor/donor domain folds that are identified by the prototype protein where they were first discovered. They are cytochrome P450BMP (BMP), cytochrome b5 (CYTB5), cytochrome c (CYTC) and the 2Fe-2S plant-type ferredoxin (FDN).

## 2.  BACKGROUND AND STRUCTURAL PROPERTIES

### 2.1  NADPH-Cytochrome P450 Reductase (CPR)

Cytochrome P450 reductase is the physiological reductant of a wide range of microsomal cytochrome P450 monooxygenase isozymes that

---

[1] The abbreviations used in this chapter are: BMP, cytochrome P450 domain of P450BM3; BMP/FMN, expressed P450BM3 heme-binding and FMN-binding domain; CPR, NADPH-cytochrome P450 reductase; CYTB5, cytochrome b5; CYTC, cytochrome c; FCB2, flavo-cytochrome b₂; FCSD, flavocytochrome c sulfide dehydrogenase; FDN, ferredoxin; FDX, flavodoxin; FNR, ferredoxin reductase; FUM, fumarate reductase; GR, glutathione reductase; PCMH, *p*-cresol methylhydroxylase; PDR, phthalate dioxygenase reductase; TIM, triosephosphate isomerase; TMADH, trimethylamine dehydrogenase.
[2] For FNR, both the NADP and FAD-binding domains are shown.

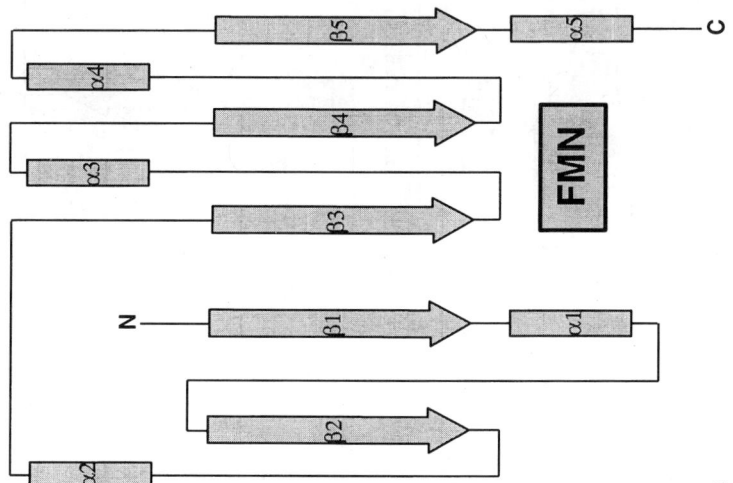

**FIGURE 2.** Folding motifs of the flavin-binding domains described in this chapter. Arrows represent β-strands and rectangles represent α-helices. The approximate locations of the flavin rings are also indicated in rectangles. (a) flavodoxin (FDX); (b) ferredoxin reductase (FNR); for this protein both the FAD- and NADP⁺-binding domains are shown. (c) $p$-cresol methylhydroxylase (PCMH); this domain consists of two subdomains each containing a mixed β-sheet with the ADP-ribityl group lying in a groove between the two subdomains; (d) glutathione reductase (GR).

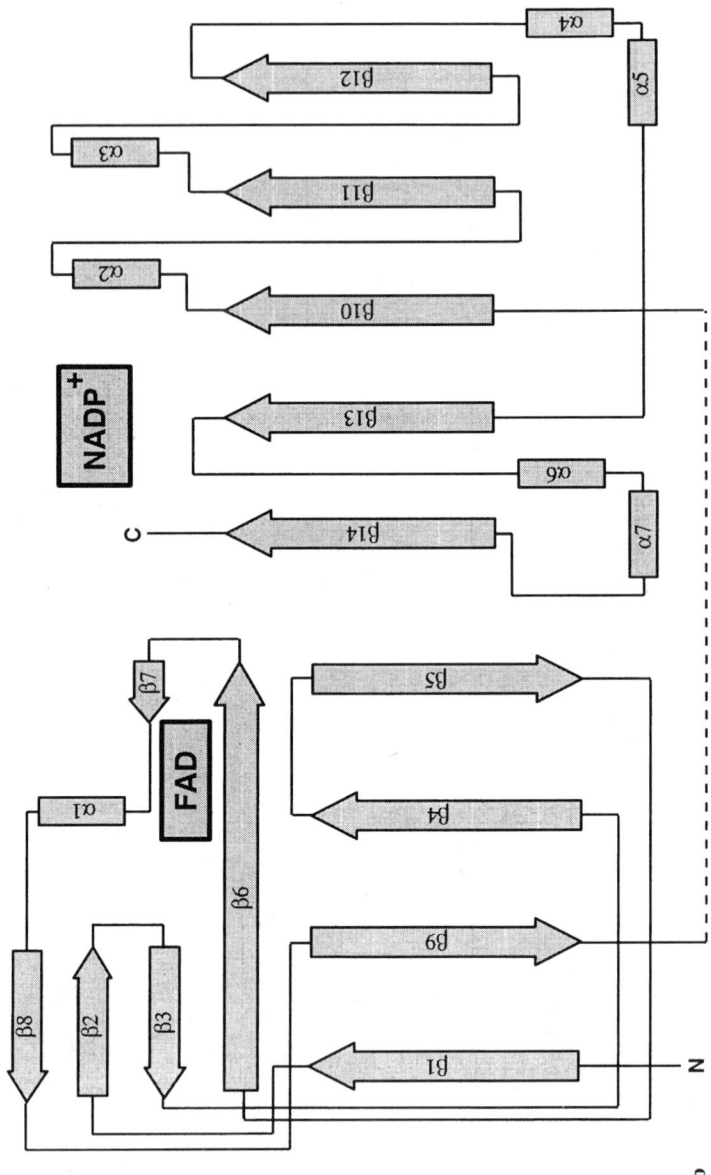

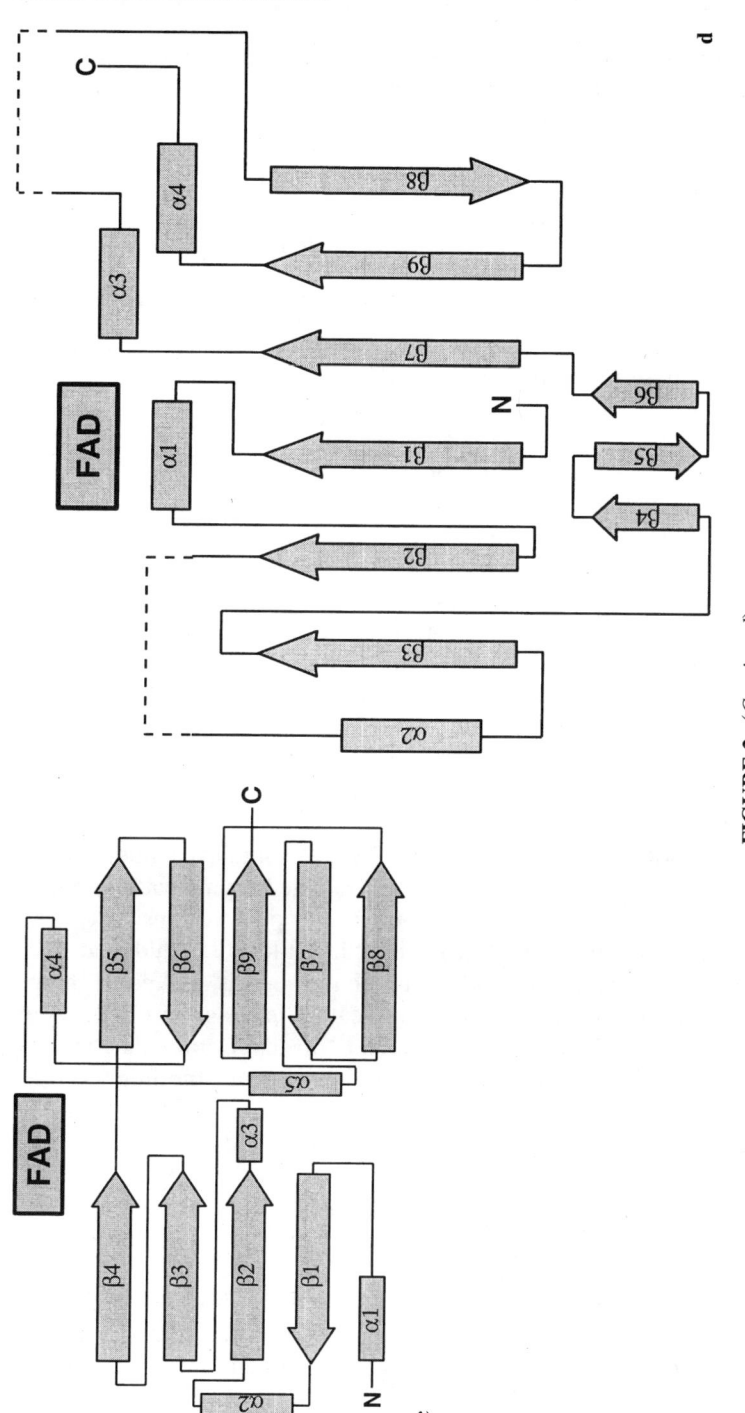

**FIGURE 2.** (*Continued*)

catalyze the oxidative metabolism of exogenous and endogenous organic compounds (Porter and Coon, 1991). Together, CPR and P450 catalyze the transfer of reducing equivalents from NADPH to molecular oxygen resulting in the insertion of an oxygen atom into the organic substrate. CPR contains both FAD and FMN as redox cofactors, a feature common to one other mammalian enzyme, nitric oxide synthase (Bredt *et al.*, 1991), and to the prokaryotic enzyme heterotetrameric sarcosine oxidase (Willie *et al.*, 1996).

During catalytic turnover, NADPH reduces FAD and the FAD subsequently reduces FMN by two electrons with the latter cycling between the hydroquinone and semiquinone states while carrying out the 1-electron reduction of the P450 heme iron (Masters *et al.*, 1966; Backes and Reker-Backes, 1988). The midpoint redox potentials of the FAD are $E'_{ox/sq}$ $\sim-290\,mV$ and $E'_{sq/red} \sim-365\,mV$ and of the FMN are $E'_{ox/sq} \sim-110\,mV$ and $E'_{sq/red} \sim-270\,mV$ (Iyanagi *et al.*, 1974). Electron transfer rates of FAD to FMN within CPR are difficult to measures and somewhat controversial and discussion of them is beyond the scope of this chapter.

Both the microsomal CPR and P450 are integral membrane proteins. CPR has two functional domains, an N-terminal 6 kDa hydrophobic domain that serves to anchor the enzyme to the endoplasmic reticulum, and a 72 kDa hydrophilic C-terminal catalytic domain (Kasper, 1971). In the absence of the N-terminal membrane anchor, which can be cleaved by proteolysis or omitted from the expression system in *E. coli*, CPR is incapable of reducing P450 but is able to reduce cytochrome c and several artificial electron acceptors.

The 2.6 Å resolution structure of the 72 kDa soluble fragment of rat liver CPR was determined by Wang *et al.* (1997). The enzyme was expressed in *E. coli* and rendered soluble by immobilized trypsin. It consists of four structural domains preceded by a short disordered loop (as deduced from N-terminal sequence analysis), indicating that there is a flexible hinge between the membrane anchor and the catalytic fragment. The four domains consist of an FMN-containing flavodoxin-like domain, a connecting domain, an FAD-binding domain and an NADP⁺-binding domain (Figure 3). The flavodoxin-like domain consists of a 5-stranded parallel β-sheet flanked by two or three α-helices on each side (Figure 2a). The NADP⁺-binding domain also contains 5 parallel β-strands flanked by α-helices (Figure 2b), but with a slightly different topology from the FMN domain. The FAD-binding domain is a mostly antiparallel β-barrel of 8 strands (Figure 2b). Together, the NADP⁺-binding and FAD-binding domains are structurally homologous to ferredoxin-NADP⁺ reductase (FNR; see Figure 2b). The FAD-binding domain is interrupted in the middle to form the majority (5/6) of the connecting domain. The remainder of the

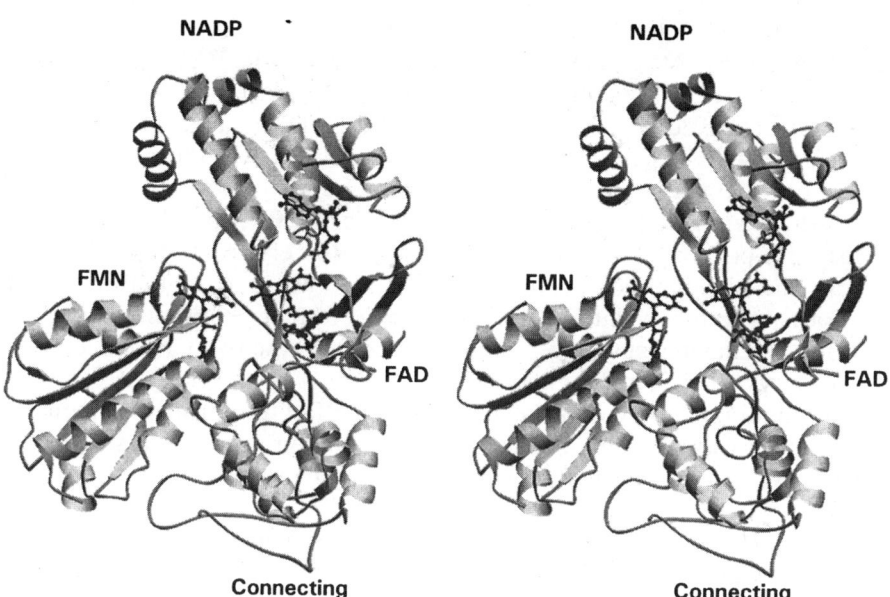

**FIGURE 3.** Stereo diagram of cytochrome P450 reductase. The Connecting, FMN-, FAD- and NADP-binding domains are indicated and the FMN, FAD, and ADP-phosphate prosthetic groups are shown in their respective domains. This and other ribbon diagrams were generated using RIBBONS (Carson, 1997).

connecting domain consists of a 30-residue segment between the FMN and the FAD domains, part of which forms a poorly ordered hinge which may enable the connecting domain to modulate electron transfer between the two flavins. The interface between the FMN and the connecting domain is largely hydrophilic and covers an area of ~880 Å² per domain. It contains 3 salt bridges and 3 hydrogen bonds. The interface between the FMN and FAD/NADPH-binding domains is smaller, ~400 Å², and is held together by 2 salt bridges. Three ordered water molecules are also present in the interface. The connecting domain and the FAD/NADPH-binding domains are partially intertwined and their interfaces are mostly hydrophobic, suggesting the latter three domains could move as a rigid unit with respect to the FMN-binding domain (Wang *et al.*, 1997).

The FMN is inserted into the flavodoxin-like domain with the pyrimidine portion of the isoalloxazine ring hydrogen bonded to the polypeptide chain and the dimethylbenzene portion exposed to the domain surface (Figure 3). The isoalloxazine ring of FAD is inserted into the boundary between the FAD and NADP$^+$ domains with the pyrimidine portion hydrogen bonded to protein side and main chain atoms of the FAD domain, and the dimethylbenzene moiety exposed near the interface between the FAD and connecting domains. Both the FMN and FAD isoalloxazine rings are sandwiched between aromatic side chains, two tyrosines in the case of FMN and a tyrosine and a tryptophan in the case of FAD.

Although NADP$^+$ is bound to the crystals of CPR, only the adenosine phosphate portion is well ordered (Figure 3), the electron density for the ribose and nicotinamide portions being weak. This suggests that multiple conformations for the latter exist. The shortest distance between the nicotinamide and flavin rings ranges from 9–14 Å within the two CPR molecules in the crystallographic asymmetric unit (Wang *et al.*, 1997). However, by reorienting the pyrophosphate group of NADP$^+$ through computer modeling, it was possible to place the nicotinamide ring in contact with the flavin ring of FAD after displacing the tryptophan ring stacked over the latter.

In CPR, FMN and FAD approach each other in an end-to-end manner through the dimethylbenzene portions of their flavin rings (Figure 3). The closest approach distance is 3.5 Å between the C-7 methyl groups and 4.5 Å between the C-8 methyl groups. The two isoalloxazine rings make an angle of ~30° to each other and together form a continuous ribbon. There are no protein side chains mediating the interaction of the two flavin rings, suggesting that electron transfer occurs directly between the two cofactors, which should be rapid considering their close contact distance.

## 2.2.   Flavocytochrome P450BM3

In contrast to mammalian microsomal P450s that require the membrane-bound CPR flavoprotein for reduction of the heme group during its catalytic cycle, most prokaryotic P450s receive electrons from small soluble iron-sulfur proteins such as putidaredoxin. However, the bacterium *Bacillus megaterium* contains a soluble, monomeric NADPH/P450 enzyme system of 119 kDa called flavocytochrome P450BM3 (Fulco, 1991). This complex flavocytochrome exists as a single intact polypeptide consisting of two major functional domains, a heme-containing P450 domain (BMP) and an FMN/FAD-containing reductase domain. The individual domains and subdomains of P450BM3 have been expressed and used to study the mechanism of domain-domain interaction and interdomain electron transfer within the system (Boddupalli *et al.*, 1992). In both the microsomal P450 system and the P450BM3 molecule, reducing equivalents from NADPH are transferred first to FAD and then to FMN within the FMN/FAD functional domain and from there to the P450 moiety via the FMN.

The midpoint potentials for the FAD, FMN and heme of P450BM3 have been measured, both in the intact flavocytochrome and in the separately-expressed domains (Daff *et al.*, 1997). For the FAD, $E'_{ox/sq}$ ~−285 mV while $E'_{sq/red}$ ~−345 mV. For the FMN, $E'_{ox/sq}$ ~−210 mV while $E'_{sq/red}$ ~−180 mV. For BMP, the midpoint potential for the unliganded heme, $E'_{Fe}$, is ~−370 mV, while the potential of the substrate-bound heme (arachadonic acid), $E'_{Fe}$, is ~−265 mV. As opposed to CPR, the FMN cycles between the oxidized and semiquinone states during catalytic turnover.

The low midpoint potential of the substrate-free P450 heme prevents electron transfer to it from the NADPH-reduced flavins in the absence of substrate, thereby avoiding wasteful cycling of electrons and loss of reducing equivalents. Only when substrate is present can electrons flow to the P450 heme and begin the cycle of heme reduction, oxygen binding, further reduction of the ferrous-oxy intermediate and oxygen incorporation into substrate.[3] This tightly coupled redox control of P450 activity is common to all P450s including the bacterial and microsomal enzymes.

The electron transfer rates in P450BM3, measured by laser flash photolysis using semicarbazide-activated 5-deazflavin semiquinone, show that no reduction of the native BMP heme occurs even though FMN could be reduced rapidly to the semiquinone (Hazard *et al.*, 1997). In the presence of carbon monoxide, which can displace water from the sixth coordination site of iron and convert the low-spin ferric iron to high spin, the intramolecular electron transfer rate is $18\,sec^{-1}$. In the presence of both CO and the substrate myristic acid, an intramolecular electron transfer rate of up to $250\,sec^{-1}$ can be obtained.

The best model for studying the intermolecular electron transfer step from FMN to heme in the P450BM3 system appears to be BMP/FMN which is the partial P450BM3 construct containing the BMP and FMN domains but lacking the BM3/FAD domain (Hazzard *et al.*, 1997; Sevrioukova *et al.*, 1997). The crystal structure of BMP/FMN has been determined at 2.03 Å resolution (Sevrioukova *et al.*, 1999a). It was found that during crystallization a 30-residue linker peptide had been cleaved, so that the structure actually consists of an intermolecular complex between BMP (residues 20–458) and BM3/FMN (residues 479–630) (Figure 4). The BMP structure is virtually identical to that of the isolated BMP molecule (Ravichandran *et al.*, 1993). The P450 BMP subunit folds into two domains, $\alpha$ and $\beta$. The $\alpha$-domain is the larger (~75% of the subunit) and is almost entirely helical. It

---

[3] Although the measured $E'_{Fe}$ for FMN is higher than the potential of the substrate-bound heme, reduction of the heme by the FMN semiquinone is observed experimentally, suggesting that some shift of relative midpoint potentials occurs during turnover.

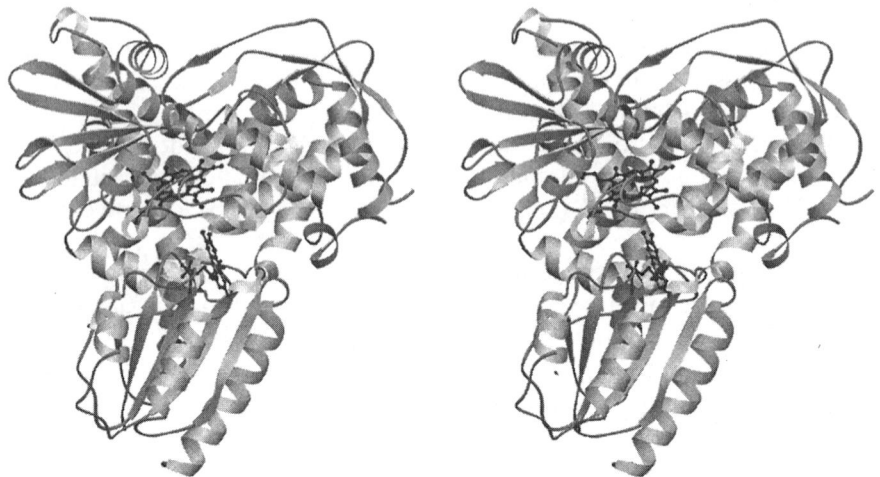

**FIGURE 4.** Global stereo diagram of the BMP/FMN complex.

contains a 4-helix bundle in the center and two pairs of helices perpendicular to the bundle at the top and the bottom. The β-domain (~25% of the subunit) is mostly β-structure with a 5-stranded mixed β-sheet and two short helices. The subunit as a whole is shaped like a disk, about 65 Å in diameter and about 35 Å thick. The heme is centered in the disk and is totally shielded from solvent, about 10 Å below the proximal surface. It is wedged between two of the helices of the 4-helix bundle and is coordinated on the proximal side by a cysteine side chain, Cys400. On the distal side, the heme group is about 20 Å from the surface and the iron is coordinated by a molecule of ethylene glycol, one oxygen of which is 2.1 Å from the iron. The entry channel for substrate, usually a lipophilic organic molecule, is contained in the β domain, and is lined with non-aromatic hydrophobic side chains.

The BM3/FMN domain is similar in fold to bacterial flavodoxins and to the FMN domain of CPR (Figure 2a). It consists of 5 parallel β-strands that form a β-sheet and four α-helices grouped in pairs on either side of the β-sheet. The FMN is bound to the flavodoxin domain so that the phosphate group of FMN is tucked into the loop between strand β1 and the following helix, α1 (Figure 2a). The FMN is oriented so that only the dimethylbenzene portion of the flavin ring (Figure 1) is exposed to solvent, the remainder being buried within the domain interior.

The interaction of the BM3/FMN domain with the BMP domain involves four segments of the former and three segments of the latter (Figure 5). The largest such segment of BM3/FMN is the β1 to α1 loop and

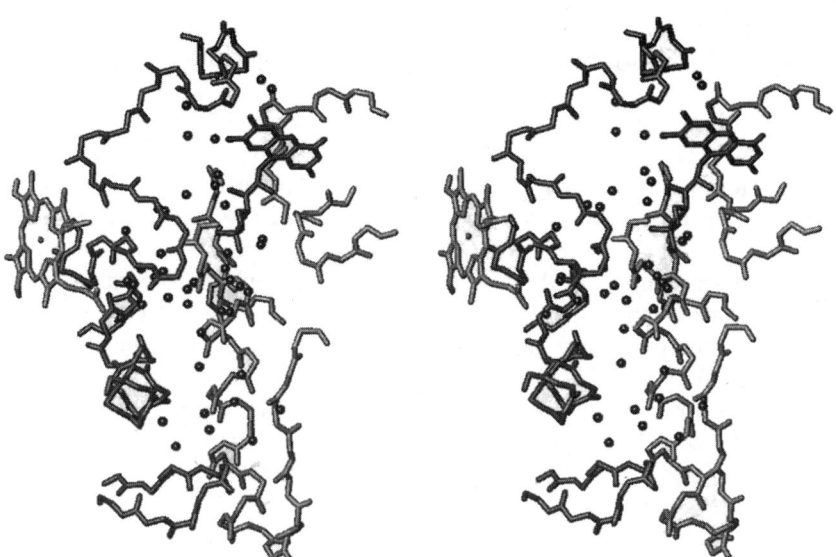

**FIGURE 5.** Stereo close-up view of the interdomain interface of the BMP/FMN complex. Helix C, the H/I loop and the helix K′-K/L-loop of BMP and the FMN of BM3/FMN are shown in darker lines. The β1-α1 helix-loop, N-terminal portion of α2, the β4/α4 and β5/α5 loops of BM3/FMN and the heme of BMP are shown in lighter lines. The locations of ~40 water molecules within the interface are shown as dark spheres.

helix (Figure 2a) which interacts with the H/I loop (between helices H and I), helix C and part of helix K′ and the K′/L loop (which includes the heme ligand, Cys400, near its C-terminus) of BMP (Ravichandran *et al.*, 1993). One end of helix $\alpha_5$ and the $\beta_4/\alpha_4$ turn of BM3/FMN interact with loop K′/L and helix K′ of BMP, respectively, while $\beta_4/\alpha_4$ of BM3/FMN interacts with part of helix C of BMP (Figure 5). The peptide segment 381–389 in the K′/L loop of BMP is characteristic of eukaryotic P450s and corresponds to an insertion in the prokaryotic P450 sequences and is found only in BMP among the prokaryotic class of P450s. The heme and flavin rings are nearly perpendicular to each other and the C7 methyl of FMN is about 18 Å from the heme iron atom. The solvent-exposed dimethylbenzene portion of FMN (Figure 1) is directed into the domain interface which contains approximately 40 water molecules (Figure 5). The two domains are linked by two direct hydrogen bonds and about ten water-mediated hydrogen bonds. The hydrogen bonds and seven of the ten bridging water molecules are located in the large helix C to (β1/α1) segment of the interface while three of the water-mediated hydrogen bonds are near the K′/L loop and the flavin ring. The surface area of the interface between the two domains is about 750 Å$^2$

per domain. Trp574 of BM3/FMN is stacked against the isoalloxazine ring of FMN at 3.3 Å and is involved in a charge-transfer interaction with the flavin, serving to shield the flavin ring from solvent.

Mutational analysis of electron transfer within BMP/FMN has been carried out using laser flash photolysis (Sevrioukova *et al.*, 1999b) to test the interaction site of the two domains identified in the crystal structure (Figure 5). Three positions in the BMP domain, Leu104, Glu372 and Gln387 were mutated to Cys and each of the new cysteine sites was then dansylated with the bulky sulfhydryl reagent 1-dimethylamino-5-sulfonate-L-cysteine. Leu104 and Gln387 are located on helix C and on the K'/L loop of BMP, respectively, and are at the domain interface in the crystal structure (Figure 5). Glu387 is distant from the interface and served as a control. The first order electron transfer rates of the wild type and dansylated Glu372Cys and Gln387Cys mutants, measured in the presence of the substrate arachidonic acid, were about $500 \, sec^{-1}$ whereas the rate for the Leu104Cys mutant was about 20-fold lower. These results are consistent with the tight contact between domains near helix C and the looser, more solvent-exposed contact near the K'/L loop of BMP within the complex (Figure 5).

The structures of CPR and P450/FMN are inconsistent with each other. This is because the FMN domain utilizes the same portion of its surface for its interaction with the FAD/NADPH domain in the CPR structure and with the BMP domain in the BMP/FMN structure. The FAD/FMN domain of P450BMP (66 kDa) is homologous to CPR (72 kDa) with ~35% sequence identity. This implies that their three dimensional structures are quite similar. If the FMN domains of CPR and BMP/FMN are superimposed, the partner domains in the two structures clash severely, indicating that the FMN domains cannot interact with both the P450 domain and the FAD/NADPH-binding domains at the same time without a major rearrangement. In fact, for the FMN domain to function catalytically in passing electrons from NADPH to the P450 heme, in either the microsomal or the BM3 system, a very large domain movement of nearly 20 Å is required. This suggests that domain mobility and large conformational changes accompany their physiological activity and indicates that the association of the FMN domain within each complex is weak.

## 2.3. Flavocytochrome b$_2$ (FCB2)

FCB2 from *Saccharomyces cerevisiae* catalyzes the oxidation of lactic acid to pyruvic acid with subsequent transfer of two electrons to cytochrome c (Lederer, 1991). The enzyme is located in the intermembrane space of yeast mitochondria and is part of an independent branch of the electron transport chain, involving cytochrome $c$ and cytochrome oxidase, which enables the organism to grow on lactate alone. FCB2 is a 230 kDa homotetramer; each 57.5 kDa monomer contains one FMN and one heme prosthetic group, both bound noncovalently.

The enzyme carries out three catalytic functions, oxidation of L-lactate to pyruvate by FMN, reoxidation of FMN by heme $b_2$ and oxidation of the $b_2$-heme by cytochrome c. The overall reaction occurs in 5 separate electron transfer steps and proceeds at a steady state turnover rate of ~100 molecules of lactate sec$^{-1}$ which corresponds to a net transfer of 200 electrons sec$^{-1}$ (Figure 6; Lederer, 1991; Daff *et al.*, 1996). The first step is the two-electron reduction of the flavin by lactate. This step occurs at a rate of ~600 sec$^{-1}$ under pre-steady state conditions and its rate is limited by $\alpha$-proton abstraction (Miles *et al.*, 1992). The second step is the 1-electron transfer from the flavin hydroquinone to the $b_2$ heme. This step occurs at a rate of ~1900 sec$^{-1}$ based on laser flash photolysis studies using 5-deazariboflavin as an oxidant of fully reduced enzyme (Hazzard *et al.*, 1994) and at a similar rate (~1500 sec$^{-1}$) based on temperature jump experiments (Tegoni *et al.*, 1998). Steps 3 and 5 involve transfer of 1-electron from the reduced $b_2$-heme to cytochrome c. This is a bimolecular reaction dependent on the rate of complex formation between FCB2 and cytochrome c and occurs within the complex at a rate greater than 1000 sec$^{-1}$ (Daff *et al.*, 1996b). Step 4, the 1-electron reduction of the $b_2$-heme by the flavin semiquinone, is the slowest step, occurring at a rate of ~120 sec$^{-1}$ according to

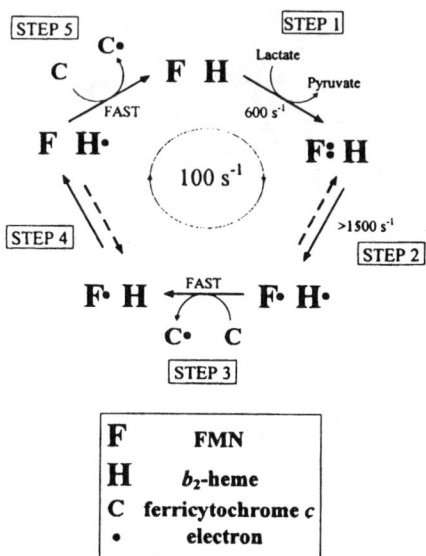

**FIGURE 6.** Schematic representation of the catalytic reaction cycle in flavocytochrome b2. Five redox intermediates of FCB2 during the oxidation of one molecule of lactate at a steady-state turnover rate of 100 sec$^{-1}$ and the reduction of two molecules of cytochrome c at the rate of 200 sec$^{-1}$ are shown. Step 4 is the rate limiting step in the steady state and the maximal rates of some of the other electron transfer steps are indicated. Reproduced from Daff *et al.*, 1996 with permission.

stopped-flow experiments (Daff *et al.*, 1996a) and ~200 sec$^{-1}$ based on temperature jump experiments (Tegoni *et al.*, 1998). Thus, the steady-state rate of lactate oxidation appears to be limited by this fourth electron transfer step. The differences in the rates of steps 2 and 4 appear to be correlated with the relative midpoint redox potentials of the flavin and heme prosthetic groups. Based on potentiometric titrations utilizing EPR (Tegoni *et al.*, 1998), these values are $E'_{ox/sq}$ ~$-135$ mV, $E'_{sq/red}$ ~$-45$ mV and $E'_{Fe}$ ~$0$ mV.

The structure of FCB2 has been determined at 2.4 Å resolution (Xia *et al.*, 1990). The four subunits are related by circular 4-fold molecular symmetry. Each subunit is composed of two domains (Figure 7). The N-terminal cytochrome domain (11 kDa) and the C-terminal flavin-binding domain (47 kDa) are connected by a short flexible hinge. In the crystals, two of the four cytochrome domains are disordered, whereas the four flavin-binding domains are well ordered. The two subunits lacking the ordered cytochrome domain each contain one molecule of pyruvate bound in the active site adjacent to the flavin ring. The flavin-binding domain is com-

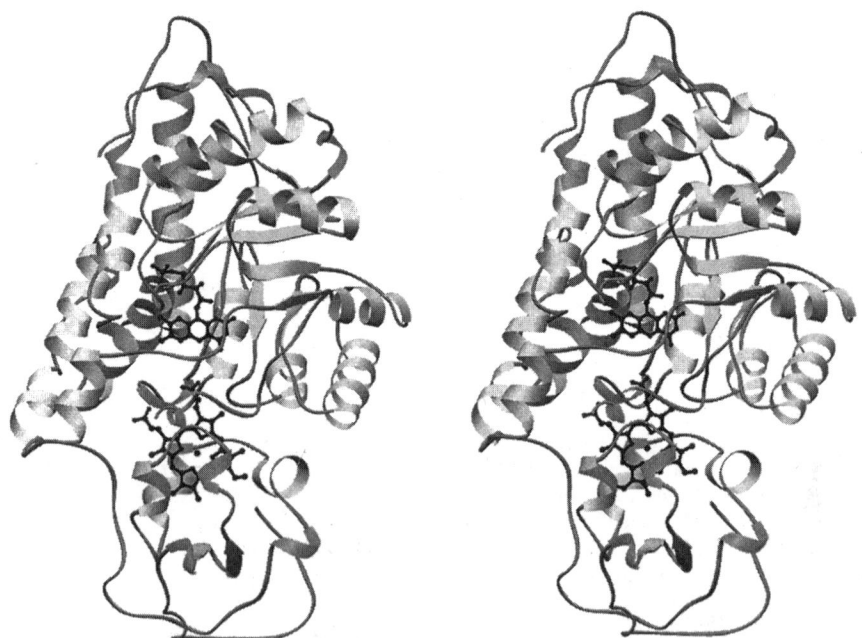

**FIGURE 7.** Stereo diagram of one subunit of flavocytochrome b₂. Only residues 1–486 are shown, the remainder being involved in intermolecular interactions. The flavin-binding domain is at the top and the cytochrome domain is at the bottom. The flavin and heme groups are shown as skeletal models.

posed of a parallel $\beta_8\alpha_8$ TIM barrel motif with the FMN located at the C terminal end of the central $\beta$-barrel. The cytochrome domain is very similar in conformation to microsomal cytochrome $b_5$.

The heme and flavin groups are nearly coplanar. The two prosthetic groups are oriented with the heme propionate groups pointing toward the N5-containing edge of the flavin. The distance from the iron atom to the center of the flavin ring is about 15 Å; the pyrrole and isoalloxazine rings are separated by about 14 Å and the heme propionate to flavin N5 distance is about 5 Å. The contact surface between the cytochrome domain and the flavin-binding domain is largely hydrophobic, although there are six direct hydrogen bonds, one salt bridge and five water molecules forming hydrogen bonding bridges between them; the interface between the two domains occupies about 900 Å$^2$ per domain. The interaction between the cytochrome and the flavin binding domains appears to be weak, however, since the flavoprotein and cytochrome domains, when separated by controlled proteolysis, no longer associate with one another (Gervais, 1983).

### 2.4.  *p*-Cresol Methylhydroxylase (PCMH)

PCMH is a flavocytochrome *c* localized in the periplasmic space of several types of Pseudomonad (Hopper and Taylor, 1977). It catalyzes the oxidation of *p*-cresol first to *p*-hydroxybenzyl alcohol and then to *p*-hydroxybenzaldehyde. Electrons are passed sequentially to the endogenous cytochrome subunit and then to an exogenous secondary electron acceptor protein, possibly an azurin or another cytochrome (Causer *et al.*, 1984).

PCMH is a 116 kDa heterotetramer (Shamala *et al.*, 1986) containing two flavoprotein subunits (49 kDa each) and two c-type cytochrome subunits (9 kDa each) (Kim *et al.*, 1994). The flavoprotein subunit binds FAD covalently through an 8-α-methyl-O-tyrosyl linkage (McIntire *et al.*, 1981). The flavoprotein and cytochrome subunits are tightly associated, having a dissociation constant of ~1 nmolar. The two types of subunits differ by 0.5 units in pI and can be resolved by isoelectric focusing (Koerber *et al.*, 1985); recombining the subunits leads to recovery of the fully active flavocytochrome. The isolated flavoprotein is dimeric and retains about 2% of the native catalytic activity towards *p*-cresol. The isolated cytochrome subunit is monomeric and its redox potential is about 70 mV lower than its value, 250 mV, in the native complex. The redox potential of the flavin is not known.

The rate of intramolecular electron transfer from the flavin to the heme in PCMH has been measured by laser flash photolysis using 5-deazariboflavin semiquinone radical as a 1-electron reductant (Bhattacharyya, 1985). Partial 1-electron reduction of heme and flavin (in different molecules) occurs rapidly as a bimolecular reaction. The reduced flavin is then reoxidized considerably more slowly concomitant with additional heme reduction, both occurring in a first order process with a rate constant of 220/s.

The structures of PCMH and of its complex with the substrate, *p*-cresol, have been determined to 2.5 Å and 2.75 Å, respectively (Cunane *et al.*, 2000). The flavoprotein subunit contains three domains, one of which comprises a newly-discovered FAD-binding motif (Figure 2c) shared by several other homologous proteins (Fraaije *et al.*, 1998).

The cytochrome subunit of PCMH is located in an indentation in the flavoprotein surface, with the vinyl-containing edge of the heme in contact with the flavoprotein surface (Figure 8). The heme iron is about 18 Å from the flavin ring and the closest edge-to-edge distance of the heme and flavin rings is about 11 Å. The benzenoid portion of the isoalloxazine ring (Figure 1) is oriented toward the thioether-containing edge of the heme group. The flavin and heme planes make an angle of about 65° to each other. The flavoprotein-cytochrome interface covers an area of approximately 1000 Å$^2$ per subunit. There are six hydrogen bonds and one salt bridge linking the two subunits as well as seven additional water-mediated hydrogen bonds. Approximately half the residues in the interface are

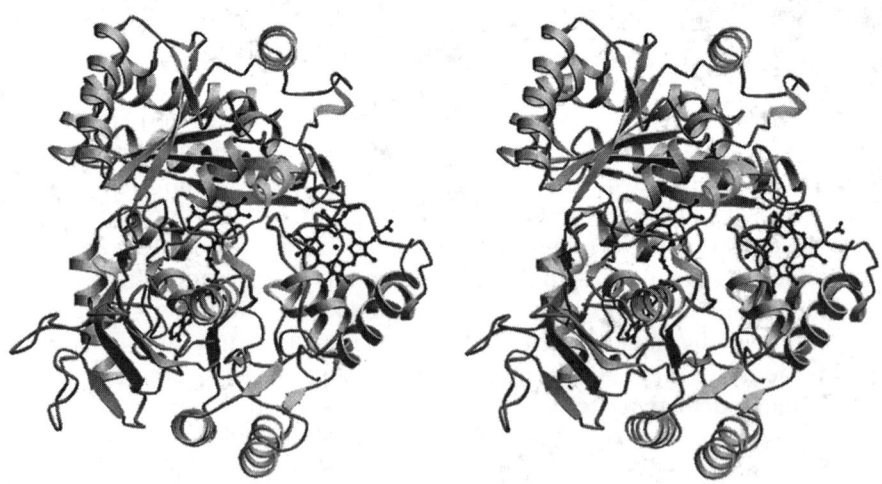

**FIGURE 8.** Stereo diagram of *p*-cresol methylhydroxylase. The flavoprotein subunit is on the left and the cytochrome subunit is on the right. The flavin-binding domain of the flavoprotein subunit is on the bottom and the catalytic domain is on the top. Skeletal models of the heme and FAD prosthetic groups are also shown.

hydrophobic and another 30% are neutral hydrophilic. This large interface and strong subunit interaction may be responsible in part for the high stability of the complex.

## 2.5. Flavocytochrome c Sulfide Dehydrogenase (FCSD)

FCSD is a periplasmic enzyme found in a number of phototrophic bacteria, as well as in *Paracoccus denitrificans*, that catalyzes the oxidation of sulfide to elemental sulfur (Cusanovich *et al.*, 1991; Wodara *et al.*, 1997). FCSD from *Chromatium vinosum* is a 67 kDa heterodimer consisting of a 46 kDa flavoprotein subunit and a 21 kDa diheme cytochrome. The secondary electron acceptor is probably a cytochrome (Gray and Knaff, 1982). The FAD is bound covalently to the flavoprotein subunit via an 8-α-methyl(S-cysteinyl)thioether linkage.

The redox potentials of the two hemes are equal and unusually low (+15 mV at pH 7.0) while the two-electron redox potential of the flavin is unusually high (−26 mV at pH 7.0) (Meyer *et al.*, 1991a). Intramolecular electron transfer in FCSD has been studied by laser flash photolysis using 5-deazriboflavin and sacrificial EDTA as the reductant. The reduction of FCSD by the 5-deazariboflavin semiquinone was second order with equal apparent reduction rates of flavin and heme, indicating that intramolecular electron transfer from flavin to heme occurs at a rapid rate, on the order of the initial second order reduction rate, *i.e.*, greater than $3 \times 10^4 \text{sec}^{-1}$. (Cusanovich *et al.*, 1985).

The structure of FCSD, shown in Figure 9, has been determined at 2.5 Å resolution (Chen *et al.*, 1994; Van Driessche *et al.*, 1996). The cytochrome subunit contains two domains and the flavoprotein subunit contains three domains. The two cytochrome domains are related by approximate 2-fold symmetry and the subunit structure closely resembles that of the diheme cytochrome $c_4$ from *Pseudomonas stutzeri* (Kadziola and Larsen, 1997). The two iron atoms are separated by 19.0 Å. The porphyrin rings are 11.4 Å apart, their planes being inclined to each other by about 30°, and they are joined by hydrogen bonding between a pair of propionic acid groups in the protein interior. The first two domains of the FCSD flavoprotein subunit each contain a pyridine nucleotide-binding motif (Rossman *et al.*, 1974) (Figure 2d) and together show a high degree of structural similarity to glutathione reductase (Karplus and Schulz, 1987). The third domain of FCSD provides the interface to the cytochrome subunit. Above the pyrimidine portion of the flavin ring is a disulfide bridge. This cystine disulfide may possibly be responsible for a charge transfer absorbance that is observed when the enzyme is treated with sodium sulfite (Meyer *et al.*, 1991b). In the intramolecular electron transfer complex of FCSD, the

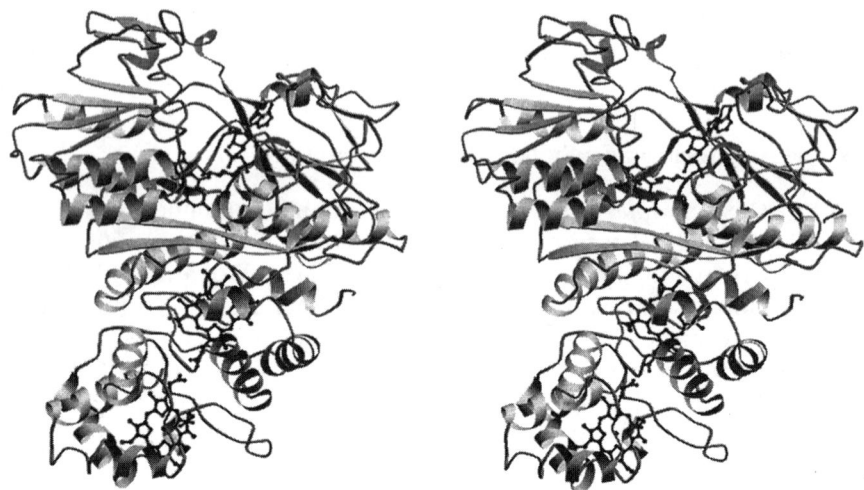

**FIGURE 9.** Stereo diagram of flavocytochrome c sulfide dehydrogenase. The flavoprotein subunit is at the top and the diheme cytochrome subunit is shown at the bottom. The flavin and two heme groups are shown as skeletal models.

pyrimidine portion of the FAD lies closest to the heme of the N-terminal cytochrome domain. The interface between the flavoprotein and cytochrome subunits covers a surface area of about $1500 \, \text{Å}^2$ per subunit. The two surfaces are complementary, with only about 5 water molecules located in the interface. There are 13 hydrogen bonds, including one salt bridge, connecting the two subunits. This sizable interface and large number of intramolecular interactions may account for the inability of FCSD to be resolved into its subunits without denaturation. The planes of the heme and the flavin rings are inclined by about $20°$ to each other. The smallest distance between the flavin and heme rings is $12.2 \, \text{Å}$. The iron is separated from the N5 atom the flavin ring by $20 \, \text{Å}$.

## 2.6.  Trimethylamine Dehydrogenase (TMADH)

Trimethylamine dehydrogenase is an iron-sulfur flavoprotein found in the methylotrophic bacterium *Methylophilus methylotrophus* W3A1. It catalyzes the oxidative N-demethylation of trimethylamine by water with formation of dimethylamine and formaldehyde (Steenkamp and Mallinson, 1976). The protein is a symmetrical dimer consisting of 166 kDa subunits (Kasprzak *et al.*, 1983; Lim *et al.*, 1982). Each subunit contains one 4Fe-4S center and one FMN cofactor. The latter is bound covalently through the 6 position of the flavin ring to a cysteine sulfur atom (Steenkamp *et al.*, 1978). The natural electron acceptor for TMADH is an FAD-containing electron transfer flavoprotein of 62 kDa molecular mass (Steenkamp and Gallup, 1978; Chen and Swenson, 1994); phenazine methosulfate and ferricenium hexafluorophosphate can serve as artificial acceptors *in vitro*.

Under stopped-flow conditions, addition of a stoichiometric amount of trimethylamine to the enzyme leads to formation of the flavin hydroquinone within a few msec while the iron-sulfur center remains oxidized (Steenkamp and Beinert, 1982). When excess substrate is used, rapid flavin reduction is followed by formation of a triplet state resulting from transfer of one electron from the reduced flavin to the 4Fe-4S center, giving rise to a very intense EPR signal which arises from strong electronic coupling between the two redox centers (Stevenson *et al.*, 1986). The latter process is biphasic with half times of 80 msec and 200 msec. When TMADH is treated with excess dithionite under anaerobic conditions, it takes up a total of three electrons per subunit to form the fully-reduced flavin and the reduced 4Fe-4S cluster. However, in the presence of tetramethylammonium chloride, a competitive inhibitor of TMADH, the enzyme takes up only two electrons per subunit and the triplet state is formed. The rates of the intramolecular electron transfer from the flavin to the iron-sulfur center in the dithionite reduced enzyme are quite fast compared with the rates in the presence of substrate, and range from about $300 \, \text{sec}^{-1}$ to $1200 \, \text{sec}^{-1}$ (Rohlfs and Hille, 1991).

These results are consistent with the behavior of the redox potential of TMADH. At pH 7, the two potentials for FMN, $E'_{ox/sq}$ and $E'_{sq/red}$ are about equal, at +40 mV, and $E'_{FeS} = +100 \, \text{mV}$ (Barber *et al.*, 1988). In the tetramethylammonium-inhibited enzyme, at pH 7, $E'_{ox/sq} = +230 \, \text{mV}$ and $E'_{sq/red} = -45 \, \text{mV}$ while $E'_{FeS}$ falls to +60 mV (Pace and Stankovich, 1991).

The crystal structure of TMADH was initially solved at 2.4 Å resolution (Lim *et al.*, 1986) and has been refined at 1.8 Å resolution (White *et al.*, 2000). Each monomer is composed of three domains. The largest domain is located at the N-terminal end of the molecule and contains the covalently bound flavin (Figure 10). The domain is folded into a parallel $\beta_8\alpha_8$ TIM barrel similar to flavocytochrome $b_2$ (Xia and Mathews, 1990) and "old yellow enzyme" (Fox and Karplus, 1994). In all of these enzymes the FMN is bound in a very similar manner, at the C-terminal end of the β barrel. The two smaller domains of TMADH (not shown) each contain an α/β motif, and are arranged geometrically and topologically like the FAD-binding and NADPH-binding domains of glutathione reductase (Karplus and Schulz, 1987). The iron-sulfur cluster is bound through four cysteine side chains located in an α/β hairpin loop situated between domains 1 and

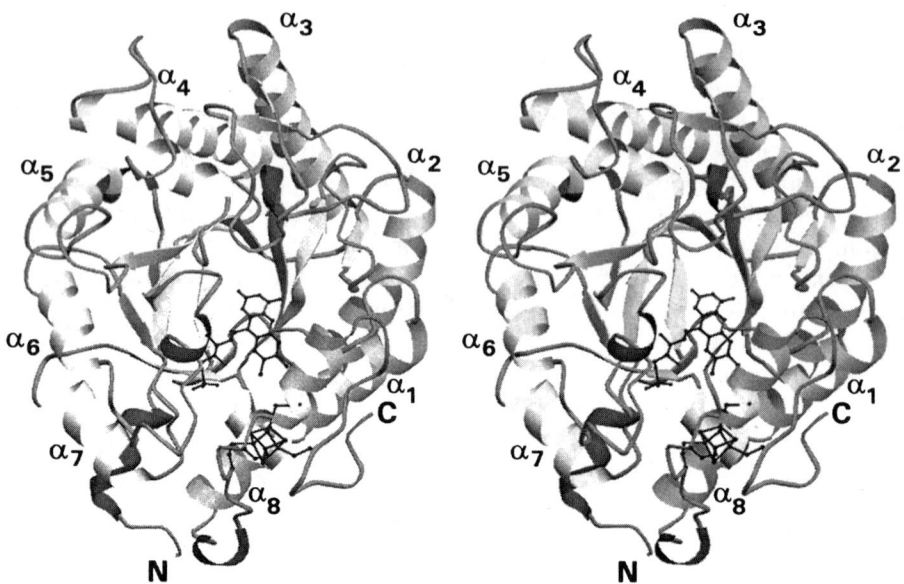

**FIGURE 10.** Stereo diagram of domain 1 of trimethylamine dehydrogenase. Residues 1–371 are shown. Helices $\alpha_1$-$\alpha_8$ of the $\beta_8\alpha_8$ TIM barrel are indicated. The iron sulfur cluster-binding loop consisting of an $\alpha$-helix and a $\beta$-strand is located at the end of helix $\alpha_8$.

2 and is attached to the C-terminus of helix 8 of the $\beta_8\alpha_8$ barrel (Figure 10). The sulfur of one of the 4Fe-4S cysteine ligands is in van der Waals contact with the 8-$\alpha$ methyl group of the flavin ring.

## 2.7.  Phthalate Dioxygenase Reductase (PDR)

In the bacterium *Pseudomonas cepacia*, the conversion of phthalate to its *cis*-dihydrodiol by molecular oxygen is catalyzed by a two-component enzyme system consisting of phthalate dioxygenase reductase (PDR), a 36 kDa monomeric protein containing both non-covalently bound FMN and a 2Fe-2S center, and phthalate dioxygenase (PDO), a tetrameric enzyme with one Rieske center, (a 2Fe2S cluster with two cysteine and two histidine ligands) and a mononuclear ferrous iron center (Gassner *et al.*, 1995). The function of PDR is to deliver reducing equivalents from NADH to PDO prior to the dioxygenase reaction. In the process, two electrons are transferred from the nicotinamide ring of NADH to the N5 position of the flavin *via* hydride transfer. The fully reduced flavin then donates the electrons, one at a time, to the 2Fe-2S cluster, temporarily existing in a metastable semiquinone state.

The midpoint redox potentials for flavin reduction in PDR are $E'_{ox/sq}$ = −174 mV for the first couple and $E'_{sq/red}$ = −274 mV for the second couple (Gassner *et al.*, 1995). The potential for the 2Fe-2S center is $E'_{FeS}$ = −174 mV. Thus, the thermodynamic driving force favors reduction of the 2Fe-2S center, especially by the first electron transfer. The rate of intramolecular electron transfer for the reduced flavin to the oxidized 2Fe-2S center has been estimated at >200 sec$^{-1}$ based on simulation of stopped flow kinetic data (Correll *et al.*, 1992).

Structures of the native oxidized PDR from *P. cepacia* at 2.0 Å resolution and of PDR in complex with reduced NADH at 2.7 Å resolution have been determined (Correll *et al.*, 1992). The enzyme folds into three domains, consisting of residues 1 to 102, 112 to 226 and 236 to 321, which bind FMN, pyridine nucleotide, and the 2Fe-2S center, respectively (Figure 11).

The flavin-binding domain is a beta barrel topologically similar to that in FNR (Figure 2b) despite the fact that the domain binds FMN rather than FAD. The FMN is bound with one side of the isoalloxazine ring facing the nicotinamide ring and the C8α-methyl group pointing in the direction of the iron-sulfur cluster (Figure 11). Strand β4 of the flavin barrel domain (Figure 2b) makes contact with the flavin N5 from the other side of the flavin ring in addition to providing hydrogen bonds with the 2Fe-2S domain. Upon flavin reduction, a serine side chain moves close to the flavin N5,

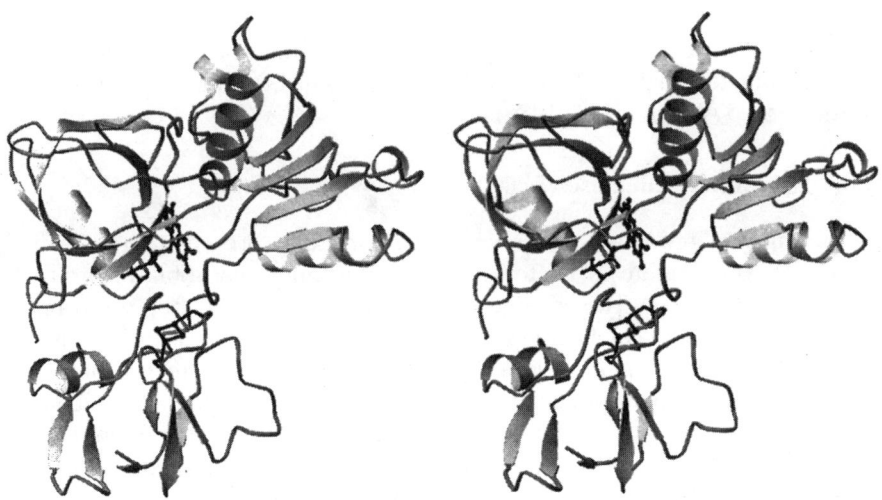

**FIGURE 11.** Stereo diagram of phthalate dioxygenase reductase. The FMN- and NADPH-binding domains are on the top and the 2Fe-2S binding domain is on the bottom. The FMN and 2Fe-2S prosthetic groups are shown as skeletal models.

forming a hydrogen bond, thus helping stabilize the semiquinone and fully reduced forms of the flavin.

The pyridine nucleotide-binding domain has a typical $\alpha/\beta$ nucleotide-binding motif consisting of a five-stranded beta sheet flanked by two helices on either side (Figure 2b). In the PDR-NADH complex the nicotinamide ring is situated close to the C-terminal ends of strands $\beta1$ and $\beta4$, and lies in a plane parallel to the flavin isoalloxazine ring. The nicotinamide ring is not stacked immediately adjacent to the flavin, being separated from it by a phenylalanine side chain from the NADH-binding domain. The iron-sulfur domain is a mixed beta sheet with a topology similar to the plant ferredoxins. The 2Fe-2S cluster is coordinated by the SG side chains of Cys272, Cys277 and Cys280 that are located in a 10-residue loop that covers the iron-sulfur group, and by Cys308 (Figure 11).

The 2Fe-2S domain interacts largely with the flavin-binding domain and, to a limited extent, with the NADPH-binding domain. The surface area of the 2Fe-2S in contact with the FAD/NADPH domains is ~650 Å². The interactions are a mixture of hydrophobic and hydrophobic, with six direct hydrogen bonds, one salt bridge and six water-mediated hydrogen bonds. The plane of the 2Fe-2S cluster is approximately perpendicular to the flavin isoalloxazine ring. The flavin C8α-methyl group is 4.1 Å from the Cys 272 SG atom which is coordinated to the 2Fe-2S center. This short distance would seem to allow for a facile electron transfer despite the small redox potential difference between acceptor and donor redox centers (Gassner *et al.*, 1995). The relative geometry of the flavin and nearest cysteine ligand of the 2Fe-2S cluster is similar to the geometry of the analogous flavin and 4Fe-4S cofactor pair of TMADH where the flavin C8α-methyl to iron separation is 4.5 Å (Lim *et al.*, 1986).

### 2.8. Fumarate Reductase (FUM)

Fumarate reductase is an integral membrane protein that converts fumarate to succinate in bacterial cells undergoing anaerobic respiration (Ackrell *et al.*, 1992). The enzyme from *E. coli* is a 121 kDa heterotetramer consisting of a 66 kDa flavoprotein subunit (Cole, 1982), a 27 kDa iron-sulfur protein (Cole *et al.*, 1982) and a pair of membrane anchor proteins each of about 14 kDa molecular mass which together bind two molecules of menaquinone, a membrane-soluble electron carrier molecule (Westenberg *et al.*, 1993). The flavoprotein subunit contains covalently-bound FAD (Walker and Singer, 1970) while the iron-sulfur protein contains three redox centers, an 2Fe-2S cluster, an 4Fe-4S cluster and a 3Fe-4S cluster (Kowal *et al.*, 1995). The complex catalyzes the reduction of fumarate using electrons derived from reduced menaquinone. These electrons are delivered to the flavin ring of FAD from the menaquinone during a reductive half reaction involving electron transfer through the iron sulfur protein. During the oxidative half reaction, the enzyme catalyzes the 2-electron reduction of fumarate to succinate. This reaction is the reverse of the reaction catalyzed by succinate dehydrogenase, a well studied enzyme of the mitochondrial respiratory chain (Ackrell *et al.*, 1992).

The 2-electron midpoint potential of the FAD in FUM is $\sim$−55 mV (Ackrell *et al.*, 1989). The 1-electron potentials for the iron-sulfur centers are $\sim$−80 mV (2Fe-2S), $\sim$−300 (4Fe-4S) and $\sim$−70 mV (3Fe-4S) (Kowal *et al.*, 1995). The electron transfer rates between the individual clusters or from the 2Fe-2S cluster to FAD are not known.

The structure of fumarate reductase has been determined at 3.3 Å resolution (Iverson *et al.*, 1999). The four subunits are arranged in a linear manner with the flavoprotein subunit at one end, the iron-sulfur subunit in the center and a tight dimer of the anchoring subunits at the other end, giving the molecule an overall length of 110 Å. The FAD-binding subunit contains a characteristic FAD-binding domain (Figure 2d) consisting of a 5-stranded parallel β-sheet flanked by three α-helices on one side and a 3-stranded antiparallel β-meander plus one α-helix on the other (Figure 12), similar to FCSD. The C-8-methyl group of FAD is covalently attached to the NE2 atom of His44 at the C-terminal end of β-strand 1. There are two small domains of 80 and 120 residues that protrude from the FAD-domain and are independently folded These domains have limited interaction with each other and with the FAD domain. In addition, there is a third domain, 160 residues in length, that is mostly helical and extends from the C-terminal end of the FAD-domain, wrapping over the latter and making substantial contact with it (Figure 12).

The iron sulfur subunit contains two domains. One domain, comprising the first 106 residues of the subunit, is characteristic of 2Fe-2S plant-type ferredoxin. The other domain, (about 125 residues in length) is similar to bacterial ferredoxins that contain 4Fe-4S and/or 3Fe-4S clusters. The three iron sulfur clusters are arranged in a linear manner and individual pairs are separated by about 12 Å. They appear to provide an electron transfer chain from the menaquinone-containing anchor subunits to the FAD subunit. The 2Fe-2S cluster is closest to the flavoprotein subunit and the 3Fe-4S cluster is closest to the membrane anchoring subunits. Each anchor subunit contains three α-helices; two helices from each subunit are packed together to form a 4-helix bundle. The remaining two helices lie approximately parallel to the helix bundle axis. Together the subunits form a thick rod about 60 Å in length that contains a menaquinone binding site at either end, one close to the 3Fe-4S cluster and the other 37 Å away.

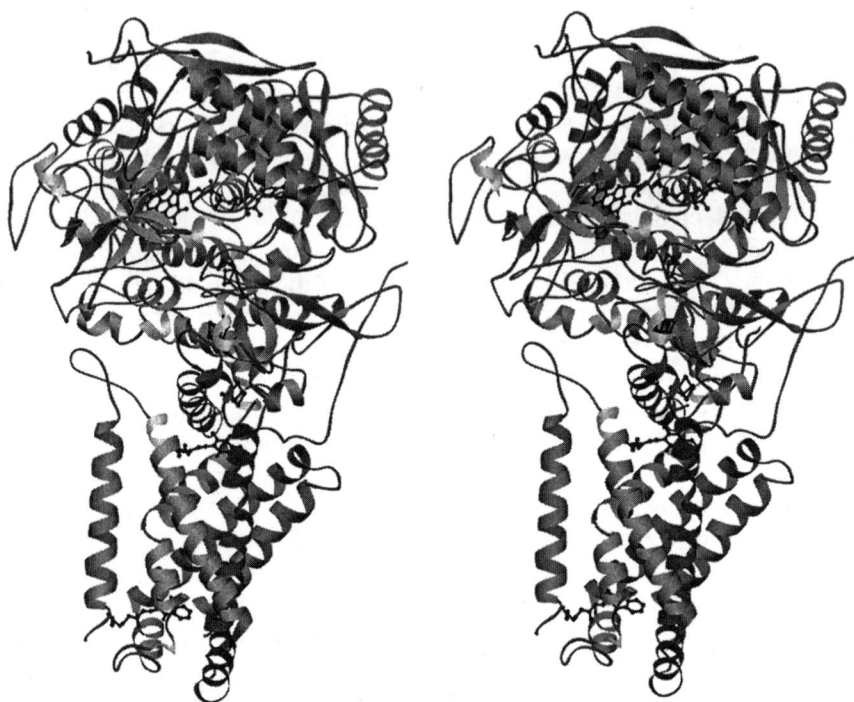

**FIGURE 12.** Stereo diagram of the complete fumarate reductase complex. The FAD-binding subunit is at the top, the iron-sulfur subunit is in the center and the two membrane anchoring subunits that provide the binding sites for two molecules of menaquinone are at the bottom. In this molecule electron transfer occurs from menaquinone at the bottom to FAD at the top during reduction of fumarate by menaquinone. Skeletal models of two molecules of menaquinone, a 3Fe-4S, a 4Fe-4S, a 2Fe-2S, an FAD molecule and one molecule of oxalate are included.

The iron-sulfur subunit interacts extensively with the flavoprotein subunit, with the 2Fe-2S domain wedged between the second and third external domains of the flavoprotein and the iron-sulfur cluster binding loop making contact with the FAD domain (Figure 13). The interface between the 2Fe-2S domain and the FAD domain covers an area of about 1450 Å$^2$ and is held together by 11 hydrogen bonds and 4 salt bridges. No water positions have been reported for the structure, possibly because of the limited resolution currently available. The planes of the 2Fe-2S and flavin rings are tilted by about 30′ to each other. The closest iron of the cluster is about 12 Å from the flavin ring and about 15 Å from the flavin N5 atom.

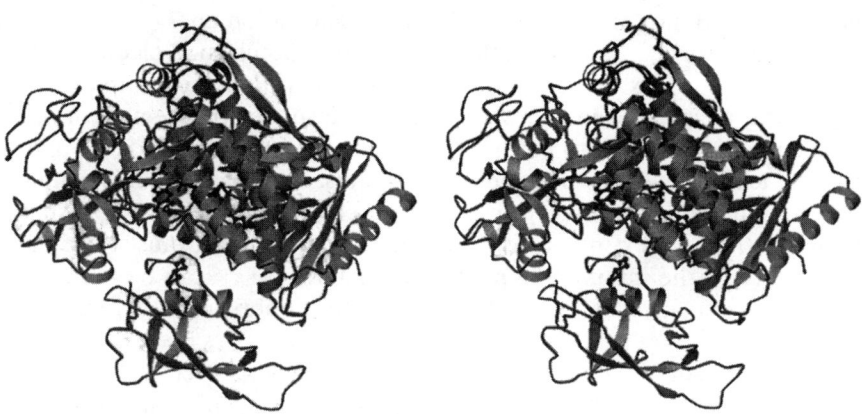

**FIGURE 13.** Stereo view of the FAD-binding subunit (top) and the 2Fe-2S domain (residues 1–106) of the iron-sulfur subunit (bottom) of fumarate reductase.

## 3.  ELECTRON TRANSFER

### 3.1.  General Aspects

Electron transfer in biological systems can be thought of as a two-step process involving formation of a complex between the donor and acceptor molecule followed by an electron transfer event. The overall electron transfer rate will depend on the rate of complex formation and stability of the complex and on the intrinsic electron transfer rate $k_{et}$ within the complex. Since the flavin electron transfer systems discussed in this chapter are all stable inter- or intramolecular complexes, the discussion will be limited to considerations of a first order intra-complex electron transfer process.

The rate of electron transfer between the donor and the acceptor redox center within the complex will depend on three main variables according to equation 1 (Marcus and Sutton, 1985).

$$k_{et} = (H_{da})^2 (2\pi/\hbar)(4\pi\lfloor k_B T)^{1/2} \exp\left(-(\Delta G^\circ - \lambda^2)/4\lambda k_B T\right) \qquad (1)$$

In this equation, $\hbar$ is Plank's constant, $k_B$ is the Boltzmann constant and T is the absolute temperature. The three variables are $\Delta G^\circ$, the difference in free energy between the donor and acceptor redox center, $\lambda$, the reorganization energy and $H_{da}$, the electronic coupling between the donor, d, and the acceptor, a. $\Delta G^\circ$ is the driving force for the electron transfer reaction and can be calculated from the difference in midpoint redox potentials between the donor and acceptor. $\lambda$, the reorganization energy, depends on conformational changes within the complex that occur upon electron transfer. Since the redox centers are generally fixed within the complexes, the main difference between the two redox states is in the arrangement of water molecules. $\lambda$ is difficult to predict, but can be measured from the temperature dependence of the electron transfer rates. $H_{da}$, the electronic coupling between the donor and acceptor redox center, is dependent on the distance between the redox centers, *i.e.*, the distance the electron must travel, and the nature of the intervening medium. The term $(H_{da})^2$ can be thought of as the conductance of the path between the donor and acceptor redox center.

The measured electron transfer rates and redox potential differences for the complexes discussed in this chapter, where clearly established, are summarized in Table 2. The flavin electron transfer rates for CPR and for fumarate reductase are not well established or are not known and the thermodynamic driving force is not known for PCMH.

The electronic coupling, $H_{da}$, can be calculated on the basis of the atomic coordinates of the flavin electron transfer complexes using the computer program GREENPATH (Regan, 1994). GREENPATH uses an algorithm developed by Beratan *et al.* (1991) to calculate the electronic coupling for electron transfer through many combinations of interacting bonds that link the donor and acceptor. Segments of each pathway are characterized as covalent, hydrogen-bonded or through space, with each segment type having a characteristic attenuation factor. The electronic coupling for a given path is then the product of the attenuation factors for the individual steps in the path and the program selects the paths of greatest coupling.

The electronic coupling has been calculated for each of the flavin electron transfer complexes described in this chapter and is described below, with the exceptions of CPR, TMADH and PDR. For these three proteins, the two redox cofactors are in direct van der Waals contact, either between the C-7 and C-8 methyl groups of two flavins (CPR), or between the flavin C-8 methyl and a cysteinyl sulfur ligand to the iron-sulfur center (TMADH and PDR). In these cases the coupling between the redox centers should be maximal and the electron transfer rates should depend only on the driving force and reorganization energy for the electron transfer processes.

For each of the five protein complexes considered, the largest electronic coupling, $H_{da}$ is computed for the structure described in Section 2 (Table 3). In all cases, the coupling has also been computed for several hypothetical "mutations" that serve to block a particular pathway in order to identify paths of lower coupling. This enables "families" of pathways to be

**Table 2**

**Structural and catalytic properties of the eight flavin electron transfer complexes**

| Complex | $\Delta E$ (mV) | Rate ($s^{-1}$) | Cofactor center-to-center distance (Å) | Cofactor edge-to-edge distance (Å)[a] | Cofactor inter-planar angle (°) | ET[b] pathway through flavin ring | Surface area of interface (Å²)[c] | H-bonds (water-mediated)[d] | Salt bridges[e] |
|---|---|---|---|---|---|---|---|---|---|
| CPR | (1) 95[f] (2) 20[g] | — | 13.4 N5–N5 | 6.3 FAD C8–FMN C7 | 30 | C8- or C7-edge (FMN + FAD) | 1,280 | 7 (0) | 7 |
| BMP/FMN | 0–50[h] | 250–500 | 23.6 N5–Fe | 18.3 FMN C8–HEM C3A | 90 | C8- or C7-edge | 740 | 2 (10) | 0 |
| FCB2 | (1) 135 (2) 45 | (1) 1,600 (2) 200 | 13.9 N5–Fe | 9.7 FMN N5–HEM C2A | 0 | O4–N5–C6 edge | 925 | 5 (5) | 1 |
| PCMH | — | 220 | 17.6 N5–Fe | 10.7 FAD C8–HEM C3C | 65 | C8- or C7-edge | 985 | 6 (7) | 1 |
| FCSD | ~40[i] | >30,000 | 19.7 N5–Fe | 12.2 FAD N3–HEM C3C | 20 | N1–O2–N3 edge | 1,470 | 12 (0) | 1 |
| TMADH | 60 | 300 | 11.0 N5–Fe | 5.3 FMN C7–Cys 351($S^\gamma$) | —[j] | C8-edge | — | — | — |
| PDR | 100 | >200 | 12.5 N5–Fe | 6.3 FMN C8–Cys 272($S^\gamma$) | 90 | C8-edge | 665 | 6 (6) | 0 |
| FUM | 25[i] | — | 14.8 N5–Fe | 10.0 FAD C7–Cys 57($S^\gamma$) | 30 | C8- or C7-edge | 1,450[k] | 11 (—) | 4 |

[a] Shortest distance between flavin ring edge atom and acceptor–cofactor edge atom.
[b] Electron transfer.
[c] Using the program GRASP (Nicholls et al., 1991), accessible surface area calculations based on the Shrake & Rupley (1973) algorithm.
[d] Hydrogen bonds in domain interfaces, with water-mediated hydrogen bonds in parentheses.
[e] Salt bridges in domain interfaces.
[f] Potential difference between $E'_{sqred}$ of FAD and $E'_{sqred}$ of FMN because the latter reduces P450 Fe.
[g] Potential difference between $E'_{oxisq}$ of FAD and $E'_{sqred}$ of FMN.
[h] Potential difference between $E'_{sqred}$ of FAD and $E'_{oxisq}$ of FMN.
[i] This range of $\Delta E$ represents a compromise between measured potentials and observed electron transfer activity.
[j] Based on measured 2-e⁻ redox potential of flavin.
[i] Not applicable to the cubane-like structure of a 4Fe4S cluster.
[k] Interface between FAD-binding subunit and the 2Fe2S domain of the iron-sulfur subunit (residues 1–106).

## Table 3

**Computed electronic coupling and relative electron transfer rates for five flavin electron transfer complexes calculated using the program GREENPATH (Regan, 1994)**

| Construct[a] | $H_{da}$[b] | $(H_{da})^2$ | Comment |
|---|---|---|---|
| **BMP/FMN** | | | |
| *No solvent:* | | | |
| Native | $7.7 \times 10^{-9}$ | $5.9 \times 10^{-17}$ | Path *via* C8M, Met 490 |
| M490G | $6.0 \times 10^{-9}$ | $3.6 \times 10^{-17}$ | Path *via* FMN phosphate, Met 490 |
| *Include solvent:* | | | |
| Native | $2.7 \times 10^{-7}$ | $7.3 \times 10^{-14}$ | Path *via* C8M, Met 490, HOH 649 |
| M490G | $2.9 \times 10^{-8}$ | $8.4 \times 10^{-16}$ | Path *via* FMN phosphate, Met 490, HOH 649 |
| M490G + FMN → isoalloxazine[c] | $1.0 \times 10^{-8}$ | $1.0 \times 10^{-16}$ | Path *via* C7M, Gln 387, HOH 437 |
| Above + Q387G | $5.6 \times 10^{-9}$ | $3.1 \times 10^{-17}$ | Path *via* C7M, HOH 945 |
| Native − FMN[d] | $1.7 \times 10^{-9}$ | $2.9 \times 10^{-18}$ | Path *via* Trp 574, HOH 32 |
| **FCB2** | | | |
| Native | $4.5 \times 10^{-5}$ | $2.0 \times 10^{-9}$ | Path *via* Tyr 143 |
| Y143F | $1.9 \times 10^{-5}$ | $3.6 \times 10^{-10}$ | Path *via* Ala 198 |
| Y143F + A198G | $1.4 \times 10^{-5}$ | $2.0 \times 10^{-10}$ | Path *via* Leu 230 |
| Native + solvent | $1.5 \times 10^{-4}$ | $2.2 \times 10^{-8}$ | Path *via* HOH 664, heme O1A |
| **PCMH** | | | |
| Native (+ substrate) | $8.8 \times 10^{-6}$ | $7.7 \times 10^{-11}$ | Path *via* Phe 381 |
| F381G | $8.7 \times 10^{-6}$ | $7.6 \times 10^{-11}$ | Path *via* Tyr 384 (covalent link to FAD) |
| F381G + Y384G | $1.6 \times 10^{-6}$ | $2.6 \times 10^{-12}$ | Path *via* Pro 155, Leu 378 |
| **FCSD** | | | |
| Native | $4.6 \times 10^{-6}$ | $2.1 \times 10^{-11}$ | Path *via* Thr 336 |
| T336G | $3.3 \times 10^{-6}$ | $1.1 \times 10^{-11}$ | Path *via* Tyr 338 |
| T336G + Y338G | $2.4 \times 10^{-6}$ | $5.8 \times 10^{-12}$ | Path *via* Tyr 306 |
| T336G + Y338G + Y306G | $1.4 \times 10^{-6}$ | $2.0 \times 10^{-12}$ | Path *via* Trp 391 |
| **FUM** | | | |
| *2Fe2S to FAD:*[e] | | | |
| Native | $2.5 \times 10^{-5}$ | $6.2 \times 10^{-10}$ | Path *via* Cys 57 ligand, Ala 47 |
| A47G | $2.3 \times 10^{-5}$ | $5.3 \times 10^{-10}$ | Path *via* Cys 57 ligand, His 44 (FAD covalent link) |
| A47G + H44G | $1.2 \times 10^{-5}$ | $1.4 \times 10^{-10}$ | Path *via* Cys 57 ligand, Ile 61 |
| Above + I61G | $1.1 \times 10^{-5}$ | $1.2 \times 10^{-10}$ | Path *via* Cys 57 ligand, Met 59 |
| Above + M59G | $5.7 \times 10^{-6}$ | $3.2 \times 10^{-11}$ | Path *via* Cys 62 ligand, Val 46 |

[a] Modified coordinates were input to the pathways calculations to simulate the effect of a structural mutation in order to search for alternative pathways.

[b] $H_{da}$ is the electronic coupling matrix element between electron donor and acceptor. The electron transfer rate is proportional to its square.

[c] FMN truncated to isoalloxazine ring to block pathway through FMN phosphate.

[d] FMN removed from coordinates, assuming a path which includes a prior jump from flavin to Trp 574, which are involved in a charge-transfer interaction.

[e] Paths from menaquinones through 3Fe4S, 4Fe4S to 2Fe2S not shown.

identified and to provide a quantitative estimate of the relative importance of various pathways. In general, water molecules that might mediate electron transfer have not been included in the calculations. Two exceptions to this are FCB2 and BMP/FMN where paths involving water are considered. The inclusion of water in the GREENPATH calculations has been found to give higher electronic coupling than when it is omitted. However, unless tightly bound to the protein, water tends to be somewhat mobile and its binding sites might not be fully occupied, leading to unreliable estimates of its importance for an electron transfer pathway.

## 3.2. Electronic Coupling

### 3.2.1. BMP/FMN

Calculation of likely electron transfer pathways in BMP/FMN using GREENPATH (Regan, 1994), indicates that the most favorable route (with solvent omitted) from the flavin N5 to the heme is *via* the C8-methyl group, and then by a through-space jump to Met 490 CE in the FMN-binding domain, along the methionine side-chain and backbone atoms to the carbonyl oxygen of Asn 489 (Figure 14a). From there the path follows a jump to the cytochrome domain at Gly 396 CA, follows a hydrogen bond from Gly 396 O to Ala 399 N and traces the protein backbone and Cys400 side chain to SG, the heme iron ligand. An M490G "mutation" results in an alternative path for electrons from the flavin N5 that follows the ribityl and phosphate segments of FMN, exiting from a phosphate oxygen atom *via* a hydrogen bond to the amide group of Gly 490, and from there following the same path as the non-mutated BMP above. The electronic coupling, $H_{da}$ for the two alternative pathways is about the same (Table 3).

The BMP structure contains a number of ordered solvent molecules in the interface between the FMN and heme-binding domains. When solvent is included in the pathways calculation the predicted best path similarly follows the C8-methyl group of flavin and Met490. However from the CB atom of Met 490, the electrons jump to a water molecule (HOH 649), make another jump to Ala 399 CB and then proceed *via* the backbone as above to Cys400 (Figure 14b). This pathway has a 30-fold larger electronic coupling (and therefore about a 1000-fold greater electron transfer rate) than the model without solvent (Table 3). In a second path with very similar coupling, electrons travel along the ribityl chain of FMN to a phosphate oxygen through a hydrogen bond to Met490N and jumps from CA to HOH 649. It then follows the same path as previously described. Alternative pathways including solvent, but with lower magnitudes of electronic coupling comparable to the paths with solvent omitted, predict that electrons will leave

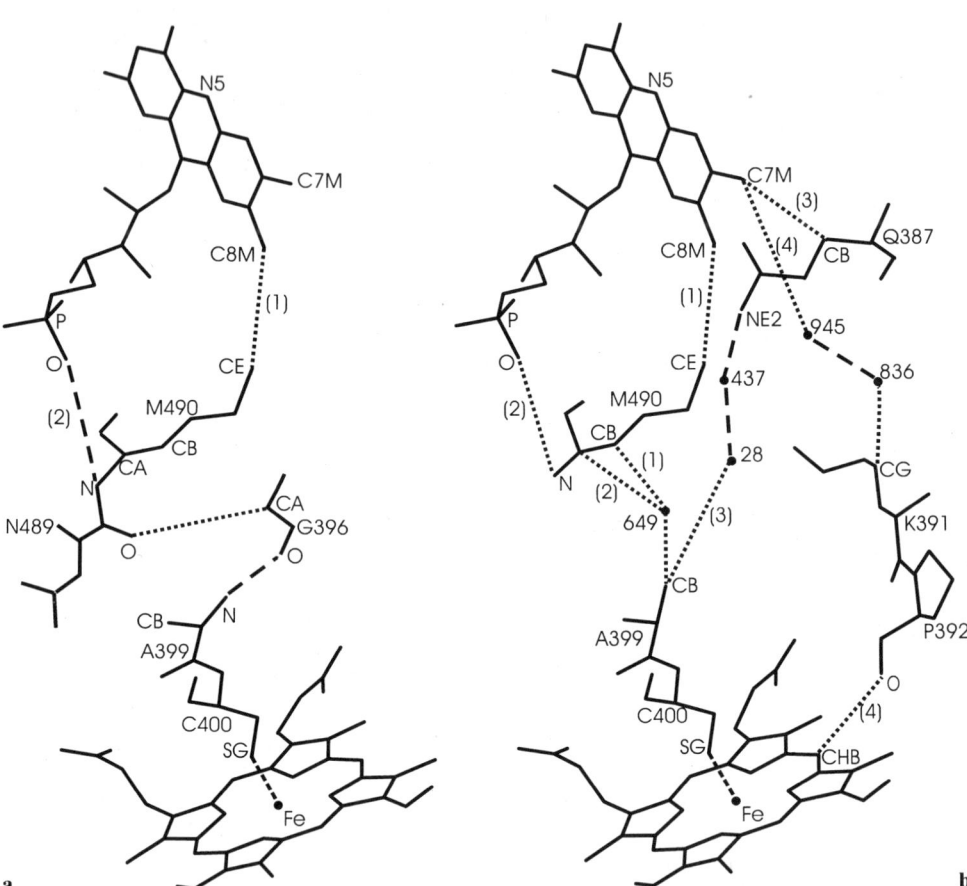

**FIGURE 14.** Electron transfer pathways in BMP/FMN computed using GREENPATH. Through-space jumps are indicated by dotted lines and paths along hydrogen bonds are indicated by dashed lines. (a) The two best paths (labeled 1 and 2) for electron transfer from the flavin N5 to the heme iron atom with no water molecules included in the calculation. (b) The four best paths (labeled 1–4) between the flavin and heme groups with water molecules included in the calculations. Paths 1 and 2 are the same at the beginning as the two paths in (a) but utilize a water molecule instead of a through-space jump between the FMN and BMP domains.

the flavin ring at the C7-methyl group and cross to the cytochrome domain. Electrons may jump from the C7-methyl group to Gln 387 CB of the heme-binding domain, go on through the Gln side-chain to tunnel *via* a hydrogen bond to HOH 437 and then travel along another hydrogen bond to HOH 28, jump to Ala 399CB, and finally tunnel through the peptide backbone to Cys 400. Alternatively, a jump from C7M to HOH 945, then along a hydrogen bond to HOH 836, followed by another jump to Lys 391 CG, backbone travel to Pro 392, and a jump from there to the CHB atom of heme, provides a possible pathway. Additional paths (not shown), for example through Trp574 (which is involved in a charge-transfer interaction with flavin) are ~4-fold lower in coupling than the best non-solvent paths (Table 3).

### 3.2.2. FCB2

The results of the GREENPATH calculation with FCB2 indicate that the most favorable route for electron flow from the flavin ring to the heme passes through atom O4 and a water molecule connecting it through hydrogen bonding to atom O1A of the heme propionate and along the propionate chain to the porphyrin ring (Figure 15). The next most favorable route follows a through-space jump from the flavin C6 to the hydroxyl of Tyr143 and then along a hydrogen bond to the heme propionate O2A. This latter route is about 5-fold lower in coupling, but is about 2 to 3-fold higher than the next two paths, one utilizing Ala198 and the other Leu 230, both involving two through-space jumps (Table 3; not shown). If the route involving the water is ignored, because of the possibilities of variable occupancy or a higher tunneling barrier, the Y143 route becomes the most important. Mutagenesis of Tyr143 to Phe experimentally reduces the electron transfer rate to the cytochrome 20-fold, consistent with about 4-fold reduction in coupling (Miles *et al.*, 1992). Measured electron transfer in the Y143F mutant to ferricyanide, an artificial electron acceptor that can interact directly with the flavin, is essentially unchanged.

### 3.2.3. PCMH

In PCMH, two equally efficient pathways for electron transfer from the flavin N5 to the heme iron were identified in GREENPATH calculations. One path follows the tyrosyl covalent link to FAD at the C8-methyl position, whereby electrons can travel from C8M *via* the Tyr 384 phenolic ether bond, tunnel through the tyrosine ring atoms and make a through-space jump across the subunit interface to the carbonyl oxygen of Ala 49 in the cytochrome (Figure 16). From there the electrons follow backbone

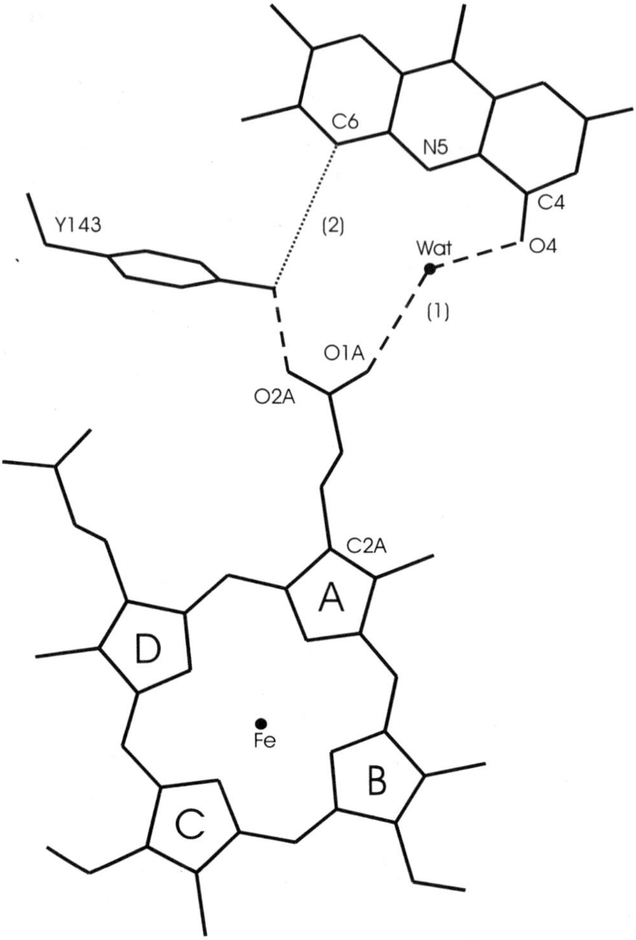

**FIGURE 15.** Electron transfer pathways in FCB2. The best path for electron transfer from flavin to heme based on GREENPATH calculations. Dashed lines represent paths along hydrogen bonds and dotted lines represent a through-space jump. Path 1 involves a water molecule while path 2 does not.

atoms to reach Met 50, one of the heme iron ligands. The other pathway consists of a through-space jump from the C7-methyl group of flavin to the aromatic ring of Phe 381, followed by a second jump directly to the heme vinyl atom CBC, and then to the porphyrin ring. A pathway with about 5 to 6-fold lower electronic coupling (corresponding to 30-fold lower rate) follows directly from flavin N5 to Pro 155, Leu 378 and to the

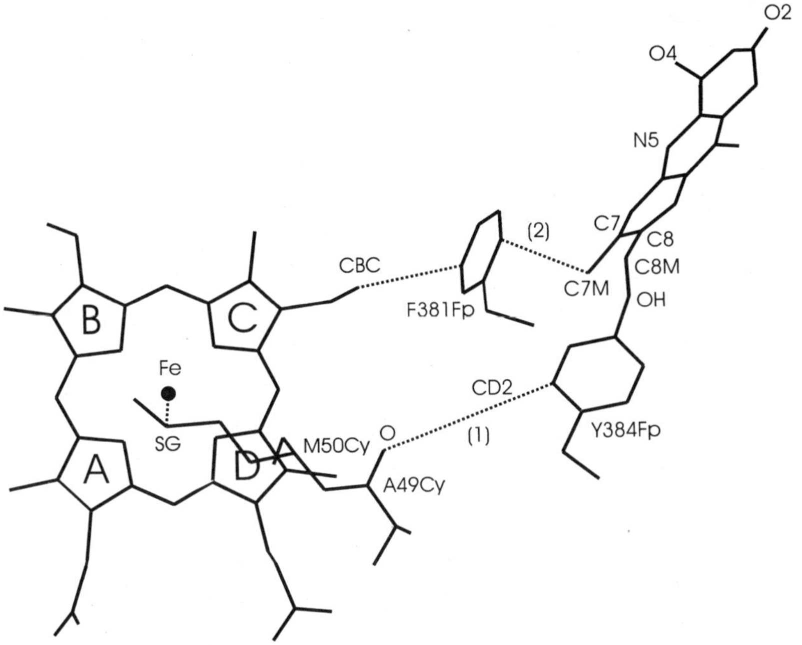

**FIGURE 16.** Electron transfer pathways in PCMH. Dotted lines represent through-space jumps. Two paths of approximately equal electronic coupling are indicated, one through Tyr384 covalently bound to the flavin ring and the other through the side chain of Phe381. Fp and Cy indicate residues in the flavoprotein and the cytochrome subunits, respectively.

heme vinyl atom CBC and involves three through-space jumps (Table 3; not shown).

### 3.2.4. FCSD

There are four side chains in the flavoprotein subunit of FCSD, Trp391, Tyr306 Tyr338 and Thr336, which provide potential pathways for electron flow from flavin to heme c (Figure 17). GREENPATH calculations show them to vary over a 3-fold range of electronic coupling (Table 3). The two best paths, with similar electronic coupling (Table 3) lead from the N3 position of the flavin ring. One involves a through-space jump to the protein backbone at Cys337 CA and thence to Thr336 CG2, followed by another jump to the heme vinyl atom CBC of the porphyrin ring. The other route follows a jump from N3 to Tyr338 N, tunneling through the tyrosine side-chain to atom CE2 followed by a jump to the heme CBC atom.

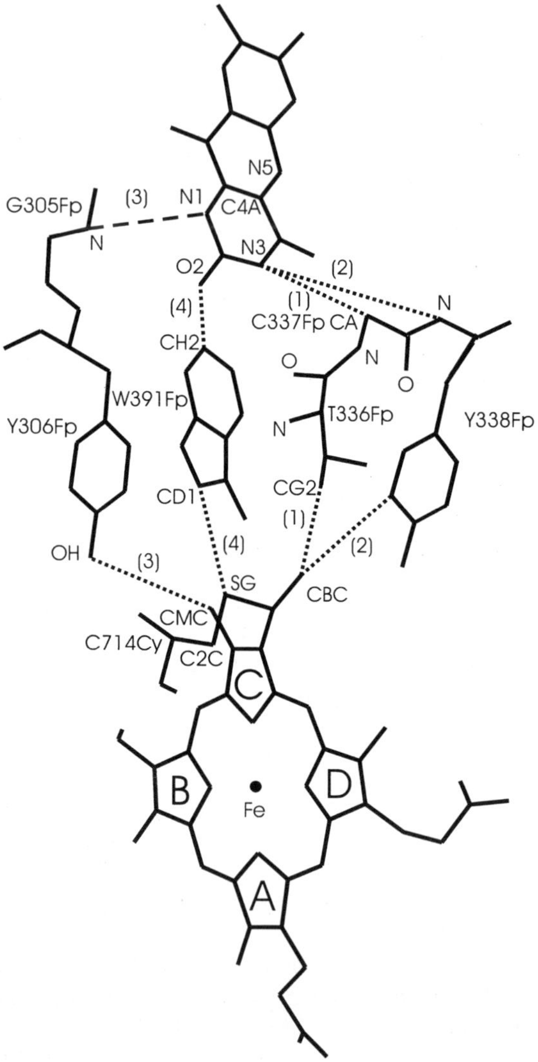

**FIGURE 17.** Electron transfer pathways in FCSD. Through-space jumps are indicated by dotted lines and paths along hydrogen bonds are indicated by dashed lines. Four paths (1–4) with decreasing electronic coupling are indicated. Fp and Cy indicate residues in the flavo-protein and the cytochrome subunits, respectively.

In a third pathway, electrons flow from the N1 position of flavin *via* a hydrogen bond to Gly 305(N), on through backbone to the Tyr 306 sidechain, followed by a jump from the tyrosine hydroxyl to the methyl CMC atom of heme. The path with weakest coupling involves a through-space jump from the flavin O2 to Trp 391, then a jump to Cys 714 SG, which is covalently linked to the heme.

### 3.2.5. FUM

The localized electron transfer from the 2Fe-2S center to FAD in fumarate reductase is of greatest relevance to this chapter on flavin electron transfer. The four best paths for this electron transfer initially involve flow from the 2Fe-2S species to the Cys 57 SG link to iron (Table 3, Figure 18). The first path involves a through-space jump to Ala 47 CB, travel along the backbone to Ala 48 CA and then a jump directly to the flavin N5 atom. Calculation with an A47G "mutation" indicates an alternative pathway, with similar coupling, involving a jump from Cys57 CA to His44 ND1 and subsequently through the His44 ring to the C8-methyl group of FAD *via* the covalent link and on to the flavin. "Mutation" of His44 to Gly in addition to the A47G "mutation" results in a path to "Gly47 CA" *via* Ile 61 from Cys57 and has a 2-fold lower coupling; an additional I61G "mutation" sends electrons instead directly from Cys57 SG to Met59 N, with similar coupling (Table 3), and then on to the flavin C7 methyl via Met59 CE (Figure 18). Further mutations lead to more drastic reduction in coupling (Table 3).

GREENPATH calculations were also carried out to predict electron transfer pathways between the two menaquinone molecules and the three iron-sulfur clusters. The results indicate that electrons might travel a distance of 40 Å from $Q_d$ to $Q_p$, using six protein residues, mostly along backbone atoms, with five intervening through-space jumps. From menaquinone $Q_p$, electrons may jump directly to Cys 204, a ligand of Fe in the 3Fe4S moiety and then to the 4Fe4S center *via* a jump between Cys210 and Cys154, ligands of the 3Fe4S and 4Fe4S centers, respectively. Electrons in the 4Fe4S cluster leave *via* the Cys 151 iron ligand, jump to Gly 63, then from its amide group to an Fe atom of the 2Fe2S.

## 4. CONCLUSIONS

### 4.1. Domain Interactions

Of the eight complexes discussed in this chapter, five involve electron transfer between covalently linked domains and three between separate

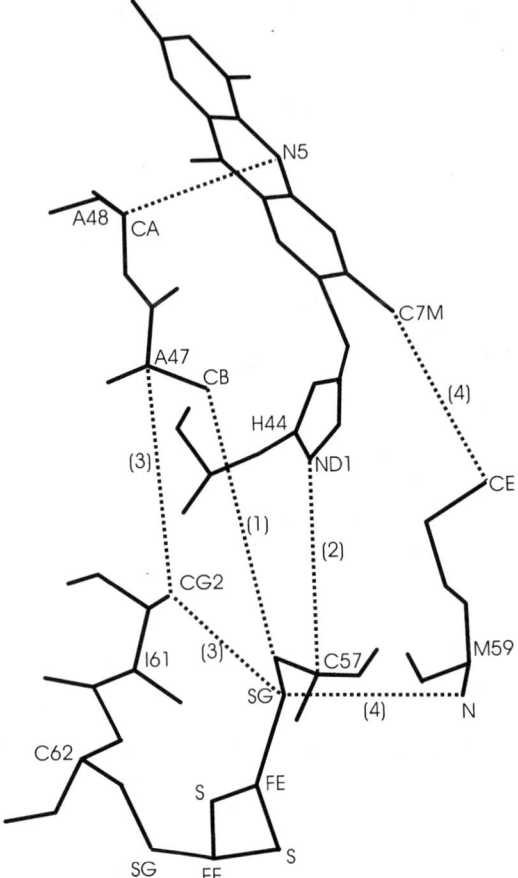

**FIGURE 18.** Electron transfer pathways from the 2Fe-2S center to the flavin ring in fumarate reductase. Dotted lines represent through-space jumps. The four best paths (1–4) are indicated, all of which involve an initial transfer of an electron from the 2Fe-2S center to the SG atom of Cys57 which forms a ligand to an iron atom.

subunits in tight complex. One of the covalent complexes, TMADH, is atypical since the flavin and 4Fe-4S redox centers are in a single domain. BMP/FMN was crystallized as a covalent complex, but suffered proteolysis during crystallization. Assuming that loss of the 21-residue covalent link did not lead to domain rearrangement, however, it should be considered as a member of the covalent class.

The surface area per interaction domain ranges from 665 Å$^2$ to 1450 Å$^2$ (Table 2) with the average for the covalent class ~900 Å$^2$ and for the non-covalent class ~1300 Å$^2$. Both classes contain hydrophilic interactions within their interfaces, with several (5–10) direct hydrogen bonds but usually only one or two salt bridges (Table 2). Thus, in most cases, there is little ionic character to the interactions in either class of complex. In general, however, there appear to be more water-mediated hydrogen bonds among the covalent complexes than the non-covalent complexes, suggesting that there is tighter surface complementarity among the latter than the former.

The stabilities of the complexes are generally consistent with the above analysis. The stabilities of the covalent complexes, where known, indicate that the isolated subunits interact more weakly than the non-covalent complexes. The dissociation constant for PCMH is in the nmolar range and FCSD cannot be resolved without denaturation. In FCB2, the separated domains do not interact significantly, indicating the covalent link provides the necessary stabilization of the complex. The large movement of the FMN domain that must occur during catalysis in NADPH cytochrome P450 reductase, inferred from its difference in orientation in the CPR structure *vs* the BMP/FMN structure, indicates weak association of the FMN domain within each covalent complex and suggests that such weak association may be important for domain mobility and the physiological activity of the covalent complexes.

## 4.2. Features of Electron Transfer

In seven of the eight complexes, the flavin interacts with a planar cofactor, either a heme, a flavin or a 2Fe2S cluster. There is considerable variability in the relative angular orientations of the flavin and its redox partner cofactor (Table 2). In the flavocytochromes, the angles between the heme and flavin planes are observed at values near 0° (FCB2), 20° (FCSD), 65° (PCMH) and 90° (BMP/FMN), while the interflavin angle in CPR is ~30° (Table 2). Likewise, the angle between the flavin and the 2Fe2S plane is ~30° (FUM) and ~90° (PDR). These observations demonstrate that the interplanar angle between redox cofactors can vary substantially in electron transfer complexes with no apparent correlation with the measured rates of electron transfer (Table 2).

The eight flavin complexes show that there is considerably less variability in the pathway for electron flow within the flavin ring to its partner cofactor. In six of the eight complexes, the electron acceptor (or donor) interacts with the dimethylbenzene ring of the flavin, electrons flowing through the C8 or C7 positions (Table 2). Only in FCB2 (C6, N5, C4 edge, Figure 1) and FCSD (N1, O2 and N3 edge) does electron transfer follow a different route through the flavin ring. Thus, the preferred route for electron transfer in the flavin ring appears to be through the dimethylbenzene portion. Since flavoenzyme catalysis seems to occur mainly at the N5 and C4A position, electron transfer through C7 and C8 would seem to be an effective way to separate the 2-electron reductive half reaction from the 1-electron oxidative half reactions within flavin electron transferases.

Electron transfer rates for six of the eight flavin electron transfer complexes have been measured and range from ~200 sec$^{-1}$ to >30,000 sec$^{-1}$, with most of the rates being below 500 sec$^{-1}$. For the limited data available, there seems to be some correlation between the electron transfer rates and the driving force, which ranges from ~25 mV (BMP/FMN) to ~100 mV (PDR), the clearest example being FCB2 where the two flavin to *b*-type heme rates (~1600 sec$^{-1}$ and ~200 sec$^{-1}$) correlate with the relative driving forces (135 mV and 45 mV, Table 2). There seems to be little correlation between the electron transfer rates and the intercofactor distance, however, which ranges from 5.3 Å (TMADH, $k_{et}$ ~ 300 sec$^{-1}$) to 23.6 Å (BMP/FMN, $k_{et}$ ~ 500 sec$^{-1}$), and with PCMH and FCSD, which have similar center to center cofactor distances (17.6 Å and 19.7 Å, respectively) but widely varying electron transfer rates (~220 sec$^{-1}$ and >30,000 sec$^{-1}$, respectively). It is clear that considerably more kinetic and thermodynamic data on flavin electron transfer complexes are needed for a better understanding of the electron transfer process, especially as new structures of such complexes become available.

ACKNOWLEDGMENTS. This work has been supported by USPHS Grants No. GM20530 and GM31611.

## 5. REFERENCES

Ackrell, B. A. C., Cochran, B., and Cecchini, G., 1989, Interactions of oxaloacetate with Escherichia coli fumarate reductase. *Arch. Biochem. Biophys.* **268**:26–34.

Ackrell, B. A. C., Johnson, M. K., Gunsalus, R. P., and Cecchini, G., 1992, Structure and function of succinate dehydrogenase and fumarate reductase, in: *Chemistry and Biochemistry of Flavoenzymes*, Volume 3 (F. Muller, ed.) CRC Press, Boca Raton, FL, pp. 229–297.

Backes, W. L., and Reker-Backes, C. E., 1988, The effect of NADPH concentration on the reduction of cytochrome P-450 LM2. *J. Biol. Chem.* **263**:247–253.

Barber, M. J., Pollock, V., and Spence, J. T. (1988) Microcoulometric analysis of trimethylamine dehydrogenase., *Biochem. J.* **256**:657–659.

Beratan, D. N., Betts, J. N., and Onuchic, J. N., 1991, Protein electron transfer rates set by the bridging secondary and tertiary structure, *Science* **252**:1285–1288.

Bhattacharyya, A., Tollin, G., McIntire, W. S., and Singer, T. P., 1985, Laser-flash-photolysis studies of p-cresol methylhydroxylase. Electron-transfer properties of the flavin and haem components, *Biochem. J.* **228**:337–345.

Boddupalli, S. S., Oster, T., Estabrook, R. W., and Peterson, J. A., 1992, Reconstitution of the fatty acid hydroxylation function of cytochrome P-450BM-3 utilizing its individual recombinant hemo- and flavoprotein domains, *J. Biol. Chem.* **267**:10375–10380.

Bredt, D. S., Hwang, P. M., Glatt, C. E., Lowenstein, C., Reed, R. R., and Snyder, S. H., 1991, Cloned and expressed nitric oxide synthase structurally resembles cytochrome P-450 reductase, *Nature,* **351**:714–718.

Carson, M, 1997, Ribbons *Methods Enzymol.* **277**:493–505.

Causer, M. J., Hopper, D. J., McIntire, W. S., and Singer, T. P., 1984, Azurin from *Pseudomonal putida*: an electron acceptor from *p*-cresol methylhydroxylase. *Biochem. Soc. Trans.* **12**:1131–1132.

Chen, D., and Swenson, R. P., 1994, Cloning, sequence analysis, and expression of the genes encoding the two subunits of the methylotrophic bacterium W3A1 electron transferring flavoprotein. *J. Biol. Chem.* **269**:32120–32130.

Chen, Z.-w., Koh, M., Van Driessche, G. V., Van Beeumen, J. J., Bartsch, R. G., Meyer, T. E., Cusanovich, M. A., and Mathews, F. S., 1994, The structure of flavocytochrome c sulfide dehydrogenase from a purple phototrophic bacterium. *Science* **266**:430–432.

Cole, S. T., 1982, Nucleotide sequence coding for the flavoprotein subunit of the fumarate reductase of Escherichia coli. *Eur. J. Biochem.* **122**:479–484.

Cole, S. T., Grundstrom, T., Jaurin, B., Robinson, J. J., and Weiner, J. H., 1982, Location and nucleotide sequence of frdB, the gene coding for the iron-sulphur protein subunit of the fumarate reductase of Escherichia coli. *Eur. J. Biochem.* **126**:211–216.

Correll, C. C., Batie, C. J., Ballou, D. P., and Ludwig, M. L., 1992, Phthalate dioxygenase reductase: a modular structure for electron transfer from pyridine nucleotides to [2Fe-2S], *Science* **258**:1604–1610.

Cunane, L. M., Chen, Z.-w., Shamala, N., Mathews, F. S., Cronin, C. N., and McIntire, W. S., 2000, Structures of the flavocytochrome p-cresol methylhydroxylase and its enzyme-substrate complex: gated substrate entry and proton relays support the proposed catalytic mechanism. *J. Mol. Biol.* **295**:357–374.

Cusanovich, M. A., Meyer, T. E., and Bartsch, R. G., 1991, Flavocytochrome c. in Chemistry and Biochemistry of the Flavoenzymes, Vol. 2 (F. Muller, ed.) CRC Press, Boca Raton, pp. 377–393.

Cusanovich, M. A., Meyer, T. E., and Tollin, G., 1985, Flavocytochromes c: transient kinetics of photoreduction by flavin analogues. *Biochemistry* **24**:1281–1287.

Daff, S. N., Chapman, S. K., Turner, K. L., Holt, R. A., Govindaraj, S., Poulos, T. L., and Munro, A. W., 1997, Redox control of the catalytic cycle of flavocytochrome P-450 BM3, *Biochemistry* **36**:13816–13823.

Daff, S., Ingledew, W. J., Reid, G. A., and Chapman, S. K., 1996a, New insights into the catalytic cycle of flavocytochrome $b_2$, *Biochemistry* **35**:6345–6350.

Daff, S., Sharp, R. E., Short, D. M., Bell, C., White, P., Manson, F. D., Reid, G. A., and Chapman, S. K., 1996b, Interaction of cytochrome c with flavocytochrome b2. *Biochemistry* **35**:6351–6357.

Fox, K. M., and Karplus, P. A., 1994, Old yellow enzyme at 2.0 Å resolution: overall structure, ligand binding, comparison with related flavoproteins. *Structure* **2**:1089–1105.

Fraaije, M. W., van Berkel, W. J. H., Benen, J. A. E., Visser, J., and Mattevi, A., 1998, A novel oxidoreductase family sharing a conserved FAD-binding motif. *Trends Biochem. Sci.* **23**:26–27.

Fulco, A. J., 1991, P450BM-3 and other inducible bacterial P450 cytochromes: biochemistry and regulation. *Annu. Rev. Pharmacol. Toxicol.* **31**:177–203.

Gassner, G. T., Ludwig, M. L., Gatti, D. L., Correll, C. C., and Ballou, D. P., 1995, Structure and mechanism of the iron-sulfur flavoprotein phthalate dioxygenase reductase. *FASEB J.* 9:1411–1418.

Gervais, M., Risler, J., and Corazzin, S., 1983, Proteolytic cleavage of Hansenula anomala flavocytochrome b2 into its two functional domains. Isolation of a highly active flavodehydrogenase and a cytochrome b2 core, *Eur. J. Biochem.* 130:253–259.

Gray, G., and Knaff, D. B., 1982, The role of a cytochrome $c_{552}$-cytochrome c complex in the oxidation of sulfide in *Chromatium vinosum. Biochem. Biophys. Acta.* 680:290–296.

Hazzard, J. T., Govindaraj, S., Poulos, T. L., and Tollin, G., 1997, Electron transfer between the FMN and heme domains of cytochrome P450BM-3. Effects of substrate and CO. *J. Biol. Chem.* 272:7922–7926.

Hazzard, J. T., McDonough, C. A., and Tollin, G., 1994, Intramolecular electron transfer in yeast flavocytochrome b2 upon one-electron photooxidation of the fully reduced enzyme: evidence for redox state control of heme-flavin communication. *Biochemistry* 33:13445–13454.

Hazzard, J. T., Govindaraj, S., Poulos, T. L., and Tollin, G., 1997, Electron transfer between the FMN and heme domains of cytochrome P450BM-3. Effects of substrate and CO. *J. Biol. Chem.* 272:7922–7926.

Hopper, D. J., and Taylor, D. G., 1977, The purification and properties of *p*-cresol-(acceptor) oxidoreductase (hydroxylating), a flavocytochrome from *Pseudomonas putida. Biochem. J.* 167:155–162.

Iverson, T. M., Luna-Chavez, C., Cecchini, G., and Rees, D. C., 1999, Structure of the Escherichia coli fumarate reductase respiratory complex. *Science* 284:1961–1966.

Iyanagi, T., Makino, N., and Mason, H. S., 1974, Redox properties of the reduced nicotinamide adenine dinucleotide phosphate-cytochrome P-450 and reduced nicotinamide adenine dinucleotide-cytochrome b5 reductases. *Biochemistry* 13:1701–1710.

Kadziola, A., and Larsen, S., 1997, Crystal structure of the dihaem cytochrome c4 from *Pseudomonas stutzeri* determined at 2.2A resolution. *Structure* 5:203–216.

Karplus, P. A., and Schulz, G. E., 1987, Refined structure of glutathione reductase at 1.54 A resolution. *J. Molec. Biol.* 195:701–729.

Kasper, C. B., 1971, Biochemical distinctions between the nuclear and microsomal membranes from rat hepatocytes. *J. Biol. Chem.* 246:577–581.

Kasprzak, A. A., Papas, E. J., and Steenkamp, D. J., 1983, Identity of the subunits and the stoichiometry of prosthetic groups in trimethylamine dehydrogenase. *Biochem, J.* 211:353–541.

Kim, J., Fuller, J. H. Cecchini, G., and McIntire, W. S. 1994, Cloning, sequencing, and expression of the structural genes for the cytochrome and flavoprotein subunits of p-cresol methylhydroxylase from two strains of Pseudomonas putida. *J. Bact.* 176:6349–6361.

Koerber, S. C., McIntire, W. S., Bohmont, C., and Singer, T. P., 1985, Resolution of the flavocytochrome *p*-cresol methylhydroxylase into subunits and reconstitution of the enzyme. *Biochemistry* 24:5276–5280.

Kowal, A. T., Werth, M. T., Manodori, A., Cecchini, G., Schroder, I., Gunsalus, R. P., and Johnson, M. K., 1995, Effect of cysteine to serine mutations on the properties of the [4Fe-4S] center in Escherichia coli fumarate reductase. *Biochemistry* 34:12284–12293.

Lederer, F., 1991, Flavocytochrome $b_2$, in *Chemistry and Biochemistry of the Flavoenzymes*, Vol. 2 (F. Muller, ed.) CRC Press, Boca Raton, pp. 153–242.

Lim, L. W., Mathews, F. S., and Steenkamp, D. J., 1982, Crystallographic study of the iron-sulfur flavoprotein trimethylamine dehydrogenase from the bacterium W3A1. *J. Mol. Biol.* 162:869–876.

Lim, L. W., Mathews, F. S., and Steenkamp, D. J., 1988, Identification of ADP in the iron-sulfur flavoprotein trimethylamine dehydrogenase. *J. Biol. Chem.* 263:3075–3078.

Lim, L. W., Shamala, N., Mathews, F. S., Steenkamp, D. J., Hamlin, R., and Xuong, N., 1986, Three-dimensional structure of the iron-sulfur flavoprotein trimethylamine dehydrogenase at 2.4 Å resolution. *J. Biol. Chem.* **261**:15140–15146.

Marcus, R. A., and Sutton, N., 1985, Electron transfer in chemistry and biology, *Biochim. et Biophys. Acta.* **811**:265–322.

Masters, B. S., Bilimoria, M. H., Kamin, H., and Gibson, Q. H., 1991, The mechanism of 1- and 2-electron transfers catalyzed by reduced triphosphopyridine nucleotide-cytochrome c reductase. *Nature* **351**:714–718.

McIntire, W., Edmondson, D. E., Hopper, D. J., and Singer, T. P., 1981, 8 alpha-(O-tyrosyl)flavin adenine dinucleotide, the prosthetic group of bacterial *p*-cresol methylhydroxylase. *Biochemistry* **20**:3068–3075.

Mathews, F. S., Chen, Z.-w., Meyer, T. E., Cusanovich, M. A., Koh, M., Cusanovich, M. A., and Van Beeumen, J. J., 1996, structural studies of flavocytochrome c sulfide dehydrogenase from the purple phototrophic bacterium *Chromatium vinosum*, in Flavins and Flavoproteins 1996 (K. J. Stephens, V. Massey, and C. H. Williams, eds.), University of Calgary Press, Calgary, 913–916.

Meyer, T. E., Bartsch, R. G., Caffrey, M. S., and Cusanovich, M. A., 1991a, Redox potentials of flavocytochrome c from the phototrophic bacteria *Chromatium vinosum* and *Chlorobium thiosulfatophilum*. *Arch. Biochem. Biophys.* **287**:128–134.

Meyer, T. E., Bartsch, R. G., and Cusanovich, M. A., 1991b, Adduct formation between sulfite and the flavin of phototrophic bacterial flavocytochromes c. Kinetics of sequential bleach, recolor, and rebleach of flavin as a function of pH. *Biochemistry* **30**:8840–8845.

Miles, C. S., Rouviere-Fourmy, N., Lederer, F., Mathews, F. S., Reid, G. A., Black, M. T., and Chapman, S. K., 1992, Tyr-143 facilitates interdomain electron transfer in flavocytochrome $b_2$. *Biochem. J.*, **285**:187–192.

Narhi, L. O., and Fulco, A. J., 1987, Identification and characterization of two functional domains in cytochrome P-450BM-3, a catalytically self-sufficient monooxygenase induced by barbiturates in Bacillus megaterium. *J. Biol. Chem.* **262**:6683–6690.

Nicholls, A., Sharp, K. A., and Honig, B., 1991, Protein folding and association: insights from the interfacial and thermodynamic properties of hydrocarbons. *PROTEINS, Structure, Function and Genetics*, **11**:281–296.

Pace, C. P., and Stankovich, M. T., 1991, Oxidation-reduction properties of trimethylamine dehydrogenase: effect of inhibitor binding. *Arch. Biochem. Biophys.* **287**:97–104.

Porter, T. D., and Coon, M. J., 1991, Cytochrome P-450: Multiplicity of isoforms, substrates, and catalytic and regulatory mechanisms. *J. Biol. Chem.* **266**:13469–13472.

Ravichandran, K. G., Boddupalli, S. S., Hasermann, C. A., Peterson, J. A., and Deisenhofer, J., 1993, Crystal structure of hemoprotein domain of P450BM-3, a prototype for microsomal P450's. *Science* **261**:731–736.

Regan, J. J., 1994, *GREENPATH Version 0.97*, San Diego.

Rohlfs, R. J., and Hille, R., 1991, Intramolecular electron transfer in trimethylamine dehydrogenase from bacterium W-3A1. *J. Biol. Chem.* **266**:15244–15252.

Rossmann, M. G., Moras, D., and Olson, K. W., 1974, Chemical and biological evolution of a nucleotide-binding protein. *Nature* **250**:194–199.

Sevrioukova *et al.*, 1999b

Sevrioukova, I. F., Li, H., Zhang, H., Peterson, J. A., and Poulos, T. L., 1999a, Structure of a cytochrome P450-redox partner electron-transfer complex. *Proc. Natl. Acad. Sci. USA.* **96**:1863–1868.

Sevrioukova, I., Truan, G., and Peterson, J. A., 1997, Reconstitution of the fatty acid hydroxylase activity of cytochrome P450BM-3 utilizing its functional domains. *Arch. Biochem. Biophys.* **340**:231–238.

Shamala, N., Lim, L. W., Mathews, F. S., McIntire, W., Singer, T. P., and Hopper, D. J. 1986, Structure of an intermolecular electron transfer complex: *p*-cresol methylhydroxylase at 6.0 Å resolution. *Proc. Natl. Acad. Sci. USA* **6**:4626–4630.

Shrake, A., Rupley, J. A., 1973, Environment and exposure to solvent of protein atoms. Lysozyme and insulin. *J. Mol. Biol.* **79**:351–371.

Steenkamp, D. J., and Beinert, H., 1982, Mechanistic studies of the dehydrogenases of methylotrophic *bacteria*. *Biochem. J.* **207**:233–239.

Steenkamp, D. J., and Gallup, M., 1978, The natural flavoprotein electron acceptor of trimethylamine dehydrogenase. *J. Biol. Chem.* **253**:4086–4089.

Steenkamp, D. J., and Mallinson, J., 1976, Trimethylamine dehydrogenase from a methylotrophic bacterium I. Isolation and steady-state kinetics. *Biochim. Biophys. Acta.* **429**:705–719.

Steenkamp, D. J., McIntire, W., and Kenney, W. C., 1978, Structure of the covalently bound coenzyme of trimethylamine dehydrogenase. Evidence for a 6-substituted flavin. *J. Biol. Chem.* **253**:2818–2824.

Stevenson, R. C., Dunham, W. R., Sands, R. H., Singer, T. P., and Beinert, H., 1986, Studies of the spin-spin interaction between flavin and iron-sulfur cluster in an iron-sulfur flavoprotein. *Biochim. et. Biophys. Acta.* **869**:81–88.

Tegoni, M., Silvestrini, M. C., Guigliarelli, B., Asso, M., Brunori, M., and Bertrand, P., 1998, Temperature-jump and potentiometric studies on recombinant wild type and Y143F and Y254F mutants of Saccharomyces cerevisiae flavocytochrome b2: role of the driving force in intramolecular electron transfer kinetics. *Biochemistry* **37**:12761–12771.

Van Driessche, G. V., Koh, M., Chen, Z.-w., Mathews, F. S., Meyer, T. E., Bartsch, R. G., Cusanovich, M. A., and Van Beeumen, J. J., 1996, Covalent structure of the flavoprotein subunit of the flavocytochrome c-sulfide dehydrogenase from the purple phototrophic bacterium *Chromatium vinosum*. *Protein Science* **5**:1753–1764.

Walker, W. H., and Singer, T. P., 1970, Identification of the covalently bound flavin of succinate dehydrogenase as 8-alpha-(histidyl) flavin adenine dinucleotide. *J. Biol. Chem.* **245**:4224–4225.

Wang, M., Roberts, D. L., Paschke, R., Shea, T. M., Masters, B. S., and Kim, J. J., 1997, Three-dimensional structure of NADPH-cytochrome P450 reductase: prototype for FMN-and FAD-containing enzymes. *Proc. Natl. Acad. Sci. USA* **94**:8411–8416.

Westenberg, D. J., Gunsalus, R. P., Ackrell, B. A., Sices, H., and Cecchini, G., 1993, Escherichia coli fumarate reductase frdC and frdD mutants. Identification of amino acid residues involved in catalytic activity with quinones. *J. Biol. Chem.* **268**:815–822.

White, S. A., Hamada, K., Veisaga, M. L., Scrutton, N. S., Hille, R., and Mathews, F. S., 2000, The refined structure of trimethylamine dehydrogenase at 1.7 Å resolution. Manuscript in preparation.

Willie, A., Edmondson, D. E., and Jorns, M. S., 1996, Sarcosine oxidase contains a novel covalently bound FMN, *Biochemistry* **35**:5292–5299.

Wodara, C., Bardischewsky, F., and Friedrich, C. G., 1997, Cloning and characterization of sulfite dehydrogenase, two c-type cytochromes, and a flavoprotein of *Paracoccus denitrificans* GB17: essential role of sulfite dehydrogenase in lithotrophic sulfur oxidation. *J. Bacteriol.* **179**:5014–5023.

Xia, Z.-x., and Mathews, F. S., 1990, Molecular structure of flavocytochrome b2 at 2.4 A resolution. *J. Mol. Biol.* **212**:837–863.

*Chapter 3*

# Methanol Dehydrogenase, a PQQ-Containing Quinoprotein Dehydrogenase

Christopher Anthony

## 1. INTRODUCTION

Methanol dehydrogenase (MDH; EC 1.1.99.8) catalyses the oxidation of methanol to formaldehyde in the periplasm of methylotrophic bacteria during growth on methanol or methane. It was first described in *Methylobacterium extorquens* (Anthony and Zatman, 1964a,b) and has subsequently been shown to be the one feature that is common to almost all methylotrophs in which it often constitutes up to 15% of their soluble protein (see Anthony, 1986 for a review of the basic enzymology of a wide range of MDHs).

MDH is a soluble quinoprotein which has pyrroloquinoline quinone (PQQ) as its prosthetic group and it uses a specific cytochrome, cytochrome $c_L$ as electron acceptor. It is usually assayed in a dye-linked system at high pH when ammonia is required as activator. It has an $\alpha_2\beta_2$ structure; each $\alpha$

**CHRISTOPHER ANTHONY**    Division of Biochemistry and Molecular Biology, School of Biological Sciences, University of Southampton, Southampton SO16 7PX.
*Subcellular Biochemistry, Volume 35: Enzyme-Catalyzed Electron and Radical Transfer*, edited by Holzenburg and Scrutton. Kluwer Academic / Plenum Publishers, New York, 2000.

subunit has one $Ca^{2+}$ ion coordinated to the PQQ. This is not covalently attached and so differs from the prosthetic groups of other quinoproteins which are derived from amino acids in the protein backbone of the enzyme and are shown in Figure 1 for comparison with PQQ: Tryptophan tryptophylquinone (TTQ) is derived from two tryptophan residues and occurs in bacterial amine dehydrogenases (McIntire *et al.*, 1991; Davidson, 1993; Davidson, this book); Topa-quinone (TPQ) is a modified tyrosine residue and is the prosthetic group of the copper-containing amine oxidases found in bacteria, yeasts, plants and animals (Janes *et al.*, 1990; Klinman, 1995); and lysyl oxidase is a special type of copper-containing amine oxidase

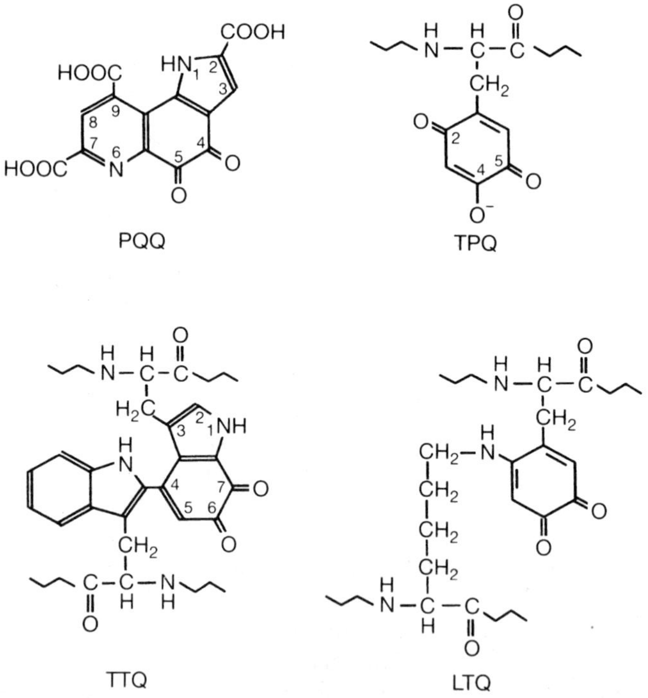

**FIGURE 1.** The prosthetic groups of quinoproteins. PQQ (pyrrolo-quinoline quinone) is the prosthetic group of dehydrogenases for methanol, higher alcohols, aldose sugars, aldehydes and polyvinyl alcohol and for hydroxylation of lupanine. TTQ (tryptophan tryptophylquinone) is the prosthetic group of amine dehydrogenases. TPQ (6-hydroxyphenylalanine or topa quinone) is the prosthetic group of the copper-containing amine oxidases in bacteria, plants and animals. LTQ (lysine tyrosylquinone) is the prosthetic group of lysyl oxidase, a specific copper-containing amine oxidase occurring in animals.

whose prosthetic group is Lysyl tyrosylquinone (LTQ) (Wang *et al.*, 1996). The structures and mechanisms of all these enzymes have been reviewed by Anthony (1996, 1998), and the biochemistry, physiology and genetics of PQQ and PQQ-containing enzymes reviewed by Goodwin and Anthony (1998).

This chapter will review the general features of the enzymology of methanol dehydrogenase, its structure as determined by X-ray crystallography, the properties of the prosthetic group (PQQ), the mechanism of its reduction by substrate, its oxidation by its specific cytochrome electron acceptor, and processes involved in its synthesis.

## 2.   GENERAL ENZYMOLOGY

### 2.1.   The Determination of MDH Activity

Although cytochrome $c_L$ is the physiological electron acceptor, it is most convenient to assay MDH in a dye-linked assay system with phenazine methosulphate or ethosulphate, linked to oxygen in an oxygen electrode, or linked to reduction of 2,6-dichlorophenol indophenol and measured spectrophotometrically (for convenient assay systems see Day and Anthony, 1990; Frank and Duine, 1990). The pH optimum in these assays is usually at least pH 9; in such conditions phenazine derivatives form free radicals and these are possibly the true electron acceptor (Duine *et al.*, 1978). This suggestion is supported by the demonstration that a better alternative to phenazine ethosulphate is Wurster's Blue which is the perchlorate salt of the cationic free radical of NNN'N'tetramethyl-*p*-phenylene diamine (Duine *et al.*, 1978; Frank and Duine, 1990). Use of Wurster's Blue avoids the problem of inactivation of MDH by phenazine dyes in the absence of protecting substrate; its only disadvantage is that it is not commercially available and must be synthesised from a commercially available precursor (Frank *et al.*, 1988; Harris and Davidson, 1994; Davidson *et al.*, 1992).

It should be noted that the dye-linked assay system is artificial in every way; the electron acceptor may also be an inhibitor; it has a high pH optimum (about pH 9); it requires ammonia as activator, but this may also inhibit; cyanide is a competitive inhibitor and may be used as a protective agent; in the absence of added substrate a high rate of dye reduction occurs which may or may not be taken into account when calculating rates of reaction. This complexity and confusion is mentioned here as an explanation for the length and complexity of some the following discussion.

## 2.2. The Substrate Specificity of MDH

MDH is able to oxidise a wide range of primary alcohols as well as methanol (the best substrate), including, for example, crotyl alcohol, bromoethanol, phenylethanol and cinnamyl alcohol. Their steric configuration is more important in determining whether or not they are oxidised than is the presence or absence of atoms or groups producing electron-displacement effects (Anthony and Zatman, 1965; Anthony, 1986). The general rule is that a second substituent on the C—2 atom appears to prevent binding, the general formula for an oxidizable substrate being $R.CH_2OH$ where R may be H, OH (hydrated aldehydes), $R'.CH_2$— or $R'.R''C{=}CH$— (Anthony and Zatman, 1965). The affinity for methanol and most other substrates is high (for methanol, $K_m = 3$–$20\,\mu M$), the affinity tending to decrease with size of the alcohol, although maximum rates are usually similar to that for methanol (Duine and Frank, 1980b; Anthony, 1986). Analysis of the reaction with $^{13}C$ deuteriated benzyl alcohols has demonstrated that the enzyme specifically removes the *pro*-S hydrogen at the C—1 carbon atom, the *pro*-R hydrogen being retained in the aldehyde product (Houck and Unkefer, cited in Anthony, 1993). Recent results using chemical models of the active site PQQ bound to $Ca^{2+}$ in organic solvents confirm that the substrate specificity of this enzyme is not due to specific aspects of the catalytic process but is dictated by the dimensions of the active site (Itoh *et al.*, 1998). When alcohols are not substrates they are unable to act as a competitive inhibitors, suggesting that if the alcohol is able to bind then it is suitable as a substrate.

MDH is able to oxidise formaldehyde (probably in the gem-diol hydrated form), although this is very unlikely to be important physiologically, as a high proportion of the formaldehyde product is assimilated into cell material. Inappropriate formaldehyde oxidation is prevented by a periplasmic modifier protein (M-protein) which decreases the affinity of MDH for formaldehyde (Page and Anthony, 1986; Long and Anthony, 1990a, 1991). M-protein has a second effect on the activity of MDH; it increases its affinity for some substrates that, in its absence, have very low affinities (e.g. 1,2-propanediol and 4-hydroxybutyrate). This protein was first observed as a result of studies of the growth of *M. extorquens* on 1,2-propanediol which ceases to be a substrate in mutants lacking the dehydrogenase, and which is oxidised only in crude extracts when M-protein is also present (Bolbot and Anthony, 1980). That the role of this protein in growth on 1,2-propanediol and 4-hydroxybutyrate is fortuitous and secondary to its main role in regulation of formaldehyde oxidation is supported by the observation that M-protein occurs in a range of methylotrophs including obligate methylotrophs which are unable to grow on multicarbon compounds such as 1,2-propanediol.

## 3.   THE KINETICS OF METHANOL DEHYDROGENASE

### 3.1.   The Reaction Cycle of MDH

It might be expected that the steady state kinetics of a dehydrogenase would be straightforward and informative with respect to the mechanism but this is not the case with MDH. When it is incubated at high pH with electron acceptor and activator (ammonia) in the absence of substrate, an unexpectedly large amount of electron acceptor becomes reduced by an unidentified "endogenous substrate" (Anthony and Zatman, 1964b; Duine *et al.*, 1978, 1980b) which is presumed to be very low concentrations of contaminating alcohols and aldehydes in buffers and reagents.

An extensive steady state kinetic analysis of MDH using phenazine ethosulphate by Duine and Frank (1980b) established many of the main features of the kinetics of MDH and solved the problem of isolating an oxidised form of the enzyme by including the competitive inhibitor cyanide in the reaction mixtures (also included at constant concentration during kinetic measurements). This study was then extended (with Wurster's Blue) using stopped flow kinetics with oxidised and reduced enzyme (Frank *et al.*, 1988). The reaction cycle proposed from this work (Figure 2) continues to be a generally valid description although modifications have been proposed by Harris and Davidson (1993a, 1994) who showed that ammonia is also an inhibitor at high concentrations (see Section 3.2). In their model, MDH binds substrate but can only form product when first activated by cyanide binding at a second (non-substrate) binding site. However, this can only be defined as an activator in the special context of demonstrating "methanol-dependent activity" which is not always seen because of the endogenous dye reduction. Cyanide, a competitive inhibitor, prevents oxidation of the very low concentration of endogenous substrate; when methanol is added then, of course, the activity resumes. In this context it should be noted that, as the endogenous substrate is oxidised by the same mechanism as added substrate which is oxidised in preference to the endogenous substrate, it is usually inappropriate to subtract the endogenous value in order to obtain a valid measure of enzyme activity (Beardmore-Gray *et al.*, 1983; Anthony, 1986).

Steady state, and stopped flow, kinetics using cytochrome $c_L$ as electron acceptor subsequently confirmed that the reaction cycle is essentially the same as with a dye electron acceptor (Figure 2) (Dijkstra *et al.*, 1989) and

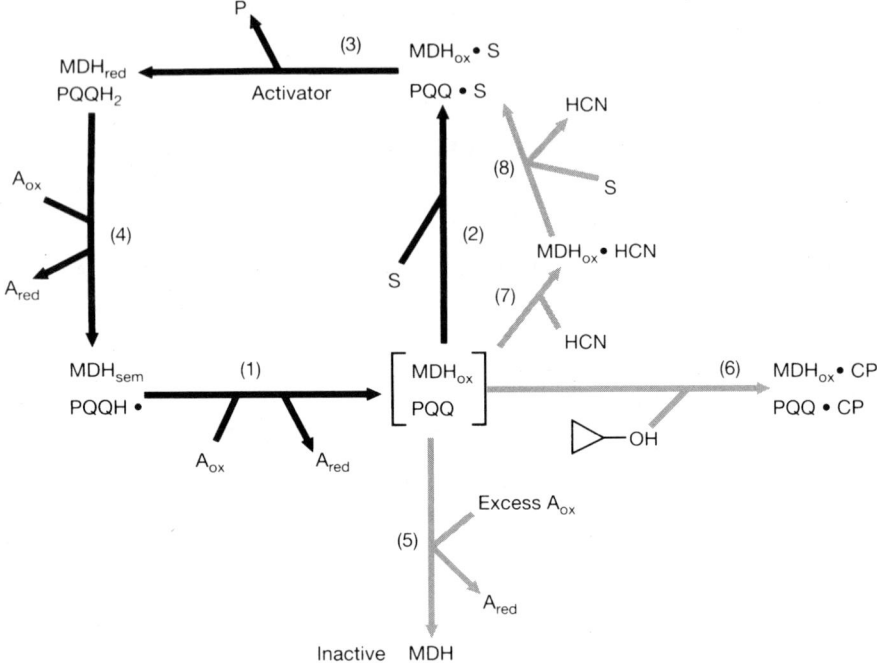

**FIGURE 2.** The reaction cycle of methanol dehydrogenase. This reaction cycle is based on the work of Frank and colleagues (Frank *et al.*, 1988; Dijkstra *et al.*, 1989). The electron acceptor (A) is either a dye such as phenazine ethosulphate or the natural electron acceptor cytochrome $c_L$. The normal cycle is outlined using dark arrows. The starting point is the semiquinone form of the enzyme as it is usually isolated $MDH_{sem}$. Reactions 1, 2 and 4 are reversible; reaction 3 is not. It should be emphasised that the enzyme does not require ammonia, or cyanide, or a high pH for full activity in the physiological reaction. Ammonia (and perhaps cyanide) may bind to MDH in a manner which either induces conformational changes or modifies amino acid residues in such a way as to mimic the in vivo form of the enzyme when it is in contact with the modifier protein, cytochrome $c_L$ or some other component of the cell.

further studies using this system have provided information for analysis by Marcus theory relating to reorganizational energy, electronic coupling and theoretical distances between redox centres (Harris and Davidson, 1993b; Harris *et al.*, 1994).

### 3.2. The Activation of MDH by Ammonia and Amines

As usually prepared, MDH has an absolute requirement for ammonium or methylammonium salts in the dye-linked assay system. The relatively higher concentrations required at lower pH values suggest that the free base is the active species (Anthony and Zatman, 1964b). This can be replaced by esters of glycine or β-alanine but not by lysine esters nor by aliphatic amines or amino acids; the affinity for the alternative activators is usually higher than for ammonium salts. The enzyme does not always need ammonia when cytochrome $c_L$ is used as electron acceptor and sometimes anaerobically-prepared enzyme has no ammonia requirement (for references to the original literature see Anthony, 1986).

Kinetic analysis of the activation by ammonia of MDH is usually difficult because of its very high affinity for methanol and the endogenous dye reduction (see above) (Frank *et al.*, 1988; Harris and Davidson, 1993a). This problem has been overcome in a modified form of enzyme in which the $Ca^{2+}$ ion in the active site has been replaced by $Ba^{2+}$, which decreases markedly the affinity for alcohol substrate and so has eradicated the "endogenous reductant" problem. These studies confirmed the previous conclusion (Frank *et al.*, 1988) that the main effect of the activator is to increase the $V_{max}$ and that the apparent affinity for methanol decreased with increasing concentrations of activator (Goodwin and Anthony, 1996). The steady state kinetic data summarised in Figure 3 show that, although methanol and ammonia are essential for activity, each binds more strongly to the enzyme in the absence of the other. This suggests that the binding site for ammonia is likely to be independent of, but very close to, that for methanol; both activator and methanol must both be bound for formation of product. These studies showed that ammonia does not activate by

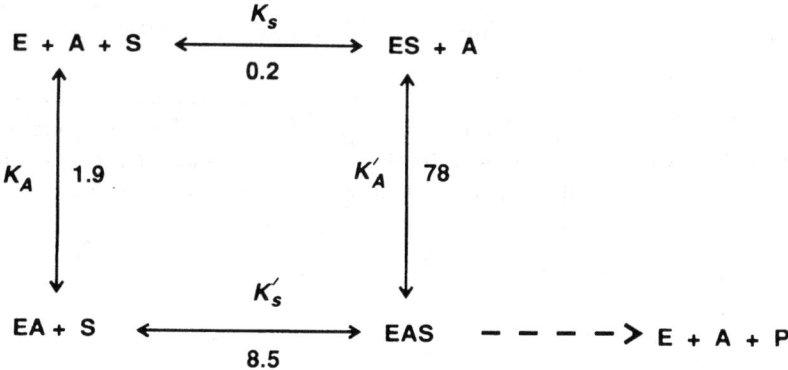

**FIGURE 3.** Kinetic scheme for the activation of Ba-MDH by ammonia. The combination of enzyme (E) with substrate (S) is not independent of its combination with activator (a), so E and EA have different affinities for the substrate. The equation for the reaction is as follows: $v = V_{max}/(1 + K'_s/S)[1 + K_A/a (1 + S/K_s)/(1 + S/K'_s)]$. The values given are for the dissociation constants for each complex (mM).

encouraging the initial binding of substrate, or by affecting the formation of any initial PQQ-methanol adduct, or by stimulating the re-oxidation of reduced PQQ. It confirmed the previous suggestion (Frank *et al.*, 1988) that ammonia activates the reduction of PQQ by methanol, the rate limiting step in this process being the transfer of the methyl hydrogen to PQQ. The high deuterium isotope effect associated with this hydrogen transfer step is affected by ammonia; increasing ammonia concentration (30-fold) decreased the measured isotope effect from 6.7 to 1.4. Using the Ba-MDH this was subsequently shown to be because the $K_A$ for deuteriated methanol is 3.5 times higher than for normal substrate (Goodwin and Anthony, 1996).

Kinetic analysis has also shown that there is a second (inhibitory) binding site for ammonia with $K_i$ of 50–100 mM; spectroscopic evidence suggests that this may be due to formation of an imino adduct with the PQQ (Harris and Davidson, 1993; Goodwin and Anthony, 1996).

### 3.3. The Low *in vitro* Rate of Cytochrome $c_L$ Reduction by MDH

Although cytochrome $c_L$ is the physiological electron acceptor for the dehydrogenase, the rate of its reduction *in vitro* is extremely slow compared with the rate measured in the artificial system with phenazine ethosulphate. This has been explained by a thorough study of the interaction of the enzyme with cytochrome $c_L$ using stopped flow kinetics, and studies of isotope effects. It appears that the differences in rates between electron acceptors are due to different rate-limiting steps in the reaction cycles (Dijkstra *et al.*, 1989). Ferricytochrome $c_L$ is an excellent oxidant of reduced enzyme at pH 7 but the substrate oxidation step is slow because the activation by ammonia is almost ineffective at this pH. At pH 9 the reverse situation exists: ferricytochrome $c_L$ is a poor oxidant of reduced enzyme; no deuterium isotope effect is observed, showing that substrate oxidation is not rate-limiting and so activation by ammonia is not seen. This conclusion emphasises or illustrates the problem with the ammonia activation. It must be presumed that an alternative (unknown) activator operates *in vivo* with cytochrome $c_L$ (see Dijkstra *et al.*, 1988) or that the dehydrogenase is altered in some way during its isolation such that the ammonia activator is now required.

### 4.  THE ABSORPTION SPECTRA OF METHANOL DEHYDROGENASE

MDH is usually coloured olive-green, the spectrum of the isolated enzyme having a characteristic absorption due to the prosthetic group at

345 nm (Anthony and Zatman, 1967); the PQQ in the isolated enzyme is usually in the semiquinone or reduced form (de Beer *et al.*, 1983; Frank *et al.*, 1988; Dijkstra *et al.*, 1989) (Figure 4). This tends to cause problems in all kinetic work and spectroscopy. For example, the effect of reduction by added substrate cannot be determined without first producing the oxidised form of the enzyme. This can only be produced by addition of small amounts of PES or Wurster's Blue in the presence of ammonia. However, the ubiquitous "endogenous substrate" rapidly re-reduces the enzyme. Incubation with an excess of PES or Wurster's Blue inactivates the enzyme unless a competitive inhibitor such as cyanide is included in the reaction mixture; as a result most spectra of the oxidised form are likely to be spectra of cyanide or ammonia adducts of the PQQ (Duine and Frank, 1980b; Frank *et al.*, 1988). Addition of ammonia at the concentration needed for activation leads to no spectral perturbation, but higher inhibitory concentrations lead to changes in the spectrum (about 400 nm) suggesting that, at these concentrations of ammonia, an imine adduct of PQQ may be formed (Harris and Davidson, 1993a).

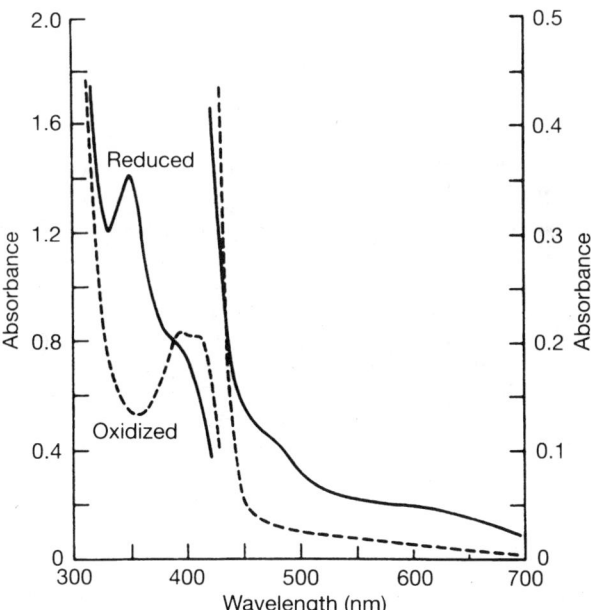

**FIGURE 4.** The absorption spectra of methanol dehydrogenase. The spectrum of the oxidised form is very difficult to determine and is often likely to be that of the cyanide or ammonia adducts. MDH is usually isolated in the reduced or half-reduced form (semiquinone).

The presence of the $Ca^{2+}$ ion in the active site has a major effect on the spectrum as shown by using a mutant (*mxaA*) of *Methylobacterium extorquens* that produces both subunits of MDH and also the PQQ, but is unable to process these in the periplasm to give the complete active enzyme. The result is an inactive enzyme containing PQQ but no calcium and having a markedly different spectrum; incubation at pH 10 with $Ca^{2+}$ led to incorporation of $Ca^{2+}$, and production of a normal spectrum together with a normally active enzyme (Figure 5) (Goodwin *et al.*, 1996). Before incorporation of $Ca^{2+}$ the PQQ was in the oxidised form but became immediately reduced by endogenous substrate as soon as $Ca^{2+}$ was incorporated. When $Ca^{2+}$ was replaced with $Ba^{2+}$ the enzyme was more active but had a hugely decreased affinity for substrate (including endogenous substrate); consequently the oxidised enzyme could easily be produced and its subsequent reduction observed (Goodwin and Anthony, 1996). This provided probably the first example of a spectrum of the oxidised MDH without adducts (Figure 6).

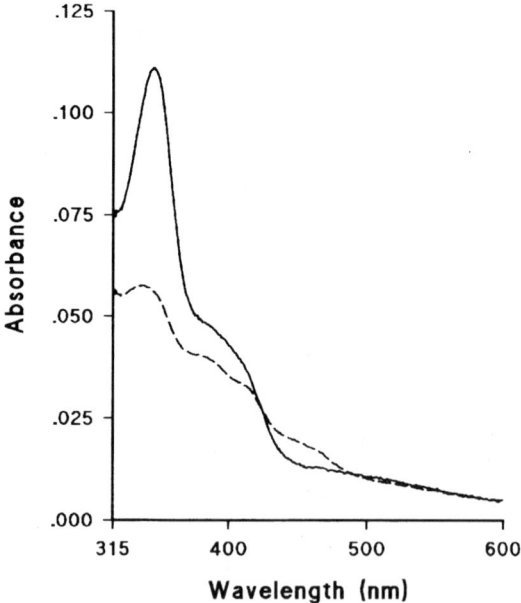

**FIGURE 5.** The absorption spectrum of MDH lacking calcium (broken line), isolated from *mxaA* mutant, and the effect of incorporation of calcium into the enzyme (solid line). Reproduced with permission from Goodwin *et al.* (1996), (*Biochemical Journal*, **319**, 839–842). © the Biochemical Society.

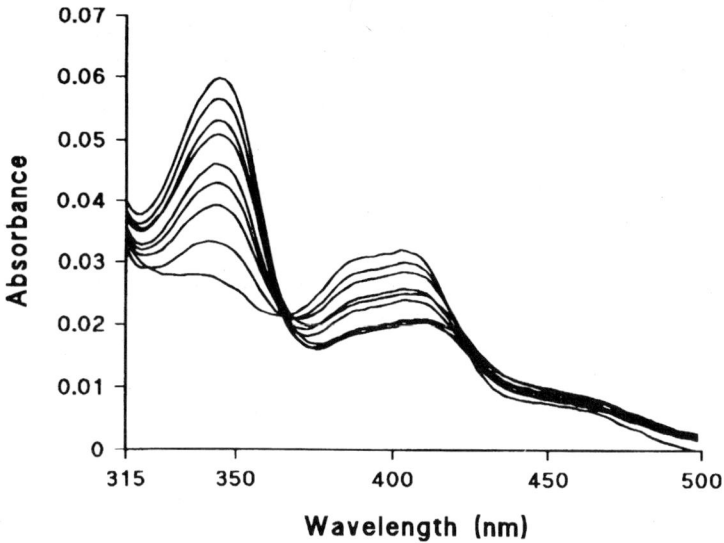

**FIGURE 6.** The reduction of oxidised Ba-MDH by endogenous substrate. MDH lacking any metal ion in its active site was produced from the *mxaA* mutant; it was incubated with $Ba^{2+}$ to produce Ba-MDH which was then oxidised with a small excess of Wurster's Blue which was then removed by rapid gel filtration and spectra recorded. Reproduced with permission from Goodwin and Anthony (1996), (*Biochemical Journal*, **318**, 673–679). © the Biochemical Society.

The oxidation state was confirmed by demonstrating that its isolated prosthetic group was predominantly in the oxidised (quinone form). Figure 6 shows the spectrum of the Ba-MDH after oxidation at 4°C with a 1.5-fold excess of Wurster's Blue at pH 9. The spectrum of the initial oxidised form of the enzyme is very similar to that seen in Figure 4.

After $Ca^{2+}$ has been incorporated into MDH it cannot usually be removed without completely denaturing the enzyme. An exceptional MDH is that isolated from the marine methylotroph *Methylophaga marina* which requires at least 50 mM NaCl for stability. Sometimes, during storage, the enzyme lost $Ca^{2+}$, and became inactive and coloured red (due to absorbance at 520 nm) (Figure 7), the spectrum being similar to that in $Ca^{2+}$-deficient MDH (see above); in this case the normal spectrum and activity was regained by incubation with $Ca^{2+}$ (Chan and Anthony, 1992b). It appears that the MDH of *M. marina* has a relatively low affinity for $Ca^{2+}$ which is probably correlated with its normal seawater environment containing about 10 mM $Ca^{2+}$.

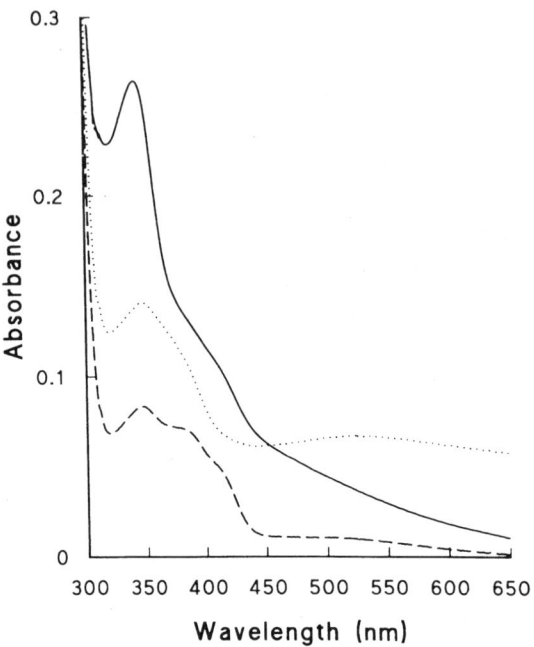

**FIGURE 7.** The spectra of the MDH from *Methylophaga marina*. The dotted line is the spectrum of the red form of the enzyme; the solid line is the active green form produced by incubation with $Ca^{2+}$, and the dashed line is the spectrum of an inactive green form produced by incubation at low pH in the absence of calcium. Reprinted from *FEMS Microbiology Letters*, **97**, Chan and Anthony, Characterisation of a red form of methanol dehydrogenase from the marine methylotroph *Methylophaga marina*, 293–298, Copyright 1992, with permission from Elsevier Science.

## 5.  PYRROLO-QUINOLINE QUINONE (PQQ): THE PROSTHETIC GROUP OF METHANOL DEHYDROGENASE

The novel prosthetic group from MDH was first isolated, purified and characterised by Anthony and Zatman (1967); it was readily removed from the enzyme by boiling or by treatment with acid or alkali, and shown to be reddish-brown in colour, highly polar and with a green fluorescence. The X-ray structure of an acetone adduct was eventually determined by Kennard and her colleagues (Salisbury *et al.*, 1979) and an extensive chemical characterisation was achieved by Frank and Duine and their co-workers (Duine and Frank, 1980a,b; Duine *et al.*, 1980, 1981; Dekker *et al.*, 1982; Duine, 1991). The structure of PQQ (full name: 2,7,9-tricarboxy-1H-pyrrolo[2,3-f]-quinoline-4,5-dione is given in Figure 1. Each tetramer of MDH has two molecules of tightly-bound PQQ which are present in the isolated enzyme in the semiquinone, free radical form (Duine *et al.*, 1981; de Beer *et al.*, 1983); this explains why addition of substrate (a two electron donor) does not lead to reduction of the PQQ.

Although PQQ is not covalently-bonded to the enzyme it impossible to reconstitute active enzyme from PQQ plus the apoenzyme of MDH, although this has been achieved with glucose dehydrogenase (Duine *et al.*, 1980; Cozier *et al.*, 1999). Figure 8 shows the absorption spectra of the quinone and quinol forms of PQQ. It has a characteristic green fluorescence which is maximal at low pH (excitation maximum, about 365 nm; emission maximum, about 460 nm) (Anthony and Zatman, 1967; Dekker *et al.*, 1982). The midpoint redox potential of the $PQQ/PQQH_2$ couple is +90 mV at pH 7.0, increasing to +419 mV at pH 2.0, indicating that it is acting at a $2e^-$ / $2H^+$ redox carrier (Duine *et al.*, 1981). Resonance Raman spectroscopy of the isolated PQQ, and other quinones, their derivatives, and quinoproteins has been extensively reviewed by Dooley and Brown (1993, 1995). Methods for measurement of PQQ, and the problems likely to be encountered are given in the review by Klinman and Mu (1994).

The crystal structures of the sodium and potassium salts of PQQ have shown that the tricyclic ring is planar and that the three carboxylate groups can be either coplanar with the ring or twisted out of the plane (van Koningsveld *et al.*, 1989). However it cannot be clear whether the observed

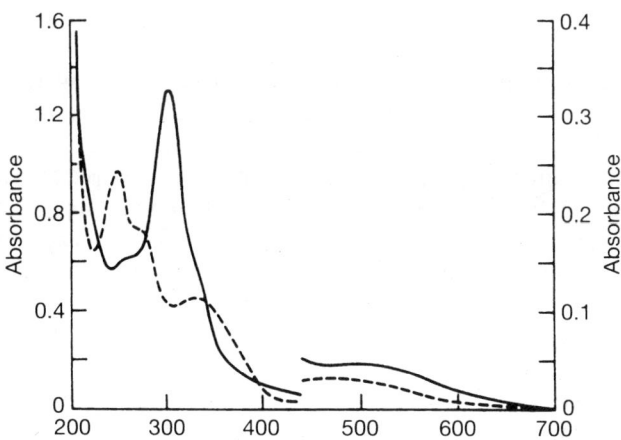

**FIGURE 8.** The absorption spectra of PQQ measured at pH 7.0. The solid line is the reduced, quinol form; the broken line is the oxidised quinone. The spectra were provided by J. A. Duine and J. Frank who have published the spectra of PQQ and those of many of its adducts (Figure 9) in Dekker *et al.* (1982).

planarity of the ring is an inherent property of PQQ itself or a manifestation of crystal packing forces. The conformation of PQQ and the role of
$Ca^{2+}$ were examined using a quantum mechanical method (Zheng and
Bruice, 1997) which indicated that, in the presence or absence of $Ca^{2+}$, the
tricyclic ring system will be essentially planar irrespective of its oxidation
state, although the C—9 carboxyl group can be coplanar or twisted out of
the plane of the ring, as seen in some X-ray structures of the enzyme (Ghosh
et al., 1995; White et al., 1996).

Extensive studies of the chemical reactivity of the isolated PQQ
demonstrated the formation, by nucleophilic attack at the C—5 position, of
a wide variety of adducts, many of which are potentially relevant to the
assay and mechanism of the enzyme, including alcohols, aldehydes, cyanide,
ammonia and amines (Dekker et al., 1982; Ohshiro and Itoh, 1993) (Figure
9). Molecular orbital calculations (Zheng and Bruice, 1997) indicated that
the two quinone carbonyl groups are not well coordinated with either the
pyrrole or the pyridine ring while the adjacency of the pyrrole ring
decreases the electron density of the quinone. As a result these two carbonyls are very reactive, the calculations supporting the experimental
observations that the C5 carbonyl is expected to be more reactive than the
C4 carbonyl. The same type of calculation indicated that the neutral and
ionic forms of the free radical semiquinone are both planar, a result that is
not consistent with the observed structures in the enzyme active site
(Section 7.4.2). It is the relative ease of formation of such adducts that can
make analysis and identification of PQQ in proteins and biological tissues
rather open to confusion, and the interpretation of spectral intermediates
in the reaction rather uncertain. For example, it has not been possible
unequivocally to demonstrate the production of a covalent reaction intermediate by spectroscopy, although it is probable that such an intermediate
occurs.

An important feature of PQQ is its ability to complex divalent metal
ions in solution. This was first shown by Mutzel and Gorisch (1991) and
exploited by Itoh et al. (1993, 1998) in their chemical model systems in which
a $Ca^{2+}$-PQQ complex is able to catalyse the oxidation of alcohols in organic
solvents. It is now generally appreciated that the PQQ-containing enzymes
probably all contain a divalent metal ion in their active sites. Itoh and
colleagues have published a series of papers on the mechanism of PQQ-
containing enzymes by using a series of PQQ model compounds in anhy-
drous organic media. The PQQ model compounds formed 1:1 complexes
with a series of alkaline earth metal ions, the site of coordination with $Ca^{2+}$,
$Sr^{2+}$ and $Ba^{2+}$ being the same as shown for $Ca^{2+}$ in the known enzyme structures (see Section 7.4.2). $Ca^{2+}$ ions bound much more tightly than the other
ions, the affinity decreasing with size of ion ($Ca^{2+} > Sr^{2+} > Ba^{2+}$). $Mg^{2+}$ binding

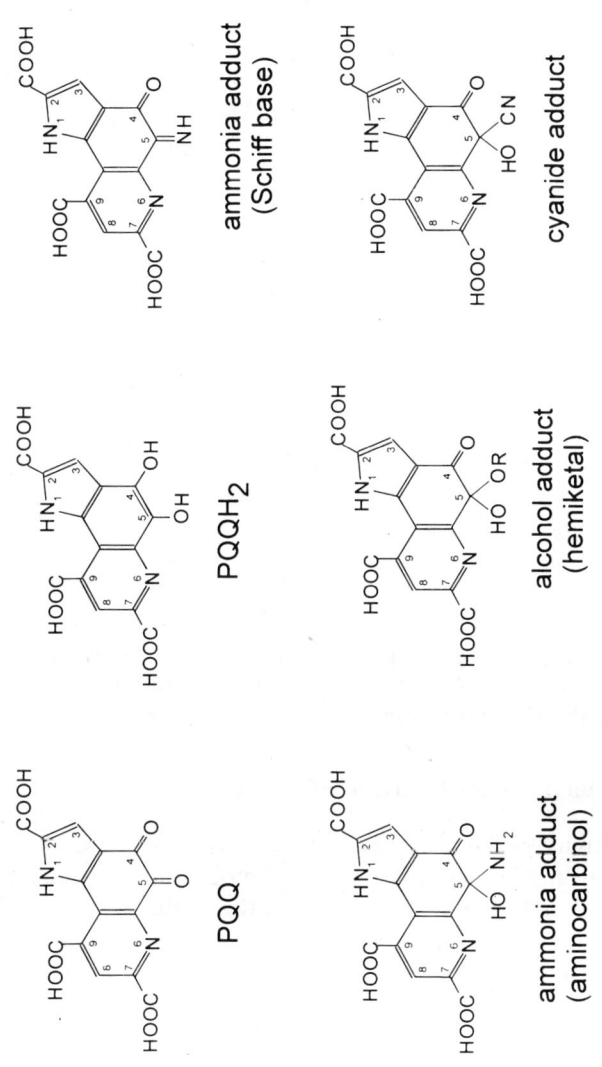

**FIGURE 9.** Adducts of PQQ. The formation and properties of these adducts are discussed extensively in Dekker *et al.* (1982).

was extremely weak, and resulted in hydration of the *o*-quinone moiety, thus deactivating the PQQ for redox reactions. Molecular orbital calculations suggested that the binding of $Mg^{2+}$ is completely different from the other alkaline earths, bonding to both the O4 and O5 atoms. This, together with the generally different properties of the $Mg^{2+}$ ion compared with other alkaline earth ions, sheds considerable doubt on the possibility that $Mg^{2+}$ replaces the active site $Ca^{2+}$ in the PQQ-containing glucose dehydrogenase, although it is usually assumed that the $Mg^{2+}$ required for reconstitution of this enzyme is essential as a constituent of the active site (Yamada *et al.*, 1998; Cozier *et al.*, 1999).

## 6.  THE REACTION MECHANISM OF METHANOL DEHYDROGENASE

Steady state kinetics showed that the enzyme catalyses a ping-pong reaction, consistent with reduction of PQQ by substrate and release of product, followed by two sequential single electron transfers to the cytochrome $c_L$, during which the $PQQH_2$ is oxidised back to the quinone by way of the free radical semiquinone (see Figure 2 for the reaction cycle) (Duine and Frank, 1980b; Frank *et al.*, 1988; Dijkstra *et al.*, 1989). The high pH optimum of the reaction with dye electron acceptors is related to the pH dependency of the oxidation of the reduced enzyme (reactions 1 and 4 in Figure 2); the pH optimum for oxidation by cytochrome $c_L$ was 7.0. The rate-limiting step is the conversion of the oxidised complex containing the substrate to the reduced enzyme plus product (reaction 3) and is the only step requiring the activator ammonia.

### 6.1.  The Reductive Half-Reaction of MDH

The reactivity of the C—5 carbonyl encourages the assumption that a covalent PQQ-substrate complex may be important in the reaction mechanism. A key piece of evidence relating to this is the demonstration of the reaction of MDH with cyclopropanol to give a C—5 propanal adduct by Frank *et al.* (1989). Because ring opening of cyclopropanol usually occurs in alkaline conditions, they suggested that the reaction is initiated by proton abstraction by a base at the active site and that this is followed by re-arrangement of the cyclopropoxy anion to a ring-opened carbanion, and attack of this on the electrophilic C—5 of PQQ. It was concluded that during oxidation of methanol a similar proton abstraction from the alcohol by an active site base must occur.

The position of Asp-303 near PQQ in the active site (see Section 7.4.2) suggests that this provides the catalytic base proposed above. Figures 10 and 11 illustrate two possible mechanisms, in which this aspartate (Asp-303) initiates the reaction by abstraction of a proton from the alcohol substrate. In these mechanisms the $Ca^{2+}$ is given a role in addition to a structural role in maintaining PQQ in an active configuration; it is proposed that the $Ca^{2+}$ acts as a Lewis acid by way of its coordination to the C—5 carbonyl oxygen of PQQ thus providing the electrophilic C—5 for attack by an oxyanion or hydride. The role of $Ca^{2+}$ in the mechanism has been given support by a study of a $Sr^{2+}$-containing MDH produced by bacteria in a high concentration of $Sr^{2+}$ (Harris and Davidson, 1994). Further evidence to support this has been provided by preparation of an active enzyme containing $Ba^{2+}$ instead of $Ca^{2+}$ (Goodwin and Anthony, 1996). This is the first example of any enzyme in which barium plays an active catalytic role; the Ba-MDH has a very low affinity for methanol ($K_m$, 3.4 mM instead of 10 μM), but its activation energy is half that of the Ca-MDH and its $V_{max}$ is doubled. It has been suggested that a further role of the $Ca^{2+}$ might be to contribute to the mechanism by coordinating to the hydroxyl oxygen of the substrate, thus decreasing its pK and facilitating the initial proton abstraction step (Zheng and Bruice, 1997; Itoh *et al.*, 1998). However, this appears unlikely as the $Ca^{2+}$ has now been shown to be too far away from the hydroxyl oxygen in the structure (Xia *et al.*, 1999).

Two general mechanisms for methanol oxidation have been proposed after the initial proton abstraction (Anthony, 1996, 1998). One is direct hydride transfer from the methanol α-methyl group to the PQQ; in the more likely mechanism, however, the oxyanion produced by the initial proton abstraction attacks the electrophilic C—5, leading to formation of a covalent hemiketal intermediate (Figure 10). The direct evidence on this point is the slight change in spectrum of a possible intermediate (half-life 2 minutes) seen during reaction with deuterated methanol in the presence of excess dye electron acceptor (Frank *et al.*, 1988; Dijkstra *et al.*, 1989). The subsequent reduction of the PQQ with release of product aldehyde is likely to require prior ionization of the C—4 carbonyl, a process facilitated by the pyrrole N atom. The involvement of this carbonyl in abstracting the methyl proton is consistent with the theoretical calculations and measurements with model PQQ systems published by Itoh *et al.* (1998). In these systems the rates of hemiketal formation from alcohols plus metal ions showed that the hemiketal stabilisation was about 6-fold greater in the presence of $Ca^{2+}$. In the $Ca^{2+}$ complex of the C5 hemiketal the $Ca^{2+}$ is coordinated by both oxygen atoms of the hemiketal (OH and OMe) and this coordination forces the methyl group of the added methanol to flip towards the C4 carbonyl

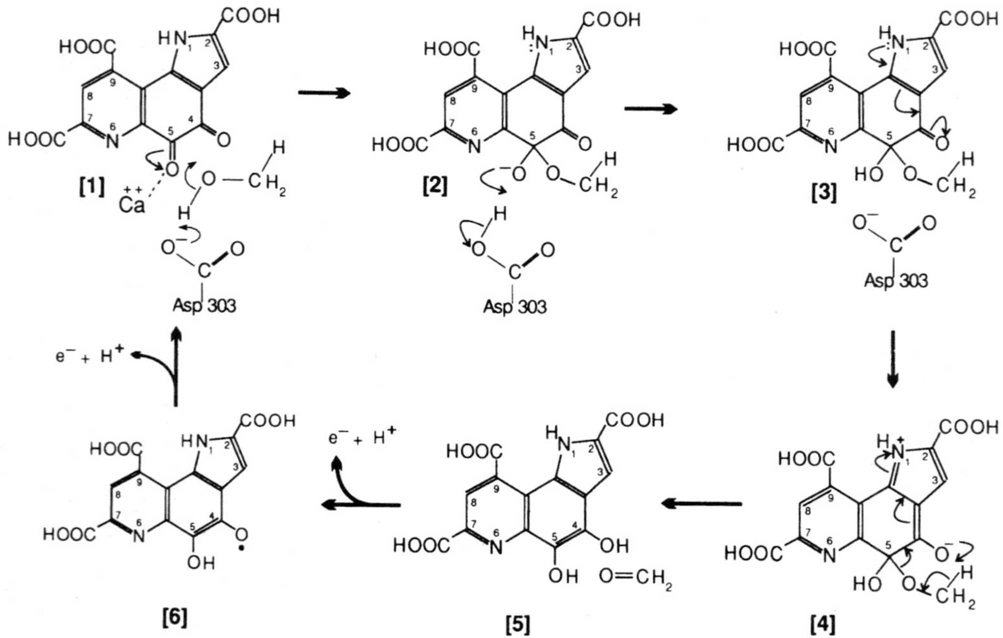

**FIGURE 10.** Proposed mechanism of MDH involving formation of hemiketal intermediate. It is suggested that $Ca^{2+}$ plays a catalytic role by acting as a Lewis acid, facilitating attack on the electrophilic C5 of PQQ. Proton abstraction by the base (Asp303) leads to an oxyanion form of the substrate which attacks C5, giving the hemiketal intermediate [3] from which the methyl proton is abstracted by the C5 carbonyl oxygen whose ionisation is facilitated by the pyrrole nitrogen atom. The oxidative phase of the reaction cycle involves electron transfer to either PES or cytochrome $c_L$ and involves the PQQ semiquinone free radical as an intermediate.

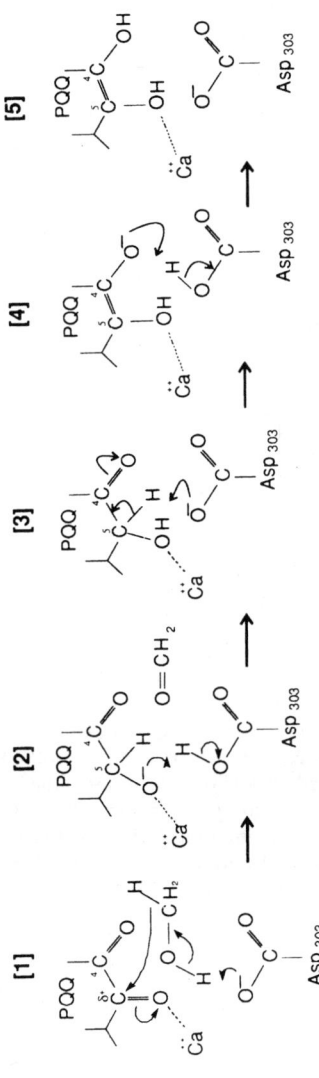

**FIGURE 11.** Alternative mechanism for MDH involving hydride transfer. In this case the initial proton abstraction is the same as in Figure 10, but the electrophilic C5 is involved directly in removal of the methyl hydrogen as a hydride. This mechanism is modified from those previously proposed (Anthony *et al.*, 1994) to emphasise the possible double involvement of the active site base (Asp303).

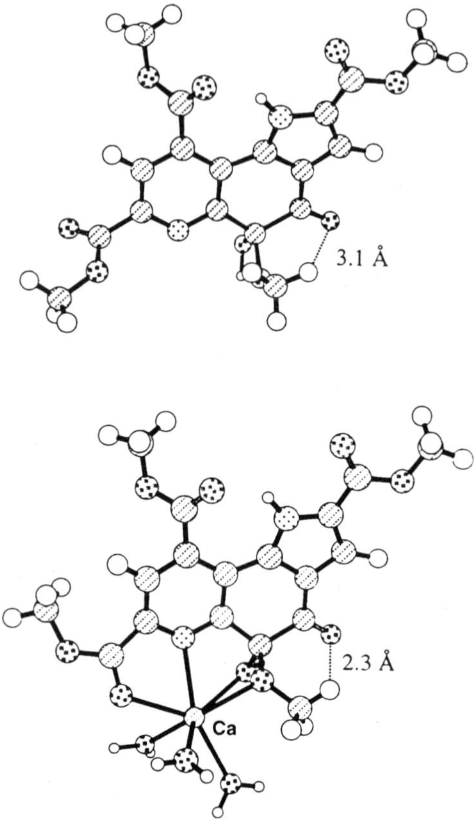

**FIGURE 12.** Computer-generated structures of the hemiketal adduct of the trimethyl ester of PQQ and its $Ca^{2+}$ complex. In the case of the $Ca^{2+}$ complex of the C5 hemiketal (the lower structure) the metal ion is coordinated by both oxygen atoms of the hemiketal (OH and Ome), and this coordination forces the methyl group to flip towards the C4 carbonyl oxygen, suggesting that this might act as a base, accepting the methyl proton from the substrate as suggested in the mechanism in Figure 10. Reprinted with permission from Itoh *et al.* (1998), (*Biochemistry*, **37**, 6562–6571). © American Chemical Society.

(Figure 12). Thus, the distance between the carbonyl oxygen at the 4-position and the methyl proton of added substrate in the $Ca^{2+}$ complex decreases by about 0.8 Å, consistent with the suggestion that the neighbouring carbonyl oxygen at the 4-position could act as a general base to accept the α-proton of the substrate. In their studies of model systems, Itoh *et al.* (1998) showed that addition of DBU (a strong base, diazabicyclo-undecene) to an acetonitrile solution containing the metal complexes of the PQQ model compounds led to reduction of the PQQ with production of the corresponding aldehyde. A large kinetic deuterium isotope effect was similar to that in the enzymic dehydrogenase reaction (about 6), clearly indicating that the base-catalyzed α-proton abstraction from the substrate is rate determining in the alcohol oxidation reaction. Divalent ions, but not monovalent cations or $Mg^{2+}$, accelerated the oxidative process, the relative effectiveness ($Ba^{2+} > Sr^{2+} > Ca^{2+}$) being the same as observed in the enzyme reaction (Goodwin and Anthony, 1996). It was suggested that this is consistent with intramolecular general base catalysis by the C4 carbonyl oxygen atom proposed above; namely, the binding of the larger $Ba^{2+}$ to the hemiketal of PQQ forces the added alcohol moiety closer to the C4 carbonyl oxygen than in the case of $Ca^{2+}$, thus facilitating the proton abstraction by the carbonyl oxygen.

A variant of the mechanism shown in Figure 10 is one in which the active site base (instead of the the O4 atom) abstracts a methyl proton from the hemiketal adduct (Itoh *et al.*, 1998) but this possibility is not consistent with the position of the methanol in the active site as seen in the X-ray structure (Section 7.4.3). This orientation of methanol lends further support to the hemiketal mechanism. The hydroxyl oxygen is close to C5 (3.2 Å) while the methyl group is considerably further away (3.9 Å). This would favour covalent bond formation between C5 and the nearby hydroxyl over hydride ion transfer from the more distant methyl group of the substrate (Xia *et al.*, 1999).

In the alternative mechanism (Figure 11), a simple acid/base-catalysed hydride transfer occurs, and no covalent bond is formed between PQQ and substrate; Asp-303 again provides the base and $Ca^{2+}$ acts again as a Lewis acid. The large deuterium isotope effect (about 6) observed during the reductive phase of the reaction is consistent with either mechanism; in both cases the step affected will be the breaking of the C—H bond, and it is this step that is affected by the activator ammonia (Frank *et al.*, 1988; Goodwin and Anthony, 1996). Although the evidence on this matter is scarce, it slightly favours the mechanism involving a hemiketal intermediate (Figure 10). Itoh *et al.* (1998) point out that their kinetics, using the model system, might be interpreted in terms of a hydride transfer mechanism in which the C5 hemiacetal intermediate is a "bystander" in a side equilibrium; however, the orders of reactivity of their various model compounds and of alkaline earth metal ions ($Ba^{2+} > Sr^{2+} > Ca^{2+}$) can only be explained by taking into account the intramolecular base catalysis by the C4 carbonyl oxygen in the hemiketal intermediate, strongly supporting the validity of the mechanism involving the covalent addition of substrate to PQQ.

## 6.2. The Role of Ammonia in the Reductive Half-Reaction Mechanism of Methanol Dehydrogenase

As mentioned above, activation of ammonia is confined to the substrate oxidation step but its mechanism of action is not known (Frank *et al.*, 1988; Dijkstra *et al.*, 1989; Goodwin and Anthony, 1996). The active form is the free base and ammonia can be replaced by methylamine or some glycine esters (Anthony, 1993a). An obvious suggestion for the involvement of ammonia would involve formation of a covalent imino-quinone adduct (Figure 9), and one of the first mechanisms proposed for this enzyme suggested that methanol binds as a methoxy group to the same carbon atom (C—4) as the ammonia and is then released as the aldehyde during reduction of the PQQ (Forrest *et al.*, 1980; Duine, 1991). However, although addition of ammonia to the C—5 position of isolated PQQ is known to occur (Dekker *et al.*, 1982; Ohshiro and Itoh, 1993), there is no convincing evidence that this occurs during the enzyme reaction. No nitrogen-containing adduct of PQQ has been isolated from the enzyme; added ammonia is easily removed by gel filtration to give inactive enzyme which must then be re-activated; in the reaction with cytochrome this activator is not always required; and some dehydrogenases for alcohol which are likely to have essentially similar mechanisms do not require ammonia as activator.

## 6.3. The Oxidative Half-Reaction of Methanol Dehydrogenase

### 6.3.1. The "Methanol Oxidase" Electron Transport Chain

In the electron transport chain for methanol oxidation, electrons are passed from MDH to cytochrome $c_L$ thence to cytochrome $c_H$ which is then oxidised by cytochrome $aa_3$ or cytochrome $co$ (Anthony, 1988, 1992a). This small electron transport chain is more like that involved in oxidation of inorganic substrates in not involving ubiquinone and the cytochrome $bc_1$ complex. It establishes a protonmotive force which drives ATP synthesis, yielding less than 1 ATP per mole of methanol oxidised. Except for the oxidase all the components of this system are in the periplasm. The concentrations of MDH and cytochrome $c_L$ in the periplasm are about 0.5 mM (Alefounder and Ferguson, 1981; Beardmore-Gray *et al.*, 1983). As these proteins are more or less globular in shape, they probably form a monolayer one molecule deep in the periplasm. Indeed if the distance between the periplasmic membrane and the outer membrane is about 70 Å then there is hardly space for the large dehydrogenase to move in more than two dimensions.

In terms of primary sequence and large size cytochromes of the cytochrome $c_L$ type constitute a novel class of acidic $c$-type cytochromes (Nunn and Anthony, 1988; Anthony, 1992a). The crystallisation of a ternary complex involving this cytochrome together with methylamine dehydrogenase and amicyanin from *Paracoccus denitrificans* has provided information on its tertiary structure (in this organism the cytochrome $c_L$ is usually referred to as cytochrome $c_{551i}$) (Chen *et al.*, 1994). It contains five α-helices, the central three of which envelope the haem group and correspond to analogous helices in most other $c$-type cytochromes which share the "cytochrome $c$ fold".

Cytochrome $c_H$ is similar in most respects to other typical bacterial cytochromes $c$ and $c_2$. However its X-ray structure (Read *et al.*, 1999) shows a number of unusual features; it bears a closer gross resemblance to mitochondrial cytochrome $c$ than to the bacterial cytochrome $c_2$ and the left hand side of the haem cleft is unique. In particular it is highly hydrophobic, the usual water is absent, and the "conserved" Tyr67 is replaced by tryptophan. A number of features of the structure demonstrate that the usual hydrogen bonding network involving water in the haem channel is not essential, and that other mechanisms for modulation of redox potentials may exist in this cytochrome. It should be emphasised that the unique character of this cytochrome does not appear to be related in any way to its special involvement in oxidising cytochrome $c_L$ in the methylotroph electron transport chain.

### 6.3.2. The Interaction of MDH with Cytochrome $c_L$

There is considerable evidence that the interaction between the dehydrogenase and cytochrome $c_L$ is electrostatic in nature, including the observation that respiration in whole bacteria is inhibited by high salt concentrations (50% inhibition by 200mM-NaCl or about 25mM-sodium phosphate) (Anthony, 1992a; Dijkstra *et al.*, 1989). Electron transfer activity between the MDH and cytochrome $c_L$ was inhibited by high ionic strength; there was an inverse linear relationship between reaction rate and the square root of the ionic strength, consistent with electrostatic interactions being involved (Chan and Anthony, 1991; Cox *et al.*, 1992).

The nature of the groups involved in the MDH / cytochrome $c_L$ interaction has been investigated by chemical modification using the proteins from a number of different classes of methylotroph including those that grow at low pH and those that grow in high salt concentrations. These studies indicated that the interaction is by way of a small number of lysyl residues on the α subunit of MDH and carboxylates on the cytochrome; modification of these residues prevents electron transfer between the proteins, without however affecting activity in the PES-linked system, showing that the active site for reaction with substrate had not been altered (Chan and Anthony, 1991; Cox et al., 1992; Anthony, 1992). Reagents which change the charge on the lysines led to inactive MDH, whereas those that modified MDH with retention of charge had relatively little effect. A range of experiments with various cross-linking reagents showed that cytochrome $c_L$ dissociates from MDH before interaction with cytochrome $c_H$; that is, the same site on cytochrome $c_L$ interacts with the electron donor and the electron acceptor, and a three component electron transfer complex is not formed. Remarkably, similar results were obtained in chemical modification studies using MDH from the marine methylotroph *Methylophaga marina*; although 1 M NaCl had no effect on activity, chemical modification of only 4 lysine residues on the MDH led to complete loss of activity.

It has been suggested that hydrophobic interactions are also important between MDH and cytochrome $c_L$ (Harris and Davidson, 1993b). PQQ is buried within an internal chamber which communicates with the exterior of the protein by way of a hydrophobic funnel in the surface; the shortest distance for electron flow between PQQ and the outside of the MDH is likely to be by way of this funnel, and so hydrophobic interactions may also be essential to hold cytochrome $c_L$ in position for electon transfer. In order to distinguish between factors affecting electron transfer and those affecting the initial binding or docking process, a fluorescence method was developed (Dales and Anthony, 1995). This confirmed the role of electrostatic interactions in the initial binding but showed a surprising result with respect to the mode of action of the inhibitor EDTA. This has no effect on the dye-linked MDH activity and is not dependent on its chelating characteristics. Studies using a fluorescent analogue of EDTA (Indo-1) showed that it acts by binding tightly to MDH ($I_{50}$, about $3 \mu M$) but not to cytochrome $c_L$; it was concluded that it inhibits methanol oxidation by binding to lysyl residues on MDH thus preventing docking with cytochrome $c_L$ (Chan and Anthony, 1992). However, the fluorescence assay for measuring the interaction of MDH and cytochrome $c_L$ showed, surprisingly, that the "docking" is not inhibited by 50 $\mu$M-EDTA which completely inhibits the overall electron transfer process. It was therefore suggested that EDTA acts by binding to nearby lysyl residues, thus preventing movement of the "docked" cytochrome to its optimal position for electron transfer, which probably involves interaction with the hydrophobic funnel in the surface of the dehydrogenase (Dales and Anthony, 1995). A further extensive kinetic investigation of the interaction of the two proteins has led to Davidson and his colleagues to a similar conclusion; that is, after a non-optimal collision there is a rearrangement of the proteins to produce the most efficient orientation for electron transfer (Harris et al., 1994). The structure of the cytochrome

$c_L$ from *P. denitrificans* shows that most of the charged residues are located on the side of the molecule away from the hydrophobic edge of the haem cleft (Chen *et al.*, 1994). This provides a challenge to the model in which these residues are involved in the docking process. The MDH / cytochrome $c_L$ complex has not been crystallized in order to determine its X-ray structure, and until this is achieved the route by which electrons pass from reduced PQQ to the haem in cytochrome $c_L$ must remain a matter for speculation.

### 6.3.3.  The Oxidation of Reduced PQQ in MDH

Very little is known about the process of electron transfer from the quinol form of PQQ to the cytochrome electron acceptor. It is reasonable to assume that this occurs in two single electron transfer steps—the semiquinone form of PQQ being produced after the first of these transfers (Figure 10). The protons are released from the reduced PQQ into the periplasmic space thus contributing to the protonmotive force (Anthony, 1988, 1993b). It was once thought that an intermediary in this process was the novel disulphide bridge between adjacent cysteines in the active site; this novel structure was very readily reduced with dithiothreitol, yielding enzyme that was inactive with cytochrome but active with the phenzine ethosulphate (PES) (Blake *et al.*, 1994). However, it has now been shown that the activity with PES is because the dye very rapidly re-oxidises the two thiols back to the original disulphide. No free thiols were ever detected during the reaction cycle and the enzyme was active with cytochrome $c_L$ even after the cysteines of the disulphide bridge had been carboxymethylated by reaction with iodoactetate, and so no longer able to undergo oxidation/reduction reactions (Avezoux *et al.*, 1995).

## 7.  THE STRUCTURE OF METHANOL DEHYDROGENASE

MDH is the only PQQ-containing enzyme for which a structure is available (Anthony and Ghosh, 1998); the structures have been determined for the MDH from the facultative methylotroph *M. extorquens* (1.94 A) (Blake *et al.*, 1994; Ghosh *et al.*, 1995) and from the obligate methylotrophs *Methylophilus methylotrophus* (2.6 A) (Xia *et al.*, 1992), and *Methylophilus* W3A1 (1.9 Å) (White *et al.*, 1993; Xia *et al.*, 1996, 1999). As expected from the similarities in protein sequence (Figure 13), all the structures are similar; the numbering system for MDH from *M. extorquens* will be used throughout this review. MDH has an $\alpha_2\beta_2$ tetrameric structure (Figure 14); the $\alpha$-subunit containing the PQQ is 66 kDa, and the $\beta$-subunit is very small

The α subunit of methanol dehydrogenase

```
AM1    1-  NDKLVELSKSDDNWVMPGKNYDSNNFSDLKQINKGNVKQLRPAWTFSTGL
W3A1   1-  DADLDKQVNTAGAWPIATGGYYSQHNSPLAQINKSNVKNVKAAWSFSTGV

AM1   51-  LNGHEGAPLVVDGKMYIHTSFPNNTFALGLDDPGTILWQDKPKQNPAARA
W3A1  51-  LNGHEGAPLVIGDMYVHSAFPNNTYALNLNDPGKIVWQHKPKQDASTKA

AM1  101-  VACCDLVNRGLAYWPGDGKTPALILKTQLDGNVAALNAETGETVWKVENS
W3A1 101-  VMCCDVVDRGLAYGAGQ------IVKKQANGELLALDAKTGKINWEVEVC

AM1  151-  DIKVGSTLTIAPYVVKDKVIIGSSGAELGVRGYLTAYDVKTGEQVWRAYA
W3A1 145-  DPKVGSTLTQAPFVAKDTVLMGCSGAELGVRGAVNAFDLKTGELKWRAFA

AM1  201-  TGPDKDLLLASDFNIKNPHYGQKGLGTGTWEGDAWKIGGGTNWGWYAYDP
W3A1 195-  TGSDDSVRLAKDFNSANPHYGQFGLGTKTWEGDAWKIGGGTNWGWYAYDP

AM1  251-  GTNLIYFGTGNPAPWNETMRPGDNKWTMTIFGRDADTGEAKFGYQKTPHD
W3A1 245-  KLNLFYYGSGNPAPWNETMRPGDNKWTMTIWGRDLDTGMAKWGYQKTPHD

AM1  301-  EWDYAGVNVMMLSEQKDKDGKARKLLTHPDRNGIVYTLDRTDGALVSANK
W3A1 295-  EWDFAGVNQMVLTDQ-PVNGKMTPLLSHRDRNGILYTLNRENGNLIVAEK

AM1  351-  LDDTVNVFKSVDLKTGQPVRDPEYGTRMDHLAKDICPSAMGYHNQGHDSY
W3A1 344-  VDPAVNVFKKVDLKTGTPVRDPEFATRMDHKGTNICPSAMGFNHQGVDSY
```

AM1   401 - DPKRELFFMGINHICMDWEPFMLPYRAGQFFVGATLNMYPGPKGDRQNYE
W3A1  394 - DPESRTLYAGLNHICMDWEPFMLPYRAGQFFVGATLAMYPGPNGPTKK--

AM1   451 - GLGQIKAYNAITGDYKWEKMERFAVWGGTMATAGDLVFYGTLDGYLKARD
W3A1  442 - EMGQIRAFDLTTGKAKWTKWEKFAAWGGTLYTKGGLVWYATLDGYLKALD

AM1   501 - SDTGDLLWKFKIPSGAIGYPMTYTHKGTQYVAIYYGVGGWPGVGLVFDLA
W3A1  492 - NKDGKELWNFKMPSGGIGSPMTYSFKGKQYIGSMYGVGGWPGVGLVFDLT

AM1   551 - DPTAGLGAVGAFKKLANYTQMGGGVVVFSLDGKGPYDDPNVGEWKSAAK
W3A1  542 - DPSAGLGAVGAFRELQNHTQMGGGLMVFSL

The β subunit of methanol dehydrogenase

AM1     1 - YDGTKCKAAGNCWEPKPGFPKKIAGSKYDPKHDPKELNKQADSIKQMEE
W3A1    1 - YDGQNCKEPQNCWENKPQYPEKIAQSKYDPKHDPVELNKQEESIKAMDA

AM1    51 - RNKKRVENFKKTGKFEYDVAKISAN
W3A1   51 - RNAKRIANAKSSGNTVFDVK

**FIGURE 13.** Amino acid sequence of the two subunits of methanol dehydrogenase. AM1 indicates the sequence for MDH from *Methylobacterium extorquens* AM1 (Anderson *et al.*, 1990; Numn *et al.*, 1989; Anthony, 1992b); W3A1 indicates the sequence for *Methylophilus* W3A1 (Xia *et al.*, 1996).

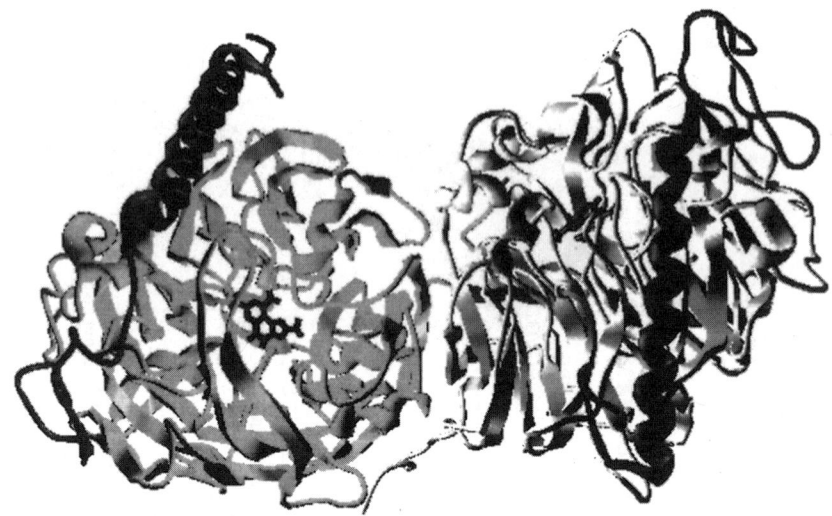

**FIGURE 14.** The $\alpha_2\beta_2$ tetrameric structure of MDH. The small $\beta$ subunit is the darker part of the structure. This Figure is based on the structure in Ghosh *et al.* (1995).

(8.5 kDa). The subunits cannot be reversibly dissociated and no function has been ascribed to the small subunit. In the tetramer the two $\alpha\beta$ subunits are arranged with their pseudo 8-fold axes approximately perpendicular to each other, and the PQQ prosthetic groups are separated by about 45 Å.

## 7.1.  The Structure of the $\alpha$-Subunit

The large $\alpha$ subunit is a superbarrel made up of eight topologically-identical four-stranded twisted antiparallel $\beta$-sheets (W-shaped), stacked radially around a pseudo eight-fold symmetry axis running through the centre of the subunit. This structure has been referred to as a propeller fold, each W motif representing a propeller blade. The 32 $\beta$-strands that make up the superbarrel structure in MDH are shown in Figure 15, the sequence of the strands being the same as in the amino acid sequence as is also seen in other superbarrel structures. The normal twist of the $\beta$-sheets enables space to be efficiently packed in the subunit and allows the large polypeptide chain to be folded in a very compact form without any other typical structural domains. There is no 'hole' along the pseudosymmetry axis, which is filled with amino acid side chains from the eight A-strands of the $\beta$-sheets; where these come closest together, at the point of contact five of the eight residues are glycine, thus facilitating close-packing.

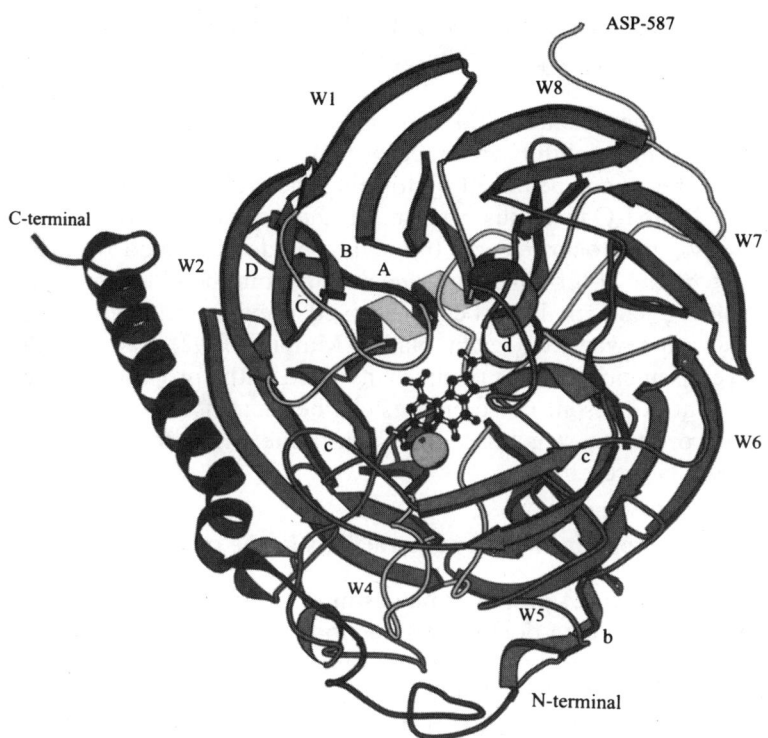

**FIGURE 15.** A drawing of an αβ unit of MDH looking down the pseudo 8-fold axis, simplified to show only the β-strands of the "W" motifs of the α-chain, and the long α-helix of the β-chain, but excluding other limited β-structures and short α-helices. The PQQ prosthetic group is in skeletal form and the calcium ion is shown as a small sphere. The outer strand of each "W" motif is the D strand, the inner strand being the A strand. The "W" motifs are arranged in this view in an anti-clockwise manner. The exceptional motif W8 is made up of strands A—C near the C-terminus plus its D strand from near the N-terminus. This Figure is based on the structure in Ghosh *et al.* (1995).

The α subunits in methanol dehydrogenase contact one another in the region containing the D-strands of the seventh and eighth W motifs, associated over a large planar interface containing many hydrophobic and hydrophilic side chain interactions. These are augmented by the ten C-terminal residues of the α-chains which form extensions which associate with the symmetry-related subunit, again through hydrophobic and hydrophilic interactions.

Cytochrome $cd_1$ (nitrite reductase) is the only other protein, whose structure has been determined, that has an eight bladed β-propeller structure. In this case the centre of the barrel contains haem $d$ instead of PQQ. Remarkably, the $\beta$-propeller domain of cytochrome $cd_1$ proves to be closely superimposable on that of MDH although the two polypeptides have no sequence identity and they use two different folding patterns to bind the first and eighth blades together (Baker *et al.*, 1997). In MDH this closure of the ring is achieved by forming the eighth blade from three $\beta$-strands from the C-terminus plus one strand provided by the N terminus. By contrast, in cytochrome $cd_1$ the eighth blade is formed from three $\beta$-strands from the N-terminus plus one strand from the C-terminus. Both of these folding patterns have been observed previously in other propeller proteins with fewer than eight blades. Murzin (1992) has published an extensive discussion of the principles involved in the propeller assembly of $\beta$ sheets, but at present no sequences can be identified which can be used to predict propeller structures or to determine which folding pattern for closure will be present.

## 7.2. The Tryptophan-Docking Interactions in the α-Subunit

The eight $\beta$ sheet propeller blades interact with each other by way of a series of tryptophan-docking interactions forming a planar, stabilising girdle around the periphery of the α subunit (Ghosh *et al.*, 1995; Xia *et al.*, 1996) (Figure 16). The interactions occur by way of 11-residue consensus sequences in the C/D region of all "W" motifs except number 8 (Figure 17). The relevant characteristics of the tryptophan residues are their planar conjugated rings, and their ability to act as a hydrogen bond donor through the indole ring NH group (Figure 18). The tryptophan at position 11 is stacked between the alanine at position 1 of the same motif ($W_n$) and the peptide bond between residue 6 and the invariant glycine at position 7 of the next motif ($W_{n+1}$). The same tryptophan is also H-bonded between its indole NH and the main chain carbonyl of residue 4 in the next motif ($W_{n+1}$). The third type of interaction involving the conserved tryptophans is a $\beta$ sheet hydrogen bond between its carbonyl oxygen and the backbone nitrogen of position 1 (usually alanine) of the same motif.

There are some exceptions to these interactions; for example, the tryptophan is replaced by Ser347 in W5 which makes a hydrogen bond with the peptide carbonyl of residue 4 in motif W6. In the other exception, Phe292 replaces tryptophan in W4 and makes only stacking interactions, with the usual glycine peptide bond of W5 on one side, and a second glycine peptide bond in position 1 of W4. Motif W8 is exceptional in having no consensus sequence (except for Trp44) and so there is no glycine to interact with the tryptophan in the previous motif (Trp508); in its place this tryptophan forms

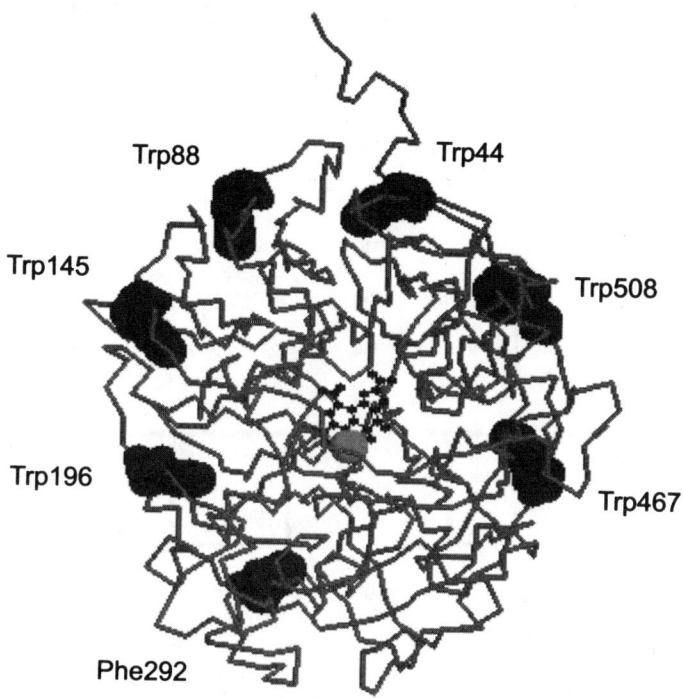

**FIGURE 16.** The girdle of tryptophan residues involved in docking the β-sheets together in the α subunit. The tryptophan residues involved in docking are shown in spacefill mode and the rest of the chain as backbone. The PQQ prosthetic group is in skeletal form and the calcium ion is shown as a small sphere. This Figure is based on the structure in Ghosh *et al.* (1995).

a hydrophobic interaction with the side chain of Leu40 and in the previous motif. The carbonyl of this leucine (position 7) replaces the usual carbonyl at position 4 in forming a hydrogen bond with the tryptophan indole NH. The 11-residue motif is extended in many cases by 2 further residues on the C and D strands which are joined by main chain hydrogen bonds (Xia *et al.*, 1996). The PQQ-containing dehydrogenases for glucose (Cozier and Anthony, 1995) and alcohols (Cozier *et al.*, 1995; Jongejan *et al.*, 1998) show the greatest similarity to the MDH in the sequences which form the W motifs which has facilitated modelling of the structure of these proteins.

No similar tryptophan docking motifs are present in the 7-bladed superbarrel proteins galactose oxidase and methylamine dehydrogenase, or in the 8-bladed nitrite reductase (cytochrome $cd_1$) (Fulop *et al.*, 1995; Baker *et al.*, 1997).

| Position | 1 | 2 | 3 | 4 | 5 | - | 6 | 7 | 8 | 9 | 10 | 11 |
|---|---|---|---|---|---|---|---|---|---|---|---|---|
| Motif | C | C | C | | | | D | D | D | D | D | D |
| W1 | Ala 77 | Leu | Gly | Leu | Asp | Asp | Pro | Gly | Thr | Ile | Leu | Trp 88 |
| W2 | Ala 135 | Leu | Asn | Ala | Glu | - | Thr | Gly | Glu | Thr | Val | Trp 145 |
| W3 | Ala 186 | Tyr | Asp | Val | Lys | - | Thr | Gly | Glu | Gln | Val | Trp 196 |
| W4 | Gly 282 | Arg | Asp | Ala | Asp | - | Thr | Gly | Glu | Ala | Lys | Phe 292 |
| W5 | Thr 337 | Leu | Asp | Arg | Thr | - | Asp | Gly | Ala | Glu | Val | Ser 347 |
| W6 | Ala 457 | Tyr | Asn | Ala | Ile | - | Thr | Gly | Asp | Tyr | Lys | Trp 467 |
| W7 | Ala 489 | Arg | Asp | Ser | Asp | - | Thr | Gly | Asp | Leu | Leu | Trp 508 |
| W8 | Val 577 | Phe | Ser | Leu | Asp 581 | - | Gln 39 | Leu | Arg | Pro | Ala | Trp 44 |
| MDH Consensus | Ala | X | Asp/ Asn | X | X | - | Thr | Gly | Asp/ Glu | X | X | Trp |
| GDH/ADH Consensus | Ala | X | Asp/ Asn | X | X | - | Thr | Gly | Lys | X | X | Trp |

**FIGURE 17.** The consensus sequences in the tryptophan docking motif (Ghosh *et al.*, 1995; Xia *et al.*, 1996). This motif occurs at the C/D corners at the end of the C strands and the beginning of the D strands of each W motif; there are no loops between these strands. The C/D corners are best characterised as 4-residue (β) turns or 5-residue turns (comprising residues 3-6 or 3-7 respectively). Consensus sequences are also included for the quinohaemoprotein alcohol dehydrogenase (ADH) and glucose dehydrogenase, which is a membrane quinoprotein (GDH).

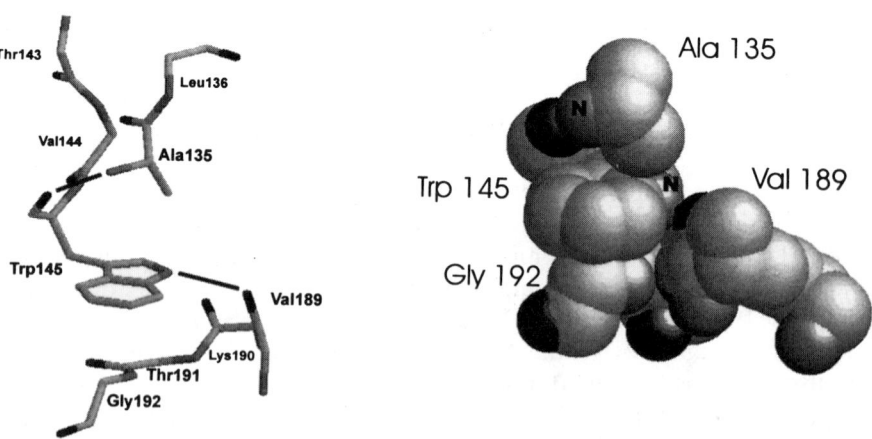

**FIGURE 18.** Part of a typical tryptophan-docking motif in stick format and spacefill format. This shows the interactions of Trp145 (in W2) with Ala135 (also in W2), and the interaction with the plane of the peptide bond between Thr191 and Gly192 (in W3). For clarity the remaining residues of the two motifs are omitted, and the side chains are omitted from all residues except for Trp145 and Ala135.

## 7.3.   Structure of the β-Subunit

The small β subunit (Figures 14 and 15) forms an unusual extended structure having no hydrophobic core. The N-terminal region (30 residues) is folded in a series of open turns and includes one intrachain disulphide bridge and a proline-rich segment. The C-terminal 54 residues, rich in charged residues, form a single straight α-helix of 30 residues in 7 turns. Overall, the β-chain forms a planar "J" shaped unit, with the long α-helix as its stem, which hooks over the globular α-subunit. Although some hydrophobic interactions occur between the α and β subunits, ion-pair interactions are predominant, as expected from the fact that 40% of the β chain residues are charged. The β chain makes contact with the edges of the W1–W4 motifs of the α chain with ion pair interactions involving glutamate, arginine and lysine residues. It makes no direct contribution to the active site and, in the absence of any other obvious function for this unusual subunit, it has been suggested that it acts to stabilise the folded form of the large α-chain. The absence of β-subunits in the other PQQ-containing quinoproteins, however, perhaps indicates that it has a more specific (unknown) function.

## 7.4.   The Active Site of Methanol Dehydrogenase

The active sites within each α subunit occur on the pseudo 8-fold symmetry axis, at the end of the superbarrel structure containing the loops which fold over the end of the superbarrel to enclose the active site chamber (Figure 15). Access is by way of a shallow funnel made up of hydrophobic surface residues (Anthony et al., 1994) leading to a narrow entrance to the chamber containing the PQQ coordinated to a $Ca^{2+}$ ion, the novel disulphide ring structure and a potential active site base (Asp303). There is no obvious interaction between the two active sites in the two α subunits.

### 7.4.1.   The Novel Disulphide Ring Structure in the Active Site

There is a remarkable novel structure within the α subunit produced by formation of a disulphide bridge between adjacent cysteine residues (Cys103–Cys104), the result being a strained 8-membered ring (Figure 19). Although such a disulphide bridge has been proposed to be present in the active site of the acetylcholine receptor (Kao and Karlin, 1986) and has been seen in the structure of an inactive, oxidised, form of mercuric ion reductase (Schiering et al., 1991), this is the first time that this structure has been seen in an active enzyme. It had been predicted that if such a

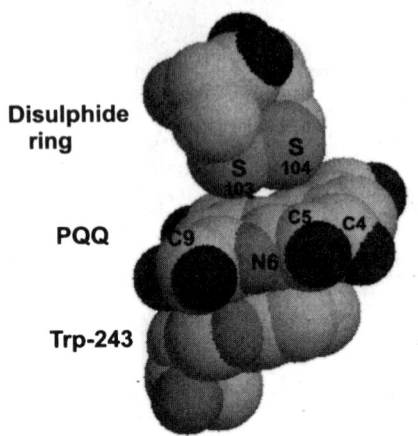

**Disulphide ring**

**PQQ**

**Trp-243**

**FIGURE 19.** The novel disulphide ring in the active site of methanol dehydrogenase (Ghosh *et al.*, 1995; Xia *et al.*, 1996). The ring is formed by disulphide bond formation between adjacent cysteine residues. The PQQ is "sandwiched" between this ring and the tryptophan that forms the floor of the active site chamber. The calcium ion is coordinated between the C—9 carboxylate, the N—6 of the PQQ ring and the carbonyl oxygen at C—5. The oxygen of the C—4 carbonyl appears to be out of the plane of the ring.

structure should unexpectedly exist in a protein then a *cis* peptide bond would be present, forming part of a strained ring structure (Ramachandran and Sasisekharan, 1968; Chandrasekaran and Balasubramanian, 1969). In our 1.94 Å structure of the *M. extorquens* MDH it is clearly seen that the peptide bond is, as predicted, non-planar, but it is in the *trans* configuration (Ghosh *et al.*, 1995). The $\omega$ angle is 145°, giving a distortion from planarity of 35°. All of the other bond lengths and bond angles in the ring are standard values, including the distance between the sulphur atoms (2.06 Å). This structure is seen in all MDHs studied (Xia *et al.*, 1996).

The rarity of this type of ring structure would suggest some special biological function. Reduction of the disulphide bond leads to loss of activity but oxidation in air or carboxymethylation of the free thiols leads to return of activity. As the carboxymethylated thiols can no longer be involved in redox activity a possible role in electron transfer appears to be ruled out (see Section 6.3.3). The disulphide ring is not present in the quinoprotein glucose dehydrogenase in which electrons are transferred to membrane ubiquinone from the quinol $PQQH_2$, and in which the semiquinone free radical is unlikely to be involved as a stable intermediate. It is possible, therefore, that this novel structure might function in the stabilization of the free radical PQQ semiquinone or its protection from solvent at the entrance to the active site in MDH (Avezoux *et al.*, 1995). Free PQQ has been reported to catalyse disulphide bond formation both in model compounds and in proteins with adjacent thiols (Park and Churchich, 1992) and it is possible that, if the two cysteines are in the reduced form during processing of MDH, then PQQ might bind at the active site and catalyse disulphide bond formation, the disulphide ring then holding the PQQ in place (White *et al.*, 1993). The disulphide ring is not essential for incorporation of $Ca^{2+}$ into the MDH purified from the processing mutants that require added $Ca^{2+}$ for production of active enzyme (Avezoux *et al.*, 1995).

### 7.4.2.   The Bonding of PQQ in the Active Site

The PQQ is sandwiched between the indole ring of Trp243 and the disulphide ring structure (Figure 19). The indole ring is within 15° of coplanarity with the PQQ ring and, on the opposite side, the two sulphur atoms of the disulphide bridge are within 3.75 Å of the plane of PQQ. In addition to these axial interactions, many amino acid residues are involved in equatorial interactions with the substituent groups of the PQQ ring system. These are exclusively hydrogen-bond and ion-pair interactions involving residues mostly on the A strands of the "W" motifs (Figure 20). Although the number of polar groups involved might indicate at first sight that the environment of the PQQ is polar, this is not the case. An oxygen of the 9-carboxyl forms a salt bridge with Arg109 and both groups are shielded from bulk solvent by the disulphide. The carboxyl group of Glu155 and a 2-carboxyl oxygen of PQQ are also shielded from solvent and it is probable

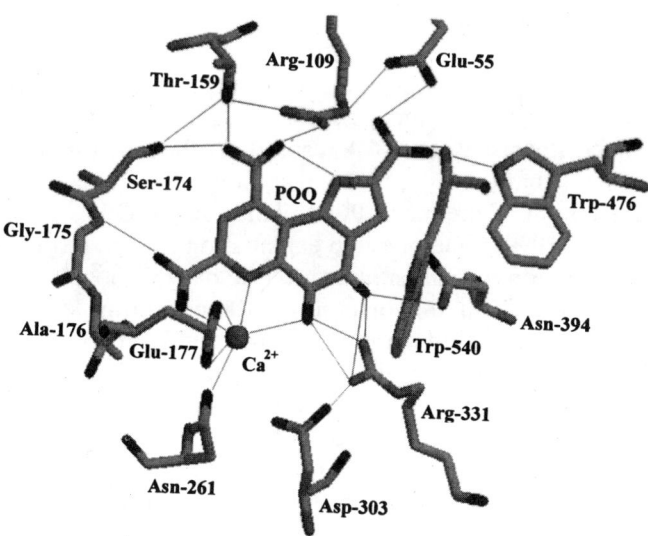

**FIGURE 20.** The equatorial interactions of PQQ and the coordination of $Ca^{2+}$ in the active site of methanol dehydrogenase (Ghosh *et al.*, 1995). This figure also shows Asp303, which is likely to act as a base, and Arg331 which may also be involved in the mechanism.

that at least one is protonated, their interaction thus being stabilised through hydrogen bond formation.

The C4 and C5 oxygens are hydrogen bonded by the $NH_1$ and $NH_2$ atoms respectively of Arg331, and in addition the C4 oxygen makes a longer hydrogen bond interaction with the amide $NH_2$ of Asn394 whose amide CO is hydrogen bonded to its own main chain NH group. It is not known if the bonding of the C4 and C5 oxygen atoms is maintained in the fully oxidised quinone and fully reduced quinol forms of the prosthetic group, in which the C4 and C5 oxygen atoms are likely to be in the plane of the ring. As well as its hydrogen bonding to the PQQ, Arg331 also makes hydrogen bonds between its $NH_2$ and the carboxylate and main-chain carbonyl of Asp303. The two side chains lie side by side, permitting free access to the carboxyl group of Asp303, which is the most likely candidate for the base required by the catalytic mechanisms previously proposed for MDH (Section 6.1).

In the fully oxidised or reduced forms of PQQ the C4 and C5 carbonyl oxygen atoms would be expected to be in the plane of the ring (Itoh *et al.*, 1993, 1998; Zheng and Bruice, 1997). In the MDH as it is usually isolated the PQQ is in the semiquinone form and it is not so obvious what structure might be expected, especially in the environment of an enzyme active site, although Zheng and Bruice (1997) calculated that the semiquinone free radical coordinated to $Ca^{2+}$ would be essentially planar. In the MDH from *M. extorquens* the O5 is in the plane of the ring whereas the O4 appears to be below the plane by about 40° (Ghosh *et al.*, 1995). The structure of the MDH from *Methylophilus* WA31 has been determined from two crystal forms. Whereas both the C4 and C5 carbonyls are in a planar configuration in the form A crystal structure (2.4 Å resolution), the structure in the B crystals (1.9 Å) shows that the C4 is trigonal and the C5 tetrahedral. As a result C5 protrudes up out of the mean PQQ plane forcing O5 to deviate downwards (Xia *et al.*, 1999). One possible explanation for the tetrahedral C5 is that adduct formation has taken place. To rule out this possibility the PQQ was modelled as the C5-methanol adduct and the resulting structure shown to give a strong disagreement with the electron density map. Xia *et al.* (1999) point out that if the PQQ is in the semiquinone state it could exist in two tautomeric forms in which either of the two quinone oxygens is in the hydroxyl form. Each tautomer can be drawn in two resonance forms, one with the unpaired electron delocalised on the ring and the other with the electron localised to the carbonyl oxygen. The presence of $Ca^{2+}$, 2.6 Å away from O5 might stabilise the negatively charged oxygen and permit distortion of C5 to a tetrahedral geometry.

The presence of the $Ca^{2+}$ ion coordinated to the PQQ in the active site is consistent with predictions that it is likely to be intimately related to the prosthetic group (Richardson and Anthony, 1992). Its co-ordination sphere contains PQQ and protein atoms (Figure 20), including both oxygens of the carboxylate of Glu177 and the amide oxygen of Asn261. The PQQ atoms include the C5 quinone oxygen, one oxygen of the C7 carboxylate and, surprisingly, the N6 ring atom which is only 2.45 Å from the metal ion. The five oxygen ligands have distances from the $Ca^{2+}$ of 2.4–2.8 Å.

### 7.4.3. The Location of Substrate in the Active Site

In the first descriptions of the active site, the position of the substrate in the active site could not be unequivocally determined; the one or two water molecules which might occupy the same space as the substrate alcohol group are in slightly different locations in the two available structures (Ghosh et al., 1995; Xia et al., 1996, 1999). In the structure of A-form crystals of MDH from Methylophilus W3A1 a water molecule makes a hydrogen bond to both O5 of PQQ and to the side chain of Asp303. In the B-form crystal structure, the elongated electron density is consistent with the binding of methanol at this site instead of water. The substrate does not react with the MDH in this form because the PQQ is in the semiquinone (not oxidised) state. The hydroxyl of the methanol appears to be hydrogen bonded to the side chain of Asp303 (at 3.1 Å), but not to the O5 of PQQ (at 3.4 Å) (Figure 21). The methyl group is in a hydrophobic cavity bounded by side chains of Trp265, Trp540 and Leu556 as well as the disulphide ring. This raises the problem that this enzyme has a broad substrate specificity, and can oxidise primary alcohols including relatively large substrates such as pentanol and cinnamyl alcohol; it is not immediately obvious how these substrates could readily gain access to the active site and

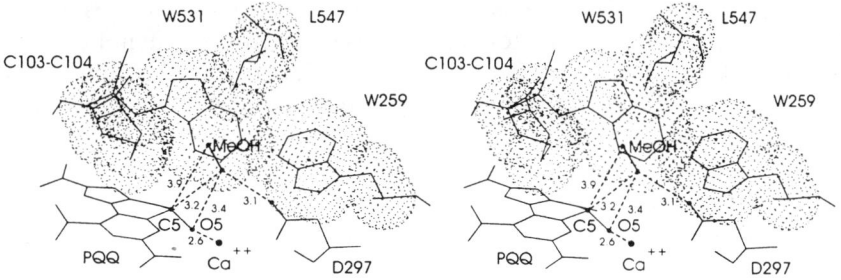

**FIGURE 21.** Stereo representation of the protein environment of the substrate methanol in the form B crystals of MDH from Methylophilus W3A1. The methyl group of methanol is above the hydroxyl group which is hydrogen bonded to Asp292 (Asp 303 in MDH from M. extroquens). The coordination bond between O5 and $Ca^{2+}$ is also shown. Reprinted with permission from Xia et al. (1999), (Biochemistry, **38**, 1214–1220). © American Chemical Society.

understanding of this awaits solution of a structure containing one of these larger substrates.

In the active site, Asp303 is well situated to serve as the catalytic base in the first step of abstracting a proton from the hydroxyl group of methanol, since one of its carboxyl oxygen atoms is 3.1 Å from the hydroxyl oxygen. This orientation of methanol in the active site lends further support to the hemiketal mechanism. The hydroxyl oxygen is close to C5 (3.2 Å) while the methyl group is considerably further away (3.9 Å). This would favour covalent bond formation between C5 and the nearby hydroxyl instead of hydride ion transfer from the more distant methyl group to the C5 atom of PQQ (see Section 6.1).

The possibility of an interaction between the methanol hydroxyl group and the $Ca^{2+}$ ion has been discussed in an extensive study of the structure of PQQ and its possible mechanism by Zheng and Bruice (1997) using *ab initio* and molecular orbital calculations. However, the hydroxyl group of methanol appears to be too distant from the $Ca^{2+}$ for coordination of the $Ca^{2+}$ ion to substrate to have any importance in the mechanism.

## 8.  PROCESSING AND ASSEMBLY OF METHANOL DEHYDROGENASE

Production of active periplasmic methanol dehydrogenase involves the following: synthesis and transport of PQQ; synthesis, transport and processing of pre-peptides for the $\alpha$ and $\beta$ subunits; folding of these proteins in the periplasm; isomerisation of prolines; insertion of disulphide bridges (including the special one at the active site); insertion of calcium and PQQ; wrapping $\beta$ chains around the $\alpha$ subunits; and association of the $\alpha\beta$ units to give the $\alpha_2\beta_2$ tetramer. Little is known about the sequence of these events. Figure 22 presents a model for the synthesis of active methanol dehydrogenase and cytochrome $c_L$ taken from a review of the functions of the various genes involved in this process (Goodwin and Anthony, 1998).

More than 25 genes (*mox* genes) are required for MDH synthesis including seven involved in PQQ biosynthesis, and these are arranged in five clusters (Figure 22). *MxaF* and *mxaI* code for the $\alpha$ and $\beta$ subunits of MDH and *mxaG* codes for its electron acceptor, cytochrome $c_L$. These three genes form part of an operon together with *mxaJ* which encodes a 30 kDa periplasmic protein which may play a role in stabilising or processing MDH (Amaratunga *et al.*, 1997ab). The sequence of genes from *mxaF* to *mxaD* form two well-defined groups. The intergenic regions in the first half of the cluster (*mxaF—mxaR*) are relatively large (105–261 bp), whereas those in the downstream half (between the putative *mxaS* and *mxaD*) are very small (0–8 bp). This draws attention to the possibility that this whole cluster might

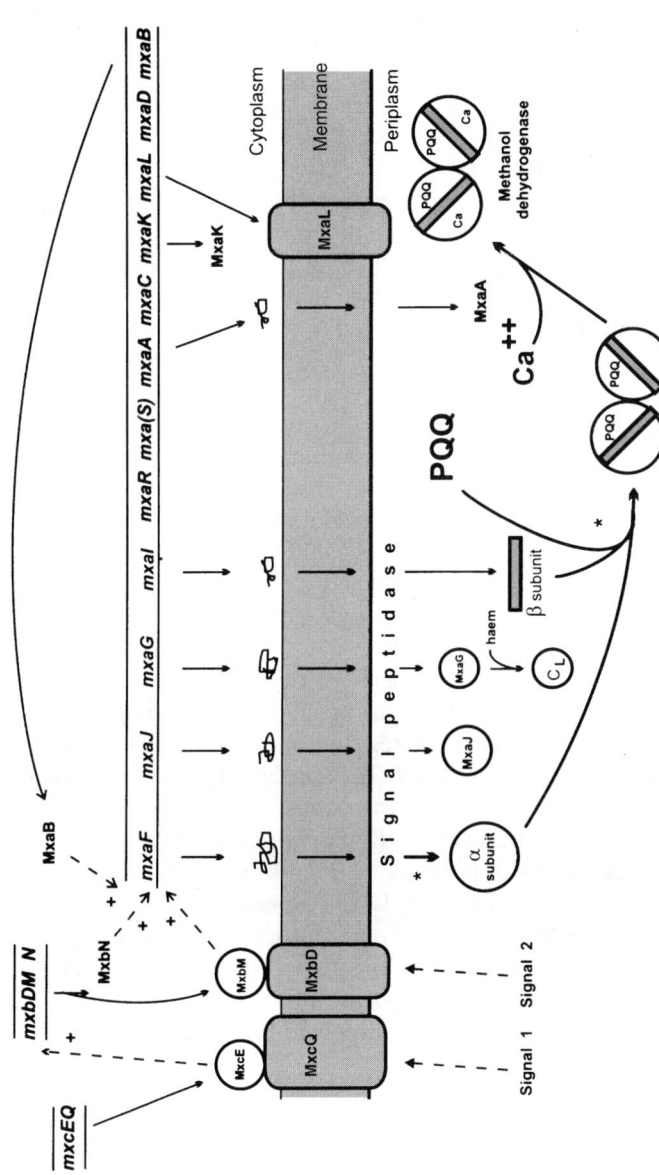

**FIGURE 22.** A model for the expression of the *mxaFJGIR(S)BACKLD* genes in *Methylobacterium extorquens*; it is essentially similar in *Paracoccus denitrificans* and related bacteria. A signal is transmitted by way of MxaX and MxaY, leading to activation of the *mxaF* promoter; 6 other genes are also involved in activating this promoter. One of the signals leading to increased expression of the *mox* genes is likely to be formaldehyde. The resulting preproteins, PQQ and haem are transported into the periplasm where they are assembled into the proteins that are specifically involved in methanol oxidation—methanol dehydrogenase and cytochrome $c_L$. The * indicates two possible steps at which PQQ may be inserted. At least three proteins (MxaA, MxaK and MxaL) are involved in incorporation of $Ca^{2+}$ into the protein and there is some evidence that MxaC and MxaD are also involved in this. Reprinted from Goodwin and Anthony (1998), (*Advances in Microbial Physiology*, **40**, 1–40) with permission from Academic Press.

be organised into two operons. However there is no obvious promoter between *mxaR* and *mxaA* and further work is required to ascertain whether or not the *mxaFJGIR(S)ACKLD* cluster constitutes more than one transcriptional unit (Amarutunga *et al.*, 1997ab).

The $\alpha$ and $\beta$ subunits of methanol dehydrogenase are presumably transported into the periplasm by either the Sec or the SRP pathway. Once in the periplasm the $\alpha$ subunits must associate with PQQ and calcium and then assemble with the $\beta$ subunits to form an active enzyme. Because the PQQ is held by non-covalent bonds and is within the centre of the super-barrel it cannot be bonded firmly to the unfolded $\alpha$ chain and it cannot be added after complete folding of this chain. Mutants lacking *mxaA*, *mxaK* or *mxaL* synthesise an inactive $\alpha_2\beta_2$ tetrameric methanol dehydrogenase, which contains PQQ and which can be converted to the active form on incubation with calcium (Richardson and Anthony, 1992; Goodwin *et al.*, 1996). This suggests that the MxaAKL proteins are required for the insertion of calcium into MDH and that $Ca^{2+}$ insertion possibly occurs after PQQ insertion; their predicted protein sequences suggest that MxaA is periplasmic, MxaK is cytoplasmic and MxaL is a membrane protein. The unusual disulphide ring structure in the active site is not essential for incorporation of $Ca^{2+}$, as active enzyme can be produced after incubation with $Ca^{2+}$ of MDH from MxaA in which the ring has been reduced and even when the free thiols have been subsequently carboxymethylated the reduced MDH (Goodwin *et al.*, 1996).

The regulation of the genes for synthesis of the methanol oxidation system is summarised in Figure 22. The *mxaF* promoter is positively regulated by MxaB which is also involved in regulation of PQQ biosynthesis (Morris and Lidstrom, 1992; Xu *et al.*, 1993). Transcription of the *mxaF* operon is also regulated by two pairs of sensor kinase—response regulator proteins, encoded by *mxbDM* and *mxcEQ* (Xu *et al.*, 1993, 1995; Springer *et al.*, 1997). MxbD and MxcQ appear to be membrane sensor kinases belonging to the histidine kinase superfamily. On detecting a signal (probably methanol or formaldehyde) these become autophosphorylated; they then react with the response regulator proteins (MxaE and MxaM) resulting in activation of transcription at the specific promoters. The *mxbDM* genes are regulated by *mxcEQ* and are also involved in regulation of PQQ synthesis.

## 9. REFERENCES

Alefounder, P. R., and Ferguson, S. J., 1981, A periplasmic location for methanol dehydroge-
    nase from Paracoccus denitrificans: implications for proton pumping by cytochrome aa3.
    *Biochem. Biophys. Res. Comm.* **98**:778–784.

Amaratunga, K., Goodwin, P. M., O'Connor, C. D., and Anthony, C., 1997, The methanol oxidation genes *mxaFJGIR(S)ACKLD* in *Methylobacterium extorquens*. *FEMS Microbiol. Lett.* **146**:31–38.

Amaratunga, K., Goodwin, P. M., O'Connor, C. D., and Anthony, C., 1997, The methanol oxidation genes *mxaFJGIR(S)ACKLD* in *Methylobacterium extorquens*. *FEMS Microbiol. Lett.* **150**:175–177. Erratum to The methanol oxidation genes *mxaFJGIR(S)ACKLD* in *Methylobacterium extorquens*. *FEMS Microbiol. Lett.* **146**:31–38.

Anderson, D. J., Morris, C. J., Nunn, D. N., Anthony, C., and Lidstrom, M. E., 1990, Nucleotide sequence of the *Methylobacterium extorquens* AM1 *moxF* and *moxJ* genes involved in methanol oxidation. *Gene* **90**:173–176.

Anthony, C., 1986, The bacterial oxidation of methane and methanol. *Advances in Microbial Physiology* **27**:113–210.

Anthony, C., 1988, Quinoproteins and energy transduction. In: *Bacterial Energy Transduction*, (C. Anthony, ed.), Academic Press, London, pp. 293–316.

Anthony, C., 1992a, The *c*-type cytochromes of methylotrophic bacteria. *Biochim. Biophys. Acta* **1099**:1–15.

Anthony, C., 1992b, The structure of bacterial quinoprotein dehydrogenases, *Int. J. Biochem.* **24**:29–39.

Anthony, C., 1993a, Methanol dehydrogenase in Gram-negative bacteria. In: *Principles and applications of quinoproteins*, (V. L. Davidson, ed.), Marcel Dekker, New York, pp. 17–45.

Anthony, C., 1993b, The role of quinoproteins in bacterial energy transduction. In: *Principles and applications of quinoproteins*, (V. L. Davidson, ed.), Marcel Dekker, New York, pp. 223–244.

Anthony, C., 1996, Quinoprotein-catalysed reactions. *Biochem. J.* **320**:697–711.

Anthony, C., 1998, Quinoprotein-catalysed reactions. In: *Comprehensive Biological Catalysis*, (M. Sinnott, ed.), Academic Press, London, pp. 155–180.

Anthony, C., and Ghosh, M., 1998, The structure and function of the PQQ-containing quinoprotein dehydrogenases. *Progress in Biophysics and Molecular Biology* **69**:1–21.

Anthony, C., and Zatman, L. J., 1964a, Isolation and properties of *Pseudomonas* sp. M27. *Biochem. J.* **92**:609–614.

Anthony, C., and Zatman, L. J., 1964b, The methanol-oxidizing enzyme of *Pseudomonas* sp. M27. *Biochem. J.* **92**:614–621.

Anthony, C., and Zatman, L. J., 1965, The alcohol dehydrogenase of *Pseudomonas* sp. M27. *Biochem. J.* **96**:808–812.

Anthony, C., and Zatman, L. J., 1967, The microbial oxidation of methanol: The prosthetic group of alcohol dehydrogenase of *Pseudomonas* sp. M27; A new oxidoreductase prosthetic group. *Biochem. J.* **104**:960–969.

Anthony, C., Ghosh, M., and Blake, C. C. F., 1994, The structure and function of methanol dehydrogenase and related PQQ-containing quinoproteins. *Biochem. J.* **304**:665–674.

Avezoux, A., Goodwin, M. G., and Anthony, C., 1995, The role of the novel disulphide ring in the active site of the quinoprotein methanol dehydrogenase from *Methylobacterium extorquens*. *Biochem. J.* **307**:735–741.

Baker, S. C., Saunders, N. F. W., Willis, A. C., Ferguson, S. J., Hajdu, J., and Fulop, V., 1997, Cytochrome $cd_1$ structure: unusual haem environments in a nitrite reductase and analysis of factors contributing to β-propeller folds. *J. Mol. Biol.* **269**:440–455.

Beardmore-Gray, M., O'Keeffe, D. T., and Anthony, C., 1983, The methanol:cytochrome c oxidoreductase activity of methylotrophs. *Journal of General Microbiology* **129**:923–933.

Blake, C. C. F., Ghosh, M., Harlos, K., Avezoux, A., and Anthony, C., 1994, The active site of methanol dehydrogenase contains a disulphide bridge between adjacent cysteine residues. *Nature, Structural Biology* **1**:102–105.

Chan, H. T. C., and Anthony, C., 1991, The interaction of methanol dehydrogenase and cytochrome $c_L$ in an acidophilic methylotroph, *Acetobacter methanolicus*. *Biochem. J.* **280**:139–146.

Chan, H. T. C., and Anthony, C., 1992a, The mechanism of inhibition by EDTA and EGTA of methanol oxidation by methylotrophic bacteria. *FEMS Microbiol Lett* **96**:231–234.

Chan, H. T. C., and Anthony, C., 1992b, Characterisation of a red form of methanol dehydrogenase from the marine methylotroph *Methylophaga marina*. *FEMS Microbiol. Lett.* **97**:293–298.

Chandrasekaran, R., and Balasubramanian, R., 1969, Stereochemical studies of cyclic peptides. VI. Energy calculations of the cyclic dipeptide cysteinyl-cysteine. *Biochim. Biophys. Acta* **188**:1–9.

Chen, L. Y., Durley, R., Poliks, B. J., Hamada, K., Chen, Z. W., Mathews, F. S., Davidson,V. L., Satow, Y., Huizinga, E., Vellieux, F. M. D., and Hol, W. G. J., 1992, Crystal structure of an electron-transfer complex between methylamine dehydrogenase and amicyanin. *Biochemistry* **31**:4959–4964.

Cox, J. M., Day, D. J., and Anthony, C., 1992, The interaction of methanol dehydrogenase and its electron acceptor, cytochrome $c_L$, in the facultative methylotroph *Methylobacterium extorquens* AM1 and in the obligate methylotroph *Methylophilus methylotrophus*. *Biochim. Biophys. Acta.* **1119**:97–106.

Cozier, G. E., Giles, I. G., and Anthony, C., 1995, The structure of the quinoprotein alcohol dehydrogenase of *Acetobacter aceti* modelled on that of methanol dehydrogenase from *Methylobacterium extorquens*. *Biochem. J.* **308**:375–379.

Cozier, G. E., and Anthony, C., 1995, The structure of the quinoprotein glucose dehydrogenase of *Escherichia coli* modelled on that of methanol dehydrogenase from *Methylobacterium extorquens*. *Biochem. J.* **312**:679–685.

Cozier, G. E., Salleh, R. A., and Anthony, C., 1999, Characterisation of the membrane glucose dehydrogenase from *Escherichia coli* and characterisation of a site directed mutant in which His262 has been changed to tyrosine. *Biochem. J.* **340**:639–647.

Dales, S. L., and Anthony, C., 1995, The interaction of methanol dehydrogenase and its cytochrome electron acceptor. *Biochem. J.* **312**:261–265.

Day, D. J., and Anthony, C., 1990, Methanol dehydrogenase from *Methylobacterium extorquens* AM1. *Methods in Enzymology* **188**:210–216.

Davidson, V. L., 1993, Methylamine dehydrogenase. In: *Principles and applications of quinoproteins*, (V. L. Davidson, ed.), Marcel Dekker, New York. pp. 73–95.

Davidson, V. L., Kumar, M. A., and Wu, J., 1992, Apparent oxygen-dependent inhibition by superoxide dismutase of the quinoprotein methanol dehydrogenase, *Biochemistry* **1992**:1504–1508.

de Beer, R., Duine, J. A. Frank, J., and Westerling, J., 1983, The role of the pyrroloquinoline semiquinone forms in the mechanism of action of methanol dehydrogenase. *Eur. J. Biochem.* **130**:105–109.

Dekker, R. H., Duine, J. A., Frank, J., Verwiel, P. E. J., and Westerling, J., 1982, Covalent addition of $H_2O$, enzyme substrates and activators to pyrroloquinoline quinone, the coenzyme of quinoproteins. *Eur. J. Biochem.* **125**:69–73.

Dijkstra, M., Frank, J., and Duine, J.A., 1988, Methanol oxidation under physiological conditions using methanol dehydrogenase and a factor isolated from *Hyphomicrobium* X. *FEBS Lett.* **227**:198–202.

Dijkstra, M., Frank, J., and Duine. J. A., 1989, Studies on electron transfer from methanol dehydrogenase to cytochrome cL, both purified from *Hyphomicrobium* X. *Biochem. J.* **257**:87–94.

Dooley, D. M., and Brown. D. E., 1993, Resonance Raman Spectroscopy of quinoproteins. In: *Principles and applications of quinoproteins*, (V. L. Davidson, ed.) pp. 132–140.

Dooley, D. M., and Brown. D. E., 1995, Resonance raman spectroscopy of quinoproteins. *Met. Enzymol.* **258**:132–140.

Duine, J. A., 1991, Quinoproteins—enzymes containing the quinonoid cofactor pyrroloquinoline quinone, topaquinone or tryptophan-tryptophan quinone. *Eur. J. Biochem.* **200**:271–284.

Duine, J. A., Frank, J., and Westerling, J., 1978, Purification and properties of methanol dehydrogenase from *Hyphomicrobium* X. *Biochim. Biophys. Acta* **524**:277–287.

Duine, J. A., and Frank, J., 1980a, The prosthetic group of methanol dehydrogenase: purification and some of its properties. *Biochem. J.* **187**:221–226.

Duine, J. A., and Frank, J., 1980b, Studies on methanol dehydrogenase from *Hyphomicrobium* X. Isolation of an oxidised form of the enzyme. *Biochem. J.* **187**:213–219.

Duine, J. A., Frank, J., and Westerling, J., 1978, Purification and properties of methanol dehydrogenase from *Hyphomicrobium* X. *Biochim. Biophys. Acta* **524**:277–287.

Duine, J. A., Frank, J., and Verwiel, P. E. J., 1980, Structure and activity of the prosthetic group of methanol dehydrogenase. *Eur. J. Biochem.* **108**:187–192.

Duine, J. A., Frank, J., and Verwiel, P. E. J., 1981, Characterization of the second prosthetic group in methanol dehydrogenase from *Hyphomicrobium* X. *Eur. J. Biochem.* **118**:395–399.

Frank, J., and Duine, J. A., 1990, Methanol dehydrogenase from *Hyphomicrobium* X. *Methods in Enzymology* **188**:202–209.

Frank, J., Dijkstra, M., Duine, J. A., and Balny, C., 1988, Kinetic and spectral studies on the redox forms of methanol dehydrogenase from *Hyphomicrobium* X. *Eur. J. Biochem.* **174**:331–338.

Frank, J., van Krimpen, S. H., Verwiel, P. E. J., Jongejan, J. A., Mulder, A. C., and Duine, J. A., 1989, On the mechanism of inhibition of methanol dehydrogenase by cyclopropane-derived inhibitors. *Eur. J. Biochem.* **184**:187–195.

Forrest, H. S., Salisbury, S. A., and Kilty, C. G., 1980, A mechanism for the enzymic oxidation of methanol involving methoxatin. *Biochem. Biophys. Res. Comm.* **97**:248–251.

Fulop, V., Moir, J. W. B., Ferguson, S. J., and Hajdu, J., 1995, The anatomy of a bifunctional enzyme: structural basis for reduction of oxygen to water and synthesis of nitric oxide by cytochrome $cd_1$. *Cell* **81**:369–377.

Ghosh, M., Anthony, C., Harlos, K., Goodwin, M. G., and Blake, C. C. F., 1995, The refined structure of the quinoprotein methanol dehydrogenase from *Methylobacterium extorquens* at 1.94 A. *Structure* **3**:177–187.

Goodwin, M. G., and Anthony, C., 1996, Characterisation of a novel methanol dehydrogenase containing barium instead of calcium. *Biochem. J.* **318**:673–679.

Goodwin, P. M., and Anthony, C., 1998, The biochemistry, physiology and genetics of PQQ and PQQ-containing enzymes. *Advances in Microbial Physiology* **40**:1–80.

Goodwin, M. G., Avezoux, A., Dales, S. L., and Anthony, C., 1996, Reconstitution of the quinoprotein methanol dehydrogenase from active $Ca^{2+}$-free enzyme with $Ca^{2+}$, $Sr^{2+}$ or $Ba^{2+}$. *Biochem. J.* **319**:839–842.

Harris, T. K., and Davidson, V. L., 1993a, A new kinetic model for the steady-state reactions of the quinoprotein methanol dehydrogenase from *Paracoccus-denitrificans*. *Biochemistry* **32**:4362–4368.

Harris, T. K., and Davidson, V. L., 1993b, Binding and electron transfer reactions between methanol dehydrogenase and its physiologic electron acceptor cytochrome $c_{551i}$—a kinetic and thermodynamic analysis. *Biochemistry* **32**:14145–14150.

Harris, T. K., and Davidson, V. L., 1994, Replacement of enzyme-bound calcium with strontium alters the kinetic properties of methanol dehydrogenase. *Biochem. J.* **300**:175–182.
Harris, T. K., Davidson, V. L., Chen, L. Y., Mathews, F. S., and Xia, Z. X., 1994, Ionic strength
dependence of the reaction between methanol dehydrogenase and cytochrome $c_{551i}$:
evidence of conformationally coupled electron transfer. *Biochemistry* **33**:12600–12608.
Itoh, S., Ogino, M., Fukui, Y., Murao, H., Komatsu, M., Ohshiro, Y., Inoue, T., Kai, Y., and Kasai,
N., 1992, C—4 and C—5 adducts of cofactor PQQ pyrroloquinolinequinone, -model
studies directed toward the action of quinoprotein methanol dehydrogenase. *J. Am. Chem.
Soc.* **115**:9960–9967.
Itoh, S., Kawakami, H., and Fukuzumi, S., 1998, Model studies on calcium-containing quino-
protein alcohol dehydrogenases. Catalytic role of $Ca^{2+}$ for the oxidation of alcohols by
coenzyme   PQQ,   4,5-dihydro-4,5-dioxo-1*H*-pyrrolo[2,3-*f*]quinoline-2,7,9-tricarboxylic
acid, *Biochemistry* **37**:6562–6571.
Janes, S. M., Mu, D., Wemmer, D., Smith, A. J., Kaur, S., Maltby, D., Buringame, A. L., and
Klinman, J. P., 1990, A new redox cofactor in eucaryotic enzymes: 6-hydroxydopa at the
active site of bovine serum amine oxidase. *Science* **248**:981–987.
Jongejan, A., Jongejan, J. A., and Duine, J. A., 1998, Homology model of quinohaemoprotein
alcohol dehydrogenase from *Comomonas testosteroni*. *Protein Engineering* **11**:185–198.
Kao, P. N., and Karlin, A., 1986, Acetylcholine receptor binding site contains a disulphide cross-
link between adjacent half-cystinyl residues. *J. Biol. Chem.* **261**:8085–8088.
Klinman, J. P., and Mu, D., 1994, Quinoenzymes in biology. *Annu. Rev. Biochem.* **63**:299–344.
Long, A. R., and Anthony, C., 1990, Modifier protein for the methanol dehydrogenase of
methylotrophs. *Methods in Enzymology* **188**:216–222.
Long, A. R., and Anthony, C., 1991, The periplasmic modifier protein for methanol dehydrogenase in the methylotrophs *Methylophilus methylotrophus* and *Paracoccus denitrificans*.
*Journal of General Microbiology* **137**:2353–2360.
McIntire, W. S., Wemmer, D. E., Chistoserdov, A., and Lidstrom M. E., 1991, A new cofactor
in a prokaryotic enzyme—tryptophan tryptophylquinone as the redox prosthetic group
in methylamine dehydrogenase. *Science* **252**:817–824.
Morris, C. J., and Lidstrom, M. E., 1992, Cloning of a methanol-inducible *moxF* promoter and
its analysis in *moxB* mutants of *Methylobacterium extorquens* AM1. *J. Bacteriol.*
**174**:4444–4449.
Mutzel, A., and Gorisch, H., 1991, Quinoprotein ethanol dehydrogenase: preparation of the
apo-form and reconstitution with pyrroloquinoline quinone and $Ca^{2+}$ or $Sr^{2+}$ ions. *Agric.
Biol. Chem.* **55**:1721–1726.
Murzin, A. G., 1992, Structural principles for the propellor assembly of β-sheets: the preference for seven-fold symmetry. *Proteins* **14**:191–201.
Nunn, D. N., and Anthony, C., 1988, The nucleotide sequence and deduced amino acid sequence
of the cytochrome $c_L$ gene of *Methylobacterium extorquens* AM1: a novel class of *c*-type
cytochromes. *Biochem. J.* **256**:673–676.
Nunn, D. N., Day, D. J., and Anthony, C., 1989, The second subunit of methanol dehydrogenase
of *Methylobacterium extorquens* AM1. *Biochem. J.* **260**:857–862.
Ohshiro, Y., and Itoh, S., 1993, The chemistry of PQQ and related compounds. In: *Principles
and applications of quinoproteins* (V. L. Davidson, ed.) Marcel Dekker, New York.
pp. 309–329.
Page, M. D., and Anthony, C., 1986, Regulation of formaldehyde oxidation by the methanol
dehydrogenase modifier proteins of *Methylophilus methylotrophus* and *Pseudomonas*
AM1. *Journal of General Microbiology* **132**:1553–1563.
Park, J., and Churchich, J. E., 1992, Pyrroloquinoline quinone coenzyme PQQ, and the xida-
tion of SH residues in proteins. *Biofactors* **3**:257–260.

Ramachandran, G. N., and Sasisekharan, V., 1968, Conformation of polypeptides and proteins. *Adv. Protein Chem.* **23**:283–437.

Read, J., Gill, R., Dales, S. D., Cooper, J. B., Wood, S. P., and Anthony, C., 1999, The molecular structure of an unusual cytochrome $c_2$ determined at 2.0 A; the cytochrome $c_H$ from *Methylobacterium extorquens*. *Protein Science* **8**:1232–1240.

Richardson, I. W., and Anthony, C., 1992, Characterization of mutant forms of the quinoprotein methanol dehydrogenase lacking an essential calcium ion. *Biochem. J.* **287**:709–715.

Salisbury, S. A., Forrest, H. S., Cruse, W. B. T., and Kennard, O., 1979, A novel coenzyme from bacterial primary alcohol dehydrogenases. *Nature* **280**:843–844.

Schiering, N., Kabsch, W., Moore, M. J., Distefano, M. D., Walsh, C. T., and Pai, E. F., 1991, Structure of the detoxification catalyst mercuric ion reductase from *Bacillus* sp. strain RC607. *Nature* **352**:168–172.

Springer, A. L., Morris, C. J., and Lidstrom, M. E., 1997, Molecular analysis of *mxbD* and *mxbM*, a putative sensor-regulator pair required for oxidation of methanol in *Methylobacterium extorquens* AM1. *Microbiology* **143**:1737–1744.

van Koningsveld, H., Jongejan, J. A., and Duine, J. A., 1989, The three dimensional structure of PQQ and related compounds. In: *PQQ and Quinoproteins* (J. A. Jongejan and J. A. Duine, eds.), Kluwer, Dordrecht, pp. 243–251.

White, S., Boyd, G., Mathews, F. S., Xia, Z. X., Dai, W. W., Zhang, Y. F., and Davidson, V. L., 1993, The active site structure of the calcium-containing quinoprotein methanol dehydrogenase. *Biochemistry* **32**:12955–12958.

Wang, S. X., Mure, M., Medzihradszky, K. F., Burlingame, A. L., Brown, D. E., Dooley, D. M., Smith, A. J., Kagan, H. M., and Klinman, J. P., 1996, A crosslinked cofactor in lysyl oxidase: a redox function for amino acid side chains. *Science* **273**:1078–1084.

Xia, Z., Dai, W., Zhang, Y., White, S. A., Boyd, G. D., and Mathews, F. S., 1996, Determination of the gene sequence and the three-dimensional structure at 2 A resolution of methanol dehydrogenae from *Methylophilus* W3A1. *J. Mol. Biol.* **259**:480–501.

Xia, Z., Dai, W., Xiong, J., Hao, Z., Davidson, V. L., White, S., and Mathews, F. S., 1992, The 3-dimensional structures of methanol dehydrogenase from 2 methylotrophic bacteria at 2.6-angstrom resolution. *J. Biol. Chem.* **267**:22289–22297.

Xia, Z., He, Y., Dai, W., White, S. A., Boyd, G. D., and Mathews, F. S., 1999, Detailed active site configuration of a new crystal form of methanol dehydrogenase from *Methylophilus* W3A1 at 1.9 A resolution. *Biochemistry* **38**:1214–1220.

Xu, H. H., Viebahn, M., and Hanson, R. S., 1993, Identification of methanol-regulated promoter sequences from the facultative methylotrophic bacterium *Methylobacterium organophilum* XX. *J. Gen. Microbiol.* **139**:743–752.

Xu, H. H., Janka, J. J., Viebahn, M., and Hanson, R. S., 1995, Nucleotide sequence of the *mxcQ* and *mxcE* genes, required for methanol dehydrogenase synthesis in *Methylobacterium organophilum* XX: a two-component regulatory system. *J. Gen. Microbiol.* **141**:2543–2551.

Yamada, M., Inbe, H., Tanaka, M., Sumi, K., Matsushita, K., and Adachi, O., 1998, Mutant isolation of *Escherichia coli* quinoprotein glucose dehydrogenase and analysis of crucial residues Asp-730 and His-775 for its function. *J. Biol. Chem.* **273**:22021–22027.

Zheng, Y-J., and Bruice, T. C., 1997, Conformation of coenzyme pyrroloquinoline quinone and role of $Ca^{2+}$ in the catalytic mechanism of quinoprotein methanol dehydrogenase. *Proc. Natl. Acad. Sci.* **94**:11881–11886.

*Chapter 4*

# Methylamine Dehydrogenase
## Structure and Function of Electron Transfer Complexes

Victor L. Davidson

## 1. METHYLAMINE DEHYDROGENASE

Methylamine dehydrogenase [MADH] is a periplasmic enzyme which has been purified from several gram negative methylotrophic and autotrophic bacteria (reviewed in Davidson, 1993; Davidson *et al.*, 1995a). It catalyzes the oxidation of methylamine to formaldehyde and ammonia, and in the process transfers two electrons from the substrate to some electron acceptor (Eq. 1). This reaction is the first step in the metabolism of methylamine, which can

$$CH_3NH_3^+ + 2A_{ox} + H_2O \rightarrow HCHO + NH_4^+ + 2A_{red} + 2H^+ \qquad (1)$$

serve as a sole source of carbon and energy for these bacteria. When MADH is assayed in vitro, small redox-active species such as phenazine

VICTOR L. DAVIDSON    Department of Biochemistry, University of Mississippi Medical Center, Jackson, MS 39214-4505
*Subcellular Biochemistry, Volume 35: Enzyme-Catalyzed Electron and Radical Transfer*, edited by Holzenburg and Scrutton. Kluwer Academic / Plenum Publishers, New York, 2000.

methosulfate are routinely used as the electron acceptor. The natural electron acceptor for most MADHs is a periplasmic type I "blue" copper protein, amicyanin, which mediates electron transfer from MADH to *c*-type cytochromes (Tobari and Harada, 1981; Husain and Davidson, 1985). This review will focus primarily on MADH from *Paracoccus denitrificans* and the redox proteins with which it interacts.

## 1.1. Structure and Function

The physical properties of the MADHs which have been characterized thus far indicate that they are a relatively well-conserved class of enzymes. Each MADH is a tetramer of two identical larger α subunits of molecular mass of 40,000–50,000, and two identical smaller β subunits of molecular mass of approximately 15,000. The smaller subunits each possess a covalently bound prosthetic group called tryptophan tryptophylquinone [TTQ] (McIntire *et al.*, 1991) (Figure 1). The complete amino acid sequences have been determined for three MADHs (Chistoserdov *et al.*, 1992; 1994; Ubbink *et al.*, 1991). Crystal structures have also been determined for these MADHs from *P. denitrificans* (Chen *et al.*, 1998), *Thiobacillus versutus* (Vellieux *et al.*, 1989) and *Methylobacterium extorquens* AMI (Labasse *et al.*, 1998). The TTQ-bearing subunits display a high level of sequence homology and structural similarity (Chen *et al.*, 1998).

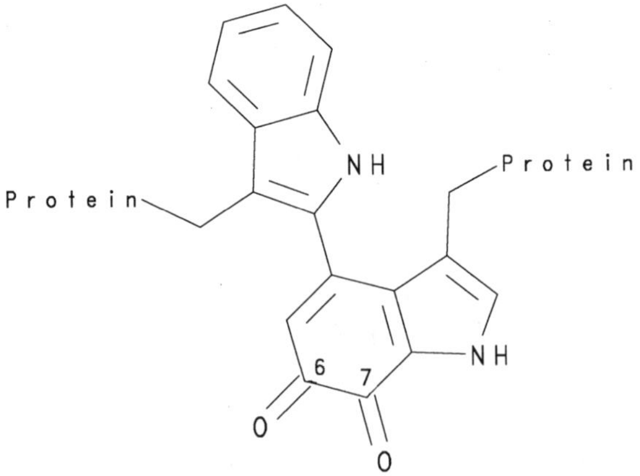

**FIGURE 1.** The structure of tryptophan tryptophylquinone. The C6 and C7 carbonyl carbons are labeled.

## 1.2.  The TTQ Prosthetic Group

The active site of MADH is relatively hydrophilic and located at the end of a hydrophobic channel between the α and β subunits. The C6 carbonyl of TTQ is exposed to solvent at the active site, and is the site of covalent adduct formation with the substrate (Huizinga et al., 1992). The two indole rings which comprise the TTQ structure are not coplanar but at a dihedral angle of approximately 38° (Chen et al., 1998). Whereas the C6 carbonyl of TTQ is present in the active site, the edge of the second indole ring which does not contain the quinone, is exposed at the surface of MADH.

TTQ is formed by a posttranslational modification of two tryptophan residues. In *P. denitrificans*, these are Trp[57] and Trp[108]. Two atoms of oxygen are incorporated into the indole ring of Trp[57] and a covalent bond is formed between the indole rings of the two tryptophan residues (see Figure 1). The mechanism by which this occurs is not known, but it does require the action of other enzymes which are subject to the same genetic regulation as the structural genes for the enzyme. The methylamine utilization gene cluster contains several genes, at least four of which encode proteins that are required for biosynthesis of MADH, in addition to the MADH structural genes (Chistoserdov et al., 1994; van der Palen et al., 1995; Graichen et al., 1999).

## 1.3.  Catalytic Reaction Mechanism

The overall oxidation-reduction reaction of MADH with methylamine and amicyanin may be divided into reductive (Eq. 2A) and oxidative (Eq. 2B) half-reactions. A detailed chemical

$$CH_3NH_3^+ + E\text{-}TTQ \rightarrow HCHO + E\text{-}TTQNH_2 + 2H^+ \qquad (2A)$$

$$E\text{-}TTQNH_2 + 2A_{ox} + H_2O \rightarrow NH_4^+ + 2A_{red} + E\text{-}TTQ \qquad (2B)$$

reaction mechanism for the overall reaction, based on results of studies of the MADH from *P. denitrificans* (Brooks et al., 1993; Davidson et al., 1995ab; Bishop et al., 1996ab; Bishop and Davidson, 1997) is shown Figure 2. The reductive and oxidative half-reactions require several steps that necessitate the presence of amino acid side chains capable of serving as either a general base or general acid. As many as 12 reaction steps involving general acids and bases (labeled $B_1$–$B_{12}$ in Figure 2) may be hypothesized. $B_1$ is required to bind and possibly deprotonate the substrate

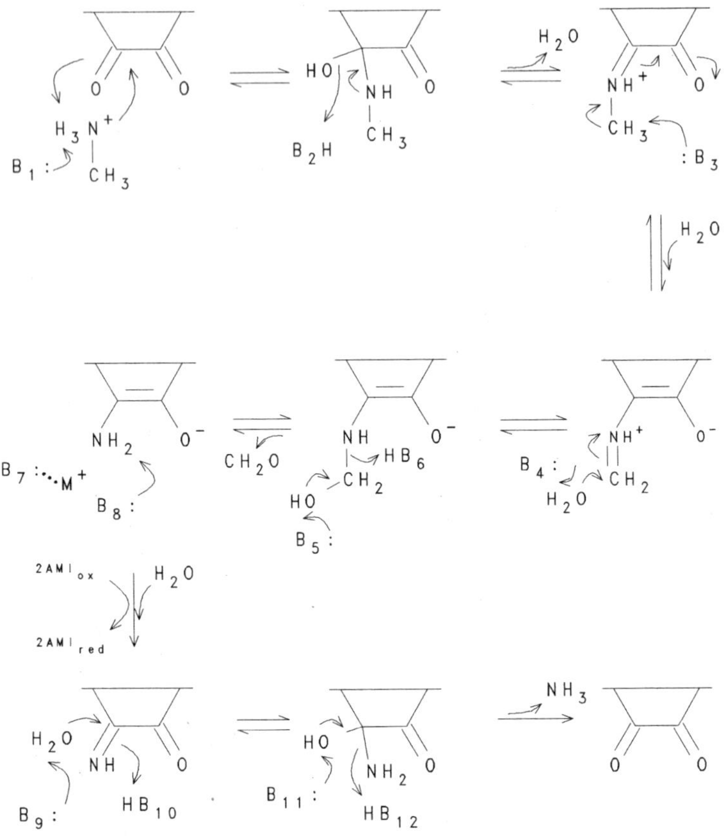

**FIGURE 2.** The reaction mechanism of MADH. Only the quinone portion of TTQ is shown. $B_1$ to $B_{12}$ represent active-site residues that may function as general acids or bases in the reaction mechanism. AMI represents amicyanin and $M^+$ is a monovalent cation. The details of the reaction mechanisms are presented in the text.

methylammonium to generate methylamine for nucleophilic attack of the C6 carbonyl carbon. $B_2H$ facilitates the dehydration of the carbinolamine intermediate to form an imine. $B_3$ abstracts a proton from the methyl carbon which leads to reduction of the TTQ cofactor (Brooks *et al.*, 1993). $B_4$ coordinates and activates water for nucleophilic attack of the imine carbon. $B_5$ and $B_6H$ facilitate cleavage of the C—N bond and release of the formaldehyde product to yield the reduced aminoquinol reaction intermediate (Bishop *et al.*, 1996b). MADH is reoxidized in two one-electron transfers to amicyanin molecules. The first electron transfer step requires the presence of a monovalent cation which is proposed to be bound to $B_7$ (Bishop and Davidson, 1997). Another general base, $B_8$, is required to deprotonate the amino nitrogen and thus activate this intermediate for electron transfer to amicyanin (Bishop and Davidson, 1997). $B_9$–$B_{12}$ facilitate the hydrolysis of the iminoquinone after electron transfer to release the ammonia product and generate the oxidized quinol. Alternatively, in the steady-state the amino nitrogen of another molecule of substrate, rather than water, may react directly with the iminoquinone to form the next enzyme-substrate adduct with concomitant release of the ammonia product (Zhu and Davidson, 1999). While 12 roles for active-site residues have been proposed, it is possible that multiple roles may be performed by a single residue so that less than 12 potentially active residues would be sufficient to catalyze the complete oxidation-reduction reaction. The crystal structure of MADH reveals the presence of four amino acid residues in the active site which could potentially participate in these reactions (Figure 3).

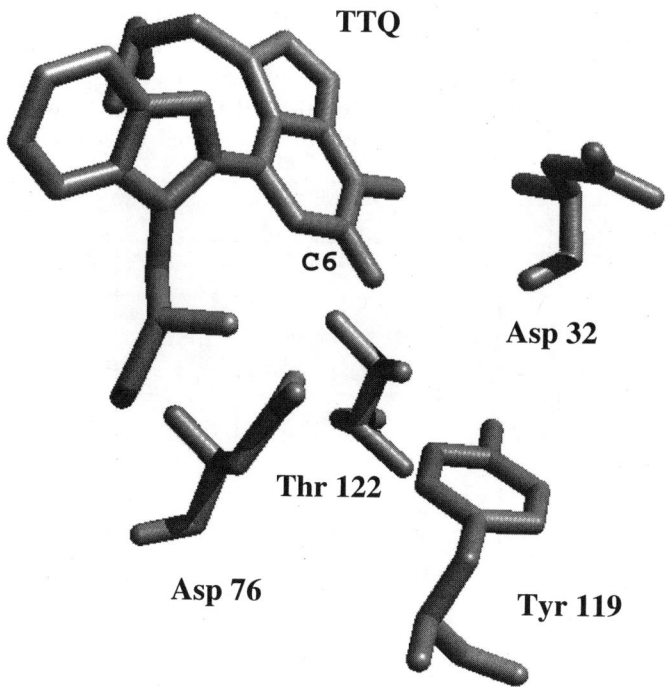

**FIGURE 3.** The active site of TTQ in MADH.

## 1.4.  Spectral and Redox Properties

The oxidized, two-electron reduced, and one-electron reduced semi-quinone redox forms of MADH exhibit very distinct visible absorption spectra (Davidson *et al.*, 1995a) (Figure 4). The oxidation-reduction mid-point potential [$E_m$] value for the two-electron oxidized/reduced couple has been determined by spectrochemical titration (Zhu and Davidson, 1998b). At pH 7.5 it is +95 mV, and over the range from pH 6.5 to 8.5 it is pH-dependent and exhibits a change of −30 mV per pH unit. This indicates that the two-electron transfer is linked to the transfer of a single proton. This result differs from what was obtained from redox studies of a TTQ model compound for which the two-electron couple is linked to the transfer of two protons (Itoh *et al.*, 1995). This result also distinguishes the redox properties of the enzyme-bound TTQ from those of the membrane-bound quinone components of respiratory and photosynthetic electron transfer chains that transfer two protons per two electrons. This difference is attributed to the accessibility of only one of the TTQ carbonyls to solvent in MADH. In the reduced form, the other quinol hydroxyl at C7 is shielded from solvent and thus is not protonated. The unusual property of TTQ enzymes of stabilizing the anionic form of the reduced quinol is important for the reaction mechanism of MADH because it allows stabilization of physiologically important aminoquinol and iminosemiquinone reaction intermediates (see Figure 2). Examination of the extent to which dispro-portionation of the MADH semiquinone occurred as a function of pH

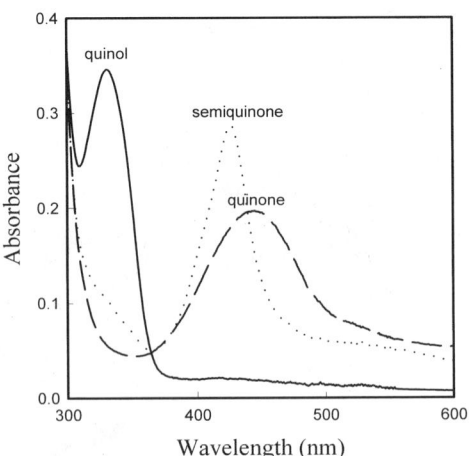

**FIGURE 4.** Absorption spectra of the different redox forms of MADH.

revealed that the equilibrium concentration of semiquinone increased with pH (Zhu and Davidson, 1998b). This indicates that the single proton transfer is linked to the semiquinone/quinol couple. Therefore, the quinol is singly protonated and the semiquinone is unprotonated and anionic. Kinetic studies have determined that the $E_m$ values at pH 7.5 for the one-electron couples are oxidized/semiquinone = +14 mV and semiquinone/reduced = +190 mV (Brooks and Davidson, 1994b). The substitution of N for O in the aminoquinol and iminosemiquinone appears to increase the $E_m$ values by approximately 41 mV (Itoh et al., 1995; Bishop and Davidson, 1998). Electronic properties of the semiquinone (Warncke et al., 1995) and aminosemiquinone (Warncke et al., 1993) forms of MADH have been characterized by electron double nuclear resonance [ENDOR] and electron spin echo envelope modulation [ESEEM] spectroscopies.

## 2. AMICYANIN

In *P. denitrificans*, amicyanin is an obligatory mediator of electron transfer from MADH to soluble *c*-type cytochromes (Husain and Davidson, 1986). Each protein is induced in this bacterium during growth on methylamine as a carbon source (Husain and Davidson, 1985; 1987). The amicyanin gene is located immediately downstream of that for MADH and inactivation of the former by gene replacement resulted in loss of the ability to grow on methylamine (Van Spanning et al., 1990). Two other structurally and functionally similar type I copper proteins, plastocyanin and azurin, did not effectively substitute for amicyanin as an electron acceptor (Gray et al., 1988; Hyun and Davidson, 1995).

### 2.1. Physical Properties

Amicyanin is a type I copper protein in which two histidines, one cysteine and one methionine provide the four ligands for the redox-active copper. The *P. denitrificans* protein is composed of 105 amino acid residues. Its primary sequence is known (Van Spanning et al., 1990) and its crystal structure has been determined (Durley et al., 1993; Cunane et al., 1996).

The redox properties of amicyanin are dependent on pH and also are affected by its association with MADH (Gray et al., 1988; Zhu et al., 1998). The pH dependence of the $E_m$ value of amicyanin free in solution correlates with that of a single protonated ligand having a $pK_a$ of 7.5. The crystal structures of oxidized amicyanin at pH 4.8 and reduced amicyanin at pH 4.4 show that the His[95] copper ligand is doubly protonated in the reduced and singly protonated in the oxidized states, and a conformational change is associated with the interconversion between the two redox states (Zhu et al., 1998). When reduced at low pH, the His[95] copper ligand rotates by 180° about the $C_\beta$-$C_\gamma$ bond relative to its position in oxidized amicyanin and is no longer in the copper coordination sphere. At pH 7.7, the conformation of His[95] in reduced amicyanin in the crystalline state appears to be that of an equilibrium mixture of the two conformers. This is what one would expect at a pH value near the $pK_a$ for this transition. It was concluded that the protonation of His[95], which is linked to the reduction of copper and to the conformational change of this residue, is responsible for the pH dependence of the $E_m$ value. When amicyanin is in complex with MADH, this rotation of His[95] that is associated with reduction at low pH is disallowed, and causes the redox potential to become pH-independent in the physiologic range of pH. Thus, in the complex with MADH, amicyanin exhibits an $E_m$ value similar to that of free amicyanin at high pH. This means that at physiologic pH, complex formation causes a decrease in the $E_m$ value of amicyanin, which makes the electron transfer reaction from reduced amicyanin to cytochromes c more thermodynamically favorable (Gray et al., 1988).

## 2.2.  Interactions with Methylamine Dehydrogenase

The specific interaction between MADH and amicyanin has been characterized by absorption spectroscopy (Gray et al., 1988), potentiometric studies (Zhu et al., 1998), steady-state kinetics (Davidson and Jones, 1991), transient kinetics (Brooks and Davidson, 1994ab; Bishop and Davidson, 1995; 1996; 1997; Bishop et al., 1996a), direct binding assays (Davidson et al., 1993), chemical cross-linking (Kumar and Davidson, 1990), resonance Raman spectroscopy (Backes et al., 1991), x-ray crystallography (Chen et al., 1992; 1994) and site-directed mutagenesis (Davidson et al., 1997). The results of these studies raise important questions concerning the relative roles of electrostatic and hydrophobic interactions in stabilizing the functional association between these two proteins. Kinetic, binding and spectroscopic studies indicated that complex formation was favored at low ionic strength (i.e. 10 mM buffer) over high ionic strength (i.e. 10 mM buffer plus 0.2 M NaCl). However, chemical cross-linking studies suggested that both hydrophobic and electrostatic forces were involved in complex formation.

A binary complex of MADH and amicyanin (Chen et al., 1992) and a ternary protein complex of these proteins plus cytochrome c-551i (Chen et al., 1994) from P. denitrificans have been crystallized and their structures have been determined. The structures of the crystallized complexes of these proteins indicate that the interface between MADH and amicyanin is stabilized largely by van der Waals' interactions, despite the presence of some charged and neutral hydrophilic residues in addition to hydrophobic residues (Chen *et al.*, 1992; 1994). One and possibly a second intermolecular salt bridge at the periphery of this hydrophobic interface may also be inferred from the structural data.

Site-directed mutagenesis was used to alter specific amino acid residues of amicyanin which appear from the crystal structure to be important for functional association with MADH (Davidson *et al.*, 1997) (Figure 5). Conversion of Arg[99] to either Asp of Leu results in a significant weakening of binding, particularly at low ionic strength. Thus, the ionic interprotein interactions involving this residue, which may be inferred from the crystal structure (Figure 5), are important for stabilizing the complex. Conversion of Phe[97] to Glu also significantly disrupts binding, demonstrating that the hydrophobic interactions involving this residue are also very important. These results demonstrate that a combination of hydrophobic and ionic interactions are required to stabilize complex formation, and that individual amino acid residues on the protein surface are able to dictate very specific interactions between soluble redox proteins.

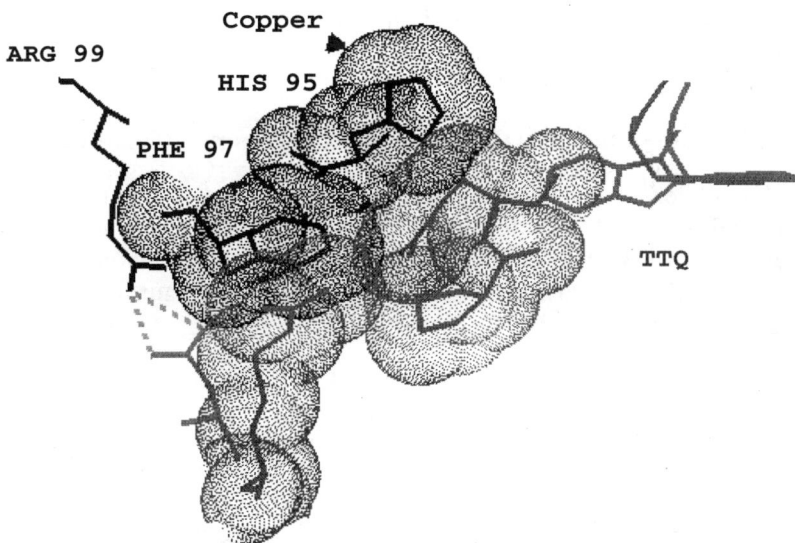

**FIGURE 5.** The amicyanin-MADH interface. Van der Waals' radii of His[95] and Phe[97] of amicyanin, and MADH residues with which they make close contact, are shown as dots. Ionic interactions between Arg[99] of amicyanin and a carboxylic residue of MADH are shown as dashed lines. MADH residues are in gray and amicyanin residues are in black. Coordinates are available in the Brookhaven data bank, entry 2MTA.

## 3.  METHYLAMINE DEHYDROGENASE-AMICYANIN-CYTOCHROME $c$-551i COMPLEX

MADH, amicyanin and cytochrome $c$-551i from *P. denitrificans* form one of the best characterized physiologic electron transfer complexes of proteins. It is the only complex of three soluble redox proteins for which a high resolution crystal structure is available (Chen *et al.*, 1994). The protein complex was shown to be functional in the crystalline state by single crystal polarized absorption spectroscopy (Merli *et al.*, 1996). In *P. denitrificans*, amicyanin is an obligatory mediator of electron transfer from MADH to soluble $c$-type cytochromes. MADH, amicyanin and cytochrome $c$-551i are isolated as individual soluble proteins, but they must form a ternary complex to catalyze methylamine-dependent cytochrome $c$-551i reduction (Husain and Davidson, 1986; Gray, *et al.*, 1986; 1988; Davidson and Jones, 1991; 1995; 1996). Although it is a thermodynamically favorable reaction, MADH does not reduce cytochrome $c$-551i in the absence of amicyanin (Husain and Davidson, 1986), probably because the proteins are unable to interact in a productive manner. Reduced amicyanin will not significantly reduce oxidized cytochrome in the absence of MADH at physiologic pH because the redox potential of free amicyanin is much more positive than that of the cytochrome (Gray *et al.*, 1988). As discussed earlier, the redox properties of amicyanin are altered on complex formation with MADH so as to facilitate the reaction (Zhu *et al.*, 1998).

### 3.1.  Structure of the MADH-Amicyanin-Cytochrome $c$-551i Complex

A complex of these proteins has been crystallized as a hetero-octamer comprised of one MADH tetramer, two amicyanins and two cytochromes (Chen *et al.*, 1994). The direct distances between redox centers are 9.4 Å from TTQ to copper, and 23 Å from copper to heme (Figure 6).

### 3.2.  Interactions Between Amicyanin and Cytochrome $c$-551i

Unlike the MADH-amicyanin protein interface which was discussed earlier, the amicyanin-cytochrome $c$-551i interface revealed in the crystal structure of the complex is relatively hydrophilic. The association between proteins appears to be stabilized by several hydrogen bond and ionic interactions (Chen *et al.*, 1994). The shortest gap between the proteins, for consideration of a possible site for interprotein electron transfer, is from the backbone O of Glu[31] of amicyanin to the backbone N of Gly[72] of cytochrome $c$-551i, a distance of less than 3 Å.

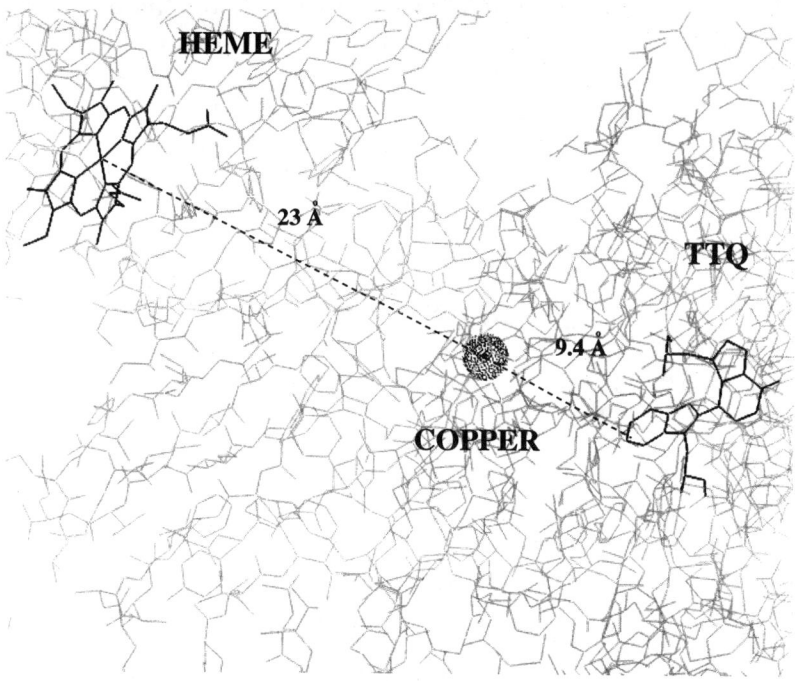

**FIGURE 6.** Orientation of redox cofactors in the MADH-amicyanin-cytochrome *c*-551i complex. A portion of the crystal structure is shown with the direct distances between the cofactors indicated. Coordinates are available in the Brookhaven Protein Data Bank, entry 2MTA.

## 3.3.  Pathways Analysis of the Electron Transfer Protein Complex

It has been suggested that long range electron transfer through proteins depends not simply on the direct distance between redox centers, but on specific pathways (Gray and Winkler, 1996). As such the structure of the ternary protein complex was analyzed to determine the most favorable predicted pathways from TTQ to copper and from copper to heme. Analysis was performed using the Greenpaths program (Regan *et al.*, 1993). This algorithm calculates the relative efficiency of electron transfer pathways according to Eq. 3 where *i* ranges over the

$$H_{AB} \propto \Pi \, \varepsilon_i \qquad (3)$$

pathway steps and $\varepsilon_i$ is a wave-function decay factor for step *i*.

For electron transfer from TTQ to copper, the most efficient predicted pathway (i.e. at least an order of magnitude more efficient than alternative pathways) involves a through space jump of 2.6 Å from a hydrogen on the surface-exposed Trp[108] indole ring of TTQ to the lone pair orbital of the carbonyl oxygen of Pro[94] of amicyanin (Figure 7). The pathway then follows six covalent bonds via the His copper ligand to the copper atom.

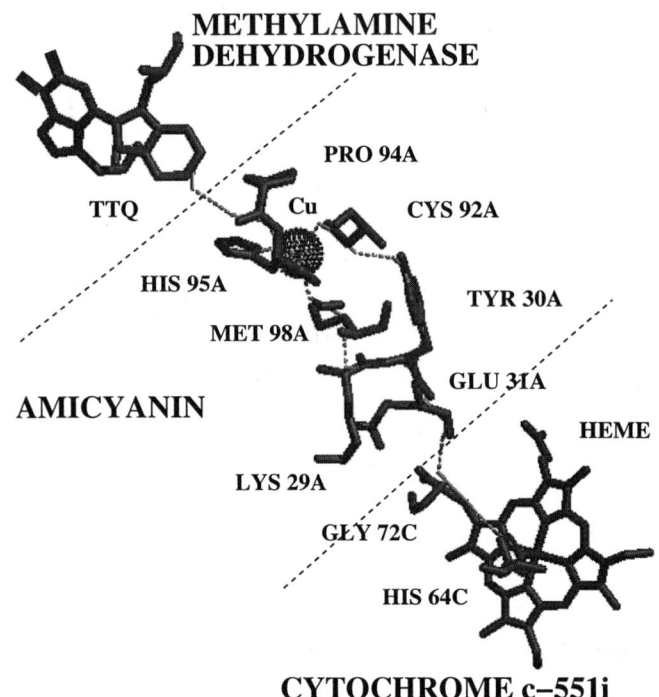

**FIGURE 7.** Predicted Electron Transfer Pathways in the MADH-Amicyanin-Cytochrome *c*-551i Complex. The most efficient predicted electron transfer pathways between redox centers are shown (i.e. at least an order of magnitude more efficient than alternative pathways). The pathway from TTQ to copper involves a through-space jump (dashed line) from the surface-exposed Trp[108] indole ring of TTQ to the carbonyl O of Pro[94] of amicyanin, followed by passage through six covalent bonds (solid line) via the His[95] ligand to copper. Two dominant sets of pathways were predicted from copper to the heme iron. In each, the point of intermolecular electron transfer was from the backbone O of Glu[31] of amicyanin to the backbone N of Gly[72] of cytochrome *c*-551i, and the entry of electrons to iron occurred either via the porphyrin ring or the His[61] ligand. In one, the exit of electrons from copper occurred via the Cys[92] copper ligand, and the phenolic side chain of Tyr[30] was an intermediate between Cys[92] to Glu[31]. In the other, the exit of electrons from copper occurred via the Met[98] copper ligand, and the backbone of Lys[29] was an intermediate between Met[98] to Glu[31].

For electron transfer from copper to heme, two dominant sets of pathways of comparable efficiency were predicted. In each set of pathways, the point of intermolecular electron transfer was from the backbone O of Glu[31] of amicyanin to the backbone N of Gly[72] of cytochrome c-551i, and the entry of electrons to iron occurred either via the porphyrin ring or the His[61] ligand. In one set of pathways the exit of electrons from copper occurred via the Cys[92] copper ligand, and the phenolic side chain of Tyr[30] was an intermediate between Cys[92] and Glu[31]. In the other set of pathways the exit of electrons from copper occurred via the Met[98] copper ligand, and the backbone of Lys[29] was an intermediate between Met[98] and Glu[31]

## 4. ELECTRON TRANSFER REACTIONS IN METHYLAMINE DEHYDROGENASE COMPLEXES

### 4.1. Electron Transfer Theory

Electron transfer theory predicts that the rate of an electron transfer reaction will vary predictably with temperature (T), $\Delta G°$, and donor-acceptor distance (r) according to the relationships given in Eq. 4 and 5 (Marcus and Sutin, 1985), where h is Planck's constant, R is

$$k_{ET} = \left\{ \left(4\pi^2 H_{AB}^2\right) \Big/ \left(h\sqrt{4\pi\lambda RT}\right)\right\} \exp\left[-(\Delta G° +\lambda)^2 \Big/ 4\lambda RT\right] \qquad (4)$$

$$k_{ET} = k_0 \exp[-\beta(r - r_0)]\exp\left[-(\Delta G° +\lambda)^2 \Big/ 4\lambda RT\right] \qquad (5)$$

the gas constant (the Boltzmann constant may alternatively be used), $k_0$ is the characteristic frequency of the nuclei which is assigned a value of $10^{13} s^{-1}$, and $r_0$ is the close contact distance and assigned a value of 3.0 Å. $H_{AB}$ is the electronic coupling between redox centers and describes the degree of wave function overlap between donor and acceptor sites. As can be seen by comparing Eqs. 4 and 5, $H_{AB}$ is related to the distance which separates the redox centers (r), and the medium which separates them ($\beta$). $\lambda$ is the reorganizational energy, which may be defined as the energy needed to deform the nuclear configuration from the reactant to the product state. Detailed discussions of the mathematical and physical meaning of $H_{AB}$ and $\lambda$ may be found in a number of excellent reviews of electron transfer theory (Marcus and Sutin, 1987; Gray and Winkler, 1996; Moser et al., 1992; McLendon and Hake, 1992).

The factor $\beta$ is related to the nature of the intervening medium between redox centers. The experimentally-derived value of r will be dependent upon the value of $\beta$ which is used in the analysis. Values which are typically used range from 0.7–1.4 $Å^{-1}$. The most appropriate value for $\beta$, and the issue of whether r describes the direct donor-acceptor distance or is pathway-dependent have been a matter of much debate (Moser *et al.*, 1992; Gray and Winkler, 1996). We are currently using the MADH-amicyanin-cytochrome *c*-551i to test whether or not specific pathways are important for long range protein electron transfer reactions. In order to correctly use this structurally well-characterized system as a model for application of theories that describe the mechanisms of non-adiabatic electron transfer, one must first have a basis for ascertaining that the experimentally-determined rate constants do in fact describe the actual electron transfer event. How to do this is discussed below.

## 4.2.  Kinetic Complexity of Protein Electron Transfer Reactions

For redox reactions involving proteins, the actual meaning of the kinetically-determined limiting first-order rate constant for the electron transfer reaction ($k_3$ in Eq. 6) must be

$$A_{ox} + B_{red} \underset{}{\overset{K_d}{\rightleftharpoons}} A_{ox}/B_{red} \underset{k_4}{\overset{k_3}{\rightleftharpoons}} A_{red}/B_{ox} \qquad (6)$$

interpreted with caution. It may not be a true electron transfer rate constant (Davidson, 1996; Hoffman and Ratner, 1987; Brunschwig and Sutin, 1989; Harris *et al.*, 1994). Protein dynamics (i.e. transient formation of unstable conformational intermediates) or catalytic events (e.g. protonation/ deprotonation) may contribute to the observed rate. In kinetic models that are used to analyze these data any spectroscopically invisible, or otherwise undetectable, events subsequent to binding and preceding the spectral change associated with the redox reaction will be reflected in this rate constant. These events could include substrate-induced, product-induced or redox-linked protein conformational changes. In this case Eq. 6 is more correctly written as Eq. 7 where $k_x$ is the forward rate constant for this prerequisite adiabatic reaction

$$A_{ox} + B_{red} \underset{}{\overset{K_d}{\rightleftharpoons}} A_{ox}/B_{red} \underset{k_{-X}}{\overset{k_X}{\rightleftharpoons}} [A_{ox}/B_{red}]^* \underset{k_{-ET}}{\overset{k_{ET}}{\rightleftharpoons}} A_{red}/B_{ox} \qquad (7)$$

step and $k_{ET}$ is the forward rate constant for the actual electron transfer event. Three situations are considered below in which the electron transfer event is preceded by some reversible prerequisite adiabatic event.

### 4.2.1.  True Electron Transfer

In a true electron transfer reaction, the electron transfer step is the slowest step in the overall redox reaction such that $k_3 = k_{ET}$. When such a reaction is analyzed by Eq. 6, the experimentally determined $\lambda$ will include no contributions from non-electron transfer reaction steps. It will be a true value that will reflect nuclear displacements in the redox centers, protein matrix and solvent. The value of $H_{AB}$ will be within the non-adiabatic limit and the electron transfer distance may be calculated from the data.

### 4.2.2. Coupled Electron Transfer

In a coupled electron transfer reaction, the preceding adiabatic reaction step influences the experimentally-determined rate constant even though the electron transfer step is the slowest for the overall redox reaction. This occurs when the relatively fast reaction step which precedes electron transfer is very unfavorable (i.e. $K_X$ ($k_X/k_{-X}$)>>1). In this case, $k_3$ will be influenced by the equilibrium constant for that non-electron transfer process such that $k_3 = k_{ET}*K_X$ (Harris et al., 1994; Davidson, 1996). It follows that the experimentally-derived $\lambda$ ($\lambda_{obs}$) may contain contributions from both the electron transfer event and the preceding reaction step (i.e. $\lambda_{obs} = f[\lambda_{ET}, \lambda_X]$). For example, $\lambda_{obs}$ for interprotein electron transfer reactions may reflect contributions from an intracomplex rearrangement of proteins after binding to achieve an optimum orientation for electron transfer. As with a true electron transfer reaction, $k_3$ will vary with $\Delta G^\circ$ since $k_3$ is proportional to $k_{ET}$, although $H_{AB}$ may also be affected by the coupling.

### 4.2.3. Gated Electron Transfer

In a gated electron transfer reaction, the adiabatic reaction step that precedes electron transfer is rate-limiting for the overall redox reaction so that the observed rate is actually that of a non-electron transfer event (i.e. $k_3 = k_X$). In contrast to true and coupled electron transfer reactions, the rate constant for a gated reaction will not exhibit a predictable dependence on the $\Delta G^\circ$ for the electron transfer (see Eqs. 4 and 5) since this reaction step is not being driven by the redox potential difference between the reactants. The reaction will still vary with temperature, but if the temperature-dependence data are analyzed by Eq. 4, the values which are obtained for $\lambda$ and $H_{AB}$ will be unrelated to the electron transfer event. When the temperature dependence of the rate constant for a gated reaction is analyzed by Eq. 4, one may expect the fitted value of $H_{AB}$ to exceed the non-adiabatic limit. In fact analysis of some gated electron transfer reactions by Eq. 5 have yielded unreasonably large values for $H_{AB}$, as well as unreasonably large values for $\lambda$, and in some cases negative fitted values for the electron transfer distance (Bishop and Davidson, 1995; Lanzilotta et al., 1998; Lee et al., 1998; Hyun et al., 1999). In this respect, analysis of the temperature dependence of the rates of electron transfer reactions by Marcus theory may be of diagnostic use in determining the kinetic mechanism of complex protein electron transfer reactions.

## 4.3.  Electron Transfer from TTQ to Copper

It is possible to study electron transfer from several different redox forms of MADH to amicyanin (Figure 8). Since TTQ is a two-electron carrier and the type I copper is a one-electron carrier, two sequential oxidations of the fully reduced TTQ by amicyanins are required to completely reoxidize MADH. O-quinol and O-semiquinone forms of MADH may be generated by reduction by dithionite (Husain *et al.*, 1987). With the O-forms of MADH it has also been possible to monitor the rates of the reverse electron transfer reactions from reduced amicyanin to the oxidized and semiquinone forms. The product of the reduction of MADH by the substrate amine is an aminoquinone (N-quinol) which retains the covalently bound substrate-derived amino group after release of the aldehyde product (Bishop *et al.*, 1996b). An iminosemiquinone (N-semiquinone) is the product of the first one-electron oxidation of the N-quinol (Bishop *et al.*, 1996a). The N-semiquinone may also be generated in vitro (Zhu and Davidson, 1998a). The rates of the electron transfer reactions from each of these four redox forms of MADH to amicyanin are different (Table 1).

### 4.3.1.  Non-adiabatic Electron Transfer Reactions

For the electron transfer reactions to amicyanin from the O-quinol, O-semiquinone and N-semiquinone, $k_{ET}$ is the slowest step in the overall reaction (Bishop and Davidson, 1998). The rates of these reactions vary predictably with driving force (i.e. the redox potential difference between the reactants) (Figure 9). Independent analyses by Eqs. 4 and 5 of the $\Delta G°$ and temperature dependencies of the electron transfer rate of each of these reactions (Brooks and Davidson, 1994ab; Bishop and Davidson, 1998) yielded identical values of $\lambda$ and $H_{AB}$, and predicted the electron transfer distance that is seen in the crystal structure. Furthermore, the rates of each of these reactions were relatively insensitive to pH and buffer composition (Table 1), as would be expected for an electron transfer reaction occurring within a protein complex.

### 4.3.2.  Mutation of Amicyanin Alters the $H_{AB}$ for Electron Transfer from MADH

Phe[97] stabilizes the MADH-amicyanin complex via van der Waals' interactions at the protein-protein interface (see Figure 5). An F97E

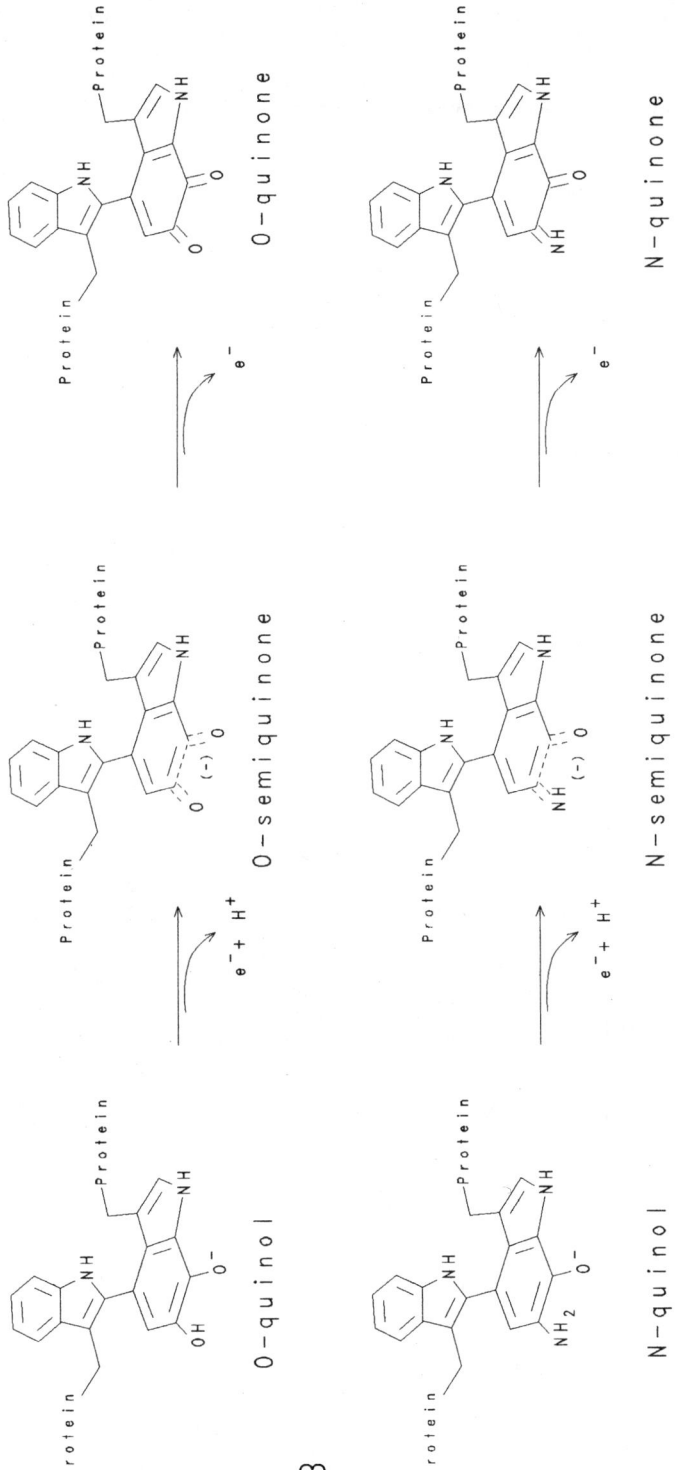

**FIGURE 8.** Sequential one-electron oxidations of dithionite-reduced (A) and substrate-reduced (B) TTQ in MADH. The protonation state of the oxygens in the quinol and semiquinone forms of TTQ was demonstrated by redox studies (Zhu and Davidson, 1998b). It was shown that the distribution of spin density in the semiquinone forms actually extends throughout the quinolated indole ring and into the second indole ring, but is asymmetric (Warncke *et al.*, 1995). The distribution of spin density should not be inferred from this figure.

## Table 1
### Reduction of Amicyanin by Different Redox Forms of MADH[a]

|  | O-quinol | O-semiquinone | N-quinol | N-semiquinone |
|---|---|---|---|---|
| $k_{ET}$ (10 mM KP$_i$, pH 7.5) |  |  |  |  |
| +0 mM KCl | $14 s^{-1}$ | $>500 s^{-1}$ | $12 s^{-1}$ | $28 s^{-1}$ |
| +200 mM KCl | $12 s^{-1}$ | $>500 s^{-1}$ | $130 s^{-1}$ | $51 s^{-1}$ |
| +400 mM KCl | $12 s^{-1}$ | $>500 s^{-1}$ | $>500 s^{-1}$ | $84 s^{-1}$ |
| $k_{ET}$ 10 mM Kp$_i$ + 0.2 M KCl |  |  |  |  |
| pH 6.5 | $12 s^{-1}$ | $>500 s^{-1}$ | $42 s^{-1}$ | $21 s^{-1}$ |
| pH 7.5 | $12 s^{-1}$ | $>500 s^{-1}$ | $130 s^{-1}$ | $51 s^{-1}$ |
| pH 9.0 | $11 s^{-1}$ | $>500 s^{-1}$ | $>500 s^{-1}$ | $110 s^{-1}$ |
| Kinetic Solvent Isotope Effect ($^{H2O}k/^{D2O}k$) at pH 7.5 | 1.5 | — | 6.5 | 1.8 |
| Marcus Parameters |  |  |  |  |
| $\lambda$ (eV) | $2.3 \pm 0.1$ | $2.3 \pm 0.1$ | $3.4 \pm 0.1$ | $2.4 \pm 0.1$ |
| $H_{AB}$ (cm$^{-1}$) | $12 \pm 7$ | $12 \pm 7$ | $>20,000$ | $13 \pm 4$ |
| r (Å, for $\beta$ = 1) | $9.6 \pm 0.7$ | $9.6 \pm 0.7$ | $<0$ | $9.4 \pm 1.2$ |
| Rate-limiting Step | Electron Transfer | Electron Transfer | Proton Transfer | Electron Transfer |

[a] Data taken from Bishop and Davidson, 1995; 1997; 1998.

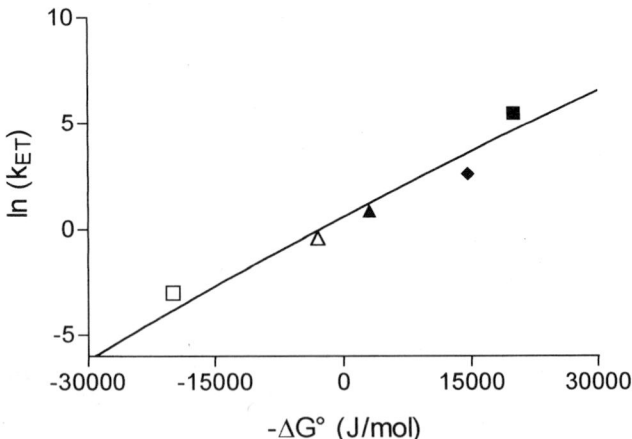

**FIGURE 9.** Marcus analysis of electron transfer reactions between MADH and amicyanin. Values of $k_{ET}$ were determined for the reactions of different redox forms of MADH with amicyanin shown in Figure 8: O-quinol (▲), O-semiquinone (■), N-semiquinone (♦). Rate constants were also obtained for the reverse reactions of the O-quinol (△) and O-semiquinone (□). The solid line represent fits to Eqs. 4 and 5, which are superimposible.

mutation increases the $K_d$ by two orders of magnitude (Davidson *et al.*, 1997). The rate constant for electron transfer from O-quinol MADH to the F97E mutant is also 24-fold slower than to the wild-type. Potentiometric and Marcus analyses of the electron transfer reaction revealed that the $\Delta G°$ and $\lambda$ associated with the electron transfer reaction are unaffected by the mutation. The decrease in $k_{ET}$ was attributable completely to a decrease in $H_{AB}$ (Davidson *et al.*, 1998). $H_{AB}$ is related to the distance or pathway which separates the redox centers. Phe$^{97}$ is not a part of the predicted pathway for electron transfer from TTQ to copper, but it is in close proximity to the point at which the electron is predicted to jump from TTQ to amicyanin. The most likely explanation for these results is that the F97E mutation at the protein-protein interface causes an increase in the interprotein distance within the protein complex. If one assumes the only segment of the electron transfer pathway that is affected by this mutation is the through-space jump from TTQ to amicyanin, then an increase of about 0.9 Å in the required through-space jump could account for the observed effects on $k_{ET}$ and $H_{AB}$ (Davidson *et al.*, 1998).

### 4.3.3. Gated Electron Transfer Reactions

Covalent incorporation of substrate-derived N into the TTQ semi-quinone altered the electron transfer rate simply by altering the redox potential of the cofactor relative to the O-semiquinone (Bishop and Davidson, 1998). However, the effects of covalent incorporation of substrate-derived N into the fully reduced TTQ were much more profound (Bishop and Davidson, 1997). The driving force for the reaction of the N-quinol has apparently been changed such that the reaction has become both extremely slow and thermodynamically unfavorable. In order for this physiologic reaction to occur, a prerequisite step is now required to activate TTQ for electron transfer. It was proposed that deprotonation of the N-quinol amino group is required to generate the "activated" intermediate from which electron transfer occurs (Bishop and Davidson, 1997). Electron transfer from this intermediate is so rapid that the redox reaction becomes rate-limited (i.e. gated) by the deprotonation event. This hypothesis is supported by the observation that only the redox reaction of the N-quinol exhibits a significant kinetic solvent isotope effect, consistent with rate-limitation by the transfer of an exchangeable proton. The rate of this redox reaction also varies markedly with pH and salt (Table 1). The salt dependence is due specifically to monovalent cations. The significant isotope effect is observed at all values of pH and salt concentrations (Bishop and Davidson, 1997). Thus, the increase in rate with pH and cations is not due to a change in the rate-limiting that "ungates" electron transfer. High pH and monovalent cations increase the rate of the redox reaction by increasing the rate of the proton transfer step that gates electron transfer from the N-quinol under all reaction conditions. To account for these observations, a detailed model has been proposed in which the pH-dependent binding of a monovalent cation to the MADH active site facilitates the rate-limiting deprotonation of the N-quinol amino group (Bishop and Davidson, 1997).

## 4.4.  Electron Transfer from Copper to Heme

The electron transfer reaction from copper to heme within the ternary protein complex was also studied in solution by stopped-flow spectroscopy. Analysis by Marcus theory of the temperature dependence of the limiting first-order rate constant for the redox reaction (Davidson and Jones, 1996) yielded values for the $\lambda$ of $1.1\,eV$ and $H_{AB}$ of $0.3\,cm^{-1}$, and predicted an electron transfer distance between redox centers which was consistent with the distance seen in the crystal structure. Thus, the electron transfer event is rate-limiting for this redox reaction. Experiments are in progress to determine the validity of the predicted pathways for electron transfer shown in Figure 7.

## 5.  APPLICATION OF MARCUS THEORY TO OTHER PROTEIN ELECTRON TRANSFER REACTIONS

In addition to the reactions described in this review, relatively few protein electron transfer reactions have been analyzed by Eqs. 4 and 5, but the results of these analyses have produced interesting results. Another TTQ-dependent enzyme, AADH (Govindaraj et al., 1994), uses the type I copper protein azurin as an electron acceptor (Edwards et al., 1995). As was observed with MADH, the oxidation of the chemically-reduced O-quinol AADH by azurin was rate-limited by electron transfer, whereas the oxidation of the substrate-reduced N-quinol AADH by azurin was gated by the transfer of an exchangeable proton (Hyun et al., 1999). The electron transfer reaction between the iron protein and molybdenum-iron protein of the nitrogenase complex was converted from one which was gated to one which was not by site-directed mutagenesis (Lanzilotta et al., 1998). The reaction of the native protein is gated by conformational events associated with either MgATP binding or hydrolysis, and yielded an $H_{AB}$ which is well in excess of the nonadiabatic limit and a negative value for the electron transfer distance. The reaction of an L127D mutant, however, yielded Marcus parameters that were characteristic of a nonadiabatic reaction and an electron transfer distance similar to what is seen in the crystal structure. Temperature dependence studies of the electron transfer reaction from methanol dehydrogenase to cytochrome $c$-551i (Harris and Davidson, 1993) yielded an electron transfer distance that correlated well with predictions based on the known structures of the individual proteins. This reaction is an example of conformationally coupled electron transfer (Harris $et$ $al.$, 1994). Temperature dependence studies of trimethylamine dehydrogenase have yielded reasonable estimates for intramolecular (Falzon and Davidson, 1996) and intermolecular (Wilson $et$ $al.$, 1995) electron transfer reactions. Temperature dependence studies have demonstrated that the electron transfer from flavin to iron in the rubredoxin reductase-rubredoxin complex is gated by an as yet undetermined process (Lee $et$ $al.$, 1998). Thus far, the reactions of the N-quinol reduced forms of the two TTQ enzymes, MADH and AADH, with their respective copper protein electron acceptors, stand out as examples of interprotein electron transfer reactions that are gated by proton transfer events.

## 6. CONCLUSIONS

Analysis by Marcus theory of the temperature dependencies of rates of biologic electron transfer reactions will provide valid estimates for $\lambda$, $H_{AB}$ and electron transfer distance, providing that the electron transfer event is rate-limiting for the observed redox reaction. It may also be useful for identifying gated electron transfer reactions when unrealistic values for these parameters are obtained (e.g. values for the N-quinol in Table 1). When an electron transfer reaction is gated, kinetic analysis including the effects of varying solution conditions on rate and kinetic isotope effect studies, can provide useful information on the nature of the reaction step which gates electron transfer. The structurally well-characterized MADH-amicyanin-cytochrome $c$-551i complex has been used as a model for developing methods to study interprotein electron transfer reactions. The application of Marcus Theory for the analysis of the temperature dependence of electron transfer reactions within this complex has yielded valuable information about both gated and ungated electron transfer reactions.

ACKNOWLEDGMENTS. Work performed in this laboratory has been supported by NIH grant GM-41574. I acknowledge the contributions of the following current and former members of this laboratory to the previously published work which was summarized herein: Limei Jones, M. Elizabeth

Graichen, Harold B. Brooks, G. Reid Bishop, Young-Lan Hyun and Zhenyu Zhu. The x-ray crystallography was performed in the laboratory of F. Scott Mathews at Washington University.

## 7. REFERENCES

Backes, G., Davidson, V. L., Huitema, F., Duine, J. A., and Sanders-Loehr, J., 1991, Character-ization of the tryptophan-derived quinone cofactor of methylamine dehydrogenase by resonance Raman spectroscopy, *Biochemistry* **30**:9201–9210.

Bishop, G. R., and Davidson, V. L., 1995, Intermolecular electron transfer from substrate-reduced methylamine dehydrogenase to amicyanin is linked to proton transfer. *Biochemistry* **34**:12082–12086.

Bishop, G. R., and Davidson, V. L., 1997, Catalytic role of monovalent cations in the mecha-nism of proton transfer which gates an interprotein electron transfer reaction, *Biochemistry* **36**:3586–13592.

Bishop, G. R., and Davidson, V. L., 1998, Electron transfer from the aminosemiquinone reac-tion intermediate of methylamine dehydrogenase to amicyanin. *Biochemistry* **37**:11026–11032.

Bishop, G. R., Brooks, H. B., and Davidson, V. L., 1996a, Evidence for a tryptophan trypto-phylquinone aminosemiquinone intermediate in the physiologic reaction between methy-lamine dehydrogenase and amicyanin, *Biochemistry* **35**:8948–8954.

Bishop, G. R., Valente, E. J., Whitehead, T. L., Brown, K. L., Hicks, R. T., and Davidson, V. L., 1996b, Direct detection by [15]N-NMR of the tryptophan tryptophylquinone aminoquinol reaction intermediate of methylamine dehydrogenase, *J. Am. Chem. Soc.* **118**:12868–12869.

Brooks, H. B., and Davidson, V. L., 1994a, Kinetic and thermodynamic analysis of a physio-logic intermolecular electron transfer reaction between methylamine dehydrogenase and amicyanin, *Biochemistry* **33**:5696–5701.

Brooks, H. B., and Davidson, V. L., 1994b, Free energy dependence of the electron transfer reaction between methylamine dehydrogenase and amicyanin. *J. Am. Chem. Soc.* **116**:11201–11202.

Brooks, H. B., Jones, L. H., and Davidson, V. L., 1993, Stopped-flow kinetic and deuterium kinetic isotope effect studies of the quinoprotein methylamine dehydrogenase from *Paracoccus denitrificans*, *Biochemistry* **32**:2725–2729.

Brunschwig, B. S., and Sutin, N., 1989, Directional electron transfer: Conformational intercon-versions and their effects on observed electron-transfer rate constants. *J. Am. Chem. Soc.* **110**:7454–7465.

Chen, L., Durley, R., Poloks, B. J., Lim, L. W., Hamada, K., Chen, Z., Mathews, F. S., Davidson, V. L., Satow, Y., Huizinga, E., Vellieux, F. M. D., and Hol, W. G. J., 1992, Interaction between a quinoprotein and a blue copper protein revealed by the crystal structure of an electron-transfer complex between methylamine dehydrogenase and amicyanin, *Biochemistry* **31**:4959–4964.

Chen, L., Durley, R., Mathews, F. S., and Davidson, V. L., 1994, Structure of an electron trans-fer complex: Methylamine dehydrogenase, amicyanin and cytochrome *c*-551i, *Science* **264**:86–90.

Chen, L., Dol, M., Durley, R. C. E., Chistoserdov, A. Y., Lidstrom, M. E., Davidson, V. L., and Mathews, F. S., 1998, Refined crystal structure of methylamine dehydrogenase from *Paracoccus denitrificans* at 1.75 Å resolution, *J. Mol. Biol.* **276**:131–149.

Chistoserdov, A. Y., Biyd, G., Mathews, F. S., and Lidstrom, M. E., 1992, The genetic organization of the *mau* gene cluster of the facultative autotroph *Paracoccus denitrificans*, *Biochim. Biophys. Res. Commun.* **184**:1226–1234.

Chistoserdov, A. Y., Chistoserdova, L. V., McIntire, W. S., and Lidstrom, M. E., 1994, Genetic organization of the *mau* gene cluster in *Methylobacterium extorquens* AM1: complete nucleotide sequence and generation and characteristics of *mau* mutants, *J. Bacteriol.* **176**:4052–4065.

Cunane L. M., Chen, Z.-W., Durley, R. C. E., and Mathews, F. S., 1996, X-ray structure of the cupredoxin amicyanin, from *Paracoccus denitrificans*, refined at 1.31 Å resolution. *Acta Cryst.* **D52**:676–686.

Davidson, V. L., 1993, Methylamine dehydrogenase, in *Principles and Applications of Quinoproteins*, (Davidson, V. L., ed.) pp. 73–95, Marcel Dekker, New York.

Davidson, V. L., 1996, Unraveling the kinetic complexity of interprotein electron transfer reactions, *Biochemistry* **35**:14035–14039.

Davidson, V. L., and Jones, L. H., 1991, Intermolecular electron transfer from quinoproteins and its relevance to biosensor technology. *Anal. Chim. Acta* **249**:235–240.

Davidson, V. L., and Jones, L. H., 1995, Complex formation with methylamine dehydrogenase affects the pathway of electron transfer from amicyanin to cytochrome *c*-551i. *J. Biol. Chem.* **270**:23941–23943.

Davidson, V. L., and Jones, L. H., 1996, Electron transfer from copper to heme within the methylamine dehydrogenase-amicyanin-cytochrome *c*-551i complex, *Biochemistry* **35**: 8120–8125.

Davidson, V. L., Graichen, M. E., and Jones, L. H., 1993, Binding constants for a physiologic electron-transfer protein complex between methylamine dehydrogenase and amicyanin. Effects of ionic strength and bound copper on binding, *Biochim. Biophys. Acta* **1144**:39–45.

Davidson, V. L., Brooks, H. B, Graichen, M. E., Jones, L. H., and Hyun, Y-L., 1995a, Detection of intermediates in TTQ enzymes, *Methods Enzymol.* **258**:176–190.

Davidson, V. L., Graichen, M. E., and Jones, L. H., 1995b, Mechanism of reaction of allylamine with the quinoprotein methylamine dehydrogenase. *Biochem. J.* **308**:487–492.

Davidson, V. L., Jones, L. H., Graichen, M. E., Mathews, F. S., and Hosler, J. P., 1997, Factors which stabilize the methylamine dehydrogenase-amicyanin electron transfer complex revealed by site-directed mutagenesis, *Biochemistry* **36**:12733–12738.

Davidson, V. L., Jones, L. H., and Zhu, Z., 1998, Site-directed mutagenesis of Phe 97 of amicyanin alters the electronic coupling for interprotein electron transfer from quinol methylamine dehydrogenase, *Biochemistry* **37**:7371–7377.

Durley, R., Chen, L., Lim, L. W., Mathews, F. S., and Davidson, V. L., 1993, The crystal structure analysis of amicyanin and apoamicyanin from *Paracoccus denitrificans* at 2.0 Å and 1.8 Å, *Protein Sci.* **2**:739–752.

Edwards, S. L., Davidson, V. L., Hyun, Y.-L., and Wingfield, P. T. 1995, Spectroscopic evidence for a common electron transfer pathway in two tryptophan tryptophylquinone enzymes. *J. Biol. Chem.* **270**:4293–4298.

Falzon, L., and Davidson V. L., 1996, Intramolecular electron transfer in trimethylamine dehydrogenase: A thermodynamic analysis, *Biochemistry* **35**:12111–12118.

Govindaraj, S., Eisenstein, E., Jones, L. H., Sanders-Loehr, J., Chistoserdov, A. Y., Davidson, V. L., and Edwards, S. L. 1994, Aromatic amine dehydrogenase, a second tryptophan tryptophylquinone enzyme. *J. Bacteriol.* **176**:2922–2929.

Graichen, M. E., Jones, L. H., Sharma, B., van Spanning, R. J. M., Hosler, J. P., and Davidson, V. L., 1999, Heterologous expression of correctly assembled methylamine dehydrogenase in *Rhodobacter sphaeroides*, *J. Bacteriol.* **181**:4210–4222.

Gray, H. B., and Winkler, J. R., 1996, Electron transfer in proteins, *Annu. Rev. Biochem.* **65**:537–561.

Gray, K. A., Knaff, D. B., Husain, M., and Davidson, V. L., 1986, Measurement of the oxidation-reduction potentials of amicyanin and *c*-type cytochromes from *Paracoccus denitrificans*, *FEBS Lett.* **207**:39–242.

Gray, K. A., Davidson, V. L., and Knaff, D. B., 1988, Complex formation between methylamine dehydrogenase and amicyanin from *Paracoccus denitrificans*, *J. Biol. Chem.* **263**:13987–13990.

Harris, T. K., and Davidson, V. L., 1993, Binding and electron transfer reactions between methanol dehydrogenase and its physiologic electron acceptor cytochrome *c*-551i. A kinetic and thermodynamic analysis, *Biochemistry* **32**:14145–14150.

Harris, T. K., Davidson, V. L, Chen, L., Mathews, F. S., and Xia, Z-H., 1994, Ionic strength dependence of the reaction between methanol dehydrogenase and cytochrome *c*-551i: Evidence of conformationally coupled electron transfer, *Biochemistry* **33**:12600–12608.

Hoffman, B. M., and Ratner, M. A., 1987, Gated electron transfer: When are observed rates controlled by conformational interconversion, *J. Am. Chem. Soc.* **109**:6237–6243.

Husain, M., and Davidson, V. L., 1985, An inducible periplasmic blue copper protein from *Paracoccus denitrificans*. Purification, properties and physiological role, *J. Biol. Chem.* **260**:14626–14629.

Husain, M., and Davidson, V. L., 1986, Characterization of two inducible periplasmic *c*-type cytochromes from *Paracoccus denitrificans*, *J. Biol. Chem.* **261**:8577–8580.

Husain, M., and Davidson, V. L., 1987, Purification and properties of methylamine dehydrogenase from *Paracoccus denitrificans*, *J. Bacteriol.* **169**:1712–1717.

Husain, M., Davidson, V. L., Gray, K. A., and Knaff, D. B., 1987, Redox properties of the quinoprotein methylamine dehydrogenase from *Paracoccus denitrificans*, *Biochemistry* **26**:4139–4143.

Huizinga, E. G., van Zanten, B. A. M., Duine, J. A., Jongejan, J. A., Huitema, F., Wilson, K. S., and Hol, W. G. J., 1992, Active site structure of methylamine dehydrogenase; hydrazines identify C6 as the reactive site of the tryptophan-derived cofactor, *Biochemistry* **31**:9789–9795.

Hyun, Y-L., and Davidson V. L., 1995, Electron transfer reactions between aromatic amine dehydrogenase and azurin, *Biochemistry* **34**:12249–12254.

Hyun, Y-L., Zhu, Z., and Davidson, V. L., 1999, Gated and ungated electron transfer reactions from aromatic amine dehydrogenase to azurin., submitted.

Itoh, S., Ogino, M., Haranou, S., Terasaka, T., Ando, T., Komatsu, M., Oshiro, Y., Fukuzumi, S., Kano, K., Takagi, K., and Ikeda, T., 1995, *J. Am. Chem. Soc.* **117**:1485–1493. TITLE

Kumar, M. A., and Davidson, V. L., 1990, Chemical cross-linking study of complex formation between methylamine dehydrogenase and amicyanin from *Paracoccus denitrificans*, *Biochemistry*, **29**:5299–5304.

Labesse, G., Ferrari, D., Chen, Z-W., Rossi, G-L., Kuusk, V., McIntire, W. S., Mathews, F. S., 1998, Crystallographic and spectroscopic studies of native, aminoquinol, and monovalent cation-bound forms of methylamine dehydrogenase from *Methylobacterium extorquens* AM1, *J. Biol. Chem.* **273**:25703–25712.

Lanzilotta, W. N., Parker, V. D., Seefeldt, L. C., 1998, Electron transfer in nitrogenase analyzed by Marcus theory: Evidence for gating by MgATP, *Biochemistry* **37**:399–407.

Lee, H. J., Basran, J., and Scrutton, N. S., 1998, Electron transfer from flavin to iron in the *Pseudomonas oleovorans* rubredoxin reductase-rubredoxin electron transfer complex. *Biochemistry* **37**:15513–15522.

Marcus, R. A., and Sutin, N., 1985, Electron transfers in chemistry and biology. *Biochim. Biophys. Acta* **811**:265–322.

McIntire, W. S., Wemmer, D. E., Christoserdov, A. Y., and Lindstrom, M. E., 1991, A new cofactor in a prokaryotic enzyme: Tryptophan tryptophylquinone as the redox prosthetic group in methylamine dehydrogenase, Science **252**:817–824.

McLendon, G., and Hake, R., 1992, Interprotein electron transfer, *Chem. Rev.* **92**:481–490.

Merli, A., Brodersen, D. E., Morini, B., Chen, Z., Durley, R. C. E., Mathews, F. S., Davidson, V. L., and Rossi, G. L., 1996, Enzymatic and electron transfer activities in crystalline protein complexes, *J. Biol. Chem.* **271**:9177–9180.

Moser, C. C., Keske, J. M., Warncke, K., Farid, R. S., and Dutton, P. L., 1992, Nature of biological electron transfer, *Nature* **355**:796–802.

Regan, J. J., Risser, S. M., Beratan, D. N., and Onuchic, J. N., 1993, Protein electron transport: Single versus multiple pathways, *J. Phys. Chem.* **97**:13083–13088.

Tobari, J., and Harada, Y., 1981, Amicyanin: an electron acceptor of methylamine dehydrogenase, *Biochim. Biophys. Res. Commun.* **101**:502–508.

Ubbink, M., van Kleef, M. A. G., Kleinjan, D.-J., Hoitink, C. W. G., Huitema, F., Beintema, J. J., Duine, J. A., and Canters, G. W., 1991, Cloning, sequencing and expression studies of the genes encoding amicyanin and the β-subunit of methylamine dehydrogenase from *Thiobacillus versutus*, *Eur. J. Biochem.* **202**:1003–1012.

Van der Palen, C. J. N. M., Slotboom, D., Jongejan, L., Reijnders, W. N. M., Harms, N., Duine, J. A., and van Spanning, R. J. M., 1995, Mutational analysis of mau genes involved in methylamine metabolism in *Paracoccus denitrificans*, *Eur. J. Biochem.* **230**:860–871.

Van Spanning, R. J. M., Wansell, C. W., Reijnders, W. N. M., Oltmann, L. F., and Stouthamer, A. H., 1990, Mutagenesis of the gene encoding amicyanin of *Paracoccus denitrificans* and theresultant effect on methylamine oxidation, *FEBS Lett.* **275**:217–220.

Vellieux, F. M. D., Huitema, F., Groendijk, H., Kalk, K. H., Frank, J. Jzn., Jongejan, J. A., Duine, J. A., Petratos, K., Drenth, J., and Hol, W. G. J., 1989, Structure of quinoprotein methylamine dehydrogenase at 2.25 Å resolution, *EMBO. J.* **8**:2171–2178.

Warncke, K., Brooks, H. B., Babcock, G. T, Davidson, V. L., and McCracken, J. L., 1993, The Nitrogen atom of substrate methylamine is incorporated into the tryptophan tryptophyl-semiquinone catalytic intermediate in methylamine dehydrogenase, *J. Am. Chem. Soc.* **115**:6464–6465.

Warncke, K., Brooks, H. B., Lee, H.-I., McCracken, J. L., Davidson, V. L., and Babcock, G. T., 1995, Structure of the dithionite-generated tryptophan tryptophylquinone cofactor radical in methylamine dehydrogenase revealed by ENDOR and ESEEM spectroscopies, *J. Am. Chem. Soc.* **117**:10063–10075.

Wilson, E. K., Mathews, F. S., Packman, L. C., and Scrutton, N. S., 1995, Electron tunneling in substrate-reduced trimethylamine dehydrogenase: Kinetics of electron transfer and analysis of the tunneling pathway, *Biochemistry* **34**:2584–2591.

Zhu, Z., and Davidson, V. L., 1998a, Methylamine dehydrogenase is a light-dependent oxidase, *Biochim. Biophys. Acta* **1364**:297–300.

Zhu, Z., and Davidson, V. L., 1998b, Redox properties of tryptophan tryptophylquinone enzymes. Correlation with structure and reactivity, *J. Biol. Chem.* **273**:14254–14260.

Zhu, Z., and Davidson, V. L., 1999, Identification of a new reaction intermediate in the oxidation of methylamine dehydrogenase by amicyanin, *Biochemistry* **38**:4862–4867.

Zhu, Z., Cunane, L. M., Chen, Z-W., Durley, R. C. E., Mathews, F. S., and Davidson, V. L., 1998, Molecular basis for interprotein complex-dependent effects on the redox potential of amicyanin, *Biochemistry* **37**:17128–17136.

*Chapter 5*

# Trimethylamine Dehydrogenase and Electron Transferring Flavoprotein

## Nigel S. Scrutton and Michael J. Sutcliffe

## 1. ELECTRON TRANSFER PROTEINS IN METHYLOTROPHIC BACTERIA

Many biological electron transfer reactions—e.g. harnessing solar energy, metabolism, defence against toxic compounds and pathogens—rely on the coexistence of protein-protein complexes in dissociation equilibrium with their constitutive reactants. The mechanism of electron transfer between weakly associating electron transfer partners has been the focus of intensive research activity in recent years. An in-depth understanding of these reactions requires knowledge of the role of protein dynamics in complex assembly, and the geometries of the complex that are compatible with inter-protein electron transfer. Knowledge of the role of electrostatics in guiding complex assembly during a diffusional encounter is central to our understanding of electron transfer processes between redox partners. The role of electrostatics, and that of other interactions (e.g. hydrophobic), in maintaining complex structure is also of central importance in maximizing electronic coupling between neighbouring redox centers. In addition to the

**NIGEL S. SCRUTTON and MICHAEL J. SUTCLIFFE**    Departments of Biochemistry and Chemistry, University of Leicester LE1 7RH UK.
*Subcellular Biochemistry, Volume 35: Enzyme-Catalyzed Electron and Radical Transfer*, edited by Holzenburg and Scrutton. Kluwer Academic / Plenum Publishers, New York, 2000.

complexities associated with diffusional encounter and complex formation, reaction rates are also governed by factors within the assembled complex. These include the varied effects of protein structure on physical parameters, such as the electronic coupling matrix element, the transfer (redox) potentials of the centers and the directionality of the transfer process (i.e. the route of electron transfer). Pathway identification is particularly complex, and invokes recent arguments about the existence of multiple pathways/pathway tubes and the role of covalent bonds/hydrogen bonds/through-space jumps in optimising the electronic coupling between redox centers. Besides these complex issues, contributions made by gating mechanisms, such as rate-limiting protonations/deprotonations, configurational gating steps (i.e. rate-limiting adjustments in protein geometry) and ligand-gating of electron transfer also need to be identified. Naively, student text books often describe electron transfer as the 'simplest of chemical reactions'—however, given the apparent complexity of electron transfers in biology and our relatively poor understanding of the process, the current, intense research activity in this area is fully justified.

Methylotrophic bacteria are a rich source of soluble electron transfer proteins required for the catabolism of organic molecules lacking carbon-carbon bonds. In recent years, these proteins have provided ideal model systems for studying the control of electron flow through assembled protein complexes. These enzymes catalyse the oxidation of methyl groups to formaldehyde and contain unusual redox cofactors. The constituent redox proteins in the pathways for methyl group oxidation are depicted in Figure 1 and have been the focus of intensive structural and mechanistic investigation. Three different enzymes capable of oxidizing trimethylamine have been identified. The majority of methylotrophs using trimethylamine as sole carbon source utilize a trimethylamine monooxygenase to generate trimethylamine-N-oxide, which is subsequently converted to dimethylamine and formaldehyde by trimethylamine-N-oxide demethylase (or aldolase) (Colby and Zatman, 1973; Colby and Zatman, 1975). On the other hand, in obligate methylotrophs and some restricted facultative methylotrophs, trimethylamine is oxidized by a nicotinamide-independent trimethylamine dehydrogenase (Colby and Zatman, 1973). This enzyme is a complex iron-sulfur flavoprotein that transfers electrons to the soluble flavoprotein known as electron transferring flavoprotein. In one case, a nicotinamide-dependent trimethylamine dehydrogenase has been reported (Loginova and Trotsenko, 1978). With dimethylamine, both a monooxygenase (Eady, Jarman and Large, 1971) and a dehydrogenase (Meiberg and Harder, 1979) are known. The latter enzyme is restricted to *Hyphomicrobium X*; most other bacteria utilize a dimethylamine monooxygenase, which is an unstable enzyme that is sensitive to CO. The oxidation of methylamine

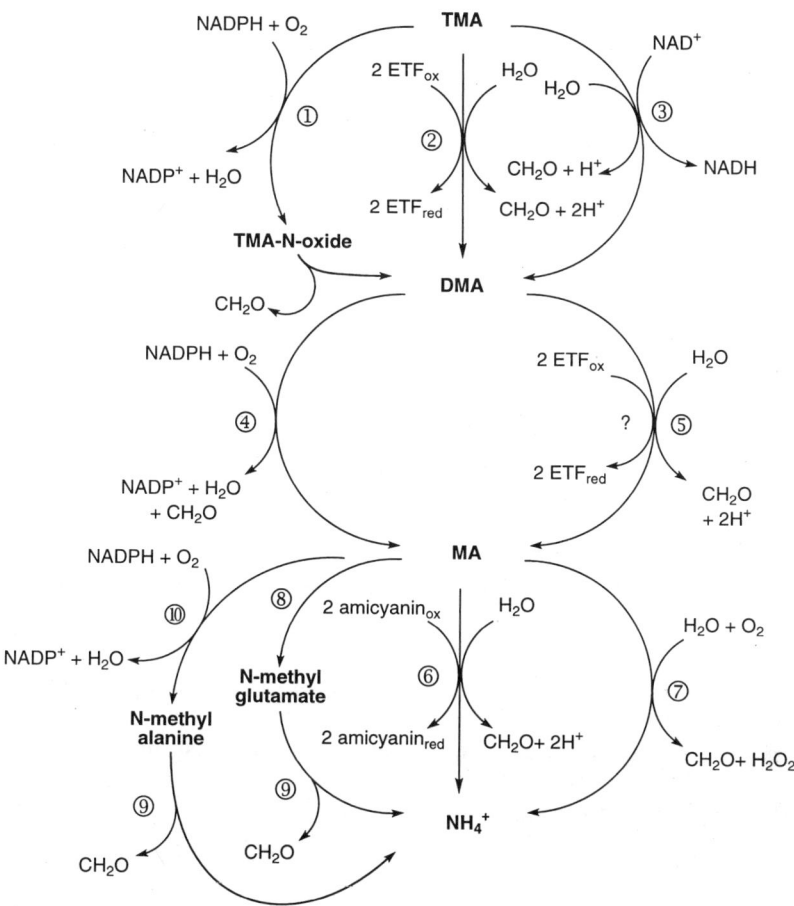

**FIGURE 1.** Pathways for methyl group oxidation in methylotrophic bacteria. The constituent proteins are denoted by numbers as follows: 1 trimethylamine monooxygenase; 2 nicotinamide-independent trimethylamine dehydrogenase; 3 nicotinamide-dependent trimethylamine dehydrogenase; 4 CO-sensitive dimethylamine monoxygenase (thought to be a cytochrome $P_{450}$); 5 dimethylamine dehydrogenase; 6 methylamine dehydrogenase (TTQ-dependent); 7 $Cu^{2+}$-containing and TPQ-dependent amino oxidase; 8 N-methyl glutamate synthetase (FMN-containing); 9 N-methyl glutamate dehydrogenase; 10 N-methyl alanine dehydrogenase.

can occur by one of three routes. Strains of *Arthrobacter* employ a copper containing amine oxidase (similar copper-containing oxidases are reviewed elsewhere in this volume) that produces ammonia, formaldehyde and hydrogen peroxide (Levering *et al.*, 1981; van Iersel, van der Meer and Duine, 1986). Perhaps the most well characterized enzyme is methylamine dehydrogenase that utilizes a tryptophan tryptophylquinone (TTQ) cofactor and transfers electrons to the blue copper protein amicyanin (Husain and Davidson, 1987). A detailed discussion of this enzyme can also be found elsewhere in this volume. An indirect method of oxidation involves the formation and reoxidation of N-methyl amino acids using the enzymes N-methylglutamate synthase and N-methylglutamate dehydrogenase [Figure 1; (Large, 1981)]. N-methylglutamate synthase is a FMN-dependent protein, but it does not catalyse a net redox reaction. N-methylglutamate dehydrogenase is FAD-dependent. Finally, methanol oxidation is accomplished via PQQ-dependent methanol dehydrogenase (Anthony, 1986), which transfers electrons to a soluble cytochrome *c*; the detailed enzymology of this redox system is also discussed elsewhere in this volume.

The focus of this chapter is the soluble electron transfer complex formed between the nicotinamide-independent trimethylamine dehydrogenase (TMADH) and electron transferring flavoprotein (ETF). Recent studies of this physiological electron transfer complex have provided invaluable insight into (i) the mechanisms of inter and intraprotein electron transfer between flavin and Fe/S centers, (ii) the role of dynamics in interprotein electron transfer and (iii) quantum mechanical mechanisms for the cleavage of substrate C-H bonds and the subsequent transfer of reducing equivalents to flavin redox centers. Brief mention is made of early structural and cofactor analyses for this redox system, but more detailed accounts of this work can be found in earlier reviews on the subject (e.g. Steenkamp and Mathews, 1992).

## 2.  PROSTHETIC GROUPS AND STRUCTURE OF TMADH

### 2.1.  Identification of the Prosthetic Groups

TMADH from *Methylophilus methylotrophus* (sp. $W_3A_1$) has been the most extensively studied, although the enzyme has also been purified from bacterium 4B6 (Colby and Zatman, 1974). The method of purification from bacterium 4B6 was adapted (and simplified) for the *M. methylotrophus* enzyme, resulting in high yields of homogeneous material suitable for structural, spectroscopic and kinetic studies (Steenkamp and Mallinson, 1976). For some time, the cofactor composition of TMADH remained unclear.

Chemical analysis revealed the presence of stoichiometric amounts of iron and acid-labile sulfur (Steenkamp and Singer, 1976) and the structure of the Fe/S center was shown to be of the conventional ferrodoxin-like 4Fe-4S type by EPR spectroscopy and by the cluster extrusion technique (Hill *et al.*, 1977). Identification of the flavin as a 6-*S*-cysteinyl FMN, however, was more problematic. Covalent linkage was demonstrated by enzyme denaturation and the flavin could be isolated as a chymotryptic-tryptic aminoacyl peptide (Kenney *et al.*, 1978). The aminoacyl flavin has been isolated by proteolysis and treatment with aminopeptidase M. The UV-visible absorption spectrum of the isolated aminoacyl flavin, however, did not resemble that of other conventional covalent flavins that are linked via the 8α methyl group of the flavin isoalloxazine ring (Mewies *et al.*, 1998). Evidence for substitution at the C6 position was gained from UV-visible spectroscopy of a flavinylated peptide (Steenkamp *et al.*, 1978b; Steenkamp and Singer, 1976), NMR data, comparison of spectroscopic data with 6-hydroxy flavins and reaction of *N*-ethylmaleimide with the product of anaerobic photolysis of the coenzyme (Steenkamp *et al.*, 1978b), and comparison with synthetic 6-*S*-cysteinylriboflavin (Ghisla *et al.*, 1980). A review of the chemical and spectroscopic evidence supporting assignment of the flavin as 6-*S*-cysteinyl FMN has been presented elsewhere (Steenkamp and Mathews, 1992). Unexpectedly, ADP was later discovered as a third prosthetic group in TMADH. This cofactor was originally discovered through analysis of the 2.4 Å resolution electron density map (see below), and was later confirmed by hplc and tlc analysis of supernatants generated following precipitation of the enzyme by perchloric acid. The ADP is tightly bound and cannot exchange with [$^{14}$C] ADP in the solvent (Lim *et al.*, 1988).

## 2.2. Structure of TMADH

### 2.2.1. Evolution of the Crystal Structure of TMADH

Early crystallographic studies of TMADH provided data from two derivatives at 6 Å resolution that revealed the domain structure and certain elements of secondary structure (Lim *et al.*, 1982; Lim *et al.*, 1984). Higher resolution data at 2.4 Å resolution have been collected and the structure solved by the multiple isomorphous replacement method with anomolous scattering (Lim *et al.*, 1986). Analysis of the diffraction pattern lead to the identification of ADP as the third cofactor in TMADH. At the time the 2.4 Å data set was analysed, there was no sequence information available for TMADH (Lim *et al.*, 1986), except for a 12 residue peptide which contained the covalently bound flavin (Kenney *et al.*, 1978). Gas-phase sequencing of isolated peptides initially provided 80% of the primary sequence of

TMADH. The chemically determined sequence was found to be approximately 80% identical with the "x-ray deduced" sequence (Barber *et al.*, 1992). Accounts of some of the earlier crystallographic studies have been reviewed (Steenkamp and Mathews, 1992). The structure of TMADH has now been solved at 2.4 Å resolution (Brookhaven code 2TMD) and more recently at 1.7 Å resolution (Mathews *et al.*, unpublished) and these studies were aided by the full determination of the primary sequence through gene sequencing methods (Boyd *et al.*, 1992).

### 2.2.2.   Domain and Quaternary Structure

TMADH is homodimeric and each subunit comprises three structural domains—a large N-terminal parallel eight-stranded β/α barrel domain, and a medium and small domain that are rich in α/β structure. The folding topology is unusual in that the medium domain is constructed around two peptide segments (residues 380 to 495 and 650 to 700) separated by an intervening segment that makes up the small domain. A short loop of the small domain extends over the large domain of the other subunit. Additionally, there is a C-terminal extension of about 30 residues that interacts with the large domains of both subunits and which cross near the two-fold axis to reach the small domain of the other subunit. Deletion of this peptide region has no affect on the ability to assemble the dimer, but removal prevents efficient flavinylation of the enzyme in the active site (Ertughrul *et al.*, 1998).

### 2.2.3.   Large Domain and Active Site Structure

The active site is contained within the large N-terminal domain. This domain is covalently linked to the FMN cofactor *via* a cysteine residue to form a 6-*S*-cysteinyl FMN. This type of linkage is unique to TMADH and the highly related DMADH (Yang *et al.*, 1995), since other covalent flavoproteins are attached *via* the 8α-methyl group of the flavin isoalloxazine ring (Mewies *et al.*, 1998). The ferredoxin-like 4Fe-4S cluster is located at the C-terminal end of the barrel domain within a α/β loop. The mode of packing within the core of the β/α barrel domain (Raine *et al.*, 1994), and sequence similarities with other member proteins, have assigned TMADH to the type I sub-class of flavin-binding β/α barrel proteins (Scrutton, 1994). The barrel domain also contains four helices external to the eight-fold β/α barrel framework. The first helix is found in an N-terminal loop that covers the bottom of the barrel. The three remaining helices extend from the C-terminal ends of the barrel and help to cover the top of the β/α barrel.

The 6-S-cysteinyl FMN is completely buried within the large domain with its *si* face exposed to the substrate-binding site at the bottom of a channel some 14 Å deep. As with other flavin-binding barrel proteins, the flavin is located at the C-terminal end of the domain. The flavin isoalloxazine ring is butterfly bent by 20° along the flavin N5 to N10 axis; the reason for this non-planar structure of the isoalloxazine is as yet unclear, but has been debated in terms of modulating flavin reduction potential (e.g. Hasford *et al.*, 1997). The ribityl side chain of the FMN is bound in an extended conformation and, as is conventional with the binding of phosphate groups in β/α barrel domains (Bork *et al.*, 1995; Wilmanns *et al.*, 1991), the phosphate is located between β-strands 7 and 8.

The active site (Figure 2) comprises the 6-S-cysteinyl FMN and the 4Fe-4S center. The flavin is relatively close to the 4Fe-4S center (6 Å from the 8α-methyl of the flavin to the closest iron atom in the cluster and almost within van der Waals contact with the cysteinyl ligand of the same iron atom). Like all type 1 flavin-binding barrel domains, an arginine residue (Arg-222) is located close to the N1/C2 carbonyl region of the flavin isoalloxazine, where it is predicted to stabilise negative charge as it develops on the flavin during enzyme reduction. Unusually, a glutamate residue (Glu-103), rather than a glutamine (as in all but one of the other members of the family), is located close to the N3/C2 carbonyl region of the flavin isoalloxazine, although the functional significance of this difference (if any) is not yet known. Arg-222 is also thought to play a major role in the flavinylation reaction of the enzyme (Mewies *et al.*, 1996). Other residues implicated in the flavinylation mechanism include Cys-30 (the residue that forms the thioether link to FMN) and His-29. The latter residue is conjectured to stabilise the thiolate of Cys-30 by forming an imidazolium-thiolate ion pair, thus enabling nucleophilic attack by the thiolate at the C6 position of the flavin (Mewies *et al.*, 1996; Packman *et al.*, 1995). The substrate-binding site is formed by the phenolic and indole rings of residues Tyr-60, Trp-264 and Trp-355. These three residues form an aromatic bowl in the active site that enable the protonated form of the substrate (the trimethylammonium cation) to bind by a process of cation-π bonding (Scrutton and Raine, 1996). Crystallographic studies at 6 Å resolution have been carried out in three oxidation states in the presence of either tetramethylammonium chloride (a competitive inhibitor), trimethylamine, or dithionite, and also a combination of dithionite and tetramethylammonium chloride (Bellamy *et al.*, 1989). Optical and EPR spectroscopic studies demonstrated that the redox state of the enzyme was identical in the crystalline and solution forms. The crystals remained intact on soaking and were isomorphous with the unliganded native crystals. The analyses revealed no major conformational changes on ligand binding or reduction. Difference density was observed

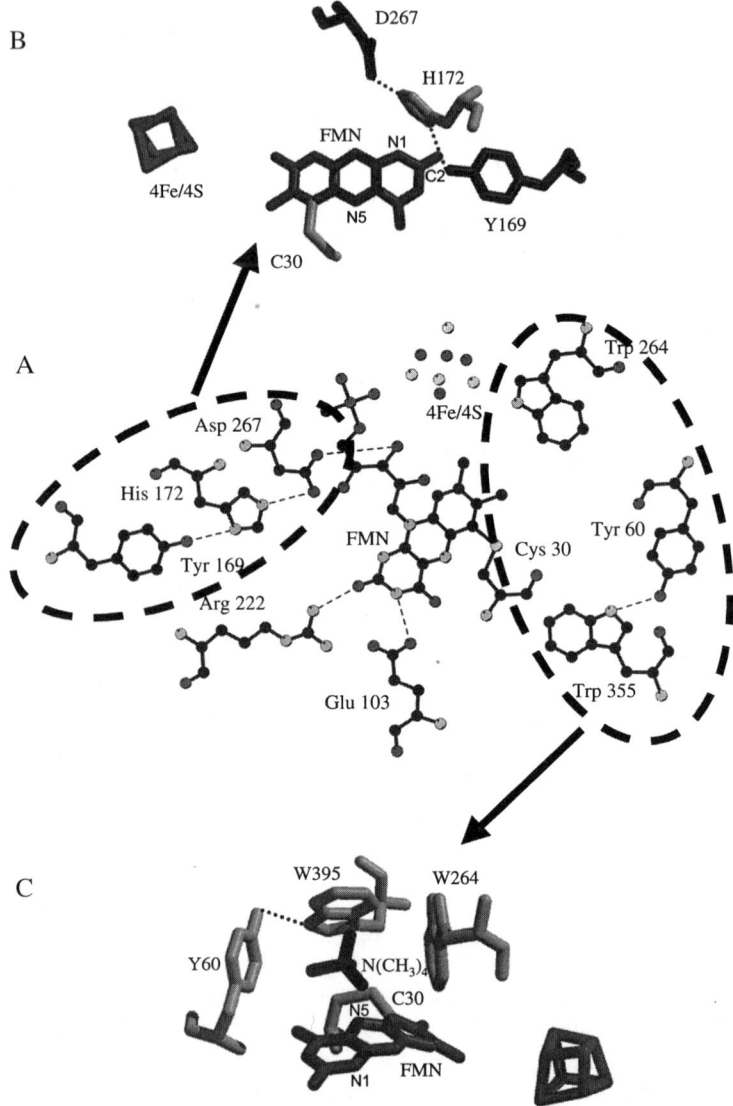

**FIGURE 2.** The active site of TMADH. A, schematic representation of the active site with carbon atoms in black, oxygen and iron in dark gray, and nitrogen and sulfur in light gray. Hydrogen bonds are shown as dashed lines. Figure generated using LIGPLOT (Wallace, 1995). B, the catalytic triad. C, the aromatic bowl; $N(CH_3)_4$ denotes the crystallographically determined binding position of the inhibitor tetramethylammonium chloride (Bellamy *et al.*, 1989).

for the soaks with tetramethylammonium chloride: a positive peak was seen in the region of the substrate-binding aromatic bowl and smaller positive and negative peaks were seen in the active site suggesting side-chain movement to accommodate the ligand. Recent site-directed mutagenesis studies have confirmed the role of the aromatic bowl in binding substrate (Basran et al., 1997). In the related DMADH of *Hyphomicrobium X*, residue Tyr-60 of TMADH is replaced by glutamine, and this exchange is thought to contribute to the difference in specificity between the two enzymes (Basran et al., 1997; Raine et al., 1995; Yang et al., 1995). An unusual feature of the active site of TMADH is the presence of a Tyr-His-Asp triad comprising residues Tyr-169, His-172 and Asp-267. The triad is located close to the pyrimidine ring of the flavin isoalloxazine ring and likely plays a major role in the mechanism of electron transfer from substrate to the flavin. Recent studies on mutants of TMADH altered in this region of the protein have been reported and their properties are beginning to cast light on the mechanism of dehydrogenation and electron transfer (see below).

### 2.2.4.  The Medium and Small Domains

The medium and small domains comprise a central five-stranded β-sheet flanked by α-helices and antiparallel β-strands (Lim et al., 1986). The folding pattern in the medium domain is similar to that seen in nicotinamide-dependent dehydrogenases and flavodoxin, having three and two α-helices, respectively, on opposite sides of the central β-sheet. The small domain resembles in fold elements of structure found in another flavoprotein family, the disulphide oxidoreductases (Williams, 1992). This family includes important enzymes such as lipoamide dehydrogenase and glutathione reductase, and the structural similarity of the small domain with the NADPH-binding domain of glutathione reductase became apparent by performing structural alignment of the relevant domains (Figure 3). Alignment of the small domain of TMADH with the NADPH-binding domain of glutathione reductase also brought the central β-sheets of the medium domain of TMADH and the FAD-binding domain of glutathione reductase into near coincidence. Moreover, the ADP molecule identified in TMADH during analysis of the crystallographic data at 2.4 Å resolution is located in the medium domain at a site that corresponds with the ADP moiety of FAD in the FAD-binding domain of glutathione reductase. The ADP in TMADH has no known physiological function; current thinking suggests that it is a vestigial remnant of an ancestral dinucleotide-binding domain recruited during the evolution of TMADH (Lim et al., 1988; Scrutton, 1994). As regards this latter point, it is interesting to note that the bile acid inducible protein baiH shows a high degree of sequence conservation with TMADH

**FIGURE 3.** Cα trace illustrating the structural similarity between the small domain of TMADH (solid line) and the NADPH-binding domain of glutathione reductase (dashed line).

and functions with NADH, two flavins, and a 4Fe-4S center (Franklund *et al.*, 1993).

## 3. ELECTRON FLOW IN TMADH—STATIC TITRATIONS, REDUCTION POTENTIALS AND INACTIVATION STUDIES

### 3.1. Static Titrations

Full reduction of TMADH requires three electrons per subunit—two for reduction of the 6-*S*-cysteinyl FMN and a third for reduction of the

4Fe-4S center. Full reduction is achieved by titration with dithionite and the titration follows a nonisosbestic course. However, during catalysis only two electrons are transferred to the enzyme as a result of breaking a C-H bond of one of the substrate methyl groups. Titration with trimethylamine under anaerobic conditions produces a characteristic two electron reduced, three-banded spectrum with a major band at 365 nm and other peaks at 440 nm and 510 nm. This spectrum is the final spectrum observed at the end of stopped-flow studies in which oxidised enzyme is mixed rapidly with substrate. Further important differences between substrate- and dithionite-reduced TMADH have become apparent from EPR spectroscopy studies. Dithionite reduction of TMADH yields a rhombic signal with g values at 2.035, 1.925 and 1.85—characteristic of a reduced iron-sulphur protein (Steenkamp et al., 1978c). In contrast, the EPR spectrum of substrate-reduced TMADH is complex and also shows an unusually strong half-field signal at g = 4, indicating spin-spin interaction between the redox centers (Steenkamp et al., 1978c). This strong spin-spin interaction is consistent with the proximity of the flavin and 4Fe-4S centers revealed from the crystallo-graphic studies of TMADH (Lim et al., 1986). Interestingly, the spin inter-acting state also forms when 2-electron-reduced TMADH is titrated with the substrate analogue tetramethylammonium chloride (TMAC) (Steenkamp and Beinert, 1982a; Steenkamp et al., 1978c). With TMAC bound in the active site, TMADH cannot be reduced to the 3-electron level with dithionite because binding of this analog perturbs the reduction poten-tial of the flavin semiquinone/hydroquinone couple (Pace and Stankovich, 1991). Formation of the spin-interacting state by the addition of TMAC to 2-electron-reduced enzyme is also consistent with the finding that excess trimethylamine is required for full development of the g = 4.0 half-field signal and the spectral changes at 365 nm during reductive titration. The development of signature at 365 nm is due to the formation of flavin anio-nic semiquinone, which is one of the paramagnetic species formed that contributes to spin-spin interaction. Binding of [14]C trimethylamine to 2-electron-reduced enzyme has been demonstrated, which is consistent with the EPR and UV-visible spectroscopy studies (Pace and Stankovich, 1991). Recent high-field EPR studies have demonstrated that in the spin-interacting state, the coupling is antiferromagnetic with $J_o = +0.72 \text{cm}^{-1}$ (Fournel et al., 1998)

## 3.2. Reduction Potentials and Selective Inactivation

The route of electron flow through the TMADH-ETF electron trans-fer complex, FMN $\rightarrow$ 4Fe-4S $\rightarrow$ ETF, is consistent with the known reduc-tion potentials of the enzyme cofactors and studies in which selective inactivation of the centers in TMADH has been achieved. The reduction potentials of the oxidised flavin/semiquinone couple (+44 mV), the semi-quinone/dihydroflavin (+36 mV) and the 4Fe-4S$^{2+}$/4Fe-4S$^{1+}$ (+102 mV) couples have been measured by microcoulometry (Barber et al., 1988) and these values are in broad agreement with values measured from UV-visible spectroelectrochemical studies (Pace and Stankovich, 1991). The reduction potential of the oxidised flavin/semiquinone couple of ETF is unusually positive (+141 mV; Wilson et al., 1997a), but is consistent with its need to accept electrons from the 4Fe-4S center of TMADH. However, recent studies indicate that the redox properties of ETF are perturbed on forming a complex with TMADH and the measured reduction potential for the oxi-dised flavin/semiquinone couple of ETF may be perturbed in the electron transfer complex (Jang et al., 1999b; see Section 6). Studies with reagents that selectively inactivate redox centers in TMADH have also helped estab-lish the route of electron flow. Selective inactivation of the 6-S-cysteinyl FMN with phenylhydrazine, by the formation of a covalent adduct con-taining a phenyl group on the flavin C4a atom, prevents reduction of the flavin and 4Fe-4S centers of TMADH with trimethylamine (Nagy et al., 1979). The 4Fe-4S center of the inactivated enzyme can, however, be reduced artificially with dithionite and is competent in transferring elec-trons to ETF (Huang et al., 1995). Conversely, inactivation of the 4Fe-4S center with ferricenium hexafluorophosphate under alkaline conditions prevents electron transfer to ETF, but the modified enzyme can be reduced by substrate at the 6-S-cysteinyl FMN (Huang et al., 1995).

## 4. SINGLE TURNOVER STOPPED-FLOW STUDIES OF ELECTRON TRANSFER

### 4.1. Early Stopped-flow Investigations

Single turnover stopped-flow studies of enzyme reduction have been particularly informative in understanding the mechanism of electron flow in the reductive half-reaction of the enzyme. Early studies demonstrated that three well-resolved kinetic phases could be identified in single turnover experiments in which enzyme was rapidly mixed with trimethylamine (Steenkamp et al., 1978a). The first phase involves a very rapid bleaching of the enzyme-bound FMN, which has been poorly characterized owing to its rapid rate (Steenkamp et al., 1978a). Subsequent steps in the reductive half-reaction are sluggish in comparison with the rate of flavin reduction, which makes the enzyme particularly amenable to stopped-flow studies since each of the three kinetic phases is well resolved (Figure 4). The spectral changes associated with the slower phases reflect intramolecular electron transfer

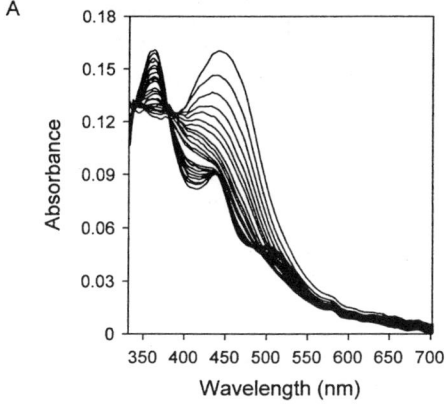

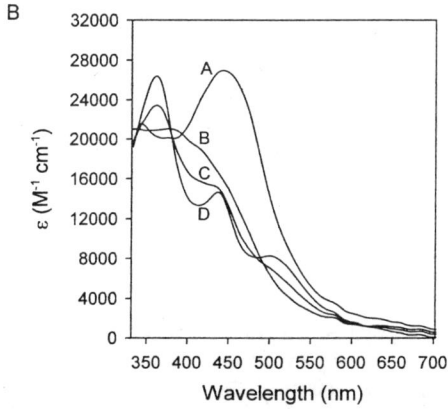

**FIGURE 4.** Spectral forms of intermediates in the reductive half-reaction of TMADH. Panel A, photodiode array analysis of the reaction of TMADH with TMA. Panel B, Denconvoluted spectra for the intermediates of the reductive half-reaction. Spectrum A, oxidised enzymes; spectrum B, reduced enzyme (dihydroflavin); spectrum C, reduced enzyme (flavin semi-quinone/reduced iron-sulfur centre); spectrum D, spin-interacting state.

from the 6-$S$-cysteinyl FMN to the 4Fe-4S center. The fact that internal electron transfer should be slow is puzzling from the viewpoint that the flavin and 4Fe-4S centers are separated by a distance of only 4 Å, measured from the flavin 8$\alpha$-methyl group to the closest cysteinyl sulphur atom that acts as a ligand to the 4Fe-4S center. This has led to the proposal that the breakdown of a substrate-flavin adduct formed during reduction of the flavin by substrate is responsible for gating intramolecular electron transfer to the 4Fe-4S center (Rohlfs and Hille, 1994; see section 5). Freeze-quench EPR studies were used to demonstrate that development of the $g = 4$ half-field signal (and the more complex signals around $g = 2$) is concomitant with the absorption increase observed at 365 nm in spectrophotometric studies (Steenkamp and Beinert, 1982b). These studies were initially complicated by the finding that internal electron transfer was found to be dependent on substrate concentration and that the chosen buffer for the studies (pyrophosphate) undergoes large changes in pH on freezing. These factors led initially to a poor correlation in the kinetics of the two processes (Steenkamp et al., 1978c). However, improvements in the experimental design eventually led to good correlations in the kinetics of formation of the EPR signals and absorption increases at 365 nm (Steenkamp and Beinert, 1982b). Single turnover stopped-flow studies demonstrated that intramolecular electron transfer showed a complex dependence on substrate concentration and that the absorption changes observed at 365 nm were in fact biphasic; a model to account for this complex behaviour was proposed for the TMADH of *Hyphomicrobium* X (Steenkamp and Beinert, 1982b). A more detailed discussion of these complex absorption changes that report on internal electron transfer in *Methylophilus methylotrophus* TMADH is discussed below (Section 4.3).

## 4.2. pH-Dependence of the Reductive Half-reaction with Diethylmethylamine

Detailed single turnover stopped-flow studies of the reductive half-reaction have been reported recently using the 'slow substrate' diethylmethylamine (Rohlfs and Hille, 1994) and the natural substrate trimethylamine (Jang et al., 1999a). These studies were conducted over the pH range 6–10 (diethylmethylamine) and pH 6–11 (trimethylamine)—combined, these provide a detailed description of the kinetic behaviour of the native enzyme. Consistent with earlier stopped-flow studies, the reaction with diethylmethylamine displayed three well-resolved kinetic phases (except at low pH where the phases become kinetically mixed). The fast phase (flavin reduction) is hyperbolically dependent on substrate concentration, revealing a two-step mechanism involving a rapid equilibrium between free enzyme and substrate with an enzyme-substrate Michaelis complex. This complex then reacts to produce the enzyme-bound dihydroflavin. The pH-dependence of the limiting rate constant ($k_{lim}$) for the fast phase exhibits a $pK_a$ value of 7.9, whereas $k_{lim}/K_d$ exhibits a $pK_a$ value of 8.7. The data thus reveal that ionizable groups are responsible for controlling the rate of flavin reduction. The intermediate and slow kinetic phases of the reductive half-reaction with diethylmethylamine were observed to be essentially independent of substrate concentration above pH 7; below this pH there is an apparent dependence due to kinetic mixing with the fast phase. The rate constant for the intermediate kinetic phase is controlled by an ionizable group of $pK_a$ value 7.3. The slow phase shows only a small dependence on pH with a shallow minimum at pH 7.7. Kinetic difference spectroscopy also demonstrated that, following completion of the intermediate kinetic phase, an additional $pK_a$ value of 6 controls the equilibrium distribution of electrons between FMN and the 4Fe-4S center (i.e. at low pH a greater fraction of the enzyme molecules contain flavin hydroquinone and oxidised 4Fe-4S center than observed at high pH). Freeze-quench EPR studies demonstrated that processes occurring in the slow kinetic phase are responsible for full development of the g = 4 signal i.e. the spin interacting state.

The observed kinetic behaviour with diethylmethylamine has led to the proposal of a mechanism for the reductive half-reaction of TMADH (Figure 5). Reduction of the flavin is envisaged to involve the formation of a flavin N5-substrate adduct, as proposed for other flavoproteins such as D-amino acid oxidase (D'Silva et al., 1987; Denu and Fitzpatrick, 1994; Porter et al., 1977), L-lactate oxidase (Ghisla and Massey, 1980) and L-lactate dehydrogenase (Ghisla and Massey, 1989). The covalent adduct was origi-nally proposed to occur through the formation of a substrate carbanion (requiring proton abstraction by an active site base; Figure 5A), but this has now been brought into question through recent mutagenesis studies of TMADH ((Basran et al., 1999a; Basran et al., 1999b); Section 4.3). It is now thought that the postulated covalent adduct may form via a radical cleav-age of the substrate C-H bond, but not by any of the mechanisms shown in Figure 5 (Section 4.3). Decay of the conjectured adduct likely accounts for the slow rate of electron transfer from the flavin to the 4Fe-4S center (i.e. the intermediate kinetic phase). The slow phase of the reductive half-reac-tion likely represents additional processes that result in full formation of the spin-interacting state subsequent to intramolecular electron transfer. In this regard, it is interesting to note that the majority of the g = 4 signal accu-mulates during this phase. These processes are proposed to be dissociation of product and/or hydrolysis of product (i.e. dialkylimminium) and the binding of a second substrate molecule to the 2-electron-reduced enzyme. Support for this part of the mechanism has recently been acquired through mutagenesis studies of residue Tyr-169 in the active site (Section 5).

## 4.3.  pH-Dependence of the Reductive Half-reaction with Trimethylamine: Native and Mutant Enzyme Studies

Stopped-flow studies with diethylmethylamine provided a conceptual framework for understanding the mechanism of electron transfer in the

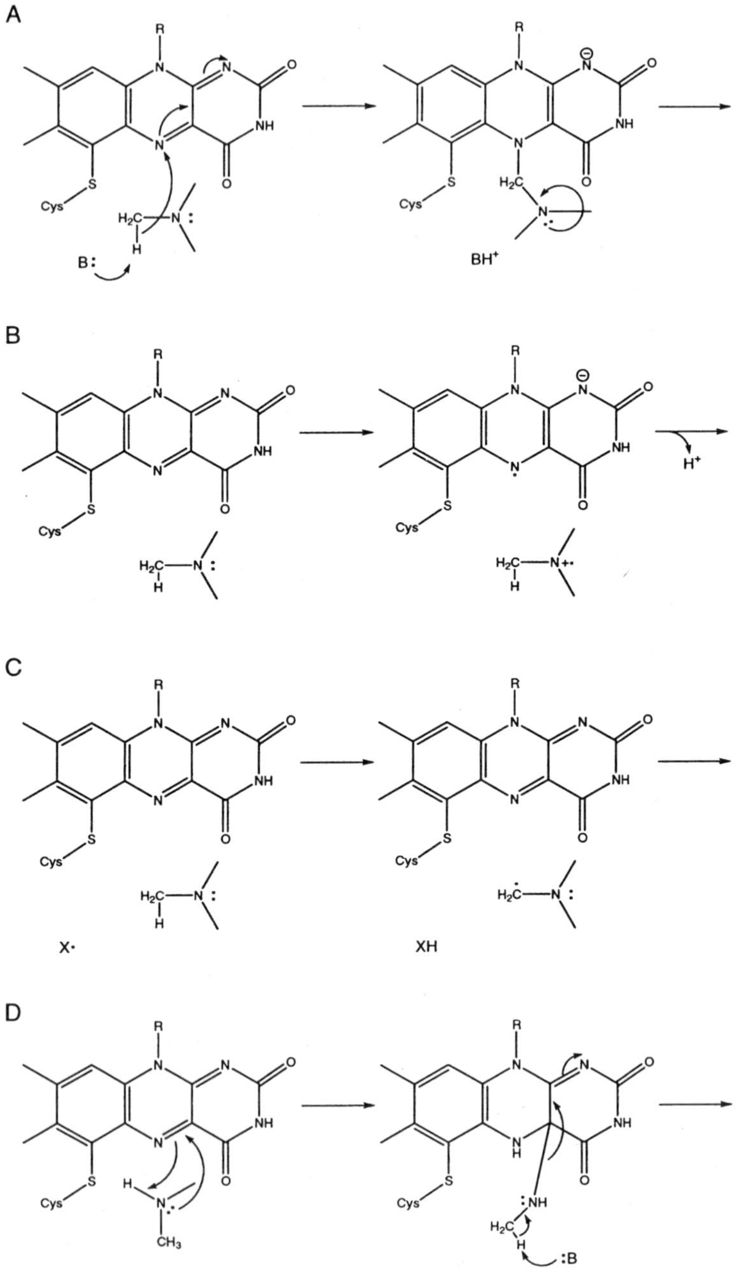

**FIGURE 5.** Proposed reaction mechanisms for flavin-catalysed oxidation of amines. A, the carbanion mechanism initially proposed for TMADH (Rohlfs and Hille, 1994). B, the amminium cation radical mechanism, as originally proposed for monoamine oxidase (Silverman, 1995); although only the pathway passing through a transient covalent intermediate is shown, several alternative pathways for breakdown of the initial flavin semiquinone/

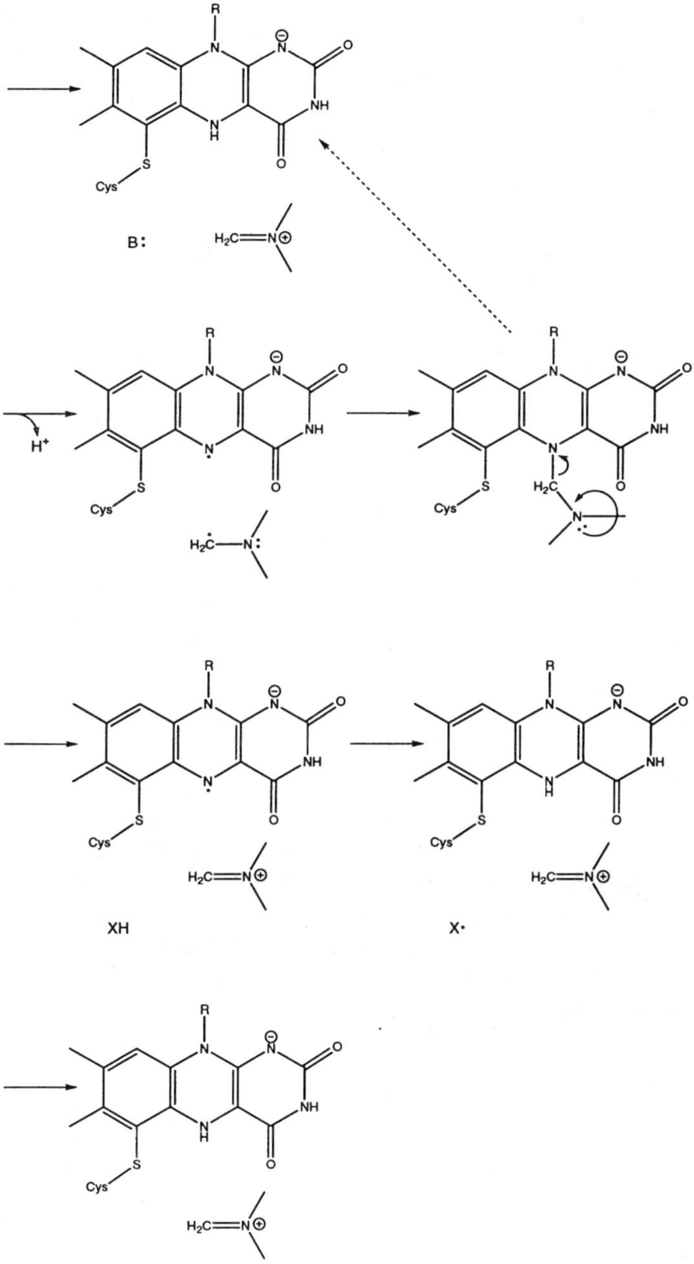

**FIGURE 5.** (*Continued*)
amminium radical intermediate have been considered. C, the hydrogen abstraction mechanism originally proposed for monoamine oxidase (Edmondson, 1995). D, the nucleophilic attack mechanism—an alternative mechanism originally proposed for monoamine oxidase (Kim, 1993).

reductive half-reaction of TMADH. Similar studies over an extended pH range (6 to 11) using the natural substrate trimethylamine, coupled with mutant studies, have now extended our understanding of the mechanism of enzyme reduction. Studies over the extended pH range have revealed an additional ionization in the $k_{lim}/K_d$ versus pH plot for the flavin reduction step of p$K_a$ value 10.0 (with trimethylamine) or 9.7 (with diethylmethylamine) (Jang et al., 1999a). Thus, the $k_{lim}/K_d$ versus pH plot has a bell-shaped dependence on solution pH; the lower ionization (p$K_a$ 9.3) may represent the ionization of Tyr-60, which is a residue involved in binding the protonated form of the substrate via cation-$\pi$ bonding (Basran et al., 1997), but this remains to be shown by mutagenesis. The value of the p$K_a$ for the ionization on the alkaline side of the bell-shaped curve suggests this ionization is the result of substrate deprotonation. Studies on the native enzyme with trimethylamine suggest (as with diethylmethylamine) a single ionization in the $k_{lim}$ versus pH plot of p$K_a$ value 7.1 (Jang et al., 1999a); these studies however, are compromised by the very fast flavin reduction rates making stopped-flow analysis difficult, especially at high pH values. However, similar studies on native TMADH with perdeuterated trimethylamine (in which flavin reduction is slowed due to an appreciable kinetic isotope effect) have revealed the presence of two ionizations of p$K_a$ values 6.5 and 8.4 (Basran et al., 1999a). The p$K_a$ value of 8.4 has been assigned to the ionization of His-172 in the Michaelis complex, since this ionization is absent in a His-172 $\rightarrow$ Gln mutant (Basran et al., 1999a). The two ionizations in the $k_{lim}$ versus pH plot are more clearly resolved in a mutant form of TMADH in which the active site residue Tyr-169 is replaced by phenylalanine (Basran et al., 1999b). Tyr-169 is H-bonded to His-172 in the wild-type enzyme, and thus the perturbation in p$K_a$ value for the more alkaline ionization (p$K_a$ = 9.5 in the Y169F mutant) is expected. The identity of the lower ionization (p$K_a$ = 6.5) in the $k_{lim}$ versus pH profile for the native enzyme is as yet unknown. All potential active site bases in TMADH have been targetted by site-directed mutagenesis and replaced by non-ionizing counterparts. In all cases, the mutant forms of the enzyme are active, which brings into serious question the validity of the carbanion mechanism (which requires an active site base for proton abstraction) for TMADH (Figure 5A).

The mechanism by which amines are oxidised by flavoproteins has been an issue of considerable debate in recent years. The debate has been particularly heated in the case of the enzyme monoamine oxidase (Silverman, 1995). Through the use of a variety of mechanism-based inhibitors and based on studies of nonenzymic mechanisms of amine oxidation, a mechanism for monoamine oxidase in which substrate is initially oxidized by single electron transfer to the enzyme flavin to give an aminium cation radical (and anionic flavin semiquinone) has been proposed (Silverman, 1995). The evidence supports a mechanism in which this radical pair first recombines to form a covalent adduct, which then decays by β-elimination, as indicated in Figure 5B. An alternative mechanism has also been considered (Edmondson, 1995) involving a hydrogen atom abstraction (Figure 5C). This was proposed principally because of (i) the absence of a significant electronic influence on reaction rate in a homologous series of benzylamine derivatives and (ii) kinetic isotope work that suggested the transition state is late rather than early in the course of the reaction. All the proposed mechanisms for monoamine oxidase begin with neutral substrate rather than the protonated form (as is likely the case here with TMADH). Thus, with TMADH it is unlikely the reaction is initiated by a single electron transfer from substrate to the enzyme-bound flavin when substrate is already positively charged. Similarly, a mechanism involving direct nucleophilic attack of the nitrogen lone-pair of substrate on the flavin 4a-carbon (Figure 5D, also considered in the case of monoamine oxidase; Kim *et al.*, 1993) can be likely eliminated for TMADH, again because the reactive form of substrate is protonated. The available evidence for TMADH does not directly address the mechanism by which C—H bond cleavage occurs, but the fact that substrate must be protonated essentially rules out two of the mechanisms shown in Figure 5. Only a hydrogen atom abstraction mechanism appears to remain fully consistent with the present results, but there is no obvious candidate for the hydrogen atom acceptor. Tyr-60 is part of the substrate-binding site and well-situated to act in this capacity, but no tyrosyl (or tryptophanyl) radical has ever been observed with the enzyme, even when treated with strong oxidants such as ferricenium ion. Clearly, additional work—along the lines of those done previously for monoamine oxidase using mechanism-based inhibitor studies—is thus required to elucidate the precise mechanism of C-H bond cleavage in TMADH.

### 4.4. Quantum Tunneling of Hydrogen

Although the precise mechanism of C-H bond breakage by TMADH remains to be elucidated, recent stopped-flow studies focused on the temperature dependence of the flavin reduction rate using trimethylamine and perdeuterated trimethylamine have revealed a role for quantum mechanical tunneling of hydrogen in the bond breakage reaction. Analysis of the temperature dependence of the rate of flavin reduction was prompted by similar work performed on *Methylophilus methylotrophus* methylamine dehydrogenase. This work revealed that a ground state hydrogen tunneling mechanism driven by the natural, thermally-activated breathing of the enzyme molecule was responsible for breaking the stable C-H bond of the substrate methylamine (Basran *et al.*, 1999c). This novel mechanism—termed vibrationally enhanced ground state tunneling theory [VEGST theory; (Basran *et al.*, 1999c; Bruno and Bialek, 1992)]—of breaking a stable C-H bond has clear energetic advantages, since the conjectured high activation energies during the breakage of a stable C-H bond need not be surmounted. Various diagnostic criteria have been described for ground state quantum tunneling during C-H/C-D bond breakage. These include parallel temperature dependence plots for C-H and C-D bond breakage when ln (k/T) is plotted against 1/T; these plots have a finite gradient, the value of which is used to calculate the energy associated with protein deformation (Scrutton *et al.*, 1999; Sutcliffe and Scrutton, 2000). In this theoretical treatment, the role of the protein is to transiently deform the active site structure into a conformation that is compatible with ground state hydrogen tunneling (i.e. the substrate is not vibrationally excited). This deformation transiently compresses the width of the activation energy barrier for bond breakage thus facilitating tunneling (tunneling probability is dependent on the width of the activation barrier). Stopped-flow studies of the temperature dependence of flavin reduction (i.e. the fast phase in the reductive half-reaction) by trimethylamine and perdeuterated trimethylamine have demonstrated that VEGST is the mechanism of C-H and C-D bond breakage in TMADH (Basran, Sutcliffe and Scrutton unpublished). Extreme tunneling is thus an attractive mechanism for cleaving stable C-H bonds—large activation energies make this energetically unfavourable for classical, over-the-barrier modes of breakage—and may be a general strategy employed by enzymes for catalyzing these difficult transformations. With regard to this latter point, it is also interesting to note that VEGST has recently been demonstrated during C-H bond cleavage by heterotetrameric sarcosine oxidase (Harris, *et al.*, 2000) and alcohol dehydrogenase (Kohen *et al.*, 1999).

## 5.  CONTROL OF INTRAMOLECULAR ELECTRON TRANSFER: PH-JUMP STOPPED-FLOW STUDIES

pH-jump stopped-flow experiments have been used to demonstrate that intramolecular electron transfer in TMADH is not intrinsically rate-limiting in catalysis and is under prototropic control (Rohlfs and Hille, 1991; Rohlfs *et al.*, 1995). This is in contrast to stopped-flow experiments in which TMADH is mixed with excess trimethylamine: in these experiments intramolecular electron transfer is slow, and may be gated by the decay of a substrate-flavin adduct that is conjectured to form during 2-electron reduction of the flavin (Section 4). Reductive optical and EPR titrations of TMADH with sodium dithionite have been performed and these studies have revealed that the equilibrium distribution of the reducing equivalents between the 6-$S$-cysteinyl FMN and 4Fe-4S center is pH-dependent. Formation of dihydroflavin with oxidised iron–sulfur center is favoured below pH 7.5, whereas above pH 8 formation of flavin semiquinone with reduced iron-sulphur center is preferred (Rohlfs and Hille, 1991). These spectral differences have been exploited in a series of pH-jump stopped-flow experiments to investigate the rate of electron transfer between the redox centers. The prototropic control of the electron transfer reaction is exerted at the level of the 6-$S$-cysteinyl FMN. For example, protonation of an anionic semiquinone is required prior to reduction thus preventing formation of the unfavourable dianionic hydroquinone. In electron transfer reactions under prototropic control, the Franck-Condon principle states that nuclear motion (such as protonation) must occur prior to electron transfer since electron transfer is much more rapid than any nuclear motion of the reactants. Identical kinetic and static difference spectra for the pH jump analyses indicated that these experiments monitored both protonation/deprotonation and electron transfer events i.e. the deprotonation/protonation reactions are not lost in the dead-time of the stopped-flow apparatus. The observed rate constants fell within the range $230\,s^{-1}$ to $1200$ $s^{-1}$ [depending on the pH-jump conditions used (Rohlfs and Hille, 1991)], and in all cases these values are significantly faster than the rates of internal electron transfer measured by stopped-flow studies with excess trimethylamine.

Although at pH 8 the electron distribution favours the formation of flavin semiquinone and reduced iron-sulfur center, the magnetic moments of the two redox centers do not interact. At pH 10, however, 2-electron-reduced TMADH exhibits the EPR spectrum diagnostic of the spin-interacting state. In a more detailed analysis using the pH-jump technique, the interconversion of three states of TMADH [state 1, dihydroflavin-oxidised 4Fe-4S center (formed at pH 6); state 2, flavin semiquinone-reduced 4Fe-4S center (formed at pH 8); state 3, spin interacting state (formed at pH 10)] were studied in both $H_2O$ and $D_2O$ (Rohlfs et al., 1995). The kinetics were found to be consistent with a reaction mechanism that involves sequential protonation/deprotonation and electron transfer events (Figure 6). Normal solvent kinetic isotope effects were observed and proton inventory analysis revealed that at least one proton is involved in the reaction between pH 6 and 8 and at least two protons are involved between pH 8 and 10. At least three protonation/

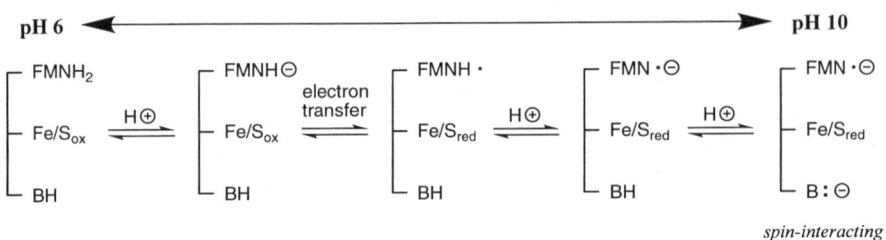

*spin-interacting*

**FIGURE 6.** Scheme illustrating the prototropic control of internal electron transfer in TMADH.

deprotonation events are thus associated with internal electron transfer with $pK_a$ values of 6, 8 and about 9.5. On moving from pH 6 to 10, Figure 6 indicates that the $pK_a$ of 6 represents deprotonation of flavin hydroquinone and the $pK_a$ of 8 the deprotonation of flavin semiquinone. The $pK_a$ of about 9.5 (which controls the formation of the spin interacting state) was assigned to an undesignated basic group on the protein molecule (Rohlfs *et al.*, 1995).

The identity of the group with $pK_a \sim 9.5$ has been realised from studies of the Y169F mutant TMADH, which is altered in one component of the triad formed by residues Tyr-169, His-172 and Asp-267 in the active site of native enzyme. EPR studies of this mutant demonstrated that its ability to form the spin-interacting state is substantially compromised when reacted with excess substrate at pH 7.0 or indeed at pH 10 (Basran *et al.*, 1999b). Interestingly, unlike for native enzyme, the Y169F TMADH can be reduced to the level of three electrons by dithionite titration in the presence of the substrate analog TMAC: 3-electron reduction of Y169F TMADH in the presence of TMAC is a direct result of perturbing the oxidation-reduction potential of the enzyme, as inferred from UV-visible spectra (Basran *et al.*, 1999b). Tyr-169 lies opposite the flavin ring from the 4Fe-4S center, but its role in formation of the spin-interacting state can be rationalised in terms of the x-ray structure of TMADH (Lim *et al.*, 1986). The phenolic side chain of Tyr-169 is positioned close to the C2 carbonyl group of the flavin isoalloxazine ring where it is in van der Waals contact with the flavin. On binding substrate (or TMAC) to 2-electron-reduced TMADH, the hydroxy group is deprotonated and the negative charge redistributes the electron density in the flavin toward the 4Fe-4S center by electrostatic repulsion. This redistribution of electron density therefore reduces the spin-spin distance and favors formation of the spin-interacting state.

## 6. THE OXIDATIVE HALF-REACTION AND TMADH-ETF COMPLEX ASSEMBLY

### 6.1. Stopped-flow Studies

Convenient stopped-flow methods for investigating electron transfer from the reduced 4Fe-4S center of TMADH to the flavin of ETF have been developed (Huang *et al.*, 1995). At the completion of the reductive half-reaction TMADH is reduced at the level of 2 electrons and the distribution of those electrons within the enzyme is determined by solution pH (Rohlfs and Hille, 1991; Rohlfs *et al.*, 1995). Rapid mixing experiments of 2 electron reduced TMADH with oxidised ETF [generated by ferricyanide treatment of the as-purified semiquinone form of ETF (Steenkamp and Gallup, 1978)] gives rise to multiphasic kinetics, thus complicating data analysis (Huang *et al.*, 1995). Studies of the oxidative half-reaction have been simplified by inactivating the 6-*S*-cysteinyl FMN by treatment with phenylhydrazine (Huang *et al.*, 1995) which places a phenyl group on the C4a atom of the flavin, thus rendering it redox inert. Enzyme treated in this way can be reduced to the level of one reducing equivalent under anaerobic conditions by dithionite and the electron is located in the 4Fe-4S center. One-electron reduced TMADH when rapidly mixed with oxidised ETF gives rise to a biphasic transient. The fast phase of these transients reports on interprotein electron transfer and is dependent on ETF concentration; the origin of the slow phase is as yet uncertain (Huang *et al.*, 1995). Reactions have been performed with 3-electron reduced TMADH with ETF; in these cases, the fast phase of the complex transients obtained are similar in rate to the fast phase of the biphasic transient obtained for phenylhydrazine inactivated TMADH. A limiting rate of $172 \, s^{-1}$ and $K_d$ for the complex of $10 \, \mu M$ have been calculated for the TMADH:ETF complex. The complex dissociation constant is similar to that measured by sedimentation equilibrium methods for oxidised TMADH and ETF (Wilson *et al.*, 1997b).

Site-directed mutagenesis studies have demonstrated that residue Tyr-442 on the surface of TMADH is involved in facilitating electron transfer from TMADH to ETF (Wilson *et al.*, 1997a). Three mutant forms of TMADH, in which Tyr-442 has been converted to Phe, Leu and Gly, have been isolated. Treatment with phenylhydrazine and subsequent stopped-flow studies revealed a substantial decline in electron transfer rate decreasing in the series native>Y442F>Y442L>Y442G. The data reveal a major role for Tyr-442 in assembly of the electron transfer complex, although the precise role (e.g. whether Tyr-442 is on a major electron transfer pathway) remains to be elucidated. Tyr 442 is located at the center of a large concave region on the surface of TMADH that is likely to be the natural docking

site for ETF. The residue is located along one side of a small surface indentation, the bottom of which is occupied by residue Val-344. This latter
residue is adjacent in the polypeptide chain to one of the cysteinyl ligands
(Cys-345) of the 4Fe-4S center. Computational analyses using the Pathways
program (Regan, 1993) predict that Val-344 is a hot spot for electron transfer out of TMADH. In recent work, a variety of mutants at position Val-
344 have been isolated and the kinetics of electron transfer investigated to
both ETF and the artificial redox acceptor ferricenium hexafluorophosphate (Basran, Sutcliffe, Hille and Scrutton unpublished work). Shortening
the side chain at position 344 (by mutation to Gly, Ala and Cys) *increases*
the rate of electron transfer to ferricenium ions compared with the rate seen
with native TMADH, whereas the introduction of larger side chains (Tyr
and Ile) substantially reduces the rate of electron transfer to ferricenium.
These data indicate that Val-344 is an important residue for coupling to ferricenium ions and simple modelling studies reveal that this redox acceptor
can approach Val-344 at the bottom of the small indentation on the surface
of TMADH. Comparable studies with ETF are equally striking. Electron
transfer to ETF is only moderately affected by mutation of Val-344 to
smaller side chains (Gly, Ala and Cys) and also to the larger Ile and Tyr side
chains. However, the largest decreases in electron transfer rate to ETF are
elicited through mutation of residue Tyr-442, but the corresponding mutant
data for reactions performed with ferricenium ions reveal only moderate
decreases in electron transfer rate. Together the data suggest that ETF
couples preferentially through residue Tyr-442, which is a greater distance
from the 4Fe-4S center than Val-344, and the reason for this mode of coupling is that ETF—unlike the ferricenium ion—cannot gain sufficient access
to Val-344 for good electronic coupling. These observations thus reveal the
existence of multiple electron transfer pathways from the 4Fe-4S center to
the surface of TMADH.

### 6.2.   A Model for the Electron Transfer Complex

A major handicap to our detailed understanding of the electron transfer reactions between TMADH and ETF is the lack of a crystallographic
structure for ETF. Crystals of ETF have been isolated (White *et al.*, 1994),
but to date no structure for the protein has been reported. A homology
model for ETF, however, has been constructed based on the crystallographic structure of human ETF (Roberts *et al.*, 1996), and this model has
been used to create a model of the electron transfer complex formed
between TMADH and ETF (Chohan *et al.*, 1998) (Figure 7). ETF comprises
two subunits, which in turn form three domains. Domain I comprises the
N-terminal region of the α-subunit, domain II comprises the C-terminal

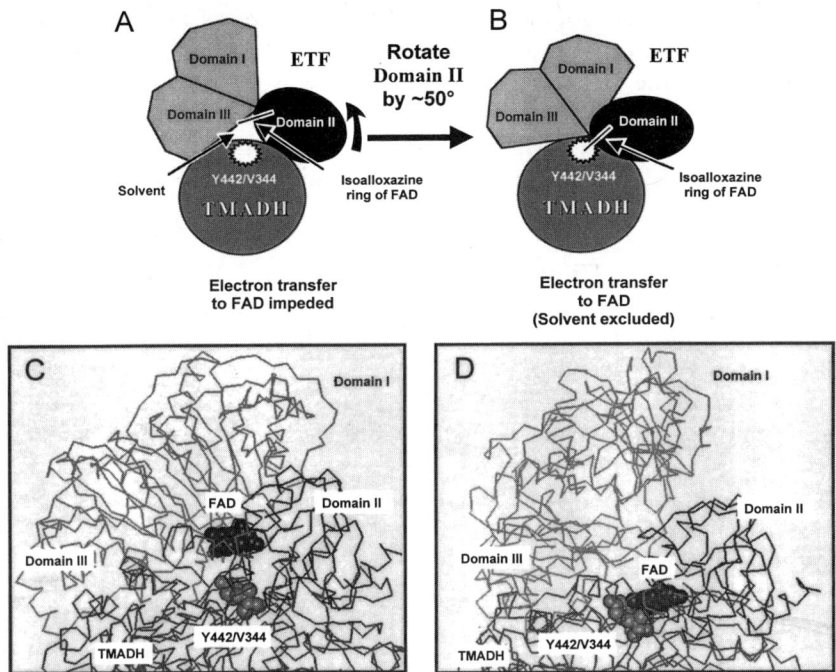

**FIGURE 7.** Proposed large-scale structural reorganisation of ETF on binding to TMADH. Schematic representations of A, the "eT-inactive" complex, and B, the "eT-active" complex. The corresponding structures (Cα trace for protein) are shown in C and D, respectively.

part of the α-subunit and a small number of residues at the C-terminus of the β-subunit, and domain III comprises the majority of the β-subunit. ETF is 'Y-shaped'. The quaternary structure is stabilised by interactions formed between domains I and III, and domain II sits in a shallow bowl formed by domains I and III. The polypeptide linking domains I and II is intrinsically flexible [[193]GGG[195] (Chen and Swenson, 1994)]. There is also a long loop linking domains III and II. ETF can thus be considered as two rigid bodies (one comprising domains I and III and the other domain II) and it is envisaged that these rigid bodies can move independently of each other. The isoalloxazine of the FAD in ETF is mainly bound by domain II, which suggests that this domain docks onto TMADH in the electron transfer complex. In modeling the complex, domain II was docked onto the surface of TMADH such that the isoalloxazine ring was positioned close to Tyr-442 and Val-344. Having docked domain II onto TMADH, domains I and III were included in the model. This required a significant rotation (~50°) of domains I and III with respect to domain II relative to their initial positions in the ETF homology model. This rotation is facilitated by the flexible hinge regions at the boundaries of domains I and II and domains III and II. In the docked conformation, the isoalloxazine ring of the FAD is "released" from ETF and interacts with the surface of TMADH in the region of the small surface groove lined by Tyr-442—a configuration that is compatible with efficient electron transfer from the 4Fe-4S center to FAD.

The structural model of the electron transfer complex formed by TMADH and ETF suggests that ETF adopts "eT-inactive" (non-complexed) and "eT-active" (complexed) conformations. Recent solution evidence supports the notion of different conformational states of ETF. For example, ETF is reduced to the semiquinone form only in the absence of TMADH, but in the presence of TMADH ETF can be reduced to the hydroquinone form (Jang *et al.*, 2000). Moreover, stopped-flow reoxidation studies of ETF by ferricenium ions have demonstrated that flavin reoxidation is slow, perhaps reflecting conformational gating by ETF of electron transfer to the artificial electron acceptor. Such large-scale structural reorganisation during complex assembly is unprecedented in interprotein electron transfer and its role therefore needs to be addressed. ETF is purified predominantly in the semiquinone (1-electron reduced) form, which illustrates how effectively the "eT-inactive" conformation of ETF protects the FAD from reoxidation by molecular oxygen. Given the exposed nature of the flavin in the "eT-active" form, in the absence of TMADH the 1-electron reduced form of this conformation is likely to be susceptible to reoxidation by molecular oxygen. A large-scale structural change following electron transfer and dissociation of the complex to produce the "eT-inactive", oxygen-insensitive conformation of reduced ETF thus provides a plausible mechanism for chaperoning electrons to membrane-bound cytochromes, thereby preventing redox piracy by molecular oxygen. Additionally, the redox potential of ETF in the "active" (complexed) conformation may be perturbed, which may lead to a fine-tuning of its redox properties—this most likely arises from movement of Arg-237, implicated in stabilising the semiquinone form, away from the isoalloxazine ring.

## 7. ENZYME OVER-REDUCTION AND SUBSTRATE INHIBITION: MULTIPLE TURNOVER STUDIES

TMADH exhibits an unusual dependence on trimethylamine concentration in steady-state reactions with either ETF or artificial electron acceptors. The enzyme is inhibited by high concentrations of substrate; the reaction rate approaching a limiting and nonzero value (Falzon and

Davidson, 1996; Steenkamp and Beinert, 1982a). The mechanism by which substrate inhibits the enzyme has been debated recently. In one model, a mechanism involving an additional inhibitory binding site was proposed (Falzon and Davidson, 1996), although there is no direct evidence for the existence of such a site from structural studies (Bellamy et al., 1989; Lim et al., 1986). An alternative model involving two linked catalytic cycles has been put forward (Huang et al., 1995) comprising an oxidised and 2-electron-reduced enzyme (0/2) cycle and a 1- and 3-electron reduced enzyme (1/3) cycle (Figure 8). The bifurcating kinetic scheme arises because trimethylamine donates two electrons to TMADH, while ETF or artificial electron acceptors remove one electron from TMADH. Also, as we have seen above, the enzyme is capable of accepting three electrons during titration with dithionite and this form of the enzyme is unique to the 1/3 catalytic cycle. The kinetic scheme predicts that at low TMA and/or high ETF concentrations, the 0/2 cycle predominates, and conversely at high substrate and/or low ETF concentrations the 1/3 cycle is more important. Moreover, turnover in the 1/3 cycle is expected to be slower than in the 0/2 cycle. This arises because the binding of substrate to 1-electron reduced enzyme is known to stabilise the semiquinone form of the flavin (Stankovich and Steenkamp, 1987). A prediction of the branching mechanism for TMADH, therefore, is that under conditions of high substrate concentration there will be a redistribution of reducing equivalents in 1-electron reduced such that the flavin center (rather than the 4Fe-4S center) becomes reduced (see boxed equilibrium in Figure 8). Since the flavin is in the semiquinone form, the substrate molecule bound at the enzyme active site cannot transfer electrons to the flavin. The effect therefore is equivalent to excess substrate inhibition. A similar mechanism of substrate inhibition in xanthine oxidase has also been described (Hille and Stewart, 1984).

The robustness of the proposed kinetic mechanism has recently been tested through the analysis of steady-state reactions using the stopped-flow technique (Roberts et al., 1999). The artificial electron acceptor ferricenium hexafluorophosphate was used in these reactions to enable a study of the major redox forms of TMADH that accumulate under steady-state conditions. The ferricenium ion is a good mimic of ETF in that it accepts electrons only from the reduced 4Fe-4S center of TMADH and not from the reduced 6-$S$-cysteinyl FMN. Photodiode array analysis of the absorption spectrum of TMADH during steady-state turnover at high substrate concentrations indicated that 1-electron reduced TMADH in the form of the flavin anionic semiquinone was accumulated. At low substrate concentrations the enzyme form that predominates under steady-state conditions is that of the oxidised enzyme. Similar studies with the slow substrate dimethylbutylamine revealed that the 1/3 cycle was poorly populated even

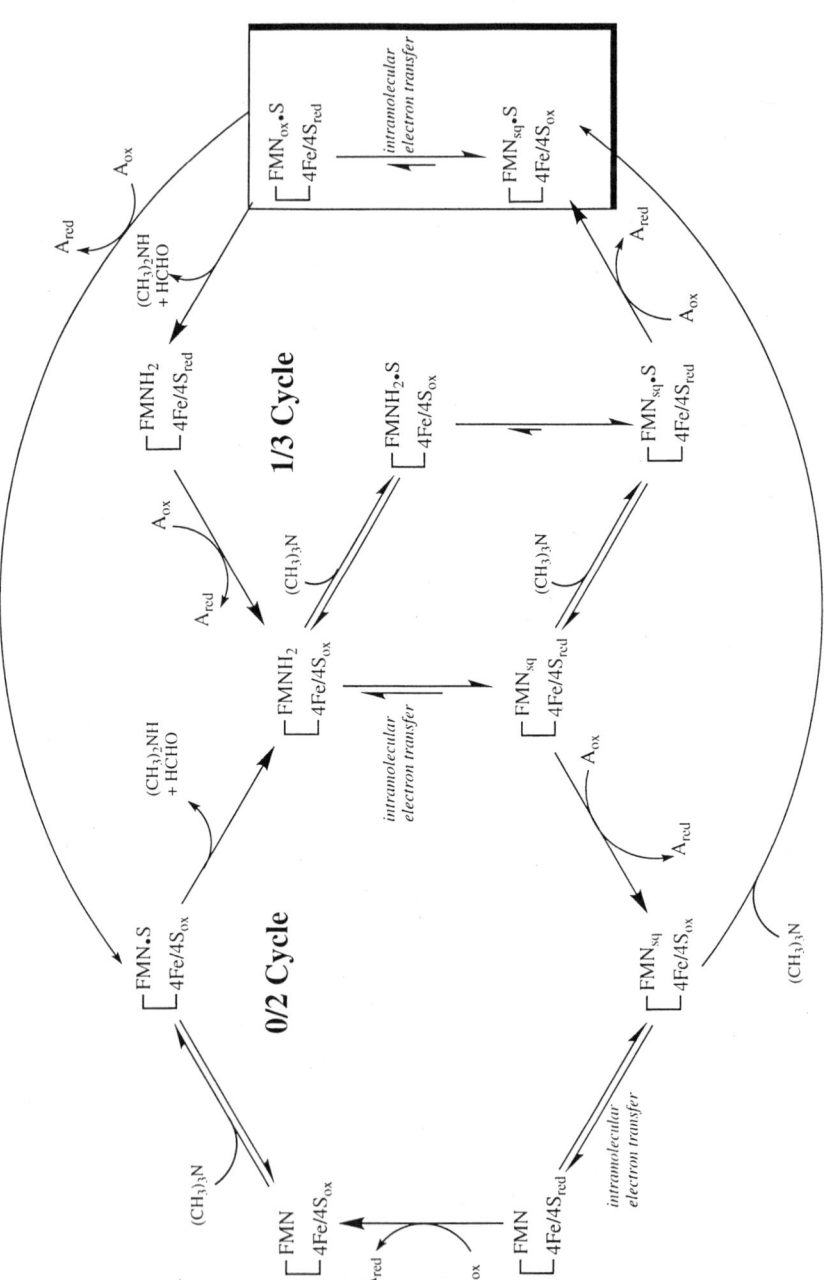

**FIGURE 8.** Branching kinetic steady-state scheme proposed for TMADH.

at high concentrations of substrate. This observation is consistent with the inability of dimethybutylamine to act as an effective inhibitor of TMADH. The enzyme-monitored turnover method was also used to demonstrate that during steady-state turnover the depletion of substrate could be correlated directly to the loss of the 1 electron reduced semiquinone form of TMADH. Together, the data provide compelling evidence for the existence of the branched kinetic scheme and thus provide an explanation for the unusual mode of substrate inhibition seen in TMADH.

## 8.  COFACTOR ASSEMBLY AND THE ROLE OF THE 6-*S*-CYSTEINYL FMN

Expression of TMADH in the heterologous host *Escherichia coli* leads to the production of recombinant protein that binds ADP and the 4Fe-4S center stoichiometrically. However, only substoichiometric amounts of FMN (linked covalently as the 6-*S*-cysteinyl FMN) are bound at the level of 30 to 50% of the subunits (Scrutton *et al.*, 1994). The remainder of the enzyme is incapable of binding FMN. The flavinylated enzyme is fully active and behaves the same as the native enzyme in steady-state and single turnover stopped-flow reactions. Covalent attachment of the flavin to the wild-type enzyme is self-catalytic (Scrutton *et al.*, 1994) and underflavinylation of TMADH when expressed in the recombinant host has been attributed to the inability of the organism to deliver FMN rapidly enough to keep up with production of the enzyme. X-ray crystallographic analyses of the recombinant wild-type enzyme (approximately 30% flavinylated) indicate a structure identical with native TMADH except for decreased flavin content (Mathews *et al.*, 1996). However, full occupancy by an anion of the ribityl phosphate-binding site of FMN is observed in the recombinant enzyme. The anion was conjectured to be inorganic phosphate; an assignment that has recently been corroborated by [31]P NMR studies.

The mechanism of flavinylation of TMADH has been investigated through the isolation of various mutant forms of the enzyme and analysis by mass spectrometry and UV-visible spectrophotometric methods (Mewies *et al.*, 1996; Packman *et al.*, 1995). Formation of the iminoquinone tautomeric state of non-covalently bound FMN is proposed in the active site of TMADH, which enhances the electrophilicity of the flavin C6 atom (Figure 9). Residue Arg-222 is thought to play a central role in formation of the tautomeric state, and this assumed role is consistent with mutagenesis experiments. Flavinylation is abolished in mutants where the positive charge at residue position 222 is removed (TMADH mutants R222V and

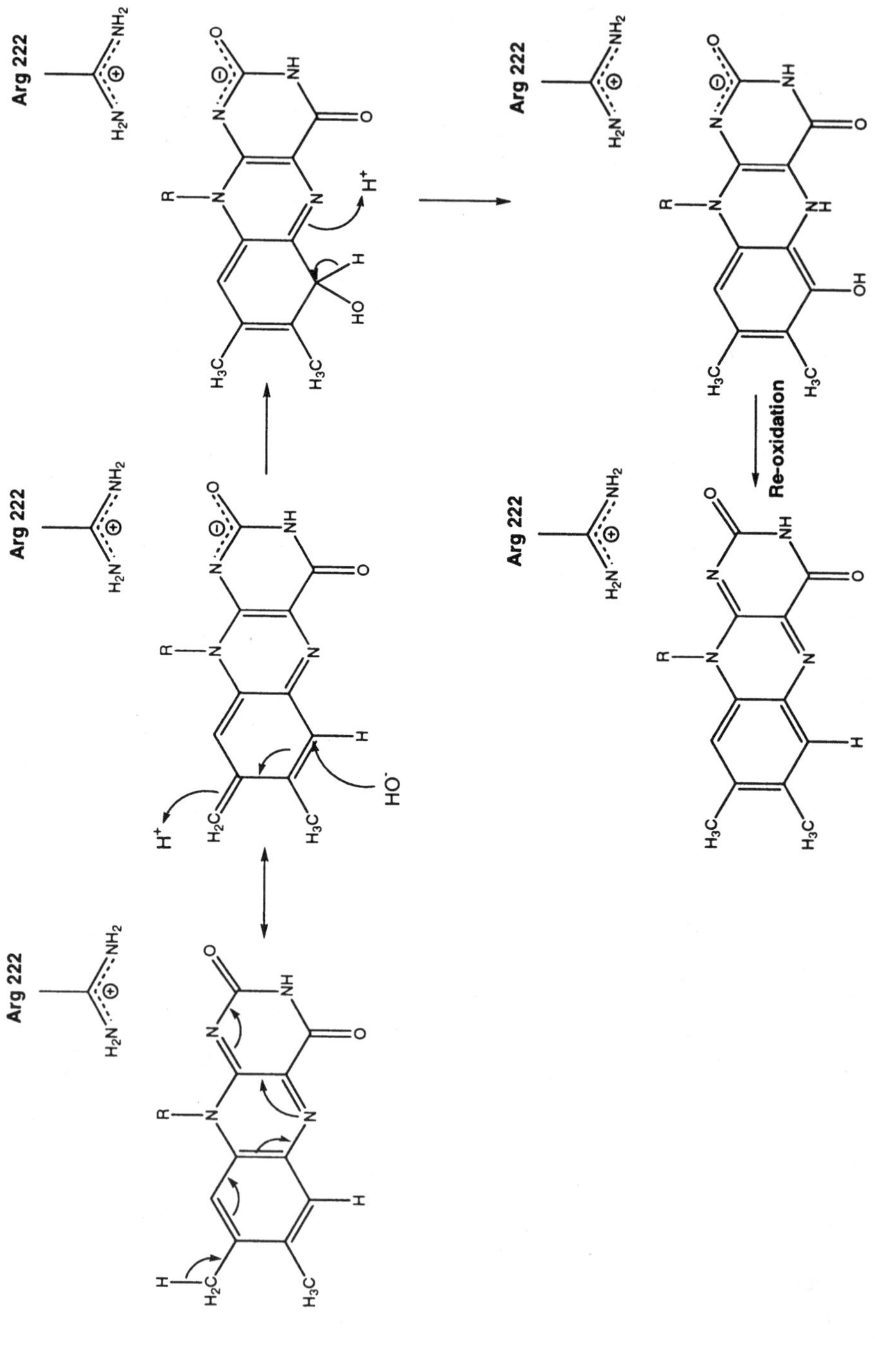

**FIGURE 9.** Mechanism of flavinylation of TMADH.

R222E), whereas flavinylation proceeds (albeit less efficiently) in a R222K mutant of TMADH where the positive charge is retained (Mewies *et al.*, 1996). Residue His-29 is also thought to play a major role in flavinylation by stabilising the thiolate form of Cys-30 (the reactive nucleophile); mutagenesis of this residue severely compromises, but does not completely abolish flavinylation of TMADH (Packman *et al.*, 1995).

The role of the 6-*S*-cysteinyl FMN in catalysis has been investigated by the isolation of a C30A mutant form of TMADH. Stopped-flow kinetic studies with the slow substrate diethylmethylamine demonstrated a reduction in the limiting rate of flavin reduction of about 6-fold compared with wild-type enzyme (Huang *et al.*, 1996). Reactions with the natural substrate, however, are more compromised; in this case there is about a 30-fold reduction in the limiting rate of flavin reduction and a 100-fold increase in the dissociation constant for the enzyme-substrate complex. The flavin recovered from the C30A enzyme following reaction with substrate is converted to the 6-hydroxy derivative, which renders the enzyme inactive. A reaction mechanism was proposed in which $OH^-$ attacks the C6 atom of the notional flavin-substrate covalent adduct that forms in the reductive half-reaction, since formation of the adduct is envisaged to increase the electrophilicity of the flavin C6 atom (Huang *et al.*, 1996). More recently, however, this mechanism has been questioned following studies of a W355L mutant of TMADH. W355 is one of three aromatic residue involved in substrate binding in the active site of the enzyme. Mutation to leucine influences the chemistry of the flavin by enabling derivatisation to the 6-hydroxy FMN form even in the absence of substrate turnover (Mewies *et al.*, 1997). This finding has led to the proposal of an alternative mechanism for hydroxylation involving the flavin iminoquinone methide tautomer to enhance the electrophilicity of the C6 atom. The mechanism is attractive since the tautomer is also an intermediate in the proposed flavinylation reaction of TMADH. Covalent linkage at the C6 atom (via Cys-30 in native TMADH) or the 8α methyl group [as with other covalently linked flavoproteins (Mewies *et al.*, 1998)] would prevent 6-hydroxylation by either direct linkage at the C6 position or by preventing formation of the reactive flavin iminoquinone methide tautomer, respectively. Thus, the major role of covalent linkage of flavin to certain flavoproteins may not be to 'fine tune' the redox properties of the flavin, but to prevent inactivation of the enzyme by hydroxylation at the flavin C6 atom. Certainly, covalent linkage in TMADH is not responsible for the high degree of butterfly bending of the flavin, since the crystal structure of the C30A TMADH reveals that the flavin is bent to the same degree as seen in the wild-type enzyme (Mathews *et al.*, 1996).

## 9. SUMMARY AND FUTURE PROSPECTS

The last decade has witnessed substantial progress in our understanding of electron transfer reactions in the TMADH:ETF redox system. Detailed kinetic studies, coupled with high resolution structural data and studies of mutant forms makes TMADH probably the best characterized iron-sulfur flavoprotein. These detailed studies have made important contributions to the field of enzyme catalyzed electron transfers. In particular, a role for large-scale structural reorganization in gating interprotein electron transfer has been demonstrated in addition to the prototropic and ligand gating reactions seen within TMADH itself. Studies with perdeuterated substrates have for the first time demonstrated the importance of a ground state nuclear tunneling mechanism involving vibrational coupling with the protein molecule in cleaving stable C-H bonds. Studies with TMADH have also provided a rationale for the evolution of covalent flavoproteins, since covalent incorporation of the flavin in TMADH prevents inactivation of the flavin through derivatisation at the C6 atom. Detailed studies of the pH dependence of the reductive half-reaction of the enzyme have provided evidence against a carbanion mechanism for C-H bond cleavage and provided indirect support for homolysis of the C-H bond.

A variety of issues remain to be elucidated with this enzyme system. Direct evidence for the mechanism of C-H bond cleavage is required, and this may involve studies with cyclopropyl radical trapping reagents as demonstrated with the flavoprotein monoamine oxidase. Evidence for the conjectured flavin-substrate adduct that is thought to gate internal electron transfer in TMADH is also required. This is a challenging area of research work, but attempts are being made to cage substrates of TMADH for use in Laue diffraction experiments; such experiments may identify a covalent flavin-substrate intermediate in the reaction mechanism. Our knowledge of the oxidative half-reaction is still fragmentary: additional studies are still required to improve our understanding of the role of protein dynamics in enhancing electronic coupling between TMADH and ETF . . . and, of course, an x-ray crystallographic structure of ETF and the its electron transfer complex with TMADH are major goals of future research.

ACKNOWLEDGMENTS. Some of the work described in this review was performed in the authors' laboratory and was funded by the BBSRC, MRC, Royal Society and Leverhulme Trust. NSS is a Lister Institute Research Fellow. MJS and NSS were Royal Society University Research Fellows. The authors wish to acknowledge Professors R. Hille and F. S. Mathews who have both made substantial contributions to studies on TMADH and with whom the authors have enjoyed productive collaborations investigating the TMADH:ETF redox system over recent years.

## 10. REFERENCES

Anthony, C., 1986, Bacterial oxidation of methane and methanol, *Adv. Microb. Physiol.* **27**:113–210.

Barber, M. J., Neame, P. J., Lim, L. W., White, S., and Matthews, F. S., 1992, Correlation of x-ray deduced and experimental amino acid sequences of trimethylamine dehydrogenase, *J. Biol. Chem.* **267**:6611–6619.

Barber, M. J., Pollock, V., and Spence, J. T., 1988, Microcoulometric analysis of trimethylamine dehydrogenase, *Biochem. J.* **256**:657–659.

Basran, J., Mewies, M., Mathews, F. S., and Scrutton, N. S., 1997, Selective modification of alkylammonium ion specificity in trimethylamine dehydrogenase by the rational engineering of cation-π bonding, *Biochemistry* **36**:1989–1998.

Basran, J., Sutcliffe, M. J., Hille, R., and Scrutton, N. S., 1999a, Reductive half-reaction of the H172Q mutant of trimethylamine dehydrogenase: evidence against a carbanion mechanism and assignment of kinetically influential ionizations in the enzyme-substrate complex, *Biochem. J.* **341**:307–314.

Basran, J., Sutcliffe, M. J., Hille, R., and Scrutton, N. S., 1999b, The role of Tyr 169 of trimethylamine dehydrogenase in substrate oxidation and magnetic interaction between FMN cofactor and the 4Fe/4S center, *J. Biol. Chem.* **274**:13155–13161.

Basran, J., Sutcliffe, M. J., and Scrutton, N. S., 1999c, Enzymatic H-transfer requires vibration-driven extreme tunneling, *Biochemistry* **38**:3218–3222.

Bellamy, H. D., Lim, L. W., Mathews, F. S., and Dunham, W. R., 1989, Studies of crystalline trimethylamine dehydrogenase in three oxidation states and in the presence of substrate and inhibitor, *J. Biol. Chem.* **264**:11887–11892.

Bork, P. J., Gellerich, H., Groth, R., Hooft, A., and Martin, F., 1995, Divergent evolution of a beta/alpha barrel subclass: detection of numerous phosphate-binding sites by motif search, *Protein Sci* **4**:268–274.

Boyd, G., Mathews, F. S., Packman, L. C., and Scrutton, N. S., 1992, Trimethylamine dehydrogenase of bacterium $W_3A_1$. Molecular cloning, sequence determination and overexpression of the gene, *FEBS Lett* **308**:271–276.

Bruno, W. J., and Bialek, W., 1992, Vibrationally enhanced tunneling as a mechanism for enzymatic hydrogen transfer, *Biophys. J.* **63**:689–699.

Chen, D. W., and Swenson, R. P., 1994, Cloning, sequence analysis, and expression of the genes encoding the two subunits of the methylotrophic bacterium $W_3A_1$ electron transfer flavoprotein, *J. Biol. Chem.* **269**:32120–32130.

Chohan, K. K., Scrutton, N. S., and Sutcliffe, M. J., 1998, Major structural reorganisation most likely accompanies the transient formation of a physiological electron transfer complex, *Prot. Pept. Lett* **5**:231–236.

Colby, J., and Zatman, L. J., 1973, Trimethylamine metabolism in obligate and facultative methylotrophs, *Biochem. J.* **132**:101–112.

Colby, J., and Zatman, L. J., 1974, Purification and properties of the trimethylamine dehydrogenase of bacterium 4B6, *Biochem. J.* **143**:555–567.

Colby, J., and Zatman, L. J., 1975, Enzymological aspects of the pathways for trimethylamine oxidation and C1 assimilation in obligate methylotrophs and restricted facultative methylotrophs, *Biochem. J.* **148**:513–520.

D'Silva, C. D., Williams, C. H., and Massey, V., 1987, Identification of methionine-110 as the residue covalently modified in the electrophilic inactivation of D-amino-acid oxidase by O-(2,4-dinitrophenyl) hydroxylamine, *Biochemistry* **26**:1717–1722.

Denu, J. M., and Fitzpatrick, P. F., 1994, Intrinsic primary, secondary, and solvent kinetic isotope effects on the reductive half-reaction of D-amino acid oxidase: evidence against a concerted mechanism, *Biochemistry* **33**:4001–4007.

Eady, R. R., Jarman, T. R., and Large, P. J., 1971, Microbial oxidation of amines. Partial purification of a mixed function secondary amine oxidase system from *Pseudomonas aminovorans* that contains an enzymically active cytochrome P-420 type hemoprotein, *Biochem. J.* **125**:449–459.

Edmondson, D. E., 1995, Aminium cation radical mechanism proposed for monoamine oxidase B catalysis: are there alternatives?, *Xenobiotica* **25**:735–753.

Ertughrul, O. W., Errington, N., Raza, S., Sutcliffe, M. J., Rowe, A. J., and Scrutton, N. S., 1998, Probing the stabilizing role of C-terminal residues in trimethylamine dehydrogenase, *Protein Eng.* **11**:447–455.

Falzon, L., and Davidson, V. L., 1996, Kinetic model for the regulation by substrate of intramolecular electron transfer in trimethylamine dehydrogenase, *Biochemistry* **35**:2445–2452.

Fournel, A., Gambarelli, S., Guigliarelli, B., More, C., Asso, M., Chouteau, G., Hille, R., and Bertrand, P., 1998, Magnetic interactions between a [4Fe-4S]$^{1+}$ cluster and a FMN radical in the enzyme trimethylamine dehydrogenase: a high-field epr study, *J. Chem. Phys.* **109**:10905–10913.

Franklund, C. F., Baron, S. F., and Hylemon, P. B., 1993, Characterisation of the *bai* H gene encoding a bile acid-inducible NADH:flavin oxidoreductase from *Eubacterium* sp. VPI 12708, *J. Bacteriol.* **175**:3002–3012.

Ghisla, S., Kenney, W. R., Knappe, W. R., McIntire, W. S., and Singer, T. P., 1980, Chemical synthesis and some properties of 6-substituted flavins, *Biochemistry* **19**:2537–2544.

Ghisla, S., and Massey, V., 1980, Studies on the catalytic mechanism of lactate oxidase. Formation of enantiomeric flavin-N(5)-glycollyl adducts via carbanion intermediates, *J. Biol. Chem.* **255**:5688–5696.

Ghisla, S., and Massey, V., 1989, Mechanisms of flavoprotein-catalyzed reactions, *Eur. J. Biochem.* **181**:1–17.

Harris, R. J., Meskys, R., Sutcliffe, M. J., and Scrutton, N. S., 2000, Kinetic studies of the mechanism of C—H bond breakage by the heterotetraneric sarcosine oxidase of *Arthrobacter* sp 1-IN, *Biochemistry* **39**:1189–1198.

Hasford, J. J., Kemnitzer, W., and Rizzo, C. J., 1997, Conformational effects on flavin redox chemistry, *J. Org. Chem.* **62**:5244–5245.

Hill, C. L., Steenkamp, D. J., Holm, R. H., and Singer, T. P., 1977, Identification of the iron-sulfur center in trimethylamine dehydrogenase, *Proc. Natl. Acad. Sci. USA* **74**:547–551.

Hille, R., and Stewart, R., 1984, The inhibition of xanthine oxidase by 8-bromoxanthine, *J. Biol. Chem.* **259**:1570–1576.

Huang, L., Rohlfs, R. J., and Hille, R., 1995, The reaction of trimethylamine dehydrogenase with electron transferring flavoprotein, *J. Biol. Chem.* **270**:23958–23965.

Huang, L., Scrutton, N. S., and Hille, R., 1996, Reaction of the C30A mutant of trimethylamine dehydrogenase with diethylmethylamine, *J. Biol. Chem.* **271**:13401–13406.

Husain, M., and Davidson, V. L., 1987, Purification and properties of methylamine dehydrogenase from *Paracoccus denitrificans, J. Bacteriol.* **169**:1712–1717.

Jang, M.-H., Basran, J., Scrutton, N. S., and Hille, R., 1999a, The reaction of trimethylamine dehydrogenase with trimethylamine, *J. Biol. Chem.* **274**:13147–13154.

Jang, M.-H., Scrutton, N. S., and Hille, R., 2000, Formation of W$_3$A$_1$ ETF hydroquinone in the TMADH/ETF protein complex, *J. Biol. Chem.* **275**:12546–12552.

Kenney, W. C., McIntire, W., and Steenkamp, D. J., 1978, Amino acid sequence of a cofactor peptide from trimethylamine dehydrogenase, *FEBS Lett* **85**:137–40.

Kim, J.-M., Bogdon, M. A., and Mariano, P. S., 1993, Mechanistic analysis of the 3-methyllumiflavin-promoted oxidative deamination of benzylamine—a potential model for monoamine oxidase catalysis, *J. Am. Chem. Soc.* **115**:10591–10595.

Kohen, A., Cannio, R., Bartolucci, S., and Klinman, J. P., 1999, Enzyme dynamics and hydrogen tunneling in a thermophilic alcohol dehydrogenase, *Nature* **399**:496–499.

Large, P. J., 1981, Microbial growth on methylated amines, in: *Microbial growth on C1 compounds: proceedings of the third international symposium* (H. Dalton, ed.), Heyden and Son, London, pp. 55–69.

Levering, P. R., van Dijken, J. P., Veenhuis, M., and Harder, W., 1981, *Arthrobacter* P1, a fast growing versatile methylotroph with amine oxidase as a key enzyme in the metabolism of methylated amines *Arch. Microbiol.* **129**:72–80.

Lim, L. W., Mathews, F. S., and Steenkamp, D. J., 1982, Crystallographic study of the iron-sulfur flavoprotein trimethylamine dehydrogenase from the bacterium W₃A₁, *J. Mol. Biol.* **162**:869–876.

Lim, L. W., Mathews, F. S., and Steenkamp, D. J., 1988, Identification of ADP in the iron-sulfur flavoprotein trimethylamine dehydrogenase, *J. Biol. Chem.* **263**:3075–3078.

Lim, L. W., Shamala, N., Mathews, F. S., and Steenkamp, D. J., 1984, Molecular structure of trimethylamine dehydrogenase from the bacterium W₃A₁ at 6.0-Å resolution, *J. Biol. Chem.* **259**:14458–14462.

Lim, L. W., Shamala, N., Mathews, F. S., Steenkamp, D. J., Hamlin, R., and Xuong, N. H., 1986, Three-dimensional structure of the iron-sulfur flavoprotein trimethylamine dehydrogenase at 2.4-Å resolution, *J. Biol. Chem.* **261**:15140–15146.

Loginova, N. V., and Trotsenko, Y. A., 1978, Carbon metabolism in methylotrophic bacteria isolated from activated sludge, *Mikrobiologiya* **47**:939–946.

Mathews, F. S., Trickey, P., Barton, J. D., and Chen, Z.-W., 1996, Crystal structures of recombinant wild-type and a C30A mutant trimethylamine dehydrogenase from *Methylophilus* W₃A₁, in: *Flavins and Flavoproteins* (K. Stevenson, V. Massey, and C. H. Williams, eds.), University of Calgary Press, Calgary, pp. 873–876.

Meiberg, J. B. M., and Harder, W., 1979, Dimethylamine dehydrogenase from *Hyphomicrobium X*: purification and properties of a new enzyme that oxidizes secondary amines, *J. Gen. Microbiol.* **115**:49–58.

Mewies, M., Basran, J., Packman, L. C., Hille, R., and Scrutton, N. S., 1997, Involvement of a flavin iminoquinone methide in the formation of 6- hydroxyflavin mononucleotide in trimethylamine dehydrogenase: a rationale for the existence of 8α-methyl and C6-linked covalent flavoproteins, *Biochemistry* **36**:7162–7168.

Mewies, M., McIntire, W. S., and Scrutton, N. S., 1998, Covalent attachment of flavin adenine dinucleotide (FAD) and flavin mononucleotide (FMN) to enzymes: the current state of affairs, *Protein Science* **7**:7–20.

Mewies, M., Packman, L. C., Mathews, F. S., and Scrutton, N. S., 1996, Flavinylation in wild-type trimethylamine dehydrogenase and differentially charged mutant enzymes: a study of the protein environment around the N1 of the flavin isoalloxazine, *Biochem. J.* **317**:267–272.

Nagy, J., Kenney, W. C., and Singer, T. P., 1979, The reaction of phenylhydrazine with trimethylamine dehydrogenase and with free flavins, *J. Biol. Chem.* **254**:2684–2688.

Pace, C. P., and Stankovich, M. T., 1991, Oxidation-reduction properties of trimethylamine dehydrogenase: effect of inhibitor binding, *Arch. Biochem. Biophys.* **287**:97–104.

Packman, L. C., Mewies, M., and Scrutton, N. S., 1995, The flavinylation reaction of trimethylamine dehydrogenase. Analysis by directed mutagenesis and electrospray mass spectrometry, *J. Biol. Chem.* **270**:13186–13191.

Porter, D. J. T., Voet, J. G., and Bright, H. J., 1977, Mechanistic features of the D-amino acid oxidase reaction studied by double stopped flow spectrophotometry, *J. Biol. Chem.* **252**:4464–4473.

Raine, A. R., Scrutton, N. S., and Mathews, F. S., 1994, On the evolution of alternate core packing in eightfold beta/alpha- barrels, *Protein Sci* **3**:1889–1892.

Raine, A. R., Yang, C. C., Packman, L. C., White, S. A., Mathews, F. S., and Scrutton, N. S., 1995, Protein recognition of ammonium cations using side-chain aromatics: a structural variation for secondary ammonium ligands, *Protein Sci* **4**:2625–2628.

Regan, J. J., 1993, PATHWAYS II, San Diego.

Roberts, P., Basran, J., Wilson, E. K., Hille, R., and Scrutton, N. S., 1999, Redox cycles in trimethylamine dehydrogenase and mechanism of substrate inhibition, *Biochemistry* **38**:14927–14940.

Roberts, D. L., Frerman, F. E., and Kim, J.-J. P., 1996, Three-dimensional structure of human electron transfer flavoprotein to 2.1 Å resolution, *Proc. Natl. Acad. Sci. USA* **93**:14355–14360.

Rohlfs, R. J., and Hille, R., 1991, Intramolecular electron transfer in trimethylamine dehydrogenase from bacterium $W_3A_1$, *J. Biol. Chem.* **266**:15244–15252.

Rohlfs, R. J., and Hille, R., 1994, The reaction of trimethylamine dehydrogenase with diethylmethylamine, *J. Biol. Chem.* **269**:30869–30879.

Rohlfs, R. J., Huang, L., and Hille, R., 1995, Prototropic control of intramolecular electron transfer in trimethylamine dehydrogenase, *J. Biol. Chem.* **270**:22196–22207.

Scrutton, N. S., 1994, α/β barrel evolution and the modular assembly of enzymes: emerging trends in the flavin dehydrogenase/oxidase family, *BioEssays* **16**:115–122.

Scrutton, N. S., Basran, J., and Sutcliffe, M. J., 1999, New insights into enzyme catalysis: ground state tunneling driven by protein dynamics, *Eur. J. Biochem.* **264**:666–671.

Scrutton, N. S., Packman, L. C., Mathews, F. S., Rohlfs, R. J., and Hille, R., 1994, Assembly of redox centers in the trimethylamine dehydrogenase of bacterium $W_3A_1$. Properties of the wild-type enzyme and a C30A mutant expressed from a cloned gene in *Escherichia coli*, *J. Biol. Chem.* **269**:13942–13950.

Scrutton, N. S., and Raine, A. R. C., 1996, Cation-π bonding and amino aromatic interactions in the biomolecular recognition of substituted ammonium ligands, *Biochem. J.* **319**:1–8.

Silverman, R. B., 1995, Radical ideas about monoamine oxidase, *Accts. Chem. Res.* **28**:335–342.

Stankovich, M. T., and Steenkamp, D. J., 1987, Redox properties of trimethylamine dehydrogenase, in: *Flavins and Flavoproteins* (D. E. Edmondson and D. B. McCormick, eds.), Walter de Gruyter, Berlin, pp. 687–690.

Steenkamp, D. J., and Beinert, H., 1982a, Mechanistic studies on the dehydrogenases of methylotrophic bacteria. 1. The influence of substrate binding to reduced trimethylamine dehydrogenase on the intramolecular electron transfer between its prosthetic groups, *Biochem. J.* **207**:233–239.

Steenkamp, D. J., and Beinert, H., 1982b, Mechanistic studies on the dehydrogenases of methylotrophic bacteria. 2. Kinetic studies on the intramolecular electron transfer in trimethylamine and dimethylamine dehydrogenase, *Biochem. J.* **207**:241–252.

Steenkamp, D. J., Beinert, H., McIntire, W. S., and Singer, T. P., 1978a, in: *Mechanisms of Oxidizing Enzymes* (T. P. Singer and R. N. Ondarza, eds.), Elsevier North-Holland Inc, New York, pp. 127–141.

Steenkamp, D. J., and Gallup, M., 1978, The natural flavorprotein electron acceptor of trimethylamine dehydrogenase, *J. Biol. Chem.* **253**:4086–4089.

Steenkamp, D. J., and Mallinson, J., 1976, Trimethylamine dehydrogenase from a methylotrophic bacterium. I. Isolation and steady-state kinetics, *Biochim. Biophys. Acta* **429**:705–719.

Steenkamp, D. J., and Mathews, F. S., 1992, The biochemical properties and structure of trimethylamine dehydrogenase, in: *Chemistry and Biochemistry of Flavoenzymes*, Volume II (F. Muller, ed.), CRC Press, Boca Raton, pp. 395–423.

Steenkamp, D. J., McIntire, W., and Kenney, W. C., 1978b, Structure of the covalently bound coenzyme of trimethylamine dehydrogenase. Evidence for a 6-substituted flavin, *J. Biol. Chem.* **253**:2818–2824.

Steenkamp, D. J., and Singer, T. P., 1976, On the presence of a novel covalently bound oxidation-reduction cofactor, iron and labile sulfur in trimethylamine dehydrogenase, *Biochem. Biophys. Res. Commun.* **71**:1289–1295.

Steenkamp, D. J., Singer, T. P., and Beinert, H., 1978c, Participation of the iron-sulphur cluster and of the covalently bound coenzyme of trimethylamine dehydrogenase in catalysis, *Biochem. J.* **169**:361–9.

Sutcliffe, M. J., and Scrutton, N. S., 2000, Enzymology takes a quantum leap forward, *Phil. Trans. Roy. Soc. Ser. A.* **358**:367–386.

van Iersel, J., van der Meer, R. A., and Duine, J. A., 1986, Methylamine oxidase from *Arthrobacter* P1. A bacterial copper-quinoprotein amine oxidase, *Eur. J. Biochem.* **161**:415–419.

Wallace, A. C., Laskowski, R. A., and Thornton, J. M., 1995. LIGPLOT: a program to generate schematic diagrams of protein-ligand interactions, *Protein Eng.* **8**:127–134

White, S. A., Mathews, F. S., Rohlfs, R. J., and Hille, R., 1994, Crystallization and preliminary crystallographic investigation of electron-transfer flavoprotein from the bacterium Methylophilus W₃A₁, *J. Mol. Biol.* **240**:265–266.

Williams, C. H. J., 1992, Lipoamide dehydrogenase, glutathione reductase, thioredoxin reductase and mercuric ion reductase family of flavoenzyme transhydrogenases, in: *Chemistry and Biochemistry of Flavoenzymes*, volume III (F. Muller, ed.), CRC Press, Boca Raton, pp. 121–211.

Wilmanns, M., Hyde, C. C., Davies, D. R., Kirschener, K., and Jansonius, J. N., 1991, Structural conservation in parallel beta/alpha barrel enzymes that catalyse three sequential reactions in the pathway of tryptophan biosynthesis, *Biochemistry* **30**:9161–9169.

Wilson, E. K., Huang, L., Sutcliffe, M. J., Mathews, F. S., Hille, R., and Scrutton, N. S., 1997a, An exposed tyrosine on the surface of trimethylamine dehydrogenase facilitates electron transfer to electron transferring flavoprotein: kinetics of transfer in wild-type and mutant complexes, *Biochemistry* **36**:41–48.

Wilson, E. K., Scrutton, N. S., Colfen, H., Harding, S. E., Jacobsen, M. P., and Winzor, D. J., 1997b, An ultracentrifugal approach to quantitative characterization of the molecular assembly of a physiological electron-transfer complex: the interaction of electron-transferring flavoprotein with trimethylamine dehydrogenase, *Eur. J. Biochem.* **243**:393–399.

Yang, C. C., Packman, L. C., and Scrutton, N. S., 1995, The primary structure of *Hyphomicrobium* X dimethylamine dehydrogenase. Relationship to trimethylamine dehydrogenase and implications for substrate recognition, *Eur. J. Biochem.* **232**:264–71.

*Chapter 6*

# Amine Oxidases and Galactose Oxidase

## Malcolm Halcrow, Simon Phillips, and Peter Knowles

## 1. INTRODUCTION

Over the past 10 years, our understanding of enzymes which effect the difficult chemical process of C—H bond cleavage has increased dramatically (Stubbe, 1989; Klinman, 1996). We know that nature employs both metal ions and reactive organic cofactors, such as radicals and quinones, derived by post-translational modification of aminoacids in the polypeptide chain of the enzyme. The two enzymes to be described in the present review are good examples: galactose oxidase employs copper and a tyrosine covalently cross-linked to a cysteine to stabilize a radical whilst amine oxidases employ copper and tyrosine-derived quinones. There is subtle interplay between the roles played by copper in the biogenesis of these novel cofactors and in the catalytic cycle of the oxidases.

Protein cross linkages are important in determining the chemical properties of these novel cofactors. In galactose oxidase, the covalent link between the cysteine and tyrosine helps lower the redox potential of the radical to a value consistent with its ready generation *in vivo*. In lysyl oxidase, however, the quinone derived from trihydroxyphenylalanine

**MALCOLM HALCROW, SIMON PHILLIPS, AND PETER KNOWLES**    Departments of Chemistry, and School of Biochemistry and Molecular Biology, University of Leeds, Leeds LS2 9J1

*Subcellular Biochemistry, Volume 35: Enzyme-Catalyzed Electron and Radical Transfer*, edited by Holzenburg and Scrutton. Kluwer Academic / Plenum Publishers, New York, 2000.

(TPQ) found in other amine oxidases is replaced by one involving a cross link between tyrosine and lysine (Wang *et al.*, 1997). There are a growing number of examples of novel cross linkages in proteins involved in redox processes and oxygen transport, including a tyrosine-histidine linkage in cytochrome oxidase (Ostermeier *et al.*, 1997), a tryptophan-tryptophan bridge in methanol dehydrogenase (McIntire *et al.*, 1991) and histidine-cysteine bridges in tyrosinases (Lerch, 1982) and some haemocyanins (Gielens *et al.*, 1991). We are only now beginning to understand the biological roles that these cross links play. A combination of detailed structural and functional studies on enzymes such as galactose oxidase, are critical to this understanding.

In this review, current understanding of the structure and function of galactose oxidase and amine oxidases will be described together with comparisons between them and future directions in this field.

## 2.   GALACTOSE OXIDASE

Galactose oxidase (GOase: EC 1.1.3.9) was first reported forty years ago (Cooper *et al.*, 1959) as an enzyme occurring in the culture medium of the fungus *Polyporus circinatus*. Initially it was suggested to be a metalloenzyme with zinc and flavin mononucleotide as prosthetic groups but later work by Horecker and coworkers (Aviqad *et al.*, 1962) established copper as the sole exogenous cofactor; the stoichiometry was one copper per enzyme molecule. These workers also showed (Amaral *et al.*, 1963) that GOase has a surprisingly low specificity for primary alcohols but is regioselective and stereo-selective; for instance D-galactose, but not L-galactose or D-glucose, is a substrate and secondary alcohol groupings are not oxidised (Figure 1).

A major problem in understanding the molecular basis for catalysis by GOase has been how an enzyme containing only a single copper and no other cofactor can catalyse the two-electron redox reaction shown in Figure 1; the normal valence states of copper are $Cu^{2+}$ and $Cu^+$ i.e. transition from

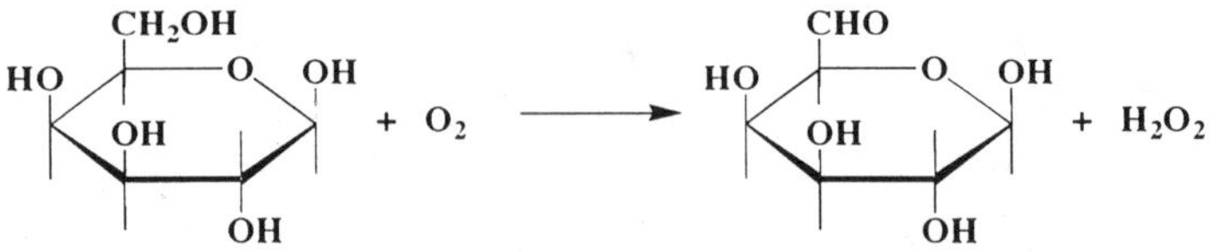

**FIGURE 1.**   Oxidation of D-galactose to aldehyde and hydrogen peroxide catalysed by GOase.

one to the other is a one electron redox process. This paradox was only finally solved when Whittaker and Whittaker (1988) suggested that the active form of the enzyme contained a protein-derived radical which acted as the second redox cofactor.

The major purpose of this review is to describe studies on the structure and function of galactose oxidase carried out over the last decade. As with all science, these advances are based on many earlier studies and it will not be possible to document with due respect how these studies advanced the field. Earlier work (up to 1984) on GOase has been reviewed (Malmstrom *et al.*, 1975; 1981; Kosman, 1984, 13–15) and more recent work is covered in reviews by Whittaker (1994) and by Knowles and Ito (1993).

## 2.1. Structure

### 2.1.1. General Structural Properties

GOase consists of a single polypeptide chain with two intramolecular disulphide bridges (Kosman *et al.*, 1974). It is reported to be fully active in 6M urea (Kosman *et al.*, 1974) attesting to its stability and potential for biotechnological applications.

### 2.1.2. Primary Structure

N-terminal amino acid sequence data from native GOase and from various protease-derived fragments were used to design primers for the polymerase chain reaction. A unique DNA fragment of 1.4 kilobases was amplified from *Dactylium dendroides* DNA and used as a probe to isolate the GOase gene from a genomic library. The DNA sequence (McPherson *et al.*, 1992) is consistent with protein sequence data and the crystal structure. Translation of the DNA sequence would produce a mature protein containing 639 amino acid residues. The primary structure of the N-terminus of the mature protein and its 41 residue leader sequence is shown in Figure 2.

The presence of a 41-residue leader sequence is indicated and protein processing events necessary to generate the mature protein will be discussed further in section 2.3.

-41                                                                    -1 ↓

MKHLLTLALCFSSINAVAVTVPHKAVGTGIPEGSLQFLSLR ASAP....

**FIGURE 2.** The 41-residue leader sequence of galactose oxidase obtained by sequencing the *gao A* gene. The arrow indicates the start site of the mature protein.

To date, GOase has only been isolated from fungi of the *Fusarium* genera. However, possible sequence homology between the GOase gene and parts of the *Arabidopsis* genome has been identified recently (McPherson, private communication); the biological significance of this observation is unknown at present.

### 2.1.3.  Secondary and Tertiary Structure

GOase from *D. dendroides* has been crystallised from acetate buffer at pH 4.5 using ammonium sulphate as precipitant. The structure has been solved by the multiple isomorphous replacement method using three heavy atom derivatives and the model refined to 1.7 Å resolution to allow detailed structural analysis (Ito *et al.*, 1991; Ito *et al.*, 1994).

The polypeptide chain is divided into three predominantly β-structural domains (Figure 3a). The high content of β-structure probably accounts for the stability of the enzyme noted above (Kosman *et al.*, 1974). Domain I (residues 1-155) has a β-sandwich structure and is linked to domain II by a well-ordered stretch of polypeptide chain. Domain II, the largest (residues 156-532), has pseudo-sevenfold symmetry with the overall appearance (Figure 3b) of a seven-bladed propeller where each blade consists of a four-stranded antiparallel β-sheet. Similar motifs have been reported for instance in methylamine dehydrogenase (Vellieux *et al.*, 1989) and G$_\beta$ proteins (Sondek *et al.*, 1996) while methanol dehydrogenase (Xia *et al.*, 1992), neuraminidase (Varghese *et al.*, 1983) and hemopexin (Faber *et al.*, 1995) are examples of 8, 6 and 4-bladed propeller domains respectively. The motif is probably widely distributed in Nature and associated with a range of protein functions (Bork and Doolittle, 1994). In GOase, the cupric ion, which is part of the active site, is located on the solvent-accessible surface of this domain, close to the pseudo-sevenfold axis. Domain III (residues 533-639) lies on the opposite side of domain II to the copper. Two of the seven β-strands in domain III form a hairpin loop that extends through the pore in the middle of domain II and provides a histidine ligand (His581) to the copper. The space between the hairpin and the walls of the pore in domain III is filled with ordered waters.

### 2.1.4.  Structure of the Active Site

In the original GOase crystals grown in the presence of acetate at pH 4.5, the copper site has square pyramidal coordination with Tyr 272, His 496, His 581 and an acetate ion as the in-plane ligands and Tyr 495, as a weakly-coordinated axial ligand (Figure 4). The site is rich in aromatic aminoacid side chains.

**FIGURE 3.** (a) Ribbon diagram of the overall structure of Galactose Oxidase with Domains I and III shaded and the copper shown as a shaded sphere. (b) View of Domain II along the propeller axis, corresponding to approaching (a) from below. Figure prepared using MOLSCRIPT (Kraulis, 1991).

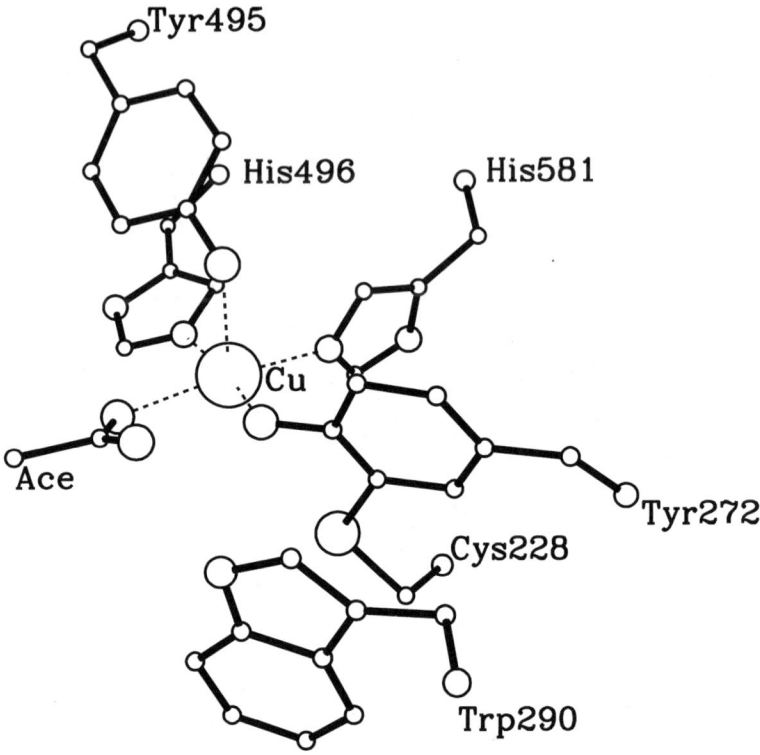

**FIGURE 4.** Structure of the Galactose Oxidase active site at pH 4. 5. Copper coordination is shown by dashed lines and "Ace" indicates the acetate ion. The larger sphere covalently bonded to $C_{el}$ of Tyr272 represents $S_\chi$ of Cys228. Figure prepared using the program NPO from the CCP4 program suite (CCP4, 1994).

A most surprising finding in the crystal structure is that one of the copper ligands, Tyr 272 is covalently bonded at $C_{el}$ to the sulphur of Cys 228 to give a thioether bond. This is consistent with a variety of independent evidence including protein chemistry (Kosman *et al.*, 1974; McPherson *et al.*, 1992; Ito *et al.*, 1991), resonance Raman spectroscopy (Whittaker *et al.*, 1989) and EPR / ENDOR data (Whittaker and Whittaker, 1990; Babcock *et al.*, 1992). These spectroscopic data, considered in conjunction with the X-ray crystal structure (Ito *et al.*, 1991), indicates that the Tyr-Cys cross-link forms part of the radical site. The cross-link creates an extended planar unit consisting of the aromatic ring, the sulphur and the cysteine β-methylene which could lower the redox potential for generation of the radical (DeFilippis *et al.*, 1991). The formation of the Tyr-Cys bond is discussed in section 2.3.

Another striking feature in the active site structure is that the indole ring of Trp 290 is parallel to the plane of the Tyr 272 ring and stacked over the Tyr-Cys bond such that the six-membered ring of the Trp is located directly above the Cys sulphur atom. The other face of the Trp indole ring is exposed to solvent and Trp 290 protects the radical site from solvent, probably stabilising the radical and "directing" the chemical reactivity of the radical to achieve catalysis.

The elegant work by Whittaker and coworkers to establish that GOase is isolated as a mixture of inactive and active forms (Whittaker and Whittaker, 1988) and the further spectroscopic studies by this group to establish a tyrosine as the site of the free radical redox cofactor needed to rationalise the 2-electron redox reaction has been reviewed recently (Whittaker, 1994; Knowles and Ito, 1993) and will not be described further. The GOase structure was originally determined for crystals grown at pH 4.5, where the enzyme is inactive, raising the question of whether it is a valid starting point for understanding the structural basis for catalysis. It has been shown, however, that soaking a crystal of GOase at pH 7.0 (where it would be active) in the presence of 50mM $K_3Fe (CN)_6$ as oxidant does not lead to significant changes in active site structure other than replacement of the coordinated acetate by water (Ito et al., 1994) suggesting that the crystal structure is representative of the activated form. EXAFS and EPR studies indicate that there is no major structural change in the active site between GOase at pH 4.5 vs 7.0 and for the activated vs inactive states, whilst NMR dispersion studies provide evidence for an equatorial water coordinated to copper, consistent with that observed in the pH 7.0 crystal structure (Knowles et al., 1995). However, direct demonstration of GOase activity in the crystals grown at pH 4.5 and transferred to pH 7.0 has not been possible since, in the crystals, the active site is very close to a symmetrically-related molecule in the crystal lattice and there is no physical space for substrates such as D-galactose to bind. This has prevented crystallographic studies of substrate turnover in the original (space groups C2) crystals (Ito et al., 1994) though computer graphics modelling of substrate binding satisfactorily explains substrate specificity. Preliminary results (Firbank et al., unpublished observations) on a new crystal form of GOase (space group $P2_12_12_1$) grown at pH 7.0 indicate that diffusion of substrate and other effector molecules into the active site should be possible.

## 2.2. Catalytic Mechanism

A simple, free radical catalytic mechanism for the reductive half cycle of GOase is shown in Figure 5.

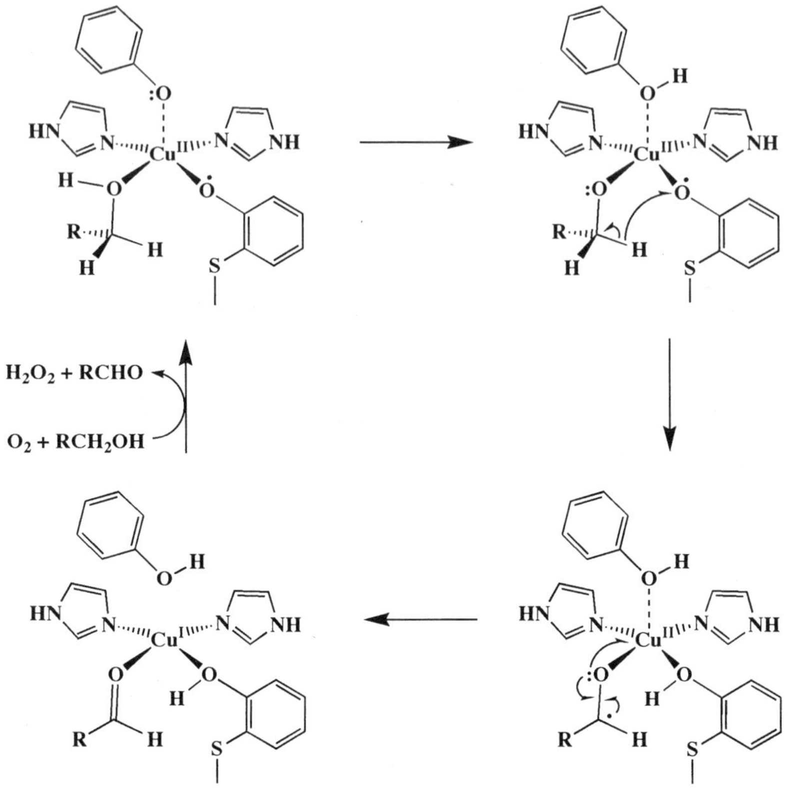

**FIGURE 5.** GOase catalytic mechanism proposed by Whittaker 1994.

The oxidative half cycle has been little studied but it is likely that dioxygen binds to $Cu^+$ and electron transfer occurs from both $Cu^+$ and from Tyr 272 to generate hydrogen peroxide and restore the activated form of the enzyme ($Cu^{2+}$ and the Tyr 272 cation radical). The source of the two hydrogens in the product peroxide is unknown (cfd amine oxidases, section 3.2 where the source of the hydrogens can be assigned with reasonable confidence).

A ping-pong mechanism for GOase was first proposed by Hamilton (Hamilton *et al.*, 1973) but discounted at the time on the grounds that $Cu^{2+}$ (rather than $Cu^+$) was still present in the substrate-reduced enzyme. The probable presence of inactive enzyme in these early studies invalidates this criticism of the ping-pong mechanism as pointed out by Whittaker *et al.* (1998). More recently, the ping-pong mechanism has been questioned on the basis of the results of steady state kinetic deuterium isotope studies for galactose substrate. A kinetic isotope effect originating from galactose in the reaction with oxygen was interpreted (Villafranca *et al.*, 1993) as evidence for a rapid equilibrium random mechanism in which both substrates are bound simultaneously in the active site. Whittaker *et al.* (1998) dispute this interpretation of the isotope effect data and cite Borman *et al.* (1997) who found in their kinetic studies that $k_{ox}$ is independent of the reducing substrate. The weight of evidence favours a ping-pong mechanism though it is difficult to eliminate definitively ternary complex mechanisms (Engel, 1977).

Assuming a ping-pong mechanism, the different steps in the reductive half cycle (Figure 5) can be described as follows.

### 2.2.1. Substrate Binding

Although it has not been established by X-ray crystallography that substrate binds to the copper by displacing the equatorial water, this is consistent with graphics modelling based on the 1.7 Å X-ray structure (Ito *et al.*, 1994). This is consistent with the water proton relaxation data (Knowles *et al.*, 1995) that show that a water proximal to copper can be titrated by the substrate dihydroxyacetone.

### 2.2.2. Substrate Activation

Whittaker and Whittaker (1993) reported that azide binding to the copper in GOase induced uptake of a proton from solution and argued that the first step following substrate binding to the copper would be transfer of the hydroxyl proton of the primary alcohol to the axial ligand, Tyr 495. Support for this proposal has come from studies with the mutational variant Y495F in which the Phe cannot coordinate to the copper and where there is no proton uptake following addition of azide (Reynolds *et al.*, 1997). The proton transfer to Tyr 495 would weaken its axial coordination to the copper, leaving it effectively 4-coordinate.

### 2.2.3. Hydrogen Abstraction from Substrate

It has been known for many years (Maradufu *et al.*, 1971) that GOase stereospecifically abstracts the *pro*-S hydrogen from the methylene group alpha to the substrate hydroxyl, and that there is a substantial kinetic isotope effect ($k_H$ / $k_D$ = 7.7) with galactose deuterated in the methylene at the 6-position. The presence of the free radical on Tyr 272 of activated GOase strongly suggests a homolytic radical hydrogen transfer from substrate to enzyme rather than heterolytic hydride ion transfer (as in alcohol dehydrogenase) or proton transfer followed by two electron transfer steps (as in amine oxidases). A recent study by Whittaker *et al.* (1998) shows that there is no evidence for a $k_{H2O} / k_{D2O}$ solvent isotope effect on the rate limiting step of the reductive half cycle, consistent with homolytic cleavage. This study further shows that reduction by substrate is rate limiting at low galactose concentrations but reoxidation of reduced enzyme becomes rate limiting at high galactose concentrations. The large kinetic isotope effect from the substrate reported first by Maradufu *et al.* (1971) is maintained under conditions where the oxygen concentration become rate limiting. This suggests that the hydrogen atom removed from the substrate alcohol is retained in the active site in the $E_{reduced}$ species and hence participates in the re-oxidation steps. Under some experimental conditions, kinetic isotope effects as high as 22 have been seen (Whittaker *et al.*, 1998) and their strong temperature-dependence is consistent with quantum mechanical tunnelling (Cha *et al.*, 1989) of the hydrogen atom in the rate limiting step.

### 2.2.4.  Formation of $E_{reduced}$

It is not known whether H atom transfer from substrate to Tyr 272 is preceded or followed by single electron transfer from substrate to $Cu^{2+}$. Whittaker *et al.* (1998) favour electron transfer being first since the 4-coordinate geometry of $Cu^{2+}$ following loss of the Tyr 495 ligand would permit fast inner sphere electron transfer. Formation of $E_{reduced}$ would release product aldehyde in a ping-pong mechanism. It has not, however, been demonstrated that product aldehyde is produced when activated GOase is reacted anaerobically with a reducing substrate.

### 2.2.5.  Reoxidation of $E_{reduced}$

It is proposed (Whittaker *et al.*, 1998) that dioxygen binds to the $Cu^+$ in $E_{reduced}$ although there is no direct evidence. Electron and proton transfer steps to the bound oxygen would generate the second product, hydrogen peroxide.

### 2.3.  Biogenesis of the Thio-Ether Bond and Other Processing Events

The identification of a thio-ether bond in the active site of GOase (Ito *et al.*, 1991) immediately encouraged speculation about its biogenesis. Is the formation of this bond autocatalytic as appears to be the case with the TPQ cofactor in amine oxidases (Dooley, 1999) or does it require chaperones as is the case in the biogenesis of the TTQ cofactor (Christoserdov *et al.*, 1991)? Does biogenesis of the thioether bond involve radicals? There are more questions than answers at present but a story is emerging (Baron et al., 1999).

The mature form of galactose oxidase is exported into the medium, with the N-terminal sequence **ASAP** (Figure 2). Sequencing of the cloned GOase gene reveals that the translated polypeptide chain has a 41 amino acid N-terminal leader sequence presumably involved in processing, including secretion events. The secretion signal targets the GOase polypeptide chain to the endoplasmic reticulum where it is cleaved leaving a pro-form with a 17-mer N-terminal extension AVGTGIPEGSLQFLSLR (**ASAP**).

Growth of *Aspergillus* expressing the wild type GOase gene under copper-depleted conditions leads to three distinct protein species in the medium all of which react with antibody against GOase. These species are clearly separated by SDS PAGE. Form I, the slowest moving species, has the N-terminal sequence corresponding to GOase with its pro-sequence intact. Form II and Form III both have the N-terminal sequence corresponding to the mature enzyme but can be identified on the basis of their SDS PAGE mobility; Form II, the intermediate species, has the same mobility as the mutational varient C228G, leading to the suggestion that, like the variant, it lacks the thioether bond. Form III, the fastest moving species, has the same mobility as mature GOase isolated from copper-enriched medium and corresponds to fully processed GOase. The anomolously high mobility is due to the "loop" in the polypeptide chain resulting from the thioether bond (Cys228-Tyr272) causing shortening of the chain and an apparently lower molecular mass in SDS PAGE.

Under rigorous copper-depleted conditions, the dominant GOase species produced is Form I and this has been isolated and purified. Aerobic incubation of Form I with $Cu^{2+}$ leads to its conversion to Form III, the visible spectrum showing this to be the activated form with the thioether bond formed. Incubation of Form I with $Cu^{2+}$ under anaerobic conditions gives a species with the same circular dichroism spectrum as inactive wild type indicating that copper has been incorporated into the protein as $Cu^{2+}$.

## 2.4.  Model Studies

A large number of copper/phenoxide complexes have been prepared as models for GOase. Several early studies were complicated by the formation of phenoxide-bridged dimeric complexes, which affects the solution chemistry of these compounds (Whittaker et al., 1996a; Zurita et al., 1996; Itoh et al., 1997). Latterly, however, several groups have succeeded in cleanly isolating monomeric square pyramidal copper (II)/phenoxide complexes, from which copper(II)/phenoxyl species can be accessed by reversible, one-electron oxidation (Halfen et al., 1997; Zurita et al., 1997;

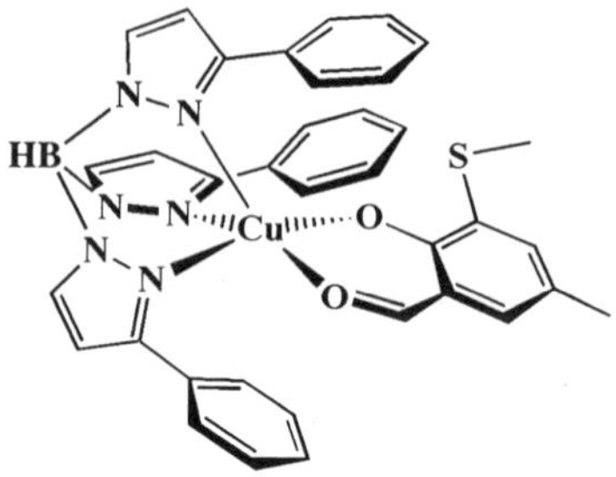

I

**FIGURE 6.** A structural model compound for galactose oxidase that forms a phenoxyl species closely resembling the oxidised form of the enzyme (Halcrow *et al.*, 1998).

Sokolowski *et al.*, 1997; Müller *et al.*, 1998; Halcrow *et al.*, 1999; Jazdzewski *et al.*, 1998). Although none of these oxidised complexes has yet been structurally characterised, their uv / vis and resonance Raman spectra are typical of phenoxyl radicals, while their magnetic properties are consistent with their formulation as biradical species. To date, only one of these compounds (**I**) (Figure 6) bears a thioether substituent at the phenoxide ring; the electronic spectrum of the phenoxyl species [**I**]$^+$ closely resembles that of GOase$_{ox}$ (Halfen *et al.*, 1997). Comparison of the ligand oxidation potentials of these various model compounds has shown that thioether substitution of a phenol ring lowers its oxidation potential by 250–500 mV (Whittaker *et al.*, 1996; Itoh *et al.*, 1997; Halfen *et al.*, 1997; Halcrow *et al.*, 1999). Recently, the first three-co-ordinate copper (I)/phenoxide complex has been prepared, as a structural model for the GOase$_{red}$ state (**I**).

The groups of Stack and Wieghardt have recently reported fully functional aerobic oxidation catalysts based on copper(II)/phenoxide complexes, which oxidise benzylic or allylic alcohols to aldehydes with the concomitant production of hydrogen peroxide. The Stack system employs the complexes **II** (Wang and Stack, 1996; Wang *et al.*, 1998; Figure 7) which are designed to afford stable and accessible phenoxyl centres, and to enforce a tetrahedral distortion at the copper ion so as to stabilise the Cu(I) oxidation state. EXAFS measurements showed that the oxidised phenoxyl species [**II**]$^+$ binds alkoxides to form a five-coordinate Cu(II) centre, which converts stoichiometrically to aldehydes and a Cu(I) species [**II**]$^-$ or [**IIH**] (Wang *et al.*, 1998). Under an O$_2$ atmosphere this reduced compound is reoxidised to [**II**]$^+$, thus completing the catalytic cycle. A kinetic isotope effect of 5.3 has been measured for this system using PhCD$_2$OH as substrate, which is consistent with a rate-limiting C—H cleavage step. In addition, a Hammet study using various *para*-substitute benzyl alcohol substrates afforded r = –0.14, typical of a radical-based mechanism. Hence, it

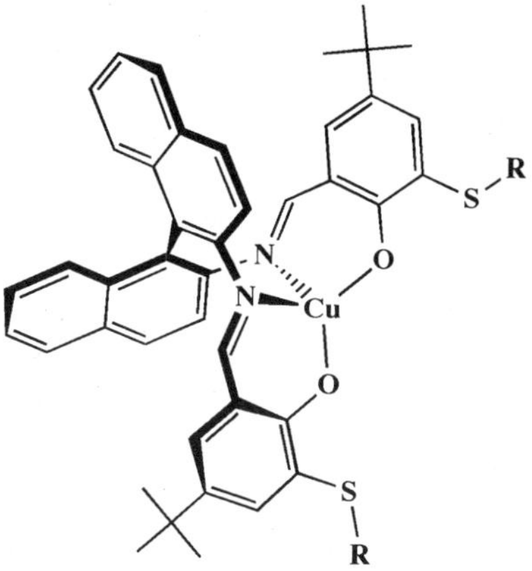

**FIGURE 7.** A functional model for galactose oxidase (Wang *et al.*, 1998).

**II**

was concluded that this catalyst operates by the same mechanism currently favoured for GOase catalysis (Wang *et al.*, 1998).

Wieghardt's group have described two generations of their catalytic system, both of which will oxidise some saturated as well as β-unsaturated alcohols. The first of these is an isolable dinuclear *bis*-phenoxyl complex **[III]**$^{2+}$, (Figure 8) which retains its structure in solution and is the active oxidation catalyst (Chaudhuri *et al.*, 1998). The second generation catalyst **IV**, (Figure 9), however, forms a mononuclear phenoxyl complex upon oxida-

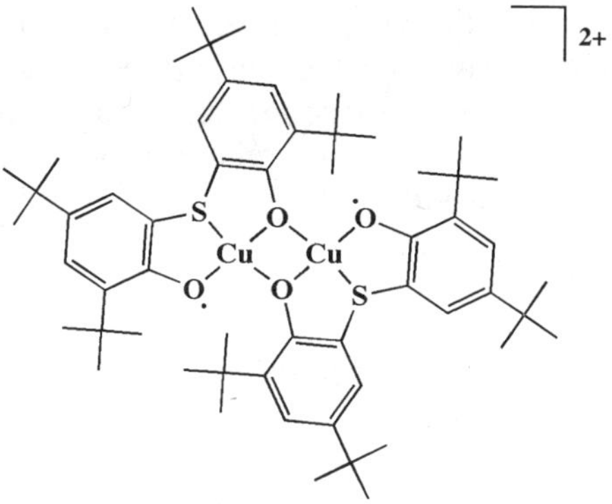

**FIGURE 8.** A functional model for galactose oxidase based on a di-copper/di-phenoxy species (Chaudhuri *et al.*, 1998).

**[III]**$^{2+}$

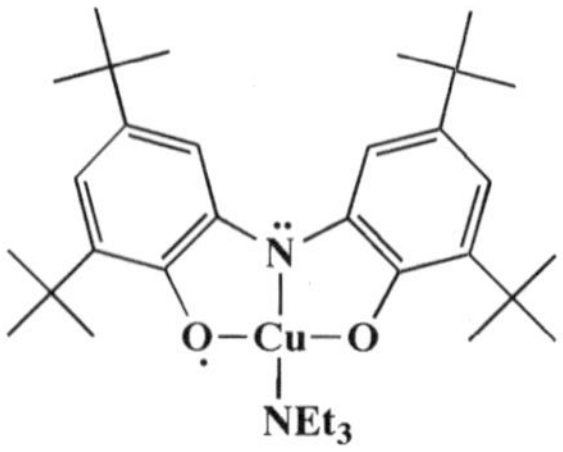

**FIGURE 9.** A functional model for galactose oxidase
(Chaudhuri *et al.*, 1999).

**IV**

tion (Chaudhuri *et al.*, 1999). Kinetic isotope effects of 8–10 were measured for both catalysts using $\alpha$-dideutero alcohols, suggesting that a similar mechanism operates here as for **II** and GOase. However, since the active catalyst $[\textbf{III}]^{2+}$ is a diphenoxyl, both electrons from the substrate can be accepted by ligand radical centres and for this compound, the Cu(I) state is not involved in catalysis.

## 2.5.  Biological Role of Galactose Oxidase

A laboratory strain of the fungus *Dactylium dendroides* has been used by most investigators of galactose oxidase. In the wild, *D dendroides* grows on rotting trees and this lead to the reasonable proposal that the physiological role of galactose oxidase, an extracellular enzyme, is to catalyse production of the hydrogen peroxide needed by the enzyme ligninase, a peroxidase involved in breakdown of the polyphenylpropanoid complex abundant in wood and plant tissue. Ligninase is secreted by other wood rotting fungi and there would be mutual benefit to the mixed population of fungi in deriving nutrients from the rotting wood. The reclassification of *D dendroides* as a *Fusarium* (Ogel *et al.*, 1994) casts some doubt on this proposal since *Fusaria* are fungal pathogens associated with cereal plants. An enzyme structurally related to galactose oxidase, glyoxal oxidase, secreted by the wood rot fungus *Phanerochetes chrysosporium* may well have the physiological role of catalysing hydrogen peroxide production for lignin degradation (Whittaker *et al.*, 1996b) since *P chrysosporium* also secretes ligninases.

What role could galactose oxidase play in *Fusarium* pathogenicity to cereal plants? It could act in concert with other fungal pathogens to help degrade the plant cell wall in a similar way to that proposed above or it could be acting more subtly by subverting the cellular signalling processes in the plant which involve hydrogen peroxide (Antonelli *et al.*, 1988). A possible role for galactose oxidase in cellular signalling is intriguing. The strict stereochemical specificity for substrates (e.g., D-galactose but not D-glucose) suggests a specific role for galactose oxidase as does the ability of galactose oxidase to oxidise the terminal galactose of oligosaccharides including ones found on the surface of macrophages (Knowles and Ito, 1993).

At present, galactose oxidase has only been found in specific classes of fungi. The completion of sequencing the genomes of *Arabidopsis*, of nematodes and in the relatively near future of humans will allow sequence data bases to be searched for a wider distribution of galactose oxidase in Nature.

## 3. AMINE OXIDASES

Amine oxidases catalyse the oxidative deamination of mono-, di- and polyamines according to the following general reactions which involve attack at either primary amine sites (1) or secondary amine sites (2)

Mono- and diamines

$$RCH_2NH_2 + O_2 + H_2O \rightarrow RCHO + H_2O_2 \tag{1}$$

Polyamines

$$RCH_2NHCH_2R' + O_2 + H_2O \rightarrow RCH_2NH_2 + R'CHO + H_2O_2 \tag{2}$$

There is a broad division of amine oxidases into two classes, flavin-containing and copper-containing. The flavin class (EC 1.4.3.4) is commonly associated with the peroxisomes and mitochondria in cells and can attack substrates at both primary and secondary amine sites, (1) and (2) above.

The copper-containing class (EC 1.4.3.6), which is the subject of the present review, can attack mono-, di- and polyamines (such as spermidine, $H_2N-(CH_2)_3NH(CH_2)_4NH_2$) but the latter only at the primary amine site. Discovery of amine oxidases in 1938 (Zeller, 1938) was followed by three decades laying the foundations for understanding the molecular basis of catalysis (Buffoni and Blaschko, 1964; Yamada *et al.*, 1963; Mondovi *et al.*, 1967; Oi *et al.*, 1970) and the pharmacology (Tabor, 1954; Buffoni, 1966). In terms of the biochemistry, this period saw the purification of amine oxidases (Buffoni and Blaschko, 1964), the identification of copper as a cofactor (Yamada *et al.*, 1963), the characterisation of the copper site by EPR (Mondovi *et al.*, 1967) and steady state kinetic studies indicating a ping-pong mechanism (Oi *et al.*, 1970). Significant advances in understanding the

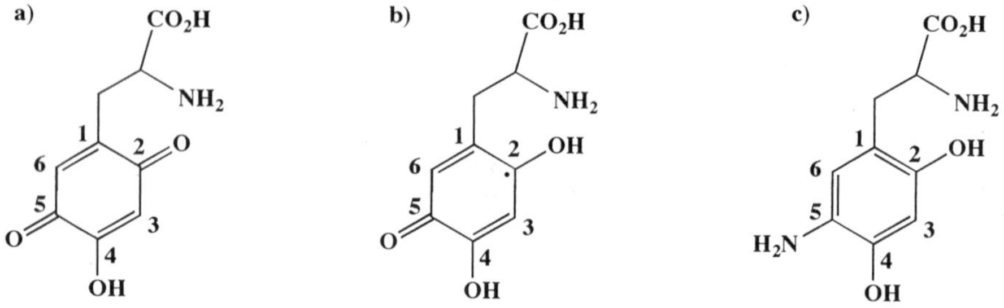

**FIGURE 10.** Structure of (a) trihydroxyphenylalanine quinone (TPQ); (b) TPQ semi-quinone; (c) aminoquinol form of TPQ present in the substrate-reduced form of amine oxidases.

catalytic mechanism and the chemistry of the copper site were made during the 70s and 80s (see Knowles and Yadav, 1984; Pettersson, 1985; Knowles *et al.*, 1983; Klinman and Mu, 1994 for reviews) but it was not until 1990 (Janes *et al.*, 1990) that the organic cofactor present in most amine oxidases was finally identified as the quinone derived from 2, 4, 5 trihydroxyphenylalanine, TPQ (Figure 10).

Copper-containing amine oxidases are widely distributed in Nature and have been identified in bacteria, yeasts and filamentous fungi, plants and animals though their precise biological roles are unclear. In prokaryotic and lower eukaryotic organisms, their role appears to be for utilising various amine substrates as nitrogen and carbon sources. In higher eukaryotic organisms, the situation is much less clear; thus in animals, amine oxidases are implicated in development and detoxification, whilst in plants they play roles in development, wound and resistance responses and in secondary metabolism (see section 3.5).

A recent review on copper-containing amine oxidases covered medical and biological aspects, as well as studies leading to an understanding of the catalytic mechanism (McIntire and Hartmann, 1993), whilst another (Knowles and Dooley, 1994) emphasised the molecular basis for catalysis and attempted to identify future challenges. Since these reviews were written, there has been an explosion in our understanding of the molecular basis of amine oxidase catalysis initiated by the first report of an atomic resolution structure in 1995 (Parsons *et al.*, 1995). The present review aims to cover this recent structural work in the context of key contributions from the earlier literature. The Cu containing protein Lysine 6-oxidase (EC 1.4.3.13), "lysyl oxidase", is a separate enzyme class and is not the main focus of this review.

## 3.1.  Structure

### 3.1.1.  General Structural Properties

Amine oxidases are homodimers with subunit molecular masses generally in the range 70 kDa to 95 kDa with one copper ion and one organic cofactor per subunit. Eukaryotic amine oxidases are glycoproteins with carbohydrate contents estimated at 3–10%. Further information on the protein chemistry of amine oxidase can be found in Knowles & Yadav (1984) and Pettersson (1985).

Lysyl oxidase, on the other hand, appears to be a monomer of molecular mass 32 kDa (Kagan and Trackman, 1992).

### 3.1.2.  Protein Structure

Primary sequence alignments for a number of amine oxidases based on cDNA sequence data are shown in Table 1.

There are 33 totally conserved residues representing only 5% of the aligned residues. The conserved residues include the TXXNYD/E motif at the active site which first identified (Janes et al., 1990) tyrosine as the precursor of the TPQ cofactor (see section 3.3). It is interesting that the sequence for human kidney diamine oxidase (Mu et al., 1992) matched that in the sequence data base for a human amiloride-binding protein believed to function as a sodium channel (Mu et al., 1994). The binding protein has now been classified as diamine oxidase. Further discussion of amine oxidases having possible dual biological functions will be resumed in section 3.5.

The first crystal of an amine oxidase giving good X-ray diffraction patterns was reported in 1993 for the pea seedling enzyme (Vignevich et al., 1993). Historically it may be noted that micro-crystals of pig plasma amine oxidase had been obtained by Buffoni and Blaschko 30 years previously (Buffoni and Blaschko, 1964) but no diffracting crystals of a mammalian amine oxidase have yet been reported, probably due to heterogeneity and disorder in the carbohydrate component which accounts for 10% of the molecular weight. This suggested that structural studies of a bacterial amine oxidase, which would not be glycosylated, would be profitable. *E coli* amine oxidase was shown to have copper and TPQ as cofactors (Cooper et al., 1992), the primary sequence (Azakami et al., 1994) and preliminary X-ray crystallographic data were reported in 1994, and the full structure to 2.0 Å resolution reported in 1995 (Parsons et al., 1995). There have subsequently been high resolution structures reported for amine oxidases from pea

Malcolm Halcrow *et al.*

## Table 1

Amino acid sequence alignments of copper oxidases. The totally conserved residues are shown in *inverse highlighting* as are the three positions corresponding to acidic positions (either aspartate or glutamate). Positions at which a residue occurs in at least 60% of the sequences are shown in *boldface*. Abbreviations: PSAO, pea seedling (Tipping and McPherson, 1995); LSAO, lentil seedling (Tipping and McPherson, 1995); ACAO, *E. coli* (Azakami *et al.*, 1992); KPAO, *Klebsiella aerogenes* (Sugino *et al.*, 1992); AGAO, *Arthrobacter globiformis* (Tanizawa *et al.*, 1994); ARAO, *Arthrobacter strain P1* (Zhang *et al.*, 1993); HPAO, *Hansenula polymorpha* (Bruinenberg *et al.*, 1989); BSAO, bovine serum (Mu *et al.*, 1992); ABPHK, human kidney (Barbry *et al.*, 1990); ABPRAT, rat kidney (Lingueglia *et al.*, 1993). From Tipping and McPherson, 1995 with permission

```
PSAO     MRLALFSVLTLLSFHAVVS----------------------------------------------------------
LSAO     KFALFSVLTLLSFHAVFS-----------------------------------------------------------
ECAO     MGSPSLYSARKYTLALVALSFAWQAPVFAHGGEAHMVPMDKTLKEFGADVQWDDYAQLFTLIKDGAYVVKPGAQTAIVNGQPLALQVPVVMKDNKAWV
KPAO     MANGLKFSPRKTALALAVAVVCAWQSPVFAHGSEAHMVPLDKTLQEFGADVQWDDYAQMFTLIKDGAYVVKVKPGAKTAIVNGKSLDLPVPVVMKEGKAWV
AGAO
ARAO
HPAO
BSAO     MFIFIFLSLMTLLVMG-------REEGGVGSEEGVGKQCHPSLPPRCPSRSPSDQPWTHPDQSQLFADLSREELTTVMSFLTQQLGPDLVDAAQARPSDNC
ABPHK                            MPALGWAVVAAILMLQTAMAEPSGTLPRKAGVFSDLSNQELKAVHSFLWSKKELRLQPSSTTMAKNT
ABPRAT                           MCLAFGWAAVILVLQTVDTASAVRTPYDKARVFADLSPQEIKAVHSFLMNREELGLQPSKEPTLAKNS

PSAO     -------------VTPLHVQHPLDPLTKEEFLAVQTTVQNKYPISNNRLAPHYIGLDDPEKDHV-LR--YETHPTL--VSIPRK--IFVVAIINSQT
LSAO     -------------FTPLHTQHPLDPITKEEFLAVQTTVQNKYPISNNRLAPHYIGVDDPEKDLV-LK--YETSPTL--ISIPRK--IFVVAIINSQT
ECAO     SDTFINGVFQSGLDQTFQVEKRPHPANALTADEIKQAVEIVKA-SADFKPWTRFTEISLLPPDKEAV--WAFAPENKP-VDQPRKADVIMLD--GKHI
KPAO     SDTFINDVFQSGLDQTFQVEKRPHPANSLSAAEISKAVTIVKA-APEFQPNVRFTEISLLHEPDKAV--WAFALQSTP--VDAPRTADVVMLD--GKHV
AGAO     MTPSTIQTASPFRLASAGEISEVQGILRT-ALGLGLRLT-ALGLVLDPARGA---------------GS--EAEDRRFRVFIHDVSGARP
ARAO     MTLNAESRALVGVSHPLDPLSRVEIARAVAILKE-GPAAAESFRFISVELREPSKDDL-------RAG--VAVAREADAVLVDRAQARS
HPAO     MERLRQIASQATAASAAPARPAHPADPLSTAEIKAATNTVKS-Y-FAGKKISFNTVTLREPARKAY-IQWK--EQGG--PLPRLAYYVILEAGKPGV
BSAO     VFSVELQLPPKAAALAHLDRGSPPAREALAIVFFGGQPQPNVTLVVGPLPQPSMRDVTVERHGGPLPYYRPVLLREYLDIDQMIFNRELPQAAGVL
ABPHK    VFLIEMLLPKKYHVLRFLDKGERHPVREARAVIFFGGQEHPNVTEFAVGPLPGPCYMRALSP-RPGYQSSWASRPISTAEY----ALLYH-TLQEATKPL
ABPRAT   VFLIEMLLPKKKHVLKFLDESGRKGPNRERARAVIFFGAQDYPNVTEFAVGPLPRPYYIRALSP-RPGHHLSWSSRPISTAEY----DLLYH-TLKRATMPL
```

```
PSAO    HEILINLRIRSIVS-DNIHNGYGFPILSVDEQSLAIKLPLKYPP-FIDS--VKKRGL-NLSEIVCSSFTMGWFGEE----KNVRTVRLDCFMKEST-VNI
LSAO    HEILIDLTIKSIVS-DNIHNGYGFPVLSAABQFLAIDLPLKYPP-FIAS--VNKRGL-NISEIVCSSFTMGWFGEE----KNSRTVRVDCFMKEST-VNI
ECAO    IEAVVDLQNKLLSWQPIKDA---HGMVLLDDFASVQNIINNSEEFAA--AVKKRGITDAKKVITTPLTVGYPDGKDLKQDARLLKIVISYL-DVGDGNY
KPAO    IRAVVDLQNKKILSWTPIKGA--HGMVLLDDFVSVQNIINYSSEFAE--VLKKHGITDRGKVVTTPLTVGFPDGKDGLQQDARLLKVVSYL-DTGDGNY
AGAO    QBVTVSVTNGTVISAVELDTAATGELPVLEBEFEVVEQLLATDERWLKA--LAARNL-DVSKVRVAPLSAGVFEYAEE--RGRRILRGLAFVQDFPEDSA
ARAO    FRAVVDLEAGTVDSWKLLAENI--QPPFMLDEFAEECDACRKDPEVIAA--LAKRGLTNLDLVCFEPWSVGYFG-EDN--EGRRLMRALVFVRDEADDSP
HPAO    KBGLVDLASLSVIETRALE---TVQPILTVEDLCSTEEVIRNDPAVIEQCVLSGIPANEMHKVYCDPWTIG--YDERWGTCGKRLQQALVYRSDEDDSQ
BSAO    HH------CCSYKQGGQKLLTM-NSAPRGVOSGDRSTWFGIYNITKGGPYNPVGLELLVDHKALD-------PADWTVQKVFFQGRYYENLAQLEEQ
ABPHK   HQFFLNTTGFSFQDCHDRCLAFTDVAPRGVASGQRRSWLII--QRYVEGYFLHPTGLELLVDHGSTD------AGHWAVEQVWYNGKFYGSPEELARK
ABPRAT  HQFFLDTTGFSFLGCDDRCLTFTDVAPRGVASGQRRSWFIV--QRYVEGYFLHPTGLEILLDHGSTD------VQDMRVEQLWYNGKFYNNPEELARK

PSAO    YVRPITGITIVADLDLMKIVEYHDRDIEAVPTAE-NTEYQVSKQ---SPPFGPKQHSLTSHQPQGPGFQING-HSVSWANWKFHIGFDVRAGIVISLASI
LSAO    YVRPITGITIVADLDLMKIVEYHDRDTEAVPTAE-NTEYQVSKQ---SPPFGPKQHSLTSHQPQGPGFQING-TSVSWANWKFHIGFDVRAGIVISLASI
ECAO    WAHPIENLVAVVDLEQKKIVKIEBGPVVPVMTA------RPFGRDRVAPAVKPMQIIEPBGKNYTTIG-DMIHWRNWMDFHLSMNSRVGPMIS-TVT
KPAO    WAHPIENLVAVVDLEAKKIIKIEBGPVIPVMEP------RPYDGRDRNAPAVKPLEITEPBGKNYTTIG-DTIHWQNWDFHLRNSRVGPILS-TVT
AGAO    WAHPVDGLVAYVPIVSKEVTRVIDTGVPFVPAEHGN--YTDPEL-TGPLRTVQKPISITQPEGRSPTTVTGNHIEWEKWSLDVGFDVREGVVLH-NIA
ARAO    YAHPIENFIVFYDLNAGKVVRLEDDQAIPVPSARGN--YL-PKY--VGEARTDLKPLNITQPEGASPTVTG-NHVTWADWSFRVGFTPREGLVLH-QLK
HPAO    YSHPLD-FCPIVDTEEKKVIFIDIPNRRRKVSKHKHANFYPKHMIEKVGAMRPEAPPINVTQPEGVSFKMTG-NRVASSLWTPSFGLGAFSGPRV-FDVR
BSAO    FEAGQVNVVIPD---DGTGGFWSLKSQVPPGPTPLQPH----------------PQGPRPSVQG-NRVACRLRLELCLPVRSSSGLQV-LNVH
ABPHK   YALGEVMVVLEDRCLGARGMTAQRSRPSSPQAPR-DFPQPHPCERP-----------PLGPAPRPSLQAG-ORCALRRLELCLPVRSSSGLQV-LNVH
ABPRAT  YAVGEVDTVVLEDPLPNG------TEKPLFSSYKPRGEFHTPVNVAGP-----------HVVQPSGPRYKLEG-NTVLYGGWCSFSYRLRSSSGLQI-FNVL
```

```
PSAO    YDLEHKKSRRVLFKGYISELFVYQDPTEEFYFKTFFDSGFGFGLSTVSLIPNRDCPPHAQFIDT-YVHSANGTPILLKNAICVFEQYGNI-MWRHTEN
LSAO    YDLEHKKSRRVLFKGYISELFVYQDPTEEFYFKTFFDSGFGFGLSTVSLIPNRDCPPHAQFIDT-YIHSADGTPIFLENAICVFEQYGNI-MWRHTET
ECAO    YN-DNGTKRKVMTEGSLGGMIVFGDPDIGWYFKAYLDSGDYGWGLTSPIARGKDAPSNAVLLNE-TIADYTGVPMEIPRDIAVFERYAGP-EYHLQEM
KPAO    YN-DNGTKRQVMEGSLGGMIVFGDPDVGWVFKAYLDSGDYGWGLTSPIVRGKDAPSNAVLLDE-VISADYTDGKPTTIPGAVAIFERYAGP-EYHLEM
AGAO    FR-DGDRLRFINRASIAEMVVFGDPSPIRSGWNYFDTGYLVGQYANSLELGCLGDITYLSP-VISDAFPNPREIRNGICMHEEDWGI-LARHSDL
ARAO    FK-DQGVDRPVINRASLSEMVVFGDTAPVQAKKNAPGBEYNIGNMANSLTLGCDCLGEIKYFDG-HSVDSHGNPWTIENAICMHEEDDSI-LWKHPDF
HPAO    YN-DHGNVRPIFHRISLSEMIVFGSPEPPHQRKHAIDICEYGAGVMTNPLSLGCDCPYLATYMD-WHFVVDSSQFKTLHDAFCVFEQNKGLPLRRH-HS
BSAO    FQGE------RLAYEISLQEAGAVVGGNTPAAMLFRYDSG-FGMGYFATPLIRYG-WGLGSVTHELAPGIDCPETATFLDTFHYY-DADDPVHYPRALCLFEMPTGVPLRRHFNS
ABPHK   FGGE------RIAYEVSVQEAVALVGGHTPAGMQTKYIDVG-WGLGSVTHELAPGIDCPETATFLDTFHYY-DADDPVHYPRALCLFEMPTGVPLRRHFNS
ABPRAT  FGGE------RVAYEVSVQEAVALVGGHTPAGMQTKYIDVG-WGLGSVTHELAPGIDCPETATFLDAFHYY-DSDGPVHYPHALCLFEMPTGVPLRRHFNS
```

## Table 1 (*Continued*)

```
PSAO    GIPN-ESIEESRTEVNLIVRTIVTVGNVDNVIDWEFKASGSIKPSIALSGLIEIKGTNIK---HKDEIKEDLHGKLVSANSIGIYHQH-FYIYYLDFDID
LSAO    GIPN-ESIEESRTEVDLAIRTVVTVGHNVLDWEFKTSGVMKPSIALSGILEIKGTNIK---HKDEIKEEIHGKLVSANSIGIYHH-FYIYYLDFDID
ECAO    G-Q----PNVSTERRELVVRWISTVGHDYIFDWIFHENGGTIGIDAGATGIEAVKGVKAKTMHDETAKDDTRYGTLIDHNIVGTHQH-IYNFRLDLDVD
KPAO    G-K----PNVSTERRELVVRWISTVGNDYIFDWHDNGGTIGIDAGATGIEAVKGVLAKTMHDPSAKEDTRYGTLIDHNIVGTHQH-IYNFRLDLDVD
AGAO    W-S----GINYTRRNRRMVISFFTIGSGFYWYLYLDGTIEFEAKATGV------VPTSAFPEGGSDNISQL--APGLGAPFHQH-IPSARLDMAID
ARAO    R-E--GTAETRRSRKLVISFIAVANTEYAFYWMHLFLDGSIEFLVKAVGI------LSTAGQLPGEKNPYGQSLNNDGLYAPIHQH-MFNVRMDFELD
HPAO    R-DN-FATSLVTRATKLVVSQIFTANNEYCLYWFPMODGHIRLDIRLTGI------LNTYILGDDEEAGPWGTRVYPNVNAHNQH-LFSLRIDPRID
BSAO    DFLS--HYFGGVAQTVLVFRSVSVI------GIEVKLHATGVISSA------FLFGAA-RRYGNQVGEHTLGPVHHS-AHYKVDLDVG
ABPHK   NFKGGPNFYAGLKGQVLVLRTTEIVNDYIWDFIPYPNGVMEAKMHATGVVHAT-------FYTPEGCARHSP--AHPDWCHIHSLVHYRVDLDVA
ABPRAT  NFKGGPNFYAGLKGVVLVLRTTSVVNDYIWDFIPYSNGVMEAKMHATGVVHAT-------FYTPEG-LRHGTRLQTHLIGNHITH-LVHYRVDVDVA

PSAO    GTHNSFEKTSLKTVRIKDGSSKRKSWTTETQTAKTESDAKITIGLAPAEL---VVVNPNIKTAV-GNEVGVRLIP------AIPAHPLLTEDDYPQIRG
LSAO    GTVQSFEKTSLKTVRIVDGGSKRKSYWTTETQTAKTESDAKITIGLAPAEL---VVVNPNIKTAV-GNEVGVRLIP------AIPAHPLLTEDDYPQIRG
ECAO    GENNSLVAMDPVVKPNTAGGP---RTSTMQVNQYNIGNEDAAQKFDPGTIRL--LSNPNKENRM-GNPVGYQIIPYAGGTHPVAAVAQFAPDEMIYHRL
KPAO    GENNTLVAMDPEVKPNTAGGP---RTSTMQVNQYTIDSEDKAAQKFDPGTIRL--LSNTSKENRM-GNPVGYQIIPYAGSTHPAATGAKFAPDEMIYHRL
AGAO    GFTNRVEEEDVV---RQTMGPGNERGNAFSRKRTVLTRESEAVREADARTGKTWIISNPESKNRLNE-PVCYKLHAHNQPTILLADPGS------SIARRA
ARAO    GVNNAVVEVDME---YPEHNPT---GTAPMAVDRILETEQKAIRKTNEAKHRFWKIANHESKNLVNE-FVAYRLIPTNGIQLLAARDDA------YVSKRA
HPAO    GQGNSAAACDAKSSPYPLGSPENMYGNAFYSEKTTPKTVKDSLTNVESATGRSWDIFWPNKVNPYSGKPPSYKLVSTQCPPLLAKEGS------LVAKRA
BSAO    GLENWWAEDMAFVPTAIPWSPEHQIQRLQVTRKQLETEEQAAPFLGGASPRYILYLASKQSNKW--GHPRGYRIQTVSFAGGPMPQNSPMERAFSW---
ABPHK   GTMNSFQTLQMKLENITNPWSPRHRVVQPTLEQTQYSWERQAAFRFKRRLPKYILLFTSPQBNFW--GHKRSYRLQIHSMADQVLPGWQEEQAITWAR--
ABPRAT  GTMNSFQTLTMKLENLTNPWSPSHSLVQPTLEQTQYSQEHQAAFRFGQTLPKYILFSSPQKNCW--GHRRSYRLQIHSMAEQVLPPGWQEERAVTWAR--

PSAO    AFT-NYNWWTAYNRTEKWAGGLIVDHSRGD--DTLAVWTKQNRE-IVNKDIVMWHVVGIIHIVPAQEDFP--IMPLLSTSFEPRPINFFERNPVLKTLSP
LSAO    AFT-NYNWWTPYNRTEKWAGGLIVDHSRGD--DTLAVWTKKNRE-IVNKDIVMWHVVGIIHIVPAQEDF---IMPLLSTSFEPRPINFFERNPVLKTLPP
ECAO    SFM-DKQLWVTRXHPGERFPGRYPNRSTHDT--GLGQY-SKDNESLDWTDAVVWMTTGTHVARABEAP--IMPTEWVHTIEKPNEFDETPTLGALKK
KPAO    SFM-DKQLWVTRYHPTERYPBGKYPNRSAHDT--GLGQY-AKDDESLTNHDDVVWITTGTHVARABEAP--IMPTEWALALEKPNAFDEETPTLGEKKK
AGAO    APA-TKDLWVTRXYADDERYPGDFVNQHSSGA--GLPSYIAQDRD-IVDTHDVVWHTTGLTHPRVVEDAP--IMPVDTVGRKLPGCFEDRSPVLDVPAN
ARAO    QFA-RNNLWVTAYDRTERFAAGEYPNQATGAD-DGLHIWTQKDRN-IVDTDLVVWYTTGHMHVVRLEDAP--VMPRQNIGFWBLPIGENQNFIANLPTS
HPAO    PWA-SHSVNVVPYKDNRLYPSGDHVPQWSGDGVRGMREWIGDGSENIDWDILFFHTFGITHPPAEDFP--LMPAEPITLMERPRHFFTNRPGLHDIQPS
BSAO    ---GRYQLAITQRKETEPSSSSVPNQNDPWTPVTVDFSDFI-NN-ETTAGKDLVAWVTGFLHIPHAEDIPNTVTGNGVGFFEPNFFDQPPSMDSADS
ABPHK   ----YPLAVTKYRESELCSSSIIYHQNDPWDPPVVFEQFL-HNNENIENPDLVAWVTGFLHIPHSEIPNTATPGNSVGFLRPEPFFPEDPSPPWHPE
ABPRAT  ----YPLAVTKYRESERYSSSLXNQNDPWDPPVVFEEFL-RNNENIEDEDLVAWVTGFLHIPHSEDVPNTATPGNSVGFLRPEPFFPEDPSLASRDT

PSAO    RDVAWPGCSN
LSAO    RDFTWPGCSN
ECAO    DK
KPAO
AGAO    PSQSGSHCHG
ARAO    TSTTQTGBADTCCHTDK
HPAO    YAMTTSEAKRAVHKETKDKTSRLAFEGSCCGK
BSAO    IYFREGQDAGSCEINPLACLPQAATCAPDLPVFSHGGYPEY
ABPHK   TL
ABPRAT  VTVWPQDKGLNRVQRWIPEDRRCLVSPPFSYNGTYKPV
```

## Table 2
### Equivalent active site residues of CuAOs;* copper ligands

| E. coli (Parsons et al., 1995) | P. sativum (Kumar et al., 1996) | A. globiformis (Wilce et al., 1997) | H. polymorpha (Li et al., 1998) |
|---|---|---|---|
| Val 367 | 284 | 382 | 303 |
| Tyr369 | 286 | 384 | 305 |
| Asp383 | 300 | 298 | 319 |
| Thr462 | 383 | 378 | 401 |
| Asn465 | 386 | 381 | 404 |
| TPQ466 | 387 | 382 | 405 |
| Asp467 | 388 | 383 | 406 |
| His524* | 442 | 431 | 456 |
| His526* | 444 | 433 | 458 |
| His689* | 603 | 592 | 624 |
| Met699 | 613 | 602 | 634 |

seedling (Kumar et al., 1996), Arthrobacter globiformis (Wilce et al., 1997) and Hansenula polymorpha (Li et al., 1998). The main structural features are present in all amine oxidases and will be presented in terms of the E coli amine oxidase structure with cross reference to the other structures where appropriate. The equivalent active site residues of amine oxidases from different sources whose structures have been determined are shown in Table 2.

The E coli amine oxidase dimer is mushroom shaped with the "stalk" comprising the first 85 amino acids of each polypeptide chain and the "cap" the remaining 640 residues (Figure 11a).

The N-terminal stalk domain(D1) is not present in all amine oxidases and consists of a five-stranded antiparallel β-sheet twisted round an α-helix. The β-sheets from the two stalk domains face towards the molecular diad axis but are not tightly packed together. The biological role of D1 is unknown.

The cap of the ECAO dimer is roughly rectangular and the bulk of the molecule, comprising the C-terminal 440 amino acids (domain D4), folds into an extensive β sandwich which contains the active site and mediates intersubunit interactions. It can be seen in Figure 11a that each subunit has a pair of small domains (D2 and D3) on the surface. These domains are closely similar to each other in both primary and tertiary structures, implying a gene duplication event. Amino acid sequence comparisons of amine oxidases from different organisms (see Table 1) shows that the domains have been conserved over a wide span of evolutionary time suggesting an important biological role at present unidentified.

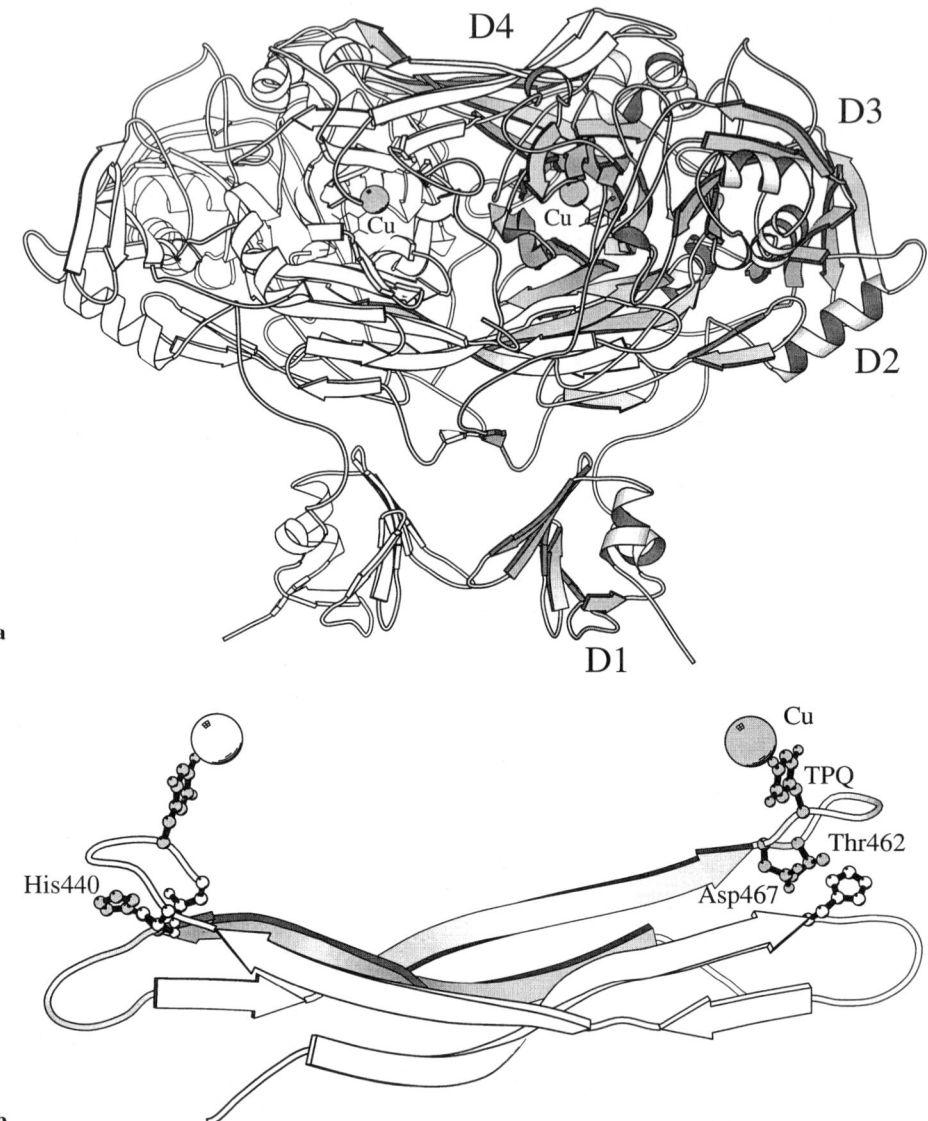

**FIGURE 11.** (a) Ribbon diagram of the overall structure of *E coli* Amine Oxidase with one subunit shaded and labelled. Domain D1 forms the stalk of the mushroom-shaped dimer, with D2 and D3 lying on the periphery and the D4 domains from both subunits forming the core. The copper is shown as a shaded sphere and TPQ is shown as stick bonds in the conformation found in the original structure determination (Parsons *et al.*, 1995). (b) Lower intersubunit hairpin shown in similar orientation to (a). Side chains are shown for residues interacting at the tips of the loops, with labels in the shaded subunit only. Thr462, TPQ and Asp467 all belong to the TXXNYD motif. Figure prepared using the program MOLSCRIPT (Kraulis, 1991).

The core D4 domain in each subunit is made up of a pair of extensive twisted β sheets that contribute to the inter-subunit interface. The area of interaction between the subunits is very large; burying some 7250 Ångstrom$^2$ of each subunit in the dimer interface.

In addition, there are two pairs of β hairpin arms which connect the subunits. One arm from each subunit lies across the upper surface of the β sandwich domain whilst the other arm from each subunit lies towards the bottom of the cap and penetrates deeply into the other subunit providing what appears to be a link between the active sites (Figure 11b).

Two fully conserved residues, His440 and Thr462 (from the TXXXNYD motif) lie at opposite ends of each arm such that Thr462 is close to the active site while His440 is close to the active site in the other subunit. Asp467, a conserved residue immediately following the active site TPQ (derived from Tyr 466) hydrogen bonds to both the adjacent Thr462 and His440 of the other subunit. It is probable that the arm linking the active sites via the hydrogen bonds described above has functional significance, perhaps for cooperativity in catalysis. This has been a controversial aspect of amine oxidases (Janes and Klinman, 1991; DiBiase *et al.*, 1996) and may be resolved when the properties of mutational variants in the active site consensus sequence are fully investigated (Schwartz *et al.*, 1998).

### 3.1.3.  Active Site Structure

**3.1.3.1.  The Copper Site.**  In a crystal form of ECAO shown to contain catalytically-active protein (Parsons *et al.*, 1995), the copper is penta-coordinated in approximate square pyramidal configuration by four basal (equatorial) ligands (His 524, His 526, His 689 and a water [We]) and an apical (axial) water (Wa). The presence of equatorial and axial waters had been first reported by Barker *et al.* (1979) from EPR, water proton relaxation and kinetic studies on pig plasma amine oxidase and the prediction of histidines and waters as the copper ligands came from EXAFS studies by Scott and Dooley (1985). The equatorial water (We) is labile and not always present. In the HPAO structure (Li *et al.*, 1998) it is present in some, but not all, of the six independent subunits in the same crystal. A comprehensive discussion of the spectroscopic properties of the copper site in amine oxidases, including the exchange rates for the equatorial and axial waters, is given in the review by Knowles and Dooley (1994).

**3.1.3.2.  The TPQ Site.**  The TPQ cofactor is not coordinated to the copper in the active, resting enzyme but is located close by. In the ECAO structure (Parsons *et al.*, 1995), the overall position of the TPQ is clear from the electron density but there is evidence for high mobility and the orientation and conformation of the ring is necessarily approximate. In fact there

**Malcolm Halcrow** *et al.*

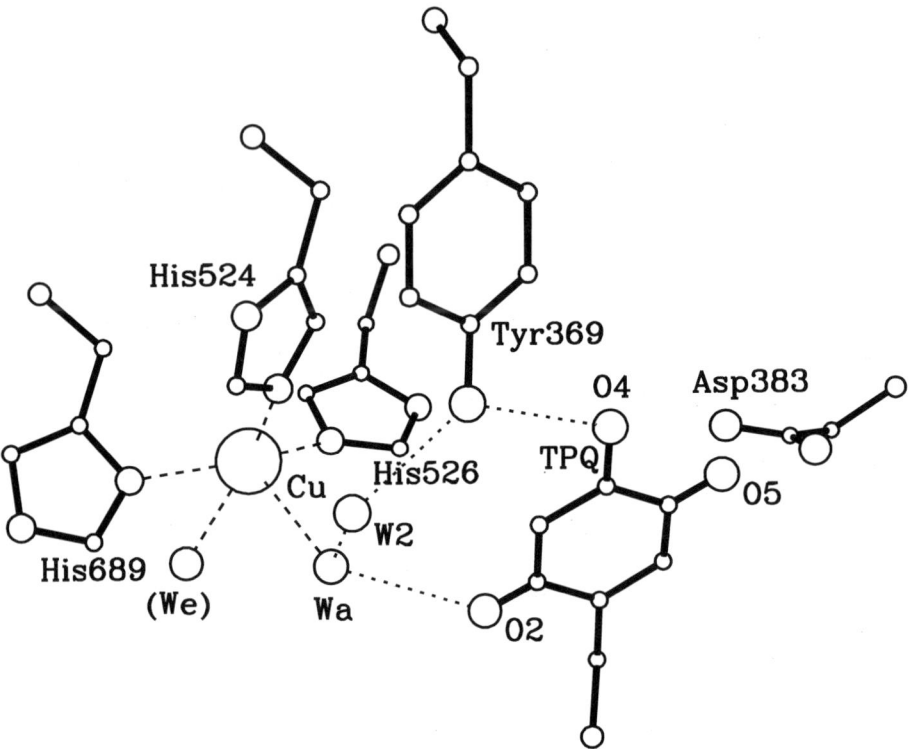

**FIGURE 12.** Structure of the *E coli* Amine Oxidase active site in the catalytically-active crystal form. Copper coordination and hydrogen bonds are shown by dashed lines and dotted lines respectively. Also shown are TPQ in its active conformation, the active site base Asp383, the conserved Tyr369 and a water molecule (W2) observed in most crystal forms of amine oxidases from various organisms. The equatorial water (We) is shown in parentheses since it is labile, variable in position and is absent in some crystal forms.

are differences in the order and orientation of the TPQ ring in the structures of amine oxidases from different sources. For PSAO, AGAO and HPAO, the TPQ ring is ordered; however whereas for PSAO (Kumar *et al.*, 1996) and AGAO (Wilce *et al.*, 1997), the oxygen at the 5 position of the TPQ is pointing *towards* the copper, in HPAO (Li *et al.*, 1998), the 5 oxygen of TPQ is pointing *away* from the copper. In PSAO, it is not clear whether this orientation of the TPQ is affected by the crystal growth conditions at pH 4, a pH at which amine oxidases are not active and where the active site aspartate is certainly protonated and forms a strong hydrogen bond to TPQ O2.

TPQ order and orientation in amine oxidases is clearly important and will be examined in later parts of this review (sections 3.2 and 3.3).

**3.1.3.3. Substrate Access Channels to the Active Site of Amine Oxidases.** An intriguing feature of the ECAO structure (Parsons *et al.*, 1995) was burial of the copper approximately 12 Ångstroms below the molecular surface of the protein, with little evidence for substrate entry channels. In the PSAO structure, there is a water-filled channel approximately 18 Ångstroms in length from the protein surface to the TPQ suggesting the route for amine entry (Kumar *et al.*, 1996) but with the difficulty that the TPQ is pointing the wrong way for substrate binding which extensive model studies (Mure and Klinman, 1993, 1995a, 1995b; Lee and Sayre, 1995) suggest occurs at the 5-position of the TPQ. The amine substrate entry channel, and the active orientation of TPQ, was clearly demonstrated in the crystal structure of the complex of ECAO with the inhibitor 2-hydrazinopyridine (Wilmot *et al.*, 1997; see section 3.2). The inhibitor is bound to TPQ O5 and remains lodged in the channel, its tip just reaching the enzyme surface. The entry path for oxygen is likely to be through a water filled cavity that runs up through the centre of the dimer between the subunits (Kumar *et al.*, 1996). This gives access to the other side of the active site away from TPQ but near the copper. A similar channel is seen in the HPAO X-ray crystallographic structure (Li *et al.*, 1998) where the TPQ is again pointing the correct way for substrate binding.

**3.1.3.4. Other Structural Features.** In addition to the copper at the catalytic centre of ECAO, the crystal structure of ECAO shows two metal ions bound per monomer. One is in a site of octahedral coordination, the other in a site of pentagonal bipyramidal coordination. These sites are both close to the molecular surface on the upper part of the β-sandwich domain and were tentatively assigned (Parsons *et al.*, 1995) as calciums on the basis of the ligand environment. This assignment has been made firmer by mass spectrometric metal analysis (Murray *et al.*, 1999). There is no evidence that these calciums have any role in catalysis or indeed any other role. However, one of these sites, the octahedral site, is almost certainly conserved in all amine oxidases as two of the carboxylate ligands (Asp533 and Asp678 in ECAO) are fully conserved and a third carboxylate (Asp 535 in ECAO) is conserved in nine out of twelve known amine oxidase sequences. This calcium site is located at the junction of two β-strands which include the three copper-binding histidines some ten residues away. It is interesting that the octahedral metal binding site is suggested to be occupied by $Mn^{2+}$ in PSAO (Kumar *et al.*, 1996) in view of an earlier report (Crabbe *et al.*, 1976) that placental diamine oxidase has bound $Mn^{2+}$ in addition to its $Cu^{2+}$.

### 3.2. Catalytic Mechanism

The catalytic mechanism of amine oxidases can be formally divided into reductive and oxidative half cycles

$$\text{Reductive half cycle } E_{ox} + RCH_2NH_2 \rightarrow E_{red} + RCHO$$

$$\text{Oxidative half cycle } E_{red} + O_2 + H_2O \rightarrow E_{ox} + H_2O_2 + NH_3$$

This mechanism suggests that amine oxidases would exhibit ping-pong steady state kinetics as has been reported from studies on the pig serum and bovine serum amine oxidases (see Petterson, 1985 and McIntire and Hartmann, 1993 for reviews). However, the mechanism is more complex than simple ping-pong suggests. For example, there is substrate inhibition at high substrate concentrations for pig serum amine oxidase (Kelly *et al.*, 1981) and this has been interpreted in terms of an enzyme form binding two molecules of substrate per active site. Again, the initial rate data for the methylamine oxidase from *Candida boidinii* (Haywood and Large, 1981) indicate an ordered rather than a ping pong mechanism; this has been discussed in terms of the low $K_m$ for oxygen (10 mM) in the case of the methylamine oxidase compared to 500 mM for the pig serum enzyme. A ping- pong mechanism for amine oxidases has recently been called into question through studies of catalysis in ECAO crystals where the structure shows product aldehyde bound when a bound dioxygen species is also present (Wilmot *et al.*, 1999).

It is often difficult, however, to distinguish unambiguously between a ping-pong and an ordered mechanism by steady state kinetics (Engel, 1977). The description as an aminotransferase mechanism is not in doubt; thus for the pig serum enzyme, quenched flow kinetic studies have shown that $NH_3$ is released following oxygen binding (Rius *et al.*, 1984) whilst for the bovine serum amine oxidase, it has been shown that $NH_3$ is not released under anaerobic conditions (Janes and Klinman, 1991). Perhaps the most compelling evidence demonstrating an aminotransferase mechanism comes from EPR and Electron Spin Echo Envelope Modulation (ESEEM) on the semi quinone radical form of TPQ generated by reduction of *Arthrobacter globiformis* amine oxidase by $[^{14}N]$- and $[^{15}N]$—methylamine which showed that the coupled nitrogen nucleus comes from substrate (Dooley *et al.*, 1990; McCracken *et al.*, 1992).

### 3.2.1. Reductive Half Cycle

The catalytic mechanism for amine oxidases is shown in Figure 13.

Pioneering kinetic studies by Pettersson and coworkers during the 1970s on pig plasma amine oxidase (Pettersson, 1985) indicated that Schiff-base intermediates were involved and that deprotonation of the α-carbon of bound substrate is rate limiting for steps between the enzyme-substrate complex and the reduced form of the enzyme. Although the carbonyl cofactor involved in these Schiff-base reactions was accepted at the time of these studies to be pyridoxal phosphate, the experimental design and logical analysis of the kinetic data in the work of Pettersson and coworkers was essentially correct in principle.

Mechanistic evidence for a quinone cofactor in amine oxidases came from reductive trapping experiments by Hartmann and Klinman (1987) which demonstrated that substrate could be irreversibly bound to enzyme during catalytic turnover by the presence of the mild reductant sodium cyanoborohydride. Using $^{14}$C- labelled substrate and tritiated cyanoborohydride, it was shown that the reductively- inactivated enzyme retained $^{14}$C but not tritium. This was rationalised by the presence as a cofactor of a *dicarbonyl* (quino-) structure which tautomerised subsequent to reduction releasing tritium to solvent. Subsequently, it was shown by Klinman and coworkers in an outstanding paper that this cofactor was a quinone formed from 2, 4, 5 trihydroxyphenylalanine, TPQ (Janes *et al.*, 1990).

**3.2.1.1.   Substrate Binding / Substrate Schiff-Base.**   Studies of TPQ model compounds by Mure and Klinman (1993, 1995a, 1995b), and by Lee and Sayre (1995) established that it is the carbonyl at position 5 of TPQ which is reactive to nucleophilic reagents and thus the site at which substrate amine would be expected to bind in the active site of amine oxidases. These model studies also showed that TPQ is essentially fully ionised at neutral pH due to its low pK (4.2 in TPQ models and ~3 in bovine serum amine oxidase) the negative charge being delocalised over the oxygens on C-2 and C-4. Recent resonance Raman studies on amine oxidases from several sources have substantiated these conclusions (Moenne-Loccoz *et al.*, 1995; Nakamura *et al.*, 1997).

The TPQ 5-position as the site of amine substrate binding is confirmed by the crystallographic structures of the complex between ECAO and the substrate-like inhibitor 2-hydrazinopyridine (Wilmot *et al.*, 1997). The complex corresponds approximately to the substrate Schiff-base (Species 2 in Figure 13) but with the CH$_2$ groups replaced by NH. The structure shows that the TPQ ring and the pyridine ring of the 2-HP are not parallel as would be the case for the product Schiff-base (Species 4 in Figure 13). This structure additionally revealed that O4 of the TPQ ring forms a short hydrogen bond with the hydroxyl of the conserved residue Tyr 369 (Figure 12). The distance between the oxygens is less than 2.5 Å consistent with one shared proton and suggesting ionisation at the O4 position of the quinone ring. It was also evident that in the 2-HP complex, the O2 position of the quinone ring is hydrogen bonded to the axial water ligand (Wa) on the copper (Figure 12). These studies indicate the importance of Tyr 369 in

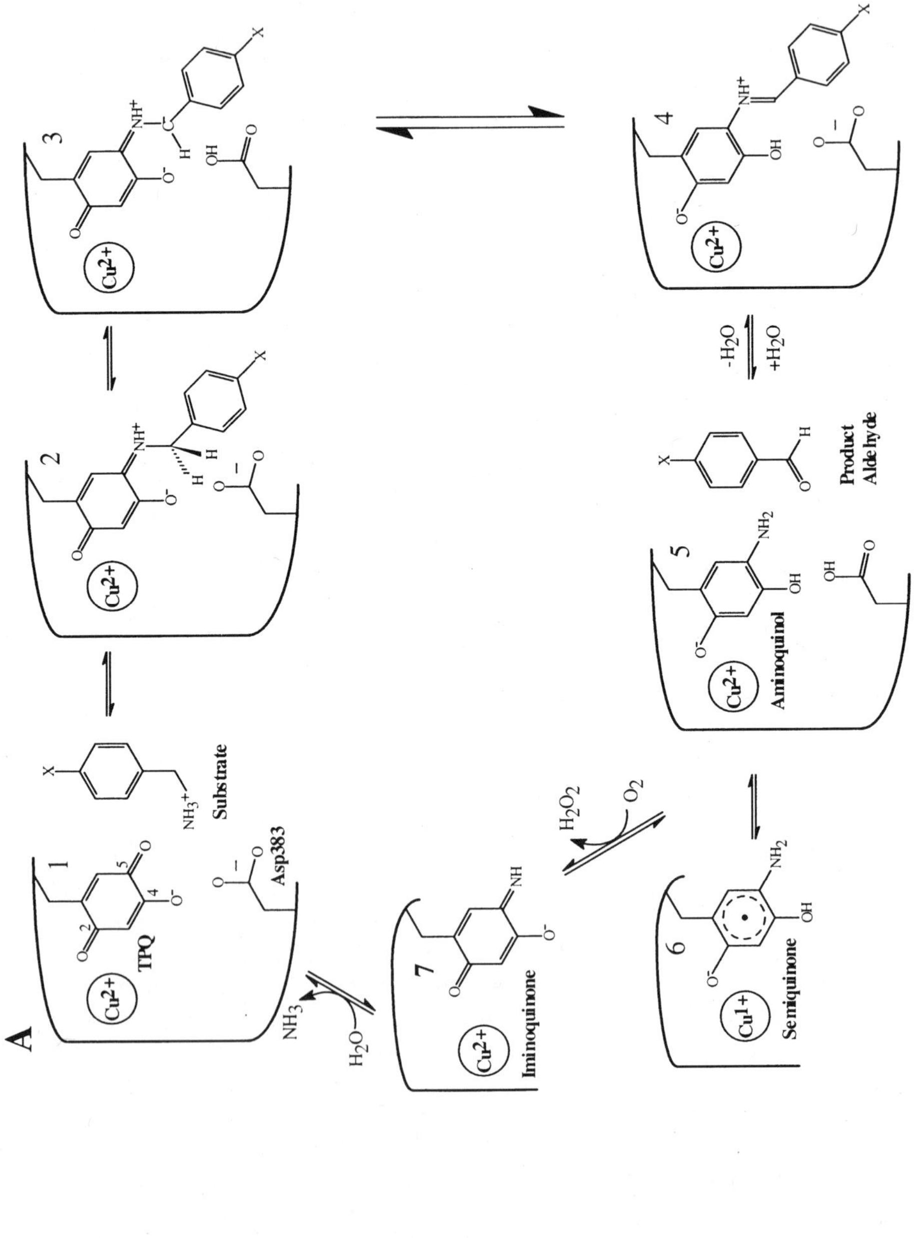

A
TPQ
Asp383
Substrate
Product Aldehyde
Aminoquinol
Semiquinone
Iminoquinone
-H2O
+H2O
H2O2
O2
NH3
H2O

stabilising the position and orientation of the quinone ring and point to hydrogen bonding to waters in the active site as a key feature of the catalytic mechanism as has been proposed by Klinman and coworkers (Su and Klinman, 1998).

Stopped flow kinetic studies by Hartmann and Klinman (1991) on bovine serum amine oxidase and by Olsson *et al.* (1976) on pig plasma amine oxidase showed that the substrate Schiff-base has an absorbance maximum at ~350 nm.

**3.2.1.2. Product Schiff-Base.** Conversion of the substrate Schiff-base to product Schiff-base involves removal of a hydrogen from the α-carbon of the substrate. This chemically-difficult step is achieved in the enzyme by a base catalysed abstraction of a proton and shows interesting stereochemical variation amongst amine oxidases from different sources (Scaman and Palcic, 1992; Coleman *et al.*, 1989, 1991; Shah *et al.*, 1993). When benzylamine is the substrate, it is always the pro-S hydrogen which is removed but when tyramine or dopamine are the substrates, removal of hydrogen from C-1 can be pro-S, pro-R or random depending upon the source of the amine oxidase (see Table 3).

There is a correlation between this proton abstraction stereochemistry and exchange of the hydrogens at C-2 of tyramine or dopamine substrates. Thus the pro-S specific amine oxidases do not catalyse exchange at C-2 whereas the pro-R specific and non-specific amine oxidases catalyse this exchange.

From the pH dependence of the BSAO-catalysed reaction, an active site base with a pK of ~5 was implicated in the proton abstraction step (Farnum *et al.*, 1986). The 2.0 Å structure of ECAO (Parsons *et al.*, 1995) indicate that Asp 383 is ideally placed to act as the active site base and this was substantiated by specific mutations at this site (Wilmot *et al.*, 1997) which eliminated catalytic activity. In addition the structures of the 2-HP complex (Wilmot *et al.*, 1997) shows Asp383 in contact with the expected site of the pro-S hydrogen. Klinman (1996) in a perceptive review presents the case for the pK of the active site aspartate being ~8 in the resting enzyme (due to its hydrophobic environment) and changing to 5.6 following formation of the protonated substrate Schiff-base due to electrostatic

---

**FIGURE 13.** Species identified in the reaction cycle of copper amine oxidases. Oxidised, resting state enzyme (**1**) reacts with substrate to form a substrate Schiff base (**2**). Proton abstraction by the active site base (Asp383 in ECAO) leads, via a carbanion intermediate (**3**) to the product Schiff base (**4**). Hydrolysis releases the product aldehyde, leaving reduced cofactor in equilibrium between aminoquinol/$Cu^{2+}$ (**5**) and semiquinone/$Cu^+$ (**6**). The reduced cofactor is reoxidised by molecular oxygen, releasing ammonium ions and hydrogen peroxide. (Modified from Wilmot *et al.*, 1999 with permission).

## Table 3
### Stereochemistry of proton abstraction at C-1 of Tyramine and Solvent Exchange Characteristics at C-2 Catalyzed by Semicarbazide-Sensitive and Copper Amine Oxidases

| Enzyme source | C-1 proton abstraction | C-2 solvent exchange | Ref. |
|---|---|---|---|
| *Semicarbazide-Sensitive Amine Oxidases* | | | |
| Porcine aorta | pro-S | Yes | Scaman and Palcic, 1992 |
| Bovine aorta | pro-S | Yes | Scaman and Palcic, 1992 |
| *Copper Amine Oxidases (EC 1. 4. 3. 6)* | | | |
| Pea seedling | pro-S | No | Coleman *et al.*, 1989 |
| Porcine kidney | pro-S | No | Coleman *et al.*, 1991 |
| Soybean seedling | pro-S | No | Coleman *et al.*, 1991 |
| Chick pea seedling | pro-S | No | Coleman *et al.*, 1991 |
| Procine plasma | pro-R | Yes | Coleman *et al.*, 1989 |
| Bovine plasma | nonstereospecific | Yes | Coleman *et al.*, 1989 |
| Sheep plasma | nonstereospecific | Yes | Coleman *et al.*, 1991 |
| Rabbit plasma | nonstereospecific | Yes | Coleman *et al.*, 1991 |
| *Lysyl oxidase* | | | |
| Bovine aorta | pro-S | Yes | Shah *et al.*, 1993 |

interactions. The transition state species between substrate and product Schiff-base species involves considerable development of negative charge, is consistent with the findings from an extensive series of kinetic studies on BSAO (Hartmann and Klinman, 1991) using ring-substituted benzylamines.

The transition state leads to the product Schiff-base where the TPQ ring has been aromatised. The TPQ ring is effectively mediating proton transfer from substrate. This will be discussed further in the section on the Oxidative Half Cycle below.

The lack of tritium incorporation during the reductive inactivation experiments with cyanoborohydride suggests that the product Schiff-base does not accumulate significantly. A short lifetime for this species is also consistent with the results of isotopic substitution during kinetic studies (Pettersson, 1985; Farnum *et al.*, 1986; Palcic and Klinman, 1983). Of particular note is that very large deuterium isotope effects (in the range 7–16) are seen in $V_{max} / K_m$ during the oxidation of benzylamine catalysed by BSAO. This indicates rate limitation in the C—H bond cleavage step. The size of these isotope effects is outside the classical range and more detailed studies of tritium isotope effects as a function of temperature indicate quantum mechanical tunnelling during amine substrate oxidation (Grant and Klinman, 1989; Kohen and Klinman, 1998).

Recent studies on HPAO (Plastino *et al.*, 1999) and ECAO (Murray *et al.*, 1999) indicate that the active site Asp performs multiple roles at different stages of the catalytic cycle. Both these studies report investigations of the properties of mutational variants (Asp to Glu and Asp to Asn) and conclude that one essential role of the active site Asp is to position the TPQ cofactor optimally for catalysis. There are, however, differences between the reported findings. The HPAO studies which focus on the kinetic properties of the variants in the oxidation of the non-aromatic substrate methylamine and, supported by resonance Raman structural data, suggest that whereas in the native enzyme the TPQ O5 is already correctly oriented towards the base, for the Asp→ Glu variant the hydrogen bonding network holding the cofactor in place is compromised, allowing it to slip into a less accessible conformation. Binding of substrate to the the Asp→ Glu variant is proposed to induce a TPQ ring orientation which brings the O5 sufficiently close to the Glu to allow the catalytic cycle to operate, though $k_{cat}$ is 200 fold decreased from wild type. It is interesting that the properties of another mutational variant of HPAO, namely the conserved Asn 404 preceding the TPQ cofactor derived from Tyr 405, also allows non-productive conformations of the TPQ ring and the consequent formation of inactive Schiff-base products when the enzyme is reacted with substrate methylamine (Schwartz *et al.*, 1998). The ECAO studies (Murray *et al.*, 1999), based mainly on comparisons of the crystallographic structures of wildtype, D383E, D383A and D383N variants, suggest that Asp 383 helps maintain the active site structure by preventing TPQ migration to the copper. In wild type ECAO, the TPQ is located in a wedge-shaped pocket which allows freedom of pivotal movement at the substrate-binding position (C5) essential for optimal catalytic activity and electron density maps clearly show this lack of order. In the D383E variant, however, the longer side chain of the Glu restricts this movement of the TPQ as shown by the clear electron density of the ordered TPQ ring and explaining the low catalytic activity of D383E. The paper concludes that Asp 383 in ECAO is not only critical in several proton transfer steps in the catalytic cycle but also plays a key role in controlling the conformation and flexibility of the TPQ cofactor.

**3.2.1.3. Reduced Forms of the Enzyme.** The hydrolysis of the product Schiff-base to release product and generate the aminoquinol form of the cofactor may involve water which has been retained in the active site. Dooley and co-workers have provided direct evidence for copper reduction during the interaction of amine oxidases with substrate under anaerobic conditions (Dooley *et al.*, 1991). By varying the temperature at which EPR spectra were recorded, it was shown for amine oxidases from several sources that there is a temperature dependent equilibrium between $Cu^{2+}$ / aminoquinol TPQ and $Cu^{+}$ / TPQ semiquinone. The $Cu^{+}$ / TPQ semiquinone form was found to be stabilised in the presence of cyanide. The temperature-dependence of the equilibrium allowed temperature jump relaxation kinetic studies to be carried out. The forward rate constant for the formation of $Cu^+$ / TPQ semiquinone was found to be $4000s^{-1}$ (Turowski *et al.*, 1993); this is much greater than the rate limiting step for catalysis by any amine oxidase suggesting, but not establishing, that the semiquinone form could be an intermediate in catalysis.

It will be seen from Figure 13 that copper is not invoked for the catalytic steps of the reductive half cycle and an unresolved issue is whether the copper-free form of amine oxidase can catalyse the reactions leading to the aminoquinol form of the TPQ cofactor and release of product aldehyde. For the lentil seedling amine oxidase (Rinaldi *et al.*, 1989), it has been reported that the copper-free form can catalyse the reactions of the reductive half cycle but that the presence of copper is required for the TPQ semiquinone species to be observed (Bellelli *et al.*, 1985). Further work is needed to establish fully the catalytic properties of apo-forms and metal substituted-forms of amine oxidases.

### 3.2.2. Oxidative Half Cycle

In chemical terms, $Cu^+$ would be a natural site for binding molecular oxygen as an intermediate in the oxidative half cycle. The $Cu^+$ / TPQ semiquinone radical described above would, therefore, be a plausible species for an important intermediate. A tentative scheme for oxygen binding at this site, and subsequent electron and proton transfer events leading to reduction to hydrogen peroxide, has been given by Klinman (1996). Su and Klinman (1998) however, in studies of BSAO examined which step is rate limiting in the oxidative half cycle. Solvent isotope studies, as well as study of the effects of solvent viscosity, argue against proton transfer or oxygen binding steps being rate limiting and the authors conclude that the initial electron transfer to dioxygen is the most probable rate limiting step. If the binding site of dioxygen is to $Cu^+$ (in the $Cu^+$ / TPQ semiquinone species), chemical precedent suggests that electron transfer from $Cu^+$ to oxygen, yielding superoxide, would be fast. Su and Klinman (1998) concluded that oxygen binds instead in a hydrophobic pocket rather than to the copper and that the rate limiting step involves electron transfer to it from the aminoquinol form of TPQ. Su and Klinman further discuss the nature of subsequent electron transfer and proton transfer events in the oxidative half cycle. They envisaged that reduced TPQ acts as a "transducer" storing two electrons and two protons for later delivery to dioxygen. Both of the electrons and one of the protons (at the O4 position on the reduced cofactor) are proposed to originate from substrate while the proton at O2 of the reduced cofactor is proposed to come from the axial water to $Cu^{2+}$. Although copper is seen as playing a role in the oxidative half cycle, it is subtle and appears not to involve $Cu^+$. Other metals might substitute at least partially for Cu as has been reported for BSAO by Suzuki and coworkers (Suzuki *et al.*, 1986).

Wilmot *et al.* (1999) have reported high resolution crystal structures of three species relevant to understanding the chemistry of the oxidative half cycle of amine oxidases, (i) anaerobic substrate-reduced ECAO, (ii) anaerobic substrate-reduced ECAO with bound nitric oxide (an oxygen mimic) and (iii) ECAO reacted aerobically with substrate to reach an equilibrium turnover state, then cryo-trapped. In all these species, product aldehyde remains bound at the back of the substrate binding pocket and this seems to be crucial in allowing build up of intermediates.

In the nitric oxide complex, electron density from a diatomic molecule is seen at the position previously occupied by the axial water (Wa) and bridging to O2 of the aminoquinol form of TPQ. The distance of the nitric oxide to copper is long ($2.4 \text{Å}$), suggesting a rather weak interaction. Thus it is possible that during the oxidative half cycle, dioxygen binds close to the aminoquinol O2 and is reduced there to superoxide as suggested by Su and Klinman (1998).

The most informative structure is the turnover complex, formed in (iii), where the electron density shows a diatomic molecule (which must be a form of dioxygen) again replacing Wa and in a very similar position to that of nitric oxide in the structure discussed above (Figure 14).

This structure allows the proton transfer pathways to the oxygen as proposed by Klinman to be visualised. In the case of the proton transfer via the aminoquinol at position O2, a direct hydrogen bonded interaction between the O2 and the dioxygen is seen. For the other proton transfer event, the cofactor O4 is in a short hydrogen bond ($2.4 \text{Å}$) interaction with the OH of Tyr 369. The OH of Tyr 369 is linked to the copper-liganded oxygen of the dioxygen by a bridging water (W2 in Figure 14) that is conserved in all the known structures of amine oxidases. It is suggested (Wilmot *et al.*, 1999) that both electron transfers from the reduced cofactor and both proton transfers have probably taken place in this trapped intermediate implying the observed diatomic molecule is hydrogen peroxide. The sequence of electron and proton transfer steps leading to this trapped intermediate remains to be determined but we now know the essential chemical features of the oxidative half cycle. It is possible also to speculate on the chemistry of the step leading to release of ammonia from the iminoquinone which probably involves activation by Asp 383 of the water designated W4 in Figure 14. The key roles played by Tyr 369 and Asp 383 in maintaining the correct location of the TPQ cofactor in the catalytic cycle as well as participating in the proton transfer steps emerges convincingly from the structural model shown in Figure 14.

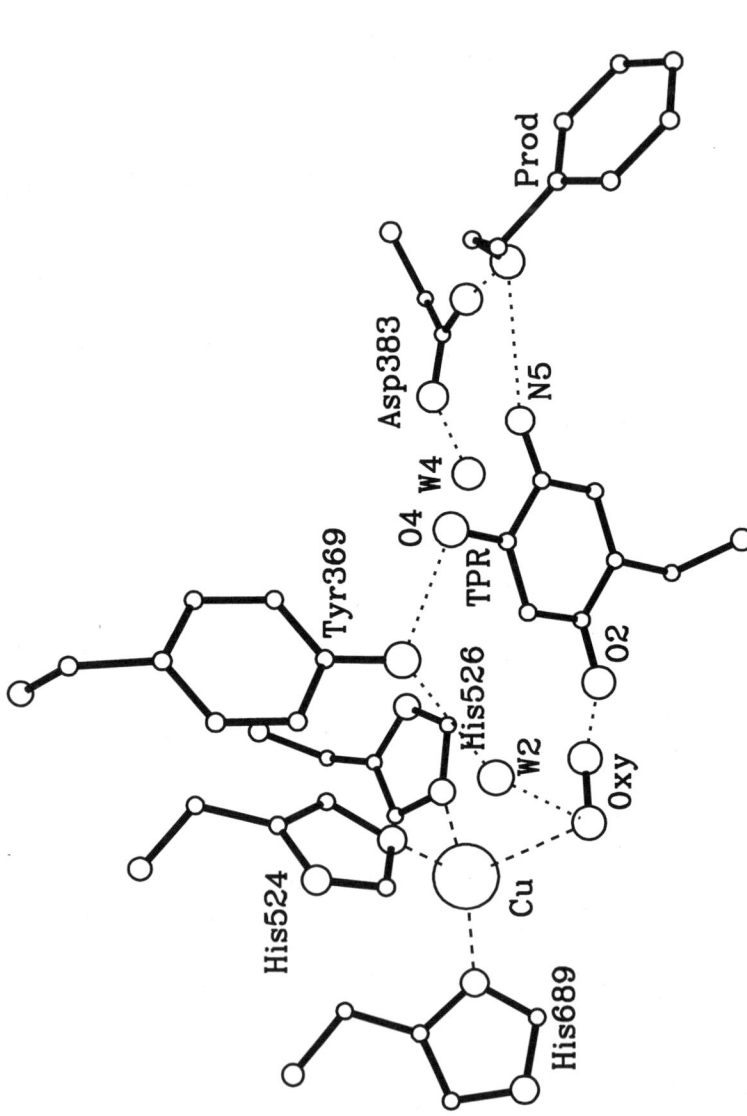

**FIGURE 14.** Structure of *E coli* amine oxidase active site in the equilibrium turnover complex. Interactions and labelling as in Figure 12 except that the reduced form of the cofactor is labelled TPR and carries a nitrogen at the 5-position. The aldehyde product, labelled Prod, is retained in the channel to the enzyme surface. Water W4 is positioned for nucleophilic attack on C5 of the cofactor and W2 is retained as a link in a proton shuttle from O4 to dioxygen.

### 3.3. Biogenesis of TPQ and Related Cofactors

The determination of the structure of TPQ by Klinman and coworkers (Janes *et al.*, 1990) was an important landmark in the field. Firstly, TPQ was the first example of a redox cofactor produced by post-translational modification of an amino acid, tyrosine. In this respect, it differs from flavin as well as from pyrroloquinolinequinone (PQQ) which had been discovered as a dissociable quinone cofactor in methanotrophic bacteria and had been suggested to be the cofactor in amine oxidases (Ameyama *et al.*, 1984; Lobenstein-Verbeek *et al.*, 1984), a suggestion which was attractive but wrong (Klinman *et al.*, 1991). Secondly, TPQ is widely-distributed in biology, found in all copper-containing amine oxidases reported to date including the so-called "semicarbazide sensitive" amine oxidases which are membrane-associated (Holt *et al.*, 1998). A related cofactor termed lysine tyrosylquinone (LTQ) has the TPQ substituted at the $O_2$ position by the ε-amino group of a lysine (Wang *et al.*, 1996, 1997) and has been found to date only in mammalian lysyl oxidase, a protein class quite distinct from the amine oxidases which are the main subject of the present review.

Reports that TPQ is derived by post-translational modification of tyrosine (Cai and Klinman, 1994; Matsuzaki *et al.*, 1994) also established that this processing is auto-catalytic within the active site of amine oxidases, rather than involving separate enzymes. It was shown (Matsuzaki *et al.*, 1994) that a precursor apoprotein lacking both TPQ and copper could be purified and that simple addition of copper in the presence of oxygen led to TPQ formation. The oxidation of tyrosine to TPQ is a six-electron redox process and the mechanism is intriguing. Dooley and coworkers (Ruggiero *et al.*, 1997; Ruggiero and Dooley, 1999) have studied the overall stoichiometry and kinetics of the reaction, showing that one equivalent of $Cu^{2+}$ per subunit is sufficient for TPQ biogenesis and excluding the possibility that unbound copper participates. The close similarity in the structures of the copper sites in the apo-form of *Arthrobacter globiformis* amine oxidase (Wilce *et al.*, 1997) and an inactive form of *E coli* amine oxidase (Parson *et al.*, 1995) in which TPQ is tetrahedrally coordinated to copper through His 524, His 526.

His 689 and the $O_4$ of TPQ support a model in which the first step in TPQ biogenesis is binding of $Cu^{2+}$ to tyrosine (Tyr 466 in ECAO) and the three His ligands. A chemical mechanism for the subsequent oxidation steps involving free radical intermediates has been proposed (Dooley, 1999).

Although this mechanism is plausible, there is currently no direct evidence for free radical involvement and an alternative mechanism has been proposed (Williams and Klinman, 1999). It is interesting that step E to F in the mechanism shown in Figure 15 which precedes introduction of the

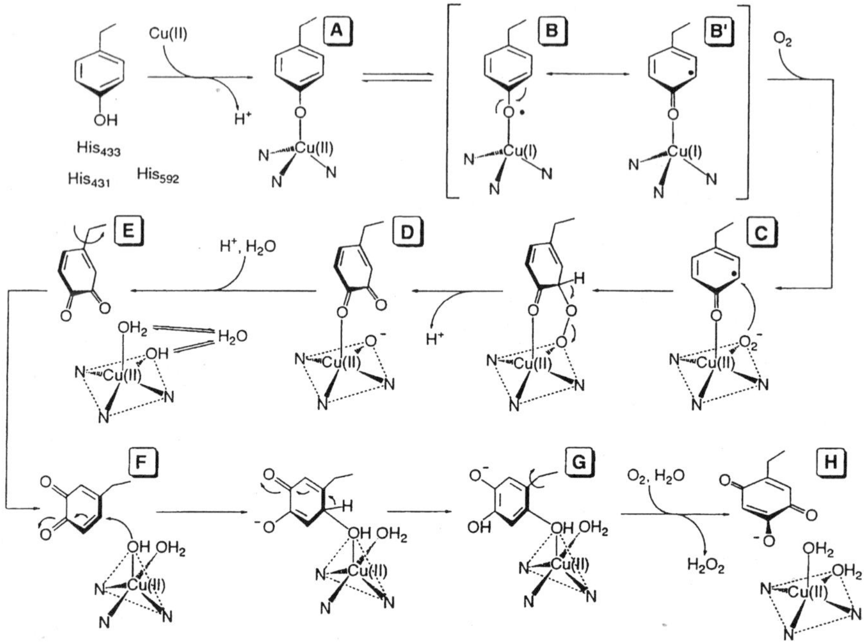

**FIGURE 15.** Possible mechanism for biogenesis of TPQ (from Dooley, 1999 with permission).

second oxygen into the ring, involves rotation of the ring of the TPQ precursor, a step bearing similarities to the reorientation of the TPQ ring in some amine oxidases during the catalytic cycle. This was first pointed out by Cai and Klinman (1994).

The biogenesis of the LTQ cofactor in lysyl oxidase may be closely similar to that of TPQ since it is also derived autocatalytically from a tyrosine precursor (Wang *et al.*, 1996). A chemically plausible mechanism has been proposed (Dooley, 1999).

Whilst the evidence is strong that the processing events to generate the TPQ and LTQ cofactors require no additional enzymes, it seems likely that insertion of copper into the apoproteins involves a chaperone protein (Harrison *et al.*, 1999). Although there is currently no direct evidence for this, it is suggested by the fact that free copper levels inside the cell are kept vanishingly small in order to minimise production of damaging oxygen radicals. A chaperone would also ensure that only copper is inserted into the oxidase; there is a growing awareness from *in vitro* studies of copper-containing enzymes that binding of zinc leads to non-functional forms and that *in vivo*, this adversely affects normal cell function (Lyons *et al.*, 1999).

## 3.4.  Model Studies

There is little inorganic chemistry in the literature specifically designed to model copper-containing amine oxidases, CAO. Several groups have shown that four-co-ordinate copper/catecholate complexes can exist either as $[Cu^{II}L_2(catecholate)]$ or $[Cu^{I}L_2(ortho\text{-semiquinonate})]$ valence tautomers, depending on the identity of the co-ligand L (Buchanan *et al.*, 1986; Speier *et al.*, 1994; Rall and Kaim, 1994). This provides some precedent for the Cu(II)/TOPA-to-Cu(I)/TPSQ interconversion in CAO, although no compound that exhibits *both* isomeric forms has yet been discovered. The only functional CAO models to have been reported to date are the complexes $[Cu(L)(OH_2)(bipy)]^+$ (LH = DL-TOPA, DL-4, 5-dihydroxyphenylalanine { DL-DOPA}; bipy = 2,2'-bipyridyl), which contain a pendant di- or tri-hydroxyphenyl moiety and catalyse the aerobic oxidation of amines with rate enhancements of up to 14 (Nakamura *et al.*, 1992; Suzuki *et al.*, 1998). No mechanistic studies of this system were possible, however, because polymerisation of the pendant quinone groups causes rapid deactivation of these catalysts (Suzuki *et al.*, 1998).

## 3.5.  Biological Roles of Amine Oxidases

### 3.5.1.  Microorganisms

Amine oxidases are induced in prokaryotic organisms such as the bacteria *Arthrobacter* P1 (Van Iersel *et al.*, 1986; McIntire, 1990) and *Escherichia coli* (Parrot *et al.*, 1987) when grown on monoamines as the sole carbon and energy source. The nutritional role of the amine oxidase in these organisms seems clear. The role of amine oxidases in eukaryotic microorganisms grown on different substrates is more complex and requires further study. For example, growth of *Penicillium chrysogenum* on glucose and spermine induces a flavin-containing polyamine oxidase (Isobe *et al.*, 1980) whereas growth on spermine as the sole carbon and nitrogen source induces an amine oxidase of broad substrate specificity (Isobe *et al.*, 1982).

### 3.5.2.  Plants

In plants, aliphatic di- and polyamines have been implicated in many processes including rapid cell division, fruit development, stress response and senescence (Evans and Malmberg, 1989; Galston and Kaur-Sawhney, 1995). Amine oxidases catalyse the catabolism of biogenic amines and hence could be involved in regulating such cellular processes. It is convenient to separate discussion of the roles of the copper-containing "diamine oxidases" in plants from the roles of "polyamine oxidases", which are probably flavoenzymes. The copper amine oxidases catalyse oxidation of the aliphatic diamine putrescine at the primary amine site to give 4-aminobutaraldehyde (which cyclises to $D^1$ pyrroline), ammonia and hydrogen peroxide. They have been characterised in many plants (Medda *et al.*, 1995) but have been most intensively studied in legumes (Tipping and McPherson, 1995) where they are associated with plant cell walls of roots, cotyledons and lignified tissue as revealed by immunolocalisation studies (Angelini *et al.*, 1990). Møller and McPherson (1998) screened the genomic library of *Arabidopsis* and isolated two cloned genes for structurally distinct copper containing amine oxidases. One of these genes, *atao1*, codes for a 668 amino acid polypeptide showing 48% homology to pea seedling amine oxidase (Tipping and McPherson, 1995). The promoter region of *atao1* has been transcriptionally fused with the reporter genes β-glucuronidase (GUS) and green fluorescent protein (GFP) and expression analysed by transformation into *Arabidopsis*. The results indicate temporally and spatially discrete patterns of gene expression in lateral root cap cells, vascular tissue of roots, developing leaves and other tissue. There is a correlation between *atao1* expression and cells destined to undergo programmed cell death. The expression patterns of the other amine oxidase gene, *atao2*, has not yet been reported. These elegant studies are a prototype for analysing the roles of amine oxidases, not only in plants, but also in mammals.

Polyamine oxidases have been studied most extensively in cereal plants and have been shown to be extracellular enzymes located in the cell wall with a probable role in cell signalling (Smith, 1985).

### 3.5.3. Mammals

The diversity of copper-containing amine oxidases in mammalian tissue suggests important cellular roles. Our knowledge of this subject is, however, fragmentary and widely spread throughout the medical and pharmacological literature (reviewed in McIntire and Hartman, 1993). Speculative suggestions of the roles of amine oxidases include regulation of levels of polyamines and other biogenic amines, and that hydrogen peroxide generation by amine oxidation acting as a cellular signal (Sundaresan *et al.*, 1995). There have been some intriguing recent findings which stem from comparisons of both gene and protein sequences. It has been found, for instance, that the sequence of the gene for human kidney diamine oxidase (Lingueglia *et al.*, 1993) is identical to that of the human amiloride-binding protein (Mu *et al.*, 1994), a protein thought to function as a $Na^+$ channel, causing the latter to be re-designated.

Human vascular adhesion protein (HVAP) is involved in targeting lymphocytes to inflamed tissue following infection (Salmi and Jalkanen, 1997). Interestingly, the sequence of HVAP cDNA is identical to that reported for the amine oxidase from human placenta (Smith *et al.*, 1998), and HVAP has been shown to have amine oxidase activity. The partial three dimensional structure of HVAP has been modelled based on homology to the known 3-D structure of ECAO (Salminen *et al.*, 1998). There is an equally intriguing sequence identity between mouse vascular adhesion protein and the membrane-bound amine oxidase associated with adipocytes (Bono *et al.*, 1998).

Finally, important advances are being made in defining the biological roles of lysyl oxidases. Much of this work has been carried out by Kagan and his collaborators; (reviewed in (Kagan and Trackman, 1992)). The gene for human lysyl oxidase, an extracellular enzyme, has been cloned and sequenced (Trackman *et al.*, 1990). Lysyl oxidase has an important role in cross-linking structural proteins such as elastin and collagen, and pathological changes in the production or regulation of lysyl oxidase are implicated in fibrosis (Kagan and Trackman, 1992; Trackman *et al.*, 1990). At the cellular level, lysyl oxidase activity is much reduced in malignantly transformed human cell lines (Kuivaniemi *et al.*, 1986). Another link to cancer has emerged recently with the discoveries that the expression of the *ras* recision gene (*rrg*) is decreased following transformation with the Ha-ras sarcoma virus and that rrg is >98% homologous with lysyl oxidase (Kenyon *et al.*, 1991). High levels of rrg expression are restored if this transformed tumorigenic cell line reverts to a normal cell line. These results indicate strongly that a protein with lysyl oxidase activity plays a role in maintaining the non-tumorigenic state of these cell lines. The molecular biology and other approaches employed in these studies point the way to future investigations of the roles of amine oxidases in mammals.

## 4. COMPARISONS BETWEEN GALACTOSE OXIDASE AND AMINE OXIDASES

Galactose oxidase and copper-containing amine oxidases catalyse similar reactions: the methylene group of a primary alcohol or primary amine is activated, a hydrogen abstracted and a product aldehyde is formed in both cases. Both enzymes contain a single copper per active site and it would have been a reasonable prediction that the role(s) played by copper would be similar. As we have seen (section 2.3 and 3.3), copper is essential for post-translational processing of tyrosine to yield novel cofactors in both oxidases and the chemistry involved in the processing events may be similar. However, the role of copper in catalytic processing of substrates seems to be different. In GOase, there is evidence that the alcohol substrate binds to copper in the active site, whereas for amine oxidases, copper appears not to involved directly (other than in helping to position the TPQ cofactor) in the reductive half cycle though it does participate in binding dioxygen species during the oxidative half cycle.

The chemistry of hydrogen abstraction from substrate is stereospecific and involves the organic cofactor in both enzymes. In galactose oxidase, however, the mechanism appears to involve hydrogen atom abstraction by the modified Tyr[272] free radical site whereas in amine oxidases, the TPQ cofactor holds the amine substrate so that a proton can be removed by an active site base (Asp[383] in ECAO). The active sites of galactose oxidase and amine oxidases are strikingly different. The GOase active site is compact and non-polar in character and the catalytic events probably do not involve large movements in active site residues. By contrast, the AO active site is polar in character harbouring several water molecules and movement in the TPQ cofactor is an essential feature of catalysis.

## 5. FUTURE DIRECTIONS

There have been dramatic advances in our understanding of the catalytic mechanism of amine oxidases over the past ten years. The availability of high resolution X-ray structures of several amine oxidases has undoubtedly been the most important factor in these advances and has stimulated research on the roles played in catalysis of the copper, the TPQ cofactor and active site amino acid residues. We already have a reasonable understanding of the reductive half cycle and there are strong indications that we will soon understand in molecular detail the chemistry of oxygen activation and reduction. The challenge of synthesising functional inorganic models based on our knowledge of the active site structure of amine oxidases is great, primarily because catalysis by the enzyme involves controlled mobility of the TPQ cofactor to mediate the chemistry.

It is safe to predict that there will be major effort in understanding the biological roles of copper-containing amine oxidases and the relationship to flavin-containing amine oxidases. The diversity of amine oxidases in all forms of life and their involvement in key cellular processes in plants and mammals underline the importance of this objective.

With galactose oxidase, our understanding of the catalytic mechanism is less advanced than for amine oxidases but all the essential foundations for continued advances are in place; high resolution X-ray structures of native and mutational variant forms, complete with advanced spectroscopic and kinetic methods. Refinement of the current functional models of galactose oxidase, and further attempts to mimic the controlled catalytic power of the free radical site in this redox enzyme, present important challenges for the future. With respect to the biological role of galactose oxidase, it would be a stimulus to future research activity if enzymes of the galactose oxidase/glyoxal oxidase class were found to be more widely distributed in Nature.

## 6. REFERENCES

Amaral, D., Bernstein, L., Morse, D., and Horecker, B. L., 1963, Galactose oxidase from *Polyporus circinatus*: a copper enzyme, *J. Biol. Chem.* **238**:2281–2284.
Ameyama, M., Hayashi, M., Matsushita, K., Shinagawa, E., and Adachi, O., 1984, Microbial production of pyrroloquinoline quinone, *Agric. Biol. Chem.* **48**:561–565.
Angelini, R., Manes, F., and Federico, R., 1990, Spatial and functional correlation between diamine-oxidase and peroxidase activities and their dependence upon de-etiolation and wounding in chick pea stems, *Planta* **182**:89–96.
Avigad, G., Amaral, D., Asensio, C., and Horecker, B. L., 1962, The D-galactose oxidase of *Polyporus circinatus*, *J. Biol. Chem.* **237**:2736–2743.
Azakami, H., Yamashita, M., Roh, J. H., Suzuki, H., Kumagai, H., and Murooka, Y., 1994, Nucleotide sequence of the gene for monoamine oxidase from *Escherichia coli*, *J. Ferment. Bio-Eng.* **77**:315–319.
Babcock, G. T., El-Deeb, M. K., Sandusky, P. O., Whittaker, M. M., and Whittaker, J. W., 1992, Electron paramagnetic resonance and electron nuclear double resonance spectroscopies of the radical site in galactose oxidase and of thioether-substituted phenol model compounds, *J. Am. Chem. Soc.* **114**:3727–3734.
Barbry, P., Champe, M., Chassande, O., Munemitsu, S., Champigny, G., Lingueglia, E., Maes, P., Frelin, C., Tartar, A., Ullrich, A., and Lazdunski, M., 1990, Human kidney amiloride binding protein: cDNA structure and functional expression, *Proc. Natl. Acad. Sci. U. S. A.* **87**:7347–7351.
Barker, R., Boden, N., Cayley, G., Charlton S. C., Henson, R., Holmes, M. C., Lelly, I. D., and Knowles, P. F., 1979, Properties of cupric ions in benzylamine oxidase from pig plasma as studied by magnetic resonance and kinetic methods, *Biochem. J.* **177**:289–302.
Baron, A. J., Rogers, M. S., Dooley, D. M., Knowles, P. F., and McPherson, M. J., 1999, Galactose oxidase pro-sequence cleavage and cofactor assembly are self-processing events, *J. Am. Chem. Soc., submitted.*
Bellelli, A., Brunori, M., Finazzi-Agro, A., Floris, G., Giartosi, A., and Rinaldi, A., 1985, Transient kinetics of copper-containing lentil (*Lens culinari*) seedling amine oxidase, *Biochem. J.* **232**:923–926.
Bono, P., Salmi, M., Smith, D. J., and Jalkanen, S., 1998, Cloning and characterisation of mouse vascular adhesion protein-1 reveals a novel molecule with enzymatic activity, *J. Imunol.* **160**:5563–5571.
Bork, P., and Doolittle, R. F., 1994, The *Drosophila* kelch motif is derived from a common enzyme fold, *J. Mol. Biol.* **236**:1277–1282.
Borman, C. D., Saysell, C. G., and Sykes, A. G., 1997, Kinetic studies on the reactions of *Fusarium* galactose oxidase with five different substrates in the presence of dioxygen, *J. Biol. Inorg. Chem.* **2**:480–487.

Bruinenberg, P. G., Evers, M., Waterham, H. R., Kuipers, J., and Arnberg, A. C., 1989, Cloning and sequencing of the peroxisomal amine oxidase gene from *Hansenula polymorpha*, *Biochem. Biophys. Acta*, **1008**:157–167.

Buchanan, R. M., Wilson-Blumenberg, C., Trapp, C., Larsen, S. K., Green, D. L., and Pierpont, C. G., 1986, Counterligand Dependence of Charge Distribution in Copper-Quinone Complexes. Structural and Magnetic properties of (3,5-Di-*tert*-butylcatecholato)(bipyridine)copper(II), *Inorg. Chem.* **25**:3070-3076.

Buffoni, F., and Blaschko, H., 1964, Benzylamine oxidase and histaminase: purification and crystallization of an enzyme from pig plasma, *Proc. Roy. Soc. B*, **161**:153.

Buffoni, F., 1966, Histaminase and related amine oxidases, *Pharmacol. Rev.* **18**:1163–1199.

Cai, D. Y., and Klinman, J. P., 1994, Copper amine oxidase: heterologous expression, purification and characterisation of an active enzyme in *Saccharomyces cerevisiae*, *Biochemistry* **33**:7647–7653.

CCP4; Collaborative Computational Project number 4 1994 *Acta Cryst* D50:760–763.

Cha, Y., Murray, C. J., and Klinman, J. P., 1989, Hydrogen tunneling in enzyme reactions, *Science* **243**:1325–1330.

Chaudhuri, P., Hess, M., Flörke, U., and Wieghardt, K., 1998, From Structural Models of Galactose Oxidase to Homogeneous Catalysis: Efficient Aerobic Oxidation of Alcohols, *Angew. Chem., Int. Ed. Engl.* **37**:2217–2220.

Chaudhuri, P., Hess, M., Weyhermüller, T., and Wieghardt, K., 1999, Aerobic Oxidation of Primary Alcohols by a New Mononuclear Cu(II)-Radical Catalyst, *Angew. Chem., Int. Ed. Engl.* **38**:1095–1098.

Christoserdov, A. Y., Tsygankov, Y. D., and Lidstrom, M. E., 1991, *J. Bact.* **173**:5901–5908.

Coleman, A. A., Hindsgaul, O., and Palcic, M. M., 1989, Stereochemistry of copper amine oxidase reactions, *J. Biol. Chem.* **264**:19500–19505.

Coleman, A. A., Scaman, C. H., Kang, Y. J., and Palcic, M. M., 1991, Stereochemical trends in copper amine oxidase reactions, *J. Biol. Chem.* **266**:6795–6800.

Cooper, J. A. D., Smith, W., Bacila, M., and Medina, H., 1959, Galactose oxidase from *Polyporus circinatus*, *J. Biol. Chem.* **234**:445–448.

Cooper, R. A., Knowles, P. F., Brown, D. E., McGuirl, M. A., and Dooley, D. M., 1992, Evidence for copper and 3,4,6 trihydroxyphenylalanine quinone cofactors in an amine oxidase from the gram-negative bacterium *Escherichia coli* K-12, *Biochem. J.* **288**:337–340.

Crabbe, M. J. C., Waight, R. D., Bardsey, W. G., Barker, R. W., Kelly, I. D., and Knowles, P. F., 1976, Human placental diamine oxidase, *Biochem. J.* **155**:679–687.

DeBiase, D., Agostinelli, E., DeMatteis, G., Mondovi, B., and Morpurgo, L., 1996, Half of the sites reactivity of bovine serum amine oxidase: reactivity and chemical identity of the second site, *Eur. J. Biochem.* **237**:727–732.

DeFilippis, M. R., Murthy, C. P., Broitman, F., Weinraub, D., Faraggi, M., and Klapper, M. H., 1991, Electrochemical Properties of Tyrosine Phenoxy and Tryptophan Indolyl Radicals in Peptides and Amino Acid Analogues, *J. Phys. Chem.* **95**:3416–3419.

Dooley, D. M., McGuirl, M. A., Brown, D. E., Turowski, P. N., McIntire, W. S., and Knowles, P. F., 1991, The Cu(I)-semiquinone state in substrate-reduced amine oxidases, *Nature* **349**:262–264.

Dooley, D. M., McIntire, W. S., McGuirl, M. A., Cote, C. E., and Bates, J. L., 1990, Characterisation of the active site of *Arthrobacter P1* methylamine oxidase: evidence for copper-quinone interactions, *J. Amer. Chem. Soc.*, **112**:2782–2789.

Dooley, D. M., 1999, Structure and biogenesis of topaquinone and related cofactors, *J. Biol. Inorg. Chem.* **4**:1–11.

Engel, P. C., 1977, *Enzyme Kinetics*, Chapman and Hall, London, p. 65.

Evans, P. T., and Malmberg, R. I., 1989, Do polyamines have roles in plant development? *Ann. Rev. Plant Physiol.* **40**:235–269.

Faber, H. R., Groom, C. R., Baker, H. M., Morgan, W. T., Smith, A., and Baker, E. N., 1995, 1. 8 Å crystal structure of the C-terminal domain of rabbit serum hemopexin, *Structure* **3**:551–559.

Farnum, M., Palcic, M., and Klinman, J. P., 1986, pH dependence of deuterium isotope effects and tritium exchange in the bovine plasma amine oxidase reaction: a role for single base catalysis in amine oxidation and imine exchange, *Biochemistry*, **25**:1898–1904.

Firbank, S., Phillips, S. E. V., Vinecombe, E., and Wilmot, C. M., unpublished.

Galston, A. W., and Kaur-Sawhney, R., 1995, Polyamines as engogenous growth regulators, in *Plant Hormones: Physiology, Biochemistry & Molecular Biology* (Davies, ed. ), Kluwer Netherlands, publishers, pp. 158–178.

Grant, K. L., and Klinman, J. P., 1989, Evidence that both protium and deuterium undergo significant tunneling in the reaction catalysed by bovine serum amine oxidase, *Biochemistry* **28**:6597–6605.

Halcrow, M. A., Chia, L. M. L., Liu, X., McInnes, E. J. L., Yellowlees, L. J., Mabbs, F. E., Scowen, I. J., McPartlin, M., and Davies, J. E., 1999, Syntheses, Structures and Electrochemistry of Copper(II) Salicylaldehyde/Tris(3-phenylpyrazolyl)borate Complexes as Models for the Radical Copper Oxidases, *J. Chem. Soc., Dalton Trans.* **1999**:1753–1762.

Halfen, J. A., Jadzdzewski, B. A., Mahaptara, S., Berreau, L. M., Wilkinson, E. C., Que, L. jr., and Tolman, W. B., 1997, Synthetic Models of the Inactive Copper(II)-Tyrosinate and Active Copper(II)-Tyrosyl Forms of Galactose and Glyoxal Oxidases, *J. Am. Chem. Soc.* **119**:8217–8227.

Hamilton, G. A., de Jersey, J., and Adolf, P. K., 1973, Galactose oxidase, in *Oxidases and Related Redox Systems*, King, T. E., Mason, H. S., and Morrison, M. editors, University Park Press, Baltimore, **1**:103–124.

Harrison, M. D., Jones, C. E., and Dameron, C. T., 1999, Copper chaperones: function, structure and copper-binding properties, *J. Biol. Inorg. Chem.* **4**:145–153.

Hartmann, C., and Klinman, J. P., 1987, Reductive trapping of substrate to bovine plasma amine oxidase, *J. Biol. Chem.* **262**:962–965.

Hartmann, C., and Klinman, J. P., 1991, Structure-function studies of substrate oxidation by bovine serum amine oxidase: relationship to cofactor structure and mechanism, *Biochemistry* **30**:4605–4611.

Haywood, G. W., and Large, P. J., 1981, Microbial oxidation of amines: distribution, purification and properties of two primary amine oxidases from the yeast *Candida boidinii* grown on amines as sole nitrogen source, *Biochem. J.* **199**:187–201.

Holt, A., Alton, G., Scaman, C. H., Loppnow, G. R., Szpacenko, A., Svendsen, I., and Palcic, M. M., 1998, Identification of the quinone cofactor in mammalian semicarbazide-sensitive amine oxidase, *Biochemistry* **37**:4946–4957.

Isobe, K., Tani, Y., and Yamada, H., 1980, Crystallization and characterisation of polyamine oxidase from *Penicillium chrysogenum*, *Agric. Biol. Chem.* **44**:2651–2658.

Isobe, K., Tani, Y., and Yamada, H., 1982, Crystallization and characterization of agmatine oxidase from *Penicillium chrysogenum*, *Agric. Biol. Chem.*, **46**:1353–1359.

Ito, N., Phillips, S. E. V., Stevens, C., Ogel, Z. B., McPherson, M. J., Keen, J. N., Yadav, K. D. S., and Knowles, P. F., 1991, Novel thioether bond revealed by a 1. 7 Å crystal structure of galactose oxidase, *Nature* **350**:87–90.

Ito, N., Phillips, S. E. V., Yadav, K. D. S., and Knowles, P. F., 1994, Crystal structure of a free radical enzyme, galactose oxidase, *J. Mol. Biol.* **238**:794–814.

Itoh, S., Takayama, S., Arakawa, R., Furuta, A., Komatsu, M., Ishida, A., Takamuku, S., and Fukuzumi, S., 1997, Active Site Models for Galactose Oxidase. Electronic Effects of the Thioether Group in the Novel Organic Cofactor, *Inorg. Chem.* **36**:1407–1416.

Janes, S. M., and Klinman, J. P., 1991, An investigation of bovine serum amine oxidase active site stoichiometry: evidence for an aminotransferase mechanism involving two carbonyl cofactors per enzyme dimer, *Biochemistry* **30**:4599–4605.

Janes, S. M., Mu, D., Wemmer, D., Smith, A. J., Kaur, S., Maltby, D., Burlingame, A. L., and Klinman, J. P., 1990, A new redox cofactor in eukaruotic enzymes: 6-hydroxydopa at the active site of bovine serum amine oxidase, *Science*, **248**:981–987.

Jazdzewski, B. A., Young, V. G. jr., and Tolman, W. B., 1998, A Three-Coordinate Copper(I)-Phenoxide Complex that Models the Reduced Form of Galactose Oxidase, *Chem. Commun.* **1998**:2521–2522.

Kagan, H. M., and Trackman, P. C., 1992, Lysyl oxidase, in *Principles and Applications of Quinoproteins*, Davidson, V. L. editor, Marcel Dekker New York publishers 1992, pp. 173–189.

Kelly, I. D., Knowles, P. F., Yadav, K. D. S., Barsley, W. G., Leff, P., and Waight, R. D., 1981, Steady state kinetic studies on benzylamine oxidase from pig plasma, *Eur. J. Biochem.* **144**:133–138.

Kenyon, K., Contente, S., Trackman, P. C., Tang, J., Kagan, H. M., and Friedman, R. F., Lysyl oxidase and *rrg* messenger RNA, 1991, *Science* 253, 802.

Klinman, J. P., and Mu, D., 1994, Quinoenzymes in biology, *Annu. Rev. Biochem.* **63**:299–344.

Klinman, J. P., 1996, Mechanisms whereby mononuclear copper proteins functionalize organic substrates, *Chem. Rev.* **96**:2541–2561.

Klinman, J. P., Dooley, D. M., Duine, J. A., Knowles, P. F., Mondovi, B., and Villafranca, J. J., 1991, Status of the cofactor identity in copper oxidative enzymes, *FEBS Lett.* **282**:1–4.

Knowles P. F., and Yadav, K. D. S., 1984, Amine oxidases in "Copper Proteins and Copper Enzymes" Vol 2 Edited by Lontie, R. CRC Press, Boca Raton, Florida Vol 2 103–129.

Knowles, P. F., and Dooley, D. M., 1994, Amine oxidases in *Metal Ions in Biological Systems* Editors Sigel, H. and Sigel, A., Publishers Marcel Dekker; **30**:361–403.

Knowles, P. F., and Ito, N., 1993, Galactose oxidase, in *Perspectives in Bioinorganic Chemistry*, Editors Hay, R. W., Dilworth, J. R., and Nolan, K. B. Jai Press, **2**:207–244.

Knowles, P. F., Brown, R. D., Koenig, S. H., Wang, S., Scott, R. A., McGuirl, M. A., Brown, D. E., and Dooley, D. M., 1995, Spectroscopic studies of the activesite of galactose oxidase, *Inorg. Chem.* **34**:3895–3902.

Knowles, P. F., Lowe, D. J., Peters, J., Thorneley, R. N. F., and Yadav, K. D. S., 1983, Kinetic and magnetic resonance studies on amine oxidases, in "The Coordination Chemistry of Metallo-Enzymes" Edited by Bertini, I., Drago, R. S., and Luchinat, C. Reidel publishers, 159–176.

Kohen, A., and Klinman, J. P., 1998, Enzyme catalysis, beyond classical paradigms, *Acc. Chem. Res.* **31**:397–404.

Kosman, D. J., 1984, Galactose oxidase, in *Copper proteins and Copper Enzymes*, Lontie, R. editor, CRC Press, **2**:1–26.

Kosman, D. J., Ettinger, M. J., Weiner, R. E., and Massaro, E. J., 1974, The molecular properties of galactose oxidase, *Arch. Biochem. Biophys.* **165**:456–467.

Kraulis, P. J., 1991, Molscript: a program to produce both detailed and schematic plots of protein structures, *J. Appl. Cryst.* **24**:946–950.

Kumar, V., Dooley, D. M., Freeman, H. C., Guss, J. M., Harvey, I., McGuirl, M. A., Wilce, M. C. J., and Zubak, V. M., 1996, Crystal structure of a eukaryotic (pea seedling) copper-containing amine oxidase at 2.2 Å resolution, *Structure*, **4**:943–955.

Kuivaniemi, H., Korhonen, R. M., Vaheri, A., and Kivirriko, K. I., 1986, Deficient production of lysyl oxidase in cultures of malgnantly transformed human cells, *FEBS Lett.* **195**:261–264.

Lee, Y., and Sayre, L. M., 1995, Model studies on the quinone-containing copper amine oxidases: unambiguous demonstration of a transamination mechanism, *J. Amer. Chem. Soc.* **117**:11823–11828.

Lerch, K., 1982, Primary structure of tyrosinase from *Neurospora crassa*: complete amino-acid sequence and chemical structure of a tripeptide containing an unusual thioether, *J. Biol. Chem.* **257**:6414–6419.

Li, R. B., Klinman, J. P., and Mathews, F. S., 1998, Copper amine oxidase from *Hansenula polymorpha*: the crystal structure determined at 2. 4 Å resolution reveals the active conformation, *Structure* 6:293–307.

Lingueglia, E., Renard, S., Voilley, N., Waldmann, R., Chassande, O., Lazdunski, M., and Barbry, P., 1993, Molecular cloning and functional expression of different molecular forms of rat amiloride binding proteins, *Eur. J. Biochem.* **216**:679–687.

Lobenstein-Verbeek C. L., Jongejan J. A., Frank J., and Duine J. A., 1984, FEBS Lett 170:305–309.

Lyons, T. J., Gralla, E. B., and Valentine, J. S., 1999, Biological chemistry of copper-zinc superoxide dismutase and its link to amytrophiclateral sclerosis in *Metal ions in biological systems*, edited by Sigel, H. and Sigel, A., Dekker, publishers, **36**:125–177.

Malmstrom, B. G., Andreasson, L. E., and Reinhammar, B., 1975, Copper oxidases and superoxide dismutase, in *The Enzymes, 3rd edition*, Boyer, P. D. editor, Academic Press, NY, **12**:507–579.

Maradufu, A., Cree, G. G. M., and Perlin, A. S., 1971, Stereochemistry of dehydrogenation by galactose oxidase, *Can. J. Chem.* 49:3429–3437.

Matsuzaki, R., Fukui, T., Sato, H., Ozaki, Y., and Tanizawa, K., 1994, Generation of the TOPA quinone cofactor in bacterial monoamine oxidase by cupric ion-dependent autooxidation of a specific tyrosyl residue, *FEBS Lett.* **351**:360–364.

McCracken, J., Peisach, J., Cote, C. E., McGuirl, M. A., and Dooley, D. M., 1992, Pulsed EPR studies of the semiquinone state of copper-containing amine oxidases, *J. Amer. Chem. Soc.* **114**:3715–3720.

McIntire, W. S., 1990, Methylamine oxidase from *Arthrobacter P1*, *Methods in Enzymology*, **188**:227–235.

McIntire, W. S., and Hartmann, C., 1993, Copper-containing amine oxidases in *Principles and Applications of Quinoproteins*, Editor Davison, V., Marcel Dekker publisher 97–171.

McIntire, W. S., Wemmer, D. E., Christoserdov, A., and Lidstrom, M. E., 1991, A new cofactor in a prokaryotic enzyme-tryptophan tryptophylquinone as the redox prosthetic group in methylamine dehydrogenase, *Science* 252:817–824.

McPherson, M. J., Ogel, Z. B., Stevens, C., Yadav, K. D. S., Keen, J. N., and Knowles, P. F., Galactose oxidase of *Dactylium dendroides*: gene cloning and sequence analysis, 1992, *J. Biol. Chem.* **267**:8146–8152.

McPherson, unpublished observations.

Medda, R., Padiglia, A., and Floris, G., 1995, Plant copper amine oxidases, *Phytochemistry* **39**:1–9.

Moenne-Loccoz, P., Nakamura, N., Steinebach, V., Duine, J. A., Mure, M., Klinman, J. P., and Sanders-Loehr, J., 1995, Characterisation of the TOPA quinone cofactor in amine oxidase from *Escherichia coli* by resonance Raman spectroscopy, 1995 *Biochemistry* **34**:7020–7026.

Møller, S. G., and McPherson, M. J., 1998, Developmental expression and biochemical analysis of the *Arabidopsis atao 1* gene encoding an $H_2O_2$-generating diamine oxidase, *The Plant Journal* **13**:781–791.

Mondovi, B., Rotilio, Costa, M. T., Finazzi-Agro, A., Chiancone, E., Hansen, R. E., and Beinert, H., 1967, Diamine oxidase from pig kidney: improved purification and properties, *J. Biol. Chem.* **242**:1160.

Mu, D., Janes, S. M., Smith, A. J., Brown, D. E., Dooley, D. M., and Klinman, J. P., 1992, Tyrosine codon corresponds to TOPA quinone at the active site of copper amine oxidases, *J. Biol. Chem.* **267**:7979–7982.

Mu, D., Medzihradszky, K. F., Adams, G. W., Mayer, P., Hines, W. M., Burlingame, A. L., Smith, A. J., Cai, D. Y., and Klinman, J. P., 1994, Primary structures for a mammalian cellular and serum copper amine oxidase, *J. Biol. Chem.*, **269**:9926–9932.

Müller, J., Weyhermüller, T., Bill, E., Hildenbracht, P., Ould-Moussa, L., Glaser, T., and Wieghardt, K., 1998, Why Does the Active Form of Galactose Oxidase Possess a Diamagnetic Ground State? *Angew. Chem., Int. Ed. Engl.* **37**:616–619.

Mure, M., and Klinman, J. P., 1995a, Model studies of topaquinone-dependent amine oxidases: oxidation of benzylamine by topaquinone analogues, *J. Amer. Chem. Soc.* **117**:8698–8706.

Mure, M., and Klinman, J. P., 1995b, Model studies of topaquinone-dependent amine oxidases: characterisation of reaction intermediates and mechanism, *J. Amer. Chem. Soc.* **117**:8707–8718.

Mure, M., and Klinman, J. P. 1993, Synthesis and spectroscopic characterisation of model compound for the active site cofactor in copper amine oxidases, *J. Amer. Chem. Soc.* **115**:7117–7127.

Murray, J. M., Wilmot, C. M., Saysell, C. G., Jaeger, J., Knowles, P. F., Phillips, S. E. V., and McPherson, M. J., 1999, The active site base controls cofactor reactivity in *Escherichia coli* amine oxidase: X-ray crystallographic studies with mutational variants, *Biochemistry*, **38**:8217–8227.

Nakamura, N., Kohzuma, T., Kuma, H., and Suzuki, S., 1992, The first topa containing copper (II) complex, $[Cu(DL\text{-}topa)(bpy)(H_2O)]BF_4$ as a model for the active site in copper-cotaining amine oxidases, *J. Am. Chem. Soc.* **114**:6550–6552.

Nakamura, N., Moenne-Loccoz, P., Tanizawa, K., Mure, M., Suzuki, S., Klinman, J. P., and Sanders-Loehr, J., 1997, Topaquinone-dependent amine oxidases: identification of reaction intermediates by Raman spectroscopy, *Biochemistry*, **36**:11479–11486.

Ogel, Z. B., Brayford, D., and McPherson, M. J., 1994, Cellulose-triggered sporulation in the galactose oxidase producing fungus *Cladobotryum (Dactylium) dendroides* NRRL-2903 and its reidentification as a species of *Fusarium*, *Mycology Research* **98**:474–480.

Oi, S., Inamasu, M., and Yasunobu, K. T., 1970, Mechanistic studies on beef plasma amine oxidase, *Biochemistry*, **9**:3378–3383.

Olsson, B., Olsson, J., and Pettersson, G., 1976, Stopped flow spectrophotometric characterisation of enzyme reaction intermediates in the anaerobic reduction of pig plasma amine oxidase by amine substrate, *Eur. J. Biochem.* **71**:375–382.

Ostermeier, C., Harrenga, A., Ermler, U., and Michel, H., 1997, Structure at 2. 7 Å resolution of the *Paracoccus denitificans* two-subunit cytochrome oxidase complexed with an antibody fragment, *Proc. Natl. Acad. Sci. USA* **94**:10547–10553.

Palcic, M. M., and Klinman, J. P., 1983, Isotopic probes yield microscopic constants: separation of binding energy from catalytic efficiency in the bovine plasma amine oxidase reaction, *Biochemistry* **22**:5957–5966.

Parrot, S., Jones, S., and Cooper, R. A., 1987, 2-Phenylethylamine catabolism by *Escherichia coli K12*, *J. Gen. Microbiol.* **133**:347–351.

Parsons, M. R., Convery, M. A., Wilmot, C. M., Yadav, K. D. S., Blakeley, V., Corner, A. S., Phillips, S. E. V., McPherson, M. J., and Knowles, P. F., 1995, Crystal structure of a quinoenzyme: copper amine oxidase of *Escherichia coli* at 2. Å resolution *Structure* **3**:1171–1184.

Pettersson, G., 1985, Amine oxidases, in *Structure and Function of Amine Oxidases*, Mondovi, B. editor, CRC Press, p. 105.

Plastino, J., Green, E. L., Sanders-Loehr, J., and Klinman, J. P., 1999, An unexpected role for the active site base in cofactor orientation and flexibility in the copper amine oxidase from *Hansenula polymorpha*, *Biochemistry*, 8204–8216D.

Rall, J., and Kaim, W., 1994, Ligand-Controlled Oxidation State Ambivalence in Copper-Quinone Complexes, *J. Chem. Soc., Faraday Trans.* **90**:2905–2908.

Reynolds, M. P., Baron, A. J., Wilmot, C. M., Vinecombe, E., Stevens, C., Phillips, S. E. V., Knowles, P. F., and McPherson, M. J., 1997, Structure and mechanism of galactose oxidase: catalytic role of tyrosine 495, *J. Biol. Inorg. Chem.* **2**:327–335.

Rinaldi, A., Giartosio, A., Floris, G., Medda, R., and Finazzi-Agro, A., 1989, Lentil seedling amine oxidase: preparation and properties of the copper-free enzyme, *Biochem. Biophys. Res. Commun.,* **120**:242–249.

Rius, F. X., Knowles, P. F., and Pettersson, G., 1984, The kinetics of ammonia release during the catalytic cycle of pig plasma amine oxidase, *Biochem. J.* **220**:767–772.

Ruggiero, C. E., and Dooley, D. M., 1999, Stoichiometry of the TOPA quinone biogenesis reaction in copper amine oxidases, *Biochemistry* **38**:2892–2898.

Ruggiero, C. E., Smith, J. A., Tanizawa, K., and Dooley, D. M., 1997, Mechanistic studies of topa quinone biogenesis in phenylethylamine oxidase, *Biochemistry* 36:1953–1959.

Salmi, M., and Jalkanen, S., 1997, How do lymphocytes know where to go: current concepts and enigmas of lymphocyte homing, *Advances in Immunology* **64**:139–218.

Salminen, T. A., Smith, D. J., Jalkanen, S., and Johnson, M. S., 1998, Structural model of the catalytic domain of an enzyme with cell adhesion activity: human vascular adhesion protein—1 (HVAP-1) D4 domain is an amine oxidase, *Protein Engineering,* **11**:1195–1204.

Scaman, C. M., and Palcic, M. M., 1992, Stereochemical course of tyramine oxidation by semicarbazide-sensitive amine oxidase, *Biochemistry* **31**:6829–6841.

Schwartz, B., Green, E. L., Sanders-Loehr, J., and Klinman, J. P., 1998, The relationship between conserved consensus site residues and the productive conformation for TPQ cofactor in a copper-containing amine oxidase from yeast, *Biochemistry* **37**:16591–16600.

Scott, R. A., and Dooley, D. M., 1985, X-ray absorption spectroscopic studies of the copper sites in bovine plasma amine oxidase, *J. Amer. Chem. Soc.* **107**:4348–4350.

Shah, M. A., Scaman, C. H., Palcic, M. M., and Kagan, H. M., 1993, Kinetics and sterospecificity of the lysyl oxidase reaction, *J. Biol. Chem.* **268**:11573–11579.

Smith, T. A., 1985, The di- and poly-amine oxidases of higher plants, *Biochem. Soc. Trans.,* **13**:319–322.

Smith, D. J., Salmi, M., Bono, P., Hellman, J., Leu, T., and Jalkanen, S., 1998, Cloning of vascular adhesion protein 1 reveals a novel multifunctional adhesion molecule, *J. Exper. Med.* **188**:17–27.

Sokolowski, A., Leutbecher, H., Weyhermüller, T., Schnepf, R., Bothe, E., Bill, E., Hildenbrandt, P., and Wieghardt, K., 1997, Phenoxyl-Copper (II) Complexes: Models for the Active Site of Galactose Oxidase, *J. Biol. Inorg. Chem.* **2**:444–453.

Sondek, J., Bohm, A., Lambright, D. G., Hamm, H. E., and Sigler, P. B., 1996, Crystal structure of a G protein βγ dimer at 2. 1 Å resolution, *Nature* **379**:369–374.

Speier, G., Tisza, S., Tyeklár, Z., Lange, C. W., and Pierpont, C. G., 1994, Coligand-Dependent Shifts in Charge Distribution for Copper Complexes Containing 3,5-Di-*tert*-butylcatecholate and 3,5-Di-*tert*-butylsemiquinonate Ligands, *Inorg. Chem.* 33:2041–2045.

Stubbe, J. A., 1989, Protein radical involvement in biological catalysis, *Annu. Rev. Biochem.,* **58**:257–285.

Su, Q., and Klinman, J. P., 1998, Probing the mechanism of proton coupled electron transfer to dioxygen: the oxidative half-reaction of bovine serum amine oxidase, *Biochemistry* 37:12513–12525.

Sugino, H., Sasaki, M., Azakami, H., Yamashita, M., and Murooka, Y., 1992, A monoamine-regulated *Klebsiella aerogenes* operon containing the monoamine oxidase (MAOA) structural gene and the MAOC gene, *J. Bact.* 174:2485–2492.

Sundaresan, M., Yu, Z. X., Ferrans, V. J., Irani, K., and Finkel, T., 1995, Requirement for generation of $H_2O_2$ for platelet-derived growth factor signal transduction, *Science* **270**:296–299.

Suzuki, S., Sakurai, T., Nakahara, A., Manabe, T., and Okuyama, T., 1986, Roles of the two copper ions in bovine serum amine oxidase, *Biochemistry* **25**:338–341.

Suzuki, S., Yamaguchi, K., Nakamura, N., Tagawa, Y., Kuma, H., and Kawamoto, T., 1998, Structures and Properties of Ternary Copper (II) Complexes with 4,5-Dihydroxyphenylalanine or 2,4,5-Trihydroxyphenylalanine and Aromatic Diamines, *Inorg. Chim. Acta* **283**: 260–267.

Tabor, H., 1954, Metabolic studies on histidine, histamine and related imidazoles, *Pharmacol. Rev.*, **6**:299–343.

Tanizawa, K., Matsuzaki, R., Shimizu, E., Yorifuji, T., and Fukui, T., 1994, Cloning and sequencing of phenylethylamine oxidase from *Arthrobacter globiformis* and implication of Tyr 382 as the precursor to its covalently bound quinone cofactor, *Biochem. Biophys. Res. Commun.* **199**:1096–1102.

Tipping, A. J., and McPherson, M. J., 1995, Cloning and molecular analysis of the pea seedling copper amine oxidase, *J. Biol. Chem.* **270**:16939–16946.

Trackman, P. C., Pratt, A. M., Wolanski, A., Tang, S.-S., Offer, G. D., Troxler, R. F., and Kagan, H. M., 1990, Cloning of rat lysyl oxidase cDNA: complete codons and predicted aminoacid sequences, *Biochemistry* **29**:4863–4870.

Turowski, P. N., McGuirl, M. A., and Dooley, D. M., 1993, Intramolecular electron transfer rate between active site copper and topa quinone in pea seedling amine oxidase, *J. Biol. Chem.* **268**:17680–17682.

VanIersel, J., Van der Meer, R. A., and Duine, J. A., 1986, Methylamine oxidase from *Arthrobacter P1*, a bacterial copper-quinoenzyme, *Eur. J. Biochem.* **161**:415–419.

Varghese, J. N., Laver, W. G., and Colman, P. M., 1983, Structure of the influenza virus glycoprotein antigen neuraminidase at 2. 9 Å resolution, *Nature* **303**:35–40.

Vellieux, F. M. D., Huitema, F., Groendijk, H., Kalk, K. H., Jzn, J. F., Jongejan, J. A., Duine, J. A., Petratos, K., Drenth, J., and Hol, W. G. J., 1989, The structure of quinoprotein methylamine dehydrogenase at 2. 25 Å resolution, *EMBO J.* **8**:2171–2178.

Vignevich, V., Dooley, D. M., Guss, J. M., Harvey, I., McGuirl, M. A., and Freeman, H. C., 1993, Crystallization and preliminary crystallographic characterization of copper-containing amine oxidase from pea seedlings, *J. Mol. Biol.* **229**:243–245.

Villafranca, J. J., Freeman, J. C., and Kotchevar, A., 1993, in *Bioinorganic Chemistry of Copper*, Karlin, K. D. and Tyeklar, Z. editors, Chapman and Hall, New York, 439–446.

Wang, S. X., Mure, M., Medzihradzky, K. F., Burlingame, A. L., Brown, D. E., Dooley, D. M., Smith, A. J., Kagan H. M., and Klinman, J. P., 1996, A cross linked cofactor in lysyl oxidase: redox function for the aminoacid side chains, *Science* **252**:1078–1084.

Wang, S. X., Nakamura, N., Klinman, J. P., and Sanders-Loehr, J., 1997, Characterization of the native lysine tyrosylquinone cofactor in lysyl oxidase by resoance Raman, *J. Biol. Chem.* **272**:28841–28844.

Wang, Y., and Stack, T. D. P., 1996, Galactose Oxidase Model Complexes: Catalytic Reactivities, *J. Am. Chem. Soc.* **118**:13097–13098.

Wang, Y., DuBois, J. L., Hedman, B., Hodgson, K. O., and Stack, T. D. P., 1998, Catalytic Galactose Oxidase Models: Biomimetic Cu(II)-Phenoxyl-Radical Reactivity, *Science* **279**:537–540.

Whittaker, J. W., 1994, The free radical-coupled copper active site in galactose oxidase, *Metal Ions in Biology*, Editors Sigel, H. and Sigel, A., Marcel Dekker publishers, **30**:315–360.

Whittaker, M. M., and Whittaker, J. W., 1988, The active site of galactose oxidase, *J. Biol. Chem.* **263**:6074–6080.

Whittaker, M. M., and Whittaker, J. W., 1993, Ligand interactions with galactose oxidase: mechanistic insights, *Biophys. J.* **64**:762–772.

Whittaker, M. M., Ballou, D. P., and Whittaker, J. W., 1998, Kinetic isotope effects as probes of the mechanism of galactose oxidase, *Biochemistry* **37**:8426–8436.

Whittaker, M. M., Duncan, W. R., and Whittaker, J. W., 1996a, Synthesis, Structure and Properties of a Model for Galactose Oxidase, *Inorg. Chem.* **35**:382–386.

Whittaker, M. M., Kersten, P. J., Nakamura, N., Sanders-Loehr, J., Schweizer, E. S., and Whittaker, J. W., 1996b, Glyoxal oxidase from *Phanerochaete chrysosporum* is a new radical-copper oxidase, *J. Biol. Chem.* **271**:681–687.

Whittaker, M. M., and Whittaker, J. W., 1990, A tyrosine-derived free radical in apogalactose oxidase, *J. Biol. Chem.* **265**:9610–9613.

Whittaker, M. M., DeVito, V. L., Asher, S. A., and Whittaker, J. W., 1989, Resonance Raman evidence for tyrosine involvement in the radical site of galactose oxidase, *J. Biol. Chem.* **264**:7104–7106.

Wilce, M. C. J., Dooley, D. M., Freeman, H. C., Guss, J. M., Matsunami, H., McIntire, W. S., Ruggiero, C. E., Tanizawa, K., and Yamaguchi, H., 1997, Crystal structures of the copper-containing amine oxidase from *Arthrobacter globiformis* in the holo and apo forms, *Biochemistry* **36**:16116–16133.

Williams, N., and Klinman, J. P., 1999, *J. Mol. Catal.* in press.

Wilmot, C. M., Hajdu, J., McPherson, M. J., Knowles, P. F., and Phillips, S. E. V., 1999, Visualisation of dioxygen bound to copper during enzyme catalysis, *Science* in press.

Wilmot, C. M., Murray, J. M., Alton, G., Parsons, M. R., Convery, M. A., Blakeley, V., Corner, A. S., Palcic, M. A., Knowles, P. F., McPherson, M. J., and Phillips, S. E. V., 1997, The catalytic mechanism of the quinoenzyme amine oxidase from *Escherichia coli*: exploring the reductive half reaction, *Biochemistry* **36**:1608–1620.

Xia, Z. X., Dai, W. W., Xiong, J. P., Hao, Z. P., Davidson, V. L., White, S., and Mathews, F. S., 1992, The 3-dimensional structures of methanol dehydrogenase from two methylotrophic bacteria at 2. 6 Å, *J. Biol. Chem.* **267**:22289–22297.

Yamada, H., Yasunobu, K., Yamano, T., and Mason, H. S., 1963, Copper in plasma amine oxidase, *Nature* **198**:1092–1093.

Zeller, E. A., 1938, Enzymatic degradation of histamine and diamines, *Helv. Chim. Acta* 21:880–890.

Zhang, X. P., Fuller, J. H., and McIntire, W. S., 1993, Cloning, sequencing, expression and regulation of the structural gene for the copper TOPA quinone-containing methylamine oxidase from *Arthrobacter P1*, a gram positive facultative methylotroph, *J. Bact.* **175**:5617–5627.

Zurita, D., Gautier-Luneau, I., Ménage, S., Pierre, J.-L., and Saint-Aman, E., 1997, A First Model for the Oxidized Active Form of the Active Site of Galactose Oxidase: a Free Radical Copper Complex, *J. Biol. Inorg. Chem.* **2**:46–55.

Zurita, D., Scheer, C., Pierre, J.-L., and Saint-Aman, E., 1996, Solution Studies of Copper(II) Complexes as Models for the Active Site of Galactose Oxidase, *J. Chem. Soc., Dalton Trans.* 4331–4336.

*Chapter 7*

# Electron Transfer and Radical Forming Reactions of Methane Monooxygenase

## Brian J. Brazeau and John D. Lipscomb

## 1. INTRODUCTION

Dioxygen is a unique reagent for biological systems being at once thermodynamically reactive and kinetically stable. Life in a sea of oxygen is possible because oxygenase and oxidase enzymes have evolved that allow the inherent reactivity of dioxygen to be realized in a controlled manner that maintains specificity. Monooxygenases are a class of such enzymes that catalyze the activation of dioxygen through reductive cleavage of the O—O bond with the incorporation of one oxygen atom into the product and the release of the second as water. Numerous mechanistic strategies are used within the monooxygenase family, but a cofactor of some type is generally required to facilitate the binding and reduction of dioxygen. Typically, hemes, metal ions, flavins, pterins, and similar redox active species are bound by the enzyme and perform this function. The cofactor selected depends to some extent on the type of

**BRIAN J. BRAZEAU and JOHN D. LIPSCOMB**    Department of Biochemistry, Molecular Biology, and Biophysics, University of Minnesota, Minneapolis, MN 55455 USA
*Subcellular Biochemistry, Volume 35: Enzyme-Catalyzed Electron and Radical Transfer*, edited by Holzenburg and Scrutton. Kluwer Academic / Plenum Publishers, New York, 2000.

substrate which will receive the oxygen atom. In particular, oxygen incorporation into unactivated hydrocarbons generally involves the well-studied monooxygenase cytochrome P-450 (P-450) (Ortiz de Montellano, 1995). P450 catalyzed reactions often entail the cleavage of a C—H bond and those with bond dissociation energies (bde) as great as 100 kcal/mol are known to be cleaved. One of the most remarkable C—H bond cleavage reactions carried out in Nature is the bacterial oxidation of methane to form methanol as part of a global carbon cycle that links aerobic and anaerobic one carbon metabolism (Dalton, 1980). Due to the high bde of methane (104 kcal/mol), this reaction would be a logical candidate for a P450 catalyzed reaction, but this is not the case. Work during the past two decades has shown that all forms of the enzyme methane monooxygenase (MMO), which catalyzes this reaction, are metalloenzymes devoid of heme. This discovery and the subsequent investigation of the novel cofactors that are employed by MMOs has led to new insights into the chemistry of oxygen activation. The current state of these studies is related here.

Methanotrophic bacteria use methane as the sole source of carbon and energy employing MMO to catalyze the first step in the methane oxidation pathway leading ultimately to $CO_2$.

$$CH_4 + O_2 + NAD(P)H + H^+ \rightarrow CH_3OH + H_2O + NAD(P)^+$$

MMO exists in both soluble (sMMO) and particulate forms and some methanotrophs can express both types depending upon the bacterial growth conditions (Stanley et al., 1983). Recently, the particulate form has been purified sufficiently to show that it is a copper containing enzyme (Chan et al., 1992; Nguyen et al., 1996; Nguyen et al., 1998) and mechanistic studies indicate that it catalyzes a direct insertion of oxygen into the C—H bond of methane (Wilkinson et al., 1996). However, it has not yet been possible to describe either the structure or the mechanism of this enzyme in great detail due to the difficulty of extracting it from the membrane in an active state. Much more is known about sMMO which can be isolated in a homogeneous and stable form. Consequently, sMMO will be the focus of this review.

Similar forms of sMMO are expressed by morphologically and metabolically distinct methanotrophs that have been classified as Type X and Type II (Lipscomb, 1994). These are typified by *Methylococcus capsulatus* Bath (MMO Bath) and *Methylosinus trichosporium* OB3b (MMO OB3b), respectively. The parallel studies of sMMO from these two organisms has facilitated the description of both the essential features of the enzyme structure and catalytic process and the species specific characteristics which might mask the more general features if viewed in isolation (Feig and Lippard, 1994; Lipscomb, 1994). Many of the studies described here have derived from our own studies of MMO OB3b, however, frequent comparison will be made to studies of MMO Bath.

As depicted in Figure 1, the sMMO system consists of three component proteins, a hydroxylase (MMOH), a reductase (MMOR), and an effector protein (MMOB). Although the fully reduced MMOH is capable of catalyzing a single turnover of methane oxidation in the absence of the other components (Fox *et al.*, 1989), multiple turnover at the maximum rate requires all three components, each playing a markedly different role in the overall process. Consequently, characterization of the components and their interactions is paramount to the understanding of the mechanism of MMO. Considerable progress has been made toward this goal through spectroscopic, crystallographic, and kinetic studies of the components and a detailed picture of the mechanism of MMO is beginning to emerge. These studies have shown that one role of MMOR is to use its inherent FAD and [Fe₂S₂] cluster cofactors to shuttle electrons from NADH to the active site of MMOH (Lund *et al.*, 1985). The chemical reaction then occurs in the α-subunit of the $(\alpha\beta\gamma)_2$ MMOH where an oxygen-bridged binuclear iron cluster is bound in a classic antiparallel 4-helix bundle structural motif (Woodland *et al.*, 1986; Ericson *et al.*, 1988; Fox *et al.*, 1988; Fox *et al.*, 1991; Rosenzweig *et al.*, 1993; Elango *et al.*, 1997). The 2-electron reduced binuclear cluster is capable of reacting with and activating molecular oxygen in preparation for attack on methane (Lee *et al.*, 1993a). Both MMOR and MMOB form specific complexes with MMOH that accelerate slow steps in the reaction cycle and maximize the concentration of the species that actually transfers oxygen to hydrocarbon substrates. Specifically, the

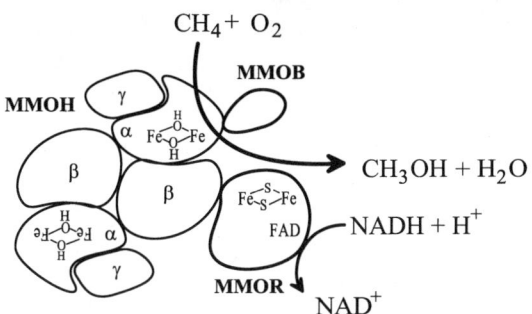

**FIGURE 1.** Reaction and protein components of the soluble form of methane monooxygenase.

cofactorless MMOB increases the overall rate of oxygen reaction with reduced MMOH by 1000-fold (Liu, Y. *et al.*, 1995). MMOR limits the formation of unreactive one-electron reduced MMOH and prevents uncoupling of product formation from NADH oxidation once two electrons have been transferred to MMOH (Liu, Y. *et al.*, 1997).

The oxidative attack on the C—H bond of methane requires a more reactive species than any described previously in a biological system. The fact that MMO is able to generate such a reagent has raised considerable interest in the details of its reaction mechanism. Through the application of a variety of transient kinetic and spectroscopic techniques a satisfying view of the reaction cycle of MMO and its unique reactive intermediates is rapidly emerging. MMOH initiates catalysis by cleaving the O—O bond of molecular oxygen as observed in many other oxygenases. However, the oxygen activation process culminates in the formation of a completely novel intermediate we have termed, compound Q (**Q**) (Lee at al., 1993a). This intermediate appears to react directly with methane and all adventitious MMO substrates. Spectroscopic studies of trapped **Q** show that this powerful oxidant contains two Fe(IV) atoms, the first such species to be described in biology (Lee *et al.*, 1993b). The structure of the iron cluster of **Q** is proposed to be a bis μ-oxo dinuclear Fe(IV) complex forming a so called "diamond core" as illustrated in Figure 2 (Shu *et al.*, 1997). Ferryl-oxo species are proposed in the mechanisms of a number of different enzymes, including P-450 and catalase and peroxidase, but **Q** is the only binuclear ferryl-oxo species. It is also the only well characterized Fe(IV) species to be trapped from an oxygenase.

The problem of determining the nature of the reaction between **Q** and substrate has proven difficult to solve, but it has also opened a new avenue of thought into the oxygen activation processes in general. For over 30 years, the mechanism of oxygen activation for attack on unactivated hydrocarbons has been assumed to be that proposed for P-450, in which a porphyrin Fe(IV)=O π cation radical is formed which subsequently abstracts

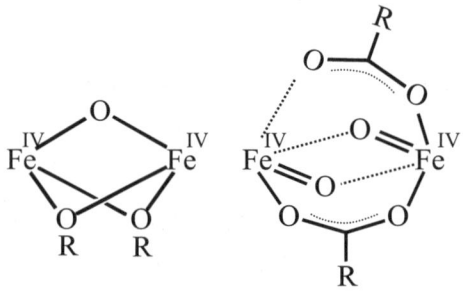

**FIGURE 2.** Postulated structures for the novel intermediate compound **Q**.

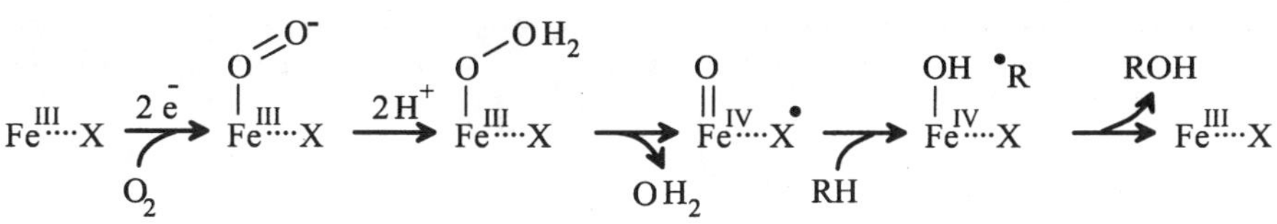

$$Fe^{III}\cdots X \xrightarrow[O_2]{2\,\bar{e}} \overset{\overset{O^{\nearrow}O^-}{\underset{|}{O}}}{Fe^{III}\cdots X} \xrightarrow{2\,H^+} \overset{\overset{O^{\nearrow}OH_2}{\underset{|}{O}}}{Fe^{III}\cdots X} \underset{OH_2}{\rightarrow} \overset{\overset{O}{\underset{||}{\phantom{O}}}}{Fe^{IV}\cdots X^{\bullet}} \underset{RH}{\xrightarrow{\phantom{x}}} \overset{\overset{OH\ ^{\bullet}R}{\underset{|}{\phantom{O}}}}{Fe^{IV}\cdots X} \xrightarrow{ROH} Fe^{III}\cdots X$$

**FIGURE 3.** Generic radical rebound mechanism of monooxygenases such as MMO and P450. X is a species that can provide one electron to stabilize the high valent oxo intermediate such as a second iron in the case of MMO or a porphyrin macrocycle in the case of P450.

a hydrogen atom from the substrate (Groves *et al.*, 1978; McMurry and Groves, 1986; Ortiz de Montellano, 1995). As illustrated generically in Figure 3, the final oxidized product is then formed by rebound of the metal bound hydroxy-radical to the substrate radical. MMO may follow a similar mechanism (Fox *et al.*, 1990; Fox *et al.*, 1991). For example, most of the experimental results are consistent with the formation of a discrete intermediate which is likely to be a radical in the case of methane because other possible intermediates would be too unstable in a biological system (Lipscomb and Que, 1998). However, most of the results pointing toward intermediates also show that the reactions occur at rates that are beyond those attainable by a conventional "rebound" reaction (Priestley *et al.*, 1992; Liu, K. *et al.*, 1993; Valentine *et al.*, 1997; Valentine *et al.*, 1999b). This suggests that the MMO reaction may be different from that of P450, at least in detail. On the other hand, MMO has allowed us to examine this important question in ways that were not possible with P450 systems in the past, and thus, a reexamination of oxygen activation processes in general is in progress.

The discussion that follows presents a summary of the results of recent studies of MMO. The complexity of the reaction and of the component interactions have mandated a multidisciplinary approach. Many principles that seem to apply globally to biological systems have frequently emerged from these studies due to the fact that MMO exhibits spectroscopic and kinetic characteristics that allow complex phenomena to be studied in unprecedented detail.

## 2. COMPONENTS

### 2.1. MMOH

#### 2.1.1. X-Ray Crystallography

The structure of the 245 kDa MMOH has been determined using X-ray crystallography for both the diferric (Rosenzweig *et al.*, 1993;

Rosenzweig *et al.*, 1995; Elango *et al.*, 1997; Rosenzweig *et al.*, 1997) and the diferrous states of the enzyme. The structures of MMOH from MMOH OB3b and MMOH Bath are remarkably similar, even more similar than expected from the degree of homology observed for the respective gene sequences (Stainthorpe *et al.*, 1989; Stainthorpe *et al.*, 1990; Cardy *et al.*, 1991a; Cardy *et al.*, 1991b). In each case as shown in Figure 4, MMOH is a dimeric protein with three different types of subunits in each protomer and a presumed active site cavity near the binuclear iron cluster in each α subunit. The crystal structure shows that the secondary structure is predominantly helical, with only three strands of β sheet found per protomer. The overall fold of the α and β subunits is very similar, and these subunits are responsible for the interactions that form the dimer, although the interactions are primarily between the β subunits. The γ subunit is somewhat elongated and interacts with both the α and β subunits on the opposite side of the protein with respect to the interface between the two protomers.

Interestingly, the detailed structure of the oxidized binuclear iron cluster environment depends to some extent on the temperature at which the data was collected and on the origin of the MMOH. The most similar structures, and those most likely to be representative of the enzyme in vivo, are derived from flash frozen MMOH Bath crystals (−160 °C) and MMOH

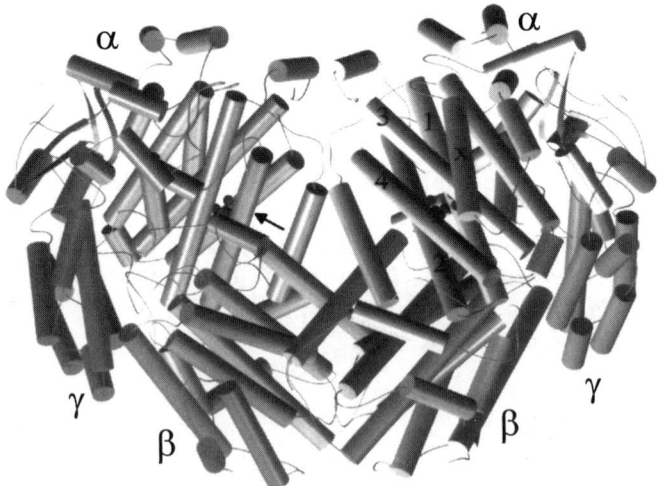

**FIGURE 4.** Backbone X-ray crystal structure of MMOH isolated from *M. trichosporium* OB3b. The cylinders represent helicies. The arrow points toward one of the two active sites in the (αβγ)₂ structure. (Elango *et al.*, 1997). The structure of the *M. capsulatus* Bath MMOH is essentially identical (Rosenzweig *et al.*, 1993).

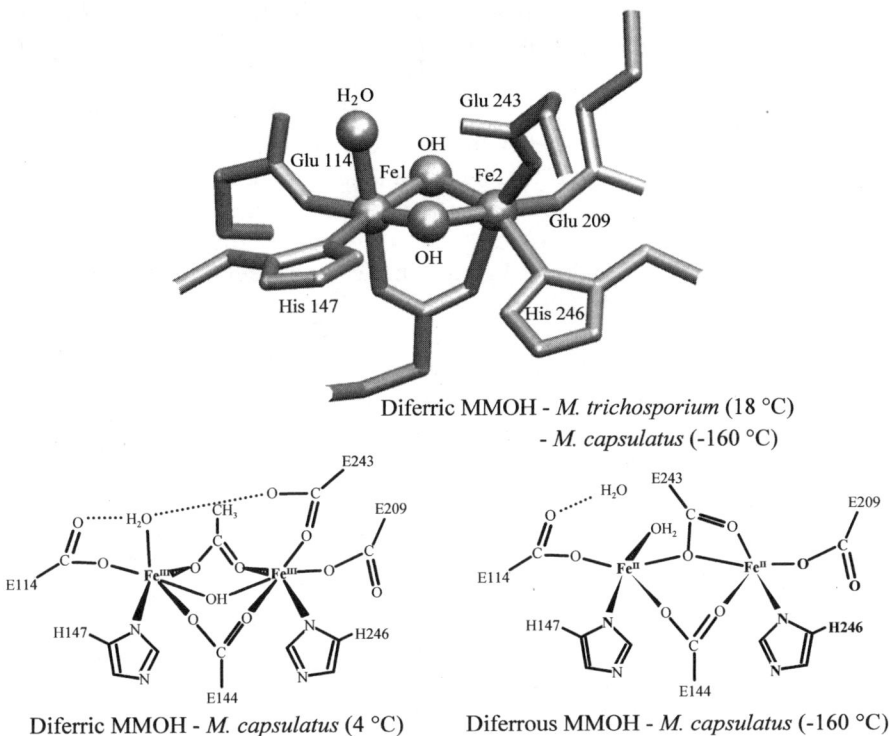

Diferric MMOH - *M. trichosporium* (18 °C)
- *M. capsulatus* (-160 °C)

Diferric MMOH - *M. capsulatus* (4 °C)          Diferrous MMOH - *M. capsulatus* (-160 °C)

**FIGURE 5.** X-ray crystal structures of the binuclear iron cluster in MMOH.

OB3b crystals at 18 °C illustrated in Figure 5 (Rosenzweig *et al.*, 1995; Elango *et al.*, 1997). In these forms, the observed Fe1-Fe2 distance is 2.99–3.1 Å, the iron atoms are each coordinated by one His ligand, and they are bridged by two hydroxo- ligands (or possibly one hydroxo- and one aquo ligand) and a glutamyl side chain carboxylate. The other protein ligands are all derived from monodentate glutamyl residues; Fe1 is coordinated by a single Glu residue, whereas Fe2 is coordinated by two Glu residues. The final ligand to Fe1 is a terminal water molecule, bringing the coordination number of each iron to six and leaving the overall charge of the cluster as neutral. The bridging aquo, hydroxo, and Glu residues in combination with the iron atoms form the "diamond core" of MMOH. This structure has been seen in model compounds of the active site of MMOH (Zang *et al.*, 1995; Que and Dong, 1996) and in another dinuclear iron enzyme, the R2 subunit of ribonucleotide reductase (Riggs-Gelasco *et al.*,

1998). This structural motif appears again in the catalytic cycle at the stage of intermediate $\mathbf{Q}$ as described below.

The first structure of diferric MMOH Bath was solved at 4 °C to 2.2 Å resolution (Rosenzweig et al., 1993). The differences between it and the structures discussed above are the absence of one bridging hydroxo ligand, the presence of a bridging acetate ligand, and an increased Fe1-Fe2 distance of 3.4 Å. The presence of acetate in the crystallization buffer accounts for the acetate ligand. However, the reason acetate is the bridging ligand instead of the a second hydroxo-ligand is not clear. It seems likely that more than one geometry can be adopted by the oxidized MMOH, and this notion is supported by spectroscopic studies (see below). Since the oxidized MMOH structures with different structures have been solved at different temperatures, a temperature dependent equilibrium may exist between the two conformations consistent with the idea that the active site cluster has considerable structural flexibility.

The structure of diferrous MMOH Bath (Rosenzweig et al., 1995) is also shown in Figure 5. Importantly, the coordination number of the iron atoms decreases from six to five upon reduction. This is significant because each iron is no longer coordinately saturated and thus has a site available to potentially bind and react with oxygen. The change in the coordination geometry arises from the replacement of both the hydroxo- and aquo-bridging ligands by a single monodentate bridge from one atom of Glu-243 that binds to Fe2 in the oxidized MMOH. This Glu undergoes a "carboxylate shift" (Rardin et al., 1991; Tolman et al., 1991) so that it is in a bridging position between the two iron atoms and chelated to Fe2. The change in the active site structure upon reduction is another example of the flexibility of the cluster in MMOH.

Adjacent to the binuclear cluster extending into the center of the protein is a cavity with a volume of 70 $Å^3$ that is lined with fifteen hydrophobic residues and only four hydrophilic residues, two Gln, and a Cys and a Thr. It has been widely postulated that this hydrophobic cavity is involved in the binding of substrate and product. Although this conclusion seems likely to be correct, no experimental evidence exists for this role and the cavity has no obvious substrate entry or product release channel. It is possible that the conformational changes known to occur upon MMOB or MMOR binding to MMOH may afford substrate access. In crystallographic studies of MMOH Bath, one of the Leu residues in the hydrophobic cavity (Leu 110) was observed to be in a different orientation in two distinct crystal forms. Therefore it was hypothesized that it could act as a gate for substrate entry or product release (Rosenzweig et al., 1997). Also, xenon has been used as a probe for cavities in the MMOH structure and at least one cavity between the surface and the putative active site has been detected in a crystal structure. Unassigned electron density was observed in the hydrophobic cavity of both diferric MMOH Bath at $-160\,^\circ$C and MMOH OB3b at $18\,^\circ$C. This was postulated to be acetate derived from the crystallization buffer in the case of the MMOH Bath and four solvent molecules in the case of MMOH OB3b, but the origin of this density has not been definitively assigned in either case.

### 2.1.2. Spectroscopy

**2.1.2.1. Diferric MMOH.** Prior to the solution of the crystal structure, MMOH had been extensively characterized in a number of spectroscopic studies. These studies initially focussed on the three stable redox states of the binuclear iron cluster of MMOH; diferric [Fe(III)Fe(III)], mixed valence [Fe(II)Fe(III)], and diferrous [Fe(II)Fe(II)]. The visible optical spectrum of diferric MMOH is very weak and exhibits no distinct features (Fox et al., 1989). This was not expected because other proteins and enzymes with binuclear iron clusters exhibit much more intense spectra with distinct features in the 300 to 510 nm range due to charge transfer interactions between the bridging oxygens and the irons (Sanders-Loehr et al., 1989). It was postulated that the weak spectrum implied substitution of the bridging oxygen(s) which is consistent with the relatively long observed bond lengths of these ligands in the crystal structure.

Due to the lack of a significant chromophore, resonance Raman (rR) studies were not feasible for resting MMOH. However, the addition of phenol to diferric MMOH OB3b resulted in a purple species that could be studied by rR (Andersson et al., 1992). Analysis of the rR spectra suggested that the purple chromophore is caused by a ligand (phenol) to metal charge transfer interaction. This conclusion indicates that relatively large molecules must have access to the diiron site, although as mentioned above, the pathway that allows access is not clear. Since the crystal structure indicates that each iron is coordinately saturated prior to phenol binding, it is likely that the phenol coordinates to the iron by displacement of the terminal water molecule on Fe1.

Mössbauer spectra of $^{57}$Fe enriched MMOH OB3b showed that each of the irons in the oxidized active site is high-spin Fe(III) ($S = 5/2$), but antiferromagnetic coupling results in a diamagnetic center ($\Delta E_Q = 1.16$ (Fe1), 0.87 (Fe(2) mm s$^{-1}$; $\delta = 0.51$ (Fe1), 0.5 (Fe2) mm s$^{-1}$) (Fox et al., 1988; DeWitt et al., 1991; Fox et al., 1993). This state of the cluster is EPR silent from the ground state, but exhibits a signal at $g = 8$ from an integer spin excited state (Fox et al., 1993). EXAFS spectra of MMOH reveal Fe-Fe distances ranging from 3.03 Å for diferric MMOH OB3b (Shu et al., 1996) to 3.42 Å for diferric MMOH Bath (DeWitt et al., 1991; DeWitt et al., 1995). The MMOH

OB3b appears to have two distances represented (3.03 and 3.32 Å) in the same sample (Shu *et al.*, 1997). This is in accord with the different Fe—Fe distances reported in the crystal structures and supports the hypothesis that the cluster in diferric MMOH can assume several conformations distinguished by the Fe-Fe separation. Rapid freeze quench EXAFS experiments using MMO OB3b revealed that the diferrous MMOH the Fe-Fe distance increases to 3.43 Å upon reduction.. Furthermore, analogous experiments showed a significantly decreased Fe-Fe distance of 2.46 Å for intermediate **Q**. Since the Fe-Fe distance changes as diferric MMOH is reduced and then again as dioxygen is activated, it follows that the flexibility of the active site geometry is important to catalysis.

**2.1.2.2.    Mixed Valence MMOH.**    The mixed-valence [Fe(II)Fe(III)] state gives a characteristic EPR spectrum from an $S = 1/2$ spin system produced by antiferromagnetic coupling of the $S = 2$ and $S = 5/2$ irons of the cluster. The spectrum is characterized by the occurrence of all of the resonances below $g = 2$ to yield a $g_{ave} = 1.85$ (Woodland *et al.*, 1986; Fox and Lipscomb, 1988; Fox *et al.*, 1988; DeWitt *et al.*, 1991). Although this state is probably not directly relevant to catalysis, the EPR resonances are very sensitive to the environment surrounding the active site and have been used be used to monitor changes caused by solvents, substrates, inhibitors, and the other MMO components (Woodland *et al.*, 1986; Hendrich *et al.*, 1990; Fox *et al.*, 1991). The EPR signal from the mixed valence state has also been used to monitor the effects of radiolytic reduction of MMOH by γ-irradiation at 77 K as a tool to probe the active site structure (Davydov *et al.*, 1997; Davydov *et al.*, 1999). Upon reduction, the active site geometry of the diferric state is maintained because molecular movement is restricted at cryogenic temperatures, thus the EPR signal observed is actually a probe of the structure of the diferric state. The EPR spectra of the radiolytically reduced sample indicates that two populations are present, in agreement with the EXAFS and crystallography experiments with diferric MMOH. The EPR signals observed from mixed valence MMOH produced in this way are different from those produced by chemical reduction in solution. However, when the radiolytically reduced sample is warmed to 230 K and then cooled to 4 K, the resultant EPR spectrum is identical to that of the mixed valence MMOH generated in solution. Therefore, it appears that a structural change occurs in the diiron site upon reduction. The addition of methanol has also been shown to effect the EPR spectra of cryogenically reduced mixed valence MMOH (Davydov *et al.*, 1999). However, the origin of the perturbation caused by the presence of this reaction product is not clear.

ENDOR studies revealed at least one nitrogen from histidine residues is present in the ligation sphere (Hendrich *et al.*, 1992) of mixed valence

MMOH, in accord with the X-ray crystal structures of diferric MMOH. In addition, the spin of the mixed valence diiron cluster was shown to interact with nine protons either through bonds or through space near the cluster. Subsequent pulsed EPR (Thomann et al., 1993) and ENDOR (DeRose et al., 1993) studies allowed the assignment of one of the protons to the terminal water ligand on one of the iron atoms and another to the proton on the bridging oxygen, both of which exchange with bulk solvent very slowly. The approach of small substrate-like molecules to the binuclear iron cluster as also detected by observing $^2$H-ENDOR from deuterium labeled derivatives (Hendrich et al., 1992). Indeed, $^2$H and $^{13}$C ENDOR experiments from MMOH Bath suggest that DMSO is O-bound to the ferric ion of the cluster (DeRose et al., 1996; Davydov et al., 1999)

**2.1.2.3. Diferrous MMOH.** The iron atoms of the diferrous state are each in the S = 2 spin state (Fox et al., 1988) and are ferromagnetically coupled to yield an S = 4 electronic ground state (Hendrich et al., 1990). This results in an EPR signal near g = 16 that is intensified when the applied magnetic field of the EPR instrument is aligned parallel with the microwave field, a so called parallel mode experiment (Fox et al., 1989). The increase in signal intensity is characteristic of an integer spin system (Hendrich and Debrunner, 1989). High-field Mössbauer studies confirmed the S = 2 spin state of each iron (Fox et al., 1993). In addition, these experiments showed that each iron of the cluster is in a similar but distinguishable environment. MCD/CD experiments showed that the ligation geometry about each iron is five coordinate square pyramidal (Pulver et al., 1993), and this was confirmed by the crystal structure of diferrous MMOH Bath (Rosenzweig et al., 1995). EXAFS experiments showed that the Fe-Fe distance is 3.43 Å, indicative of progressive increase in this distance with reduction as indicated by the crystal structure (Shu et al., 1997).

## 2.2. MMOB

The B component (MMOB), is a 15.8 kDa protein that has no cofactors. It has been termed the gating protein (Liu, Y. et al., 1995) and the coupling protein (Rosenzweig et al., 1993) by different research groups. In the MMO Bath system, MMOB appears to be necessary to couple NADH oxidation to hydroxylated product formation (Green and Dalton, 1985; Lund et al., 1985) which is the origin of the coupling protein nomenclature. However, the MMO OB3b system remains reasonably well coupled in the absence of MMOB, suggesting that this is not its main role. Extensive studies of both MMO systems have uncovered a common role for MMOB as an effector which gates reactivity with molecular oxygen giving rise to the gating protein nomenclature. Many multicomponent oxygenase systems include a small "effector" component with a regulatory function. The regulation takes many forms and the gating function of MMOB represents a novel manifestation of this phenomenon. Kinetic and spectroscopic experiments have shown that MMOB actually exerts many effects on aspects of MMOH structure and catalysis (Fox *et al.*, 1993). For example, it has been shown that MMOB OB3b alters the regiospecificity of hydroxylation on substrates that are more complex than methane (Froland *et al.*, 1992). Thus, during turnover of propane in the absence of MMOB, MMO yields 93% of the product as 2-propanol and 7% as the primary alcohol. When MMOB is added, even in substoichiometric amounts, the product distribution shifts such that 2-propanol is favored in only a 2:1 ratio. A similar change in product distribution is observed for butane and other small hydrocarbons. One interpretation of these results is that MMOB changes the binding orientation of the substrate in the active site so that a different carbon is presented to the reactive oxygen species.

### 2.2.1. NMR Solution Structure

Since MMOB has no cofactors, the spectroscopic tools used to study MMOH are of no help in determining its structure, and all attempts to crystallize the protein have failed. However, the small size of the protein allowed the solution structure to be determined by NMR methods. The first solution structure of this protein was determined for MMOB from MMO OB3b with a mean distribution for the 50 best structures of 1.1 Å for all backbone atoms and 1.6 Å for all non-hydrogen atoms excluding the apparently disordered N- and C-terminal regions (Chang *et al.*, 1999). As shown in Figure 6, the structure is an α/β fold that has two domains. Three α-helices and six antiparallel β-strands associate to form a βαββ and a βααββ domains. The structure of MMOB Bath has also been recently determined and was found to be similar in its fold, the major difference being the presence of seven β-strands (Walters *et al.*, 1999). The overall dimensions of MMOB would allow it to fit in a grove formed by the intersection of the two MMOH αβγ protomers which would bring it close to the binuclear iron cluster in the MMOH α-subunit.

### 2.3. MMOR

The reductase (MMOR) is a 39.7 kDa protein that contains both FAD and a $[Fe_2S_2]$ cluster in a single polypeptide chain (Lund and Dalton, 1985; Prince and Patel, 1986; Fox *et al.*, 1989; Gassner and Lippard, 1999). The two cofactors are commonly found in proteins that are involved in electron transfer in other oxygenase systems, therefore MMOR is proposed to

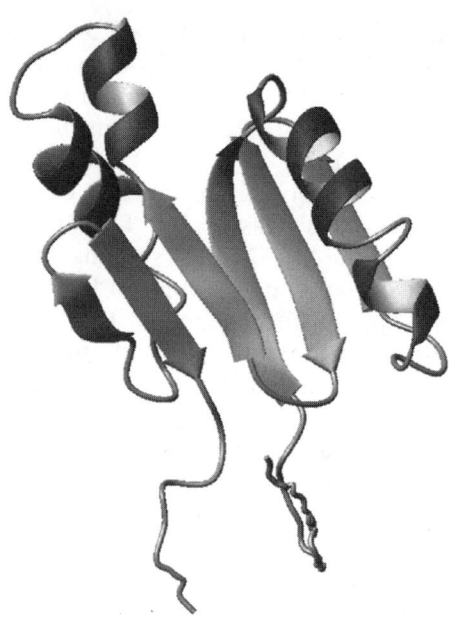

**FIGURE 6.** NMR solution structure of MMOB isolated from *M. trichosporium* OB3b. (Adapted from Chang *et al.*, 1999).

accept and store electrons from NADH and transfer them to the diiron site of MMOH. After reduction by one equivalent, an EPR signal from the FAD semiquinone is observed at g = 2.0. The iron-sulfur cluster has been characterized by EPR and Mössbauer spectroscopies (Fox *et al.*, 1993). Upon reduction of MMOR by a second equivalent, the characteristic EPR signal of a $[Fe_2S_2]^{1+}$ cluster at g = 1.94 is observed. Finally, after adding a third reducing equivalent, the g = 2.0 signal is lost as the FAD becomes fully reduced.

## 3. COMPONENT COMPLEXES

All chemical, spectroscopic, and kinetic studies of the MMO systems conducted thus far are consistent with the formation of specific, high affinity complexes between the components. The points of interactions between MMOH and the other two components were first defined using chemical cross-linking procedures. MMOB is readily cross-linked to the MMOH α-subunit while MMOR cross-links to the β-subunit when a water-soluble carbodiimide is reacted with a stoichiometric mixture of components in solution (Fox *et al.*, 1991). The endogenous tryptophan fluorescence of MMOH and the concentration dependence of the MMO components in kinetic experiments can used to determine their binding affinity and stoichiometry. For the MMO OB3b system it was found that MMOB binds to diferric MMOH with high affinity $K_D = 8–26$ nM (Fox et al., 1991). MMOR also binds tightly, but it exhibits different $K_D$ values for binding in two sites ($K_{D1} = 10$ nM, $K_{D2} = 25$ µM). Recently, similar values determined through the use of isothermal titration calorimetry were reported for the MMO Bath system (Gassner and Lippard, 1999).

In the MMO OB3b system, the $K_D$ values could be used to predict the concentration dependence of the MMOB enhancement on the rate of the multiple turnover reaction. The fit to the experimental data predicts that the maximum rate is attained when a stoichiometric ternary complex (based on active site concentration) is established. Excess MMOB is inhibitory, apparently due to the formation of inactive MMOB-MMOR and MMOB-MMOB complexes, or perhaps binding of MMOB in the MMOR binding site. Cross-linking experiments were used to demonstrate the formation of each of these inhibitory complexes. Component complexes also play a significant role during the single turnover reaction as described below.

There are several indications that complex formation between MMOB and MMOH affects the structure of MMOH. For example, when the mixed valence state of MMOH is formed in the presence of MMOB the g-values and magnitude of antiferromagnetic coupling observed for the binuclear iron cluster of MMOH are significantly changed (Fox et al., 1991; Hendrich et al., 1992). Similarly, the g = 16 EPR signal of diferrous MMOH is sharpened considerably in the presence of MMOB and both the MCD and CD spectra are perturbed (Pulver et al., 1997). Interestingly, the alternative substrate trans-1,2-dichloroethylene and the inhibitor tetrachloroethylene perturb the CD spectra only in the presence of MMOB, suggesting that MMOB allows these small molecules to have access to the diiron site by changing the active site conformation. Recently, X-ray scattering data has been used to shown that MMOB induces large conformational changes in MMOH (Gallagher et al., 1999). Collectively, these observations suggest that the formation of the MMOH:MMOB complex affects the diiron cluster. The nature of the effect appears to be a change in the active site conformation.

## 4.  OXIDATION-REDUCTION POTENTIALS

The oxidation reduction potentials of MMOR have been measured for two MMO systems by monitoring the changes in the optical and EPR spectra. In MMO Bath the potentials for the FAD/FADH• $[Fe_2S_2]^{2+}/[Fe_2S_2]^{1+}$ and FADH•/FADH$_2$ couples were found to be −150, −220, and

−260 mV, respectively (Lund and Dalton, 1985). For MMO from *Methylobacterium* species CR-26 (Type II, similar to OB3b) the corresponding potentials were found to be −195, −247 and −250 mV (Prince and Patel, 1986). For each of these types of MMOR, as well as the MMOR from MMO OB3b (Zhang and Lipscomb, 1999), the stepwise addition of electrons appears to reduce the FAD to the semiquinone state first, then the iron sulfur cluster to the +1 state, and finally the FAD to the two electron reduced state. In the case of the MMOR OB3b, isosbestic points are observed throughout each stage of the spectroscopic titration showing that each stage is nearly complete before the next begins. Consequently, the three potentials appear to differ by at least 60 mV similar to those reported for the MMOR Bath.

The two redox potential values for binuclear iron cluster of MMOH OB3b have been determined by monitoring the appearance of the $g_{ave}$ = 1.85 and g = 16 EPR signals from the mixed valence and diferrous states, respectively (Paulsen *et al.*, 1994). The quantitations of these spectra were also determined using Mössbauer spectroscopy to directly measure the relative concentration of iron in each state. The potentials at 4 °C and pH 7.0 were found to be $E_1^{o'}$ = +76 mV and $E_2^{o'}$ = +21 mV. Quite different results were reported in three separate studies using MMOH Bath. In an early study, $E_1^{o'}$ = +350 mV and $E_2^{o'}$ = −25 mV were determined (Woodland *et al.*, 1986). Later, values of $E_1^{o'}$ = +48 mV and $E_2^{o'}$ = −135 mV (Liu and Lippard, 1991) and finally $E^{o'}$ = −4 mV and $E_2^{o'}$ = −386 mV (Kazlauskaite *et al.*, 1996) were determined using different approaches. The origin of the wide range of values for the MMO Bath system is unclear, but the values found for the MMO OB3b are reproducible and account well for the observed state of reduction under all conditions examined.

Formation of complexes between MMOH and MMOB and/or MMOR cause significant shifts in the redox potential. In the presence of a stoichiometric concentration of MMOB, the potentials of MMOH were determined to be $E_1^{o'}$ = −52 mV and $E_2^{o'}$ = −115 mV (Paulsen *et al.*, 1994). Despite this large negative shift relative to the potentials of the uncomplexed MMOH, the separation of potentials is changed only slightly, so the maximal formation of mixed valence MMOH during the redox titration is unchanged. However, the redox potential is tightly coupled with the free energy of MMOB binding, so the $K_D$ for the MMOH-MMOB complex increases by about 5 orders of magnitude when MMOH is reduced. This change in affinity was observed by direct fluorescence titration (Fox *et al.*, 1991).

Formation of the MMOH-MMOR complex with or without MMOB has the opposite effect on redox potential from that described for the binding of MMOB. In this case, a stoichiometric concentration of MMOR causes little change in $E_1^{o'}$, but $E_2^{o'}$ shifts positively to +125 mV (Paulsen *et al.*, 1994; Liu, Y. *et al.*, 1997). Thus, when MMOR binds, the second electron transfer is less energetic than the first, so the diferrous state is favored over the mixed valence state. Because the two electron reduced MMOH is necessary for reaction with $O_2$, this experiment reveals an important regulatory function of MMOR. The binding of MMOR not only allows transfer of the electrons necessary for the reaction, but it also assures that the form of MMOH that can proceed with the reaction is favored.

One possible problem with a regulatory scheme based on MMOH-MMOR complexation is that MMOR is found in only about 1/20 of the MMOH concentration in vivo (Fox *et al.*, 1989). Remarkably, when the MMOH potential was measured in a solution containing a 1:20 ratio of MMOR to MMOH (sites), the positive shift in the MMOH redox potential was still observed (Liu *et al.*, 1997). This suggests that there is substantial hysteresis in the structural changes of MMOH that accompany MMOR binding so that a small amount of MMOR can maintain a dynamic equilibrium of the modified form of MMOH. Similar hysteretic effects are observed for the MMOH-MMOB complex (Froland *et al.*, 1992; Liu, Y. *et al.*, 1995).

## 5.  ELECTRON TRANSFER KINETICS

Recent transient kinetic studies have provided insight into the reaction between NADH and MMOH as illustrated in Figure 7 (Gassner and Lippard, 1999). These experiments indicate that NADH binds to MMOR and forms a complex with oxidized FAD that is distinguished by a pyridine nucleotide-flavin charge transfer absorbance that is observed at 725 nm. The rate of formation of this complex at 4 °C is $397 s^{-1}$ which far exceeds the turnover rate for methane ($1–5 s^{-1}$ depending on the system). In the next step, two electrons are delivered to FAD via hydride transfer from NADH, forming $FAD^-$ and releasing $NAD^+$. The rates of formation and decay for this complex are $180 s^{-1}$ and $106 s^{-1}$, respectively. The rate of $FAD^-$ decay is equal to the formation of FAD semiquinone, observed as the increase in absorbance at 625 nm. The semiquinone is formed as a result of intramolecular electron transfer to the $[Fe_2S_2]$ cluster.

The addition of MMOH does not affect the rates of electron transfer between NADH and oxidized FAD. However, in the MMO Bath system, the presence of MMOH results in less semiquinone formation, and the reduction of the $[Fe_2S_2]$ cluster is delayed. This suggests that when the iron-sulfur cluster of MMOR is reduced, intermolecular electron transfer to the diiron center of MMOH is very fast. Subsequently, the second electron is

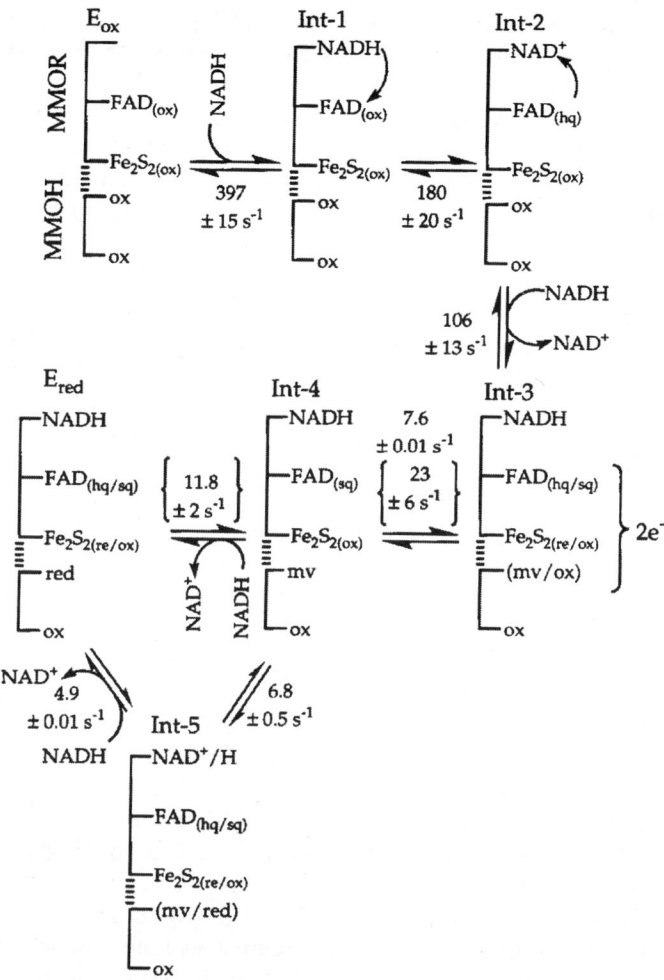

**FIGURE 7.** Electron transfer pathway and rates from NADH through MMOR to MMOH isolated from *M. capsulatus* Bath. (From Gassner *et al.*, 1999).

thought to be transferred from FAD semiquinone through the oxidized [Fe$_2$S$_2$] cluster to the diiron center of MMOH. This final electron transfer results in oxidation of both cofactors of the reductase and the two electron reduction of the diiron center of MMOH. However, the oxidized FAD is not observed due to the rapid formation of a complex with NADH followed by another hydride transfer yielding NAD$^+$ and FADH$_2$.

The effects of MMOB on the rates of intra- and intermolecular transfer were also studied in the MMO Bath system and it was found that MMOB does not effect the rate of hydride transfer to oxidized FAD or electron transfer to oxidized ferredoxin (Gassner and Lippard, 1999). However, the rate of electron transfer out of the reduced ferredoxin was shown to be increased as the concentration of MMOB increases. This result is in contrast with earlier results that suggested MMOB inhibits the rates of intermolecular electron transfer (Green and Dalton, 1989). Saturating methane was shown to have no effect on electron transfer kinetics, as opposed to previous studies which showed that substrate could increase the rate of intermolecular electron transfer in the presence of MMOB (Green and Dalton, 1989). Interestingly, when the reaction product methanol was present at saturating levels, it was shown to slow the intermolecular electron transfer steps (Gassner and Lippard, 1999). Methanol has been shown to form a complex with oxidized hydroxylase that is thought to be similar to the product complex observed when the substrate nitrobenzene is hydroxylated (Davydov et al., 1999). It has been proposed that the reduction of this complex by MMOR to reform the diferrous active site decreases the affinity of the diiron center for methanol, possibly initiating product release.

Electron transfer in the MMO OB3b system has been monitored by using optical and EPR spectroscopies to observe both the rate of departure of the electrons from MMOR and the rate of arrival of electrons at MMOH (Zhang and Lipscomb, 1999). In contrast, to earlier reports for the MMO Bath system, it was found that the rate of transfer is not rate limiting under any conditions including the presence of MMOB and/or methane. The presence of MMOB caused the rate to decrease slightly in the MMO OB3b system, but the observed rate was still greatly in excess of the turnover number for the complete system. The MMO Bath system has been reported to uncouple the system from oxygenase chemistry in the absence of MMOB converting it into an oxidase independent of the presence or absence of methane (Green and Dalton, 1989). However, the addition of MMOB in the absence of methane halted NADH oxidation. The current results from both MMO systems indicate that the omission of MMOB does not cause any major changes in electron transfer to MMOH. NADH turnover is slowed, but this is probably due to the gating function of MMOB on oxygen reactivity occurring after MMOH reduction (Liu, Y. et al., 1995).

## 6. TURNOVER SYSTEMS

In vivo, MMO functions with three protein components and both the rate and efficiency of catalysis are maximized when all three components are present. However, catalysis is possible without MMOR and MMOB under certain conditions (Fox *et al.*, 1989; Andersson *et al.*, 1991; Froland *et al.*, 1992). These alternative turnover systems have been very useful in defining the roles of MMOR and MMOB in addition to the characterization of the transient intermediates of the reaction cycle. The complete system consists of all three protein components with NADH as the source of electrons. In studies with MMO OB3b, the most efficient catalysis is observed when MMOH (active sites), MMOB and MMOR are present in approximately $1:1:1$ ratio depending on the absolute concentration of the proteins relative to the $K_D$ values for complexes with MMOH (Fox *et al.*, 1989). At higher concentrations of MMOB, the reaction is inhibited, and excess MMOR causes uncoupling. This is expected because MMOR has been shown to catalyze the oxidation of NADH and reduction of $O_2$ in the absence of the other two components. Recent studies with MMO Bath show that the oxidation of NADH is significantly uncoupled from methane oxidation at MMOR concentrations only 20% of MMOH concentrations, yet remains tightly coupled to propylene epoxidation even at equimolar amounts of the components (Gassner and Lippard, 1999). This may be due to uncoupling later in the reaction cycle, although a similar substrate dependent phenomenon has not been observed for the MMO OB3b system.

A second catalytic system consists of only diferric MMOH plus $H_2O_2$ and substrate, a so called "peroxide shunt" (Andersson *et al.*, 1991; Froland *et al.*, 1992; Jiang *et al.*, 1993). This system does not require MMOR, MMOB, NADH or $O_2$ to yield the expected products from any MMO substrate that has been examined, including methane. The normal catalytic cycle begins with two electron reduction of MMOH followed by addition of $O_2$. Since $H_2O_2$ is $O_2$ reduced by two electrons, it is reasonable that it should support catalysis if it can gain access to the binuclear iron cluster. High concentrations of $H_2O_2$ are required for rapid catalysis by the peroxide shunt, so the protein evidently limits access to $H_2O_2$ in some manner. A similar peroxide shunt has also been observed to support catalysis for P450 (Ortiz de Montellano, 1995), validating mechanistic comparisons between the two monooxygenases.

A third catalytic system consists of chemically or electrochemically reduced MMOH, substrate and $O_2$ (Fox *et al.*, 1989; Froland *et al.*, 1992). In this system, diferrous MMOH reacts with $O_2$ and turns over a single time to yield the expected products. This system is ideally constituted to search for intermediates in the reaction cycle and to determine the rate constants for the formation and decay of these intermediates.

Using these catalytic systems it has been relatively straight forward to demonstrate that interactions between the components affect the course of the chemical reaction but probably not the overall mechanistic strategy. For example, the turnover of isopentane by the complete reconstituted MMO system or in the single turnover system with added MMOB forms all possible products, but the primary carbons are preferentially oxidized in rough accord with their greater relative abundance in the molecule (Froland et al., 1992). This is unexpected based on lower bond strengths of the secondary and tertiary carbons which should facilitate C—H bond cleavage and/or stabilize intermediate radicals in these positions. However, this result can readily be rationalized in an enzyme catalyzed reaction by invoking accessibility or kinetic arguments as long as the reagent generated in the active site is capable of reacting with the stronger primary C—H bonds. Interestingly, when MMOB is omitted from any of the systems, the same products are observed from isopentane turnover but the distribution shifts toward that expected based on the relative bond strengths. The precise distribution observed is unique for each combination of MMOH, MMOB and MMOR and also depends on the initial redox state of MMOH. This suggests that the component complexes and the redox state of MMOH control the manner in which the substrate is presented to the reactive oxygen species generated on the binuclear iron cluster. The manner in which this occurs is unknown but may involve structural changes as suggested by changes in the spectroscopic properties of MMOH as the result of complex formation described above.

## 7.  REACTION CYCLE INTERMEDIATES

### 7.1.  Transient Intermediates of the Reaction Cycle of MMOH

Using the single turnover catalytic system, it is possible to initiate the reaction by exposing a high concentration of the diferrous enzyme with or without the other MMO components and substrate to $O_2$. As illustrated in Figure 8, the reaction will then proceed through the cycle, release product and stop because there is no means to begin the cycle again at the diferrous state. The inability to cycle greatly simplifies both the detection of intermediates and the determination of rate constants for the process. As discussed above, diferrous MMOH is nearly colorless, so traditional optically detected stopped flow techniques cannot be used to monitor the loss of this species. However, through the use of freeze quench procedures, it is possible to monitor the loss of the characteristic $g = 16$ EPR signal from this state. Upon exposure to $O_2$, the $g = 16$ signal was found to decay with a rate constant of $22 s^{-1}$ at $4 °C$, $pH = 7.7$ in the presence of stoichiometric MMOB (Lee et al., 1993a). The reaction itself was found to occur only in the presence of oxygen, but its rate was independent of $O_2$ concentration. Remarkably, the same value for the rate constant was determined for this reaction

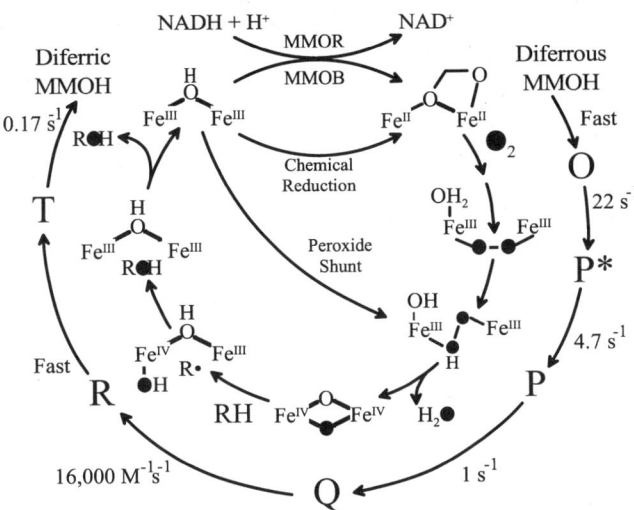

**FIGURE 8.** Postulated mechanism for MMO. The inner cycle are postulated intermediates in the catalytic cycle (only the binuclear iron cluster of the MMOH component is shown). The outer cycle represents the intermediates detected during a single turnover beginning with diferrous MMOH and ending with diferric MMOH. The rate constants shown are for 4 °C and pH 7.7. The rate shown for the substrate reaction RH with **Q** is that for methane. The alignment of the two cycles shows the postulated structures for the intermediates.

catalyzed by the MMOH Bath supporting the idea that the mechanisms of these enzymes are very similar (Liu, K. *et al.*, 1994; Liu, K. *et al.*, 1995). When MMOB was omitted from the MMOH OB3b reaction, the rate of reaction with oxygen decreased by about 1000-fold revealing, perhaps, the major role of MMOB in gating oxygen reactivity as discussed above (Liu, Y. *et al.*, 1995). In contrast to the diferrous MMOH, some of the subsequent intermediates in the reaction cycle are chromophoric, and thus their formation and decay can be followed optically. The most easily detected intermediate exhibits a bright yellow chromophore ($\lambda_{max}$ = 430 and 330 nm, $\varepsilon_{430,330}$ = 7500 $M^{-1} cm^{-1}$) and appears with a rate constant of 1 $s^{-1}$ as shown in Figure 9 (Lee *et al.*, 1993a). Because the rate of **Q** formation is less than that of the decay of diferrous MMOH, at least one intermediate, which we termed compound P (**P**), must occur between **Q** and diferrous MMOH in the reaction cycle. Subsequent studies using MMOH Bath revealed a weakly chromophoric species ($\lambda_{max}$ = 650 nm) that was assigned to **P** (termed $H_{Peroxo}$ in the MMO Bath system) (Liu, K. *et al.*, 1995). Recent studies with MMOH OB3b show that a similar chromophoric species forms with an absorbance maximum around 700 nm ($\varepsilon_{700}$ = 2500 $M^{-1} cm^{-1}$) (Lee and Lipscomb, 1999).

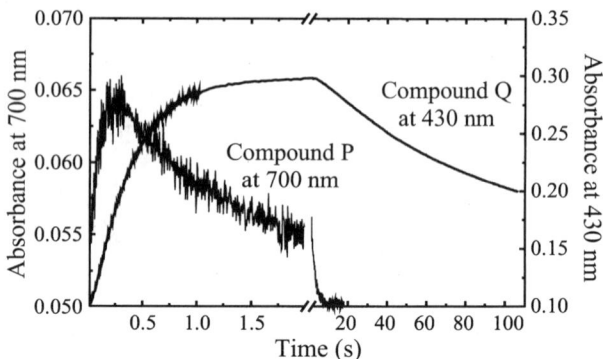

**FIGURE 9.** Formation and decay kinetics of the reaction cycle intermediates **P** and **Q**. (Adapted from Lee and Lipscomb, 1999).

The optical studies with MMOH OB3b show that the rate of **P** formation is $4.7 s^{-1}$ and its decay to **Q** is $1 s^{-1}$, at pH 7.7 (Lee and Lipscomb, 1999). Thus, it is reasonable to propose that **P** is the direct precursor of **Q**, but it cannot be the direct product of the reaction of diferrous MMOH with $O_2$ which occurs at $22 s^{-1}$. In fact, our data strongly suggests that there are at least two intermediates between diferrous MMOH and **P**. Since the rate of decay of the $g = 16$ EPR signal is independent of oxygen concentration (Lee *et al.*, 1993a), there must be an irreversible oxygen binding step before the loss of this signal. This first step would require that oxygen binding does not result in oxidation or perturbation of the coupling of the binuclear iron cluster so that the $g = 16$ signal is retained. We call this intermediate compound O (**O**) for the oxygen complex, and it is envisioned as resulting from binding of oxygen to the enzyme prior to binding to the cluster (Liu, Y. *et al.*, 1995). In the next step, a reaction must occur between oxygen and MMOH that causes the decay of the EPR signal, either by oxidation of one or both irons in the diiron site or by a perturbation of the ferromagnetic coupling between them. Direct coordination of the $O_2$ to the binuclear iron cluster with subsequent transfer of electron density to the oxygen would be expected to cause the loss of the $g = 16$ signal. Consequently, because the $g = 16$ is retained in **O**, the decay rate of $22 s^{-1}$ from the freeze quench EPR studies must be the rate of decay of **O**, not diferrous MMOH. Because the first observable intermediate **P** forms at $4.7 s^{-1}$, yet another intermediate(s), **P***, must form at $22 s^{-1}$ and decay at $4.7 s^{-1}$ (Lee and Lipscomb, 1999). The nature of **P*** is not known, but it seems likely that it is more like **P** than diferrous MMOH since it lacks the $g = 16$ signal. Recent studies from MMOH OB3b have provided additional evidence for the existence of **P***.

The rate of **P** formation was decreased when the pH was increased, whereas the rate of diferrous decay was not affected (Lee and Lipscomb, 1999). The different effects of pH indicate that there must be an intermediate between diferrous MMOH and **P**.

Neither methane nor oxygen concentration has been shown to effect the formation or decay rate of any of the intermediates up to and including the formation of **Q**. Conversely, the rate of **Q** decay increases linearly with substrate concentration, implicating **Q** as the intermediate that first reacts with substrate (Lee *et al.*, 1993a). All of the MMO substrates that have been tested increase the rate of **Q** decay but the magnitude of the increase varies widely. This fact has been exploited to identify another intermediate following **Q** (Lee *et al.*, 1993a). MMO reacts with nitrobenzene to form *p*—nitrophenol (as well as the *o*- and *m*-adducts) that is retained in the nonpolar active site to yield a characteristic optical spectrum which changes upon slow release into the polar solvent. The time course of this reaction is consistent with the formation of a single intermediate which is envisioned as the enzyme substrate product complex, compound T (**T**). Accordingly, comparison of chemical quench data collected during the formation of **T** with the rate of the optical change during its breakdown shows that it forms at the same rate as **Q** decays and breaks down at the same rate as the steady state turnover number for the enzyme with nitrobenzene as a substrate. Thus, product release from **T** is the rate determining step in nitrobenzene oxidation. Importantly, the correlation of the product formation and **Q** decay rates strongly supports the notion that **Q** is the intermediate that reacts directly with substrates to transfer oxygen into a C—H bond. Following decay of **T**, the enzyme is returned to the resting diferric state. The intermediates observed thus far and their possible relationship to the proposed catalytic cycle are illustrated above in Figure 8.

The mechanism of **Q** decay has been thought of as a single step process in which the Fe(IV)=O reacts with substrate. However, recent kinetic and thermodynamic studies suggest that the mechanism is more complex and that temperature mediates the rate-determining step in the decay of **Q**. In recent studies with MMO Bath, Eyring plots of the rate of **Q** decay in the presence of methane and acetylene suggest that the rate-determining step of **Q** decay changes at 17 °C, from **Q** formation in the low temperature region to **Q** decay in the high temperature region (Valentine *et al.*, 1999a). At pH 7 and 4 °C, the rate of formation of **Q** from MMO Bath is much slower than that of MMO OB3b, $0.45 \, s^{-1}$ and $2.5 \, s^{-1}$, respectively. In contrast, the decay rates are similar. The greater formation rate in MMO OB3b allows **Q** to form maximally and therefore allows the decay reaction to be monitored over the entire temperature range. In similar experiments to those of MMO Bath described above, nonlinear Arrhenius plots were also observed in the presence methane and other straight chain alkanes (Brazeau and Lipscomb, 1999). However, it was found that the temperature of the inflection point is dependent upon substrate concentration. Since the rate of **Q** formation is independent of substrate concentration, these data suggest that there are at least two steps involved in the reaction of **Q** with substrate.

## 7.2.  Structures of the Intermediates

The determination of the structure of several of the intermediates has been possible through the use of spectroscopic and synthetic model studies. Intermediate **P** from both MMO Bath and MMO OB3b have been studied using Mössbauer spectroscopy (Liu, K. *et al.*, 1994; Shu *et al.*, 1997). The spectrum shown in Figure 10 consists of a single quadrupole doublet with an isomer shift ($\delta$) of 0.67 mm/s showing that the irons are in the ferric oxidation state and reside in very similar electronic environments. This would be expected if the $O_2$ is bound symmetrically between the irons in this species and electron density moves onto the oxygen from the irons. The fact that $\delta$ is slightly higher than that observed for the resting diferric state is

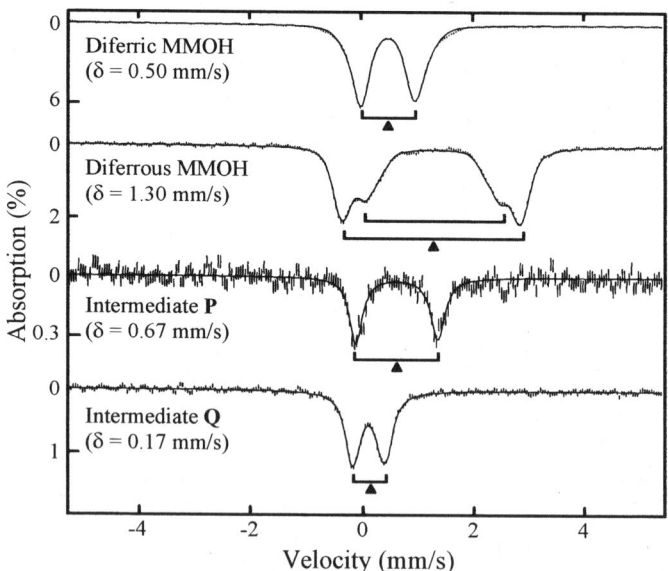

**FIGURE 10.** Mössbauer spectra of MMOH reaction cycle intermediates. The isomer shift $\delta$ (triangles) is determined primarily by the oxidation state of the iron. The larger the value of $\delta$, the lower the oxidation state of the iron. (Shu *et al.*, 1997).

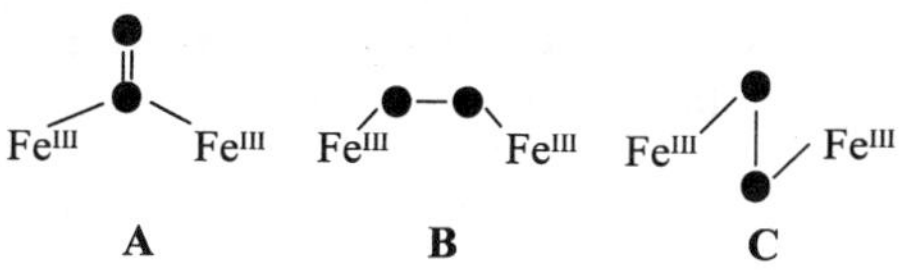

**FIGURE 11.** Hypothetical structures for the intermediate compound **P**.

consistent with incomplete delocalization of electron density away from the irons of the cluster.

There are several possible structures for **P**, some of which are illustrated in Figure 11. In principle, it is possible to distinguish between these structures using resonance Raman (rR) spectroscopy. Although results from rR studies on **P** have been inconclusive, studies of similar species recently found in the reaction cycles of ferritin and stearoyl acyl carrier protein $\Delta^9$ desaturase have been informative (Broadwater et al., 1998; Moenne-Loccoz et al., 1999). These studies are most consistent with a trans-μ peroxo binding mode similar to **C** in Figure 11. Additional support for such a structure comes from an inorganic chelate model compound with a *trans*-μ-peroxo bridge as in **C** which has Mössbauer characteristics similar to those observed for **P** (Kim and Lippard, 1996). In contrast, other peroxo-adducts of binuclear iron clusters differ markedly in either optical or Mössbauer spectroscopic characteristics (Que and Dong, 1996).

Mössbauer studies of intermediate **Q** revealed some intriguing properties. Both iron atoms in the intermediate from MMOH OB3b have identical $\delta = 0.17$ mm/s (Figure 10) (Lee et al., 1993b; Shu et al., 1997). This isomer shift is too low to be in the range for high spin ferric iron, but it is in the range of Fe(IV) or possibly low spin Fe(III). The low spin Fe(III) possibility has largely been discounted by subsequent Mössbauer studies. Ferryl-oxo species have been proposed in the mechanism of many heme and non-heme iron enzymes, but have not previously been observed in non-heme systems. Indeed, the ferryl-oxo species of cytochrome P450 has been proposed to be the intermediate that can react with hydrocarbons but it has proven impossible to isolate for spectroscopic characterization; exciting recent results suggest that it may be possible to trap this species in the crystal for structural studies (Schlichting et al., 1999). Thus, **Q** was the first reactive ferryl-oxo species to be isolated in any oxygenase enzyme and the first binuclear ferryl-oxo species to have been isolated and characterized in any biological system. High-field Mössbauer experiments indicate that the diiron center of **Q** is antiferromagnetically coupled resulting in a diamagnetic state (Lee et al., 1993b). **Q** can also be trapped from the catalytic cycle of MMOH Bath and exhibits similar spectroscopic properties except that the irons are in slightly different environments as revealed by the presence of two distinguishable quadrupole doublets from Fe(IV) (Liu, K. et al., 1995).

The likely structure of **Q** was revealed through EXAFS and Mössbauer studies of both the intermediate itself and of novel inorganic chelate complexes designed to model the high valent core (Shu et al., 1997). Two features in the EXAFS spectrum indicated that the structure of the binuclear iron cluster in **Q** differs substantially from those in the diferric and diferrous MMOH. First, a single short 1.77 Å Fe—O bond was observed for each iron, and second, a very short Fe-Fe distance of 2.46 Å was determined. The Fe—O bond is too long to be assigned as a terminal Fe(IV)=O, however, it is in the range of a bridging (μ) oxo ligand in a diiron cluster. The Fe-Fe distance is the shortest to be reported for any diiron-oxo protein and is too short to be explained by the bridging ligand geometries of binuclear iron clusters we have encountered in the intermediates discussed in this review thus far. Synthetic model studies have recently yielded in a family of tripodal ligands that can be stabilized in the high valent Fe(III)Fe(IV) state (Dong et al., 1995a; Dong et al., 1995b; Que and Dong, 1996; Hsu et al., 1999). Although this is one electron more reduced than **Q**, the model compounds can react with some hydrocarbons to yield oxidized products equivalent to those that would be produced by MMO turnover (Kim et al., 1997). Extensive spectroscopic characterizations of these model compounds have been carried out (Dong et al., 1995b). Recently, the crystal structure for an alkylated derivative of the ligand TPA, $[Fe_2(\mu\text{-O})_2(5\text{-Et}_3\text{-TPA})_2]$ $(ClO_4)$ was solved (Hsu et al., 1999). The structure shown in Figure 12 revealed a

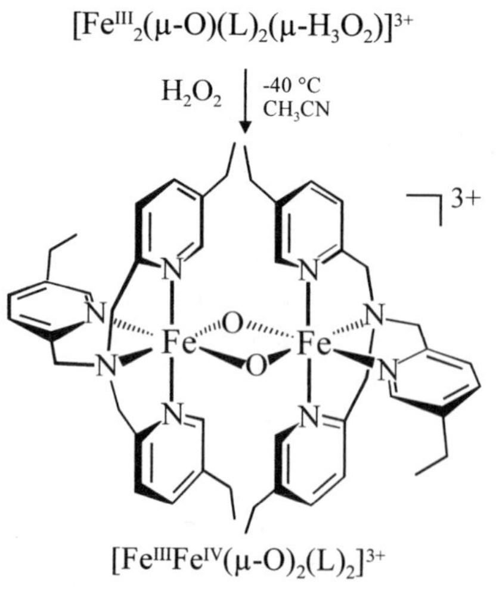

$$[Fe^{III}_2(\mu\text{-O})(L)_2(\mu\text{-H}_3O_2)]^{3+}$$

FIGURE 12. Structure of a binuclear Fe(III)Fe(IV) chelate complex used as a model for **Q**. L=tris(5-ethyl-2-pyridylmethyl)amine. The structure shown is based on the X-ray crystal structure of the complex (Hsu et al., 1999).

"diamond core" structure in which two single oxygen atoms bridge the irons. This is the first structure for a non-heme Fe(IV)-O species that is catalytically competent. As in **Q**, the best characterized model compound contained only one short Fe—$\mu$—O bond of 1.80 Å per Fe and exhibited a short Fe-Fe distance of 2.8 Å. Many other types of binuclear iron model complexes have been synthesized, but these diamond core complexes are the only ones with Fe-Fe distances under 2.7 Å, implying that the short Fe—Fe bond in **Q** can only result from the presence of at least two single atom bridges between the irons. The simplest such structure is shown above in Figure 2, although other structures such as those incorporating single atom carboxylate bridges cannot be ruled out. The presence of the single short Fe—O bond in both the model complexes and **Q** is an unusual feature that derives from a distortion in the diamond core structure. One way to view the **Q** structure is as two Fe(IV) oxo adducts bound head to tail with a weaker bond linking each oxo group to the neighboring Fe(IV). The additional stabilization that this structure would lend to an Fe(IV)-oxo moiety may account for the significantly enhanced stability of **Q** over the equivalent heme oxygenase ferryl-oxo intermediates. It is likely that one or more of the usual bridging carboxylate ligands are also present in the **Q** structure because this type of bridge is present in both the diferric and diferrous clusters and the additional bridge would account for the shorter Fe-Fe distance in **Q** relative to that observed in the models. **Q** also exhibits an unusually intense X-ray absorption pre-edge feature suggesting that the coordination geometry of **Q** is highly strained and that each iron has no more than five ligands (Shu *et al.*, 1997). This aspect of the **Q** may be an important for catalysis as described below.

## 7.3.  The Mechanism of Oxygen Cleavage

The stoichiometry of a monooxygenase reaction requires that one oxygen atom of $O_2$ be incorporated into product and the other into a water molecule via the addition of two protons. The step, or steps, in which the protons are donated has not been clear until recently. Investigation of the pH dependence of the rates of formation of intermediates **P** and **Q** in the MMO OB3b system showed that both were highly pH dependent (Lee and Lipscomb, 1999). The rates were both maximized at about pH 6 and approached zero as the pH was raised above 8.5 (Figure 13), suggesting that the input a proton is required to form each intermediate. This was examined further through the use of proton inventory solvent isotope techniques in which the rate is determined as a function of the fractional concentration of $D_2O$ in the reaction solvent. It was found that there is a substantial deuterium solvent isotope effect in the formation reactions of both **P** and

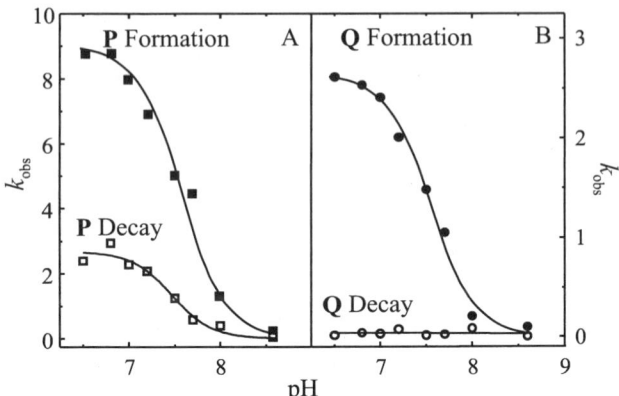

**FIGURE 13.** Rate of formation and decay of intermediates **P** and **Q** as a function of pH. (From Lee and Lipscomb, 1999).

**Q** and that the proton inventory for each reaction is linear. Linear plots generally indicate that a single proton is delivered in a single step during the reaction. Thus, the most direct interpretation of these observations is that the two protons required for the overall stoichiometry are delivered during the **P\*** to **P** and **P** to **Q** steps of the reaction cycle. The decay of **P** shows the same pH dependence as the formation of **Q** as expected if these are successive steps. In contrast, the decay of **Q** and the steps leading to **P\*** do not exhibit a pH dependence, suggesting that proton is only transferred in the two steps that have been identified.

Based on the crystal structure of MMOH OB3b, the most likely proton donor candidates were postulated to be Thr 213, Cys 151, or solvent molecules in the active site. Cys 151 was discounted as a donor because it is not conserved among the other diiron enzymes that activate oxygen. Moreover, the results of the proton inventory plots suggest that the donor must have a fractionation factor near unity, whereas the value for Cys is typically 0.5. A clue as the proton donor derived from the observation that the $pK_a$ of pH dependence for the formation of both **P** and **Q** is 7.6 (Figure 13) which may represent the $pK_a$ of the direct donor. Despite being highly conserved, Thr 213 would be expected to have a $pK_a$ significantly higher than 7.6 unless it is activated by nearby base which is not evident in the crystal structure. Also, such an interaction would lower its fractionation factor to 0.7, contrary to what is actually observed. A more likely candidate for the proton donor would be a solvent bound to Fe(III) in the **P\*** and **P** intermediates as illustrated in Figure 14. The two protons required for the reaction might be donated to the bound $O_2$ molecule either asymmetrically if both go to a single oxygen or symmetrically if both oxygens are protonated during the

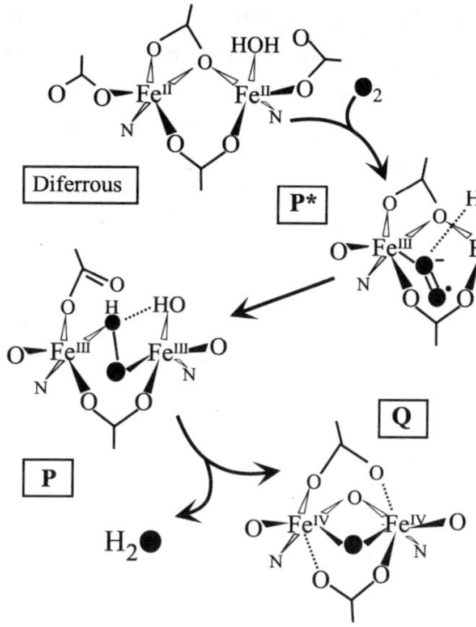

**FIGURE 14.** Hypothetical mechanism for proton delivery during MMO catalysis. (From Lee and Lipscomb, 1999).

steps leading to **Q**. However, it is likely that a symmetrically protonated peroxide would simply dissociate from the diiron site as $H_2O_2$ rather than undergoing bond cleavage, therefore asymmetric protonation of **P** is favored. Asymmetric protonation would probably lead to heterolytic bond cleavage and the loss of a water molecule prior to the formation of **Q**. The oxygen atom from the solvent that served as the proton donor would then presumably become one of the bridging oxygens of the bis $\mu$-oxo binuclear Fe(IV) diamond core proposed for **Q**. Alternatively, binding of a single bridging atom from an active site carboxylate residue might fulfill this role. One attractive aspect of this hypothesis is that because the proton donor is part of the cluster, the overall charge of the cluster does not change at any step during the formation of **Q** from resting enzyme. This would minimize the free energy changes required to form **Q** and promote its formation in high yield as observed.

## 8.  MECHANISM OF C—H BOND CLEAVAGE AND OXYGEN INSERTION

MMO presents a unique opportunity to study monooxygenase chemistry because it forms a series of intermediates that buildup in sufficient yields to allow direct study. In addition, MMO turns over a wide variety substrates. This fact has been put to use in the form of diagnostic substrates such as radical clocks, isotopically labeled substrates, and chiral molecules that can reveal the nature of the chemistry catalyzed by MMO. Based on these experiments, many possible mechanisms for MMO have been proposed (Feig and Lippard, 1994; Wallar and Lipscomb, 1996). Generally these varied proposals fall into two groups. One type of mechanism is based on the occurrence of an intermediate substrate species such as a substrate radical during the oxygen insertion process, while in the second group of mechanisms it is proposed that the oxygen insertion reaction is concerted.

## 8.1. Radical Rebound Mechanism

Our original proposal for the mechanism of MMO (Figure 8) is representative of the radical intermediate group of potential MMO mechanisms (Fox et al., 1989; Fox et al., 1990). It was based in part on the mechanism proposed for the only other type of enzyme known to catalyze hydroxylation of unactivated hydrocarbon substrates, cytochrome P-450 (McMurry and Groves, 1986; Ortiz de Montellano, 1995). In the case of P-450, it is widely accepted that the reactive species is an Fe(IV)-oxo $\pi$ cation radical in which the radical resides on the heme. This powerful oxidant is proposed to abstract a hydrogen atom from the hydrocarbon, forming Fe(IV)-OH and a substrate radical. The substrate radical and hydroxy radical recombine to form product. In the case of MMO, the analogous species to the reactive high valent intermediate in P450 appears to be **Q**. The second iron of the active site cluster of **Q** performs the same function as the heme in P-450, donating an electron to complete the valence octet of the reactive oxygen atom resulting from heterolytic O—O bond cleavage. The general aspects of this mechanism are supported by several types of experiments described in the following sections.

## 8.2. Isotope Effects

The proposal of hydrogen atom abstraction from methane by **Q** suggests that if C—H bond breaking contributes to the rate limiting step, then a kinetic isotope effect (KIE) should be observed upon substituting a deuterium atom for each protium in methane. In previous steady state experiments from MMOH OB3b and MMOH Bath, only a small isotope effect on $V_{max}$ was observed (Rataj et al., 1991; Wilkins et al., 1994). Based on the transient kinetic studies described above, we now know that product release is the rate limiting step in steady state turnover, so isotope effects would not be expected. Interestingly, the steady state kinetics show that there is a very large V/K deuterium isotope effect of approximately 20 (Rataj *et al.*, 1991). V/K is sensitive to the steps in catalysis through the first irreversible step (Cleland, 1975), which is presumably the C—H bond breaking step. This is the step that involves the decay of **Q** in our mechanism, and it can be directly monitored by stopped flow techniques. As shown in Figure 15, a deuterium isotope effect of 50–100 per C—H bond is observed for the reaction of methane with **Q** (Nesheim and Lipscomb, 1996). This is one of the largest deuterium KIE's ever observed in biology, strongly suggesting that C—H bond breaking is the rate limiting step in the decay of **Q** when methane is the substrate. Because the observed isotope effect increases linearly with the number of C—D bonds in the methane isotopomer used, this large KIE value is due almost entirely to a primary effect. Recently, this work was confirmed through the observation of a similarly large KIE (= 28) for the analogous reaction using MMOH Bath (Valentine *et al.*, 1999a). Interestingly, when ethane or $d_6$-ethane is the substrate in otherwise identical experiments, essentially no isotope effect is observed for **Q** decay, although a small V/K isotope effect of approximately 4 is observed for steady state turnover (Rataj *et al.*, 1991; Priestley *et al.*, 1992). Presently, the basis for this difference is not understood, but it points to differences in the interaction of **Q** with methane versus adventitious substrates.

The KIE of 50–100 far exceeds the classical limits derived from the zero point energy difference between protium and deuterium; the maximum expected $k_H/k_D$ is 6–7 (Bell, 1973). Larger isotope effects can be accounted for by mechanisms that involve branched reaction pathways,

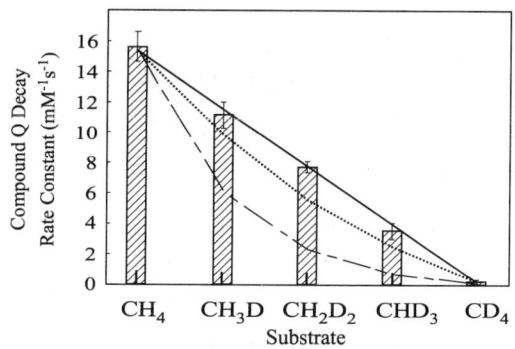

**FIGURE 15.** Effect of substrate deuteration on the rate of methane reaction with **Q**. The solid line is the expected relationship if the isotope effect is predominantly primary. The dotted and dashed lines show the expected relationship of the rate constants if there were a secondary isotope effect of 1.2 or 2, respectively. (From Nesheim and Lipscomb, 1996).

multiple steps that result in a cumulative KIE, magnetic isotope effects, or quantum mechanical tunneling (Hwang and Grissom, 1994). The first two explanations have been ruled out based on mechanistic arguments (Nesheim and Lipscomb, 1996), and we have not observed magnetic isotope effects in preliminary experiments, suggesting that quantum mechanical tunneling of hydrogen may be involved. Similar tunneling reactions have been described recently for a variety of biological reactions involving the movement of protons, hydrogen atoms, and hydride ions (Kohen and Klinman, 1999; Kohen et al., 1999; Kohen and Klinman, 1998).

## 8.3.   Radical Clock Substrates

Substrate radical intermediates can sometimes be detected using radical clock substrates that reorganize on a well defined time scale when a radical occurs in their structure. Figure 16 illustrates some of the radical clock substrates utilized in studies of MMO from both MMO OB3b and MMO Bath, and the potential products that result from radical pair rebound or rearrangement. The known rate of rearrangement in each case allows the rate of radical recombination to be calculated. Surprisingly, the product distribution resulting from some types of radical clock experiments

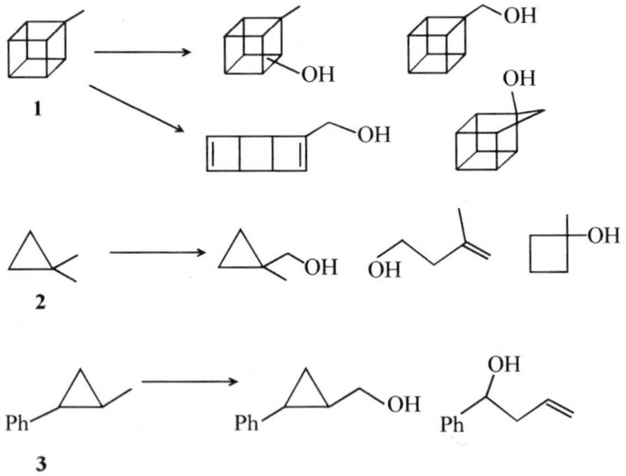

**FIGURE 16.** Radical clock substrates used in the study of MMO mechanism. **1** = methyl-cubane. The top row of products are unrearranged, while the second row are rearranged products resulting from intermediate radical or cation formation, respectively. **2** = 1,2 dimethylcyclopropane. The first product is unrearranged while the second and third are rearranged products resulting from radical and cationic intermediates, respectively. **3** = 1-phenyl-2-methylcyclopropane with following structures of the unrearranged and rearranged products.

varies depending on the bacterial source of MMO. Also, a wide range of rebound rates are found for MMO OB3b but have not been reported for MMO Bath. For example, in one study turnover of methylcubane (**1**) by MMO Bath yielded only cubylmethanol (Choi *et al.*, 1996) whereas turnover by MMO OB3b yields cubylmethanol, three different methyl-cubanols, and probably a high yield of rearranged product (Jin and Lipscomb, 1999). The expected products for a mechanism with a radical intermediate are largely the cubylmethanols due to stabilization engendered by the cubyl structure. If the radical is formed on the methyl group, rearranged products may be observed if the radical lifetime is on the order of $10^{-10}$ s or longer. Cubyl methanol would be expected as the only product in two scenarios: 1) only the methyl group can approach the reactive oxygen and the lifetime of the radical intermediate is very short, or 2) the reaction mechanism is actually concerted insertion of oxygen, in which case insertion into a methyl C—H is favored over a cubyl C—H (Choi *et al.*, 1996). Consequently, the data suggest that a radical mechanisms is possible for MMO OB3b, but they are inconclusive for MMO Bath. Recently, the methylcubane experiments for MMO Bath were repeated in light of the results from MMO OB3b (Choi *et al.*, 1999). In this set of experiments, all the possible hydroxylation products were found as observed for MMO OB3b, but the rearranged product was identified as 1-homocubanol. This product would derive from cationic rather than radical rearrangement. Additional evidence for a cationic intermediate was presented using a radical clock that can differentiate between radical and cationic intermediates. A minor product was shown to be the species resulting from cationic rearrangement. The mechanistic significance of cation formation is not clear because it is unlikely that a cation could be formed as an intermediate in methane oxidation by an enzyme. The route by which the cation is formed is also unclear. It is possible that it may be formed by two very rapid one electron abstractions, in which case a radical intermediate would be generated.

If it is assumed that the intermediate in methylcubane oxidation is a radical, the yield of rearranged product allows a radical lifetime of about $10^{-10}$ s to be estimated. A similar lifetime was estimated for turnover of 1,1-dimethylcyclopropane (**2**) which gave both radical and cationic products (Ruzicka *et al.*, 1990). In contrast, MMO Bath showed no rearranged product when 1-methyl-2-phenylcyclopropane (**3**) was used (Valentine *et al.*, 1999b). For this substrate, MMO OB3b gave only 3% rearranged product, suggesting that if a radical is formed at all, it has a lifetime of less than $10^{-12}$ s.

The two MMO systems give very similar results for turnover of (R) and (S)-1-[$^1$H,$^2$H,$^3$H]-ethane (Priestley *et al.*, 1992; Valentine *et al.*, 1997). In these experiments, MMO catalyzed conversion of ethane to ethanol as the only product. Formation of an adduct of the ethanol with a chiral reagent gave a diasteromeric compound. The extent of inversion during the MMO catalyzed reaction could then be assessed using tritium NMR because carrier free tritium was used in the synthesis of the chiral ethane. The degree of chirality was expected to change if a radical intermediate formed and lived long enough to allow rotation about the ethyl radical C—C bond. This is, perhaps, the best experiment to evaluate the MMO mechanism because ethane is similar in size, structure, and reactivity to methane. It was found that both stereoisomers are oxidized to ethanol with 30–35% inversion stereochemistry, supporting the occurrence of an intermediate in this reaction.

The observation of a substantial decrease in the chirality of chiral ethane both strongly supports the existence of a substrate radical and brings into question the pure P450-like radical rebound mechanism for MMO. The extent of racemization from the chiral ethane experiments indicates that the rebound of the ethyl radical and hydroxy radical would have to occur at a rate of about $5 \times 10^{12} \text{s}^{-1}$, similar to that computed for some of radical clock studies for MMO from both bacterial sources. This rate is about that for bond vibrations, representing the limit for a chemical reaction. The rate seems to high for rebound chemistry as currently conceived, which requires physical migration of radical species in the active site.

## 8.4. Modified Radical Rebound Mechanism

Several mechanisms have been suggested to account for the chiral ethane and radical clock results. It may be that the underlying assumptions that the radical clocks and ethane have free access to the reactive oxygen species and unencumbered rotation around single bonds is flawed as is often the case in the active sites of enzymes. Consequently, the apparent variable lifetime of the radical intermediates for MMO OB3b could be explained on the basis of access to the reactive species. Bulky substrates such as 1 and 2 which cannot approach the cluster closely may react by the usual radical rebound mechanism giving the intermediate radical a relatively long lifetime. On the other hand, small substrates such as methane, ethane, and 3, may approach the cluster closely so that rebound chemistry occurs very rapidly (Jin and Lipscomb, 1999). One possibility suggested by quantum mechanical calculations is that an Fe—C bond is formed rapidly after the substrate radical is generated for small substrates as shown in Figure 17 (Siegbahn and Crabtree, 1997). In a sense, the rebound chemistry occurs by movement of oxygen in one iron ligand position into the Fe—C bond formed using another ligand position. This would have the effect of

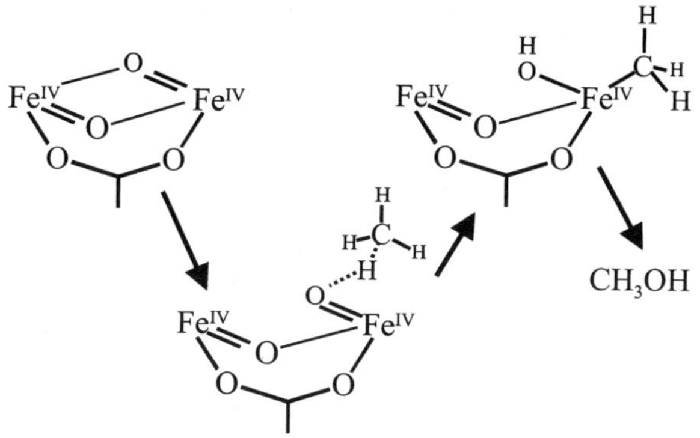

**FIGURE 17.** Postulated mechanism for **Q** reaction with methane based on Density Functional Theory calculations. (Siegbahn and Crabtree, 1997).

halting single bond rotation and give an artificially short lifetime for the radical intermediate. The close proximity of the substrate and reactive oxygen species in this mechanism may also account for the large deuterium KIE observed in the case of methane turnover if hydrogen tunneling is involved in the hydrogen atom abstraction process. This would be very sensitive to distance and bond strength so that similar molecules that are slightly more bulky or have weaker bonds such as ethane, propane and hexane may not exhibit large isotope effects as observed (Nesheim and Lipscomb, 1996).

## 8.5. Concerted Oxygen Insertion Mechanisms

An intriguing alternative to the radical rebound mechanism that retains the concept of radical character on the substrate at some point during the oxygen insertion process has been termed the nonsynchronous concerted oxygen insertion mechanism (Newcomb *et al.*, 1995; Choi *et al.*, 1996). In this mechanism, the substrate is activated by both iron-bound oxygen atoms, as shown in Figure 18. The O—H bond formation, C—H bond breaking, and C—O bond formation steps occur in a concerted manner, but the different bond vibration rates allow a very short-lived substrate radical to form as part of the transition state ensemble. It is estimated that the lifetime of this species is approximately 70 fs which is similar to the lifetime of radical species observed in MMO catalyzed reactions with non-bulky substrates. If C—H bond breaking is the rate determining step, an isotope effect may be observed for this mechanism but the magnitude of the effect is unlikely to be large because the mechanism lacks a linear

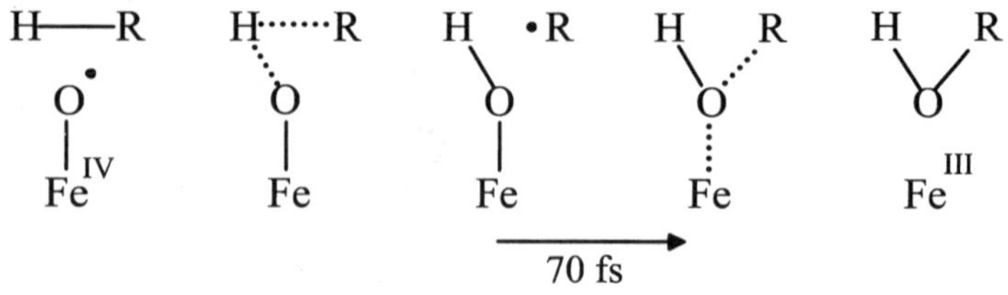

**FIGURE 18.** Nonsynchronous concerted mechanism for hydrocarbon oxidation. (From Newcomb *et al.*, 1995).

transition state. The mechanism also depends on a precise alignment of reactive groups which is not consistent with the exceptional substrate range of MMO. It might be predicted that primary carbons would be hydroxylated preferentially over secondary and tertiary carbon atoms because the former would generate a transition state most like that envisioned for methane. However, it is known that in steady-state experiments with MMO OB3b secondary and tertiary hydroxylation products are favored for longer straight chain hydrocarbons (Froland *et al.*, 1992). Furthermore, the regiospecificity changes as a consequence of the turnover system employed (see above). Together, these experimental results argue against a nonsynchronous concerted oxygen insertion mechanism.

Another type of concerted mechanism involves the formation of a five-coordinate carbon intermediate (Shteinman, 1995; Basch *et al.*, 1999; Shilov and Shteinman, 1999). As shown in Figure 19, the electrophilic oxygen atom bound to the iron adds to the carbon, forming a five-coordinate carbon intermediate that is proposed to be stabilized by the second bridging oxygen atom of **Q**. The oxygen atom then migrates into the C—H bond to form the product, methanol. This mechanism is consistent with the absence of rearranged product found in some of the radical clock experiments since no radical intermediate is formed. It can also account for the outcome of the chiral ethane studies because five coordinate carbon can undergo pseudorotation resulting in the loss of chirality. However, as in the case of the nonsynchronous concerted mechanism, it is difficult to rationalize with the large observed deuterium KIE.

### 8.6. Mechanistic Theory based on Calculations

The theoretical treatments of the MMO mechanism are based on high level quantum mechanical methods that form models of the diiron site in which the ligand geometry is built from X-ray crystallographic studies and other biophysical techniques. The calculations of Siegbahn and coworkers based on density functional theory that were described briefly above led to

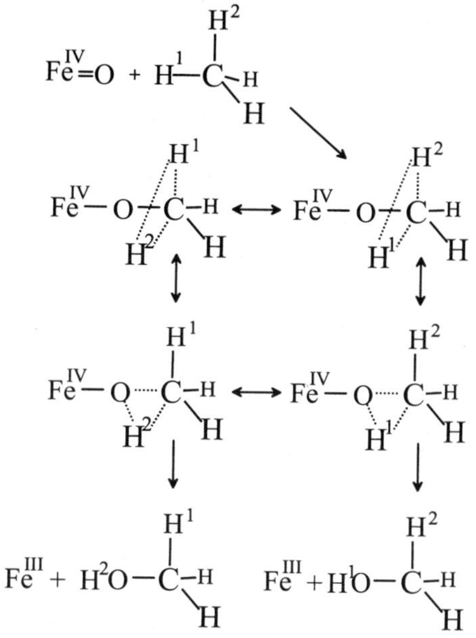

$$Fe^{IV}=O + H^1-C\overset{H^2}{\underset{H}{-}}H$$

$$Fe^{IV}-O\overset{H^1}{\cdots}C\overset{H^2}{\underset{H}{-}}H \longleftrightarrow Fe^{IV}-O\overset{H^2}{\cdots}C\overset{}{\underset{H}{-}}H$$

$$Fe^{IV}-O\cdots C\overset{H^1}{\underset{H}{-}}H \longleftrightarrow Fe^{IV}-O\cdots C\overset{H^2}{\underset{H}{-}}H$$

$$Fe^{III} + H^2O-C\overset{H^1}{\underset{H}{-}}H \qquad Fe^{III}+H^1O-C\overset{H^2}{\underset{H}{-}}H$$

FIGURE 19. Mechanism of hydrocarbon oxidation and racemization by formation of a 5-coordinate carbon intermediate. (Adapted from Shilov and Shteinman, 1999).

the mechanism of methane hydroxylation shown in Figure 17 (Siegbahn and Crabtree, 1997). Interestingly, these calculations successfully predicted the structure of **Q** without knowledge of the spectroscopic studies that were being conducted simultaneously (Shu *et al.*, 1997). This mechanism supports the proposal for a radical intermediate and further defines the attacking species. In the first calculations, the attacking species was proposed to be a radical resonance form of **Q** in which the bridge is opened. A more advanced study however, found that that **Q** in the diamond core form was sufficiently reactive to attack methane (Siegbahn *et al.*, 1998; Siegbahn, 1999). Similar conclusions were reached by Basch *et al.* (1996). Other theoretical studies support a two step concerted mechanism in which the substrate coordinates to the Fe and weakens the C—H bond (Yoshizawa, 1998). A pentavalent carbon is formed on the way to an HO-Fe-CH$_3$ intermediate that leads to product via methyl migration. The results form the theoretical calculations are providing a new means to test and extend oxygen activation theory. They suggest experiments that can be performed with MMO, P450, and a large range of model compounds that either exist or can be prepared with the required characteristics to evaluate the predictions from the calculations.

## 9.  CONCLUSIONS

MMO has provided a excellent system in which to evaluate the mechanism of oxygen activation. It complements the cytochrome P450 systems, in the sense that the same or similar chemical reactions are catalyzed with a completely different cofactor. Consequently, the fundamental aspects of the underlying chemistry can be resolved from the specific chemical characteristics of a given metallo-cofactor. It may be that in the end that a single mechanism will not suffice even when the same substrate is the target of oxidation. On the other hand, some common characteristics are now apparent. First, a high valent intermediate that is one electron equivalent more oxidized that $Fe(IV)=O$ appears to be required. The nature of this intermediate is different in MMO and P450 systems, but the net result in each case is a powerful oxidizing species at the same formal oxidation state. Second, each of the hydrocarbon oxidizing enzymes regulate aspects of the processes in which electrons are added to initiate catalysis and in which substrates are added to allow turnover. This is done through effector proteins (or other types of effector biomolecules in some cases) that form complexes with the oxygenase component and change its structure through coupling of binding free energies. The structural changes lead to other changes such as shifts in redox potential or access to the reactive metal center. These fundamental aspects of catalysis are related in that the high valent species must be protected from both reaction with adventitious substrates and destruction by reduction at the wrong stage in the catalytic cycle. A third potentially common aspect of catalysis, the mechanism by which unactivated C—H bonds are broken, is the current point of controversy. Substantial evidence of the long postulated radical abstraction and rebound chemistry has been accumulated for both P450 and MMO. On the other hand, the catalytic power of MMO, the advent of new probes, and the stunning recent advances in the ability to calculate reaction profiles has allowed the introduction of alternatives that have not been seriously considered or tested previously. The near future may bring rapid advances in our understanding of this ubiquitous and essential chemistry.

## 10.  REFERENCES

Andersson, K. K., Elgren, T. E., Que, L., Jr., and Lipscomb, J. D., 1992, Accessibility in the active site of methane monooxygenase: the first demonstration of exogenous ligand binding to the diiron center, *J. Am. Chem. Soc.* **114**:8711–8713.

Andersson, K. K., Froland, W. A., Lee, S.-K., and Lipscomb, J. D., 1991, Dioxygen independent oxygenation of hydrocarbons by methane monooxygenase hydroxylase component, *New J. Chem.* **15**:411–415.

Basch, H., Mogi, K., Musaev, G. D., and Morokuma, K., 1999, Mechanism of the methane to methanol conversion reaction catalyzed by methane monooxygenase: A density functional study, *J. Am. Chem. Soc.* **121**:7249–7256.

Bell, R. P., 1973, *The Proton in Chemistry, 2nd Edition*, Cornell University Press, Ithaca, NY.

Brazeau, B. J., and Lipscomb, J. D., 1999, Effect of temperature on the methane monooxygenase compound Q formation and decay processes, *J. Inorg. Biochem.* **74**:81.

Broadwater, J. A., Ai, J., Loehr, T. M., Sanders-Loehr, J., and Fox, B. G., 1998, Peroxodiferric intermediate of stearoyl-acyl carrier protein Δ9 desaturase: oxidase reactivity during single turnover and implications for the mechanism of desaturation, *Biochemistry* **37**:14664–14671.

Cardy, D. L., Laidler, V., Salmond, G. P., and Murrell, J. C., 1991a, The methane monooxygenase gene cluster of *Methylosinus trichosporium*: cloning and sequencing of the mmoC gene, *Arch. Microbiol.* **156**:477–483.

Cardy, D. L., Laidler, V., Salmond, G. P., and Murrell, J. C., 1991b, Molecular analysis of the methane monooxygenase (MMO) gene cluster of *Methylosinus trichosporium* OB3b, *Mol. Microbiol.* **5**:335–342.

Chan, S. I., Nguyen, H.-H. T., Shiemke, A. K., and Lidstrom, M. E., 1992, Biochemical and biophysical studies toward characterization of the membrane-associated methane monooxygenase. *7th Intern. Symp. on Microbial Growth on C1 Compounds.* J. C. Murrell, and D. P. Kelly. Andover UK, Intercept Ltd., 93–107.

Chang, S. L., Wallar, B. J., Lipscomb, J. D., and Mayo, K. H., 1999, Solution structure of component B from methane monooxygenase derived through heteronuclear NMR and molecular modeling, *Biochemistry* **38**:5799–5812.

Choi, S.-Y., Eaton, P. E., Hollenberg, P. F., Liu, K. E., Lippard, S. J., Newcomb, M., Putt, D. A., Upadhyaya, S. P., and Xiong, Y., 1996, Regiochemical variations in reactions of methylcubane with *tert*-butoxyl radical, cytochrome P-450 enzymes, and a methane monooxygenase system, *J. Am. Chem. Soc.* **118**:6547–6555.

Choi, S.-Y., Eaton, P. E., Kopp, D. A., Lippard, S. J., Newcomb, M., and Shen, R., 1999, Cationic species can be produced in soluble methane monooxygenase-catalyzed hydroxylation reactions; radical intermediates are not formed, *J. Am. Chem. Soc.*, in press.

Dalton, H., 1980, Oxidation of hydrocarbons by methane monooxygenase from a variety of microbes, *Adv. Appl. Microbiol.* **26**:71–87.

Davydov, A., Davydov, R., Gräslund, A., Lipscomb, J. D., and Andersson, K. K., 1997, Radiolytic reduction of methane monooxygenase dinuclear iron cluster At 77 K—EPR evidence for conformational change upon reduction or binding of component B to the diferric state, *J. Biol. Chem.* **272**:7022–7026.

Davydov, R., Valentine, A. M., Komar-Panicucci, S., Hoffman, B. M., and Lippard, S. J., 1999, An EPR study of the dinuclear iron site in the soluble methane monooxygenase from *Methylococcus capsulatus* (Bath) reduced by one electron at 77 K: the effects of component interactions and the binding of small molecules to the diiron(III) center, *Biochemistry* **38**:4188–4197.

DeRose, V. J., Liu, K. E., Kurtz, J., D. M., Hoffman, B. M., and Lippard, S. J., 1993, Proton ENDOR identification of bridging hydroxide ligands in mixed-valent diiron centers of proteins: methane monooxygenase and semimet azidohemerythrin, *J. Am. Chem. Soc.* **115**:6440–6441.

DeRose, V. J., Liu, K. E., Lippard, S. J., and Hoffman, B. M., 1996, Investigation of the dinuclear Fe center of methane monooxygenase by advanced paramagnetic resonance techniques: on the geometry of DMSO binding, *J. Am. Chem. Soc.* **118**:121–134.

DeWitt, J. G., Bentsen, J. G., Rosenzweig, A. C., Hedman, B., Green, J., Pilkington, S., Papaefthymiou, G. C., Dalton, H., Hodgson, K. O., and Lippard, S. J., 1991, X-ray absorption, Mössbauer, and EPR studies of the dinuclear iron center in the hydroxylase component of methane monooxygenase, *J. Am. Chem. Soc.* **113**:9219–9233.

DeWitt, J. G., Rosenzweig, A. C., Salifoglou, A., Hedman, B., Lippard, S. J., and Hodgson, K. O., 1995, X-ray absorption spectroscopic studies of the diiron center in methane monooxygenase in the presence of substrate and the coupling protein of the enzyme system, *Inorg. Chem.* **34**:2505–2515.

Dong, Y., Fujii, H., Hendrich, M. P., Leising, R. A., Pan, G., Randall, C. R., Wilkinson, E. C., Zang, Y., Que, L., Jr., Fox, B. G., Kauffmann, K., and Münck, E., 1995b, A high-valent nonheme iron intermediate. Structure and properties of $[Fe_2(\mu\text{-}O)_2(5\text{-Me-TPA})_2](ClO_4)_3$, *J. Am. Chem. Soc.* **117**:2778–2792.

Dong, Y., Kauffmann, K., Münck, E., and Que, L., Jr., 1995a, An exchange-coupled complex with localized high-spin $Fe^{IV}$ and $Fe^{III}$ sites of relevance to cluster **X** of *Escherichia coli* ribonucleotide reductase, *J. Am. Chem. Soc.* **117**:11377–11378.

Elango, N., Radhakrishnan, R., Froland, W. A., Wallar, B. J., Earhart, C. A., Lipscomb, J. D., and Ohlendorf, D. H., 1997, Crystal structure of the hydroxylase component of methane monooxygenase from *Methylosinus trichosporium* OB3b, *Protein Sci.* **6**:556–568.

Ericson, A., Hedman, B., Green, J., Bentsen, J. G., Beer, R. H., Lippard, S. J., Dalton, H., and Hodgson, K. O., 1988, Structural characterization by EXAFS spectroscopy of the binuclear iron center in protein A of methane monooxygenase from *Methylococcus capsulatus* (Bath), *J. Am. Chem. Soc.* **110**:2330–2332.

Feig, A. L., and Lippard, S. J., 1994, Reactions of non-heme iron(II) centers with dioxygen in biology and chemistry, *Chem. Rev.* **94**:759–805.

Fox, B. G., and Lipscomb, J. D., 1988, Purification of a high specific activity methane monooxygenase hydroxylase component from a type II methanotroph, *Biochem. Biophys. Res. Commun.* **154**:165–170.

Fox, B. G., Surerus, K. K., Münck, E., and Lipscomb, J. D., 1988, Evidence for a μ-oxo-bridged binuclear iron cluster in the hydroxylase component of methane monooxygenase. Mössbauer and EPR studies, *J. Biol. Chem.* **263**:10553–10556.

Fox, B. G., Froland, W. A., Dege, J. E., and Lipscomb, J. D., 1989, Methane monooxygenase from *Methylosinus trichosporium* OB3b. Purification and properties of a three-component system with high specific activity from a type II methanotroph, *J. Biol. Chem.* **264**:10023–10033.

Fox, B. G., Borneman, J. G., Wackett, L. P., and Lipscomb, J. D., 1990, Haloalkene oxidation by the soluble methane monooxygenase from *Methylosinus trichosporium* OB3b: mechanistic and environmental implications, *Biochemistry* **29**:6419–6427.

Fox, B. G., Liu, Y., Dege, J. E., and Lipscomb, J. D., 1991, Complex formation between the protein components of methane monooxygenase from *Methylosinus trichosporium* OB3b. Identification of sites of component interaction, *J. Biol. Chem.* **266**:540–550.

Fox, B. G., Hendrich, M. P., Surerus, K. K., Andersson, K. K., Froland, W. A., Lipscomb, J. D., and Münck, E., 1993, Mössbauer, EPR, and ENDOR studies of the hydroxylase and reductase components of methane monooxygenase from *Methylosinus trichosporium* OB3b, *J. Am. Chem. Soc.* **115**:3688–3701.

Froland, W. A., Andersson, K. K., Lee, S.-K., Liu, Y., and Lipscomb, J. D., 1992, Methane monooxygenase component B and reductase alter the regioselectivity of the hydroxylase component-catalyzed reactions. A novel role for protein-protein interactions in an oxygenase mechanism, *J. Biol. Chem.* **267**:17588–17597.

Gallagher, S. C., Callaghan, A. J., Zhao, J., Dalton, H., and Trewhella, J., 1999, Global Conformational Changes Control the Reactivity of Methane Monooxygenase, *Biochemistry* **38**:6572–6760.

Gassner, G. T., and Lippard, S. J., 1999, Component interactions in the soluble methane monooxygenase system from *Methylococcus capsulatus* (Bath), *Biochemistry* **38**: 12768–12785.

Green, J., and Dalton, H., 1985, Protein B of soluble methane monooxygenase from *Methylococcus capsulatus* (Bath). A novel regulatory protein of enzyme activity, *J. Biol. Chem.* **260**:15795–15801.

Green, J., and Dalton, H., 1989, A stopped-flow kinetic study of the soluble methane monooxygenase from *Methylococcus capsulatus* (Bath), *Biochem J.* **259**:167–172.

Groves, J. T., McClusky, G. A., White, R. E., and Coon, M. J., 1978, Aliphatic hydroxylation by highly purified liver microsomal cytochrome P-450. Evidence for a carbon radical intermediate, *Biochem. Biophys. Res. Commun.* **81**:154–160.

Hendrich, M. P., and Debrunner, P. G., 1989, Integer-spin electron paramagnetic resonance of iron proteins, *Biophys. J.* **56**:489–506.

Hendrich, M. P., Münck, E., Fox, B. G., and Lipscomb, J. D., 1990, Integer-spin EPR studies of the fully reduced methane monooxygenase hydroxylase component, *J. Am. Chem. Soc.* **112**:5861–5865.

Hendrich, M. P., Fox, B. G., Andersson, K. K., Debrunner, P. G., and Lipscomb, J. D., 1992, Ligation of the diiron site of the hydroxylase component of methane monooxygenase. An electron nuclear double resonance study, *J. Biol. Chem.* **267**:261–269.

Hsu, H., Dong, Y., Shu, L., Young, J., V. G., and Que, J., L., 1999, Crystal structure of a synthetic high-valent complex with an $Fe_2(\mu\text{-}O)_2$ diamond core. Implications for the core structures of methane monooxygenase intermediate Q and ribonucleotide reductase intermediate X, *J. Am. Chem. Soc.* **121**:5230–5237.

Hwang, C.-C., and Grissom, C. B., 1994, *J. Am. Chem. Soc.* **116**:795–796.

Jiang, Y., Wilkins, P. C., and Dalton, H., 1993, Activation of the hydroxylase of sMMO from *Methylococcus capsulatus* (Bath) by hydrogen peroxide, *Biochim. Biophys. Acta* **1163**:105–112.

Jin, Y., and Lipscomb, J. D., 1999, Probing the mechanism of C—H activation: oxidation of methylcubane by soluble methane monooxygenase from *Methylosinus trichosporium* OB3b, *Biochemistry* **38**:6178–6186.

Kazlauskaite, J., Hill, H. A., Wilkins, P. C., and Dalton, H., 1996, Direct electrochemistry of the hydroxylase of soluble methane monooxygenase from *Methylococcus capsulatus* (Bath), *Eur. J. Biochem.* **241**:552–556.

Kim, C., Dong, Y. H., and Que, L., Jr., 1997, Modeling nonheme diiron enzymes—Hydrocarbon hydroxylation and desaturation by a high-valent $Fe_2O_2$ diamond core, *J. Am. Chem. Soc.* **119**:3635–3636.

Kim, K., and Lippard, S. J., 1996, Structure and Mössbauer spectrum of a ($\mu$-1,2-peroxo)-bis($\mu$-carboxylato)diiron(III) model for the peroxo intermediate in the methane monooxygenase hydroxylase reaction cycle, *J. Am. Chem. Soc.* **118**:4914–4915.

Kohen, A., and Klinman, J. P., 1998, Enzyme catalysis: beyond classical paradigms, *Acc. Chem. Res.* **31**:397–404.

Kohen, A., and Klinman, J. P., 1999, Hydrogen tunneling in biology, *Chem. Biol.* **6**:191–197.

Kohen, A., Cannio, R., Bartolucci, S., and Klinman, J. P., 1999, Enzyme dynamics and hydrogen tunneling in a thermophilic alcohol dehydrogenase, *Nature* **399**:496–499.

Lee, S. K., and Lipscomb, J. D., 1999, Oxygen activation catalyzed by methane monooxygenase hydroxylase component: proton delivery during the O—O bond cleavage steps, *Biochemistry* **38**:4423–4432.

Lee, S.-K., Nesheim, J. C., and Lipscomb, J. D., 1993a, Transient intermediates of the methane monooxygenase catalytic cycle, *J. Biol. Chem.* **268**:21569–21577.

Lee, S.-K., Fox, B. G., Froland, W. A., Lipscomb, J. D., and Münck, E., 1993b, A transient intermediate of the methane monooxygenase catalytic cycle containing a $Fe^{IV}Fe^{IV}$ cluster, *J. Am. Chem. Soc.* **115**:6450–6451.

Lipscomb, J. D., 1994, Biochemistry of the soluble methane monooxygenase. *Ann. Rev. Microbiol.* **48**:371–399.

Lipscomb, J. D., and Que, L., Jr., 1998, MMO—P450 in wolf's clothing?, *JBIC* **3**:331–336.

Liu, K. E., and Lippard, S. J., 1991, Redox properties of the hydroxylase component of methane monooxygenase from *Methylococcus capsulatus* (Bath). Effects of protein B, reductase, and substrate [published erratum appears in *J. Biol. Chem.* 1991, **266**:24859], *J. Biol. Chem.* **266**:12836–12839.

Liu, K. E., Johnson, C. C., Newcomb, M., and Lippard, S. J., 1993, Radical clock substrate probes and kinetic isotope effect studies of the hydroxylation of hydrocarbons by methane monooxygenase, *J. Am. Chem. Soc.* **115**:939–947.

Liu, K. E., Wang, D., Huynh, B. H., Edmondson, D. E., Salifoglou, A., and Lippard, S. J., 1994, Spectroscopic detection of intermediates in the reaction of dioxygen with the reduced methane monooxygenase hydroxylase from *Methylococcus capsulatus* (Bath), *J. Am. Chem. Soc.* **116**:7465–7466.

Liu, K. E., Valentine, A. M., Wang, D. L., Huynh, B. H., Edmondson, D. E., Salifoglou, A., and Lippard, S. J., 1995, Kinetic and spectroscopic characterization of intermediates and component interactions in reactions of methane monooxygenase from *Methylococcus capsulatus* (Bath), *J. Am. Chem. Soc.* **117**:10174–10185.

Liu, Y., Nesheim, J. C., Lee, S.-K., and Lipscomb, J. D., 1995, Gating effects of component B on oxygen activation by the methane monooxygenase hydroxylase component, *J. Biol. Chem.* **270**:24662–24665.

Liu, Y., Nesheim, J. C., Paulsen, K. E., Stankovich, M. T., and Lipscomb, J. D., 1997, Roles of the methane monooxygenase reductase component in the regulation of catalysis, *Biochemistry* **36**:5223–5233.

Lund, J., and Dalton, H., 1985, Further characterisation of the FAD and $Fe_2S_2$ redox centres of component C, the NADH:acceptor reductase of the soluble methane monooxygenase of *Methylococcus capsulatus* (Bath), *Eur. J. Biochem.* **147**:291–296.

Lund, J., Woodland, M. P., and Dalton, H., 1985, Electron transfer reactions in the soluble methane monooxygenase of *Methylococcus capsulatus* (Bath), *Eur. J. Biochem.* **47**:297–305.

McMurry, T. J., and Groves, J. T., 1986, Metalloporphyrin models for cytochrome P-450. *Cytochrome P-450 Structure, Mechanism, and Biochemistry.* P. R. Ortiz de Montellano. New York, Plenum Press, 1–28.

Moenne-Loccoz, P., Krebs, C., Herlihy, K., Edmondson, D. E., Theil, E. C., Huynh, B. H., and Loehr, T., 1999, The ferroxidase reaction of ferritin reveals a diferric μ-1,2 bridging peroxide intermediate in common with other $O_2$-activating non-heme diiron proteins, *Biochemistry* **38**:5290–5295.

Nesheim, J. C., and Lipscomb, J. D., 1996, Large isotope effects in methane oxidation catalyzed by methane monooxygenase: evidence for C—H bond cleavage in a reaction cycle intermediate, *Biochemistry* **35**:10240–10247.

Newcomb, M., Tadic-Biadatti, M.-H. L., Chestney, D. L., Roberts, E. S., and Hollenberg, P. F., 1995, A nonsynchronous concerted mechanism for cytochrome P-450 catalyzed hydroxylation, *J. Am. Chem. Soc.* **117**:12085–12091.

Nguyen, H. H., Nakagawa, K. H., Hedman, B., Elliott, S. J., Lidstrom, M. E., Hodgson, K. O., and Chan, S. I., 1996, X-ray absorption and EPR studies on the copper ions associated with the particulate methane monooxygenase from *Methylococcus capsulatus* (Bath)—Cu(I) Ions and their implications, *J. Am. Chem. Soc.* **118**:12766–12776.

Nguyen, H. H., Elliott, S. J., Yip, J. H., and Chan, S. I., 1998, The particulate methane monooxy-genase from *Methylococcus capsulatus* (Bath) is a novel copper-containing three-subunit enzyme. Isolation and characterization, *J. Biol. Chem.* **273**:7957–7966.

Ortiz de Montellano, P. R., 1995, *Cytochrome P450: structure, mechanism, and biochemistry*, Plenum Press, New York.

Paulsen, K. E., Liu, Y., Fox, B. G., Lipscomb, J. D., Münck, E., and Stankovich, M. T., 1994, Oxidation-reduction potentials of the methane monooxygenase hydroxylase component from *Methylosinus trichosporium* OB3b, *Biochemistry* **33**:713–722.

Priestley, N. D., Floss, H. G., Froland, W. A., Lipscomb, J. D., Williams, P. G., and Morimoto, H., 1992, Cryptic stereospecificity of methane monooxygenase, *J. Am. Chem. Soc.* **114**:7561–7562.

Prince, R. C., and Patel, R. N., 1986, Redox properties of the flavoprotein of methane monooxy-genase, *FEBS Lett.* **203**:127–130.

Pulver, S., Froland, W. A., Fox, B. G., Lipscomb, J. D., and Solomon, E. I., 1993, Spectroscopic studies of the coupled binuclear non-heme iron active site in the fully reduced hydroxy-lase component of methane monooxygenase: Comparison to deoxy and deoxy-azide hemerythrin, *J. Am. Chem. Soc.* **115**:12409–12422.

Pulver, S. C., Froland, W. A., Lipscomb, J. D., and Solomon, E. I., 1997, Ligand field circular dichroism and magnetic circular dichroism studies of component B and substrate binding to the hydroxylase component of methane monooxygenase, *J. Am. Chem. Soc.* **19**:387–395.

Que, L., Jr., and Dong, Y., 1996, Modeling the oxygen activation chemistry of methane monooxygenase and ribonucleotide reductase, *Acc. Chem. Res.* **29**:190–196.

Rardin, R. L., Tolman, W. B., and Lippard, S. J., 1991, *New J. Chem.* **15**:417–430.

Rataj, M. J., Kauth, J. E., and Donnelly, M. I., 1991, Oxidation of deuterated compounds by high specific activity methane monooxygenase from *Methylosinus trichosporium*. Mechanistic implications, *J. Biol. Chem.* **266**:18684–18690.

Riggs-Gelasco, P. J., Shu, L. J., Chen, S. X., Burdi, D., Huynh, B. H., Que, L., and Stubbe, J., 1998, Exafs Characterization of the intermediate X generated during the assembly of the *Escherichia coli* ribonucleotide reductase R2 diferric tyrosyl radical cofactor, *J. Am. Chem. Soc.* **120**:849–860.

Rosenzweig, A. C., Frederick, C. A., Lippard, S. J., and Nordlund, P., 1993, Crystal structure of a bacterial non-haem iron hydroxylase that catalyses the biological oxidation of methane, *Nature* **366**:537–543.

Rosenzweig, A. C., Nordlund, P., Takahara, P. M., Frederick, C. A., and Lippard, S. J., 1995, Geometry of the soluble methane monooxygenase catalytic diiron center in two oxida-tion states, *Chem. Biol.* **2**:409–418.

Rosenzweig, A. C., Brandstetter, H., Whittington, D. A., Nordlund, P., Lippard, S. J., and Frederick, C. A., 1997, Crystal structures of the methane monooxygenase hydroxylase from *Methylococcus capsulatus* (Bath): implications for substrate gating and component interactions, *Proteins* **29**:141–152.

Ruzicka, F., Huang, D. S., Donnelly, M. I., and Frey, P. A., 1990, Methane monooxygenase cat-alyzed oxygenation of 1,1-dimethylcyclopropane. Evidence for radical and carbocationic intermediates, *Biochemistry* **29**:1696–1700.

Sanders-Loehr, J., Wheeler, W. D., Shiemke, A. K., Averill, B. A., and Loehr, T. M., 1989, Electronic and Raman spectroscopic properties of oxo-bridge dinuclear iron centers in proteins and model compounds, *J. Am. Chem. Soc.* **111**:8084–8093.

Schlichting, I., Berendzen, J., Chu, K., Stock, A. M., Sweet, R. M., Ringe, D., Petsko, G. A., Davies, M., Gerber, N. C., Mueller, E. J., Benson, D., Vidakovic, M., and Sligar, S. G., 1999, Crystal structures of intermediates occurring along the reaction coordinate of cytochrome P450cam, *J. Inorg. Biochem.* **74**:49.

Shilov, A. E., and Shteinman, A. A., 1999, Oxygen atom transfer into C—H bond in biological and model chemical systems. mechanistic aspects, *Acc. Chem. Res.* **32**:763–771.

Shteinman, A. A., 1995, The mechanism of methane and dioxygen activation in the catalytic cycle of methane monooxygenase, *FEBS Lett.* **362**:5–9.

Shu, L., Liu, Y., Lipscomb, J. D., and Que, L., Jr., 1996, EXAFS studies of the methane monooxygenase hydroxylase component from *Methylosinus trichosporium* OB3b, *JBIC* **1**:297–304.

Shu, L., Nesheim, J. C., Kauffmann, K., Münck, E., Lipscomb, J. D., and Que, L., Jr., 1997, An $Fe_2^{IV}O_2$ diamond core structure for the key intermediate Q of methane monooxygenase, *Science* **275**:515–518.

Siegbahn, P. E. M., and Crabtree, R. H., 1997, Mechanism of C—H activation by diiron methane monooxygenases—quantum chemical studies, *J. Am. Chem. Soc.* **119**:3103–3113.

Siegbahn, P. E. M., Crabtree, R. H., and Nordlund, P., 1998, Mechanism of methane monooxygenase—a structural and quantum chemical perspective, *JBIC* **3**:314–317.

Siegbahn, P. E. M., 1999, Theoretical model studies of the iron dimer complex of MMO and RNR, *Inorg. Chem.* **38**:2880–2889.

Stainthorpe, A. C., Murrell, J. C., Salmond, G. P., Dalton, H., and Lees, V., 1989, Molecular analysis of methane monooxygenase from *Methylococcus capsulatus* (Bath), *Arch. Microbiol.* **152**:154–159.

Stainthorpe, A. C., Lees, V., Salmond, G. P., Dalton, H., and Murrell, J. C., 1990, The methane monooxygenase gene cluster of *Methylococcus capsulatus* (Bath), *Gene* **91**:27–34.

Stanley, S. H., Prior, S. D., Leak, D. J., and Dalton, H., 1983, Copper stress underlies the fundamental change in intracellular location of methane monooxygenase in methane-oxidizing organisms: Studies in batch and continuous cultures, *Biotech. Lett.* **55**:487–492.

Thomann, H., Bernardo, M., McCormick, J. M., Pulver, S., Andersson, K. K., Lipscomb, J. D., and Solomon, E. I., 1993, Pulsed EPR studies of mixed valence [Fe(II)Fe(III)] forms of hemerythrin and methane monooxygenase: Evidence for a hydroxide bridge, *J. Am. Chem. Soc.* **115**:8881–8882.

Tolman, W. B., Liu, S., Bentsen, J. G., and Lippard, S. J., 1991, Models of the reduced forms of polyiron-oxo proteins: An asymmetric, triply carboxylate bridged diiron(II) complex and its reactions with dioxygen, *J. Am. Chem. Soc.* **113**:152–164.

Valentine, A. M., Wilkinson, B., Liu, K. E., Komarpanicucci, S., Priestley, N. D., Williams, P. G., Morimoto, H., Floss, H. G., and Lippard, S. J., 1997, Tritiated chiral alkanes as substrates for soluble methane monooxygenase from *Methylococcus capsulatus* (Bath)—probes for the mechanism of hydroxylation, *J. Am. Chem. Soc.* **119**:1818–1827.

Valentine, A. M., Stahl, S. S., and Lippard, S. J., 1999a, Mechanistic studies of the reaction of reduced methane monooxygenase hydroxylase with dioxygen and substrates, *J. Am. Chem. Soc.* **121**:3876–3887.

Valentine, A. M., LeTadic-Biadatti, M. H., Toy, P. H., Newcomb, M., and Lippard, S. J., 1999b, Oxidation of ultrafast radical clock substrate probes by the soluble methane monooxygenase from *Methylococcus capsulatus* (Bath), *J. Biol. Chem.* **274**:10771–10776.

Wallar, B. J., and Lipscomb, J. D., 1996, Dioxygen activation by enzymes containing binuclear non-heme iron clusters, *Chem. Rev.* **96**:2625–2657.

Walters, K. J., Gassner, G. T., Lippard, S. J., and Wagner, G., 1999, Structure of the soluble methane monooxygenase regulatory protein B, *Proc. Natl. Acad. Sci. USA* **96**:7877–7882.

Wilkins, P. C., Dalton, H., Samuel, C. J., and Green, J., 1994, Further evidence for multiple pathways in soluble methane-monooxygenase-catalysed oxidations from the measurement of deuterium kinetic isotope effects, *Eur. J. Biochem.* **226**:555–560.

Wilkinson, B., Zhu, M., Priestley, N. D., Nguyen, H.-H. T., Morimoto, H., Williams, P. G., Chan, S. I., and Floss, H. G., 1996, A concerted mechanism for ethane hydroxylation by the particulate methane monooxygenase from *Methylococcus capsulatus* (Bath), *J. Am. Chem. Soc.* **118**:921–922.

Woodland, M. P., Patil, D. S., Cammack, R., and Dalton, H., 1986, ESR studies of protein A of the soluble methane monooxygenase from *Methylococcus capsulatus* (Bath), *Biochim. Biophys. Acta* **873**:237–242.

Yoshizawa, K., 1998, Two-step concerted mechanism for alkane hydroxylation on the ferryl active site of methane monooxygenase, *JBIC* **3**:318–324.

Zang, Y., Dong, Y., Kauffmann, K., Münck, E., and Que, L., Jr., 1995, The first bis(μ-oxo)-diiron(III) complex. Structure and magnetic properties of $[Fe_2(\mu\text{-}O)_2(6TLA)_2](ClO_4)_2$, *J. Am. Chem. Soc.* **117**:1169–1170.

Zhang, X.-Y., and Lipscomb, J. D., 1999, Kinetic studies on electron transfer reactions in methane monooxygenase from *M. trichosporium* OB3b, *J. Inorg. Biochem.* **74**:349.

*Chapter 8*

# Flavocytochrome $b_2$

## Christopher G. Mowat and Stephen K. Chapman

## 1. INTRODUCTION

Flavocytochromes $b_2$ are 2-hydroxyacid dehydrogenases found in the inter-membrane space of yeast mitochondria where they couple oxidation of the substrate to reduction of cytochrome $c$. Examples include the enzymes from *Saccharomyces cerevisiae* and *Hansenula anomala*, both of which are L-lactate dehydrogenases (Chapman *et al.*, 1998), and the enzyme from *Rhodotorula graminis* which is a L-mandelate dehydrogenase (Ilias *et al.*, 1998). This article will concentrate on the flavocytochrome $b_2$ (L-lactate:cytochrome $c$ oxidoreductase) from *S. cerevisiae* (Bakers' yeast), since this is by far the most studied of these enzymes (Chapman *et al.*, 1991). Therefore, throughout this article, the term flavocytochrome $b_2$ will refer specifically to the enzyme from *S. cerevisiae* unless otherwise stated.

As a respiratory enzyme, the production of flavocytochrome $b_2$ is induced by the presence of oxygen and, more specifically, L-lactate. In addition to providing pyruvate (the product of lactate oxidation) for the Krebs cycle, flavocytochrome $b_2$ also participates in a shorter respiratory chain that ultimately directs the energy gained from L-lactate dehydrogenation

**CHRISTOPHER G. MOWAT and STEPHEN K. CHAPMAN**   Department of Chemistry, University of Edinburgh, Mayfield Road, Edinburgh EH9 3JJ, Scotland, U.K.
*Subcellular Biochemistry, Volume 35: Enzyme-Catalyzed Electron and Radical Transfer*, edited by Holzenburg and Scrutton. Kluwer Academic / Plenum Publishers, New York, 2000.

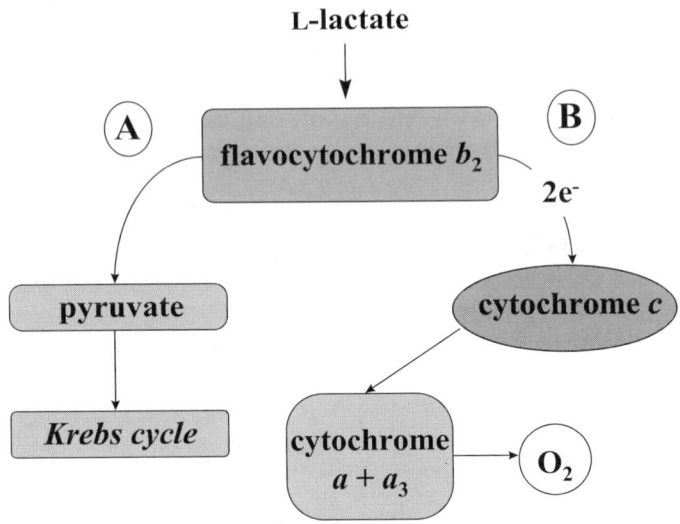

**FIGURE 1.** The Physiological Role of *S. cerevisiae* Flavocytochrome $b_2$. Flavocytochrome $b_2$ acts as a pyruvate source for the Krebs cycle (A) and also transfers electrons directly to cytochrome $c$ (B).

to oxygen, bypassing the Krebs cycle (Figure 1). This process yields one molecule of ATP for every L-lactate molecule consumed.

Three-dimensional structures are available for flavocytochrome $b_2$ in both the native (Xia and Mathews, 1990) and recombinant (Tegoni and Cambillau, 1994) forms. The enzyme is a homotetramer with subunit $M_r$ of 57.5. Each subunit is composed of two structurally and functionally distinct domains: a $N$-terminal cytochrome domain (100-residues) containing protoheme IX; and a $C$-terminal flavodehydrogenase domain (400-residues) containing FMN. The two domains are connected via a short region of peptide thought to function as a hinge (Sharp *et al.*, 1996a). The domain structure of flavocytochrome $b_2$ is illustrated in Figure 2.

Electron flow through flavocytochrome $b_2$ has been extensively studied in both the *S. cerevisiae* (Tegoni *et al.*, 1998; Daff *et al.*, 1996a; Chapman *et al.*, 1994; Pompon, 1980) and *H. anomala* (Capeillére-Blandin *et al.*, 1975) enzymes. The catalytic cycle is shown in Figure 3. Firstly, the flavin is reduced by L-lactate; a carbanion mechanism has been proposed for this redox step (Lederer, 1991). Complete (two-electron) reduction of the flavin is followed by intra-molecular electron transfer from fully-reduced flavin to heme, generating flavin semiquinone and reduced heme (Daff *et al.*,

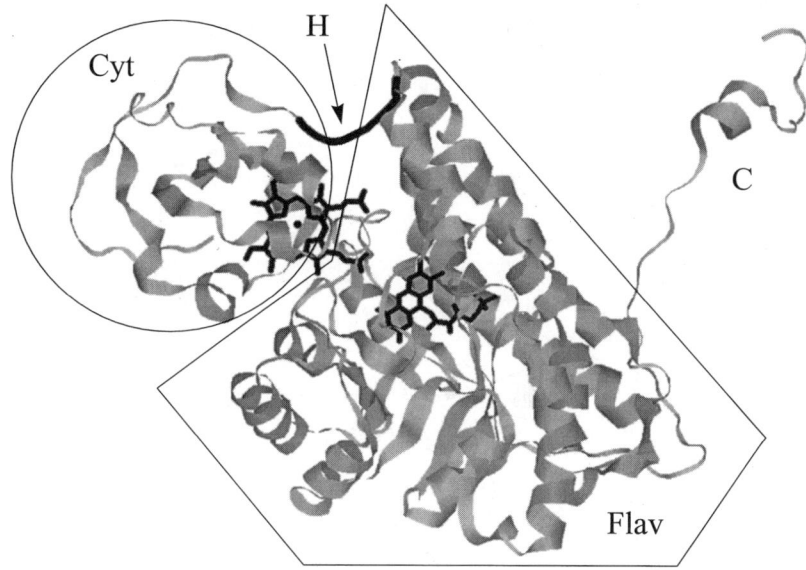

**FIGURE 2.** A Subunit of *S. cerevisiae* Flavocytochrome $b_2$. The protein is shown as a ribbon diagram with the heme and flavin cofactors in a stick representation. The two domains are clearly delineated. Cyt, cytochrome domain; Flav, flavin domain; H, interdomain hinge peptide; C, C-terminal tail.

1996a; Capeillére-Blandin, 1975). There then occurs the first of two intermolecular electron transfers from $b_2$-heme to cytochrome *c* (Short *et al.*, 1998). This results in an oxidised $b_2$-heme which is then re-reduced by the flavin semiquinone. Finally, the second electron is transferred from the $b_2$-heme to cytochrome *c* (Daff *et al.*, 1996b).

It would appear then that the redox properties of flavocytochrome $b_2$ are well understood. While this is generally true, there are a number of aspects which remain controversial and it is these that will form the main focus of this article. There are three major questions which will be addressed: (i) Does the transfer of redox equivalents from lactate to flavin really involve a carbanion intermediate? (ii) What controls the intramolecular electron transfers from flavin to heme? (iii) Where, on the surface of flavocytochrome $b_2$, does cytochrome *c* bind prior to inter-molecular electron transfer?

In addition to these topics, the factors controlling substrate specificity in flavocytochrome $b_2$ will also be discussed since considerable progress has been made in this area over recent years. This discussion will also

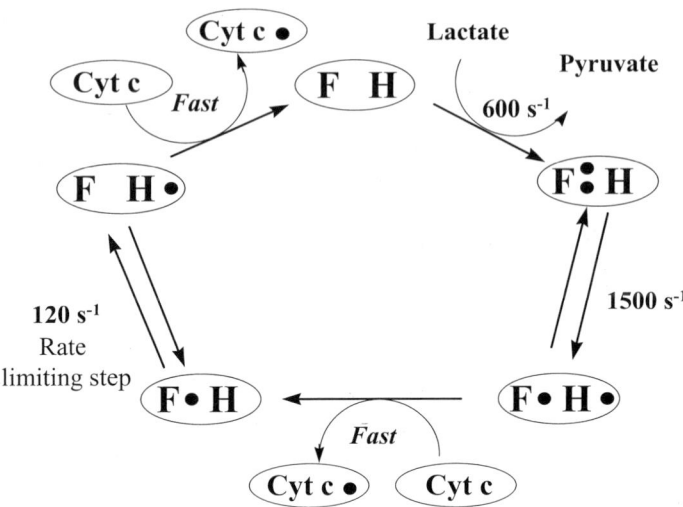

**FIGURE 3.** The Catalytic Cycle for Flavocytochrome $b_2$. F, flavin; H, heme; Cyt c, cytochrome c. Electrons are shown as black dots and are used to indicate: the two-electron reduced flavin (hydroquinone), F with two dots; one-electron reduced flavin (semiquinone), F with one dot; reduced heme, H with one dot; reduced cytochrome c, Cyt c with one dot. The rate constants shown are for *S. cerevisiae* flavocytochrome $b_2$ at 25°C, pH 7.5, I = 0.10 M. The whole catalytic cycle turns over at approximately $100\,s^{-1}$. The details of the cycle are described in the main text.

address the question of how the substrate specificity of the enzyme can be re-engineered.

## 2.  FLAVIN REDUCTION AND SUBSTRATE OXIDATION

The first step in the catalytic cycle of flavocytochrome $b_2$ is the oxidation of L-lactate to pyruvate and the reduction of the flavin. Our understanding of how this occurs has been dominated by what can only be described as the dogma of the carbanion mechanism. Although this mechanism for flavoprotein catalysed substrate oxidations is accepted by many, doubts remain, and the alternative hydride transfer process cannot be ruled out. The carbanion mechanism has been extensively surveyed in the past, reviews by Lederer (1997 and 1991) and Ghisla and Massey (1989) are recommended, and for this reason there is little point in covering the same ground in the present article in any great detail.

The fundamentals on which the carbanion mechanism is founded are the early studies on the related enzymes D-amino acid oxidase (Walsh *et al.*, 1972 and 1971), lactate oxidase (Walsh *et al.*, 1973) and flavocytochrome $b_2$ (Urban and Lederer, 1985; Pompon and Lederer, 1985). It is interesting that all of the key work, which established the carbanion mechanism, was done before any 3-dimensional structures were available on the respective enzymes.

The core requirement for the carbanion mechanism to operate is that an active-site base must abstract the $\alpha$-carbon hydrogen of the substrate, as a proton, forming a carbanion intermediate (Lederer, 1991). This would then require the equivalent of two electrons to be transferred to the flavin either with or without the formation of a covalent intermediate between the $\alpha$-carbon and the flavin N-5 (Ghisla and Massey, 1989). With this in mind, it is intriguing to find that the crystal structure of D-amino acid oxidase reveals that there is no residue correctly located to act as the active-site base required for the carbanion mechanism (Mattevi *et al.*, 1996; Mizutani *et al.*, 1996). In fact, the crystallographic information available is far more consistent with this enzyme operating a hydride transfer mechanism (Mattevi *et al.*, 1996). If this is correct then the earlier experiments on D-amino acid oxidase, which were claimed to be diagnostic of a carbanion mechanism, are called into question. It is important to note that similar experiments were used to provide support for a carbanion mechanism in the case of flavocytochrome $b_2$.

Doubts on the carbanion mechanism have also been raised in the case of L-lactate oxidase (Yorita *et al.*, 1997). Studies on this enzyme, using *para*-substituted L-mandelates as substrates, gave results compatible with any mechanism in which the transition state develops little charge. The authors concluded that this would be consistent with a hydride transfer or a synchronous mechanism in which the C2 hydrogen is abstracted as a proton (as in the classical carbanion mechanism) but that any developing charge is simultaneously neutralised by other events (Yorita *et al.*, 1997).

So what about flavocytochrome $b_2$? In this case the crystal structure (Xia and Mathews, 1990) does indicate the presence of an active-site base (His373) that might act to abstract the $\alpha$-carbon hydrogen of the substrate, as a proton. However, such a base is equally likely to remove the hydroxyl proton of lactate, thereby promoting hydride transfer to the flavin. The two contrasting roles for this active-site base are illustrated in Figure 4.

The high-level expression of recombinant flavocytochrome $b_2$ in *E. coli* (Black *et al.*, 1989) has allowed the active site of the enzyme to be probed using site-directed mutagenesis. Two particular residues, His373 and Tyr254, have been examined in detail, since they have important roles in catalysis. The substitution of His373 by glutamine resulted in an enzyme with some

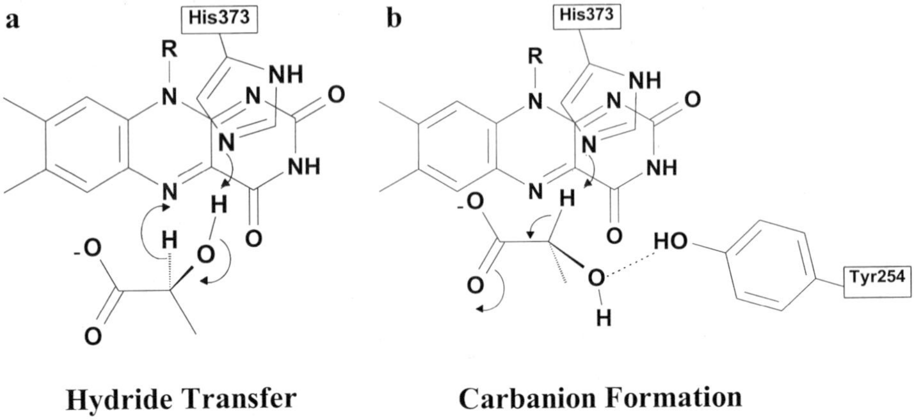

Hydride Transfer                           Carbanion Formation

FIGURE 4. Mechanisms for lactate dehydrogenation. **a.** By hydride transfer: His373 abstracts the hydroxyl proton and promotes hydride transfer to the N-5 of the flavin. **b.** By carbanion formation: His373 removes the α-carbon hydrogen as a proton forming a carbanion. A subsequent transfer of two-electrons to the flavin is required.

$10^5$-fold less activity than that seen for the wild-type enzyme (Gaume *et al.*, 1995), consistent with this residue acting as the active-site base, whichever mechanism operates. Of more interest was the effect of mutations at position 254. In the carbanion mechanism this residue is proposed to hydrogen bond to the hydroxyl group of L-lactate (Figure 4b). One would therefore expect substitution of this residue to significantly affect the $K_M$ value for L-lactate. In fact, the replacement of Tyr254 by phenylalanine (Dubois *et al.*, 1990; Reid *et al.*, 1988) and by leucine (Gondry *et al.*, 1995) had no effect on $K_M$ but did result in a decreased $k_{cat}$ value. The equivalent residue in the related flavoenzyme, glycolate oxidase (Tyr129) has also been substituted by phenylalanine and again shows little effect on substrate $K_M$ (Macheroux *et al.*, 1993). The inevitable conclusion is that this residue plays a role in transition state stabilisation and is definitely not involved in Michaelis complex formation. The depiction in the literature of this residue forming a hydrogen bond to the substrate hydroxyl, as in Figure 4b, should therefore be discouraged.

An examination of the flavocytochrome $b_2$ active site, as defined by the crystal structure (Xia and Mathews, 1990), clearly shows that if His373 formed a hydrogen bond to the substrate hydroxyl, then the hydrogen on the α-carbon would be ideally placed for hydride transfer to flavin N-5 as shown in Figure 4a. Tyr254 would then come into play, stabilising the transition state (in which the α-carbon would have less $sp^3$ and more $sp^2$ character). Thus a hydride transfer mechanism is certainly consistent with recent crystallographic and mutagenesis studies. It should therefore be clear that neither the carbanion nor the hydride transfer mechanism can be said to have been conclusively proved to operate in flavocytochrome $b_2$. As such, keeping an open mind on the subject is certainly to be recommended.

Whichever mechanism operates, it is clear that the rate of reduction of the flavin group is totally limited by the cleavage of the $\alpha$C-H bond since the deuterium kinetic isotope effect for this step is around 8 (Miles *et al.*, 1992; Pompon *et al.*, 1980). However, in flavocytochrome $b_2$ the rate of flavin reduction is some 6-fold faster than the overall steady-state turnover rate (Daff *et al.*, 1996a). As a consequence the flavin reduction step contributes little to the rate limitation of the overall catalytic cycle (Figure 3). In fact it is electron transfer from flavin-semiquinone to $b_2$-heme that is the major rate-determining step and this is discussed in the following section.

## 3. FLAVIN TO HEME ELECTRON TRANSFER

Intra-molecular electron transfer from fully-reduced flavin (flavohydroquinone) to $b_2$-heme has been studied extensively by stopped-flow spectrophotometry (Daff *et al.*, 1996a; Chapman *et al.*, 1994; Pompon, 1980; Capeillère-Blandin, 1975), laser flash photolysis (Hazzard *et al.*, 1994) and temperature jump experiments (Tegoni *et al.*, 1998). All of these studies confirm that this electron transfer is several times faster than catalytic turnover and at 25°C, pH 7.5, the rate constant for this step is likely to be between 1500 and 2000$s^{-1}$ (Daff *et al.*, 1996a). The second intra-molecular electron transfer, from flavin-semiquinone to $b_2$-heme, has also been investigated and it is clear from both stopped-flow (Daff *et al.*, 1996a) and temperature jump (Tegoni *et al.*, 1998) studies that it is this step which rate-limits the overall catalytic cycle in flavocytochrome $b_2$ (Figure 3). Confirmation of this was provided by quenched-flow EPR studies (Daff *et al.*, 1996a). However, the key question is—what controls these intra-molecular electron transfers from flavin to heme? There are two contrasting explanations. One is that mobility between the two domains, i.e. protein dynamics, is the main factor limiting electron transfer. The other is that these electron transfers are limited purely by thermodynamics.

The suggestion that the cytochrome domain of flavocytochrome $b_2$ might be mobile with respect to the flavodehydrogenase domain was originally based on observations from NMR studies (Labeyrie *et al.*, 1988). This NMR approach has recently been challenged, at least in the case of the *S. cerevisiae* enzyme (Bell *et al.*, 1997). However, crystallographic evidence also lends support to domain mobility and allows the identification of a region of polypeptide, linking the two domains, which might function as a hinge (Xia and Mathews, 1990). This region of the protein is highlighted in Figure 2. The role of this hinge and its effect on inter-domain electron transfer has been examined in detail (Sharp et al., 1996a). Deletions and insertions in the hinge region have been shown to have dramatic effects on the rate of electron transfer from flavohydroquinone to $b_2$-heme (Sharp et al., 1996b and 1994) consistent with this step being controlled by protein dynamics.

It has also been demonstrated, by site-directed mutagenesis, that Tyr143 plays a key role in flavin to heme electron transfer (Miles et al., 1992). In the crystal structure of flavocytochrome $b_2$ Tyr143 is shown to have two different conformations (Xia and Mathews, 1990). In subunit 1 Tyr143 is shown to hydrogen bond directly to one of the $b_2$-heme propionate groups forming an interdomain contact. However in subunit 2, which has pyruvate bound and in which the cytochrome domain is positionally disordered, Tyr143 forms a hydrogen bond to the carboxylate of the product. Substitution of this tyrosine by phenylalanine produced a mutant enzyme in which the rate of electron transfer from flavohydroquinone to $b_2$-heme was dramatically decreased (Miles et al., 1992) indicating that Tyr143 plays a crucial role in the first inter-domain electron transfer.

However, the studies on the hinge and Tyr143 provided no information on the electron transfer from flavin-semiquinone to $b_2$-heme, the step which rate limits the catalytic cycle. In some elegant experiments, Tegoni et al. (1998) showed that the variation of the rate constant (ln $k_{et}$) for this step correlates well with variations in the driving force ($\Delta G$). Thus, this slow intra-molecular electron transfer is almost certainly controlled by thermodynamics. Tegoni et al. (1998) also provided some evidence that the faster, flavohydroquinone $\rightarrow$ $b_2$-heme, electron transfer may also be under thermodynamic control although this remains to be conclusively demonstrated.

## 4.  THE FLAVOCYTOCHROME B₂: CYTOCHROME C INTERACTION

The characterisation of the complexation between flavocytochrome $b_2$ and cytochrome $c$ has been the subject of many studies (see for example: Short et al., 1998; Daff et al., 1996b; and Capeillère-Blandin, 1995). Work on the H. anomala flavocytochrome $b_2$, for which there is no crystal structure, led to the conclusions that the cytochrome $c$ binding site involved both the flavodehydrogenase and cytochrome domains (Capeillère-Blandin and Albani, 1987) and that the complex was stabilised by electrostatic interactions (Capeillère-Blandin, 1982). It is clear that similar conclusions hold true for the S. cerevisiae enzyme (Daff et al., 1996b) for which the crystal structure is known (Xia and Mathews, 1990). It is also clear that the rate of inter-protein electron transfer between flavocytochrome $b_2$ and cytochrome $c$ is controlled by the rate of complex formation, which must occur to permit efficient electron transfer (Short *et al.*, 1998; Daff *et al.*, 1996b).

Since attempts to obtain the flavocytochrome $b_2$:cytochrome $c$ complex in crystalline form have failed, molecular modelling has been used to try and provide a 3-dimensional understanding of the interaction. A hypothetical complex between flavocytochrome $b_2$ and cytochrome $c$ was reported by Tegoni *et al.* (1993). These authors considered several general areas as plausible sites on the flavocytochrome $b_2$ surface for cytochrome $c$ to bind. They decided to restrict themselves to examining the surface of flavocytochrome $b_2$ in the area of the interface between the two domains. Further restrictions placed on the model resulted in the discarding of binding sites that were not compatible with the crystal packing of the flavocytochrome $b_2$ tetramer. Such restrictions may, however, rule out perfectly reasonable sites that would be accessible in solution and this highlights the care needed when proposing a hypothetical complex based on crystal structural data. The final model selected by the authors predicted several important residues for the formation of a complex. The interaction area is described as extensive, however a key point is that the interacting surfaces are not perfectly complementary, a fact noted by the authors (Tegoni *et al.*, 1993). The stoichiometry is four cytochromes $c$ per tetramer, and a proposed $\sigma$-tunnelling pathway from the flavocytochrome $b_2$ heme, through the backbones of Ile50 and Lys51 and through the aromatic ring of Phe52 on to the cytochrome $c$ heme, looked convincing but has subsequently been called into question (Short *et al.*, 1997). In addition to the pathway, the model made a number of other predictions including the residues involved in complex formation. To test these predictions a number of mutant flavocytochromes $b_2$ were generated and analysed (Daff *et al.*, 1996b; Short *et al.*, 1997 and 1998). The results of these studies indicated that the model of the flavocytochrome $b_2$: cytochrome $c$ interaction reported by Tegoni *et al.* (1993) was not a realistic representation of a catalytically competent complex.

The failure of the Tegoni model provided the impetus for a new hypothesis of how flavocytochrome $b_2$ and cytochrome $c$ interact with one another. An examination of the crystal structure of flavocytochrome $b_2$ (Xia and Mathews, 1990) reveals a reasonably uniform distribution of charge, however, several patches of surface acidity can be found. These areas were considered as possible locations for binding sites for cytochrome $c$ by Short *et al.* (1998). The prime concern of these authors in locating such a binding site was that it should allow the $b_2$-heme and the $c$-heme to be as close together as possible. In addition, they considered the sequence conservation between the flavocytochrome $b_2$-heme domain and cytochrome $b_5$ in the context of the known cytochrome $b_5$:cytochrome $c$ models (Guillemette *et al.*, 1994). From this analysis two residues in flavocytochrome $b_2$ (Glu63 and Asp72), conserved in the cytochrome $b_5$ family, were implicated in the binding of cytochrome $c$. Using this information, Short *et al.* (1998) generated a number of possible docked structures which centred around these two residues and which also maintained a reasonably short heme-to-heme distance. Most of the hypothetical complexes produced shared a similar Fe-Fe vector but differed in the orientation about this vector. Thus there is no single *best-configuration* describing the interaction between the two molecules, but rather a set in which the heme group and docking-face of cytochrome $c$ can approach the heme of flavocytochrome $b_2$ in a variety of orientations. The proposed cytochrome $c$ docking site on flavocytochrome $b_2$ involved the acidic residues Glu63, Asp72 and Glu237 forming an acidic triangular recognition site for cytochrome $c$. The distance between the C$\alpha$ atoms of Glu237, Asp72 and Glu63 is between 15.1 and 15.7 Å.

Having indicated that there were a number of possible complexes with sterically acceptable docked structures, Short *et al.* (1998) went on to give detailed intermolecular interactions for two of these. In the "best" model the two protein molecules are bound together by three salt bridges Lys27..Glu63, Lys79..Glu237 and Lys13..Asp72 (where the lysine residues are from cytochrome $c$). Through this triangle of interactions runs a possible σ-tunnelling pathway for electron transfer. This pathway begins with the imidazole ring of His66 (a $b_2$-heme-iron ligand) and then passes through the backbone of Ala67 and the ring of Pro68 which is in van der Waals contact with the heme of cytochrome $c$. The direct edge-to-edge distance from the imidazole of His66 to the C3C pyrrole ring of the cytochrome $c$ heme is around 13 Å (Figure 5).

The flavocytochrome $b_2$ molecular surface which is buried and hidden from solvent on binding cytochrome $c$ has a total area of 657 Å$^2$. The model shows a number of significant general features. Firstly, four cytochromes $c$ can be accommodated on the flavocytochrome $b_2$ tetramer. However, each cytochrome $c$ forms interactions with only one flavocytochrome $b_2$ subunit. This is in stark contrast to the model of Tegoni *et al.* (1993) in which each cytochrome $c$ makes interactions with three of the four subunits of the flavocytochrome $b_2$ tetramer. Another important difference between the opposing models is in the relationship of the heme groups. In the model from Tegoni *et al.* (1993) the heme groups are essentially co-planar, whereas Short *et al.* (1998) indicate that they are perpendicular to each other. The interaction of a cytochrome $c$ molecule with a single flavocytochrome $b_2$ subunit as predicted by Short *et al.* (1998) is shown in Figure 6.

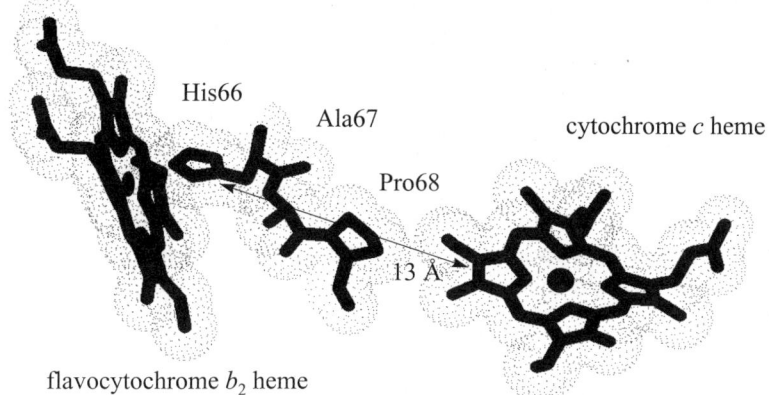

**FIGURE 5.** The Disposition of the Heme Groups in the Model Complex Proposed by Short *et al.* (1998). The figure shows the approximate perpendicular relationship between the flavo-cytochrome $b_2$ and cytochrome $c$ hemes. A possible σ-tunnelling pathway for electron transfer between the hemes is highlighted. The shortest edge-to-edge distance between the imidazole ring of His66 and a pyrrole ring of the cytochrome $c$ heme is indicated by the arrowed line.

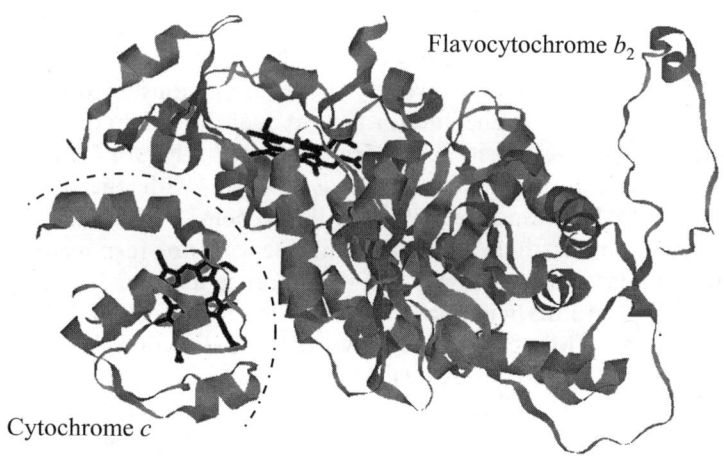

**FIGURE 6.** The Interaction of a Cytochrome $c$ Molecule with a Single Flavocytochrome $b_2$ Subunit as Predicted by Short *et al.* (1998). The two proteins are shown as ribbon diagrams. The heme groups are shown in stick representation. The interface region between the two proteins is indicated by the dotted line.

The triangular nature of the new model does, as explained above, have the potential to allow cytochrome $c$ to "sample" a number of different yet similar binding modes. In the context of electron transfer it is intriguing that in each of these binding modes the pathway and edge-to-edge distance for electron transfer remain essentially the same.

The model proposed by Short $et\ al.$ (1998) has been rigorously tested by site-directed mutagenesis and kinetic analysis of the mutant enzymes. It has been shown to be consistent with all of the results from these studies and therefore it may indeed represent the physiological complex formed between flavocytochrome $b_2$ and cytochrome $c$.

## 5.  ENGINEERING SUBSTRATE SPECIFICITY IN FLAVOCYTOCHROME $B_2$

Although flavocytochrome $b_2$ acts physiologically as a L-lactate dehydrogenase, it is able to catalyse the oxidation of a number of other 2-hydroxy acids (Chapman $et\ al.$, 1991). The X-ray crystal structure of the enzyme (Xia and Mathews, 1990) has allowed the active site for 2-hydroxyacid dehydrogenation to be identified. In fact, in one of the two crystallographically distinguishable subunits in the asymmetric unit, a molecule of pyruvate (the product of the reaction) can be found. The carboxylate end of pyruvate is bound by Arg376 and Tyr143, while the pyruvate carbonyl is held by His373 and Tyr254. These residues clearly control the group-specific recognition of substrate by the enzyme. However, they are obviously not responsible for the selectivity of the enzyme for lactate over other 2-hydroxy acids. An examination of the active site of flavocytochrome $b_2$ (Figure 7) indicates that the methyl group of pyruvate makes contact with the hydrophobic side-chains of Ala198 and Leu230, with Ile326 being a little further away. It would appear likely then, that these interactions are responsible for the selection of L-lactate by the enzyme in preference to other 2-hydroxyacids.

Flavocytochrome $b_2$ is related both structurally and by sequence to a number of other FMN-containing 2-hydroxyacid dehydrogenases/oxidases, including spinach glycolate oxidase (Lindqvist $et\ al.$, 1991) and $R.\ graminis$ L-mandelate dehydrogenase (Ilias $et\ al.$, 1998). However, these enzymes exhibit a kinetic preference for alternative substrates, glycolate and mandelate respectively. Sequence comparisons indicate that residues Ala198 and Leu230, highlighted in Figure 7, are not well conserved in these enzymes adding support to the idea that they are important in substrate selection. A shift in the selectivity of flavocytochrome $b_2$, away from lactate and towards longer chain substrates, has been achieved by making substitutions of these residues by site-directed mutagenesis (Daff $et\ al.$, 1994a and 1994b). The

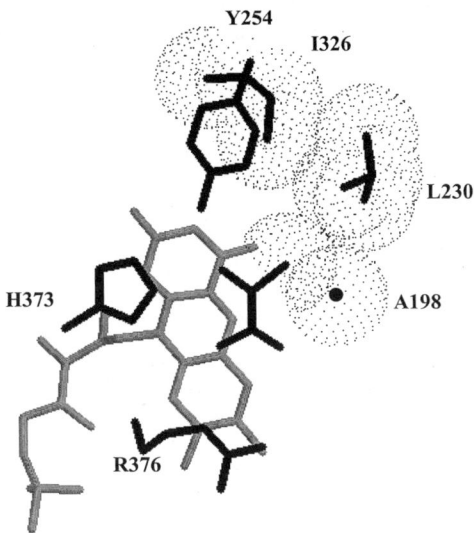

**FIGURE 7.** The Active Site in *S. cerevisiae* Flavocytochrome $b_2$. The dotted surfaces represent van der Waals radii. Hydrophobic contacts are shown between the methyl group of pyruvate and a number of alkyl amino acid side-chains.

substitution of Leu230 by alanine had a significant effect, causing the selectivity of the enzyme for 2-hydroxyoctanoate over lactate to increase by a factor of 80 (Daff *et al.*, 1994a). The substitution of Ile326 by alanine, which removes substantial bulk from a position directly adjacent to Leu230, also shifted the selectivity of the enzyme towards longer chain substrates and particularly 2-hydroxyoctanoate (Daff *et al.*, 1994b). A rather interesting observation was that, although the enzyme works well on straight-chain substrates and substrates which branch at the C-3 position, no activity was seen with substrates branched at C-4 (Daff *et al.*, 1994a and 1994b).

The activity of flavocytochrome $b_2$ towards the aromatic 2-hydroxy acid, L-mandelate, is near the limits of reliable measurement with a $k_{cat}$ value (25°C) of only $0.02\,s^{-1}$ compared to a value of $400\,s^{-1}$ seen with L-lactate (Sinclair *et al.*, 1998). This is somewhat surprising when one considers the extensive similarity between *S. cerevisiae* flavocytochrome $b_2$ and the L-mandelate dehydrogenase from *R. graminis* (Ilias *et al.*, 1998). In order to engineer a shift in activity towards L-mandelate and away from L-lactate, a number of mutant enzymes were generated and tested including Ala198→Gly, Leu230→Ala, Leu230→Gly, Ile326→Ala and the double mutation Ala198→Gly:Leu230→Ala (Sinclair *et al.*, 1998). Of these, the double mutation had the greatest effect resulting in an enzyme with a $k_{cat}$ value for L-mandelate of $8.5\,s^{-1}$, which represents an increase of greater than 400-fold over the wild-type enzyme. In fact this mutant enzyme was shown to be more efficient with L-mandelate than it was with L-lactate. Thus, Sinclair *et al.* (1998) had succeeded in engineering the enzyme into a better mandelate dehydrogenase than it was a lactate dehydrogenase.

## 6. CONCLUSIONS

There are a number of factors which make flavocytochrome $b_2$ an ideal model system for studying both intra- and inter-molecular electron transfer. Reasons include: (i) the fact that it has been expressed at a high level in *E. coli* (Black *et al.*, 1989) and is soluble and easily obtained; (ii) crystal structures of the native (Xia and Mathews, 1990) and recombinant (Tegoni and Cambillau, 1994) enzymes are available; (iii) a hypothetical structure for the flavocytochrome $b_2$:cytochrome $c$ complex has been proposed (Short *et al.*, 1998); (iv) many mutant forms of the enzyme have been constructed; (v) there is a wealth of data on the mechanism of action of the enzyme (Chapman *et al.*, 1991; Lederer, 1991).

In this article the key mechanistic questions have been discussed, giving the reader an understanding of our current state of knowledge concerning the catalytic properties of flavocytochrome $b_2$. In addition the potential for re-engineering flavocytochrome $b_2$ has been described and this may lead to a number of future applications in, for example, bio-sensing and bio-transformations. Hopefully, the reader will conclude that flavocytochrome $b_2$ is a truly fascinating enzyme

ACKNOWLEDGEMENTS. The authors acknowledge C. Miles for exceptionally helpful discussions. CGM thanks the Caledonian Research Foundation for a scholarship. SKC is grateful to the Biotechnology and Biological Sciences Research Council (U.K.) for funding.

## 7. REFERENCES

Bell, C., Uhrinova, S., Barlow, P. N., Chapman, S. K., and Reid, G. A., 1997, Domain mobility in flavocytochrome $b_2$: Fact or fiction, in *Flavins and Flavoproteins 1996* (K. J. Stevenson, V. Massey, and C. H. Williams, eds.), University of Calgary Press, Calgary, Alberta, pp. 555–558.
Black, M. T., White, S. A., Reid, G. A., and Chapman, S. K., 1989, High-level expression of fully active yeast flavocytochrome $b_2$ in *Escherichia coli*, *Biochem. J.* **258**:255–259.

Capeillère-Blandin, C., 1975, Flavocytochrome $b_2$: Simulation studies of the electron transfer reactions among prosthetic groups, *Eur. J. Biochem.* **56**:91–101.

Capeillére-Blandin, C., Bray, R. C., Iwatsubo, M., and Labeyrie, F., 1975, Flavocytochrome $b_2$: Kinetic studies by absorbance and electron paramagnetic resonance spectroscopy of electron distribution among prosthetic groups, *Eur. J. Biochem.* **54**:549–566.

Capeillère-Blandin, C., 1982, Transient kinetics of the one electron transfer reaction between reduced flavocytochrome $b_2$ and oxidised cytochrome c. Evidence for the existence of a protein complex in the reaction, *Eur. J. Biochem.* **128**:533–542.

Capeillère-Blandin, C., 1995, Flavocytochrome $b_2$-cytochrome c interactions: The electron transfer reaction revisited, *Biochimie* **77**:516–530.

Capeillère-Blandin, C., and Albani, J., 1987, Cytochrome $b_2$, an electron carrier between flavocytochrome $b_2$ and cytochrome c. Rapid kinetic characterisation of the electron transfer parameters with ionic strength dependence, *Biochem. J.* **245**:159–165.

Chapman, S. K., White, S. A., and Reid, G. A., 1991, Flavocytochrome $b_2$, in: *Advances in Inorganic Chemistry* Volume 36 (A. G. Sykes, ed.), Academic Press, San Diego, pp. 257–301.

Chapman, S. K., Reid, G. A., Daff, S., Sharp, R. E., White, P., Manson, F. D., and Lederer, F., 1994, Flavin to heme electron transfer in flavocytochrome $b_2$, *Biochem. Soc. Trans.* **22**:713–718.

Chapman, S. K., Reid, G. A., and Munro, A. W., 1998, Flavocytochromes: Nature's electrical transformers, in: *Biological Electron Transfer Chains: Genetics, Composition and Mode of Operation* (G. W. Canters and E. Vijgenboom, eds.), Kluwer Academic Publishers, Dordrecht, The Netherlands, pp. 165–184.

Daff, S., Manson, F. D. C., Reid, G. A., and Chapman, S. K., 1994a, Strategic manipulation of the substrate specificity of *Saccharomyces cerevisiae* flavocytochrome $b_2$, *Biochem. J.* **301**:829–834.

Daff, S., Manson, F. D .C., Reid, G. A., and Chapman, S. K., 1994b, Manipulation of the substrate specificity of flavocytochrome $b_2$, *Biochem. Soc. Trans.* **22**:282.

Daff, S., Ingledew, W. J., Reid, G. A., and Chapman, S. K., 1996a, New insights into the catalytic cycle of flavocytochrome $b_2$, *Biochemistry* **35**:6345–6350.

Daff, S., Sharp, R. E., Short, D. M., Bell, C., White, P., Manson, F. D., Reid, G. A., and Chapman, S. K., 1996b, Interaction of cytochrome c with flavocytochrome $b_2$, *Biochemistry* **35**:6351–6357.

Dubois, J., Chapman, S. K., Mathews, F. S., Reid, G. A., and Lederer, F., 1990, Substitution of tyr254 with phe at the active site of flavocytochrome $b_2$, *Biochemistry* **29**:6393–6400.

Gaume, B., Sharp, R. E., Manson, F. D. C, Chapman, S. K., Reid, G. A., and Lederer, F., 1995, Mutation to glutamine of histidine 373, the catalytic base of flavocytochrome $b_2$, *Biochimie* **77**:621–630.

Ghisla, S., and Massey, V., 1989, Mechanisms of flavoprotein-catalysed reactions, *Eur. J. Biochem.* **181**:1–17.

Gondry, M., Lê, K. H. D., Manson, F. D. C., Chapman, S. K., Mathews, F. S., Reid, G. A., and Lederer, F., 1995, On the lack of coordination between protein folding and flavin insertion in *Escherichia coli* for flavocytochrome $b_2$ mutant forms Y254L and D282N, *Protein Science* **4**:925–935.

Guillemette, J. G., Barker, P. D., Eltis, L. D. Lo, T. P., Smith, M., Brayer G. D., and Mauk, A. G., 1994, Analysis of the bimolecular reduction of ferricytochrome c by ferrocytochrome $b_5$, *Biochimie* **76**:592–604

Hazzard, J. T., McDonough, C. A., and Tollin, G., 1994, Intramolecular electron transfer in yeast flavocytochrome $b_2$ upon one-electron photooxidation of the fully reduced enzyme: Evidence for redox state control of heme-flavin communication, *Biochemistry* **33**:13445–13454.

Ilias, R. M., Sinclair, R., Robertson, D., Neu, A., Chapman, S. K., and Reid, G. A., 1998, L-Mandelate dehydrogenase from *Rhodotorula graminis*, *Biochem. J.* **333**:107–115.

Labeyrie, F., Beloeil, J. C., and Thomas, M. A., 1988, Evidence by NMR for mobility of the cytochrome domain of flavocytochrome $b_2$, *Biochim. Biophys. Acta.* **953**:134–141.

Lederer, F., 1991, Flavocytochrome $b_2$, in: *Chemistry and Biochemistry of Flavoenzymes* Volume 2 (F. Muller, ed.), CRC Press, Boca Raton, Fl., pp. 153–242.

Lederer, F., 1997, The mechanism of flavoprotein-catalyzed α-hydroxy acid dehydrogenation, in *Flavins and Flavoproteins 1996* (K. J. Stevenson, V. Massey, and C. H. Williams, eds.), University of Calgary Press, Calgary, Alberta, pp. 545–554.

Lindqvist, Y., Brändén, C-I., Mathews, F. S., and Lederer, F., 1991, Spinach glycolate oxidase and yeast flavocytochrome $b_2$ are structurally homologous and evolutionarily related enzymes wioth distinctly different function and FMN binding, *J. Biol. Chem.* **266**:3198–3207.

Macheroux, P., Kieweg, V., Massey, V., Söderland, E., Stenberg, K., and Lindqvist, Y., 1993, Role of tyrosine 129 in the active site of spinach glycolate oxidase, *Eur. J. Biochem.* **213**:1047–1054.

Mattevi, A., Vanoni, M. A., Todone, F., Rizzi, M., Teplyakov, A., Coda, A., Bolognesi, M., and Curti, B., 1996, Crystal structure of D-amino acid oxidase: a case of active site mirror-image convergent evolution with flavocytochrome $b_2$, *Proc. Natl. Acad. Sci. USA.* **93**:7496–7501.

Mattevi, A., Vanoni, M. A., and Curti, B., 1997, Structure of D-amino acid oxidase: New insights from an old enzyme, *Curr. Opin. Struc. Biol.* **7**:804–810.

Miles, C. S., Rouvière-Fourmy, N., Lederer, F., Mathews, F. S., Reid, G. A., Black, M. T., and Chapman, S. K., 1992, Tyr-143 facilitates interdomain electron transfer in flavocytochrome $b_2$, *Biochem. J.* **285**:187–192.

Mizutani, H., Miyahara, I., Hirotsu, K., Nishima, Y., Shiga, K., Setoyama, C., and Miura R., 1996, Three-dimensional structure of porcine kidney D-amino acid oxidase at 3 Å resolution, *J. Biochem.* **120**:14–17.

Pompon, D., 1980, Flavocytochrome $b_2$ from baker's yeast. Computer simulation studies of a new kinetic scheme for intramolecular electron transfer, *Eur. J. Biochem.* **106**:151–159.

Pompon, D., Iwatsubo, M., and Lederer, F., 1980, Flavocytochrome $b_2$ (baker's yeast). Deuterium isotope effect studied by rapid kinetic methods as a probe for the mechanism of electron transfer, *Eur. J. Biochem.* **104**:479–488.

Pompon, D., and Lederer, F., 1985, On the mechanism of flavin modification during inactivation of flavocytochrome $b_2$ from baker's yeast by acetylenic substrates, *Eur. J. Biochem.* **148**:145–154.

Reid, G. A., White, S., Black, M. T., Lederer, F., Mathews, F. S., and Chapman S. K., 1988, Probing the active site of flavocytochrome $b_2$ by site-directed mutagenesis, *Eur. J. Biochem.* **178**:329–333.

Sharp, R. E., Chapman, S. K., and Reid, G. A., 1994, Role of the interdomain hinge of flavocytochrome $b_2$ in intra- and inter-protein electron transfer, *Biochemistry* **33**:5115–5120.

Sharp, R. E., Chapman, S. K., and Reid, G. A., 1996a, Modulation of flavocytochrome $b_2$ intraprotein electron transfer via an interdomain hinge, *Biochem. J.* **316**:507–513.

Sharp, R. E., Chapman, S. K., and Reid, G. A., 1996b, Deletions in the interdomain hinge region of flavocytochrome $b_2$: Effects on intraprotein electron transfer, *Biochemistry* **35**:891–899.

Short, D. M., Walkinshaw, M. D., Taylor, P., Reid, G. A., and Chapman, S. K., 1997, The cytochrome *c* recognition site on flavocytochrome $b_2$, in *Flavins and Flavoproteins 1996* (K. J. Stevenson, V. Massey, and C. H. Williams, eds.), University of Calgary Press, Calgary, Alberta, pp. 575–578.

Short, D. M., Walkinshaw, M. D., Taylor, P., Reid, G. A., and Chapman, S. K., 1998, Location of a cytochrome $c$ binding site on the surface of flavocytochrome $b_2$, *J. Biol. Inorg. Chem.* **3**:246–252.

Sinclair, R. Reid, G., and Chapman, S. K., 1998, Re-design of *Saccharomyces cerevisiae* flavocytochrome $b_2$: introduction of L-mandelate dehydrogenase activity, *Biochem. J.* **333**:117–120.

Tegoni, M., and Cambillau, C., 1994. The 2.6 Å refined structure of the *Escherichia coli* recombinant *Saccharomyces cerevisiae* flavocytochrome $b_2$ sulfite complex, *Prot. Sci.* **3**:303–313.

Tegoni, M., Silvestrini, M. C., Guigliarelli, B., Asso, M., Brunori, M., and Bertrand, P., 1998, Temperature jump and potentiometric studies on recombinant wild-type and Y143F and Y254F mutants of *Saccharomyces cerevisiae* flavocytochrome $b_2$: Role of the driving force in intramolecular electron transfer kinetics, *Biochemistry* **37**:12761–12771.

Tegoni, M., White, S. A., Roussel, A., Mathews, F. S., and Cambillau, C., 1993, A hypothetical complex between crystalline flavocytochrome $b_2$ and cytochrome $c$, *Proteins: Structure, Function, and Genetics* **16**:408–422

Urban, P., and Lederer, F., 1985, Intermolecular hydrogen transfer catalysed by a flavodehydrogenase, baker's yeast flavocytochrome $b_2$, *J. Biol. Chem.* **260**:11115–11122.

Walsh, C. T., Schonbrunn, A., and Abeles, R., 1971, Studies on the mechanism of action of D-amino acid oxidase. Evidence for removal of substrate α-hydrogen as a proton, *J. Biol. Chem.* **246**:6855–6866.

Walsh, C. T., Schonbrunn, A., Lockridge, O., Massey, V., and Abeles, R., 1972, Inactivation of a flavoprotein, lactate oxidase, by an acetylenic substrate, *J. Biol. Chem.* **247**:6004–6006.

Walsh, C. T., Lockridge, O., Massey, V., and Abeles, R., 1973, Studies on the mechanism of action of the flavoenzyme lactate oxidase. Oxidation and elimination with b-chlorolactate, *J. Biol. Chem.* **248**:7049–7054.

Xia, Z. X., and Mathews, F. S., 1990, Molecular structure of flavocytochrome $b_2$ at 2.4 Å resolution, *J. Mol. Biol.* **212**:837–863.

Yorita, K., Janko, K., Aki, K., Ghisla, S., Palfey, B. A., and Massey, V., 1997, On the reaction mechanism of L-lactate oxidase, *Proc. Natl. Acad. Sci. USA.* **94**:9590–9595.

*Chapter 9*

# Flavocytochrome P450 BM3
## Substrate Selectivity and Electron Transfer in a Model Cytochrome P450

Andrew W. Munro, Michael A. Noble,
Tobias W. B. Ost, Amanda J. Green, Kirsty J. McLean,
Laura Robledo, Caroline S. Miles, Jane Murdoch,
and Stephen K. Chapman

## 1. INTRODUCTION

The cytochromes P450 (P450s) are haem *b*-containing monooxygenase enzymes that play crucial roles in the processes of drug metabolism, steroid and fatty acid metabolism in mammals (Munro and Lindsay, 1996). They were first recognised as carbon monoxide-binding haem pigments in mammalian liver microsomes. The fact that treatment of reduced microsomes with carbon monoxide led to loss of drug-metabolising activity and that this

**ANDREW W. MUNRO and KIRSTY J. MACLEAN**    Department of Pure & Applied Chemistry, University of Strathclyde, The Royal College, 204 George St., Glasgow, G1 1XL. UK.    **MICHAEL A. NOBLE, TOBIAS W. B. OST, AMANDA J. GREEN, LAURA ROBLEDO, CAROLINE S. MILES, JANE MURDOCH, and STEPHEN K. CHAPMAN** Department of Chemistry, University of Edinburgh, The King's Buildings, West Mains Rd., Edinburgh, EH9 3JJ. UK.

*Subcellular Biochemistry, Volume 35: Enzyme-Catalyzed Electron and Radical Transfer*, edited by Holzenburg and Scrutton. Kluwer Academic / Plenum Publishers, New York, 2000.

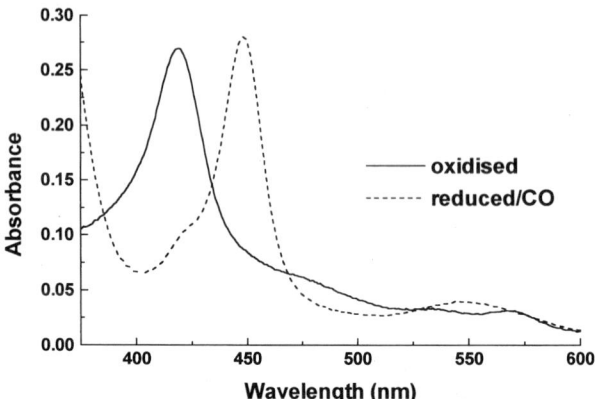

**FIGURE 1.** The visible spectral properties of flavocytochrome P450 BM3. The oxidised (solid line) and reduced/carbon monoxide-bound (dotted line) forms of the haemoprotein are shown. The protein concentration is 2.7 µM. The major (Soret) band of P450 BM3 is at 419 nm in the oxidised form, and shifts to 450 nm in the sodium dithionite-reduced/CO-bound form.

could be correlated with the formation of a new spectral species that had a visible absorbance maximum at 450 nm led to the realisation of the importance of the species in drug and xenobiotic oxidation, and to the title of this enzyme superfamily (P450 = Pigment with absorption at 450 nm in the reduced/carbon monoxide-bound form) (Figure 1) (Omura and Sato, 1964). Great interest has surrounded the P450s due not only to their involvement in critical metabolic processes and their implication in mammalian carcinogenesis, but also because of their ability to catalyse the cleavage of molecular oxygen ($O_2$) and to insert one of the derived atoms into the substrate (which may be e.g. a fatty acid, steroid or eicosanoid). The second atom of oxygen is reduced to water. The P450s catalyse regio- and stereospecific oxidations of organic substrates, a process of great interest to industry. A point of particular interest is the mechanism by which the P450s bind $O_2$ and control its reduction to achieve productive scission of oxygen and the oxidation of substrate, avoiding the damaging effects of oxygen radicals on protein and substrate (Munro *et al.*, 1999). Due to their very short lifetimes, a number of iron-oxygen intermediates that occur after oxygen binds to the ferrous P450 haem remain to be positively identified. However, a popular mechanism for the P450-dependent activation of oxygen exists (Figure 2).

The P450s are found in all life forms and have been postulated to have evolved to catalyse fatty acid oxidations required for the formation of prim-

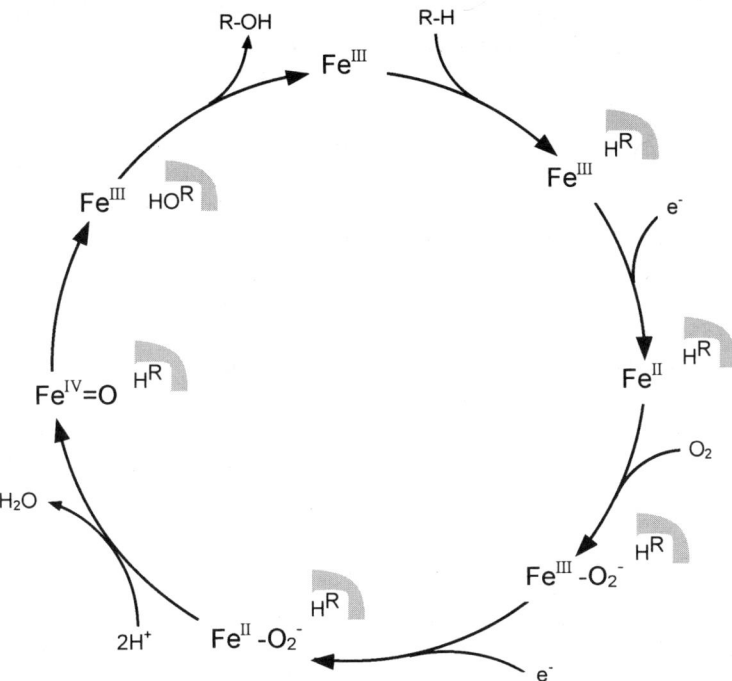

**FIGURE 2.** The catalytic cycle of cytochrome P450. RH denotes substrate (fatty acid in the case of P450 BM3). Substrate-induced increase in haem iron reduction potential triggers electron transfer to the ferric haem. Dioxygen binds to the ferrous form and a second electron is delivered to form the oxyferrous intermediate. Following the second electron transfer are a series of very rapid steps which culminate in the scission of dioxygen. A single atom of oxygen is inserted into the substrate, with the second forming water. Oxygenated product (ROH) dissociates from the active site to regenerate the resting ferric form.

itive cell membranes (Munro and Lindsay, 1996). Several hundred types of P450 are now known, with eukaryotes having many distinct forms that catalyse the mono-oxygenation of numerous different endogenous metabolites and exogenous compounds that must be targetted for excretion from the body. Many bacteria are also known to contain P450s, although usually rather fewer than found in higher organisms. For instance, only one P450 form has been reported from the Gram negative bacterium *Salmonella typhimurium*, with four evident from the determination of the entire genome sequence of *Bacillus subtilis* (Kunst *et al.*, 1997). *Escherichia coli* does not appear to encode a P450. However, this does not prevent the use of *E. coli* as a heterologous expression system for P450s. Several eukaryotic and prokaryotic P450 genes have been overexpressed in *E. coli* (e.g. Miles *et al.*, 1992; Gillam *et al.*, 1994). It is very interesting to note that *Mycobacterium tuberculosis*, the causative agent of TB, possesses a very large number of P450-encoding genes, more than 20 in all (Cole *et al.*, 1998). This information is also derived from the recent determination of the *M. tuberculosis* genome sequence, and may provide the stimulus for generation of new compounds as anti-TB drugs. It has been shown that P450 substrate analogues complexed to haem-ligating groups such as imidazole can produce very potent inhibitors of P450. Compounds of this type have proved very effective as anti-fungal treatments. Steroid-linked imidazoles such as clotrimazole, saperconazole and ketoconazole target and inactivate the lanosterol 14$\alpha$-demethylase enzyme which catalyses sterol conversions critical for membrane fluidity and integrity (Ortiz de Montellano and Correia, 1995). A similar approach to combat *M. tuberculosis* is an attractive strategy, once the substrate specificity properties of the various P450s have been determined.

   In general, it is true that the most significant advances in our understanding of the structure and mechanism of cytochromes P450 have come through the study of bacterial forms. This results from the fact that bacterial P450s have historically proven simpler to overexpress, purify in active form, study spectroscopically and enzymatically, and crystallise for structural determination by X-ray diffraction. There are obvious reasons why this has proven to be the case. Almost without exception, eukaryotic P450s are membrane-bound. They are integral membrane proteins, tethered to the membrane by a short segment of hydrophobic amino acids at the N-terminal of the P450. Eukaryotic P450s must also interact with membrane-bound reductase systems in order to receive electrons required for function. By contrast, prokaryotic P450s and their redox partners are soluble enzymes (Figure 3). Until relatively recently, problems associated with obtaining large quantities of pure, membranous eukaryotic P450s for biophysical studies meant that the bacterial P450s and their redox partners were far more attractive systems for study. However, in recent years there have been considerable technological advances in expression systems for eukaryotic P450s, including the development of good expression systems in yeast, insect cell (baculovirus) and *E. coli* systems (Sakaki *et al.*, 1990; Gonzalez *et al.*, 1991; Sandhu *et al.*, 1994). An important development has been the demonstration that active, soluble forms of P450 can be generated after the genetic removal of the membrane-spanning segment. Recently, similar advances in expression technology for NADPH-cytochrome P450 reductase (CPR), the diflavin (FAD- and FMN-containing) redox partner of non-

## Class I

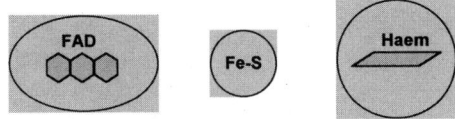

## Class II

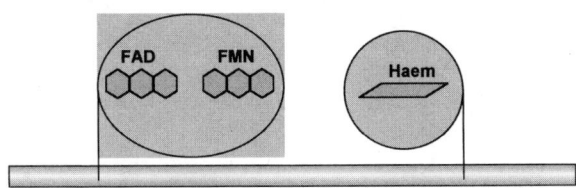

## Class III

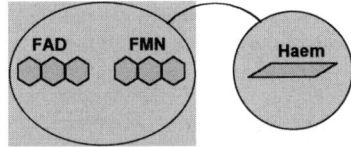

**FIGURE 3.** A diagrammatic representation of the different classes of cytochrome P450 systems. Class I systems are typified by cytosolic bacterial P450s and membrane associated eukaryotic mitochondrial forms. Electrons are delivered from NAD(P)H via an FAD-containing ferredoxin reductase and a small iron-sulfur ferredoxin. Class II systems are typified by mammalian hepatic forms, where electrons from NADPH are shuttled through a diflavin reductase to the P450. Both enzymes are membrane-bound. P450 BM3 is a class III P450, in which the redox components of a class II system are fused in a single polypeptide. The flavin isoalloxazine ring system is indicated by the fused hexagons, with the rhombus indicating the haem plane.

mitochondrial forms, led to the determination of the atomic structure of a soluble fragment of the rat CPR (Figure 4) (Wang *et al.*, 1997). The first atomic structure of a microsomal cytochrome P450 is also expected very soon. Notwithstanding these recent breakthroughs, the bulk of structural and mechanistic information on cytochrome P450 has been collected by

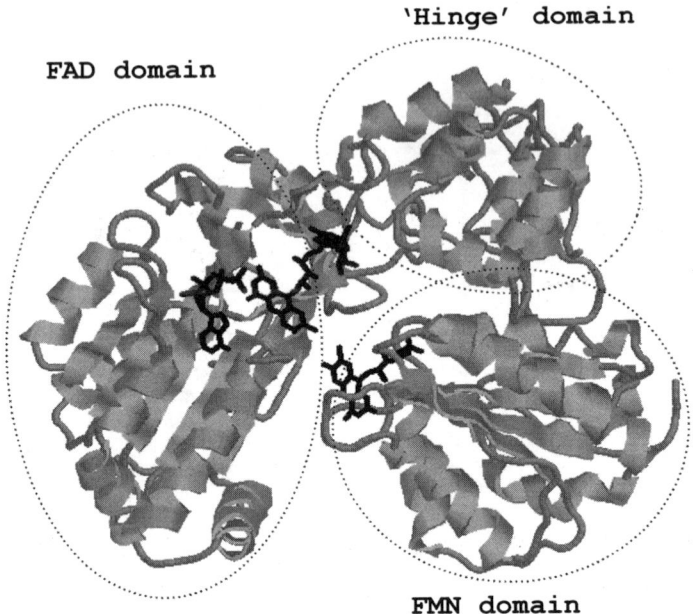

**FIGURE 4.** The structure of rat cytochrome P450 reductase (Wang *et al.*, 1997). The main secondary structural features are shown (helices and sheets) and the three major "domains" of the structure (FAD/NADPH, FMN and "hinge") are highlighted. The cofactors are shown in black, and the tightly arranged FAD and FMN isoalloxazine ring systems indicate that electron transfer from FAD to FMN occurs without requirement for intervening amino acids.

study of prokaryotic forms. Two systems in particular, the P450 cam camphor hydroxylase (Mueller *et al.*, 1995) and P450 BM3 fatty acid hydroxylase (Munro *et al.*, 1999), have proven the most useful model systems.

## 2. BACTERIAL MODEL P450 SYSTEMS

The P450 cam camphor hydroxylase from *Pseudomonas putida* is one of the best characterised of all enzymes. It catalyses the 5-exo hydroxylation of camphor, the first step in the breakdown of the compound as an energy source (Sligar and Gunsalus, 1976). The atomic structure of the P450 has been solved in a variety of different forms (substrate-free, camphor-bound, inhibitor-bound, mutant forms) and was the first P450 for which a structure was determined (e.g. Poulos *et al.*, 1987; Poulos and Howard, 1987; Raag *et al.*, 1991). It is a "typical" bacterial P450 in that it receives electrons for its function from NADH via a class I (or Bacterial (B) class) reductase system. This consists of an FAD-containing NADH-ferredoxin reductase (putidaredoxin reductase) and a small iron-sulfur centre ferredoxin (puti-daredoxin) (Mueller *et al.*, 1995, Figure 3). The ferredoxin reductase receives two reducing equivalents from NADH in the form of a hydride ion, and shuttles electrons one at a time to the haem via the ferredoxin. The P450 cam class I system is analogous to that used by mammalian mito-chondrial P450 systems, perhaps reflecting the fact that mitochondria are considered to have evolved from bacteria. However, mammalian class I P450s and their reductases are membrane-associated, whereas bacterial systems are soluble. This has greatly aided the expression and characteri-sation of wild-type and mutant forms of bacterial P450s, yielding important spectroscopic data on structure/function. For instance, X-ray crystallogra-phy of P450 cam confirmed that a cysteine (C400) provides the proximal sulfhydryl bond to the haem iron, and suggested roles for Tyr 96 and Thr 252 in binding of substrate and in dioxygen scission, respectively. These roles have subsequently been confirmed by kinetic and spectroscopic studies of P450 cam variants mutated at these amino acids (Atkins and Sligar, 1988; Imai *et al.*, 1989). For several years, the atomic structure of P450 cam was the only P450 structure available, and provided the only model on which the structural and mechanistic properties of mammalian P450s could be based. This raised one particular problem, in that the majority of eukary-otic P450s use a different electron transfer apparatus to the bacterial enzymes, a class II (or Eukaryotic (E)) system (Figure 3). For instance, mammalian hepatic P450s receive electrons from NADPH (rather than NADH) and these are delivered by a single enzyme—a diflavin (FAD and FMN) NADPH-cytochrome P450 reductase, or CPR. This enzyme is thought to have evolved from a fusion of ancestral genes encoding a ferre-doxin/flavodoxin reductase-like enzyme (as used in the P450 cam system) and a small FMN-containing flavodoxin. This theory is supported by the recently solved atomic structure of a rat CPR, in which boundaries between the component domains of the enzyme are obvious (Wang *et al.*, 1997, Figure 4). It would be expected that the redox partner binding sites on P450s would differ according to which class of reductase system is used. For this reason, a class II model P450 enzyme is desirable. This requirement was met after the discovery and characterisation of flavocytochrome P450 BM3 from *Bacillus megaterium* by Fulco and co-workers (Narhi and Fulco, 1986). This soluble fatty acid hydroxylase P450 not only utilises a class II CPR system, but has the reductase fused to the P450 in a single polypeptide (Narhi and Fulco, 1987). P450 BM3 is a special example of the class II system—often referred to as a "class III" P450. The structural arrangement of haem oxygenase and flavoprotein reductase is similar to those of the various forms of nitric oxide reductase (Marletta, 1994). The last decade has seen enormous progress made on the structural and mechanistic properties of P450 BM3, and these developments provide the basis for the remainder of the review.

## 3.  P450 BM3 STRUCTURE AND MECHANISM

P450 BM3 (or CYP 102A1, according to systematic nomenclature) (Nelson *et al.*, 1996) has many intriguing properties. 119 kDa P450 BM3 has the highest monooxygenase activity reported for any P450, with the oxidation of arachidonate catalysed at >15,000 $min^{-1}$ (Noble *et al.*, 1999). The main reason underlying its high catalytic centre activity seems to be the rapid flavin-to-haem electron transfer possible due to the fusion of the P450 (N-terminal) to the reductase (C-terminal) by a short peptide linker region. P450 BM3 catalyses the NADPH-dependent hydroxylation (usually at $\varpi$-1 to $\omega$-3 positions) of numerous saturated and unsaturated fatty acids, alcohols and amides ($\sim C_{12}$–$C_{20}$) (Narhi and Fulco, 1986). Expression of the gene (*cyp*102A1) is inducible by barbiturates, and is controlled at the transcriptional level by both positive and negative regulatory DNA-binding proteins (Liang and Fulco, 1995). Again, this situation is similar to several mammalian hepatic P450s, which show inducibility by barbiturates and related compounds (Whitlock and Denison, 1995). More recent studies have demonstrated that the *cyp102*A1 gene can also be induced by fatty acids, and it may be the case that these are the true physiological inducers (Palmer *et al.*, 1998). The gene has been cloned and overexpressed in *E. coli*, and the flavocytochrome purified to homogeneity. Using PCR, the sub-genes encoding the P450 (haem) and reductase (diflavin) domains were also expressed and the proteins purified (Miles *et al.*, 1992). More recently, the FAD/NADPH and FMN sub-domains of P450 BM3 were expressed and purified from *E. coli* (Govindaraj and Poulos, 1997). The availability of domains containing single coloured cofactors (flavin or haem) greatly simplified the deconvolution of the redox properties of the enzyme (Daff *et al.*, 1997), and have allowed the kinetic and spectroscopic properties of the enzyme to be studied in more detail. For instance, stopped-flow absorption spectroscopy studies with the diflavin domain and intact flavocytochrome have indicated that reduction by NADPH and inter-flavin electron transfer is very rapid (biphasic, with a "fast" phase of $758 s^{-1}$ and a "slow" second phase of $118 s^{-1}$ at 25°C) and that the first flavin-to-haem electron transfer is an important rate determining step in catalysis (Munro *et al.*, 1996).

Mutations in the active site of the enzyme can decrease the flavin-to-haem electron transfer rate to an extent that it is almost identical to the steady-state turnover rate of the enzyme (Noble *et al.*, 1999). Also, problems with flavin fluorescence can be circumvented by studies of the haem domain of P450 BM3, and this has proven of great value for resonance Raman characterisation of the haem site. One important finding from such studies has been that the active site of P450 BM3 is large enough to accommodate both a fatty acid and a large inhibitor molecule (metyrapone) simultaneously (Macdonald *et al.*, 1996).

Key to the understanding of the mechanism of P450 BM3 has been the solution of the atomic structures of substrate-free and palmitoleate-bound forms of the haem domain (Ravichandran *et al.*, 1993; Li and Poulos, 1997) (Figure 5). The overall topology of the molecules are similar to that of P450 cam, resembling a triagonal prism. P450 BM3 haem domain has a long

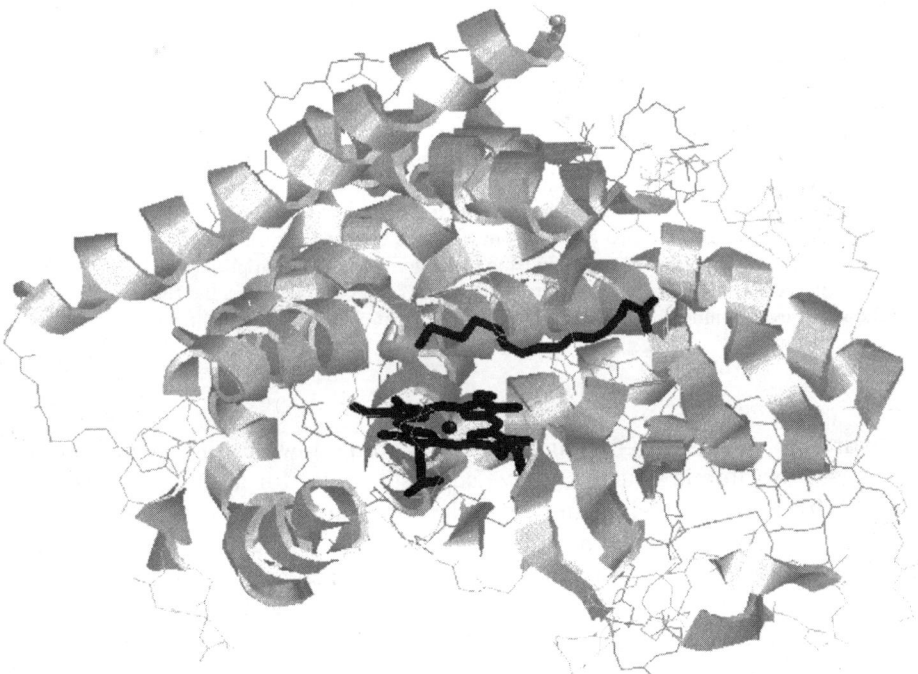

**FIGURE 5.** The atomic structure of the haem domain of substrate-bound P450 BM3 (Li and Poulos, 1997). The major helices and sheets are shown, along with the haem macrocycle and active site-bound palmitoleic acid in black. The overall shape of the molecule approximates a triangular prism.

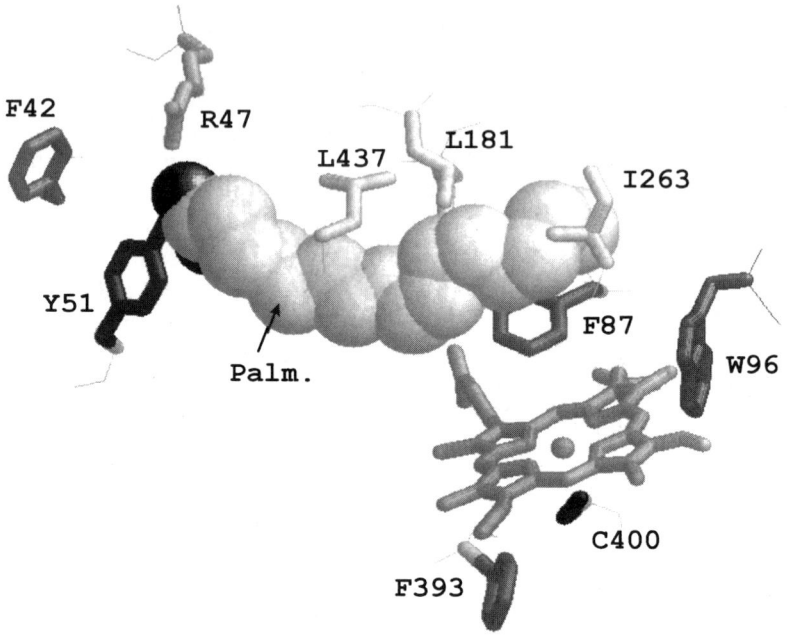

**FIGURE 6.** The active site of the P450 BM3 haem domain, with the haem group, palmitoleic acid substrate ("palm." in spacefill) and selected amino acids. At the mouth of the active site, F42 places a phenyl "cap" on the active site, with R47 and Y51 interacting with fatty acid carboxylate. Residues L437, I263 and L181 make close contacts with the central section of the fatty acid, while W96 provides a stabilising hydrogen bond to haem propionate. Residue F87 protects the ϖ-end of the fatty acid from oxidation, and C400 provides the thiolate ligand to the haem iron. Amino acid F393 is a potential electron transfer mediator from flavin (FMN) to the haem.

hydrophobic substrate-binding channel leading down to the haem (Figure 6), and the atomic structures have suggested roles for a number of amino acids in substrate binding and catalysis. Important residues include Cys 400 (provides the sulfur ligand to the haem iron), Arg 47 and Tyr51 (fatty acid carboxylate-tethering motif at the mouth of the active site) and Phe87 (protects ω-end of fatty acid from attack). A large structural change is clear between the two molecules, and this is presumed to occur on substrate binding in the active site. The finding (from the substrate-bound atomic structure) that palmitoleate binds at too great a distance from the haem iron for oxygenation to occur is consistent with NMR measurements of haem methyl proton relaxation times. The two methods predict that the ω-end of a fatty acid molecule is located approximately 0.8 nm from the haem iron (Li and Poulos, 1997; Modi *et al.*, 1995). However, further NMR measurements taken during haem reduction indicate that substrate moves down the channel by 0.6 nm when the first electron transfer to the haem occurs (Modi *et al.*, 1996). Thus, P450 BM3 undergoes a number of major conformational changes during catalysis. These findings are consistent with molecular dynamics simulations of P450 BM3, which suggest that the channel can undergo large opening and closing motions (Paulsen and Ornstein, 1995). Further clues to the complex nature of structural changes induced by substrates are the findings that fatty acid-linked imidazole inhibitors of P450 BM3 bind more tightly when $NADP^+$ is present. This indicates that there is conformational communication between the diflavin and haem domains of P450 BM3. Moreover, this phenomenon is not observed in a FMN-deficient mutant, suggesting that the flavin is important in the transmission of structural change induced by $NADP^+$ (Noble *et al.*, 1998). Clearly, the enzyme has evolved elegant conformational switching mechanisms to control its catalytic activity. As detailed below, thermodynamic control is also important to ensure the efficient use of electrons from NADPH.

## 4. ELECTRON TRANSFER AND ITS CONTROL

As with P450 cam, fatty acid substrate-binding to P450 BM3 induces a shift in the iron spin-state equilibrium from low-spin ferric ($S = 1/2$) towards high-spin ferric ($S = 5/2$) (Sligar, 1976). The spin-state shift is accompanied by a change in the visible spectrum of the P450, with the major (Soret) band moving from 419 nm to 390 nm (Miles *et al.*, 1992) (Figure 7). The $K_d$ for different substrates can be determined by plotting the absorbance shift induced at known concentration of substrate *versus* that substrate concentration, and fitting the data to the Michaelis function. Such analysis indicates that the polyunsaturated $C_{20}$ fatty acid arachidonate is an excellent substrate, with a $K_d$ of ~2 μM. There is minimal electron transfer to the haem from the flavin in the absence of fatty acid. A major reason for this is that substrate association triggers a change in the reduction potential of the P450 haem iron. The spin-state change (and associated spectral perturbation) are a reflection of the displacement of a water axial ligand from the haem iron, leaving it in a 5-coordinate state. The substrate-bound, high-spin iron has a much more positive reduction potential than that in the resting form, and electron transfer from the flavin becomes thermodynamically feasible (Figure 8). This redox switch prevents wasteful electron transfer from the reductase to the P450 if substrate is not present. This would result in the production of damaging radicals through reduction of oxygen at the

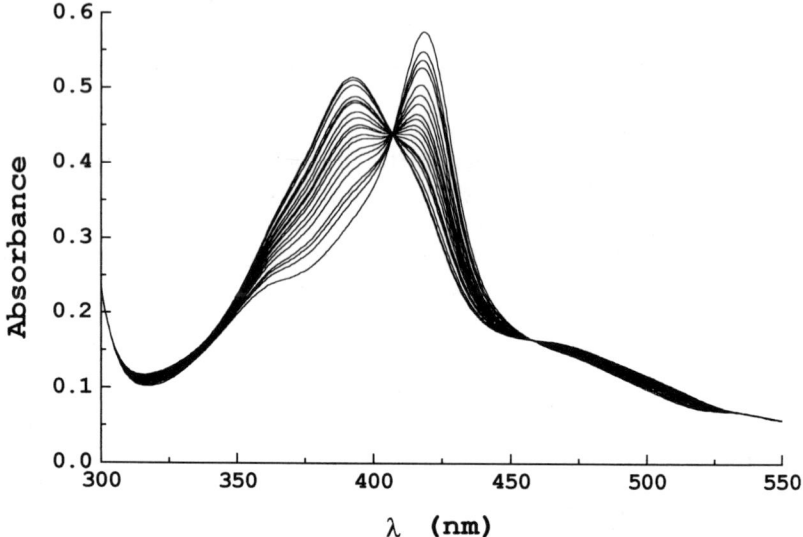

**FIGURE 7.** Spectral titration of flavocytochrome P450 BM3 (6 µM) with arachidonic acid (0–40 µM). The tight-binding substrate displaces the $6^{th}$ (aqua) ligand to the haem iron and induces a shift in the spin-state equilibrium of the haem iron towards the high-spin form. This is manifest as a shift in absorption of the Soret band from 419 nm (low-spin) to 390 nm (high-spin).

haem iron (Daff *et al.*, 1997). With saturating arachidonate, the reduction potential of the haem iron is increased from −368 mV to −239 mV. However, slow wastage of reducing equivalents is still possible through oxygen reduction from the flavins in the reductase domain of P450 BM3 if NADPH is in excess (Munro *et al.*, 1995). Remarkably, P450 BM3 seems to have a further trick up its sleeve in response to such an excess of NADPH. Under these conditions, P450 BM3 converts slowly to an inactive fatty acid hydroxylase. However, complete activity can be recovered by simple dialysis of the enzyme into fresh buffer. On the basis of redox potentiometric measurements, the reductase domain can be reduced to a three electron form (FAD semiquinone, FMN hydroquinone) by NADPH, and it is presumed that this form can undergo a slow, reversible change into the inactive conformer.

It is becoming increasingly obvious that the activity of P450 BM3 is controlled both thermodynamically and by structural changes triggered by substrate (fatty acid or NADP(H)) or redox state of the enzyme. The availability of the crystal structures of substrate-free and bound forms allows investigation of the roles of various amino acids in the processes of substrate binding, electron transfer and oxidative catalysis. This can be achieved

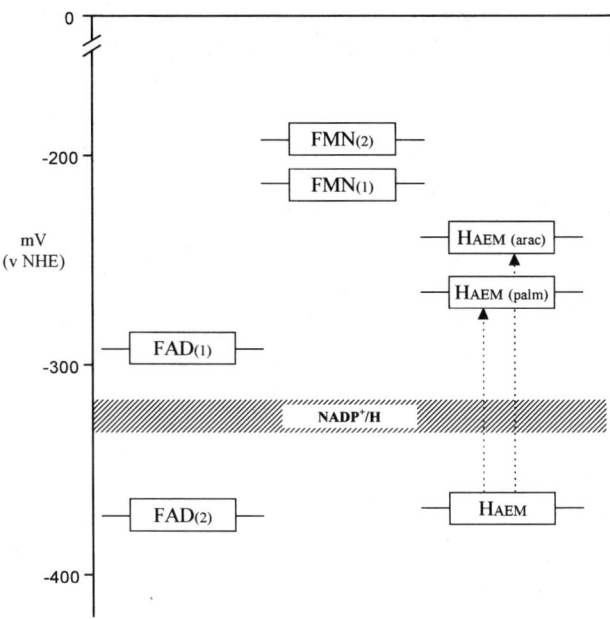

**FIGURE 8.** A comparison of individual reduction potentials for the FMN, FAD (in the diflavin domain) and haem iron (in the P450 domain) of flavocytochrome P450 BM3. Values are relative to the standard hydrogen electrode. The midpoint potential for NADP$^+$/H (striped bar: −320 mV) is close to that for the two electron reduction of the FAD at −332 mV (FAD$_{(1)}$ [ox/sq] = −292 mV, FAD$_{(2)}$ [sq/hq] = −372 mV). However, after the transfer of two electrons (as a hydride ion) to the FAD, electrons can be readily transferred to the FMN. Both the redox couples for the FMN are much more positive than those for FAD (FMN$_{(1)}$ = −213 mV, FMN$_{(2)}$ = −193 mV). In the absence of fatty acid substrates, the enzyme can be reduced to only a 3-electron form—since the FAD sq/hq and substrate-free haem Fe$^{3+}$/Fe$^{2+}$ couples are much more negative than that of NADP$^+$/H. Binding of substrates palmitate (palm) or arachidonate (arac) to the P450 increases the reduction potential of the haem iron by ~130 mV, permitting the equilibration of electrons from the reduced FMN and initiating the P450 catalytic cycle.

by rational site-directed mutagenesis, and there have already been many important mechanistic features of the enzyme elucidated by characterisation of mutant forms.

## 5. SITE-DIRECTED MUTAGENESIS IN THE STUDY OF SUBSTRATE SELECTIVITY AND ELECTRON TRANSFER

Several point mutants of flavocytochrome P450 BM3 have yielded important information on amino acids involved in substrate binding in the active site. One obvious question arising from the structure of the palmitoleate-bound form of P450 BM3 was the relative importance of amino acids Arg 47 and Tyr 51 in the interaction with the carboxylate of the fatty acid. It is these interactions that orientate the fatty acid with its carboxylate near the mouth of the active site, and the ω-end closer to the haem (Figure 9). The problem has been resolved by study of Y51F and R47A, G mutants. The R47 mutants are more badly affected in efficiency of catalysis than is Y51F ($k_{cat}/K_m$ for laurate hydroxylation = 4.6/4.2 μM$^{-1}$ min$^{-1}$ for R47A/G against 14.2 μM$^{-1}$ min$^{-1}$ for Y51F and 17.8 μM$^{-1}$ min$^{-1}$ for wild-type (Noble *et al.*, 1999). The electrostatic interaction between the Arg 47 guanidinium group and the fatty acid carboxy group is involved in both initial binding and transition state stabilisation, whereas Tyr 51 is important only for the binding. However, it should be noted that the Y51F mutant

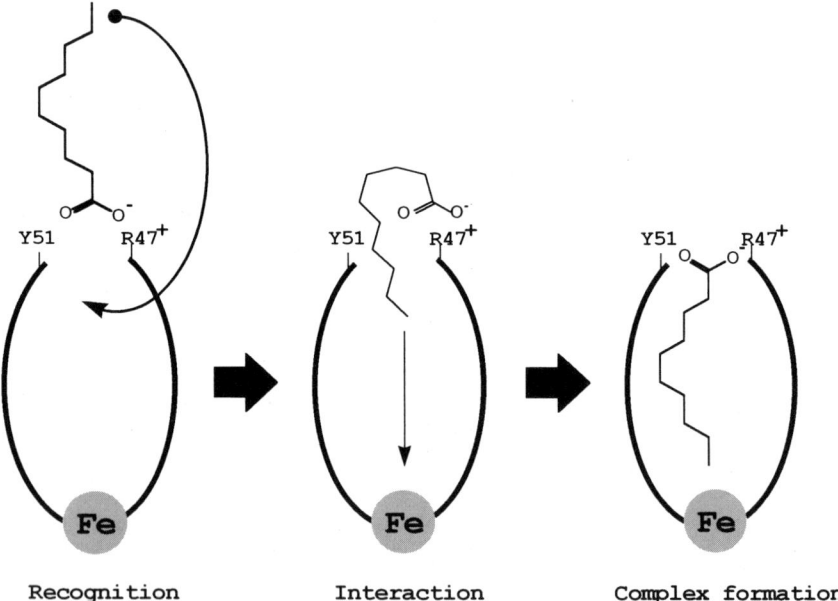

**FIGURE 9.** A scheme for the roles of residues arginine 47 and tyrosine 51 in the interaction and binding with fatty acid substrate. Substrate first interacts with the arginine guanidinium group through electrostatic attraction with the fatty acid carboxylate (recognition). The carbon chain of the substrate is pulled into the active site, leaving the carboxylate at the mouth, tethered by electrostatic and hydrogen bonding interactions with both Arg 47 and Tyr 51 (interaction). Finally, the substrate is positioned correctly in the active site to induce structural change and loss of the haem water ligand, so triggering electron transfer and catalysis (complex formation).

does not undergo fatty acid-induced spin-state shift to the same extent as the wild-type enzyme, perhaps indicating that Try 51 is important for the correct orientation of the fatty acid in the active site (Noble *et al.*, 1999).

Another key residue is Phe 87, which lies close to and perpendicular with the haem ring in the substrate-free form, and rotates to be almost parallel when substrate binds (Li and Poulos, 1997) (Figure 6). The residue has been shown to protect the $\omega$-end of lauric acid from hydroxylation, a reaction which occurs in mutant F87A (Oliver *et al.*, 1997). In addition, kinetic studies on mutants F87Y and F87G show that these enzymes undergo a catalysis-dependent inactivation during turnover, converting from a "fast" form to a "slow" form during fatty acid hydroxylation. It is possible that these mutant enzymes are disrupted in a key conformational step, and that product release becomes an important rate-determining step in the "slow" form (Noble *et al.*, 1999). Also important is Trp 96, a residue postulated to mediate electron transfer to the haem iron (Baldwin *et al.*, 1991). Mutants at this site (W96A/Y/F) exhibit similar properties of low haem content and residual haem of mixed spin-state in the resting form (Munro *et al.*, 1994). Trp 96 does not appear to be an essential electron transfer mediating amino acid, but is in hydrogen bonding contact with a haem propionate (Figure 6) and is more likely to have roles in the stabilisation of haem binding and in the maintenance of the low-spin resting state.

One amino acid with a large effect not predicted from examination of the atomic structures is Phe 42, which appears to place a phenyl "cap" over the mouth of the active site, above Tyr 51 and Arg 47 (Figure 6). Mutant F42A (removal of the phenyl group) shows very little substrate-dependent modulation of the haem iron spin-state, and has a 10-fold increased $K_m$, 2.5-fold-decreased $k_{cat}$ for laurate oxidation compared with wild-type. The large effects on the catalytic parameters can be rationalised by considering that Phe 42 act as the lid of the active site, preventing the entry of water during the catalytic cycle. The small movement of the phenyl ring of Phe 42 seen between the substrate-free and palmitoleate-bound structures is consistent with it moving to close the channel in the substrate-bound form (Noble *et al.*, 1999).

Other amino acids found to be important for catalysis include Thr 268 (involved in oxygen activation; Yeom *et al.*, 1995), Pro 25 (possibly involved in an initial docking region for fatty acid substrates; Maves *et al.*, 1997) and Gly 570 and Trp 574 (flavin-binding residues in the reductase domain, with possible roles in the mediation of electron transfer; Munro *et al.*, 1996). However, many potentially important mutants remain to be generated and characterised. The pathway of electron transfer from flavin (FMN) to haem is not defined, and the recent structure of the fused FMN-haem domains of the protein did not clarify the route, since the linker region between the domains was found to be proteolysed in the crystal (Sevrioukova *et al.*, 1999). One strong candidate for involvement in this process is Phe 393, located close to the exposed cysteine ligand to the haem iron. Other amino acids of interest include Leu 181 and Leu 437, which make close contacts with central carbon atoms of palmitoleate in the substrate-bound structure (Li and Poulos, 1997) (Figure 6). It is possible that selected mutations at these sites may result in modification of the substrate selectivity of the enzyme. In cytochrome P450 2A4, mutant L209F is converted efficiently from a testosterone hydroxylase to a coumarin hydroxylase (Lindberg and Negishi, 1989). Leucine 209 in P450 2A4 is the residue corresponding to Leu 181 in P450 BM3.

## 6. CONCLUSIONS

Flavocytochrome P450 BM3 has proved a crucial model system for our comprehension of structure and mechanism in the cytochromes P450. The future holds further exciting prospects. Spectroscopic studies of the P450 will doubtless provide deeper insight into the way in which the P450 catalyses oxygen activation, electron and proton transfer. Rational mutagenesis may allow us to understand exactly the requirements for binding substrates of choice, so that chiral compounds of value can be generated. In the shorter term, the most important breakthrough would be the determination of the electron transfer pathway through the intact flavocytochrome, which might allow us to understand the roles of inter-domain interactions in the enzyme, and the pathway of electron transfer from FMN-to haem.

ACKNOWLEDGMENTS. The authors would like to thank the Leverhulme Trust, Royal Society of Edinburgh, Royal Society and BBSRC for their generous support of research staff involved in the preparation of this review.

## 7. REFERENCES

Atkins, W. M., and Sligar, S. G., 1988, The roles of active site hydrogen bonding in cytochrome P450 cam as revealed by site-directed mutagenesis, *J. Biol. Chem.*, **263**:18842–18849.

Baldwin, J. E., Morris, G. M., and Richards, W. G., 1991, Electron transport in cytochrome P450 by covalent switching, *Proc. Roy. Soc. London Series B* **245**:43–51.

Cole, S. T., Brosch, R., Parkhill, J., Garnier, T., Churcher, C. *et al.*, 1998, Deciphering the biology of *Mycobacterium tuberculosis* from the complete genome sequence, *Nature (London)* **393**:537–559.

Daff, S. N., Chapman, S. K., Turner, K. L., Holt, R. A., Govindaraj, S., Poulos, T. L., and Munro, A. W., 1997, Redox control of the catalytic cycle of flavocytochrome P450 BM3, *Biochemistry* **36**:13816–13823.

Gillam, E. M. J., Guo, Z., and Guengerich, F. P., 1994, Expression of modified human cytochrome P450 2E1 in *Escherichia coli*, purification, and spectral and catalytic properties, *Arch. Biochem. Biophys.* **312**:59–66.

Gonzalez, F. J., Kimura, S., Tamura, S., and Gelboin, H. V., 1991, Expression of mammalian cytochrome P450 using baculovirus, *Methods Enzymol.* **206**:93–99.

Govindaraj, S., and Poulos, T. L., 1997, The domain architecture of cytochrome P450 BM3, *J. Biol. Chem.* **272**:7915–7921.

Imai, M., Shimada, H., Watanabe, Y., Matsushima-Hibiya, Y., Makino, R., Koga, H., Horiuchi, T., and Ishimura, Y., 1989, Uncoupling of the cytochrome P450 cam monooxygenase reaction by a single mutation, threonine-252 to alanine or valine: A possible role of the hydroxy amino acid in oxygen activation, *Proc. Natl. Acad. Sci. USA* **86**:7823–7827.

Kunst, N., Ogasawara, N., Moszer, I., Albertine, A. M., Alloni, G., Azevedo, V. *et al.*, 1997, The complete genome sequence of the Gram positive bacterium *Bacillus subtilis, Nature (London)* **390**:249–256.

Li, H. Y., and Poulos, T. L., 1997, The structure of the cytochrome P450 BM3 haem domain complexed with fatty acid substrate palmitoleic acid, *Nature Struct. Biol.* **4**:140–146.

Lindberg, R. L. P., and Negishi, M., 1989, Alteration of mouse cytochrome P450$_{coh}$ substrate specificity by mutation of a single amino acid residue, *Nature (London)* **339**:632–634.

Liang, Q. W., and Fulco, A. J., 1995, Transcriptional regulation of the genes encoding cytochrome P450 BM1 and P450 BM3 in *Bacillus megaterium* by the binding of BM3R1 repressor to barbie box elements and operator sites, *J. Biol. Chem.* **270**:18606–18614.

Macdonald, I. D. G., Smith, W. E., and Munro, A. W., 1996, Inhibitor/fatty acid interactions with cytochrome P450 BM3, *FEBS Lett.* **396**:196–200.

Marletta, M. A., 1994, Nitric oxide synthase—aspects concerning structure and catalysis, *Cell* **78**:927–930.

Maves, S. A., Yeom, H., Mclean, M. A., and Sligar, S. G., 1997, Decreased substrate affinity upon alteration of the substrate-docking region in cytochrome P450 BM3, *FEBS Lett.* **414**:213–218.

Miles, J. S., Munro, A. W., Rospendowski, B. N., Smith, W. E., McKnight, J., and Thomson, A. J., 1992, Domains of the catalytically self-sufficient cytochrome P450 BM3, *Biochem. J.* **288**:503–509.

Modi, S., Primrose, W. U., Boyle, J. M. B., Gibson, C. F., Lian, L. Y., and Roberts, G. C. K., 1995, NMR studies of substrate-binding to cytochrome P450 BM3—comparisons to cytochrome P450 cam, *Biochemistry* **34**:8982–8988.

Modi, S., Sutcliffe, M. J., Primrose, W. U., Lian, L. Y., and Roberts, G. C. K., 1996, The catalytic mechanism of cytochrome P450 BM3 involves a 6 Angstrom movement of the bound substrate on reduction, *Nature Struct. Biol.* **3**:414–417.

Mueller, E. J., Loida, P. J., and Sligar, S. G., 1995, Twenty-five years of P450 cam research, in *Cytochrome P450: Structure, Mechanism and Biochemistry* (P. R. Ortiz de Montellano, ed.), Plenum Press, New York, pp. 83–124.

Munro, A. W., Daff, S., Coggins, J. R., Lindsay, J. G., and Chapman, S. K., 1996, Probing electron transfer in flavocytochrome P450 BM3 and its component domains, *Eur. J. Biochem.* **239**:403–409.

Munro, A. W., Lindsay, J. G., Coggins, J. R., Kelly, S. M., and Price, N. C., 1995, NADPH oxidase activity of cytochrome P450 BM3 and its constituent reductase domain, *Biochim. Biophys. Acta* **1231**:255–264.

Munro, A. W., and Lindsay, J. G., 1996, Bacterial cytochromes P450, *Mol. Microbiol.* **20**:1115–1125.

Munro, A. W., Malarkey, K., McKnight, J., Thomson, A. J., Kelly, S. M., Price, N. C., Lindsay, J. G., Coggins, J. R., and Miles, J. S., 1994, The role of tryptophan 96 of cytochrome P450 BM3 from *Bacillus megaterium* in catalytic function, *Biochem. J.* **303**:423–428.

Munro, A. W., Noble, M. A., Miles, C. S., Daff, S. N., Green, A. J., Quaroni, L., Rivers, S., Ost, T. W. B., Reid, G. A., and Chapman, S. K., 1999, Flavocytochrome P450 BM3: A paradigm for the analysis of electron transfer and its control in the P450s, *Biochem. Soc. Trans.* **27**:190–196.

Narhi, L. O., and Fulco, A. J., 1986, Characterization of a catalytically self-sufficient 119,000 Dalton cytochrome P450 monooxygenase induced by barbiturates in *Bacillus megaterium*, *J. Biol. Chem.* **261**:7160–7169.

Narhi, L. O., and Fulco, A. J., 1987, Identification and characterization of two functional domains in cytochrome P450 BM3, a catalytically self-sufficient monooxygenase induced by barbiturates in *Bacillus megaterium*, *J. Biol. Chem.* **262**:6683–6690.

Nelson, D. R., Koymans, L., Kamataki, K., Stegeman, J. J., Feyereisen, R., Waxman, D. J., Waterman, M. R., Gotoh, O., Coon, M. J., Estabrook, R. W., Gunsalus, I. C., and Nebert, D. W., 1996, The P450 superfamily: Update on new sequences, gene mapping, accession numbers and nomenclature, *Pharmacogenetics* **6**:1–42.

Noble, M. A., Miles, C. S., Chapman, S. K., Lysek, D. A., Mackay, A. C., Reid, G. A., Hanzlik, R. P., and Munro, A. W., 1999, Roles of key active site residues in flavocytochrome P450 BM3, *Biochem. J.* **339**:371–379.

Noble, M. A., Quaroni, L., Chumanov, G. D., Turner, K. L., Chapman, S. K., Hanzlik, R. P., and Munro, A. W., 1998, *Biochemistry* **37**:15799–15807.

Omura, T., and Sato, 1964, The carbon monoxide-binding pigment of liver microsomes. I. Evidence for its hemoprotein nature, *J. Biol. Chem.* **239**:2370–2378.

Ortiz de Montellano, P. R., and Correia, M. A., 1995, Inhibition of cytochrome P450 enzymes, in *Cytochrome P450: Structure, Mechanism and Biochemistry* (P. R. Ortiz de Montellano, ed.), Plenum Press, New York, pp. 305–364.

Palmer, C. N. A., Axen, E., Hughes, V., and Wolf, C. R., 1998, The repressor protein, Bm3R1, mediates an adaptive response to toxic fatty acids in *Bacillus megaterium*, *J. Biol. Chem.* **273**:18109–18116.

Paulsen, M. D., and Ornstein, R. L., 1995, Dramatic differences in the motions of the mouth of open and closed cytochrome P450 BM3 by molecular dynamics simulations, *Proteins: Structure, Function and Genetics* **21**:237–243.

Poulos, T. L., Finzel, B. C., and Howard, A. J., 1987, High resolution crystal structure of cytochrome P450 cam, *J. Mol. Biol.* **195**:697–700.

Poulos, T. L., and Howard, A. J., 1987, Crystal structures of the metyrapone and phenylimidazole inhibited complexes of cytochrome P450 cam, *Biochemistry* **26**:8165–8174.

Raag, R., Martinis, S. A., Sligar, S. G., and Poulos, T. L., 1991, Crystal structure of the cytochrome P450 cam active site mutant Thr252Ala, *Biochemistry* **30**:11420–11429.

Ravichandran, K. G., Boddupalli, S. S., Hasemann, C. A., Peterson, J. A., and Deisenhofer, J., 1993, Crystal structure of hemoprotein domain of P450 BM3: A prototype for microsomal P450s, *Science* **261**:170–176.

Sakaki, T., Shibata, M., Yabusaki, Y., Murakami, H., and Ohkawa, H., 1990, Expression of bovine cytochrome P450c21 and its fused enzymes with yeast NADPH-cytochrome P450 reductase in *Saccharomyces cerevisiae*, *DNA Cell Biol.* **9**:603–614.

Sandhu, P., Guo, Z., Baba, T., Martin, M. V., Tukey, R. H., and Guengerich, F. P., 1994, Expression of modified human cytochrome P450 1A2 in *Escherichia coli*: stabilization, purification, spectral characterization and catalytic activities of the enzyme, *Arch. Biochem. Biophys.* **309**:168–177.

Sevrioukova, I. F., Li, H. Y., Zhang, H., Peterson, J. A., and Poulos, T. L., 1999, Structure of a cytochrome P450-redox partner electron transfer complex, *Proc. Natl. Acad. Sci. USA* **96**:1863–1868.

Sligar, S. G., 1976, Coupling of spin, substrate and redox equilibria in cytochrome P450, *Biochemistry* **15**:5399–5406.

Sligar, S. G., and Gunsalus, I. C., 1976, A thermodynamic model of regulation: Modulation of redox equilibria in camphor monooxygenase, *Proc. Natl. Acad. Sci. USA* **73**:1078–1082.

Wang, M., Roberts, D. L., Paschke, R., Shea, T. M., Masters, B. S. S., and Kim, J.-J. P., 1997, Three dimensional structure of NADPH-cytochrome P450 reductase: Prototype for FMN- and FAD-containing enzymes, *Proc. Natl. Acad. Sci. USA*, **94**:8411–8416.

Whitlock, Jr., J. P., and Denison, M. S., 1995, Induction of cytochrome P450 enzymes that metabolize xenobiotics, in *Cytochrome P450: Structure, Mechanism and Biochemistry* (P. R. Ortiz de Montellano, ed.), Plenum Press, New York, pp. 367–390.

Yeom, H., Sligar, S. G., Li, H. Y., Poulos, T. L., and Fulco, A. J., 1995, The role of Thr268 in oxygen activation of cytochrome P450 BM3, *Biochemistry* **34**:14733–14740.

*Chapter 10*

# Peroxidase-Catalyzed Oxidation of Ascorbate

## Structural, Spectroscopic and Mechanistic Correlations in Ascorbate Peroxidase

Emma Lloyd Raven

Ascorbate-dependent peroxidase activity was first reported in 1979 (Groden and Beck, 1979; Kelly and Latzko, 1979) and ascorbate peroxidase (APX) is, therefore, a relative newcomer to the peroxidase field—horseradish (HRP) and cytochrome *c* (C*c*P) peroxidases were, for example, first identified in 1903 (Bach and Chodat, 1903) and 1940 (Altschul *et al.*,

---

Abbreviations: APX, ascorbate peroxidase; pAPX, pea cytosolic APX; C*c*P, cytochrome *c* peroxidase; HRP, horseradish peroxidase; LiP, lignin peroxidase; MnP, manganese peroxidase; PNP, peanut peroxidase; ARP, *Arthromyces ramosus* peroxidase; 5-c/HS, 5-coordinate, high-spin heme; 6-c/HS, 6-coordinate, high-spin heme; 6-c/LS, 6-coordinate, low-spin heme; SHE, standard hydrogen electrode.

---

**EMMA LLOYD RAVEN**     Department of Chemistry, University of Leicester, University Road, Leicester, LE1 7RH, England, UK.

*Subcellular Biochemistry, Volume 35: Enzyme-Catalyzed Electron and Radical Transfer*, edited by Holzenburg and Scrutton. Kluwer Academic / Plenum Publishers, New York, 2000.

1940) respectively. The APX area was reviewed by Dalton in 1991 (Dalton, 1991): at that time, there was very little detailed kinetic, spectroscopic or functional information available and no structural information had been published. Since 1991, there have been some major advances in the field, most notably with the publication, in 1995, of the first crystal structure for an APX enzyme (Patterson and Poulos, 1995). This information, together with the availability of new recombinant expression systems (Yoshimura et al., 1998; Caldwell et al., 1998; Dalton et al., 1996; Patterson and Poulos, 1994), served as a catalyst for the publication of new functional and spectroscopic data and has meant these data could be sensibly rationalized at the molecular level. The aim of this review is to summarize the more recent advances in the APX area and, as far as possible, to draw comparisons with other, more well-characterized peroxidases. The review will concentrate on the ways in which structural, spectroscopic and mechanistic information have been used in a complementary way to provide a more detailed picture of APX catalysis. The more biological and physiological aspects of APX enzymes have been previously covered in a comprehensive manner (Dalton, 1991) and will not, therefore, be dealt with in detail here.

## 1.  INTRODUCTION

Ascorbate peroxidase is a heme-containing enzyme that catalyses the ascorbate-dependent reduction of potentially damaging hydrogen peroxide in plants and algae, Eq. (1),

$$2 \text{ ascorbate} + H_2O_2 + 2H^+ \rightarrow 2 \text{ monodehydroascorbate} + 2H_2O \quad (1)$$

and an ascorbate peroxidase can therefore be defined as such if the physiological substrate is ascorbate. This, in itself, is not a particularly enlightening definition however, since some APXs are rather indiscriminate in their choice of redox partner and are able to catalyze the oxidation of non-physiological, substrates (in some cases at rates comparable to that of ascorbate itself). It is, therefore, often difficult to identify a clearly perceptible physiological substrate amidst the catalytic confusion of all other, non-physiological, substrates. An alternative and more sensible definition is to classify a peroxidase as an APX when the specific activity of the enzyme for ascorbate is higher than that for other substrates. On the whole, this has been adopted (although there are exceptions, section 4) and has the advantage that it is able to discriminate between bona fide APXs and other, classical peroxidases[1] that are, by coincidence alone, able to catalyze oxidation of ascorbate.

In terms of a more rational classification within the plant peroxidase superfamily, APXs are defined as belonging to the class I group of peroxidases (Welinder, 1992). This classification, based on sequence homology, contains three major categories: class I contains the enzymes of prokaryotic origin, class II contains the fungal enzymes (e.g. MnP, LiP) and class III contains the classical secretory peroxidases (e.g. HRP). Cytosolic and chloroplastic APXs, together with CcP (pAPX shows 33% sequence homology with CcP (Mittler and Zilinskas, 1991a)) and the gene-duplicated bacterial catalase-peroxidases, fall under the class I peroxidase umbrella. It is possible, however, that this general classification may need reviewing: recent experiments (Kvaratskhelia *et al.*, 1999; Kvaratskhelia *et al.*, 1997b) on tea leaves have identified two APX enzymes that have closer sequence homology to classical class III peroxidases than to class I.

## 2. cDNA SEQUENCES AND BACTERIAL EXPRESSION OF RECOMBINANT APXs

Since the publication of the first cDNA sequence for an ascorbate peroxidase, namely that of pAPX (Mittler and Zilinskas, 1991a), a plethora of cDNA sequences for APXs from various sources have appeared (Jespersen *et al.*, 1997). These include: (a) the soluble cytosolic enzymes from soybean (Caldwell *et al.*, 1998; Chatfield and Dalton, 1993), spinach (Webb and Allen, 1995), *Arabidopsis* (Kubo *et al.*, 1992), carrot (Jespersen *et al.*, 1997), radish (Lopez *et al.*, 1996), maize (van Breusegem *et al.*, 1995), tobacco (Orvar and Ellis, 1995), rice (Morita *et al.*, 1997), strawberry (Kim and Chung, 1998) and bell pepper (Schantz *et al.*, 1995), (b) the stromal (chloroplastic) enzymes from *Arabidopsis* (Jespersen *et al.*, 1997; Newman *et al.*, 1996), spinach (Ishikawa *et al.*, 1996a) and pumpkin (Mano *et al.*, 1997), (c) the thylakoid-bound (chloroplastic) enzymes from *Arabidopsis* (Jespersen *et al.*, 1997; Newman *et al.*, 1996), spinach (Ishikawa *et al.*, 1996a) and pumpkin (Mano *et al.*, 1997; Yamaguchi *et al.*, 1996) and (d) the glyoxysomal/peroxisomal (microbody-bound) enzymes from *Arabidopsis* (Jespersen *et al.*, 1997; Zhang *et al.*, 1997; Newman *et al.*, 1996), cotton (Bunkelmann

---

[1] The classical peroxidases oxidize phenolic molecules much more efficiently than other substrates.

and Trelease, 1996), *M. crystallinum* (Jespersen *et al.*, 1997) and spinach (Ishikawa *et al.*, 1995).

Whilst a variety of cloned genes for various APXs are now available, expression of these genes in *E. coli* (for structure/function studies) has been less successful. In fact, recombinant expression of any kind is rarely straightforward, and the development of an efficient bacterial expression system still remains a largely empirical, and very often serendipitous, affair. Recombinant expression systems for APX have, nevertheless, been reported in a few cases: the first was for a cytosolic APX from pea (Patterson and Poulos, 1994), although this expression system was subsequently modified (Cheek *et al.*, 1999). Expression of cytosolic enzymes from soybean (Caldwell *et al.*, 1998; Dalton *et al.*, 1996) and spinach (Yoshimura *et al.*, 1998) has also been reported and the authenticity of the recombinant wild type pea and soybean enzymes has been confirmed using a variety of spectroscopic and crystallographic techniques (Nissum *et al.*, 1998; Jones *et al.*, 1998; Patterson *et al.*, 1995; Patterson and Poulos, 1995). Recently, bacterial expression of two chloroplastic (stromal and thylakoid-bound) enzymes (Yoshimura *et al.*, 1998) and a glyoxysomal enzyme (Ishikawa *et al.*, 1998) from spinach have been reported.

## 3. ISOLATION AND CHARACTERISATION OF APXs

Cytosolic APX enzymes have now been identified and purified from pea (Mittler and Zilinskas, 1991b; Gerbling *et al.*, 1984; Groden and Beck, 1979; Kelly and Latzko, 1979), Japanese radish (Ohya *et al.*, 1997), soybean (Dalton *et al.*, 1987), wheat (De Gara *et al.*, 1997; De Gara *et al.*, 1996), potato tubers (Elia *et al.*, 1992), maize (Koshiba, 1993), komatsuna (Ishikawa *et al.*, 1996c), tea (Kvaratskhelia *et al.*, 1997b; Chen and Asada, 1989), spinach (Yoshimura *et al.*, 1998; Tanaka *et al.*, 1991) and rice (Ushimara *et al.*, 1997). Chloroplastic enzymes have been isolated and purified from spinach (Yoshimura *et al.*, 1998; Miyake *et al.*, 1993; Tanaka *et al.*, 1991; Nakano and Asada, 1987), pea (Jimenez *et al.*, 1998; Jimenez *et al.*, 1997; Gillham and Dodge, 1986; Jablonski and Anderson, 1982), wheat (Meneguzzo *et al.*, 1998) and tea (Kvaratskhelia *et al.*, 1999; Kvaratskhelia *et al.*, 1997b; Chen *et al.*, 1992; Chen and Asada, 1989). Glyoxysomal enzymes have also been identified in cotton (Bunkelmann and Trelease, 1997), pumpkin (Yamaguchi *et al.*, 1995) and spinach (Ishikawa *et al.*, 1998; Ishikawa *et al.*, 1995); peroxisomal enzymes have been identified in cucumber (Corpas and Trelease, 1998), pea (Jimenez *et al.*, 1998; Jimenez *et al.*, 1997) and pumpkin (Yamaguchi *et al.*, 1995). APXs have also been identified from several algal sources, namely *Galdieria partita* (Sano *et al.*, 1996),

*Selenastrum capricornutum* (Sauser *et al.*, 1997), *Chlorella vulgaris* (Takeda *et al.*, 1998) and *Euglena gracilis* (Ishikawa *et al.*, 1996b; Shigeoka *et al.*, 1980), in cyanobacteria (Miyake *et al.*, 1991; Tel-Or *et al.*, 1986) and in insects (Mathews *et al.*, 1997). An APX enzyme has also been identified in *Trypanosoma cruzi* (Boveris *et al.*, 1980). Recently, an APX enzyme has been isolated and purified from bovine eye (Wada *et al.*, 1998), raising the intriguing possibility that APX enzymes may have a functional role in mammalian systems.

## 4.  GENERAL PROPERTIES

A summary of selected physical properties for various cytosolic, chloroplastic, algal and mammalian APXs, for cases where more than a bare minimum of functional data have been published, is presented in Tables 1 and 2. Wavelength maxima for the ferric enzymes, which range from 401–409 nm, are consistent in all cases with the presence of a high spin heme iron (the details of the heme coordination geometry are discussed in section 6). Values of $K_m$ for ascorbate fall within a relatively narrow range (55–770 μM) and for $H_2O_2$ are in the range 3–188 μM. In terms of substrate specificity, APXs are able to catalyze the oxidation of a variety of non-physiological substrates, some of which are included in Tables 1 and 2. In fact, the cytosolic enzymes can, in some cases, catalyze oxidation of pyrogallol at rates which are better than that for ascorbate itself (although this is not true for the chloroplastic enzymes). None of the enzymes isolated so far exhibit any activity against NADH or NADPH. In addition, none of the enzymes in Tables 1 and 2 have been found to exhibit activity against cytochrome *c*. In view of the structural similarities between APX and CcP (section 5), it seems curious that APX is unable to catalyze oxidation of cytochrome *c* whereas CcP is able to catalyze, albeit slowly (Yonetani and Ray, 1965), oxidation of ascorbate. This difference may reflect the unique substrate binding properties of CcP and underlines the subtle functional differences that can exist between structurally related enzymes (a more detailed discussion of the substrate binding properties of APX is presented in section 9).

## 5.  STRUCTURAL STUDIES

The publication, in 1995, of the crystal structure of recombinant pAPX (Patterson and Poulos, 1995) was arguably the single most important contribution to the area in recent years. The structure provided a dichotomy of

Emma Lloyd Raven

## Table 1
### Selected physical data for various cytosolic APXs

| Source | $M_r^a$ | $K_m$ ascorbate$^b$ | $K_m$ H$_2$O$_2$$^b$ | Specific activity$^c$ | $\lambda_{max}$ Fe$^{III}$ | Fe$^{II}$ | Fe$^{III}$CN | Fe$^{II}$CN | pH optimum | Relativity activity$^d$ | Reference |
|---|---|---|---|---|---|---|---|---|---|---|---|
| Spinach$^e$ | 31 | 600 | 10 | 293 | 401 | 403, 532 | 419, 534 | — | 7.0 | 1:3.7:0.04 | v, w |
| Komatsuna$^f$ | 28 | 402 ± 11 | 24 ± 1.5 | 56 | — | — | — | — | 6.5 | 1:2.5:0 | x |
| Maize$^g$ | 28 | 380 | 35 | 408 | 403, 502, 638 | 435, 556, 585 | 429, 534, 562 | — | 5–6 | 1:2.4:0.05 | y |
| Potato tubers$^h$ | ~30 | 55 | 30 | 233 | — | — | — | — | ~7 | 1:nd:0 | z |
| Japanese radish$^i$ | 28 | 770 | 130 | 561 | 403, 498, 636 | 436, 556, 590 | — | — | 6 | 1:0.06:nd | aa |
| Pea$^{j,k}$ | 29.5 | 325$^l$ | 20 | 411 | 403 | 435, 550, 585 | — | 425, 530, 560 | 5–8 | 1:0.28$^m$:0.23$^m$ | bb |
| Soybean$^{n,o}$ | 30 | 70$^m$ | 3 | 34 | 407 | 434, 554 | — | 431, 529, 558 | — | 1:0.37:1 | cc |
| Tea$^{p,q}$ | 57 | 416 | 30 | 100 | — | — | — | — | 6 | 1:4.97:0.25 | dd |
| Tea$^{q,r,s}$ | 34.660 ± 0.01$^t$ | 470 | 188 | 150 | 406, 500, 640$^u$ | 438, 560 | — | — | 4.5–5.0 | 1:1.5:0.82 | ee |

$^a$ Molecular weight given per monomer (kDa). Errors given where published.

$^b$ $K_m$ in µM. Errors given where published.

$^c$ Given in µmol/min/mg.

$^d$ Given as ascorbate:pyrogallol:guaiacol (ascorbate activity defined as 1; nd = not determined).

$^e$ Spinacia oleracea.

$^f$ Brassica rapa.

$^g$ Zea mays.

$^h$ Solanum tuberosum.

$^i$ Raphanus sativus.

$^j$ Pisum sativum.

$^k$ Dimer molecular weight = 57,500 ± 500 (by gel filtration).

$^l$ Non-Michaelis kinetics observed. Value for $K_m$ defined where rate = ½$V_{max}$.

$^m$ Values reported at 0.1 mM H$_2$O$_2$. At 0.5 mM H$_2$O$_2$ relative activities were 1:1.74:0.4.

$^n$ Glycine max.

$^o$ Dimer molecular weight = 47,000 (by gel filtration).

$^p$ Camellia sinensis.

$^q$ The two entries for tea APX represent different isozymes: the first is acidic, the second is basic.

$^r$ Camellia sinensis (designated isozyme I).

$^s$ Though to be a class III, not a class I, peroxidase.

$^t$ From electrospray mass spectrometry.

$^u$ $\lambda_{max}$ of Compound I: 407, 553, 650 nm.

$^v$ Yoshimura et al., 1998.

$^w$ Tanaka et al., 1991.

$^x$ Ishikawa et al., 1996c.

$^y$ Koshiba, 1993.

$^z$ Elia et al., 1992.

$^{aa}$ Ohya et al., 1997.

$^{bb}$ Mittler and Zilinskas, 1991b.

$^{cc}$ Dalton et al., 1987.

$^{dd}$ Chen and Asada, 1989.

$^{ee}$ Kvaratskhelia et al., 1997b.

## Table 2
### Selected physical data for various chloroplastic,[a] algal[b] and mammalian[c] APXs

| Source | $M_r$[d,e] | $K_m$ ascorbate[e] | $K_m$ H$_2$O$_2$[e] | Specific activity[f] | $\lambda_{max}$ Fe$^{III}$ | $\lambda_{max}$ Fe$^{II}$ | $\lambda_{max}$ Fe$^{III}$CN | $\lambda_{max}$ Fe$^{II}$CN | pH optimum | Relativity activity[g] | Reference |
|---|---|---|---|---|---|---|---|---|---|---|---|
| Tea[a,h] | 34 | 220 | 80 | 580 | 407, 478, 535 | 420, 556 | 416, 535 | — | 7.0 | 1:0.41:0.05 | r |
| Tea[a,j,k] | 33.698 ± 0.01[l] | — | — | 1500[m] | 403, 501, 640[n] | 438, 560 | — | — | 4.5 | 1:0.62:0.01 | s |
| Spinach[a,i] | 40 ± 2 | 500 | 87 | 1056 | 403, 620 | 433, 455 | 420, 534 | — | — | 1:0.06:nd | t |
| Spinach[a,o] | 30 ± 1 | 300 | 30 | 580 | 403 | 420, 560 | — | 535, 560 | 7.2 | 1:0.01:0.03 | u |
| Pea[a,p] | — | 0.6 | — | — | — | — | — | — | 8.2 | 1:0:nd | v, w, x |
| Chlorella vulgaris[b] | 32 | 111 ± 9 | 20 ± 2.5 | 1307 | 403 | 427, 560 | 419, 538 | — | 6.1 | 1:0.62:nd | y |
| Galdieria partita[b] | 32–36 | — | — | 160 | — | — | — | — | — | 1:2.9:0.19 | z |
| Euglena gracilis Z[b] | 58 | 410 | 56 | 476 | 407[q] | — | 421 | — | 6.2 | 1:2:0 | aa, bb |
| Bovine eye[c] | 43 ± 2 | 130 | 290 | 43 | 401, 620 | 435, 562 | 418, 535 | — | 7.2 | 1:0.51:nd | cc |

[a] Chloroplastic.
[b] Algal.
[c] Mammalian.
[d] Molecular weight given per monomer (kDa). Errors given where published.
[e] $K_m$ in μM. Errors given where published.
[f] Given in μmol/min/mg.
[g] Given as ascorbate:pyrogallol:guaiacol (ascorbate activity defined as 1; nd = not determined).
[h] Camellia sinensis.
[i] Thylakoid-bound.
[j] Though to be a class III, not a class I, peroxidase.
[k] Camellia sinensis (designated isozyme II). See footnote r, Table 1.
[l] From electrospray mass spectrometry.
[m] Highest reported for an APX.
[n] $\lambda_{max}$ of Compound I: 403, 553, 650 nm.
[o] Stromal.

[p] Pisum sativum.
[q] $\lambda_{max}$ of Compound I: 414 nm.
[r] Chen and Asada, 1989.
[s] Kvaratskhelia et al., 1999.
[t] Miyake et al., 1993.
[u] Nakano and Asada, 1987.
[v] Jimenez et al., 1998.
[w] Jablonski and Anderson, 1982.
[x] Mittler and Zilinskas, 1991b.
[y] Takeda et al., 1998.
[z] Sano et al., 1996.
[aa] Ishikawa et al., 1996b.
[bb] Shigeoka et al., 1980.
[cc] Wada et al., 1998.

ideas: on the one hand, it revealed similarities with other peroxidase enzymes that were not altogether unexpected and, on the other, it highlighted some rather surprising structural differences, the functional consequences of which have yet to be fully understood. The early publication of a crystal structure provided an important stimulus: it meant that spectroscopic and functional data, as it appeared, could be sensibly rationalized in a way that would not have been possible in the absence of structural information. By comparison, forty years elapsed between the first identification of CcP (Altschul et al., 1940) and the publication of a structure (Finzel et al., 1984; Poulos et al., 1980) and almost 100 years (Gajhede et al., 1997; Bach and Chodat, 1903) in the case of HRP—although, in view of the known difficulty in obtaining crystals for HRP, the latter probably represents an extreme example.

One of the major incentives for solving the structure of pAPX was related to the known sequence homology with CcP: pAPX was known from sequence comparisons (Mittler and Zilinskas, 1991a) to contain the same active site Trp residue (Trp179) that in CcP (Trp191) was, uniquely, known to be the site of a protein-based radical in Compound I (Erman et al., 1989; Sivaraja et al., 1989; Scholes et al., 1989) and essential for enzyme activity (Mauro et al., 1988). Other peroxidases have a phenylalanine residue at this position and use a porphyrin-based, rather than a protein-based, radical in Compound I. It was therefore of interest to establish whether CcP, which for years had been regarded as a benchmark for peroxidase catalysis, was representative of class I peroxidases as a whole—that is, was Trp191 a compulsory catalytic requirement for all class I peroxidases or, on the other hand, was the unusual specificity of CcP for a large, macromolecular substrate (cytochrome $c$) somehow responsible for the exceptional behavior?

The recombinant pAPX enzyme crystallized as a pair of homodimers, in keeping with the known dimeric behavior of the enzyme (Mittler and Zilinskas, 1991b). The monomeric structure of the enzyme was almost identical to CcP: it comprised of 12 helices (as for CcP) but contained less β-sheet structure than CcP. A schematic diagram of the monomeric structure is shown in Figure 1. The dimer interface was found to consist of a series of electrostatic interactions between Lys18 and Asp229, Glu112 and Arg24/Arg21, and Lys22 and Glu228. The active site was, as expected, very similar to CcP and is shown in Figure 2. On the proximal side, the catalytic triad of residues (His163-Asp208-Trp179) was identified: the hydrogen bond distance for the Asp-His ($3.2 \pm 0.1$ Å) bond in pAPX is longer than for CcP (3.0 Å), but the Asp-Trp distance is the same in both cases (2.8 Å). Although close to the limits of crystallographic detection, pAPX also contains a longer Fe-His bond (2.15 Å) than CcP (2.0 Å), which is reflected in

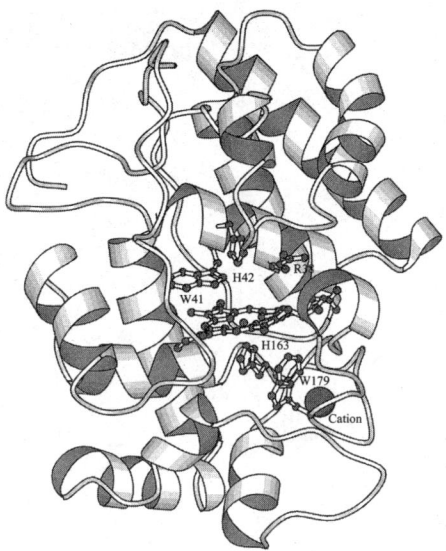

**FIGURE 1.** Schematic stereo representation of pAPX monomer 1. Active-site amino acids and the proximal metal cation are indicated. Taken from Patterson and Poulos, 1995.

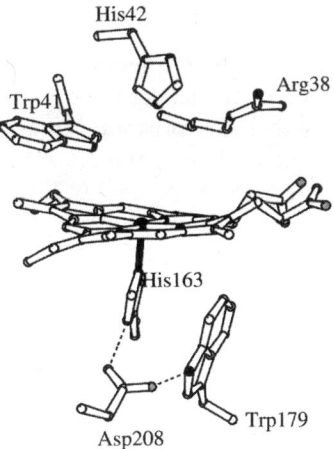

**FIGURE 2.** Stereo representation of the proximal and distal pAPX monomer active sites. Carbon atoms are open circles, nitrogen atoms and the heme iron are black circles. Dashed lines indicated hydrogen bonds. Taken from Patterson and Poulos, 1995.

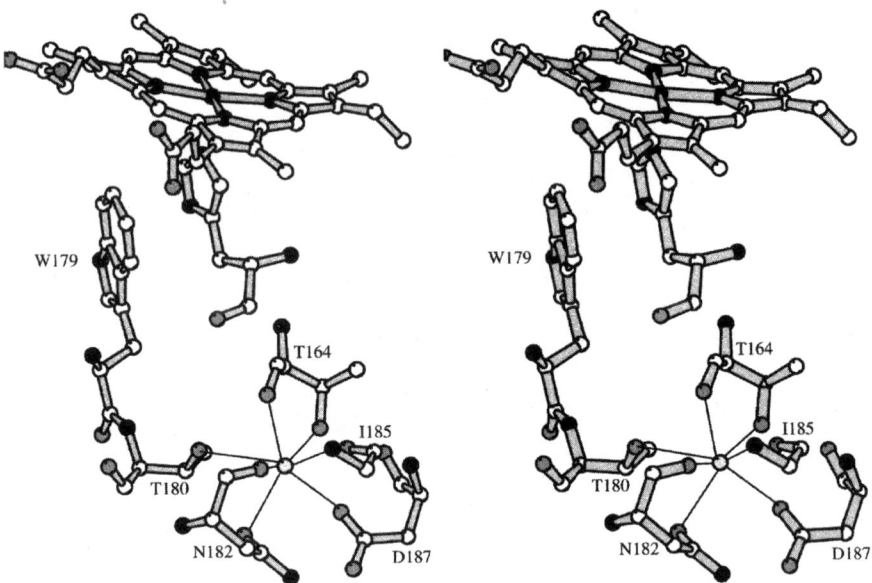

**FIGURE 3.** Stereo model of the pAPX cation site. Oxygen atoms and the cation are gray, and nitrogen atoms are black. The thin lines highlight ligand-cation interactions. Reprinted with permission from Bonagura et al., 1996.

the resonance Raman stretching frequency (Nissum et al., 1998) for this bond (section 6). On the distal side, the catalytic distal histidine and arginine residues, known to be important for activity in CcP and HRP (reviewed in Smith and Veitch, 1998; Erman, 1998), were identified, together with Trp41 and four ordered water molecules, Figure 2. The corresponding Trp residue in CcP was initially, and incorrectly, thought to be the site of the radical species in Compound I (Finzel et al., 1984; Poulos and Kraut, 1980).

A completely unexpected discovery was the identification of a metal binding site close to the heme in pAPX. The metal was found to be 7-coordinate, with ligands provided from Asp187 ($O^\delta$), Thr164 ($O^\gamma$), Thr180 ($O^\gamma$), Asn182 ($O^\delta$) and three main chain carbonyl oxygens (Asn182, Ile185 and Thr164), Figure 3. The metal was assigned as a monovalent $K^+$ ion, rather than a $Ca^{2+}$ ion as found in class III peroxidases, on the basis of electroneutrality arguments and bond distances (X-ray diffraction techniques alone are unable to distinguish between $K^+$ and $Ca^{2+}$ since they both contain the same number of electrons). Hence, only one charged ligand (Asp187) is bound to the metal ion in pAPX, whereas two ligands are normally found for divalent ions, and the average metal-ligand bond distance for pAPX (2.8 Å) was much longer than that expected for divalent ions ($\approx 2.4$ Å). In fact, the metal ion on the proximal side is linked, through an extensive array of hydrogen bonds, to the distal histidine itself and is only $\approx 8$ Å from the $\alpha$-carbon of the Trp179 residue. It is likely that this metal site plays a structural role, in a similar way to that previously observed in other class II (e.g. MnP (Sutherland and Aust, 1997; Sutherland *et al.*, 1997), LiP (Nie and Aust, 1997; Doyle and Smith, 1996)) and class III (e.g. HRP (Smith *et al.*, 1990; Morishima *et al.*, 1986; Ogawa *et al.*, 1979), PNP (Maranon *et al.*, 1993)) peroxidases. A functional role for the metal ion has also been proposed (Patterson *et al.*, 1995). These features are discussed in more detail in section 8.

In the absence of extensive functional information for the entire range of APXs from different sources, Welinder and co-workers (Jespersen *et al.*, 1997) have carried out a detailed comparative analysis of all APX sequences and have been able to establish some general principles which are consistent with the published structure of pAPX and which can provide workable hypotheses for future experimental strategies on new enzymes. Perhaps not surprisingly, the catalytic residues His42, Arg38, His163 and Asp208 are conserved in all APXs identified so far. Interestingly, the proximal Trp179 residue is also fully conserved except for the membrane-bound enzymes from spinach and *M. crystallinum*, both of which have a Phe residue instead (the Trp41 position, also invariant in all other APXs examined, is also replaced with a Phe in these two enzymes). In terms of the quaternary structure of APXs, the charged residues (Arg21/24, Lys18/22, Glu112/228, Asp229) involved in dimer formation (Patterson and Poulos, 1995) are, with one or two exceptions (i.e. Gln21/Ala112 in rice APX, Thr110 in glyoxysomal cotton APX), conserved in all the cytosolic enzymes but are not conserved in the chloroplastic APXs. Even so, the structural basis for dimer formation is, at present, incompletely understood: whilst the chloroplastic enzymes, as predicted, are uniformly monomeric, the cytosolic enzymes are not always dimeric as might be predicted from the sequence. Of the residues acting as ligands to the bound potassium ion, Thr180, Thr164 and Asp187 are all invariant (except for the thylakoid-bound APX enzyme from pumpkin which has a Asn residue at position 187), strongly suggestive of a structural and/or functional role (discussed in section 8), but Ile185 (which provides one main chain carbonyl bond) and Asn182 are not.

## 6.  SPECTROSCOPIC AND SPIN-STATE CONSIDERATIONS

As noted earlier (section 4, Tables 1 and 2), Soret maxima for ferric APXs from different sources are observed between 401 and 409 nm, which, whilst broadly consistent with the presence of high-spin iron, is nevertheless indicative of a variation in absolute quantities of individual spin states from source to source. The most detailed spectroscopic information available to date is for the recombinant cytosolic enzyme from pea. Hence, at neutral pH, this enzyme shows features in the visible region that are consistent with the presence of a high-spin heme iron ($\lambda_{max} = 404, 507, 640$ nm) (Nissum et al., 1998)). Interestingly, the Soret maximum is slightly different from that of the wild type ($\lambda_{max} = 403$ nm (Mittler and Zilinskas, 1991b)). Indeed, the exact position of the Soret band is rather variable: maxima at 403 (Turner et al., 1999; Patterson and Poulos, 1994) and 406 (Hill et al., 1997) have also been reported for the same recombinant enzyme (these maxima shift to $\lambda_{max} = 405, 508, 532, 637$ nm for the same pea enzyme expressed in a different vector (Cheek et al., 1999)). It is very likely that these variations merely reflect differing populations of individual heme spin states (5-c/HS, 6-c/HS, 6-c/LS), although a detailed quantitative analysis is not possible from electronic spectroscopy. Indeed, the exact spin distribution in ferric CcP is also known to be acutely, and rather unpredictably, sensitive to a range of variables, including pH, buffer composition, temperature and the age of the sample (Bosshard et al., 1991). For recombinant pAPX, resonance Raman experiments (Nissum et al., 1998) have provided some clarification and have shown that the enzyme does contain a mixture of 5-c/HS, 6-c/HS and 6-c/LS species at neutral pH, although the 6-coordinate species represent only fairly minor contributions. The exact nature of the sixth ligand is not known with certainty in either case but, for the 6-c/HS species at least, is almost certainly derived from coordination of one of the distal water molecules to the heme iron. MCD spectra of recombinant ferric pAPX are largely consistent with this spin distribution (Turner et al., 1999; Cheek et al., 1999). EPR spectroscopy has also identified mixtures of spin states in APX samples. Hence, signals characteristic of high-spin ($g = 5.96$, 5.18, 1.98 (Patterson et al., 1995); $g = 6.04$, 5.27 (Jones et al., 1998)) and low-spin ($g = 2.67$, 2.21, 1.78 (Patterson et al., 1995); $g = 2.69$, 2.21, 1.79 (Jones et al., 1998)) heme have been identified independently for recombinant pAPX, and similar features have been observed for recombinant soybean APX ($g = 5.96$, 5.23 and $g = 2.68$, 2.21, 1.78 for the high-spin and low-spin components respectively (Jones et al., 1998)). EPR spectra for the pea and soybean enzymes are shown in Figure 4.

For the recombinant ferrous pea enzyme, electronic spectra ($\lambda_{max} = 434$, 556, 587 nm (Turner and Lloyd Raven, 1999)) are in satisfactory agreement

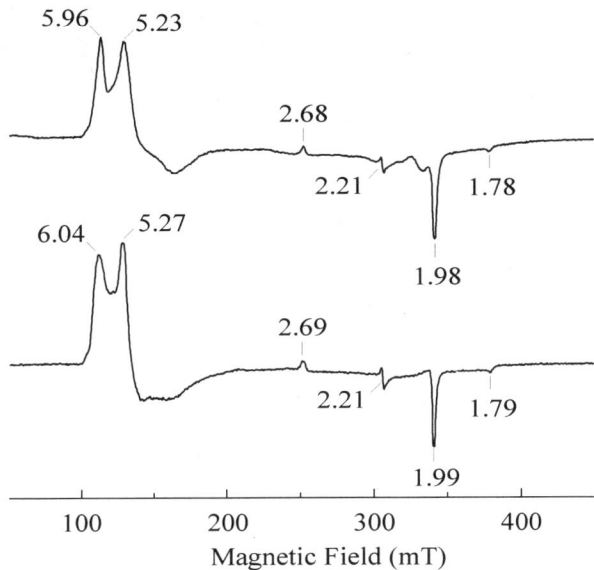

**FIGURE 4.** Low-temperature (7 K) EPR spectra of soybean cytosolic APX (top) and pAPX (bottom) (25 mM sodium phosphate, pH 6.99, 50% glycerol). Reprinted with permission from Jones *et al.*, 1998.

with the early data on the wild type enzyme ($\lambda_{max} = 435, 550, 585$ nm (Mittler and Zilinskas, 1991b)) and with the existence of a high-spin heme in the reduced enzyme. Resonance Raman spectra of recombinant ferrous pAPX (Nissum *et al.*, 1998) have also identified differences in the Fe-His163 stretching frequency compared to CcP, that are consistent with the longer Fe-His163 bonding and His163-Asp208 hydrogen bonding interactions in pAPX identified crystallographically (Patterson and Poulos, 1995).

## 7. CATALYTIC MECHANISM

### 7.1. Steady-state Kinetics

Whilst detailed spectroscopic and kinetic data for APX have been very slow to emerge, largely as a result of relatively poor yields and (in some cases) instability of the purified enzyme from the early isolations (section 11), steady state data, on the other hand, have been a fairly prominent feature of most of the early literature on APX. Values of $K_m$ for $H_2O_2$ have been reported for a number of purified and partially purified APXs and are summarized in Tables 1 and 2. Plots of rate *versus* [$H_2O_2$] show a hyperbolic dependence, consistent with a normal Michaelis-Menten treatment. For ascorbate, the situation is less clear-cut. Hence, in cases where the rate *versus* [ascorbate] dependence has been examined, the data fall into two categories: (a) APXs for which a normal hyperbolic dependence is observed and (b) APXs for which the steady state analyses show a non-hyperbolic (sigmoidal) dependence. The latter category includes the cytosolic APXs from pea (Mandelman *et al.*, 1998a; Mandelman *et al.*, 1998b; Mittler and Zilinskas, 1991b) and soybean (Dalton *et al.*, 1987). Since the pea and soybean enzymes are known to exist as homodimers (Patterson and Poulos, 1995; Dalton *et al.*, 1987), whilst the other enzymes are monomeric, it would be tempting to conclude that that source of the sigmoidal kinetics derives from a cooperative (allosteric) interaction between the two subunits. Indeed, arguments of this kind have been previously proposed (Mandelman *et al.*, 1998b; Dalton *et al.*, 1996). Recent experiments (Mandelman *et al.*, 1998b) on recombinant pAPX aimed at unravelling these complexities do not, however, support this conclusion. Hence, replacement of Glu112, which forms a salt bridge with Arg24 in the other monomeric half of the homodimer, would be expected to destabilize the dimer interaction and therefore influence the kinetics. In fact, replacement of Glu112 with a lysine residue results in a monomeric enzyme below 0.25 μM that, contrary to expectation, does not exhibit hyperbolic kinetics.[2] Hence, it seems unlikely that the sigmoidal kinetics derive from a cooperative interaction between the two subunits of the dimer.

These conclusions are supported by a series of related experiments on the same recombinant pea cytosolic enzyme in which the steady state oxidation of *p*-cresol has also been found to exhibit sigmoidal kinetics (Celik *et al.*, 1999). In this case, the data were satisfactorily fitted to the Hill equation, Eq. (2),

$$\frac{v}{V_{max}} = \frac{[S]^n}{K^n + [S]^n} \qquad (2)$$

(v = initial rate, n = number of substrate binding sites, $K$ = substrate concentration at which the velocity is half-maximal, $V_{max}$ = maximum velocity)

---

[2] Replacement of the Glu112 residue with an alanine group (E112A variant), which would similarly be expected to destabilize the dimer interaction, provides a less conclusive and slightly puzzling scenario. Hence, in contrast to the E122K variant, E112A is not monomeric under any of the experimental conditions examined.

and a value for n of 2.01 ± 0.15 was derived (pH 7.0, I = 2.2 mM). Whilst Eq. (2) is consistent with a mechanism involving allosteric interactions between the two monomers of a homodimer (n = 2 per dimer), the fact that identical oxidation kinetics are observed under conditions of high (I = 500 mM) ionic strength, that would be expected to disfavor dimer formation on electrostatic grounds, argues against this type of cooperative mechanism. An alternative interpretation, also consistent with Eq. (2), is to invoke non-allosteric behavior where n now represents the number of substrate binding sites for each non-interacting monomer of the dimer, and a model in which two binding sites are involved in $p$-cresol oxidation has therefore been proposed (Celik *et al.*, 1999). The possible existence of a second binding site for $p$-cresol is curious but, when viewed in the context of other data on the same enzyme, not altogether unexpected. In fact, a consensus of data for pAPX is beginning to appear that is suggestive of a binding interaction with some substrates that is more complex than the expected 1:1 ratio (section 9). Hence, NMR data for binding of ascorbate to recombinant pAPX are consistent with the existence of two binding sites for the substrate (Hill *et al.*, 1997) and experiments on site-directed variants and chemically modified derivatives of the same pAPX enzyme (Mandelman *et al.*, 1998a) have established that the enzyme likely utilizes different sites for aromatic and non-aromatic substrates. When viewed more generally in the context of other peroxidases, the existence of a second binding site is not unprecedented. Hence, binding stoichiometries of greater than one have been identified under certain conditions for the cytochrome $c$/CcP (Erman, 1998 and references therein) and manganese (II)/MnP (Mauk *et al.*, 1998) interactions, and discrete binding sites for different substrates have been proposed in LiP (Doyle *et al.*, 1998) and CcP (DePillis *et al.*, 1991). In general, however, the kinetic and functional competence of these multiple sites is more difficult to establish in an unambiguous manner and, for pAPX, remains unclear.

## 7.2. Pre-Steady State Kinetics

The first information on the mechanism of catalysis in wild type pAPX was published in 1996 (Marquez *et al.*, 1996). The kinetic data are consistent with a scheme in which the ferric enzyme is oxidized by two electrons to a so-called Compound I intermediate with concomitant release of one mole of water, followed by two successive single electron reductions of the intermediate by ascorbate (S) to regenerate ferric enzyme, Eqs. (3)–(5).

$$APX + H_2O_2 \xrightarrow{k_1} Compound\ I + H_2O \qquad (3)$$

$$\text{Compound I} + S_{red} \xrightarrow{k_2} \text{Compound II} + S_{ox} \qquad (4)$$

$$\text{Compound II} + S_{red} + 2H^+ \xrightarrow{k_3} \text{APX} + S_{ox} \qquad (5)$$

Rate constants (pH 7.8, 20°C) for Eqs. (3) and (4) have been defined ($k_1 = 8.0 \times 10^7 \, M^{-1} s^{-1}$; $k_2 = 8.2 \times 10^7 \, M^{-1} s^{-1}$; $k_3 = 3.8 \times 10^3 \, M^{-1} s^{-1}$) (Marquez et al., 1996) and have been largely confirmed in the recombinant pAPX enzyme ($k_1 = 8.3 \times 10^7 \, M^{-1} s^{-1}$ (Mandelman et al., 1998a) (pH 7.0 20°C), $k_1 = 9 \times 10^7 \, M^{-1} s^{-1}$ (Pappa et al., 1996); $k_2 = 3.4 \times 10^7 \, M^{-1} s^{-1}$ (Mandelman et al., 1998a) (pH 7.0, 20°C)). Hence, reduction of Compound I is competitive with its formation, and Compound II reduction represents the rate-limiting step. In contrast, while the rate constant for reaction of CcP with $H_2O_2$ is also fast ($k_1 = 4.5 \times 10^7 \, M^{-1} s^{-1}$ (Erman, 1998)), the activity with ascorbate (Yonetani and Ray, 1965) is very poor ($k_{cat} = 3 \, s^{-1}$), although HRP is more reactive with ascorbate ($k_2 = 2.3 \times 10^5 \, M^{-1} s^{-1}$). Pre-steady state kinetic data are also available for an APX isozyme from tea (Kvaratskhelia et al., 1999): rate constants (pH 7.0, 23°C) were found to be $k_1 = 8.9 \times 10^6 \, M^{-1} s^{-1}$, $k_2 = 4.5 \times 10^6 \, M^{-1} s^{-1}$, $k_3 = 3.7 \times 10^4 \, M^{-1} s^{-1}$. Whilst there is a multitude of kinetic data on the effects of active-site amino acids on the various rate constants for other peroxidases (notably CcP and HRP (reviewed in (Dunford, 1999; Smith and Veitch, 1998; Erman, 1998), there is currently no such information for APX.

## 8.   RADICAL CHEMISTRY

### 8.1.   Fate of the Monodehydroascorbate Radical

The immediate product of the oxidative reaction, the monodehydroascorbate radical (Eq. [1]), is a fairly reactive and unstable species which, in the presence of a suitable reductase system, is reduced back to ascorbate. Monodehydroascorbate reductases have been identified and purified in a few cases (Ushimara et al., 1997; Dalton et al., 1992; Shigeoka et al., 1987; Borraccino et al., 1986; Hossain et al., 1984) and cDNA sequences have been published for the pea (Murthy and Zilinskas, 1994) and cucumber enzymes (Sano et al., 1995). A bacterial expression system is also available for the cucumber enzyme (Sano et al., 1995). In the absence of a suitably efficient reductase system, the monodehydroascorbate radicals disproportionate to dehydroascorbate and ascorbate: in this case, ascorbate is regenerated using a glutathione-dependent dehydroascorbate reductase enzyme (Foyer and Mullineaux, 1998 and references therein). Under non-physiological conditions, disproportionation also occurs and steady-state data need to be corrected for the regeneration of ascorbate.

## 8.2. Nature of the Intermediates

### 8.2.1. Compound I

There is general agreement that the initial product (Compound I) of the reaction of pAPX with $H_2O_2$ is a porphyrin $\pi$-cation intermediate (as found in HRP), and not a protein-based radical species as found in CcP. Evidence for this comes from the similarity of the spectra observed immediately after the addition of $H_2O_2$ to both wild type pAPX ($\lambda_{max}$ = 404 nm (Marquez et al., 1996)) and recombinant pAPX ($\lambda_{max}$ = 404 nm (Lad et al., 1999)) to those previously published for HRP (Dunford, 1999). Recently, spectra for the Compound I of two APX enzymes from tea have been published (Kvaratskhelia et al., 1999; Kvaratskhelia et al., 1997b), both of which are consistent with the formation of a porphyrin $\pi$-cation, and not a protein-based, radical ($\lambda_{max}$ (isozyme I) = 407, 553, 650 nm; $\lambda_{max}$ (isozyme II) = 403, 553, 650 nm). Rapid freeze-quench EPR data for recombinant pAPX (Patterson et al., 1995) are also consistent with the assignment of Compound I as an oxidized porphyrin $\pi$-cation species. Hence, a weak axial resonance extending from g = 3.27 to g ≈ 2 is observed and no stable radical signal (g ≈ 2) analogous to that reported in CcP Compound I is detected.

### 8.2.2. Compound II

APX Compound I is very unstable and decays ($k_{decay}$ = 0.811 s$^{-1}$ (Lad et al., 1999)), in the absence of substrate, to a stable Compound II-like intermediate. Indeed, instability of Compound I is a characteristic feature of APX that distinguishes it from the classical peroxidases, which generally have stable Compound I intermediates. The spectrum of this intermediate is consistent with reduction of the porphyrin $\pi$-cation radical of Compound I to give a ferryl-containing Compound II. Wavelength maxima for the wild type ($\lambda_{max}$ = 413, 528, 558 nm (Marquez et al., 1996)) and recombinant wild type ($\lambda_{max}$ = 413, 532, 560 nm (Mandelman et al., 1998a)) have been reported. Spectra for Compound II of isozyme I ($\lambda_{max}$ = 420, 531 and 552 nm) and isozyme II ($\lambda_{max}$ = 419, 530 and 558 nm) of tea APX have also been published (Kvaratskhelia et al., 1999; Kvaratskhelia et al., 1997b). In contrast to Compound I, Compound II of pAPX is stable for up to an hour after preparation.

### 8.2.3.  Role of the Proximal Metal Ion

Whilst it has been established that Trp191 in CcP is the site of a protein-based radical in Compound I (Sivaraja *et al.*, 1989; Scholes *et al.*, 1989; Erman *et al.*, 1989) and that formation of a radical at this position is essential for function (Mauro *et al.*, 1988), all other peroxidases are known to form porphyrin-based, rather than protein-based, radicals during their catalytic cycle (Dunford, 1999). The absence a Trp amino acid at the position corresponding to Trp191 in CcP has provided a satisfactory rationalization of these data. The subsequent identification (Patterson *et al.*, 1995) of a porphyrin-based, rather than a protein-based, radical in Compound I of pAPX therefore provided an unexpected surprise, particularly when one considers the similarity in the crystal structures in the proximal region. Various theoretical arguments have, with the considerable benefit of hindsight, been able to provide data in support of the experimental observations (Menyhard and Naray-Szabo, 1999; Jensen, 1998; Naray-Szabo, 1997). Initially, it was proposed (Patterson *et al.*, 1995) that the proximal potassium binding site in pAPX (section 5, Figure 3), located at a distance of ≈8 Å from the α-carbon of Trp179, might partially account for these differences on electrostatic grounds. Hence, whilst CcP contains the same polypeptide conformation around the metal site, the ligating amino acids (Asp187, Thr164, Thr180, Asn182), and therefore the metal ion itself, are missing (replaced by Thr187, Ala164, Gly180 and Ala182 respectively). Cation radical formation at Trp179 would therefore be disfavored in APX but not in CcP. Attempts to confirm this hypothesis in an experimental manner have, however, been rather more troublesome. Hence, it has been possible to engineer the appropriate metal ligands into CcP and the presence of a metal ion, analogous to that found in APX, has been confirmed by X-ray crystallography (Bonagura *et al.*, 1999; Bonagura *et al.*, 1996). Unfortunately, these structural alterations, whilst clearly destabilising radical formation at Trp191, were not able to support oxidation of the porphyrin as an alternative (interestingly, incorporation of a manganese (II) binding site into CcP does not preclude formation of a radical at Trp191 (Wilcox *et al.*, 1998)). The analogous experiments in pAPX, aimed at removal of the metal site to generate a CcP-like enzyme, were thwarted by major conformational changes induced by the mutations. Hence, mutagenic removal of the metal ligands (Cheek *et al.*, 1999) resulted in the formation of a new low-spin form of the enzyme that had electronic and MCD spectroscopic properties similar to those observed for bis-histidine ligated proteins (e.g. cytochrome $b_5$). Although the exact nature of the axial ligands in this low-spin derivative have not been unambiguously assigned, it is clear that the potassium site itself plays an important role in stabilization of the overall structure in

APX. Interestingly, the same spectroscopic changes are observed in a low pH form of pAPX (Turner and Lloyd Raven, 1999), suggesting that the same sort of conformational rearrangement, perhaps involving titration of one of the ligands to the potassium ion, might have occurred. In terms of the influence of the metal ion itself on the actual location of the radical in Compound I, there is clearly no single structural feature responsible for the differences between APX and CcP, and further work will be necessary in order to define and understand these features in detail. In this context, it has recently been proposed (Menyhard and Naray-Szabo, 1999; Naray-Szabo, 1997) that differences in the protonation state of the Trp residue are important in controlling the location of the radical species in CcP and APX, although this proposal is inconsistent with ENDOR experiments on CcP (Huyett et al., 1995).

### 8.2.4. Role of Trp179

An equally intriguing feature of APX catalysis is the observation (Pappa et al., 1996) that Trp179 itself is not a necessary requirement for oxidation of ascorbate. In fact, whilst removal of Trp179 does not affect turnover of ascorbate in recombinant pAPX (Pappa et al., 1996), the corresponding mutation in CcP almost eliminates cytochrome c oxidation (Mauro et al., 1988). Although, at first sight, this comparison is difficult to rationalize, it is consistent with the view that a protein-based radical at Trp179 is not utilized in pAPX and may, instead, reflect differences in substrate specificity between the two enzymes. Hence, whilst Trp191 is clearly an essential component of the electron transfer pathway for at least one productive substrate binding interaction in CcP, this is certainly not the case for APX and may provide an important clue to the exact location of the ascorbate binding site in APX.

## 9. SUBSTRATE RECOGNITION

As mentioned above (section 7), binding of some substrates[3] to APX appears to be more complex than a simple 1:1 interaction involving a single binding site. As yet, though, there is no definitive structural information on

---

[3] Whilst steady-state kinetics for pAPX with ascorbate are clearly not hyperbolic, this is not the case for all substrates. In fact, steady state kinetics of pAPX guaiacol have been shown (Mandelman et al., 1998a) to exhibit normal Michaelis behavior. The reasons for these differences between substrates are not yet clear.

the location of the ascorbate binding site for an APX. The best information available at present comes from two sources. First, NMR-derived distance constraints indicate that the ascorbate binds between 9.0 and 11.2 Å from the heme iron (Hill *et al.*, 1997). This would be in keeping with the general mechanism of peroxidase catalysis (Ortiz de Montellano, 1987) in which substrates are bound and oxidized at the heme edge rather than close to the heme iron (as is the case for cytochrome P450). Initial molecular modeling studies, carried out using the NMR data and the crystal structure coordinates, indicate that there are two distinct binding sites, both of which are consistent with the NMR information: one close to the 6-propionate (γ-meso position) and one close to the δ meso position of the heme, Figures 5A and B. Second, a series of experiments utilizing a combined site-directed mutagenesis/chemical modification approach have provided additional information on the nature of the ascorbate binding site (Mandelman *et al.*, 1998a). Hence, site-specific removal or chemical modification of the single Cys32 residue in pAPX compromise the activity against ascorbate by a factors of ≈3 and ≈1000 respectively. Whilst the former might be regarded as a relatively modest kinetic effect, the latter certainly might not and the data are therefore indicate of an ascorbate interaction at the heme edge in the region of Cys32 and the heme propionates (which is close to the γ-meso position implicated by NMR (Hill *et al.*, 1997)). Interestingly, oxidation of guaiacol, a classical

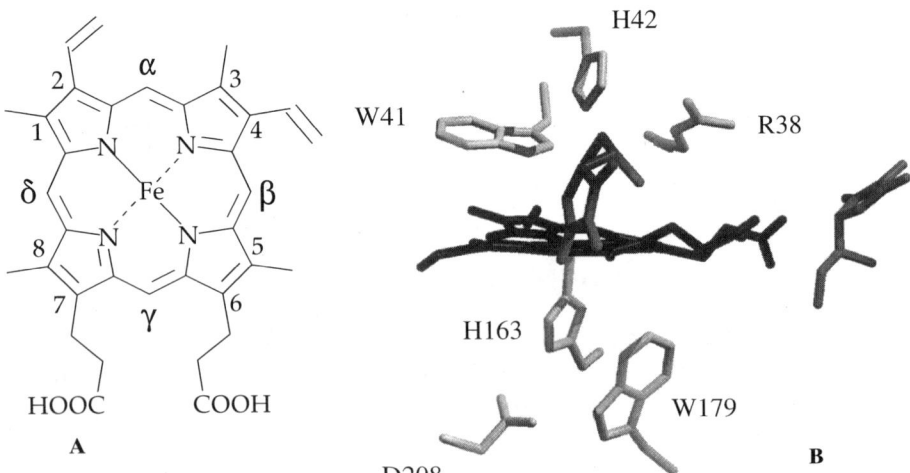

**FIGURE 5A.** Structure of iron protoporphyrin IX, showing the nomenclature used in this review.
**FIGURE 5B.** Diagram showing the proposed binding location of ascorbate (dark gray) at the δ-meso (center) and 6-propionate (right) positions in APX (Hill *et al.*, 1997). The heme is shown in black and the proximal and distal amino acids in light gray.

aromatic substrate, is unaffected by either removal or modification of Cys32, which is consistent with the idea that aromatic substrates bind at an alternative location, probably close to the δ-meso position, and with chemical modification experiments using the aromatic substrate phenylhydrazine (Hill et al., 1997), which have shown that δ-meso-modified heme is the dominant product. As mentioned earlier (section 7), when considered in a slightly more global context, the idea that substrates of different kind can utilize different binding locations is not inconsistent with the known behavior of CcP (DePillis et al., 1991) and of other, more well-characterized peroxidase enzymes (Mauk et al., 1998; Erman, 1998; Doyle et al., 1998).

In terms of the actual accessibility of the heme itself to substrate, several lines of evidence indicate that the δ-meso position is more exposed in the class I peroxidases than in the class II and III. First, as mentioned above, chemical modification experiments (Hill et al., 1997) using phenylhydrazine reveal that a much more substantial fraction of the δ-modified heme derivative is formed for pAPX (≈60%) than for HRP (≈8%). Examination of the proposed binding region and amino acids sequence comparisons of a number of class I, II and III peroxidases provide some rationalization of these data. Hence, Ala134 in pAPX is located close to the exposed δ-meso position and sequence comparisons with other class II and III peroxidases (e.g. HRP, MnP, LiP and ARP) indicate that this position is usually occupied by a more bulky proline residue. Similar conclusions have been reached from chemical modification (DePillis et al., 1991) and site-directed mutagenesis (Wilcox et al., 1996) experiments on CcP: CcP also contains an alanine at the corresponding position (Ala147) and a more exposed heme edge has been proposed. Interestingly, alignment of all known APX sequences (Jespersen et al., 1997) indicates that Ala134 is conserved throughout, with the exception of the membrane-bound enzymes from spinach and *M. crystallinum* which contain a proline. In fact, these membrane-bound APXs also contain Phe residues, analogous to those found in the class II and III peroxidases, at the otherwise conserved Trp179 and Trp41 positions (section 5)—functional studies on these enzymes should therefore provide interesting comparisons with other class I enzymes as well as the class II and III peroxidases.

## 10. REDOX PROPERTIES

There is, at present, very little redox information available for APX, which is as much a reflection on the difficulties associated with electrochemical measurements on complex metalloproteins as a whole (Armstrong et al., 1997; Armstrong et al., 1993) as on any special difficulty with peroxidases (or APX) itself. Hence, detailed redox information for peroxidases in general has been rather slow to emerge when compared to the impressive range of kinetic, structural and spectroscopic data that dominate the peroxidase literature (although some of these deficiencies are now being overcome (Mondal *et al.*, 1998; Mondal *et al.*, 1996; Armstrong and Lannon, 1987).

The reduction potential for the Fe(III)/Fe(II) couple of recombinant soybean APX has been measured by mediated spectroelectrochemistry (Jones *et al.*, 1998), an approach which has been successfully applied to CcP (Bujons *et al.*, 1997). The ferric/ferrous couple for this APX has been determined as $-159 \pm 2\,\mathrm{mV}$ (pH 7.0, 25.0°C, $\mu = 0.10\,\mathrm{M}$), Figure 6, and is in a similar range as reported values for other peroxidases (all *vs* SHE): HRP ($-261\,\mathrm{mV}$ (Tanaka *et al.*, 1998)), $-258\,\mathrm{mV}$ (Yamada *et al.*, 1975) and $-270\,\mathrm{mV}$ (Harbury, 1957)), CcP ($-194\,\mathrm{mV}$ (Conroy *et al.*, 1978)) and $-182\,\mathrm{mV}$ (Goodin and McRee, 1993)), MnP ($-93\,\mathrm{mV}$ (Millis *et al.*, 1989)) and LiP ($-130\,\mathrm{mV}$ (Millis *et al.*, 1989)). The observed potential for soybean APX is therefore consistent with the generally accepted mechanism of peroxidase catalysis, Eqs. [3]–[5], in which stabilization of high iron oxidation states is required. The corresponding potentials for the globins, where stabilization of low oxidation states is required for oxygen binding, are

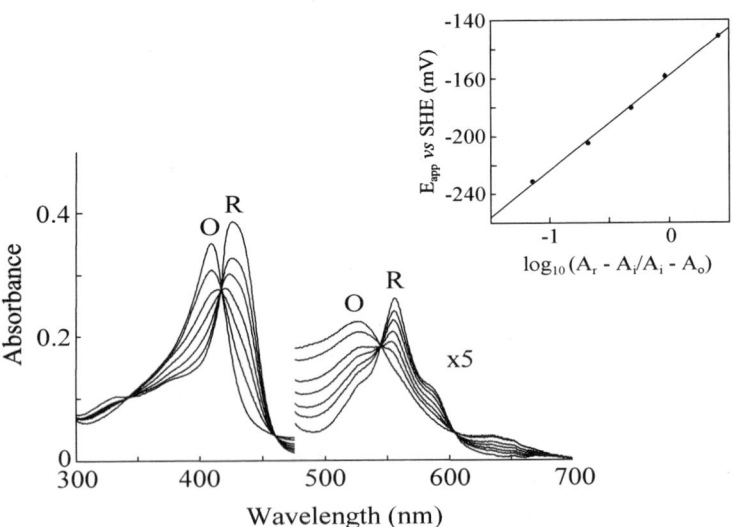

**FIGURE 6.** Thin-layer spectra (sodium phosphate, pH 7.0, 25.0°C, I = 0.10M) and corresponding Nernst plot of soybean APX at various applied potentials, $E_{app}$ (mV vs SHE). The fully oxidized (O) and fully reduced (R) spectra are indicated. For clarity, the visible region has been expanded. Reprinted with permission from Jones *et al.*, 1998.

much lower (e.g. for horse heart myoglobin the observed potential *vs* SHE is 64 mV (pH 6.0) (Lim, 1989)). In terms of a more general comparison with CcP, it is worth pointing out that the reduction potential of APX is slightly higher than that of CcP. In this context, the cation binding site, not present in CcP and within ≈8 Å of the α-carbon atom of Trp 179 in APX (Patterson and Poulos, 1995), might have a destabilising influence on high-oxidation-state intermediates and this possibility remains to be explored.

Although one electron reduction potentials for the Compound I/Compound II and Compound II/ferric couples for *A. ramosus* (915 and 982 mV respectively, pH 7.0), (Farhangrazi *et al.*, 1994)) and HRP (878 and 869 mV respectively (Farhangrazi *et al.*, 1995), 879 and 903 respectively (He *et al.*, 1996), and 880 and 900 respectively (Hayashi and Yamazaki, 1979) (all pH 7.0)) have been reported, these are not yet available for APX.

## 11. INACTIVATION

Should one wish to pursue it and given favorable conditions, one can, with varying amounts of success, inactivate one's enzyme, although the mechanistic details are not clear in all cases.

### 11.1. Inactivation by Cyanide

In all cases where the effects of cyanide have been investigated, inactivation occurs (Ohya *et al.*, 1997; Kvaratskhelia *et al.*, 1997b; Ishikawa *et al.*, 1996c; Sano *et al.*, 1996; Miyake *et al.*, 1993; Koshiba, 1993; Elia *et al.*, 1992; Mittler and Zilinskas, 1991b; Chen and Asada, 1989; Dalton *et al.*, 1987; Gerbling *et al.*, 1984; Jablonski and Anderson, 1982). This can be understood in terms of simple binding interaction between the heme and the powerful cyanide ligand that precludes reaction of the metal with $H_2O_2$. Azide, for the same reasons, also inhibits APX activity, albeit less effectively, which is merely a reflection of the increased ligand field strength for cyanide compared to azide. Binding constants, $K_d$, have not been reported for any wild type APX, but are available for recombinant pAPX ($K_{d(cyanide)} = 11.6 \pm 0.4 \mu M$ (pH 7.00, I = 0.10 M, 25.0°C) (Hill *et al.*, 1997); ($K_{d(azide)} = 300 \pm 75 \mu M$ (pH 7.00, I = 0.10 M, 25.0°C) (Turner and Lloyd Raven, 1999)).

### 11.2. Inactivation by Sulphydryl Reagents

It was known from quite early on (Gerbling *et al.*, 1984) that APX was sensitive to thiol-modifying reagents (in this case 2-mercaptoethanol and dithiothreitol). In fact, inhibition by thiol reagents is one of the distinguishing characteristics of (class I) APXs—the classical (class III) gua-iacol peroxidases are not affected by sulphydryl reagents.[4] Subsequently, other thiol-specific reagents (e.g. *p*-chloromercuribenzoate, *p*-chloromercuriphenylsulphonate and 5,5'-dithiobis-2-nitrobenzoic acid (DTNB)) were found to inhibit APX activity (Ohya *et al.*, 1997; Sano *et al.*, 1996; Miyake *et al.*, 1993; Elia *et al.*, 1992; Mittler and Zilinskas, 1991b; Chen and Asada, 1989). It was initially proposed (Asada *et al.*, 1993) that the source of the inhibition might lie in the participation of a thiol group in the formation of Compound I. However, a comprehensive series of chemical modification, kinetic and structural experiments has recently indicated this not to be the case (Mandelman *et al.*, 1998a). Hence, Compound I formation in recombinant pAPX has been shown to be unaffected by DTNB-modification (at the single Cys32 residue), but the activity of the modified enzyme with ascorbate is only $\approx 1\%$ of the wild type value. Replacement of the single cysteine amino acid with serine (Cys32Ser variant) has similar, albeit less dramatic, kinetic effects. These data are consistent with a mechanism in which binding of ascorbate occurs close to Cys32 and inhibition of enzyme activity therefore occurs as a result of a lower affinity of the substrate for the modified enzyme.

## 11.3.  Inactivation in Ascorbate-Depleted Media

Certain APXs are known to be inactivated in ascorbate-depleted media: this was first reported for APX from *Euglena gracilis* (Shigeoka *et al.*, 1980) and later for other peroxidases from different sources (Yoshimura *et al.*, 1998; Takeda *et al.*, 1998; Kvaratskhelia *et al.*, 1997b; Asada *et al.*, 1996; Tanaka *et al.*, 1991; Chen and Asada, 1989; Nakano and Asada, 1987; Hossain and Asada, 1984; Shigeoka *et al.*, 1980). The chloroplastic enzymes appear to be much more susceptible to inactivation than the cytosolic APXs (Asada *et al.*, 1996; Chen and Asada, 1989), which partially accounts for the failure to detect APX activity in early isolation procedures from chloroplasts. It was later realised that addition of ascorbate to the medium during purification largely prevents inactivation and that, once purified, both the cytosolic and chloroplastic enzymes are stable. Ascorbate appears to be acting simply as a reducing agent and other electron donors will substitute for it and also prevent inactivation. The origin of the sensitivity to ascorbate-depleted media is believed to arise from

---

[4] Interestingly, an APX isozyme from tea has been shown (Kvaratskhelia *et al.*, 1997b) to be insensitive to thiol reagents, consistent with the idea that this APX is more reminiscent of a class III enzyme than a class I (Table 1, section 1).

decomposition of Compound I by (excess) hydrogen peroxide when no reducing substrate is present to maintain enzyme turnover (Miyake and Asada, 1996). This mechanism, although plausible and consistent with known behavior of other peroxidases in the presence of excess $H_2O_2$ (Dunford, 1999; Frew and Jones, 1984; Dunford, 1982; Nicholls and Schonbaum, 1963), does not, however, account for the special sensitivity of the chloroplastic enzymes compared to the cytosolic.

### 11.4. Inactivation by Hydrogen Peroxide

The sensitivity of APX to high concentrations of hydrogen peroxide was recognized at an early stage (Hossain and Asada, 1984). Whilst this has been reported to derive from the instability of Compound I to excess peroxide in the absence of reducing substrate (Miyake and Asada, 1996), the mechanism has not been defined in detail. For other peroxidases, the inactivation mechanism, which has been investigated in detail (Dunford, 1991; Frew and Jones, 1984; Dunford, 1982), involves formation of a so-called Compound III intermediate that does not participate in the catalytic cycle. The only information for an APX comes from recent experiments on two tea enzymes (Kvaratskhelia et al., 1999; Kvaratskhelia et al., 1997b). Whilst a Compound III derivative ($\lambda_{max}$ = 417, 547, 581 nm) similar to that seen in HRP was observed after reaction of isozyme I with excess hydrogen peroxide, isozyme II did not form an analogous Compound III-like intermediate. Instead, a stable Compound II was observed for isozyme II that, at very high concentrations of peroxide, eventually decayed to an inactive species having a spectrum reminiscent of verdoheme ($\lambda_{max}$ = 670 nm). The reasons for the increased resistance of isozyme II to peroxide-induced inactivation are not currently understood.

### 11.5. Inactivation by Salicylic Acid

Initial reports (Durner and Klessig, 1995) indicating that salicylic acid acts as an inhibitor of APX activity were later questioned (Kvaratskhelia et al., 1997a). In fact, it seems more likely that salicylic acid acts as a reducing substrate for Compound I and Compound II (second order rate constants, $k_2$ and $k_3$, of $4.0 \times 10^2$ and $1.5 \times 10^2 M^{-1}s^{-1}$ respectively), although the nature of the radical products has not been identified (Kvaratskhelia et al., 1997a).

ACKNOWLEDGMENTS. The author's own work has been supported by grants from the Royal Society, the EPSRC, the Nuffield foundation, the BBSRC and an Association of Commonwealth Universities Development Fellowship. Mrs Sangita Lad is gratefully acknowledged for excellent secretarial assistance.

## 12. REFERENCES

Altschul, A. M., Abrams, R., and Hogness, T. R., 1940, Cytochrome c peroxidase, *J. Biol. Chem.* **136**:777–794.

Armstrong, F. A., and Lannon, A. M., 1987, Fast interfacial electron-transfer between cytochrome-c peroxidase and graphite-electrodes promoted by aminoglycosides-novel electroenzymic catalysis of H₂O₂ reduction, *J. Am. Chem. Soc.* **109**:7211–7212.

Armstrong, F. A., Butt, J. N., and Sucheta, A., 1993, Voltammetric studies of redox-active centres in metalloproteins adsorbed on electrodes, *Meth. Enzymol.* **227**:479–500.

Armstrong, F. A., Heering, H. A., and Hirst, J., 1997, Reactions of complex metalloproteins studied by protein-film voltammetry, *Chem. Soc. Rev.* **26**:169–179.

Asada, K., Miyake, C., Ogawa, K., and Hossain, M. A., 1996, Microcompartmentation of ascorbate peroxidase and regeneration of ascorbate from ascorbate radical: its dual role in chloroplasts, in: *Plant Peroxidases: Biochemistry and Physiology* (C. Obinger, U. Burner, R. Ebermann, C. Penel, and H. Greppin, eds.), University of Geneva, pp. 163–167.

Asada, K., Miyake, C., Sano, S., and Amako, K., 1993, *Plant Peroxidases: Biochemistry and Physiology* (K. Welinder, S. Ramusen, H. Penel, and H. Greppin, eds.), University of Geneva, Geneva, Switzerland, pp. 243–250.

Bach, A. N., and Chodat, R., 1903, Untersuchungen uber die rolle der peroxyde in der chemie der lebenden zelle. IV. Uber peroxydase, *Ber* **36**:600.

Bonagura, C. A., Sundaramoorthy, M., Bhaskar, B., and Poulos, T. L., 1999, The effects of an engineered cation site on the structure, activity and EPR properties of cytochrome c peroxidase, *Biochemistry* **38**:5538–5545.

Bonagura, C. A., Sundaramoorthy, M., Pappa, H. S., Patterson, W. R., and Poulos, T. L., 1996, An engineered cation site in cytochrome c peroxidase alters the reactivity of the redox active tryptophan, *Biochemistry* **35**:6107–6115.

Borraccino, G., Dipierro, S., and Arrigoni, O., 1986, Purification and properties of ascorbate free-radical reductase from potato tubers, *Planta* **167**:521–526.

Bosshard, H. R., Anni, H., and Yonetani, T., 1991, Yeast cytochrome c peroxidase, in: *Peroxidases in Chemistry and Biology*, Vol. 2 (J. Everse, K. E. Everse, and M. B. Grisham, eds.), CRC Press, Boca Raton, pp. 51–84.

Boveris, A., Sies, H., Martino, E. E., Docampo, R., Turrens, J. F., and Stoppani, A. O. M., 1980, Deficient metabolic utilisation of hydrogen peroxide in *Trypanosoma cruzi*, *Biochem. J.* **188**:643–648.

Bujons, J., Dikiy, A., Ferrer, J. C., Banci, L., and Mauk, A. G., 1997, Charge reversal of a critical active-site residue of cytochrome c peroxidase: characterization of the Arg48Glu variant, *FEBS Lett.* **243**:72–84.

Bunkelmann, J. R., and Trelease, R. N., 1996, Ascorbate peroxidase: a prominent membrane protein in oilseed glyoxysomes, *Plant Physiol.* **110**:589–598.

Bunkelmann, J. R., and Trelease, R. N., 1997, Expression of glyoxysomal ascorbate peroxidase in cotton seedlings during postgerminative growth, *Plant Science* **122**:209–216.

Caldwell, C. R., Turano, F. J., and McMahon, M. B., 1998, Identification of two cytosolic ascorbate peroxidase cDNAs from soybean leaves and characterisation of their products by functional expression in *E. coli*, *Planta* **204**:120–126.

Celik, A., Cullis, P. M., and Lloyd Raven, E., 2000, *Arch. Biochem. Biophys.* **373**:175–181.

Chatfield, M., and Dalton, D. A., 1993, Ascorbate peroxidase from soybean root nodules, *Plant Physiol.* **103**:661–662.

Cheek, J., Mandelman, D., Poulos, T. L., and Dawson, J. H., 1999, A study of the K⁺-site mutant of ascorbate peroxidase: mutations of the protein on the proximal side of the heme cause changes in iron ligation on the distal side, *J. Biol. Inorg. Chem.* **4**:64–72.

Chen, G.-X., and Asada, K., 1989, Ascorbate peroxidase in tea leaves: occurrence of two isozymes and the differences in their enzymic and molecular properties, *Plant Cell Physiol.* **30**:987–998.

Chen, G.-X., Sano, S., and Asada, K., 1992, The amino acid sequence of ascorbate peroxidase from tea has a high degree of homology to that of cytochrome *c* peroxidase from yeast, *Plant Cell Physiol.* **33**:109–116.

Conroy, C. W., Tyma, P., Daum, P. H., and Erman, J. E., 1978, Oxidation-reduction potential measurements of cytochrome *c* peroxidase and pH dependent spectral transitions in the ferrous enzyme, *Biochim. Biophys. Acta* **537**:62–69.

Corpas, F. J., and Trelease, R. N., 1998, Differential expression of ascorbate peroxidase and a putative molecular chaperone in the boundary membrane of differentiating cucumber seedling peroxisomes, *J. Plant Physiol.* **153**:332–338.

Dalton, D. A., 1991, Ascorbate peroxidase, in: *Peroxidases in Chemistry and Biology*, Vol. 2 (J. Everse, K. E. Everse, and M. B. Grisham, eds.), CRC Press, Boca Raton, pp. 139–154.

Dalton, D. A., Diaz del Castillo, L., Kahn, M. L., Joyner, S. L., and Chatfield, J. M., 1996, Heterologous expression and characterization of soybean cytosolic ascorbate peroxidase, *Arch. Biochem. Biophys.* **328**:1–8.

Dalton, D. A., Hanus, F. J., Russell, S. A., and Evans, H. J., 1987, Purification, properties and distribution of ascorbate peroxidase in legume root nodules, *Plant Physiol.* **83**:789–794.

Dalton, D. A., Langeberg, L., and Robbins, M., 1992, Purification and characterisation of mondehydroascorbate reductase from soybean root nodules, *Arch. Biochem. Biophys.* **292**:281–286.

De Gara, L., de Pinto, M. C., and Arrigoni, O., 1997, Ascorbate synthesis and ascorbate peroxidase activity during the early stage of wheat development, *Physiol. Plant.* **100**:894–900.

De Gara, L., de Pinton, M. C., Paciolla, C., Cappetti, V., and Arrigoni, O., 1996, Is ascorbate peroxidase only a scavenger of hydrogen peroxide? in: *Plant Peroxidases: Biochemistry and Physiology* (C. Obinger, U. Burner, R. Ebermann, C. Penel, and H. Greppin, eds.), University of Geneva, pp. 157–162.

DePillis, G. D., Sishta, B. P., Mauk, A. G., and Ortiz de Montellano, P. R., 1991, Small substrates and cytochrome *c* are oxidized at different sites of cytochrome *c* peroxidase, *J. Biol. Chem.* **266**:19334–19341.

Doyle, W. A., Blodig, W., Veitch, N. C., Piontek, K., and Smith, A. T., 1998, Two substrate binding sites in lignin peroxidase revealed by site-directed mutagenesis, *Biochemistry* **37**:15097–15105.

Doyle, W. A., and Smith, A. T., 1996, Expression of lignin peroxidase H8 in *Escherichia coli*: folding and activation of the recombinant enzyme with $Ca^{2+}$ and haem, *Biochem. J.* **315**:15–19.

Dunford, H. B., 1982, Peroxidases, *Adv. Inorg. Biochem.* **4**:41.

Dunford, H. B., 1991, Horseradish peroxidase: structure and kinetic properties, in: *Peroxidases in Chemistry and Biology*, Vol. 2 (J. Everse, K. E. Everse, and M. B. Grisham, eds.), CRC Press, Boca Raton, pp. 1–24.

Dunford, H. B., 1999, *Heme peroxidases*, John Wiley, Chichester.

Durner, J., and Klessig, D. F., 1995, Inhibition of ascorbate peroxidase by salicylic acid and 2,6-dichloroisonicotinic acid, two inducers of plant defense responses, *Proc. Natl. Acad. Sci. USA* **92**:11312–11316.

Elia, M. R., Borracino, G., and Dipierro, S., 1992, Soluble ascorbate peroxidase from potato tubers, *Plant Science* **85**:17–21.

Erman, J. E., 1998, Cytochrome *c* peroxidase: a model heme protein, *J. Biochem. Mol. Biol.* **31**:307–327.

Erman, J. E., Vitello, L. B., Mauro, J. M., and Kraut, J., 1989, Detection of an oxy-ferryl porphyrin π-cation radical in the reaction between hydrogen peroxide and a mutant yeast cytochrome *c* peroxidase. Evidence for tryptophan-191 involvement in the radical site of compound I, *Biochemistry* **28**:7992–7995.

Farhangrazi, Z. S., Copeland, B. R., Nakayama, T., Yamazaki, I., and Powers, L. S., 1994, Oxidation-reduction properties of compounds I and II of *Arthromyces ramosus* peroxidase, *Biochemistry* **33**:5647–5652.

Farhangrazi, Z. S., Fossett, M. E., Powers, L. S., and Ellis, W. R., 1995, Variable-temperature spectroelectrochemical study of horseradish peroxidase, *Biochemistry* **34**:2866–2871.

Finzel, B. C., Poulos, T. L., and Kraut, J., 1984, Crystal structure of cytochrome *c* peroxidase refined at 1.7 Å resolution, *J. Biol. Chem.* **259**:13027–13036.

Foyer, C. H., and Mullineaux, P. M., 1998, The presence of dehydroascorbate and dehydroascorbate reductase in plant tissues, *FEBS Lett.* **425**:528–529.

Frew, J. E., and Jones, P., 1984, Structure and functional properties of peroxidases and catalases, in: *Advances in Inorganic Chemistry* (A. Sykes, ed.), Academic Press, New York, pp. 175.

Gajhede, M., Schuller, D. J., Henricksen, A., Smith, A. T., and Poulos, T. L., 1997, Crystal structure of horseradish peroxidase C at 2.15 angstrom resolution, *Nature Struct. Biol.* **4**:1032–1038.

Gerbling, K.-P., Kelly, G. J., Fischer, K.-H., and Latzko, E., 1984, Partial purification and properties of soluble ascorbate peroxidases from pea leaves, *J. Plant Physiol.* **115**:59–67.

Gillham, D. J., and Dodge, A. D., 1986, Hydrogen-peroxide-scavenging systems within pea chloroplasts, *Planta* **167**:246–251.

Goodin, D. B., and McRee, D. E., 1993, The Asp-His-Fe triad of cytochrome *c* peroxidase controls the reduction potential, electronic-structure and coupling of the tryptophan free-radical to the heme, *Biochemistry* **32**:3313–3324.

Groden, D., and Beck, E., 1979, $H_2O_2$ destruction by ascorbate-dependent systems from chloroplasts, *Biochim. Biophys. Acta* **546**:426–435.

Harbury, H. A., 1957, Oxidation-reduction potentials of horseradish peroxidase, *J. Biol. Chem.* **225**:1009–1024.

Hayashi, Y., and Yamazaki, I., 1979, The oxidation-reduction potentials of compound I/II and compound II/ferric couples of horseradish peroxidase $A_2$ and C, *J. Biol. Chem.* **254**:9101–9106.

He, B., Sinclair, R., Copeland, B. R., Makino, R., and Powers, L. S., 1996, The structure/function relationship and reduction potentials of high oxidation states of myoglobin and peroxidase, *Biochemistry* **35**:2413–2420.

Hill, A. P., Modi, S., Sutcliffe, M. J., Turner, D. D., Gilfoyle, D. J., Smith, A. T., Tam, B. M., and Lloyd, E., 1997, Chemical, spectroscopic and structural investigation of the substrate-binding site in ascorbate peroxidase, *Eur. J. Biochem.* **248**:347–354.

Hossain, M. A., and Asada, K., 1984, Inactivation of ascorbate peroxidase in spinach chloroplasts on dark addition of hydrogen peroxide: its protection by ascorbate, *Plant Cell Physiol.* **25**:1285–1295.

Hossain, M. A., Nakano, Y., and Asada, K., 1984, Monodehydroascorbate reductase in spinach chloroplasts and its participation in regeneration of ascorbate for scavenging hydrogen peroxide, *Plant Cell Physiol.* **25**:385–395.

Huyett, J. E., Doan, P. E., Gurbiel, R., Houseman, A. L. P., Sivaraja, M., Goodin, D. B., and Hoffman, B. M., 1995, Compound ES of cytochrome $c$ peroxidase contains a Trp π-cation radical—characterisation by CW and pulsed Q-band ENDOR spectroscopy, *J. Am. Chem. Soc.* **117**:9033–9041.

Ishikawa, T., Sakai, K., Takeda, T., and Shigeoka, S., 1995, Cloning and expression of a cDNA encoding a new type of ascorbate peroxidase from spinach, *FEBS Lett.* **367**:28–32.

Ishikawa, T., Sakai, K., Yoshimura, K., Takeda, T., and Shigeoka, S., 1996a, cDNAs encoding spinach stromal and thylakod-bound ascorbate peroxidase, differing in the presence or absence of their 3′-coding regions, *FEBS Lett.* **384**:289–293.

Ishikawa, T., Takeda, T., Kohno, H., and Shigeoka, S., 1996b, Molecular characterisation of *Euglena* ascorbate peroxidase using monoclonal antibody, *Biochim. Biophys. Acta* **1290**:69–75.

Ishikawa, T., Takeda, T., and Shigeoka, S., 1996c, Purification and characterisation of cytosolic asorbate peroxidase from komatsuna (*Brassica rapa*), *Plant Science* **120**:11–18.

Ishikawa, T., Yoshimura, K., Sakai, K., Tamoi, M., Takeda, T., and Shigeoka, S., 1998, Molecular characterisation and physiological role of a glyoxysome-bound ascorbate peroxidase from spinach, *Plant Cell Physiol.* **39**:23–34.

Jablonski, P. P., and Anderson, J. W., 1982, Light-dependent reduction of hydrogen peroxide by ruptured pea chloroplasts, *Plant Physiol.* **69**:1407–1413.

Jensen, G. M., 1998, Energetics of cation radical formation at the proximal active site tryptophan of cytochrome $c$ peroxidase and ascorbate peroxidase, *J. Phys. Chem.* **102**:8221–8228.

Jespersen, H. M., Kjaersgard, I. V. H., Ostergaard, L., and Welinder, K. G., 1997, From sequence analysis of three novel ascorbate peroxidases from *Arabidopsis thaliana* to structure, function and evolution of seven types of ascorbate peroxidase, *Eur. J. Biochem.* **326**:305–310.

Jimenez, A., Hernandez, J. A., del Rio, L. A., and Sevilla, F., 1997, Evidence for the presence of the ascorbate-glutathione cycle in mitochondria and peroxisomes of pea leaves, *Plant Physiol.* **114**:275–284.

Jimenez, A., Jimenez, J. A., Barcelo, A. R., Sandalio, L. M., del Rio, L. A., and Sevilla, F., 1998, Mitochondrial and peroxisomal ascorbate peroxidase of pea leaves, *Physiol. Plant.* **104**:687–692.

Jones, D. K., Dalton, D. A., Rosell, F. I., and Lloyd Raven, E., 1998, Class I heme peroxidases: characterisation of soybean ascorbate peroxidase, *Arch. Biochem. Biophys.* **360**:173–178.

Kelly, G. J., and Latzko, E., 1979, Soluble ascorbate peroxidase, *Naturewissenschaften* **66**:617–618.

Kim, I. J., and Chung, W. I., 1998, Molecular characterisation of a cytosolic ascorbate peroxidase in strawberry fruit, *Plant Science* **133**:69–77.

Koshiba, T., 1993, Cytosolic ascorbate peroxidase in seedlings and leaves of maize (*Zea mays*), *Plant Cell Physiol.* **34**:713–721.

Kubo, A., Saji, H., Tanaka, K., Tanaka, K., and Kondo, N., 1992, Cloning and sequencing of a cDNA encoding ascorbate peroxidase from *Arabidopsis thaliana*, *Plant Mol. Biol.* **18**:691–701.

Kvaratskhelia, M., George, S. J., and Thorneley, R. N. F., 1997a, Salicylic acid is a reducing substrate and not an effective inhibitor of ascorbate peroxidase, *J. Biol. Chem.* **272**:20998–21001.

Kvaratskhelia, M., Winkel, C., Naldrett, M. T., and Thorneley, R. N. F., 1999, A novel high activity cationic ascorbate from tea (*Camellia sinensis*)—a class III peroxidase with unusual substrate specificity, *J. Plant Physiol.* **154**:273–282.

Kvaratskhelia, M., Winkel, C., and Thorneley, R. N. F., 1997b, Purification and characterisation of a novel class III peroxidase isozyme from tea leaves, *Plant Physiol.* **114**:1237–1245.

Lad, L., Basran, J., Scrutton, N. S., and Lloyd Raven, E., 1999, unpublished.

Lim, A. R., 1989, PhD dissertation, University of British Columbia.

Lopez, F., Vansuyt, G., CasseDelbart, F., and Fourcroy, P., 1996, Ascorbate peroxidase activity, not the mRNA level, is enhanced in salt-stressed *Raphanus sativus* plants, *Physiol. Plant.* **97**:13–20.

Mandelman, D., Jamal, J., and Poulos, T. L., 1998a, Identification of two-electron transfer site in ascorbate peroxidase using chemical modification, enzyme kinetics, and crystallography, *Biochemistry* **37**:17610–17617.

Mandelman, D., Schwarz, F. P., Li, H., and Poulos, T. L., 1998b, The role of quaternary interactions on the stability and activity of ascorbate peroxidase, *Protein Science* **7**:2089–2098.

Mano, S., Yamaguchi, K., Hayashi, M., and Nishimura, M., 1997, Stromal and thylakoid-bound ascorbate peroxidase are produced by alternative splicing in pumpkin, *FEBS Lett.* **413**:21–26.

Maranon, M. J. R., Stillman, M. J., and van Huystee, R. B., 1993, CD analysis of co-dependency of calcium and porphyrin for the integrate molecular structure of peanut peroxidase, *Biochem. Biophys. Res. Comm.* **194**:326–333.

Marquez, L. A., Quitoriano, M., Zilinskas, B. A., and Dunford, H. B., 1996, Kinetic and spectral properties of pea cytosolic ascorbate peroxidase, *FEBS Lett.* **389**:153–156.

Mathews, M. C., Summers, C. B., and Felton, G. W., 1997, Ascorbate peroxidase: a novel antioxidant enzyme in insects, *Arch. Insect Biochem. Physiol.* **34**:57–68.

Mauk, M. R., Kishi, K., Gold, M. H., and Mauk, A. G., 1998, pH-linked binding of Mn(II) to manganese peroxidase, *Biochemistry* **37**:6767–677.

Mauro, J. M., Fishel, L. A., Hazzard, J. T., Meyer, T. E., Tollin, G., Cusanovich, M. A., and Kraut, J., 1988, Tryptophan-191-phenylalanine, a proximal-side mutation in yeast cytochrome *c* peroxidase that strongly affects the kinetics of ferrocytochrome *c* oxidation, *Biochemistry* **27**:6243–6256.

Meneguzzo, S., Sgherri, C. L. M., Navari-Izzo, F., and Izzo, R., 1998, Stromal and thylakoid-bound ascorbate peroxidases in NaCl-treated wheat, *Physiol. Plant.* **104**:735–740.

Menyhard, D. K., and Naray-Szabo, G., 1999, Electrostatic effect on electron transfer at the active site of heme peroxidases: a comparative molecular orbital study on cytochrome *c* peroxidase and ascorbate peroxidase, *J. Phys. Chem.* **103**:227–233.

Millis, C. D., Cai, D., Stankovich, M. T., and Tien, M., 1989, Oxidation-reduction potentials and ionization states of extracellular peroxidases from the lignin-degrading fungus *Phanerochaete crysoporium*, *Biochemistry* **28**:8484–8489.

Mittler, R., and Zilinskas, B. A., 1991a, Molecular cloning and nucleotide sequence analysis of a cDNA encoding pea cytosolic ascorbate peroxidase, *FEBS Lett.* **289**:257–259.

Mittler, R., and Zilinskas, B. A., 1991b, Purification and characterisation of pea cytosolic ascorbate peroxidase, *Plant Physiol.* **97**:962–968.

Miyake, C., and Asada, K., 1996, Inactivation mechanism of ascorbate peroxidase at low concentrations of ascorbate: hydrogen peroxide decomposes Compound I of ascorbate peroxidase, *Plant Cell Physiol.* **37**:423–430.

Miyake, C., Cao, W.-H., and Asada, K., 1993, Purification and molecular properties of the thylakoid-bound ascorbate peroxidase in spinach chloroplasts, *Plant Cell Physiol.* **34**:3881–3889.

Miyake, C., Michihata, F., and Asada, K., 1991, Scavenging of hydrogen peroxide in prokaryotic and eukaryotic algae: acquisition of ascorbate peroxidase during the evolution of cyanobacteria, *Plant Cell Physiol.* **32**:33–43.

Mondal, M. S., Goodin, D. B., and Armstrong, F. A., 1998, Simultaneous voltammetric comparisons of reduction potentials, reactivities, and stabilities of the high-potential catalytic states of wild-type and distal-pocket mutant (W51F) yeast cytochrome c peroxidase, *J. Am. Chem. Soc.* **120**:6270–6276.

Mondal, M. S., Fuller, H. A., and Armstrong, F. A., 1996, Direct measurement of the reduction potential of catalytically active cytochrome c peroxidase compound I: voltammetric detection of a reversible, cooperative two-electron transfer reaction, *J. Am. Chem. Soc.* **118**:263–264

Morishima, I., Kurono, M., and Shiro, Y., 1986, Presence of endogenous calcium-ion in horseradish peroxidase—elucidation of metal-binding site by substitutions of divalent and lanthanide ions for calcium and use of metal-induced NMR $H^1$ and $Cd^{113}$ resonances, *J. Biol. Chem.* **261**:9391–9399.

Morita, S., Kaminaka, H., Yokoi, H., Masumara, T., and Tanaka, K., 1997, Differential responses of two cytosolic superoxide dismutase genes and two cytosolic ascorbate peroxidase genes in rice to environmental stresses, *Plant Physiol.* **114**:102.

Murthy, S. S., and Zilinskas, B. A., 1994, Molecular cloning and characterisation of a cDNA encoding pea monodehydroascorbate reductase, *J. Biol. Chem.* **269**:31129–31133.

Nakano, Y., and Asada, K., 1987, Purification of ascorbate peroxidase in spinach chloroplasts: its inactivation in ascorbate-depleted medium and reactivation by monodehydroascorbate radical, *Plant Cell Physiol.* **28**:131–140.

Naray-Szabo, G., 1997, Electrostatic modulation of electron transfer in the active site of heme peroxidases, *J. Biol. Inorg. Chem.* **2**:135–138.

Newman, T., De Bruijin, F. J., Green, P., Keegstra, K., Kende, H., McIntish, L., Ohlrogge, J., Raikhel, N., Somerville, S., Thomashow, M., Retzel, E., and Somerville, C., 1996, Genes galore: a summary of methods for accessing results from large-scale partial sequencing of anonymous *Arabidopsis* cDNA clones, *Plant Physiol.* **106**:1241–1255.

Nicholls, P., and Schonbaum, G. R., 1963, in: *The Enzymes*, Vol. 8 (P. D. Boyer, H. Lardy, and K. Myrbach, eds.), Academic Press, New York, pp. 147–225.

Nie, G., and Aust, S. D., 1997, Effect of calcium on the reversible thermal activation of lignin peroxidase, *Arch. Biochem. Biophys.* **337**:225–231.

Nissum, M., Neri, F., Mandelman, D., Poulos, T. L., and Smulevich, G., 1998, Spectroscopic characterisation of recombinant pea cytosolic ascorbate peroxidase: similarities and differences with cytochrome c peroxidase, *Biochemistry* **37**:8080–8087.

Ogawa, S., Shino, Y., and Morishima, I., 1979, Calcium binding by horseradish peroxidase C and the heme environmental structure, *Biochem. Biophys. Res. Comm.* **90**:674–678.

Ohya, T., Morimura, Y., Saji, H., Mihara, T., and Ikawa, T., 1997, Purification and characterisation of ascorbate peroxidase in roots of Japanese radish, *Plant Science* **125**:137–145.

Ortiz de Montellano, P. R., 1987, Control of the catalytic activity of prosthetic heme by the structure of hemoproteins, *Acc. Chem. Res.* **20**:289–294.

Orvar, B. L., and Ellis, B. E., 1995, Isolation of a cDNA encoding cytosolic ascorbate peroxidase from tobacco, *Plant Physiol.* **108**:839–840.

Pappa, H., Patterson, W. R., and Poulos, T. L., 1996, The homologous tryptophan critical for cytochrome c peroxidase function is not essential for ascorbate peroxidase activity, *J. Biol. Chem.* **1**:61–66.

Patterson, W. R., and Poulos, T. L., 1994, Characterization and crystallization of recombinant pea cytosolic ascorbate peroxidase, *J. Biol. Chem.* **269**:17020–17024.

Patterson, W. R., and Poulos, T. L., 1995, Crystal structure of recombinant pea cytosolic ascorbate peroxidase, *Biochemistry* **34**:4331–4341.

Patterson, W. R., Poulos, T. L., and Goodin, D. B., 1995, Identification of a porphyrin π-Cation radical in ascorbate peroxidase Compound I, *Biochemistry* **34**:4342–4345.

Poulos, T. L., Freer, S. T., Alden, R. A., Edwards, S. L., Skogland, U., Takio, K., Eriksson, B., Xuong, N., Yonetani, T., and Kraut, J., 1980, The crystal structure of cytochrome $c$ peroxidase, *J. Biol. Chem.* **255**:575–580.

Poulos, T. L., and Kraut, J., 1980, The stereochemistry of peroxidase catalysis, *J. Biol. Chem.* **255**:8199–8205.

Sano, S., Miyake, C., Mikami, B., and Asada, K., 1995, Molecular characterisation of monodehydroascorbate radical reductase from cucumber highly expressed in *Escherichia coli*, *J. Biol. Chem.* **270**:21354–21361.

Sano, S., Ueda, M., Kurano, N., Miyachi, S., and Yokota, A., 1996, Ascorbate peroxidase from a red alga, *Galdiera partita*, in: *Plant Peroxidases: Biochemistry and Physiology* (C. Obinger, U. Burner, R. Ebermann, C. Penel, and H. Greppin, eds.), University of Geneva, pp. 168–172.

Sauser, K. R., Liu, J. K., and Wong, T.-Y., 1997, Identification of a copper-sensitive ascorbate peroxidase in the unicellular green alga *Selenastrum capricornutum*, *BioMetals* **10**:163–168.

Schantz, M.-L., Schreiber, H., Guillemaut, P., and Schantz, R., 1995, Changes in ascorbate peroxidase activities during fruit ripening in *Capiscum annuum*, *FEBS Lett.* **358**:149.

Scholes, C. P., Liu, Y., Fishel, L. A., Farnum, M. F., Mauro, J. M., and Kraut, J., 1989, Recent ENDOR and pulsed electron paramagnetic resonance studies of cytochrome $c$ peroxidase Compound I and its site-directed mutants, *Israel J. Chem.* **29**:85–92.

Shigeoka, S., Nakano, Y., and Kitaoka, S., 1980, Purification and some properties of L-ascorbic acid-specific peroxidase in *Euglena gracilis* Z, *Arch. Biochem. Biophys.* **201**:121–127.

Shigeoka, S., Yasumoto, R., Onishi, T., Nakano, Y., and Kitaoka, S., 1987, Properties of monodehydroascorbate reductase and dehydroascorbate reductase and their participation in the regeneration of ascorbate in *Euglena gracilis*, *J. Gen. Microbiol.* **133**:227–232.

Sivaraja, M., Goodin, D. B., Smith, M., and Hoffman, B. M., 1989, Identification by ENDOR of Trp[191] as the free-radical site in cytochrome $c$ peroxidase Compound ES, *Science* **245**:738–740.

Smith, A. T., Santama, N., Dacey, S., Edwards, M., Bray, R. C., Thorneley, R. N. F., and Burke, J. F., 1990, Expression of a synthetic gene for horseradish peroxidase C in *Escherichia coli* and folding and activation of the recombinant enzyme with $Ca^{2+}$ and heme, *J. Biol. Chem.* **265**:13335–13343.

Smith, A. T., and Veitch, N. C., 1998, Substrate binding and catalysis in heme peroxidases, *Curr. Opin. Chem. Biol.* **2**:269–278.

Sutherland, G. R. J., and Aust, S. D., 1997, Thermodynamics of binding of the distal calcium to manganese peroxidase, *Biochemistry* **36**:8567–8573.

Sutherland, G. R. J., Zapanta, S., Tien, M., and Aust, S. D., 1997, Role of calcium in maintaining the heme environment of manganese peroxidase, *Biochemistry* **36**:3654–3662.

Takeda, T., Yoshimura, K., Ishikawa, T., and Shigeoka, S., 1998, Purification and characterisation of ascorbate peroxidase in *Chlorella vulgaris*, *Biochimie* **80**:295–301.

Tanaka, K., Takeuchi, E., Kubo, A., Sakaki, T., Haraguchi, K., and Kawamura, Y., 1991, Two immunologically different isozymes of ascorbate peroxidase from spinach leaves, *Arch. Biochem. Biophys.* **286**:371–375.

Tanaka, M., Ishimori, K., and Morishima, I., 1998, Structural roles of the highly conserved Glu residue in the heme distal site of peroxidases, *Biochemistry* **37**:2629–2638.

Tel-Or, E., Huflejt, M. E., and Packer, L., 1986, Hydroperoxide metabolism in cyanobacteria, *Arch. Biochem. Biophys.* **246**:396–402.

Turner, D. D., Andrew, C. R., Cheesman, M. R., Thomson, A. J., and Lloyd Raven, E., 1999, submitted.

Turner, D. D., and Lloyd Raven, E., 1999, unpublished.

Ushimara, T., Maki, Y., Sano, S., Koshiba, K., Asada, K., and Tsuji, H., 1997, Induction of enzymes involved in the ascorbate-dependent antioxidative systems, namely ascorbate peroxidase, monodehydroascorbate reductase and dehydroascorbate reductase, after exposure to air of rice (*Oryza sativa*) seedlings germinated under water, *Plant Cell Physiol.* **38**:541–549.

van Breusegem, F., Villarroel, R., van Montagu, M., and Inze, D., 1995, Ascorbate peroxidase cDNA from maize, *Plant Physiol.* **107**:649–650.

Wada, N., Kinoshita, S., Matsuo, M., Amako, K., Miyake, C., and Asada, K., 1998, Purification and molecular properties of ascorbate peroxidase from bovine eye, *Biochem. Biophys. Res. Comm.* **242**:256–261.

Webb, R. P., and Allen, R. D., 1995, Isolation and characterisation of a cDNA for spinach cytosolic ascorbate peroxidase, *Plant Physiol.* **108**:1325.

Welinder, K. G., 1992, Superfamily of plant, fungal and bacterial peroxidases, *Curr. Opin. Struct. Biol* **2**:388–393.

Wilcox, S. K., Jensen, G. M., Fitzgerald, M. M., McRee, D. E., and Goodin, D. B., 1996, Altering substrate specificity at the heme edge of cytochrome *c* peroxidase, *Biochemistry* **35**:4858–4866.

Wilcox, S. K., Putnam, C. D., Sastry, M., Blankenship, J., Chazin, W. J., McRee, D. E., and Goodin, D. B., 1998, Rational design of a functional metalloenzyme: introduction of a site for manganese binding and oxidation into a heme peroxidase, *Biochemistry* **37**:16853–16862.

Yamada, H., Makino, R., and Yamazaki, I., 1975, Effects of 2,4-substituents of deuteroheme upon redox potentials of horseradish peroxidases, *Arch. Biochem. Biophys.* **169**:344–353.

Yamaguchi, K., Hayashi, M., and Nishimura, M., 1996, cDNA cloning of thylakoid-bound ascorbate peroxidase in pumpkin and its characterisation, *Plant Cell Physiol.* **37**:405–409.

Yamaguchi, K., Hori, H., and Nishimura, M., 1995, A novel isoenzyme of ascorbate peroxidase localized on glyoxysomal and leaf peroxisomal membranes in pumpkin, *Plant Cell Physiol.* **36**:1157–1162.

Yonetani, T., and Ray, G. S., 1965, Studies on cytochrome *c* peroxidase. Purification and some properties, *J. Biol. Chem.* **240**:4503–4514.

Yoshimura, K., Ishikawa, T., Nakumura, Y., Tamoi, M., Takeda, T., Tada, T., Nishimura, K., and Shigeoka, S., 1998, Comparative study on recombinant chloroplastic and cytosolic ascorbate peroxidase isozymes of spinach, *Arch. Biochem. Biophys.* **353**:55–63.

Zhang, H., Wang, J., Nickel, U., Allen, R. D., and Goodman, H. M., 1997, Cloning and expression of an *Arabidopsis* gene encoding a putative peroxisomal ascorbate peroxidase, *Plant Mol. Biol.* **34**:967–971.

*Chapter 11*

# Adenosylcobalamin-Dependent Enzymes

E. Neil G. Marsh and Daniel E. Holloway

## 1. INTRODUCTION

This chapter describes recent developments in our understanding of adenosylcobalamin[1] (AdoCbl, coenzyme $B_{12}$)-dependent enzymes. We will focus on the ten $B_{12}$ enzymes known that catalyze unusual rearrangement reactions. These reactions involve the interchange of an electron-withdrawing group, X, on one carbon with a hydrogen atom on an adjacent carbon, as shown in Figure 1. This review will place greatest emphasis on those enzymes that catalyze carbon skeleton rearrangements, for which a significant amount of new structural and mechanistic information is now emerging. The eleventh AdoCbl-dependent enzyme, class II ribonucleotide reductase, will be discussed only in relation to the function of the AdoCbl coenzyme, which serves the same basic role as in the

---

[1] The abbreviations used are: AdoCbl, adenosylcobalamin; CNCbl, cyanocobalamin; Cbl(II) cob(II)alamin; $AdoCH_3$, 5′-deoxyadenosine; $AdoCH_2^{\bullet}$, 5′-deoxyadenosyl radical.

---

**E. NEIL G. MARSH**    Department of Chemistry, University of Michigan, Ann Arbor MI 48109, United States of America.    **DANIEL E. HOLLOWAY**    Department of Biology and Biochemistry, University of Bath, Claverton Down, Bath BA2 7AY, United Kingdom.

*Subcellular Biochemistry, Volume 35: Enzyme-Catalyzed Electron and Radical Transfer*, edited by Holzenburg and Scrutton. Kluwer Academic / Plenum Publishers, New York, 2000.

**FIGURE 1.** The general rearrangement catalyzed by adenosylcobalamin-dependent enzymes.

isomerases. Ribonucleotide reductase enzymes are reviewed more com-
prehensively elsewhere in this volume.

## 1.1.   Structure and Reactivity of Cobalamins

Few biological cofactors are as large or as structurally complex
as cobalamins. Vitamin $B_{12}$ (cyanocobalamin), originally identified as the
anti-pernicious anemia factor present in liver, was the first cobalamin to
be isolated and purified (Rickes *et al.*, 1948; West, 1948). However, even
when the complete three-dimensional structure was determined by X-ray
crystallography (Hodgkin *et al.*, 1955), its biochemical function was not
obvious. A major clue was provided by H. Albert Barker and associates who
discovered an unusual isomerisation of L-glutamate to L-*threo*-3-methy-
laspartate that occurs during fermentation of glutamate by certain anaerobic bacteria (Barker *et al.*, 1958). The reaction required a cofactor which
was reminiscent of, but chemically distinct from the $B_{12}$ vitamin previously
described. Its structure was determined, again by X ray-crystallography
(Lenhert and Hodgkin, 1961), providing the first view of AdoCbl or coenzyme $B_{12}$ as it is also known (Figure 2).

AdoCbl is one of two biologically active $B_{12}$ cofactors known, the
other being methylcobalamin (MeCbl). Common to both is the corrin ring
(IUPAC-IUB Commission on Biochemical Nomenclature, 1974), a par-
tially-reduced tetrapyrrole which differs from the closely-related porphyrin
macrocycle in that the A and D rings are directly connected, giving a characteristic puckered conformation (Gruber *et al.*, 1998). The corrin ring is
liberally functionalized around its periphery: acetamide side chains project
"upwards" from the β-face, while propionamide side chains project "downwards" from the α-face. Under most conditions, the central cobalt atom is
hexacoordinate. In both $B_{12}$ cofactors, the four equatorial ligands are
pyrrole nitrogens and the "lower" (Co$_\alpha$-) axial ligand is the N-3 atom of
5,6-dimethylbenzimidazole. This nitrogenous base forms part of a pseudo-
ribonucleotide loop, the other end of which is esterified to the ring periphery at C-17. In AdoCbl, the remaining (Co$_\beta$-) coordination position is
occupied by a 5'-deoxyadenosine moiety bonded to the cobalt through the
5'-carbon of the nucleoside; in MeCbl, the Co$_\beta$-substituent is a methyl
group. It is the unusual properties of these Co—C bonds that give the

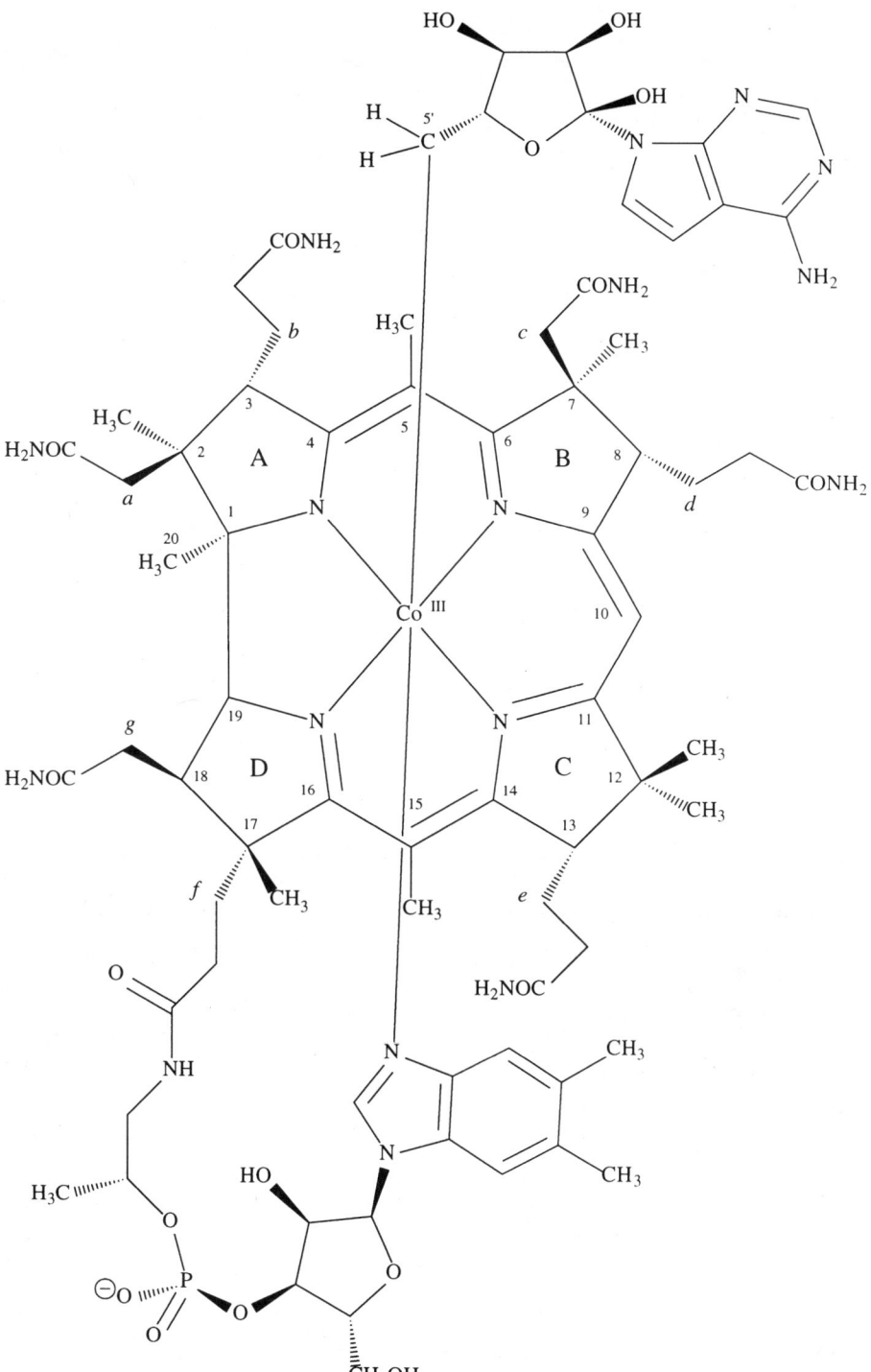

**FIGURE 2.** Structure of adenosylcobalamin (coenzyme $B_{12}$). Carbons 1–20 of the corrin ring are labeled, as are the four pyrroles (A–D) and the corrin side chains (*a–g*). The α-face lies below the plane of the corrin ring and the β-face above.

cofactor its unique chemical reactivity. The Co—C bond is essentially covalent in character (Wirt *et al.*, 1991), yet owing to the variety of oxidation states that cobalt can assume, its cleavage may yield $(Co^I + R^+)$, $(Co^{II} + R^{\bullet})$, or $(Co^{III} + R^-)$ (Pratt, 1972). Such metal-carbon bonds are particularly rare in Nature and it is the making and breaking of this bond which is central to all known reactions catalyzed by $B_{12}$-dependent enzymes. Although structurally similar, MeCbl and AdoCbl catalyze strikingly different types of chemical reactions.

## 1.2. Methylcobalamin Dependent Enzymes

MeCbl serves as the intermediate methyl carrier in a number of methyl transfer reactions. Cobalamin-dependent methyltransferases are distributed among prokaryotes and eukaryotes but mainly are restricted to bacteria which possess a distinct $C_1$ metabolism, where they participate in both catabolic and anabolic pathways (Stupperich, 1993). Of note are the prokaryotic methionine biosynthesis and the eukaryotic methionine salvage pathways which employ a cobalamin-dependent methionine synthase (EC 2.1.1.13), that catalyzes the transfer of a methyl group from methyltetrahydrofolate to homocysteine (Figure 3). The enzyme from *E. coli* is by far the most extensively studied (Drennan *et al.*, 1998), although the enzyme is also present in mammals. The resting form of the enzyme contains MeCbl at the active site; during catalysis, transfer of the methyl group to homocysteine generates the highly nucleophilic $Co^I$ species which accepts a methyl group from methyltetrahydrofolate in a second half reaction (Banerjee *et al.*, 1990). One of the consequences of human cobalamin deficiency is decreased activity of methionine synthase, and it is the resulting changes in cellular tetrahydrofolate pools which appear to be the

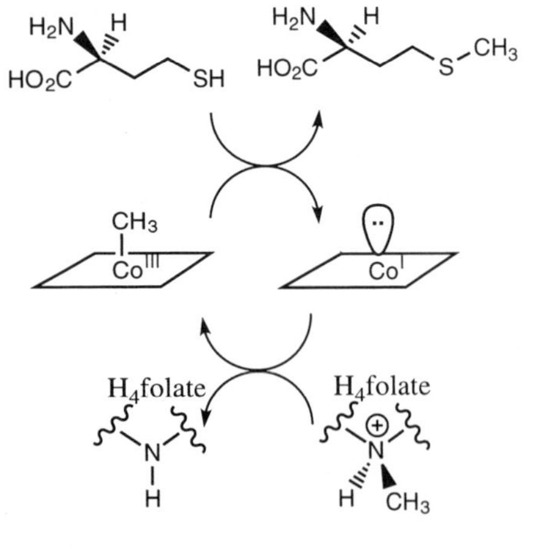

FIGURE 3. The synthesis of methionine from methyltetrahydrofolate and homocysteine catalyzed by methionine synthase. $H_4$-folate: tetrahydrofolate.

primary causes of the megaloblastic anemia associated with either cobalamin or folate deficiency.

## 1.3.  Adenosylcobalamin-Dependent Enzymes

AdoCbl functions as a masked form of carbon-centered free radicals that are generated by homolysis of the Co—C bond (Halpern, 1985). The study of AdoCbl-dependent enzymes and complementary model compounds has placed these enzyme reactions among an expanding group of remarkable biological reactions for which protein-bound organic radicals are not only key intermediates but true catalysts (Marsh, 1995; Retey, 1990; Stubbe, 1988). AdoCbl mediates a number of unusual and chemically difficult reactions; the rates of the corresponding uncatalyzed reactions are close to zero and hence the rate enhancement provided by these enzymes is virtually infinite.

Eleven adenosylcobalamin-dependent enzymes have been documented (Burkhardt et al., 1998; Golding and Rao, 1987; Babior and Krouwer, 1979). Ten of these catalyze rearrangement reactions (Figure 4) that involve the interchange of hydrogen on one carbon with an electron-withdrawing group on an adjacent carbon, so that a 1,2-migration is effected. These migrations can be conveniently divided into three groups. Firstly, there are those that catalyze the migration of hydroxyl or primary amine groups in vicinal diols or amino alcohols to yield aldehydes after subsequent dehydration or deamination. There are three known examples, diol dehydrase (EC 4.2.1.28), glycerol dehydrase (EC 4.2.1.30) (Toraya, 1994), and ethanolamine ammonia-lyase (EC 4.3.1.7) (Babior, 1982). Secondly, there are the aminomutases which catalyze the 1,2-migration of a primary amine group within an amino acid. There are also three known examples, D-$\alpha$-lysine (L-$\beta$-lysine) 5,6-aminomutase (EC 5.4.3.3/EC 5.4.3.4), D-ornithine 4,5-aminomutase (EC 5.4.3.5) and L-leucine 2,3-aminomutase (EC 5.4.3.6) (Baker and Stadtman, 1982). Ethanolamine ammonia-lyase could perhaps be placed in this group since it catalyzes the migration of a primary amine group, but it seems to have much in common with the dehydrase reactions. Thirdly, and perhaps most interestingly, there are those which catalyze a carbon-skeleton arrangement. The four known examples are glutamate mutase (EC 5.4.99.1) (Marsh et al., 1998; Switzer, 1982), methylmalonyl-CoA mutase (EC 5.4.99.2) (Francalanci et al., 1986), 2-methyleneglutarate mutase (EC 5.4.99.4) (Chemaly, 1994), and the more recently discovered isobutyryl-CoA mutase (EC 5.4.99.13) (Burkhardt et al., 1998).

The 1,2-rearrangement of a substrate molecule is now recognized as a common feature of several bacterial fermentation pathways. In general, the rearrangement enables the substrate to be readily assimilated into

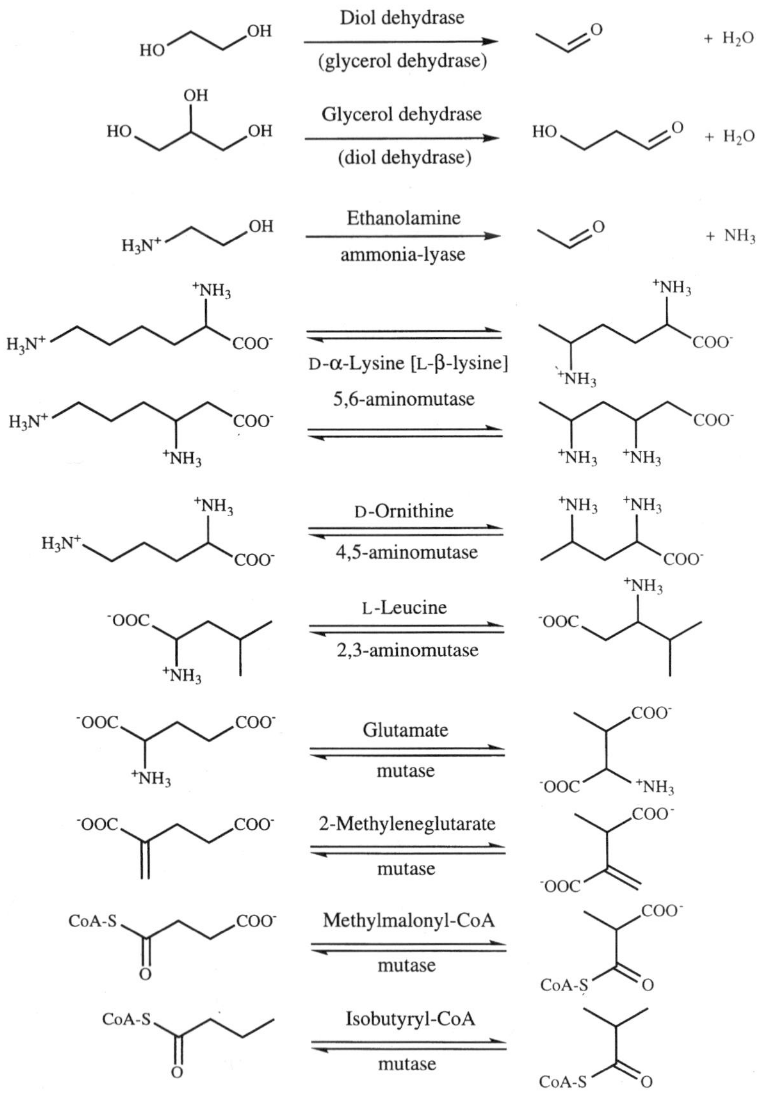

**FIGURE 4.** The eleven adenosylcobalamin-dependent rearrangements so far described. Notes: D-α-lysine [L-β-lysine] 5,6-aminomutase catalyzes similar rearrangements on two substrates; diol dehydrase and glycerol dehydrase have over-lapping substrate specificity.

common metabolic pathways. Mammals contain one such enzyme, methyl-malonyl-CoA mutase, which generates succinyl-CoA in the pathway by which odd-chain and branch-chain fatty acids are converted to Krebs Cycle intermediates (Rosenberg and Fenton, 1989). Notably, no cobalamin-independent enzymes have yet been discovered that catalyze analogous carbon skeleton rearrangements.

## 1.4.  Mechanism of Adenosylcobalamin-Dependent Rearrangements

Early studies using isotopically-labeled substrates and cofactors demonstrated the transfer of the migrating hydrogen atom from the 5′-carbon of AdoCbl to both the substrate and product in several AdoCbl-dependent isomerisations (Frey *et al.*, 1967; Retey, 1966). With those that have been tested, it has also been shown that the migrating hydrogen is "scrambled" (i.e. becomes equivalent) with the two methylene hydrogens at the 5′-position of AdoCbl (Miller and Richards, 1969; Frey *et al.*, 1967). The only reasonable way for this to occur is through formation of an intermediate methyl group at the 5′-carbon, implying that the Co—C bond is cleaved during catalysis. The radical nature of the reaction has been demonstrated by EPR spectroscopy of a number AdoCbl-dependent enzymes. Signals dependent on the presence of substrate or suitable analogue have been detected (Leutbecher *et al.*, 1992; Zhao *et al.*, 1992; Valinsky *et al.*, 1974; Babior *et al.*, 1974), indicating that these enzymes induce homolytic fission of the $Co_\beta$-alkyl bond to give cob(II)alamin and the 5′-deoxyadenosyl radical. Based on these observations the minimal mechanistic scheme shown in Figure 5 was proposed (Halpern, 1985) in which the rearrangement is envisaged to occur by the following sequence of events:

(i) Enzyme-induced homolytic cleavage of the AdoCbl Co—C bond to generate cob(II)alamin and a 5′-deoxyadenosyl radical ($AdoCH_2^{\bullet}$).

(ii) Abstraction of a hydrogen atom from the substrate by $AdoCH_2^{\bullet}$ to generate a substrate radical and 5′-deoxyadenosine ($AdoCH_3$).

(iii) Rearrangement of the substrate radical to the corresponding product radical by way of a 1,2-migration of group X.

(iv) Abstraction of a hydrogen atom back from $AdoCH_3$ by the product radical to complete the rearrangement.

A shortcoming of this mechanism is that it provides no explicit role for the protein, nor does it address the issue of how the substrate and product radicals interconvert (this is discussed in more detail in Section 3.2). Later on (Section 3.1) we shall see that even the participation of $AdoCH_2^{\bullet}$ in

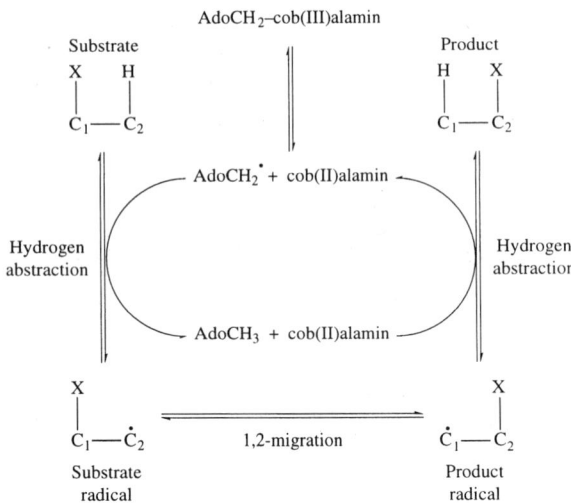

**FIGURE 5.** A minimal scheme for the mechanism of AdoCbl-dependent rearrangements. Adapted from Halpern (1985).

the mechanism may need to be reconsidered in the light of recent experimental findings.

## 1.5.  Ribonucleotide Reductase

The remaining AdoCbl-dependent enzyme is the class II ribonucleotide reductase (EC 1.17.4.2) of the type isolated from the bacterium *Lactobacillus leichmannii* (Lawrence and Stubbe, 1998). This ubiquitous enzyme plays a central role in DNA biosynthesis, catalyzing the reduction of ribonucleotides to give 2′-deoxyribonucleotides (Figure 6). Reduction occurs through the intermediacy of protein thiols that are oxidized concomitant with substrate reduction (Jordan and Reichard, 1998). Reducing equivalents are first transferred from NADPH to oxidized thioredoxin by thioredoxin reductase, and then on to the thiol groups of ribonucleotide reductase. The role of AdoCbl in the reduction process is to initiate the formation of a *protein* thiyl radical. The thiyl radical activates the substrate towards reduction by abstracting the 3′-hydrogen atom of the ribonucleotide (Licht *et al.*, 1996) which serves to labilize the 2′-hydroxyl towards leaving. Although no transfer of hydrogen (tritium) between substrate and cofactor occurs during turnover, it was established early on that the enzyme catalyzes exchange of $^3$H between the 5′-position of AdoCbl and solvent (Hogenkamp *et al.*, 1968; Abeles and Beck, 1967), implying that Co—C bond cleavage occurs during catalysis. This has proved a useful reaction with

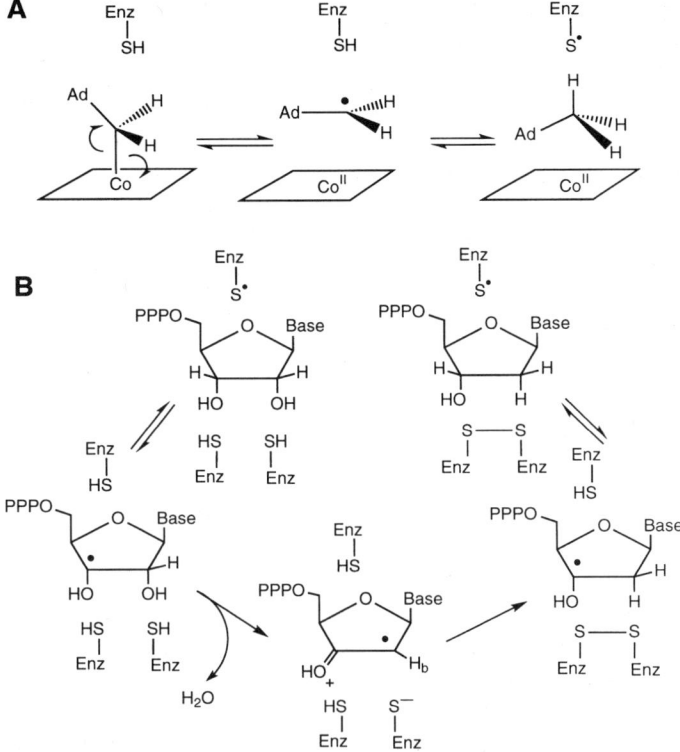

**FIGURE 6.** The role of free radicals in the reduction of ribonucleotides to 2'-deoxyribonu-cleotides catalyzed by AdoCbl-dependent ribonucleotide triphosphate reductase.

which to examine the mechanism of AdoCbl homolysis (Section 3.2). Interestingly, AdoCbl is just one of several radical-generating cofactors that are used by various ribonucleotide reductases (Reichard, 1993; Marsh, 1995) from different organisms; the AdoCbl-dependent class are almost exclusively limited to prokaryotes. Although the radical-generating system differs between enzymes, the mechanism of ribonucleotide reduction appears invariant and always involves free radicals.

## 1.6.  Carbon-Centered Radicals in Enzyme Reactions

Free radicals provide the only feasible means of activating certain key C—C, C—O or C—N bonds in otherwise inert substrate molecules; this is achieved by abstraction of a hydrogen atom from an *unactivated* carbon (Frey, 1990). The resulting unpaired electron leaves the carbon atom highly reactive and amenable to various further reactions. In the case of AdoCbl, the homolytic cleavage of a relatively weak organocobalt bond provides, in a most elegant manner, a source of the highly reactive AdoCH$_2^{\bullet}$ radical. From a practical point of view, AdoCbl is admirable in that it is conveniently soluble in aqueous solution and kinetically stable within the cell. Furthermore, within the active site of an enzyme, homolysis is entirely reversible and relatively independent of oxygen, an avid radical scavenger.

Adenosylcobalamin is not the only source of AdoCH$_2^{\bullet}$ in Nature. The enzymes lysine 2,3-aminomutase from *Clostridium subterminale* (EC 5.4.3.2) (Frey, 1993), pyruvate formate-lyase from *Escherichia coli* (EC 2.3.1.54) (Knappe *et al.*, 1993), and the (class III) anaerobic ribonucleotide reductase from *Escherichia coli* (Reichard, 1993) all rely on the cleavage of *S*-adenosylmethionine to provide AdoCH$_2^{\bullet}$. All three enzymes are acutely sensitive to oxygen, and each requires an activation step which is not readily reversible. As a consequence of these shortcomings, adenosylmethionine has been likened to a "poor man's adenosylcobalamin" (Frey, 1993). Indeed, it is likely that *S*-adenosylmethionine-dependent radical systems reflect an early point in evolutionary history when only primitive AdoCH$_2^{\bullet}$ cofactors were present on Earth; due to their oxygen sensitivity, the advent of photosynthesis would have severely limited the breadth of their use. At the other extreme, (class I) iron-tyrosyl radical ribonucleotide reductases have a radical-generating system whose assembly is absolutely *dependent* upon oxygen (Stubbe and RiggsGelasco, 1998). An entire subunit of this enzyme is devoted to maintenance of a stable radical which resides on a deeply-buried tyrosine residue. The oxygen-dependence of radical initiation establishes this mechanism as relatively young in evolutionary terms, and demonstrates that radical chemistry is important enough for Nature to have developed new solutions to the radical generation problem over the course of time.

## 1.7. Long-Standing Questions

Having presented AdoCbl as an elegant initiator of radical catalysis, it must be stressed that the cofactor does not operate without considerable assistance from its protein counterpart, i.e. the apoenzyme. The homolytic bond dissociation energy of the Co—C bond has been estimated to be 30 kcal/mol (Halpern *et al.*, 1984; Finke and Hay, 1984). Although weak compared with typical covalent bonds in organic molecules, $t_{1/2}$ for thermally-induced homolysis is a matter of years at room temperature. The Co—C bond is clearly kinetically very stable, yet turnover rate ($k_{cat}$) values as high as 370 s$^{-1}$ (i.e. $t_{1/2} = 2.5$ ms) have been reported for diol dehydrase (Toraya *et al.*, 1976). If Co—C bond cleavage occurs once every turnover, this represents acceleration of bond cleavage by the massive factor of $10^{12}$;

in other cases, the rate enhancement would be at least $10^{10}$. A major task for the protein must therefore be to displace the Co—C bond cleavage equilibrium accordingly towards cleavage.

Determining precisely how the enzyme accelerates Co—C bond cleavage is a long-standing goal of those studying AdoCbl-dependent catalysis. It has been thought that the enzyme may use the intrinsic AdoCbl-protein binding energy to trigger homolysis (Hay and Finke, 1987). This hypothesis can now be examined in detail due to the recent determination of several high-resolution holoenzyme structures. In addition, the parallel study of cobalamin-dependent methyltransferases offers the means to identify how homolysis can be promoted in preference to heterolysis and *vice-versa*. Three-dimensional structures have also shed light on how AdoCbl-dependent enzymes are able to stabilize and protect highly-reactive radical intermediates during the catalytic cycle. Finally, a host of spectroscopic and kinetic techniques have been combined with recombinant DNA technology to probe the mechanism of catalysis in ever greater detail. The following sections describe significant developments in these fields.

## 2. STRUCTURAL FEATURES OF ADENOSYLCOBALAMIN-DEPENDENT ENZYMES

The genes encoding a number of cobalamin-dependent enzymes have now been cloned and sequenced. Although differing considerably in molecular mass and quaternary structure, these enzymes can be placed in two groups, which as discussed below, reflect differences in the way they bind cobalamins (summarized in Table 1). The first group exhibits weak but clear sequence homology (Figure 7) that is characterized by the "**D-x-H-x-x-G**" sequence motif. Among the homologous sequences are representatives from both the AdoCbl-dependent carbon-skeleton mutases and the MeCbl-dependent methyltransferases (Marsh and Holloway, 1992). However, the second group that includes enzymes such as ethanolamine ammonia-lyase from *Salmonella typhimurium* (Faust *et al.*, 1990), diol dehydrases and glycerol dehydrases from various sources (Tobimatsu *et al.*, 1998) and AdoCbl-dependent ribonucleotide reductase (Booker and Stubbe, 1993) do not contain this motif and (apart from diol dehydrases and glycerol dehydrases) bear no obvious similarity to each other. Recently, X-ray structures have been obtained for representatives from both groups of AdoCbl-dependent enzymes (Shibata *et al.*, 1999; Reitzer *et al.*, 1999; Mancia *et al.*, 1996) that reveal both remarkable similarities in the global folds of each enzyme and striking differences in the way that the coenzyme is bound.

**Table 1**
**Corrinoid-dependent enzymes for which the cofactor-binding mode is known.**
**Data compiled from: A. Padmakumar et al., 1995; Mancia et al., 1996. B. Zelder**
**et al., 1995; Reitzer et al., 1999. C. Drennan et al., 1994a. D. Sauer and Thauer,**
**1998b. E. Sauer and Thauer, 1998a. F. Abend et al., 1998; Yamanishi et al., 1998.**
**G. Lawrence et al., 1999. H. Ragsdale et al., 1987; Jablonski et al., 1993; Grahame**
**and DeMoll, 1996**

|  | $Co_\alpha$ coordination state | | |
|---|---|---|---|
|  | Base-off / His-on | Base-on | Base-off |
| DxHxxG motif present | Methylmalonyl-CoA mutase[A] Glutamate mutase[B] Methionine synthase[C] Methanol:coenzyme M methyltransferase[D] | | |
| DxHxxG motif absent | $N^5$-methyl-$H_4$-methanopterin: coenzyme M methyltransferase[E] | Diol dehydrase[F] Ribonucleotide reductase[G] | Corrinoid/Fe-S protein of CO dehydrogenase[H] |

## 2.1. Cobalamin Binding by Enzymes Containing the D-x-H-x-x-G Motif

This group of enzymes share a conserved domain comprising about 100 amino acid residues that contains the following invariant "finger print" residues, although there are several additional positions where the side-chain character is conserved (Figure 7).

$$D\text{-}X\text{-}H\text{-}X\text{-}X\text{-}G\text{-}X_{41-42}\text{-}S/T\text{-}X\text{-}L\text{-}X_{24-28}\text{-}G\text{-}G$$

In the case of the MeCbl-dependent methyltransferases, physical techniques have established that the aligned region falls within the corrinoid-binding subunit or domain of the protein (Banerjee et al., 1989; Sauer et al., 1997; Kaufmann et al., 1998), suggesting that the coenzyme binding process could be fundamentally similar in all these enzymes. This domain encompasses most of the small (15 kDa) S subunit of glutamate mutase but constitutes a smaller fraction of the other proteins. Thus, the aligned portions of 2-methyleneglutarate mutase, the methylmalonyl-CoA mutases and the selected methyltransferases of M. barkeri and A. dehalogenans are located at the C-termini of subunits ranging from 21 to 80 kDa in mass, whereas in methionine synthases the conserved region falls in the middle of a particularly long (136 kDa) polypeptide chain. Since the homologous sequences include representatives from archaebacteria, eubacteria, invertebrates and

```
Cte-MutS   MEKKTIVLGVGIGSDCHAVGNKILDHSFTNAGENVVNIGVLSSQEDFINAA      50
Cco-GlmS   MEKKTIVLGVGIGSDCHAVGNKILDHAFTNAGENVVNIGVLSPQEVFIKAA      50
Cba-Mgm    TRPEKIVLATVGADAHVNGINVIREAFQDAGKDVVYLRGMNLPESVAEVA      519
Psh-MutB   GRRPRILLAKMGQDGHDRGQKVIATAYADLGEDVDVGPLFQTPEETARQA      644
Cel-MutA   GRQPRIVVAKMGQDGHDRGAKVIATGFADLGEDVDVGPLFQTPLEAAQQA      652
Hsa-MutA   GRRPRLVVAKMGQDGHDRGAKVIATGFADLGEDVDIGPLFQTPREVAQQA      661
Hsa-MTR    PYQGTIVLATVKGDVHDIGKNIVGVVLGCNNERVIDLGVMTPCDKILKAA      819
Cel-MTR    PYQGTVVIATVKGDVHDIGKNIVSVVLGCNNEKVVDLGVMTPCENIIKAA      809
Eco-MetH   KTNGKMVIATVKGDVHDIGKNIVGVVLQCNNYELVDLGVMVPAEKILRTA      793
Ade-OdmA   ASLGTCVIGTVAGDLHDIGKNLVSMMIESAGEDMVDLGVDVPADTFVQAV      133
Mba-MtaC   KTKGTVVCHVAEGDVHDIGKNIVTALLRANGMNVVDLGRDVPAEEVLAAV      170
Mba-MtsB   PSQGKVVSLVIVGDLHDIGKNIVAAILRANGEEVIDLGRDVTVEAAVEAV      192


Cte-MutS   IE-TKADLICVSSLYGQGEIDCKGLREKCDEAGL-KGIKLFVGGNIVVGK      98
Cco-GlmS   IE-TKADALLSSLYGQGEIDCKGLRQKCDEAGL-EGILLYVGGNIVVGK      98
Cba-Mgm    AE-VGADAVGVSNLLGLGMELFPRVSKRLEELGLRDKWVVCAGGRIAEKE      568
Psh-MutB   VE-ADVHVVGVSSLAGGHLTLVPALRKELDKLGR-PDILITVGG--VIPE      690
Cel-MutA   VD-ADVHVIGASSLAAGHLTLIPQLIGELKKLGR-PDILVVAGG--VIPP      698
Hsa-MutA   VD-ADVHAVGVSTLAAGHKTLVPELIKELNSLGR-PDILVMCGG--VIPP      707
Hsa-MTR    LD-HKADIIGLSGLITPSLDEMIFVAKEMERLAI--RIPLLIGGATTSKT      866
Cel-MTR    IE-EKADFIGLSGLITPSLDEMVYVAKEMNRVGL--NIPLLIGGATTSKT      856
Eco-MetH   KE-VNADLIGLSGLITPSLDEMVNVAKEMERQGF--TIPLLIGGATTSKA      840
Ade-OdmA   KDNTNVKLVACSGLLTTTMPALKEAVQTIKAAYP--DIKVIVGGAPVTPE      181
Mba-MtaC   QK-EKPIMITGTALMTTTMYAFKEVNDMLLENGI--KIPFACGGAVNQD      217
Mba-MtsB   KS-TKANLVTGTTLMSTTKGGLKALANALEPE----GVPLACGGAAVDRR      237


Cte-MutS   QNWPDVEQRFKA-----MGFDRVYPPGTSPETTIADMKEVLGVE*-----    137
Cco-GlmS   QHWPDVEKRFKD-----MGYDRVYAPGTPPEVGIADLKKDLNIE*-----    137
Cba-Mgm    EEHRQFEEKIQKEGSAFMGMDGFFGPGSSPEDCVKIIGDMINAKKAMMSD    618
Psh-MutB   QDFDELRK--------DGAVEIYTPGTVIPESAISLVVKKLRASLDA*--    728
Cel-MutA   QDYKELYD--------AGVALVFGPGTRLPACANQILEKLEANLPEAPG    739
Hsa-MutA   QDYEFLFE--------VGVSNVFGPGTRIPKAAVQVLDDIEKCLEKKQQ    748
Hsa-MTR    HTAVKIAPRYS------APVIHVLDASKSVVVCSQLLDENLKDEYFEEIM    910
Cel-MTR    HTAVKISPRYP------HPVVHCLDASKSVVVCSSLSDMSVRDAFLQDLN    900
Eco-MetH   HTAVKIEQNYS------GPTVYVQNASRTVGVVAALLSDTQRDDFVARTR    884
Ade-OdmA   YAAEVGADG----------YAPDAG--SAAVKARELATA*----------    208
Mba-MtaC   FVSQFALGVYG------EEAADAPKIADAIIAGTTDVTEL-REKFHKH*-    258
Mba-MtsB   FVDTFGNSVYG------RTPLDAVKIAKEICEGKSWEEAR-NELY*----    275
```

**FIGURE 7.** Sequence similarities between the conserved cobalamin-binding domains of various MeCbl- and AdoCbl-dependent enzymes. Invariant residues are outlined in black; strongly conserved residues in gray. Cte-MutS, *Clostridium tetanomorphum* glutamate mutase S subunit; Cco-GlmS, *Clostridium cochlearium* glutamate mutase S subunit; Cba-Mgm, *Clostridium barkerii* 2-methyleneglutarate mutase; Psh-MutB, *Propionibacterium shermanii* methylmalonyl-CoA mutase α-subunit; *C. elegans* methylmalonyl-CoA mutase α-subunit; *Homo sapiens* methylmalonyl-CoA mutase α-subunit; Hsa-TR, *Homo sapiens* methyl-tetrahydrofolate reductase; Cel-MTR, *C. elegans* methyl-tetrahydrofolate reductase; Eco-MetH, *E. coli* methionine synthase; Ade-OdmA *Acetobacterium dehalogenans* vanillate specific O-demethylase corrinoid protein; Mba-MtaC, *Methanosarcina barkeri* methanol:coenzyme M methyltransferase; Mba-MtsB, *Methanosarcina barkeri* methylthiol:coenzyme M methyltransferase 30 kDa subunit.

mammals, this transposable domain appears to have been conserved over much of the course of evolution.

High-resolution structural data have revealed that the finger print residues are key components of a conserved corrinoid-binding domain. In the crystal structures of *P. shermanii* methylmalonyl-CoA mutase (Mancia and Evans, 1996; Mancia *et al.*, 1998), *C. cochlearium* glutamate mutase

(Reitzer *et al.*, 1999) and a MeCbl-binding fragment of *E. coli* methionine synthase (Drennan *et al.*, 1994), the cofactor is sandwiched between two domains (Figure 8). The conserved domain possesses an $\alpha/\beta$ structure reminiscent of the Rossmann fold of nucleotide-binding proteins (Rossmann *et al.*, 1974) and consists of a twisted $\beta$-sheet of five parallel strands encased by five $\alpha$-helices. It binds the lower, $\alpha$-face of the corrin macrocycle and the substituents projecting "down" from this face, notably the dimethylbenzimidazole ribofuranosyl nucleotide loop.

It is clear that remarkable changes in both cofactor and protein structure take place when cobalamin binds to each of these enzymes. A comparison of the crystal structure of free AdoCbl with that of AdoCbl bound in the methylmalonyl-CoA mutase holoenzyme permits an assessment of enzyme-induced changes in cofactor conformation (Figure 9). The bacterial enzyme is a heterodimer of 150 kDa, but only the 80 kDa $\alpha$-chain forms an active complex with cobalamin. Upon binding to this subunit, cobalamin undergoes a major conformational change whereby the dimethylbenzimidazole moiety dissociates from the cobalt and becomes buried in an elongated hydrophobic pocket deep in the Rossmann domain. Extensive intermolecular contacts are made in this region, with strands $\beta3$ and $\beta4$ forming one side of the nucleotide-binding pocket and helices $\alpha1$ and $\alpha5$ forming the other. The conserved serine (Ser655) and glycines (Gly613, Gly685 and Gly686) are among those residues lining the pocket, all of which contribute substantially to the total binding surface and serve to anchor the cofactor in place. Comparison with the nucleotide-binding domain of lactate dehydrogenase shows that the tail is accommodated partly by a reduction in side-chain volume (conserved glycines replace valine at several key positions of close approach) and partly by substitution of segments of helix with loops (Drennan *et al.*, 1994).

Other changes in the conformation of AdoCbl have potential bearing on the process of Co—C bond labilization. Two protein loops pass under the corrin ring, providing binding interactions with the macrocycle and its side-chains (Figure 9). From one of these loops, the side-chain of a conserved leucine (Leu657) makes contact with a methyl group protruding from corrin ring A. The other loop makes intimate contact with the cobalt atom itself, taking up the vacant $Co_\alpha$ coordination position *via* the N$\epsilon$2 atom of a histidyl residue in the protein. This residue (His610) is the histidine found in the motif identified by sequence comparisons and is held firmly in place by a hydrogen-bond network involving the side-chain of Asp608 (conserved) and extending to the backbone nitrogen of Leu657 (conserved) and the side-chain of Lys604. Similar, though not identical, networks are found in glutamate mutase and methionine synthase. The length of the Co—N(His) bond has been difficult to determine accurately from methyl-

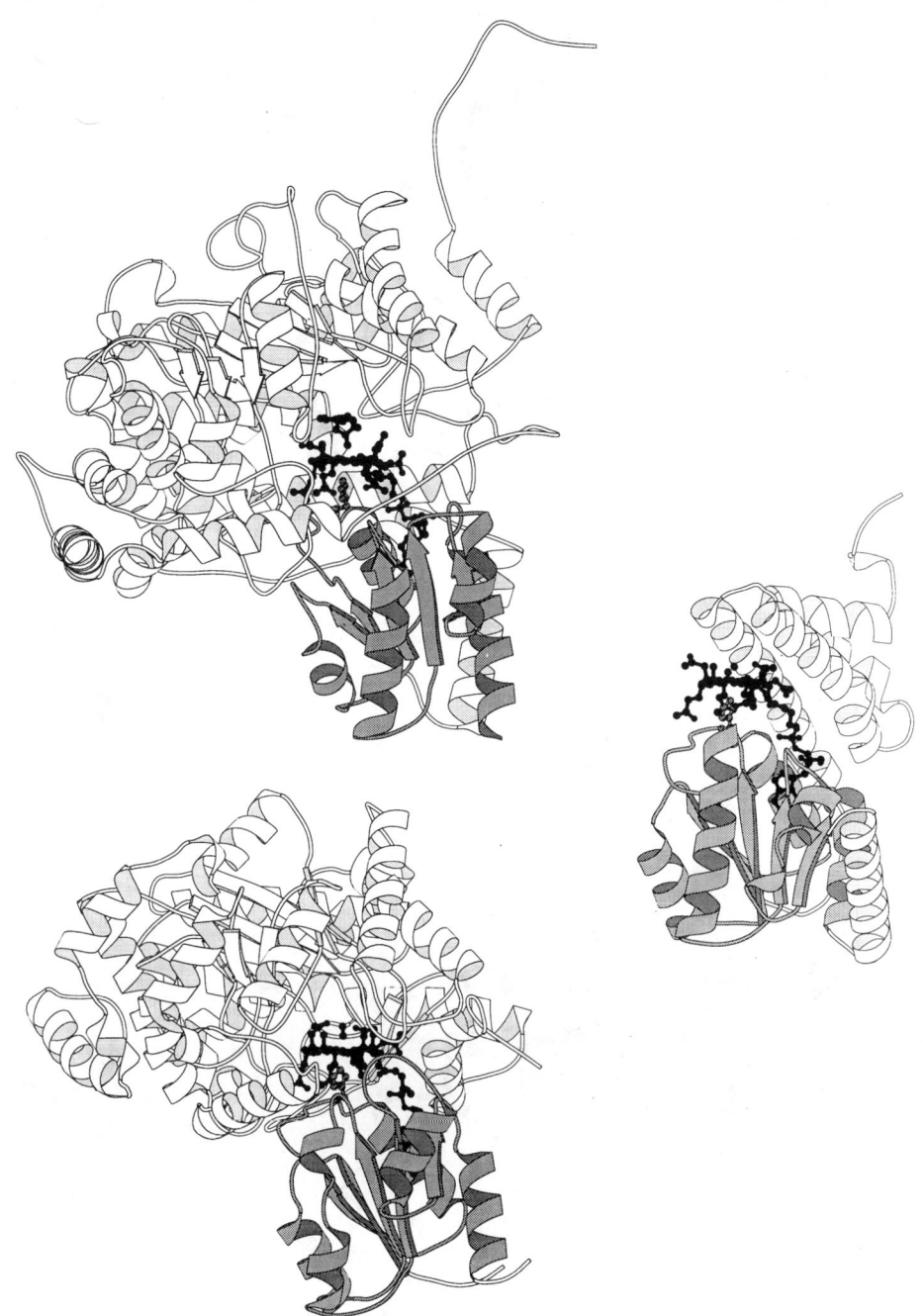

**FIGURE 8.** Three-dimensional structures of cobalamin-dependent enzymes. Schematic representations of the protein and cobalamin components of *Top*, α-subunit of *P. shermanii* methylmalonyl-CoA mutase holoenzyme reconstituted with AdoCbl (Mancia & Evans, 1998); *Middle*, fragment of *E. coli* methionine synthase reconstituted with MeCbl (Drennan *et al.*, 1994); *Bottom*, *C. cochlearium* glutamate mutase-tartrate complex reconstituted with CNCbl (Reitzer *et al.*, 1999). Secondary structure elements comprising the conserved $\alpha_5/\beta_5$ cobalamin-binding domain (residues 599–727 of methylmalonyl-CoA mutase, residues 748–871 of methionine synthase and residues 5–134 of glutamate mutase component S) are shown in gray. Drawn in ball-and-stick representation are bound cobalamin (black) and the protein-derived $Co_\alpha$-ligand (gray). Figures drawn with the program MOLSCRIPT (Kraulis, 1991).

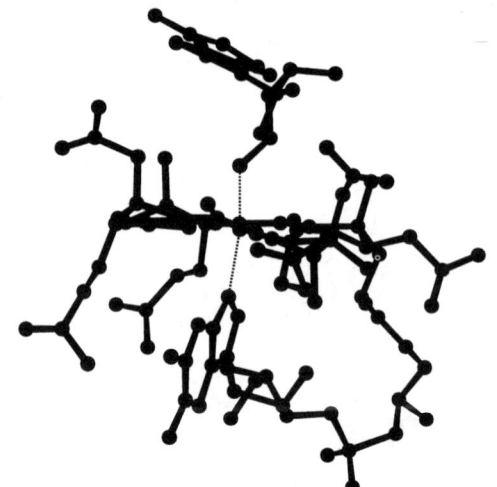

A

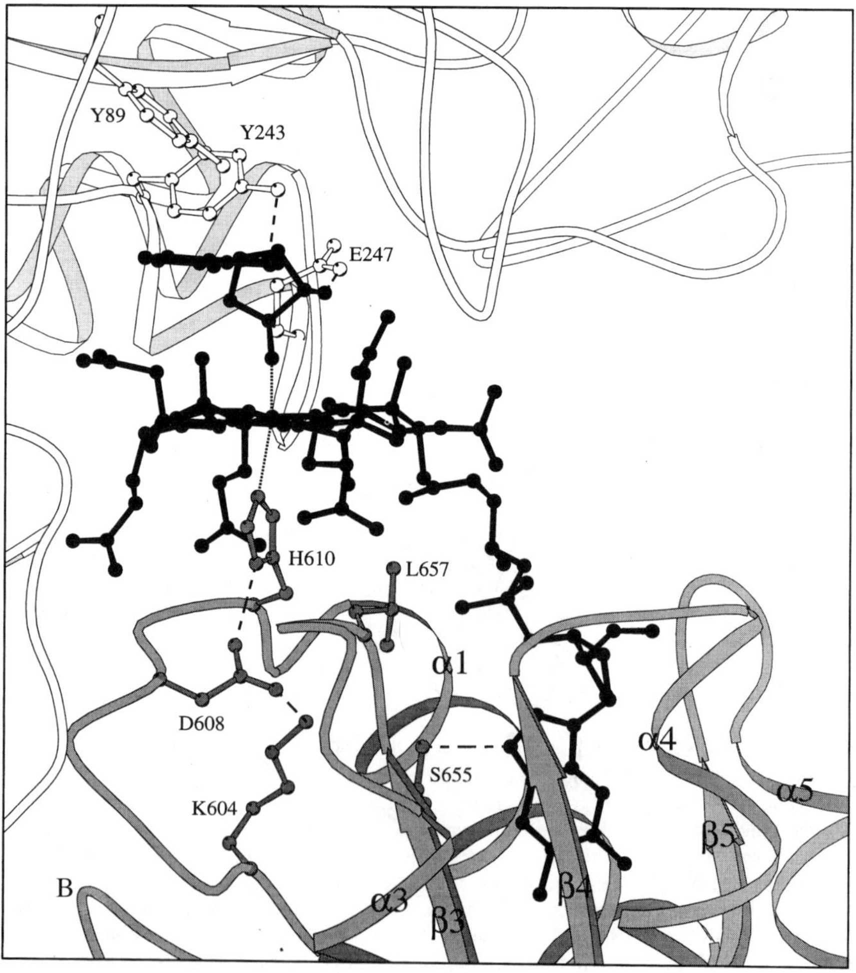

B

Y89    Y243

E247

H610    L657

α1

D608

S655    α4

K604    α5

β5

B    α3    β3    β4

malonyl-CoA mutase crystal structures, but the weight of evidence from X-ray crystallographic and EXAFS analyses suggests that it has a length greater than that expected on the basis of data from isolated Co(III) cofactors, more closely resembling that of cob(II)alamin (Kräutler et al., 1987; Mancia et al., 1996; Scheuring, 1997; Mancia and Evans, 1998). Also, the corrin ring is somewhat "flatter" in the enzyme-bound state than in free solution, as indicated by a reduction of $-6°$ in the fold angle about the Co—C-10 axis. However, the structures of methionine synthase and glutamate mutase reveal a normal length for the Co—N(His) bond (i.e. the length is similar to the axial Co—N bond found in the free coenzyme). Therefore, the mechanistic significance of these findings, if any, remains unclear.

The "upper" β-face of the corrin macrocycle is recognized by a separate, substrate-binding domain or subunit. In methylmalonyl-CoA mutase, this takes the form of an eight-stranded α/β triose phosphate isomerase (TIM) barrel (Section 2.2). Upon binding, the Co-bound 5'-deoxyadenosyl group is repositioned over ring B by virtue of a 90° rotation relative to the corrin ring. In its new position, the adenine ring stacks between the side-chain of Tyr89 and the corrin side-chain attached to C-7, necessitating a $-10°$ reduction in the twist of corrin ring B about the C-7 C-8 bond. The orientation is stabilized by hydrogen bonds between the ribose ring and two residues, Tyr243 and Glu247. A change in the position of the Co-bound C-5' atom and flattening of the macrocycle have also been inferred from data obtained by resonance Raman spectroscopy (Dong et al., 1998).

Substantial changes in protein structure also accompany cobalamin-binding, snapshots of which have been provided by high-resolution structural studies on glutamate mutase component S. The apo-form of C. tetanomorphum component S (Tollinger et al., 1998) and a CNCbl-

---

**FIGURE 9.** Binding of adenosylcobalamin to methylmalonyl-CoA mutase **(A)** Three-dimensional structure of AdoCbl as determined by X-ray crystallography (Lenhert, 1968). Molecule viewed approximately from the plane of the corrin macrocycle and drawn in ball-and-stick representation. **(B)** Detail of the cobalamin-binding site of P. shermanii methyl-malonyl-CoA mutase holoenzyme (Mancia & Evans, 1998) in schematic form. The $\alpha_5/\beta_5$ cobalamin-binding domain is shown in gray, the $(\alpha/\beta)_8$ substrate-binding domain in white and AdoCbl in black. The side-chains of selected conserved and/or key residues are drawn in ball-and-stick representation: His610 coordinates with the cobalt atom of AdoCbl and participates in a hydrogen-bonded network involving Asp608 and Lys604; the conserved serine (Ser655) forms a hydrogen bond with N3 of dimethybenzimidazole; the conserved leucine (Leu657) makes hydrophobic contacts with the corrin C20 methyl group and the beginning of the pseudo-nucleotide tail; Tyr243 and Glu247 form hydrogen bonds with O2' and O3' of the 5'-deoxyribose moiety, and Tyr89 participates in stacking interactions with the adenine ring. Secondary structure elements in the cobalamin-binding domain are labeled using the designations given to the homologous domains of methionine synthase (Drennan et al., 1994) and glutamate mutase (Reitzer et al., 1999; Tollinger et al., 1998).

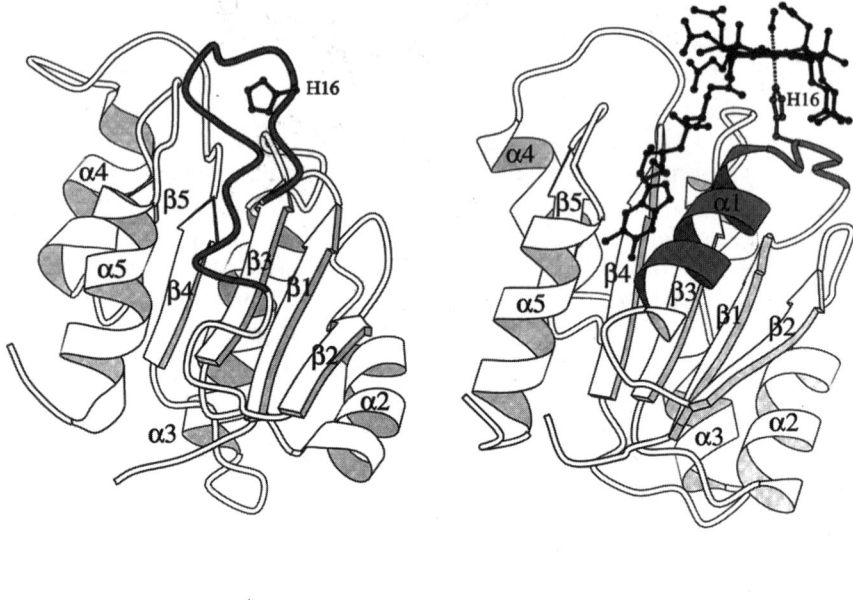

A                                                              B

**FIGURE 10.** Conformational changes accompanying the binding of cyanocobalamin to glutamate mutase component S. Schematic representations of **(A)** *C. tetanomorphum* glutamate mutase component S apoprotein (Tollinger *et al.*, 1998); **(B)** *C. cochlearium* glutamate mutase component S and CNCbl as observed in the CNCbl-glutamate mutase-tartrate complex (Reitzer *et al.*, 1999). Protein secondary structure elements present in both the apoprotein and the ternary enzyme complex are colored white. Upon binding CNCbl, the segment encompassing residues Ser13-Phe27 (gray) is dramatically restructured to form helix α1 and to position the side-chain of His16 for ligation to cobalt. Shown in ball-and-stick form are CNCbl (black) and the side-chain of His16 (gray), with dotted lines denoting the axial bonds to cobalt. Figure drawn with the program MOLSCRIPT (Kraulis, 1991).

glutamate mutase-tartrate complex involving the *C. cochlearium* enzyme (Reitzer *et al.*, 1999) have been characterized in detail (Figure 10). The component S proteins from the two species are extremely close homologues, sharing 83% amino acid sequence identity and being able to substitute for one another in formation of an active complex with component E from *C. tetanomorphum* (Zelder *et al.*, 1994). In the apoprotein, the five-stranded β-sheet core is intact but surrounded by just four α-helices. NMR relaxation measurements revealed considerable dynamic behavior in several parts of the protein, particularly between sheets β1 and β2 that are spanned by a poorly-structured loop containing a highly mobile set of residues (Ser13-

Phe27) including the **D-x-H-x-x-G** motif. However, the structure of the CNCbl-glutamate mutase-tartrate complex reveals that, upon binding CNCbl, the mobile loop is radically restructured to provide the ligand to cobalt and to form helix α1, lining one side of the nucleotide-binding pocket. The hydrogen-bonded network around His16 is also not formed until binding has taken place, and hence one can conclude that only part of the cobalamin-binding surface is pre-formed, with the remainder awaiting an initial encounter with the cofactor.

Solution studies have also demonstrated histidine ligation, as revealed by electron paramagnetic resonance (EPR) spectra obtained from holoenzyme complexes reconstituted from isotopically-substituted apoenzyme and cofactor (Padmakumar et al., 1995; Zelder et al., 1995; Drennan et al., 1994a). In these experiments, protein, cofactor or both are $^{15}$N-labeled and the cobalt resonance is analyzed. Application of the same technique to other corrinoid-dependent enzymes has made it possible to examine the wider occurrence of the "base-off/His-on" cofactor binding mode (Table 1). Interestingly, histidine ligation is not exclusive to those cobalamin-dependent enzymes carrying the **D-x-H-x-x-G** motif, but neither does it appear to be essential for the participation of enzyme-bound corrinoids in either radical initiation or methyl transfer chemistry. Thus, it can be concluded that different enzymes can initiate and control the same aspects of corrinoid chemistry in a variety of different ways.

## 2.2. Substrate Binding and Initiation of Catalysis

Both methylmalonyl-CoA mutase and glutamate mutase share strikingly similar global folds (Figure 8), even though sequence similarity is limited to the small α/β domain and their quaternary structures are quite different. The "catalytic" domains of both enzymes take the form of a $(\alpha/\beta)_8$ TIM-barrel; the $C_\alpha$ atoms of the two structures can be superimposed with an r.m.s. deviation of only 2 Å (Reitzer et al., 1999). However, the active site residues of these enzymes (other than those involved in binding the lower face of the coenzyme) do not seem to be conserved and the substrates are bound very differently.

The TIM-barrel domain of methylmalonyl-CoA mutase is connected to the conserved α/β domain through a long linker region (Mancia et al., 1996; Mancia and Evans, 1998). Unexpectedly, in the absence of substrate the barrel is incomplete (Mancia et al., 1996; Mancia and Evans, 1998), lacking hydrogen bonds between strands 1 and 2 and between strands 5 and 6. This novel "split" barrel conformation is employed in an equally novel mode of substrate binding (Figure 11). When desulpho-CoA is added, the barrel is induced to close up along the length of the pantetheine chain,

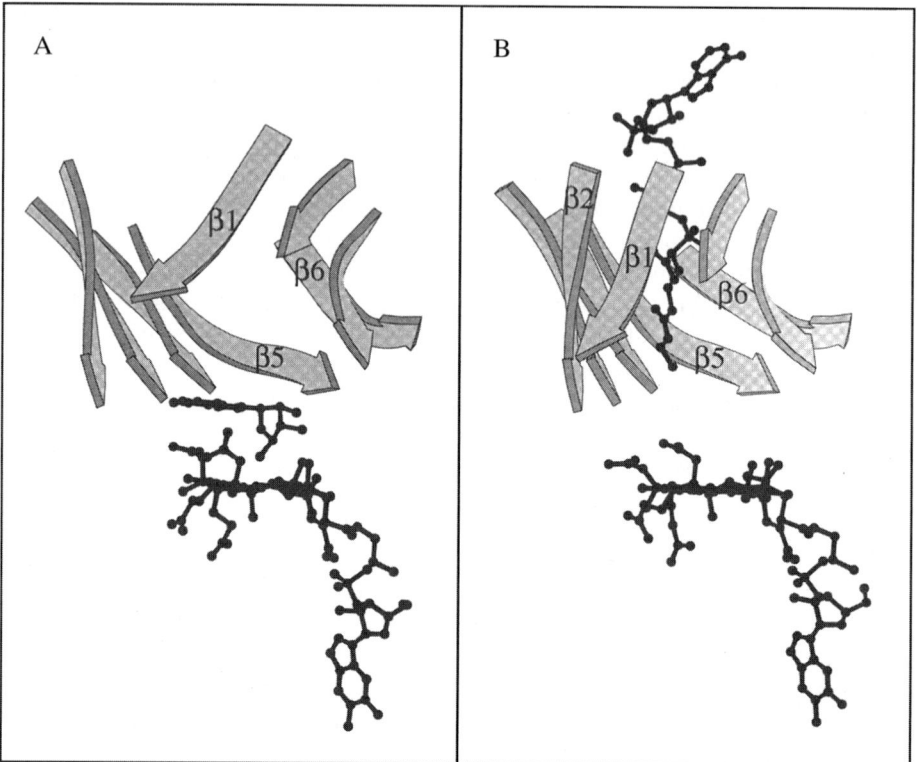

**FIGURE 11.** Binding of desulpho-CoA to methylmalonyl-CoA mutase. Detail from the structure of *P. shermanii* methylmalonyl-CoA mutase, **(A)** in the absence of substrate (Mancia & Evans, 1998), and **(B)** in the presence of the substrate analogue desulpho-CoA (Mancia *et al.*, 1996). The β-strands of the substrate-binding TIM-barrel are shown in gray. Bound cobalamin and desulpho-CoA are drawn as ball-and-stick models with the cobalamin located below the TIM-barrel. In the holoenzyme, the TIM-barrel is split apart, lacking hydrogen bonds between strands β1 and β2 and between strands β5 and β6. However, when desulpho-CoA binds, the barrel strands close up around the substrate analogue to form a sealed channel. The binding event also induces the loss of 5′-deoxyadenosine from the structure. Figure drawn with the program MOLSCRIPT (Kraulis, 1991).

leaving the terminal ADP-3′-phosphate moiety partly protruding into solvent. Closure can be considered as the movement of two four-stranded β-sheets about a hinge region between strands 5 and 6, bringing a collection of small and/or hydrophilic side-chains into contact with the pantetheine chain. This mode of binding is so far unique among TIM-barrel proteins and it is the only case known where the substrate spans the entire barrel;

in all other barrel proteins the central channel is instead occluded by large hydrophobic side-chains, giving rigidity to the fold.

Desulpho-CoA also appears to induce Co—C homolysis and the loss of AdoCH$_3$ during crystallization of methylmalonyl-CoA mutase (Mancia *et al.*, 1996). This leaves a vacant cavity bounded by loops located at the C-terminal ends of the barrel strands and by the corrin ring and some of its amidated side-chains. Its volume is substantially smaller than that of the equivalent cavity in the holoenzyme, forcing the side-chain of Tyr89 to rotate and move into the area previously occupied by the 5′-deoxyadenosine moiety of AdoCbl. Thus, one hypothesis is that substrate binding energy is harnessed to drive a conformational change which displaces the AdoCH$_2$-group to initiate catalysis. Furthermore, in the "open" enzyme conformation, the Co-linked 5′-deoxyadenosyl group is accessible to solvent through the central channel. If Co—C homolysis were allowed to occur in the absence of substrate, solvent-associated radical scavenging would doubtless result in inactivation of the enzyme. In contrast, the active site cavity in the ternary complex is effectively screened from solvent, providing an ideal environment for the protection of highly-reactive radical intermediates. Similar, shielded environments are provided by other free radical enzymes including galactose oxidase (Baron *et al.*, 1994) and prostaglandin H$_2$ synthase (Picot *et al.*, 1994).

The structure of glutamate mutase was determined in tartrate buffer and contains one tartrate ion (which may be considered a good mimic of methylaspartate) bound in the barrel domain about 5 Å from the coenzyme (Reitzer *et al.*, 1999). A number of residues make hydrogen bonds to the tartrate (Figure 12): these include Arg66 and Arg149 that bind what would be the α-carboxylate of glutamate, Arg100 and Tyr181 that bind the ω-carboxylate, and Glu171 that binds the α-carboxylate of tartrate but might well interact with the positively-charged amino group of glutamate. The large number of hydrogen bonds to the substrate analogue provides an explanation for the strict substrate specificity observed for the enzyme. The structure has so far only been determined with MeCbl and CNCbl bound, so the location of the adenosyl moiety is not known. It seems unlikely that a major reorganization of the barrel domain (c.f. methylmalonyl-CoA mutase) is necessary for the substrate to gain access to the active site: the (partial) dissociation of the two protein subunits would provide a simple mechanism for the substrates to enter and leave.

## 2.3. Structure of Diol Dehydrase

Diol dehydrase comprises three subunits and adopts a $(\alpha\beta\gamma)_2$ structure. The α and γ subunits are tightly associated with each other as an $(\alpha\gamma)_2$

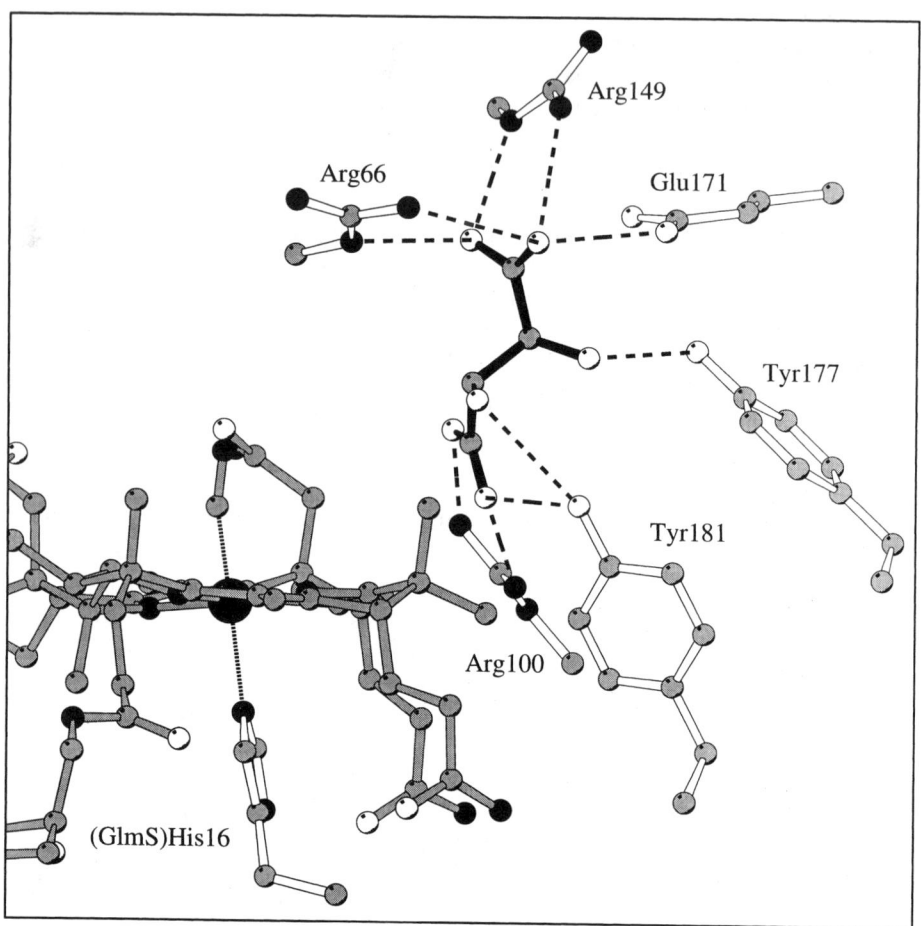

**FIGURE 12.** The active site of glutamate mutase as observed in the CNCbl-glutamate mutase-tartrate complex crystallized by Reitzer *et al.* (1999). Shown in ball-and-stick representation are the positions of the cofactor analogue CNCbl (gray sticks), the substrate analogue tartrate (black sticks) and key residues of the protein (white sticks). Atoms are colored according to type: carbon gray, oxygen white and nitrogen black, with the cobalt atom drawn as a larger black sphere. Dotted lines denote the axial bonds to cobalt and dashed lines denote hydrogen bonds. The tartrate ion is the centerpiece of an extensive hydrogen bond network, each of its oxygen atoms participating in at least one hydrogen bond. Figure drawn with the program MOLSCRIPT (Kraulis, 1991).

complex but the β subunit can be separated from the $(\alpha\gamma)_2$ complex by ion exchange chromatography (Tobimatsu *et al.*, 1997). In this respect, the enzyme resembles glutamate mutase where assembly of the E and S subunits is required for $B_{12}$ binding (Holloway and Marsh, 1994). Despite the lack of sequence similarity to either methylmalonyl-CoA or glutamate mutase and the different subunit structure, the overall structure of diol dehydrase is strikingly similar (Shibata *et al.*, 1999).

Like the mutase enzymes the core of the α subunit is made up of an $(\alpha\beta)_8$ TIM barrel which harbors the active site and binds the reactive face of cobalamin (Figure 13) (for technical reasons the enzyme was crystallized with cyanocobalamin rather than AdoCbl bound). Additional α-helixes located in the N- and C-terminal regions of the α-subunit surround the barrel domain. The β subunit recognizes the lower portion of the cofactor and like the mutases the central part of this subunit adopts a Rossmann fold-like structure. However, the corrin ring is oriented quite differently with respect to the β-sheet of the Rossmann fold and in this case the dimethylbenzimidazole ligand remains coordinated to cobalt. The coenzyme is predominantly surrounded by hydrophilic residues from the α and β subunits and many of the hydrogen-bonding contacts are made by backbone amides: this may explain the absence of a $B_{12}$-binding motif for this class of AdoCbl enzymes. The small γ subunit is predominantly β-helical and binds to the side of the TIM barrel formed by the α subunit. It does not make any contacts with the coenzyme or substrate and its role therefore may be purely structural.

The active site of diol dehydrase is formed within the TIM barrel of the α-subunit. The enzyme was crystallized in the presence of potassium ion, which is known to be required for activity, and racemic propane-1,2-diol; both of these are found bound to the active site (it was assumed that the *S*-isomer was bound by the protein). The potassium ion is deeply buried within the protein (Figure 13) and coordinated by oxygen atoms from Gln141, Glu170, Glu221, Gln296 and Ser362, most interestingly, the sixth and seventh coordination sites are occupied by the oxygen atoms of propanediol. The potassium ion may be both structural, serving to rigidify the barrel and facilitate substrate binding (Shibata *et al.*, 1999) and, as discussed in section 3.2, play an important catalytic role. Because the protein was crystallized with cyanocobalamin bound, it is not known how the adenosyl group of the coenzyme interacts with the protein, however C-1 and C-2 of the substrate are 8.4 and 9.0 Å respectively from the cobalt which would certainly allow room for the 5′-carbon of $AdoCH_3$ to occupy its expected position between the substrate and corrin ring.

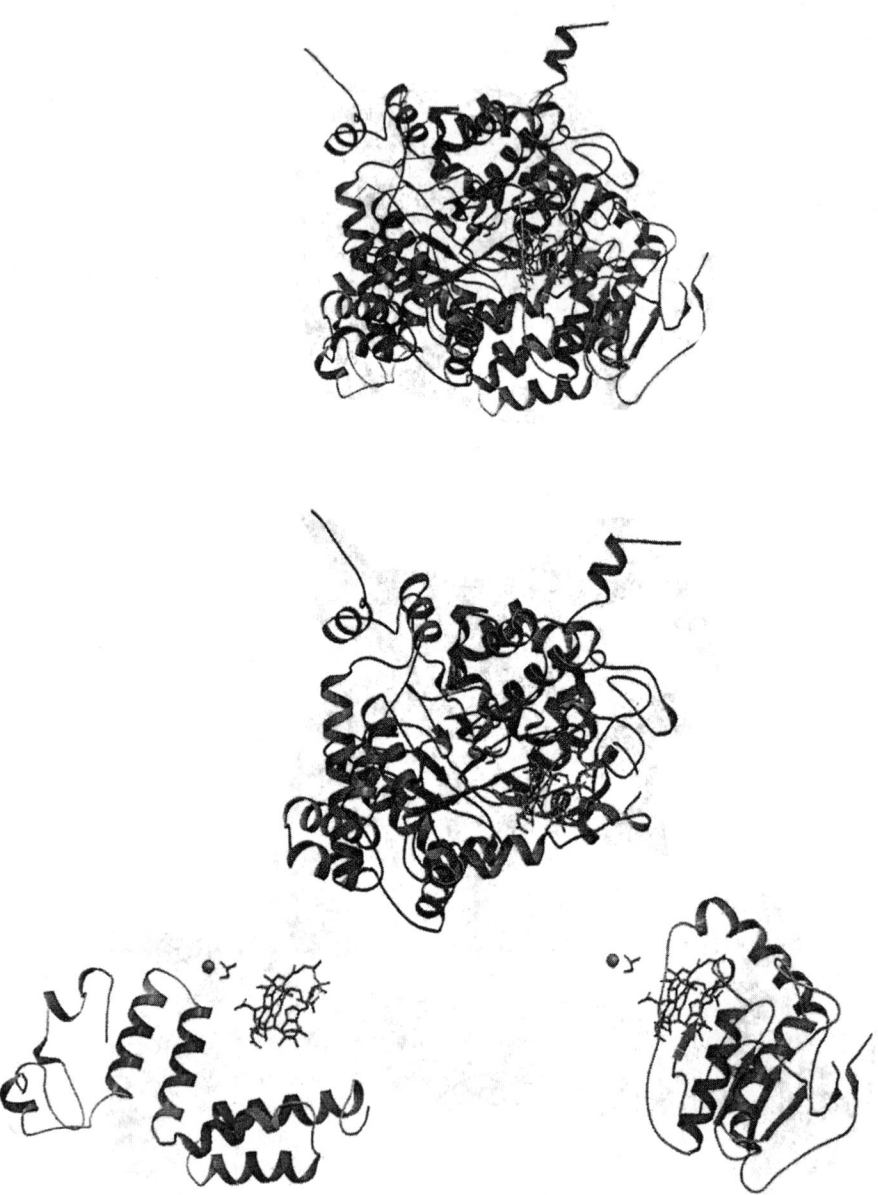

**FIGURE 13.** The structure of diol dehydrase-cyanocobalamin complex. The top figure shows the complete αβγ trimer; the lower figures show each subunit separately and its spatial relationship to the coenzyme, substrate and essential potassium ion. Reprinted from Shibata *et al.*, 1999, with permission from Elsevier Science.

## 3. MECHANISTIC ASPECTS OF ADENOSYLCOBALAMIN-MEDIATED CATALYSIS

### 3.1. Homolysis of Adenosylcobalamin and the Formation of Substrate Radicals

#### 3.1.1. EPR Studies on Adenosylcobalamin Enzymes

Electron paramagnetic resonance spectroscopy has proved a valuable tool in the study of AdoCbl-dependent enzymes. AdoCbl itself is EPR-silent, but upon homolysis to form Cbl(II), two spins are formed, one on the cobalt (which now has low-spin $d^7$ configuration) and one on the organic radical. Typically, the two unpaired electrons remain close enough in the enzyme's active site that they interact with one another to give complex, but informative, EPR spectra.

Among the first enzymes studied by EPR were ethanolamine deaminase and AdoCbl-dependent ribonucleotide reductase (Babior et al., 1974; Orme-Johnson et al., 1974). The EPR spectrum of ethanolamine deaminase was extensively characterized in the presence of 2-aminopropanol, which is a slow substrate for the enzyme. It exhibits a broad feature at g = 2.34 attributed to Cbl(II), and a sharp doublet at g = 2.01 attributed to an organic radical. Using isotopically-labeled substrates, the organic radical was identified as the C-1 radical of 2-aminopropanol (Babior et al., 1974). The doublet splitting was attributed to dipolar coupling of the Co(II) spin with the substrate radical and the distance between cobalt and the substrate radical was calculated to be about 6 Å. Thus, these experiments yielded the first structural information on the active site of a $B_{12}$ enzyme.

Rapid-freeze quench experiments on ribonucleotide reductase demonstrated the kinetic competence of radical formation when the enzyme, allosteric activator and AdoCbl were mixed (Orme-Johnson et al., 1974). The EPR spectra of the active enzyme-AdoCbl complex differ significantly from those observed simply with Cbl(II) and $AdoCH_3$ bound to the enzyme: whereas Cbl(II) has a g-value of 2.23 and a cobalt hyperfine splitting of 110 Gauss, the intermediate has g = 2.12 and cobalt hyperfine splitting of 50 Gauss. Although it was originally thought that the signal originated from the interaction of unpaired electrons on Co(II) and $AdoCH_2^•$, the signal was unchanged when either $5'-(^{13}C)$-AdoCbl or $5'-(^2H_2)$-AdoCbl were used in the experiment. More recently, evidence has accumulated that a thiyl radical on Cys-408 is involved in catalysis (Mao et al., 1992). In experiments employing ribonucleotide reductase containing cysteine deuterated at the β-carbon, Stubbe's laboratory demonstrated that the EPR spectrum (Figure 14) arises from the interaction of Cbl(II) and the thiyl radical (Licht

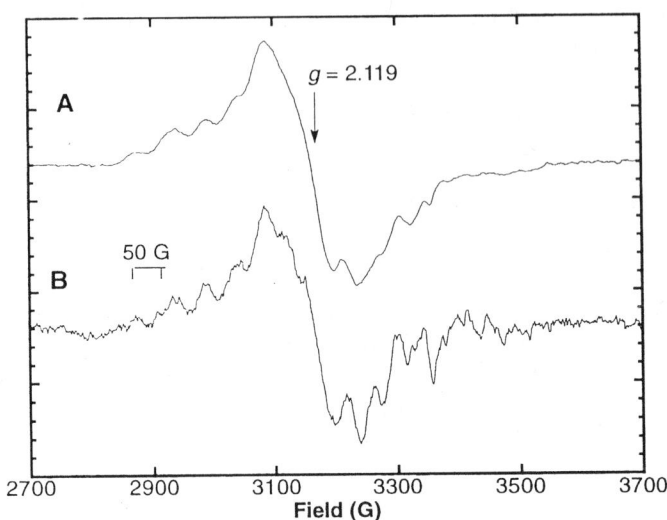

**FIGURE 14.** EPR spectra of AdoCbl-dependent ribonucleotide triphosphate reductase. The top spectrum exhibits complex features arising from the coupling of the Co(II) signal to an organic radical. The lower spectrum was obtained from a sample of enzyme that incorporated [β-$^2$H$_2$]-cysteine; the deuteration results in a sharpening of many of the spectral features and provides evidence that the organic radical is situated on a cysteine residue. Reprinted with permission from Licht *et al.*, 1996, copyright 1996 American Association for the Advancement of Science.

*et al.*, 1996). They were able to model the spectrum and hence estimated the distance between Co(II) and the cysteine residue to be 5.5–6.5 Å.

EPR spectroscopy has been used to characterize several other AdoCbl enzymes, among them diol dehydrase, glutamate mutase, methylmalonyl-CoA mutase and 2-methyleneglutarate mutase (Yamanishi *et al.*, 1998a; Zelder *et al.*, 1995; Zhao *et al.*, 1994; Michel *et al.*, 1992). These enzymes all give spectra that are consistent with a Co(II) species interacting with an organic radical. In particular, a recent study on glutamate mutase using $^{13}$C- and $^2$H-labeled glutamates has identified the C-4 radical of glutamate as the organic partner of Co(II), separated from each other by about 6 Å (Bothe *et al.*, 1998). This is in excellent agreement with the recently determined crystal structure of glutamate mutase (Reitzer *et al.*, 1999). It is noteworthy that in every case where active enzymes have been examined (there have also been some studies on inactive enzyme-cobinamide complexes that show Cbl(II)-like spectra (Michel *et al.*, 1992)), EPR-active species are only generated in the presence of the substrate or an analog. Although these EPR spectra have often been interpreted as arising from AdoCH$_2^{\bullet}$

interacting with Cbl(II) on the enzyme, AdoCH$_2$• has never been unambiguously observed by EPR. In all cases where it has been investigated the organic radical is associated with the substrate. As discussed below, it is likely that AdoCH$_2$• is far too unstable to exist in any significant concentration and indeed may not actually be a true intermediate in AdoCbl-mediated catalysis.

### 3.1.2. Stopped Flow Studies of Adenosylcobalamin Homolysis

The u.v.-visible spectrum of the coenzyme changes significantly upon homolysis of the Co—C bond: AdoCbl is pink, a characteristic of the 6-coordinate Co(III) atom, whereas Cbl(II) is yellow-brown due to the presence of the 5-coordinate Co(II) species. This property provides a convenient and sensitive spectroscopic probe with which to monitor the enzyme-catalyzed homolysis of AdoCbl, the key step which results in the generation of free radicals. Thus, for several enzymes the rate of AdoCbl homolysis has been measured by stopped flow spectroscopy.

B$_{12}$-dependent ribonucleotide reductase was the first enzyme for which the rate of AdoCbl homolysis was examined. Pioneering studies by Blakely and co-workers established that a homolytic mechanism was indeed operating, as opposed to heterolytic cleavage to give Co(I), and that the rate of homolysis ($k = 38\,\text{s}^{-1}$) was kinetically competent (Tamao and Blakely, 1973). Subsequently, the kinetic competency of Co—C cleavage in ethanolamine deaminase was established (Hollaway *et al.*, 1978), and, as discussed below, homolysis has been found to be far more rapid than turnover for all the enzymes so far examined.

Recently, more extensive measurements employing deuterated substrates and coenzyme have been made on several enzymes, and these have provided further insights into the mechanism of homolysis. For glutamate mutase, the kinetics of AdoCbl homolysis have been investigated with both substrates, L-glutamate and L-*threo*-3-methylaspartate, in their proteo-forms and with the abstractable hydrogen substituted with deuterium (Marsh and Ballou, 1998). The homolysis reaction occurred in two phases, each of similar amplitude, an effect that is most likely due to negative cooperativity arising from the dimeric structure of the enzyme (Figure 15). Although this phenomenon probably has no physiological significance it does illustrate the sensitivity of this reaction to what are most likely very subtle changes in protein structure. The cobalt centers of the two active sites are separated by ca. 45 Å and it is not apparent from the structure how the two sites communicate with each other. When protiated substrates were reacted with the enzyme the first phase was almost complete within the dead time of the spectrometer, whereas the slower phase

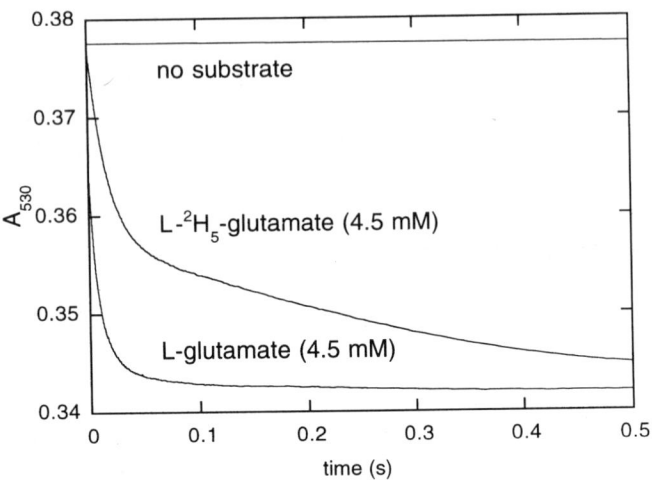

**FIGURE 15.** The effect of substrate deuteration on AdoCbl homolysis by glutamate mutase. Co—C bond cleavage was monitored by the decrease in absorbance at 530 nm after mixing the holo-enzyme with substrate in the stopped flow spectrophotometer. With deuterated glutamate homolysis is much slower and is clearly biphasic (most likely a result of negative cooperativity arising from the dimeric subunit structure of the enzyme); protiated glutamate also exhibits biphasic kinetics, but most of the more rapid phase occurs within the deadtime of the instrument.

occurred with observed rate constants of $97\,s^{-1}$ with glutamate and $80\,s^{-1}$ with methylaspartate (substrates at saturating concentrations). For comparison, $k_{cat}$ is about $5\,s^{-1}$ with either substrate.

Measurements with deuterated substrates provided an unexpected result: the rate of homolysis was much slower, by factors of 28 and 35 with glutamate and methylaspartate respectively. For the mechanism discussed in Section 1 and Figure 5, isotopic substitution is not expected to affect the rate of substrate-triggered homolysis because this step does not formally involve hydrogen abstraction from the substrate. The isotope effect could arise in two ways (Figure 16). One explanation is that Co—C bond cleavage and homolysis are concerted, so that adenosyl radical is never formed as an intermediate. The other is that adenosyl radical and Cbl(II) are formed at very low concentrations in a rapidly established but unfavorable equilibrium

**FIGURE 16.** Concerted and kinetically coupled mechanisms for Co—C bond homolysis and hydrogen atom abstraction, illustrated for the reaction catalyzed by glutamate mutase. Either mechanism could give rise to the deuterium isotope effects observed in pre-steady state stopped flow experiments.

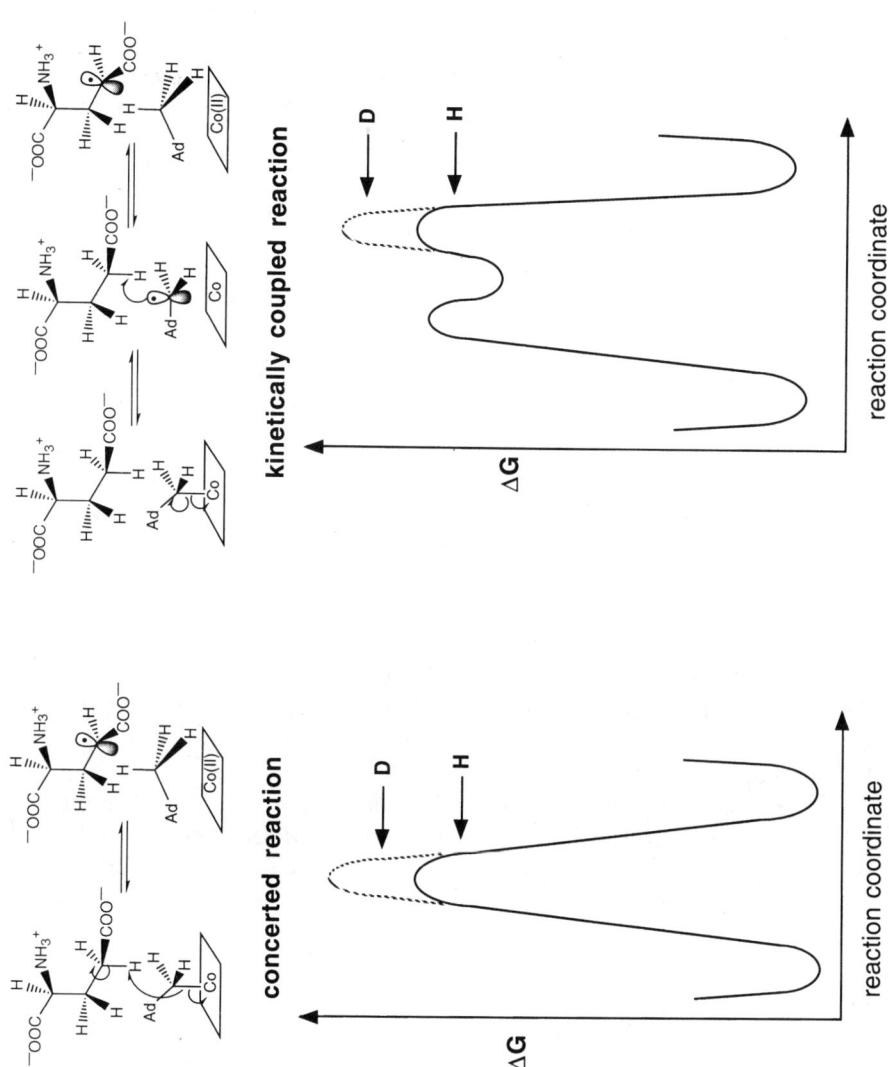

with AdoCbl, and that the subsequent (favorable) reaction with substrate displaces the equilibrium towards homolysis. An important point is that in neither case can 5′-deoxyadenosyl radical accumulate on the enzyme.

A similar result has been observed with methylmalonyl-CoA mutase (Padmakumar and Banerjee, 1997). When this enzyme was reacted with *proteo*-methylmalonyl-CoA, homolysis of AdoCbl was almost complete within the deadtime of the stopped flow spectrometer. However, with ($d_3$-methyl)-methylmalonyl-CoA homolysis is much slower and the isotope effect is estimated to be at least 20. This suggests that coupling between Co—C bond cleavage and substrate hydrogen abstraction is likely to be a general phenomenon.

The magnitude of these isotope effects is also noteworthy: they are much larger than would be expected simply from differences in zero-point vibrational energies of the reactive bonds. Most likely, this indicates that quantum tunneling of hydrogen plays a significant role in these reactions. (The very large isotope effects associated with quantum tunneling arise from the fact that the de Broglie wavelengths of deuterium and tritium are much shorter than that of hydrogen and so they tunnel much less easily (Bahnson and Klinman, 1995).) Hydrogen tunneling is seen to be important in the mechanisms of an increasing number of enzymes that catalyze proton and hydride transfers, and, in particular, hydrogen atom transfers (Bahnson and Klinman, 1995). Indeed, it has been speculated that enzymes may have evolved to catalyze hydrogen transfer by introducing compression along the reaction coordinate (Bahnson *et al.*, 1997; Kohen and Klinman, 1998). This represents a novel approach to catalysis, in which the energy barrier is *narrowed* rather than lowered to speed up the reaction! Such a mechanism has recently been demonstrated in bacterial trimethylamine dehydrogenase and methylamine dehydrogenase (see chapter by Scrutton and Sutcliffe).

The most comprehensive set of kinetic studies on AdoCbl homolysis have been performed with AdoCbl-dependent ribonucleotide reductase by Stubbe and coworkers (Licht *et al.*, 1999a; Licht *et al.*, 1999b). In this enzyme, AdoCbl is used to generate a thiyl radical on a cysteine residue; it is this thiyl radical that abstracts hydrogen from the 3′-position of the ribonucleotide to facilitate reduction at C-2′. In the presence of dGTP, an allosteric activator of the enzyme, the enzyme catalyzes the reversible cleavage of AdoCbl and formation of thiyl radical in the absence of substrate. This partial reaction proceeds rapidly enough for it to be mechanistically relevant and provides a system to study AdoCbl homolysis which is both simple and amenable to detailed kinetic analysis.

For ribonucleotide reductase, the rate constants, and hence kinetic isotope effects, were measured for the pre-steady state homolysis of

AdoCbl when a) the coenzyme was deuterated at the 5'-position: $k_H/k_D = 1.7$, b) the thiol was deuterated (by performing the experiment in $D_2O$): $k_H/k_D = 1.6$, and c) both coenzyme and thiol were deuterated: $k_H/k_D = 2.7$. By assessing the fit of these data to kinetic models in which homolysis and hydrogen transfer reactions are presumed to occur either as stepwise or concerted processes, and by employing some assumptions based on chemical precedents, the authors argue that a concerted mechanism is more likely to be operating (Licht *et al.*, 1999a).

In a further study, the temperature dependence of AdoCbl homolysis was examined and thermodynamic parameters determined for formation of thiyl radical, assuming a concerted mechanism (Licht *et al.*, 1999b). $\Delta G$ for the formation of Cbl(II), $AdoCH_3$ and Enz-S$^•$ from the Enz-SH-AdoCbl complex is 0.5 kcal/mol, i.e. close to zero. For comparison, the free energy change associated with the non-enzymatic reaction is estimated to be about 25 kcal/mol. $\Delta H$ for the enzymatic reaction is unfavorable by ca. 20 kcal/mol but is compensated for by a large increase in entropy, $\Delta S = 70$ cal/mol/K. The activation energy for thiyl radical formation, $\Delta G^{\ddagger}$, is 15 kcal/mol, and again this associated with a large favorable increase in entropy in the transition state, $\Delta S^{\ddagger} = 96$ kcal/mol/K. The enthalpy of activation, $\Delta H^{\ddagger} = 46$ kcal/mol, is in fact *greater* than that measured in free solution [$\Delta H^{\ddagger} = 35$ kcal/mol (Hay and Finke, 1988)].

This important observation argues against stabilization of the transition state by enthalpically favorable protein-coenzyme interactions, as has been proposed by many workers in this field. It has been proposed that changes in solvation between the ground state and the transition states, possibly as a result of a conformational change in the enzyme, could be responsible for the large increase in entropy. It should be noted that Brown and Li (1998) have also investigated the energetics of AdoCbl homolysis in ribonucleotide reductase and have reached different conclusions, i.e. that enthalpic factors play the major role in catalysis. The two sets of experiments were performed under slightly different conditions and the data were also analyzed differently, so it is presently unclear why the two laboratories obtained different results.

### 3.1.3. Magnetic Field Effects on Adenosylcobalamin-Dependent Reactions

Harkins and Grissom have made the interesting observation that catalysis by ethanolamine deaminase is affected by external magnetic fields (Harkins and Grissom, 1994). It is expected that reactions involving free radicals may be sensitive to magnetic fields, and there are prior examples in the chemical literature (for a review see Steiner and Ulrich, 1989), but this is the first time that an enzyme-catalyzed reaction has been shown to be affected. The authors found that although $V_{max}$ is unchanged with increasing magnetic field strength, $V_{max}/K_m$ decreases by up to 25% at 0.1 T. The insensitivity of $V_{max}$ to the applied magnetic field is explained by the fact that product release is rate-limiting in the enzyme. However, $V_{max}/K_m$ reflects the steps up to and including the first irreversible step, and these include formation of free radicals on the enzyme.

The magnetic field can alter the rate of a chemical reaction involving a biradical intermediate by either increasing or decreasing the rate of intersystem crossing between singlet and triplet spin-correlated states (Figure 17). Homolysis of the AdoCbl Co—C bond is expected to result initially in a singlet state (electron spins paired), and in this state the radicals can readily recombine to reform the Co—C bond. However, if intersystem crossing to the triplet state (spins unpaired) occurs, recombination is forbidden by the Pauli exclusion principle. If the magnetic field decreases the rate of intersystem crossing, the probability of recombination will be increased, and as a result there will be a decrease in the forward commitment to catalysis. Assuming this occurs before the first irreversible step, this will be manifested as a decrease in $V_{max}/K_m$, which is what was observed.

Interestingly, when perdeuterated ethanolamine was the substrate, a magnetic field-dependent isotope effect on $V_{max}/K_m$ was observed. $V_{max}/K_m$ was decreased by 60% with $d_4$-ethanolamine, reaching a minimum value at 0.15 T (c.f. 0.1 T for protiated substrate) (Figure 17). The different dependency of $V_{max}/K_m$ on field strength may arise because protium and deuterium have different nuclear spin and nuclear magnetic moments. The isotope effect again suggests coupling between homolysis and hydrogen abstraction, as has been seen with other AdoCbl enzymes (Section 3.1.2.). Since molecular motion rapidly randomizes the spin states of correlated radical pairs on a time-scale of $10^{-10}$–$10^{-6}$ s, events that happen slower than this are unlikely to be affected by the magnetic field. This implies that hydrogen transfer between 5′-deoxyadenosine and substrate occurs within 1 µs of the Co—C bond being broken.

### 3.1.4. Resonance Raman Experiments

Early experiments using simple alkyl cobalt compounds as models for AdoCbl fostered the notion that steric crowding around the Co—C bond might be an important factor that enzymes could exploit to effect the remarkable rate enhancements for homolysis of AdoCbl (Ng et al., 1983). These studies demonstrated that the more bulky the alkyl group attached to cobalt, the faster the rate of homolysis, leading to the idea that the enzymes might distort the coenzyme and thereby weaken the Co—C bond.

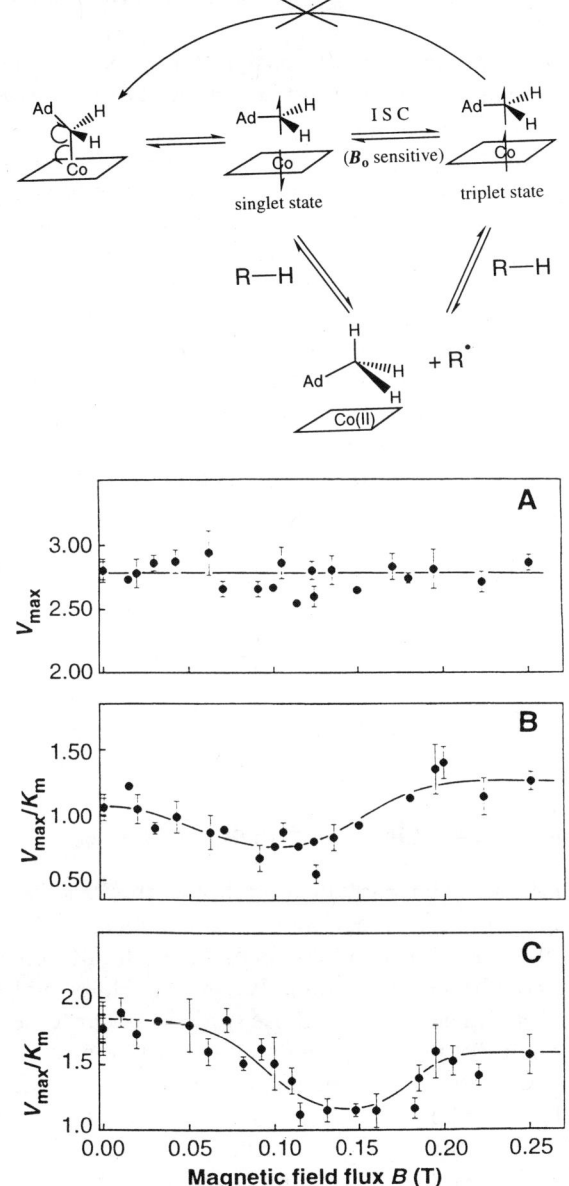

**FIGURE 17.** Magnetic field effects on ethanolamine ammonia-lyase. The top scheme shows how a magnetic field can affect the rates of radical processes by altering the relative rates of intersystem crossing (ISC), thereby altering the commitment to catalysis. Bottom, the effect of magnetic field strength on **(A)** $V_{max}$ and **(B)** $V_{max}/K_m$ for unlabeled ethanolamine; **(C)** $V_{max}/K_m$ for [1,1,2,2-$^2$H$_4$]-ethanolamine. Reprinted with permission from Harkins and Grissom, 1994, copyright 1994 American Association for the Advancement of Science.

It has recently been possible to test this hypothesis using resonance Raman spectroscopy to measure the vibrational frequency of the Co—C bond of AdoCbl while bound to the enzymes themselves.

Resonance Raman spectra have been recorded for AdoCbl both in free solution and when bound to methylmalonyl-CoA mutase (Dong *et al.*, 1998). Using 5'-$^{13}$C-labelled AdoCbl it was possible to assign the Co—C stretching frequencies to signals at 428 and 441 cm$^{-1}$ for the free coenzyme; the presence of two frequencies is attributed to the adenosyl group occupying two different conformations relative to the corrin ring. When AdoCbl is bound to methylmalonyl-CoA mutase, extensive changes in the intensity and frequency of many of the vibrational modes are seen, but significantly the frequencies associated with the Co—C stretch are only lowered by 6 cm$^{-1}$ (Figure 18). This translates to only about 0.5 kcal of additional bond weakening by the enzyme in the ground state. Similar results have been obtained with glutamate mutase. In this case, the resonance Raman spectrum of the coenzyme appears even less perturbed when bound by the protein and virtually no change in the Co—C stretching frequency is observed (E. N. G. Marsh and C. R. Hille, unpublished results). Both these results argue against an enzyme-induced ground state distortion of the coenzyme as a means of activating the coenzyme towards homolysis. This is consistent with the crystal structures of methylmalonyl-CoA mutase and glutamate mutase in which only modest changes in corrin ring conformation occur upon binding to the proteins.

### 3.1.5.   Role of the Axial Ligand to Cobalt in Catalysis

The role of the axial base in modulating the reactivity of AdoCbl remains unclear. There have been numerous chemical studies on the thermolysis of alkyl cobaloximes and cobinamides with different axial ligands coordinated to cobalt; see for example (Geno and Halpern, 1987; Ng *et al.*, 1982; Ng *et al.*, 1983). The results of these studies appear to depend to some extent on the choice of model system and will not be discussed here [for a discussion of this aspect of B$_{12}$ chemistry see (Sirovatka and Finke, 1997) and references therein]. Recently it has been possible to address this question more directly by studying the enzymes themselves. The structures of methylmalonyl-CoA mutase and glutamate mutase immediately suggested that the cobalt-coordinating histidine-aspartate pair may play an important role in both binding cobalamin and controlling the reactivity of the cobalt-carbon bond.

For glutamate mutase, site specific mutagenesis of the conserved histidine and aspartate residues has been performed (Chen and Marsh, 1997). Mutation of either the histidine or the aspartate results in a dramatic

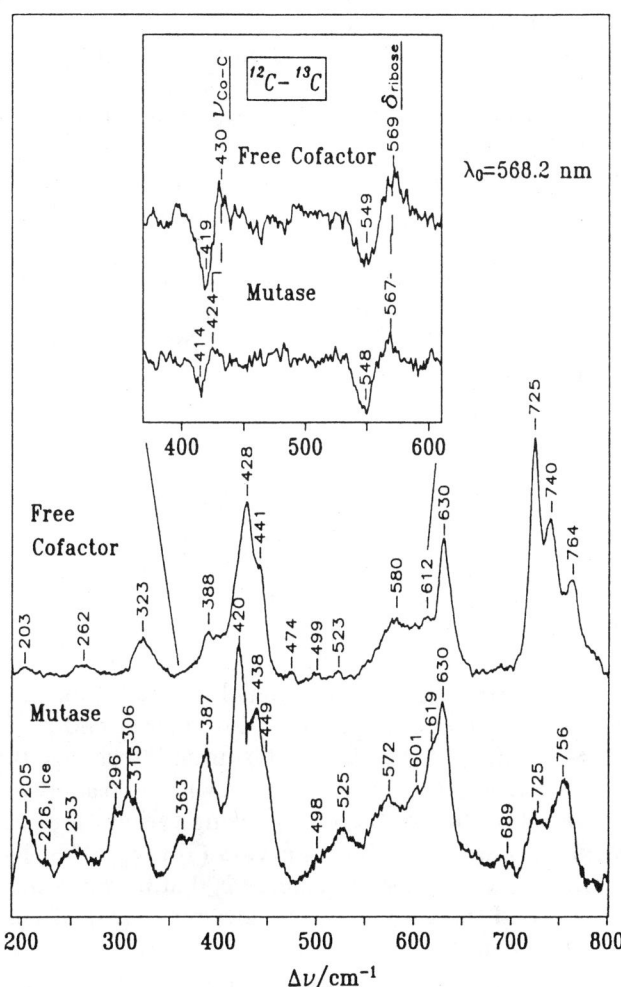

**FIGURE 18.** Low frequency resonance Raman spectra of AdoCbl in free solution and bound to methylmalonyl-CoA mutase. Inset are spectra obtained with [5′-$^{13}$C]-AdoCbl that identify the Co—C stretch and the δ-ribose vibrations. It is evident that neither of these vibrational modes in changed greatly when the coenzyme is bound by the protein. Reprinted with permission from Dong *et al.*, 1998, copyright 1998 American Chemical Society.

decrease in $k_{cat}$ of ca. 1000-fold. The binding of AdoCbl is also significantly weakened; deletion of the histidine raises the apparent $K_d$ for AdoCbl by about 50-fold, as might be expected since the histidine supplies a bond to cobalt. Mutation of the aspartate also raises $K_d$, although only by 5- to 10-fold, depending upon the choice of mutation. Interestingly, the aspartate mutations appear to affect how strongly the histidine coordinates the cobalt. Whereas mutation to asparagine results in cobalt still being predominantly "base-on" (as judged by the u.v.-visible spectrum) mutation to either alanine or glutamate results in a substantial proportion of the coenzyme being bound "base-off", i.e. the histidine is no longer strongly coordinated to cobalt.

These experiments serve to demonstrate the importance of these residues in catalysis, but their exact role is still not fully understood. Since the mutant enzymes no longer accumulate Cbl(II) during steady state turnover, it is reasonable to assume that the mutations affect the ability of the enzyme to break the Co—C bond. However, the decrease in $k_{cat}$, although large, represents no more than $10^3$ of the $10^{12}$-fold increase in the rate of AdoCbl homolysis required to explain the observed rate of turnover (Hay and Finke, 1987; Marsh and Ballou, 1998). This observation, and the fact that enzymes such as diol dehydrase bind AdoCbl with the coenzyme nucleotide tail still coordinated to cobalt (Yamanishi et al., 1998b), suggest that the histidine-aspartate pair exerts a relatively small effect on the reactivity of the coenzyme and serves only to fine-tune catalysis.

A different approach has been taken to investigate the importance of the heterocyclic base in the diol dehydrase reaction. Since the coenzyme "tail" remains coordinated to cobalt when bound to this enzyme it has been possible to use "chemical mutagenesis" to change the structure of the axial ligand. In the modified coenzyme, a trimethylene spacer replaces the ribosyl portion of the tail, allowing various heterocyclic bases to be easily incorporated into the coenzyme (Toraya, 1998; Toraya et al., 1994). An interesting correlation to emerge is that between the size of the heterocyclic base and the activity of the coenzyme analog. The dimethyl-benzimidazole analogue has 59% of the activity of AdoCbl whereas the imidazole analogue has only 8% of the activity. This suggests that the steric bulk of the axial base, rather than the $pK_a$, is important for enzyme activity.

## 3.2.  Rearrangement of Substrate Radicals

As discussed above, homolysis of AdoCbl and hydrogen abstraction from the substrate to form a substrate radical appear to be closely coupled or concerted reactions. This leaves isomerisation of the substrate radical to product radical as the remaining step, the mechanism of which depends upon the nature of the substrate.

An interesting question is, to what extent does the protein catalyze the rearrangement of substrate radical to product radical? Attempts to develop mechanism-based inhibitors of $B_{12}$-dependent isomerases (by incorporating radical-trapping functionality such as cyclopropyl groups into the substrate) that might label the active site have been largely unsuccessful because the enzymes tend to exhibit very high substrate specificity. One view, which gains some support from chemical models, is that substrate radicals, once formed, are sufficiently reactive that their rearrangement occurs spontaneously (Dowd et al., 1992; Wollowitz and Halpern, 1988; Wollowitz and Halpern, 1984). In this view, the main function of the protein is to exclude reactants such as water and oxygen from the vicinity of the radical and to prevent unwanted side reactions; this strategy has been termed "negative catalysis" (Retey, 1990). Whereas the protein undoubtedly does serve these functions, evidence is now accumulating that the enzyme also applies conventional "positive catalysis" to promote radical rearrangements. The availability of crystal structures for methylmalonyl-CoA mutase and, more recently for glutamate mutase, has allowed the catalytic role of the protein to be investigated through a combination of mutagenesis and kinetic analyses, as discussed in sections 3.2.1. and 3.2.2.

For diol dehydrase, elegant experiments employing isotopically labeled propanediol demonstrated that the dehydration reaction proceeds through the stereospecific formation of a gem-diol intermediate (Zagalak et al., 1966). This implies that a 1,2-OH shift occurs, and it was proposed some time ago that protonation of the migrating hydroxy could facilitate this migration (Golding and Rao, 1986). Most interestingly, the structure of diol dehydrase reveals that a potassium ion, essential for catalysis, is tightly bound at the active site and that the hydroxyl groups of the substrate diol are coordinated to this ion (Shibata et al., 1999). It is plausible that the potassium ion could provide electrophilic catalysis to facilitate the migration of the hydroxyl group, functioning in the same way as a proton, as shown in Figure 19. The mechanism of ethanolamine deaminase is less clearly defined; it is not known whether direct elimination of the amino group occurs (through protonation to give ammonia) or whether a 1,1-aminoalcohol is formed in a manner analogous to that of the diol dehydrase reaction (Tan et al., 1986).

The amino mutases require pyridoxal phosphate for activity, and although little is known about them mechanistically, Frey's work on lysine 2,3-aminomutase has proved very informative (Lieder et al., 1998; Frey, 1997). Lysine 2,3-aminomutase is not a $B_{12}$ enzyme but it functions very similarly (Section 1). The enzyme uses pyridoxal phosphate to facilitate the 1,2

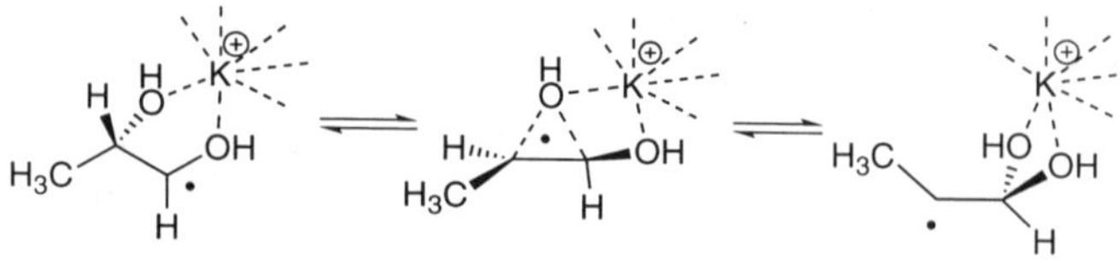

**FIGURE 19.** Mechanism for the rearrangement of propane-1,2-diol radical by diol dehydrase in which the migration of the hydroxyl group if facilitated through electrostatic catalysis by a tightly-bound potassium ion at the active site (Shibata *et al.*, 1999).

migration of the amino group, forming a Schiff base with the coenzyme so that the migrating nitrogen is sp$^2$ hybridized. This allows the reaction to proceed *via* a cyclic aziridine intermediate in which the unpaired electron is delocalized into the $\pi$-system of the pyridoxal ring (Figure 20). The stabilization of radical intermediates by pyridoxal phosphate represents a further novel role for this remarkably versatile coenzyme.

Since they have no ready counterparts in conventional organic chemistry, much attention has centered upon rearrangements catalyzed by the carbon skeleton isomerases. In methylmalonyl-CoA, isobutyryl-CoA and 2-methyleneglutarate mutases, the migrating carbon is sp$^2$ hybridized, and interconversion of substrate and product radicals can occur through a cyclopropyl radical intermediate as shown in Figure 21. This mechanism is supported by model studies which show that radicals generated chemically on compounds designed to mimic the substrate radicals produced in the 2-methyleneglutarate and methylmalonyl-CoA mutase

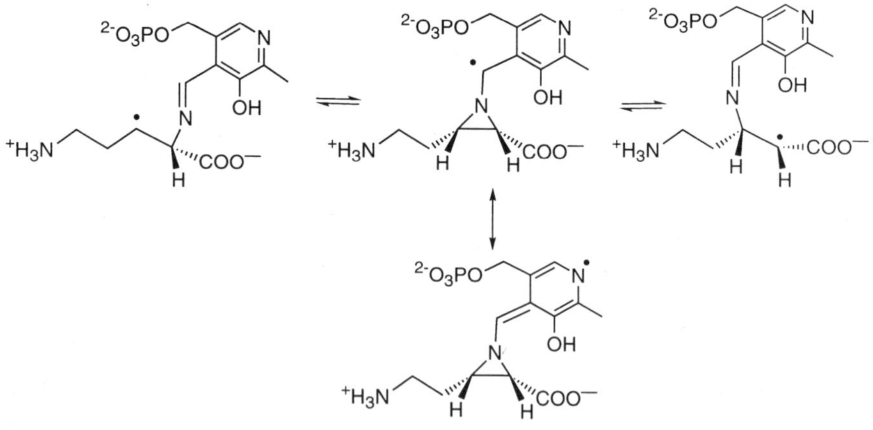

**FIGURE 20.** Mechanism for the migration of the amino group in the reaction catalyzed by lysine-2,3-aminomutase, for discussion of the mechanism see the text.

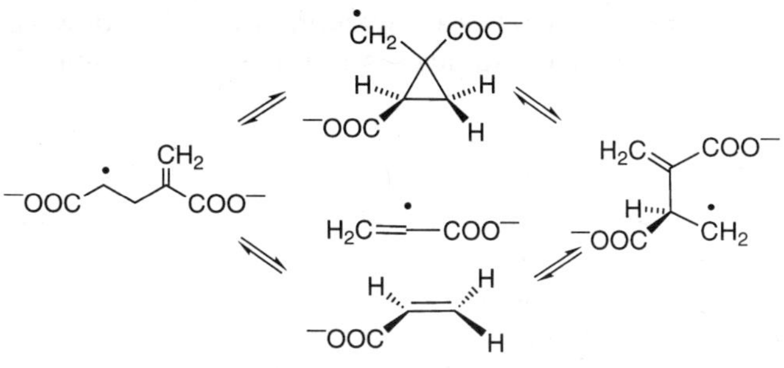

2-methyleneglutarate mutase

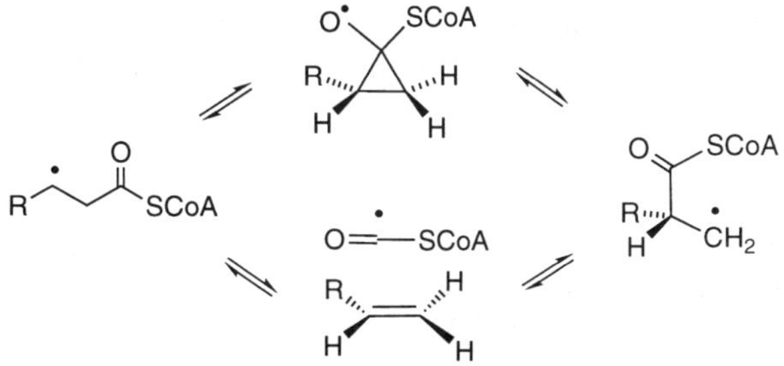

Acyl-CoA mutases ($R = CH_3, COO^-$)

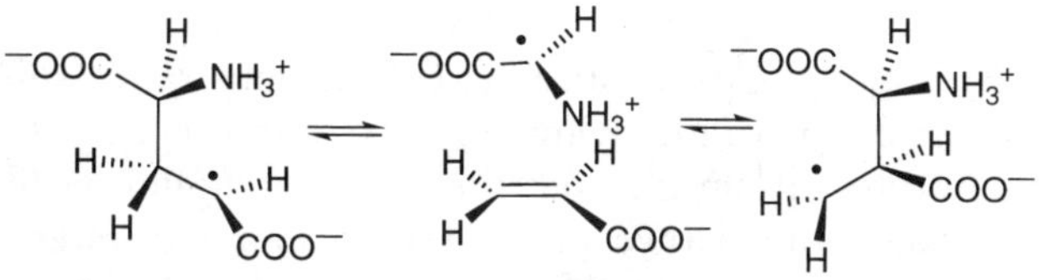

Glutamate mutase

**FIGURE 21.** Mechanisms for the rearrangement of substrate radicals in the reactions catalyzed by carbon skeleton mutases. For 2-methyleneglutarate mutase and the acyl-CoA mutases both associative (upper pathway) and dissociative (lower pathway) mechanisms have been proposed (Halpern, 1985; Buckel & Golding, 1996), whereas for glutamate only a dissociative mechanism appears feasible.

reactions can undergo spontaneous rearrangement (Dowd *et al.*, 1992; Ashwell *et al.*, 1989).

The reaction catalyzed by glutamate mutase stands out as the only case in which the migrating carbon is $sp^3$ hybridized, and hence rearrangement through a cyclic intermediate is not possible (pyridoxal phosphate is not a cofactor). A plausible mechanism for the rearrangement of glutamyl radical to methylaspartyl radical involves a fragmentation and recombination reaction with acrylate and glycyl radical as intermediates (Figure 21). If this occurs it would explain the lack of success in modeling this rearrangement in free solution, although model studies employing hydrophobic $B_{12}$ analogs buried in lipid micelles provide some evidence that the rearrangement may take place when the reactive species are enclosed within the lipid bilayer (Murakami et al., 1992). Recently, it has been proposed that all the carbon skeleton rearrangement reactions may proceed through this type of fragmentation/recombination mechanism (Buckel and Golding, 1996). Whereas this proposal would unify the mechanisms of these four enzymes, as yet there is no firm evidence in its favor. Indeed, there is no chemical imperative for the mechanisms to be the same, and in the case of 2-methyleneglutamate mutase in particular, chemical model studies provide strong chemical precedence for the reaction proceeding through a cyclopropyl carbenyl radical intermediate (Ashwell et al., 1989).

### 3.2.1.  Mechanistic Studies on Methylmalonyl-CoA Mutase

Tritium partitioning experiments have been used to compare the rate with which hydrogen is transferred from coenzyme to substrate and the rate with which the substrate and product radicals interconvert in the methylmalonyl-CoA mutase reaction. Using AdoCbl tritiated at the exchangeable 5′-position, Leadlay and coworkers observed that the distribution of tritium between substrate and product varied when the enzyme was incubated with either methylmalonyl-CoA or succinyl-CoA as substrate (Meier et al., 1996). If the interconversion of methylmalonyl-CoA and succinyl-CoA radicals is slow compared to tritium transfer, then tritium should partition preferentially into the substrate. However, if interconversion of radicals is fast, then the ratio of tritium in methylmalonyl-CoA to tritium in succinyl-CoA should be the same regardless of which direction the reaction was run in. In practice, tritium partitions between methylmalonyl-CoA and succinyl-CoA in a constant $1:3$ ratio, irrespective of the direction of reaction, indicating that substrate rearrangement is relatively rapid. From such experiments the tritium isotope effect on hydrogen transfer from coenzyme to product can also be calculated: with methylmalonyl-CoA as substrate, $k_H/k_T$ is only 4.9. These results suggest that hydrogen transfer is only partially rate limiting and that a subsequent step such as product release contributes significantly to the overall reaction velocity. This would seem quite plausible in light of the large conformational changes in the protein that appear to occur when the substrate is bound (Section 2.2).

The tritium partitioning technique has been used to investigate the role of a conserved tyrosine in methylmalonyl-CoA mutase, Tyr89 (Thoma et al.,

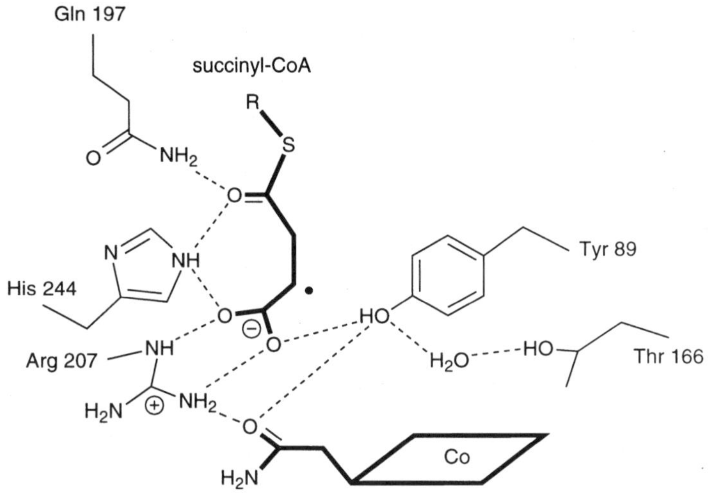

**FIGURE 22.** Detail of the active site of methylmalonyl-CoA mutase showing the interactions of Tyr-89 with substrate and coenzyme.

1998). This residue makes hydrogen bonds with both the terminal carboxyl group of the substrate and a side chain of the corrin ring (Figure 22). When this residue is mutated to phenylalanine, $k_{cat}$ is decreased by about 580-fold whereas $K_m$ is little altered. Importantly, the crystal structure of the mutant shows that no changes in protein structure have occurred. In the mutant enzyme the partitioning ratios are now different in each direction, indicating that the rearrangement step is now slower than tritium transfer. Furthermore, tritium is lost from AdoCbl much faster when succinyl-CoA is the substrate than when methylmalonyl-CoA is the substrate. The tritium isotope effect is also increased to ca. 30, indicating that hydrogen transfer is more rate limiting in the mutant. The changes in the free energy profile caused by the mutation are summarized in Figure 23. The catalytic role of the tyrosine hydroxyl is not entirely clear from these experiments; it may serve to anchor the substrate and coenzyme in the optimal conformation for hydrogen transfer. The experiments do, however, provide a clear demonstration that the protein provides more than just an inert environment to contain free radicals but plays a specific role in controlling their reactions.

### 3.2.2. Mechanistic Studies on Glutamate Mutase

An EPR study designed to identify the radical species present in the glutamate mutase reaction under steady state conditions was recently

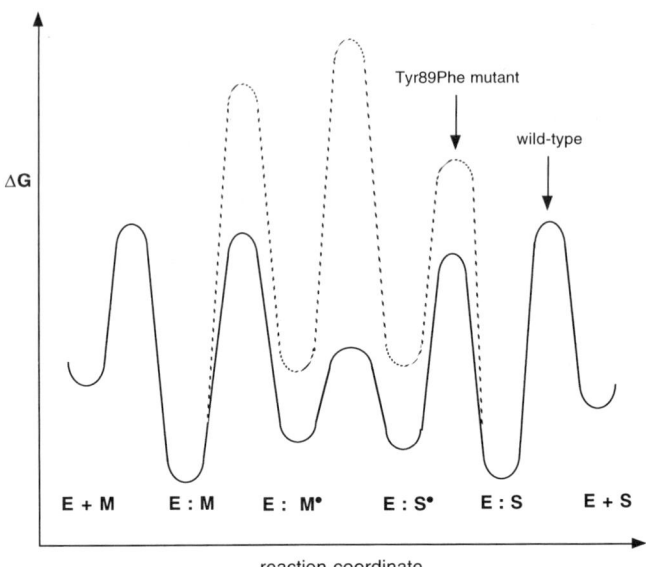

**FIGURE 23.** The effects of mutating Tyr-89 on the free energy profile of the isomerisation catalyzed by methylmalonyl-CoA mutase. **E**, enzyme; **M**, methylmalonyl-CoA, **S**, succinyl-CoA.

undertaken (Bothe *et al.*, 1998). Glutamate molecules, isotopically labeled with either $^{13}$C or $^{2}$H at C-2, C-3, or C-4, were incubated with the holoenzyme and rapidly frozen in liquid nitrogen. EPR spectra were recorded at 50 K and from changes in the hyperfine coupling interaction resulting from isotopic substitution the identity of the radicals were determined. The major organic radical was assigned as the C-4 radical of glutamate, and further modeling of the spectrum indicated that it was coupled to Co(II) and separated by a distance of ca. 6 Å—in good agreement with expectations from the crystal structure. Glutamate $^{13}$C- or $^{2}$H-labelled at C-2 produced a slight perturbation of the spectrum which was ascribed to a very low concentration of glycyl radical. However, with C-3-labeled glutamate no perturbation of the EPR spectrum was evident, suggesting that methylaspartyl radical is present at significant concentration. This is not surprising, as chemical reasoning would predict this primary methylene radical to be as unstable as 5'-deoxyadenosyl radical, which, as discussed above, cannot be formed in any significant concentration on the enzyme.

Pre-steady state kinetic studies have provided information on the free energy profile of the glutamate mutase reaction (Figure 24) (Chih and Marsh, 1999; Marsh and Ballou, 1998). In contrast to MMCM, rapid quench

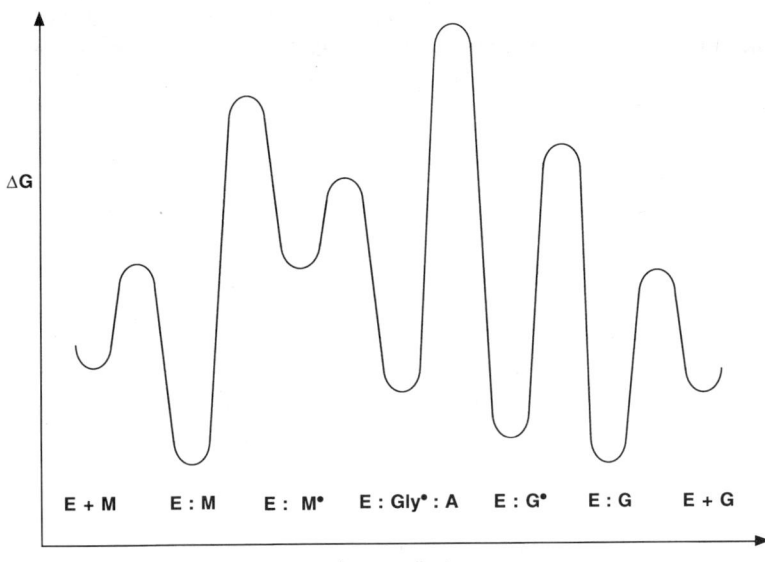

$\Delta G$

reaction coordinate

**FIGURE 24.** Free energy profile for the reaction catalyzed by glutamate mutase. **E**, enzyme; **M**, methylaspartate, **Gly**, glycine; **A**, acrylate; **G**, glutamate.

flow studies have demonstrated that product release is not rate limiting in either direction for the wild-type enzyme. Measurements of the deuterium isotope effect for hydrogen transfer between substrate and coenzyme under both steady state and pre-steady state conditions indicated that a step other than hydrogen transfer was partially rate limiting in the overall reaction. Furthermore, with methylaspartate as the substrate a lag phase is observed, which suggests the accumulation of an intermediate. The identity of this intermediate has not been unambiguously established, but likely it represents either the formation of the presumptive glycyl radical and acrylate, or the accumulation of glutamyl radical on the enzyme.

Recently it has been found that the enzyme will tolerate sterically conservative substitutions at C-2 of glutamate, which is interesting because the substituent should affect the stability of any radical at C-2. Depending upon the substrate analog, either partial reaction, isomerisation, or irreversible inhibition is observed (Table 2). Thus 2-ketoglutarate will undergo tritium exchange with $5'$-$^3$H-AdoCbl and generates a u.v.-visible spectrum similar to that observed for the steady state turn-over of glutamate (Roymoulik et al., 1999). This suggests the formation of a 2-ketoglutaryl radical, but this radical does not appear to react further, as the expected product, 3-

**Table 2**
**Reactions of various substrate analogues of glutamate with glutamate mutase.**
**Data from: A, Roymoulik et al., 1999; B, Roymoulik and Marsh, 1999; C, Pickett**
**and Marsh, unpublished results**

| Substrate analogue | Tritium exchange with AdoCbl | Isomerisation of radicals | Mechanism-based activation |
|---|---|---|---|
| 2-ketoglutarate[A] | Yes | No | No |
| L-2-hydroxy-glutarate[B] | Yes | Yes | No |
| 2-methylene-glutarate[C] | Yes | (Yes) | Yes |

methyloxalacetate is not observed. L-2-hydroxyglutarate, however, is a substrate for the enzyme, although $k_{cat}$ is only about 1% of that with L-glutamate (Roymoulik and Marsh, 1999). Most likely, the slow turnover reflects the difficulty in forming glyoxal radical, the intermediate analogous to the glycyl radical that is assumed to be formed in the reaction of glutamate. Glyoxal radical is expected to be less stable than glycyl radical, because oxygen stabilizes carbon-based radicals less effectively than nitrogen.

Lastly, 2-methyleneglutarate, which is a substrate for the related AdoCbl enzyme 2-methyleneglutarate mutase, interestingly appears to be a mechanism-based inhibitor of glutamate mutase (E. N. G. Marsh and J. Pickett, unpublished results) (previously it was reported that this molecule was a competitive inhibitor of the enzyme (Leutbecher et al., 1992)). The molecule undergoes tritium transfer between substrate and coenzyme, suggesting that 2-methyleneglutaryl radical can be formed reversibly on the enzyme. One possibility is that inactivation results from further reaction of the 2-methyleneglutaryl radical with amino acid residues at the active site.

## 4. PERSPECTIVE

Early studies on $B_{12}$ enzymes were in many cases hampered by difficulties in obtaining pure preparations of enzyme in large enough quantities for them to be studied by physical techniques such as stopped flow spectroscopy and X-ray crystallography. The complex subunit structure, oxygen sensitivity and requirement for cofactors other than AdoCbl that some enzymes display also posed serious technical challenges. Molecular genetics has, as with many areas of enzymology, revitalized the study of $B_{12}$ enzymes by facilitating the cloning, sequencing and heterologous expression of many of these enzymes in the last decade. The ability to express these enzymes in a corrin-free background that allows pure apo-enzyme to be obtained is a particular advantage, as preparations from the natural host often included large amounts of inactive enzyme that contained tightly bound corrinoids.

Sequencing studies initially revealed clear structural similarities, characterized by the **D-x-H-x-x-G** motif, between one sub-family of the AdoCbl enzymes, the carbon skeleton isomerases, which interestingly are shared by some MeCbl-dependent enzymes. The structure of the $B_{12}$-binding domain of methionine synthase first revealed the quite unexpected mode of $B_{12}$ binding involving the coordination of cobalt by the conserved histidine residue. [Although histidine coordination was presaged by EPR studies of Stupperich *et al.* (1990) on corrinoid-dependent methyl-transfer in *Sporomusa ovata*] However, further studies revealed that other enzymes which lack this motif, such as ribonucleotide reductase and diol dehydrase, keep the endogenous dimethylbenzimidazole ligand coordinated to cobalt.

Until very recently, this difference in the mode of $B_{12}$-binding, combined with the lack of sequence similarity and the diverse quaternary structures of the AdoCbl-dependent enzymes, would have led one to think that this group of enzymes would be very diverse structurally. But the structures of methylmalonyl-CoA mutase, glutamate mutase and diol dehydrase, three enzymes with very different quaternary structures and little sequence similarity, reveal an extraordinary unity of design. In each case the "lower" portion of AdoCbl including the nucleotide "tail" is recognized, albeit in different ways, by a Rossmann fold-like structure, which is a ubiquitous nucleotide-binding motif. The active site of each enzyme is deeply buried within a β-barrel, another extremely common protein fold, which sits atop the reactive face of AdoCbl: this appears to provide a highly screened environment in which radical chemistry can occur. It is tempting to speculate that this arrangement in which AdoCbl is sandwiched between β-barrel and Rossmann fold-like domains will be common to all AdoCbl-dependent enzymes.

The crystal structures have also revealed the details of the substrate-binding site for methylmalonyl-CoA mutase, glutamate mutase and diol dehydrase, and has highlighted amino acid residues and metal ions that may be important in catalysis. This will allow more sophisticated mechanistic hypotheses relating to the proteins' role in catalyzing these unusual chemical reactions to be formulated and tested.

Most hypotheses concerning the remarkable enzymatic activation of the AdoCbl Co—C bond towards homolysis have been based on studies of small molecules in which simple ligands to cobalt replaced the complex corrin ring. From such studies arose the idea that a distortion of the Co—

C bond, applied by the enzyme, could sufficiently lower the bond dissociation energy to achieve the substantial rates of catalysis seen. Studies on the
enzymes themselves have failed to find any evidence that binding AdoCbl
results in destabilization of the Co—C bond: the stretching frequency of the
Co—C bond is changed very little on binding to the protein; crystal structures reveal no great perturbation of the conformationally flexible corrin
ring; and the products of homolysis, Cbl(II) and $AdoCH_3$, are not very tight-
binding competitive inhibitors of the enzyme, as might be expected if
AdoCbl was being strained towards homolysis. Models that point towards
the $pK_a$ and steric bulk of the axial ligand to cobalt being important have
found more support. Mutation of the histidine ligand is indeed deleterious
to catalysis, but imidazole is clearly smaller than dimethylbenzimidazole,
contrary to the idea that large ligands labilize the Co—C bond through distortion of the corrin ring. Furthermore, the correlation between increasing
$pK_a$ of the axial ligand and increasing Co—C bond dissociation energy
evident in the cobaloxime model system is not reproduced in studies that
use the true corrin-ring structure.

So how *do* enzymes catalyze AdoCbl-dependent reactions? The
most significant mechanistic finding to emerge is the coupling of Co—
C bond homolysis to hydrogen abstraction from the substrate (or protein).
This makes sound mechanistic sense in that reactive radicals cannot be
generated in the absence of substrate, thereby reducing the possibility of
deleterious side reactions. It is quite possible, if the reaction is truly
concerted, that adenosyl radical, a species central to mechanistic thinking
about these enzymes, is in fact never formed! A concerted reaction would
require a highly ordered transition state, but optimizing reaction geometry
is an art that enzymes excel in. The transition state itself appears to be
unusual, as evidenced by the very large kinetic isotope effects observed for
several enzymes. This suggests that quantum tunneling of hydrogen, a phenomenon that is increasingly coming to light in the study of enzymes that
use free radicals, is probably an important component in AdoCbl-mediated
catalysis.

What is still not well understood is the remarkable stabilization of free
radicals that these enzymes are able to achieve. The bond dissociation
energy of the AdoCbl Co—C bond is ca. 30 kcal/mol, and in many cases, at
saturating substrate concentrations, organic radical species accumulate to
comprise 20–50% of the enzyme concentration under steady state conditions. This would seem to imply that the enzyme is able to destabilize the
Co—C covalent bond by about 30 kcal/mol. Put another way, the equilibrium constant for homolysis is shifted by a factor of $10^{22}$ in going from
free solution to the enzyme active site, a truly impressive feat even for
an enzyme! This, however, is an overestimation of the enzyme's prowess

because it is a *substrate*-based radical that accumulates on the enzyme, not adenosyl radical, and these are expected to be chemically much more stable. For example, the estimated free energy of forming thiyl radical from AdoCbl and cysteine, as occurs in ribonucleotide reductase (Licht *et al.*, 1999a), is only unfavorable by about 15 kcal/mol, which is much closer to the stabilization energies often seen in enzymatic reactions.

The AdoCbl enzymes have challenged our understanding of enzyme catalysis and enriched our knowledge of organic chemistry. Although neither widely distributed in nature, nor implicated in major diseases, the enzymes do provide excellent systems with which to study a difficult and fundamental problem—how to generate and control highly reactive free radical species—in the context of a simple biological reaction. What we learn from these enzymes should provide insight into the mechanisms of more complex radical-requiring enzymes and broaden our understanding of biological catalysis.

ACKNOWLEDGMENTS. We thank Prof. Christoph Kratky, University of Graz, for kindly making the structure of glutamate mutase available to us ahead of publication.

## 5. REFERENCES

Abeles, R. H., and Beck, W. S., 1967, The mechanism of action of cobamide coenzyme in the ribonucleotide reductase reaction. *J. Biol. Chem.* **242**:3589–3593.

Ashwell, S., Davies, A. G., Golding, B. T., Haymotherwell, R., and Mwesigyekibende, S, 1989, Model experiments pertaining to the mechanism of action of vitamin $B_{12}$-dependent α-methyleneglutarate mutase. *J. Chem. Soc. Chem. Comm.* **19**:1483–1485.

Babior, B. M., 1982, Ethanolamine ammonia-lyase. In $B_{12}$ (D. Dolphin, ed.), Vol. 2, John Wiley and Sons, New York, pp. 263–287.

Babior, B. M., and Krouwer, J. S., 1979, The mechanism of adenosylcobalamin-dependent reactions. *C.R.C. Crit. Rev. Biochem.* **6**:35–102.

Babior, B. M., Moss, T. H., Orme-Johnson, W. H., and Beinert, H., 1974, The mechanism of action of ethanolamine ammonia-lyase, a $B_{12}$-dependent enzyme. The participation of paramagnetic species in the catalytic deamination of 2-aminopropanol. *J. Biol. Chem.* **249**:4537–4544.

Bahnson, B. J., and Klinman, J. P., 1995, Hydrogen tunneling in enzyme catalysis. In *Enzyme Kinetics and Mechanism, Pt D.* Methods in Enzymology, **249**:373–397.

Bahnson, B. J., Colby, T. D., Chin, J. K., Goldstein, B. M., and Klinman, J. P., 1997, A link between protein structure and enzyme catalyzed hydrogen tunneling. *Proc. Natl. Acad. Sci. U.S.A.* **94**:12797–12802.

Baker, J. J., and Stadtman, T. C., 1982, Amino mutases. In $B_{12}$ (D. Dolphin, ed.), Vol. 2, John Wiley and Sons, New York, pp. 203–232.

Banerjee, R. V., Johnston, N. L., Sobeski, J. K., Datta, P., and Matthews, R. G., 1989, Cloning and sequence analysis of the *Escherichia coli metH* gene encoding cobalamin-dependent methionine synthase and isolation of a tryptic fragment containing the cobalamin-binding domain. *J. Biol. Chem.* **264**:13888–13895.

Banerjee, R. V., Frasca, V., Ballou, D. P., and Matthews, R. G., 1990, Participation of cob(I)alamin in the reaction catalyzed by methionine synthase from *Escherichia coli*: a steady-state and rapid kinetic analysis. *Biochemistry* **29**:11101–11109.

Barker, H. A., Weissbach, H., and Smyth, R. D., 1958, A coenzyme containing pseudovitamin $B_{12}$. *Proc. Natl. Acad. Sci. U.S.A.* **44**:1093–1097.

Baron, A. J., Stevens, C., Wilmot, C., Seneviratne, K. D., Blakeley, V., Dooley, D. M., Phillips, S. E. V., Knowles, P. F., and McPherson, M. J., 1994, Structure and mechanism of galactose oxidase. The free radical site. *J. Biol. Chem.* **269**:25095–25105.

Booker, S., and Stubbe, J., 1993, Cloning, sequencing, and expression of the adenosylcobalamin-dependent ribonucleotide reductase from *Lactobacillus leichmannii*. *Proc. Natl. Acad. Sci. U.S.A.* **90**:8352–8356.

Bothe, H., Darley, D. J., Albracht, S. P., Gerfen, G. J., Golding, B. T., and Buckel, W., 1998, Identification of the 4-glutamyl radical as an intermediate in the carbon skeleton rearrangement catalyzed by coenzyme $B_{12}$-dependent glutamate mutase from *Clostridium cochlearium*. *Biochemistry* **37**:4105–4113.

Brown, K. L., and Li, J., 1998, Activation parameters for the carbon-cobalt bond homolysis of coenzyme $B_{12}$ induced by the $B_{12}$-dependent ribonucleotide reductase from *Lactobacillus leichmannii*. *J. Am. Chem. Soc.* **120**:9466–9474.

Buckel, W., and Golding, B. T., 1996, Glutamate and 2-methyleneglutrate mutase: from microbial curiosities to paradigms for coenzyme $B_{12}$-dependent enzymes. *Chem. Soc. Rev.* 329–337.

Burkhardt, K., Phillipon, N., and Robinson, J. A., 1998, Isobutyryl-CoA mutase from Streptomycetes. In *Vitamin $B_{12}$ and $B_{12}$-proteins* (B. Krautler, D. Arigoni, and B. T. Golding, eds.), Wiley-VCH, Weinheim, pp. 265–272.

Chemaly, S. M., 1994, α-Methyleneglutarate mutase: an adenosylcobalamin-dependent enzyme, *S. Afr. J. Chem.* **47**:37–47.

Chen, H. P., and Marsh, E. N. G., 1997, How enzymes control the reactivity of adenosylcobalamin: effect on coenzyme binding and catalysis of mutations in the conserved histidine-aspartate pair of glutamate mutase. *Biochemistry* **36**:7884–7889.

Chih, H.-W., and Marsh, E. N. G., 1999, Pre-steady state investigation of intermediates in the reaction catalyzed by adenosylcobalamin-dependent glutamate mutase, *Biochemistry* **38**: in press.

Dong, S. L., Padmakumar, R., Maiti, N., Banerjee, R., and Spiro, T. G., 1998, Resonance raman spectra show that coenzyme $B_{12}$ binding to methylmalonyl-coenzyme A mutase changes the corrin ring conformation but leaves the Co—C bond essentially unaffected. *J. Am. Chem. Soc.* **120**:9947–9948.

Dowd, P., Wilk, B., and Wilk, B. K., 1992, 1st Hydrogen abstraction rearrangement model for the coenzyme $B_{12}$-dependent methylmalonyl-CoA to succinyl-CoA carbon skeleton rearrangement reaction. *J. Am. Chem. Soc.* **114**:7949–7951.

Drennan, C. L., Huang, S., Drummond, J. T., Matthews, R. G., and Ludwig, M. L., 1994a, How a protein binds $B_{12}$: a 3.0 Å structure of $B_{12}$-binding domains of methionine synthase. *Science* **266**:1669–1674.

Drennan, C. L., Matthews, R. G., and Ludwig, M. L., 1994b, Cobalamin-dependent methionine synthase: the structure of a methylcobalamin-binding fragment and implications for other $B_{12}$-dependent enzymes. *Curr. Opin. Struct. Biol.* **4**:919–929.

Drennan, C. L., Dixon, M. M., Hoover, D. M., Jarrett, J. T., Goulding, C. W., Matthews, R. G., and Ludwig, M. L., 1998, Cobalamin-dependent methionine synthase from *Escherichia coli*: structure and reactivity. In *Vitamin $B_{12}$ and $B_{12}$-proteins* (B. Krautler, D. Arigoni, and B. T. Golding, eds.), Wiley-VCH, Weinheim, pp. 133–156.

Faust, L. P., Connor, J. A., Roof, D. M., Hoch, J. A., and Babior, B. M., 1990, Cloning, sequencing, and expression of the genes encoding the adenosylcobalamin-dependent ethanolamine ammonia-lyase of *Salmonella typhimurium*. *J. Biol. Chem.* **265**:12462–12466.

Finke, R. G., and Hay, B. P., 1984, Thermolysis of adenosylcobalamin: a product, kinetic, and Co—C5′ bond dissociation study. *Inorg. Chem.* **23**:3041–3043.

Francalanci, F., Davis, N. K., Fuller, J. Q., Murfitt, D., and Leadlay, P. F., 1986, The subunit structure of methylmalonyl-CoA mutase from *Propionibacterium shermanii*. *Biochem. J.* **236**:489–494.

Frey, P. A., 1990, Importance of organic radicals in enzymatic cleavage of unactivated C—H bonds. *Chem. Rev.* **90**:1343–1357.

Frey, P. A., 1993, Lysine 2,3-aminomutase: is adenosylmethionine a poor man's adenosylcobalamin? *FASEB J.* **7**:662–670.

Frey, P. A., 1997, Radicals in enzymatic reactions. *Current Opinion in Chemical Biology* **1**:347–356.

Frey, P. A., Essenberg, M. K., and Abeles, R. H., 1967, Studies on the mechanism of hydrogen transfer in the cobamide coenzyme-dependent dioldehydrase reaction. *J. Biol. Chem.* **242**:5369–5377.

Geno, M. K., and Halpern, J., 1987, Why does nature not use the porphyrin ligand in vitamin $B_{12}$?. *J. Am. Chem. Soc.* **109**:1238–1240.

Golding, B. T., and Rao, D. N. R., 1986, Adenosylcobalamin-dependent enzymic reactions. In *Enzyme Mechanisms* (M. L. Page and A. Williams, eds.), London: Royal Society of Chemistry, pp. 404–428.

Gruber, K., Jogl, G., Klintschar, G., and Kratky, C., 1998, High-resolution crystal structures of cobalamins. In *Vitamin $B_{12}$ and $B_{12}$-proteins* (B. Krautler, D. Arigoni, and B. T. Golding, eds.), Wiley-VCH, Weinheim, pp. 335–348.

Halpern, J., 1985, Mechanisms of coenzyme $B_{12}$-dependent rearrangements. *Science* **227**:869–875.

Halpern, J., Kim, S.-H., and Leung, T. W., 1984, Cobalt-carbon bond dissociation energy of coenzyme $B_{12}$. *J. Am. Chem. Soc.* **106**:8317–8319.

Harkins, T. T., and Grissom, C. B., 1994, Magnetic field effects on $B_{12}$ ethanolamine ammonia lyase: evidence for a radical mechanism. *Science* **263**:958–960.

Hay, B. P., and Finke, R. G., 1987, Thermolysis of the Co—C bond in adenosylcorrins .3. quantification of the axial base effect in adenosylcobalamin by the synthesis and thermolysis of axial base-free adenosylcobinamide—insights into the energetics of enzyme-assisted cobalt carbon bond homolysis. *J. Am. Chem. Soc.* **109**:8012–8018.

Hay, B. P., and Finke, R. G., 1988, Thermolysis of the Co—C bond in adenosylcobalamin (coenzyme B12) .4. products, kinetics and Co—C bond-dissociation energy studies in ethylene-glycol. *Polyhedron* **7**:1469–1481.

Hodgkin, D. C., Pickworth, J., Robertson, J. H., Trueblood, K. N., Prosen, R. J., and White, J. G., 1955, Structure of vitamin $B_{12}$. The crystal structure of the hexacarboxylic acid derived from $B_{12}$ and the molecular structure of the vitamin. *Nature* **176**:325–328.

Hogenkamp, H. P. C., Ghambeer, R. K., Brownson, C., Blakley, R. L., and Vitols, E., 1968, Cobamides and ribonucleotide reduction. VI. Enzyme-catalyzed hydrogen exchange between water and deoxyadenosylcobalamin. *J. Biol. Chem.* **243**:799–808.

Holloway, D. E., and Marsh, E. N. G., 1994, Adenosylcobalamin-dependent glutamate mutase from *Clostridium tetanomorphum*. Overexpression in *Escherichia coli*, purification, and characterization of the recombinant enzyme. *J. Biol. Chem.* **269**:20425–20430.

Hollaway, M. R., White, H. A., Joblin, K. N., Johnson, A. W., Lappert, M. F., and Wallis, O. C., 1978, A spectrophotometric rapid kinetic study of reactions catalysed by coenzyme-$B_{12}$-dependent ethanolamine ammonia-lyase. *Eur. J. Biochem.* **82**:143–154.

IUPAC-IUB Commision on Biochemical Nomenclature, 1974, The nomenclature of corrinoids. *Eur. J. Biochem.* **45**:7–12.

Jordan, A., and Reichard, P., 1998, Ribonucleotide reductases. *Ann. Rev. Biochem.* **67**:71–98.

Kaufmann, F., Wohlfarth, G., and Diekert, G., 1998, O-Demethylase from *Acetobacterium dehalogenans*. Cloning, sequencing, and active expression of the gene encoding the corrinoid protein. *Eur. J. Biochem.* **257**:515–521.

Knappe, J., Elbert, S., Frey, M., and Wagner, A. F. V., 1993, Pyruvate formate-lyase mechanism involving the protein-based glycyl radical. *Biochem. Soc. Trans.* **21**:731–734.

Kohen, A., and Klinman, J. P., 1998, Enzyme catalysis: Beyond classical paradigms. *Acc. Chem. Res.* **31**:397–404.

Kraulis, P. J., 1991, MOLSCRIPT: a program to produce both detailed and schematic structures. *J. Appl. Cryst.* **24**:946–950.

Kräutler, B., Keller, W., and Kratky, C., 1987, Coenzyme $B_{12}$ chemistry: the crystal and molecular structure of cob(II)alamin. *J. Am. Chem. Soc.* **111**:8936–8938.

Lawrence, C. C., and Stubbe, J., 1998, The function of adenosylcobalamin in the mechanism of ribonucleoside triphosphate reductase from *Lactobacillus leichmannii. Curr. Opin. Chem. Biol.* **2**:650–655.

Lawrence, C. C., Gerfen, G. J., Samano, V., Nitsche, R., Robins, M. J., and Stubbe, J., 1999, Binding of Cob(II)alamin to the adenosylcobalamin-dependent ribonucleotide reductase from *Lactobacillus leichmannii*—Identification of dimethylbenzimidazole as the axial ligand, *J. Biol. Chem.* **274**:7039–7042.

Lenhert, P. G., and Hodgkin, D. C., 1961, Structure of the 5,6-dimethylbenzimidazolylcobamide coenzyme. *Nature* **192**:937–938.

Leutbecher, U., Albracht, S. P. J., and Buckel, W., 1992, Identification of a paramagnetic species as an early intermediate in the coenzyme $B_{12}$-dependent glutamate mutase reaction: a cob(II)amide?, *FEBS Lett.* **307**(2):144–146.

Leutbecher, U., Bocher, R., Linder, D., and Buckel, W., 1992, Glutamate mutase from *Clostridium cochlearium*. Purification, cobamide content and stereospecific inhibitors, *Eur. J. Biochem.* **205**:759–765.

Licht, S., Gerfen, G. J., and Stubbe, J., 1996, Thiyl radicals in ribonucleotide reductases, *Science* **271**:477–481.

Licht, S. S., Booker, S., and Stubbe, J., 1999a, Studies on the catalysis of carbon-cobalt bond homolysis by ribonucleoside triphosphate reductase: evidence for concerted carbon-cobalt bond homolysis and thiyl radical formation, *Biochemistry* **38**:1221–1233.

Licht, S. S., Lawrence, C. C., and Stubbe, J., 1999b, Thermodynamic and kinetic studies on cobalt-carbon bond homolysis by ribonucleotide triphosphate reductase: The importance of entropy in catalysis, *Biochemistry* **34**:1234–1242.

Lieder, K. W., Booker, S., Ruzicka, F. J., Beinert, H., Reed, G. H., and Frey, P. A., 1998, S-adenosylmethionine-dependent reduction of lysine 2,3-aminomutase and observation of the catalytically functional iron-sulfur centers by electron paramagnetic resonance, *Biochemistry* **37**:2578–2585.

Mancia, F., and Evans, P. R., 1998, Conformational changes on substrate binding to methylmalonyl CoA mutase and new insights into the free radical mechanism, *Structure* **6**:711–720.

Mancia, F., Keep, N. H., Nakagawa, A., Leadlay, P. F., McSweeney, S., Rasmussen, B., Bösecke, P., Diat, O., and Evans, P. R., 1996, How coenzyme $B_{12}$ radicals are generated: the crystal structure of methylmalonyl-coenzyme A mutase at 2 Å resolution. *Structure* **4**:339–350.

Mao, S. S., Yu, G. X., Chalfoun, D., and Stubbe, J., 1992, Characterization of C439SR1, a mutant of Escherichia coli ribonucleotide diphosphate reductase: evidence that C439 is a residue essential for nucleotide reduction and C439SR1 is a protein possessing novel thioredoxin-like activity, *Biochemistry* **31**:9752–9759.

Marsh, E. N. G., 1995, A radical approach to enzyme catalysis, *Bioessays* **17**:431–441.

Marsh, E. N. G., and Holloway, D. E., 1992, Cloning and sequencing of glutamate mutase component S from *Clostridium tetanomorphum*. Homologies with other cobalamin-dependent enzymes, *FEBS Lett.* **310**:167–170.

Marsh, E. N. G., and Ballou, D. P., 1998, Coupling of cobalt-carbon bond homolysis and hydrogen atom abstraction in adenosylcobalamin-dependent glutamate mutase. *Biochemistry* **37**:11864–11872.

Marsh, E. N. G., Holloway, D. E., and Chen, H.-P., 1998, Glutamate mutase. In *Vitamin B₁₂ and B₁₂-proteins* (B. Krautler, D. Arigoni, and B. T. Golding, eds.), Wiley-VCH, Weinheim, pp. 253–264.

Meier, T. W., Thoma, N. H., and Leadlay, P. F., 1996, Tritium isotope effects in adenosylcobalamin-dependent methylmalonyl-CoA mutase. *Biochemistry* **35**:11791–11796.

Michel, C., Albracht, S. P., and Buckel, W., 1992, Adenosylcobalamin and cob(II)alamin as prosthetic groups of 2-methyleneglutarate mutase from *Clostridium barkeri*, *Eur. J. Biochem.* **205**:767–773.

Miller, W. W., and Richards, J. H., 1969, Mechanism of action of coenzyme B₁₂. Hydrogen transfer in the isomerization of methylmalonyl coenzyme A to succinyl coenzyme A, *J. Am. Chem. Soc.* **91**:1458–1507.

Murakami, Y., Hisaeda, Y., Song, X.-M., and Ohno, T., 1992, A migrating group in a glutamate mutase model reaction mediated by a functionalised bilayer membrane. *J. Chem. Soc. Perkin Trans. II*:1527–1528.

Ng, F. T. T., Rempel, G. L., and Halpern, J., 1982, Ligand effects on transition-metal alkyl bond-dissociation energies, *J. Am. Chem. Soc.* **104**:621–623.

Ng, F. T. T., Rempel, G. L., and Halpern, J., 1983, Steric influences on cobalt-alkyl bond-dissociation energies, *Inorganica Chimica Acta-Letters* **77**:L165–L166.

Orme-Johnson, W. H., Beinert, H., and Blakley, R. L., 1974, Cobamides and ribonucleotide reduction. XII. The electron paramagnetic resonance spectrum of "active coenzyme B₁₂", *J. Biol. Chem.* **249**:2338–2343.

Padmakumar, R., and Banerjee, R., 1997, Evidence that cobalt-carbon bond homolysis is coupled to hydrogen atom abstraction from substrate in methylmalonyl-CoA mutase. *Biochemistry* **36**:3713–3718.

Padmakumar, R., Taoka, S., Padmakumar, R., and Banerjee, R., 1995, Coenzyme B₁₂ is coordinated by histidine and not dimethylbenzimidazole on methylmalonyl-CoA mutase, *J. Am. Chem. Soc.* **117**(26):7033–7034.

Picot, D., Loll, P. J., and Garavito, R. M., 1994, The X-ray structure of the membrane protein prostaglandin H₂ synthase-1. *Nature* **367**:243–249.

Pratt, J. M., 1972, *Inorganic chemistry of vitamin B₁₂*, Academic Press, London.

Reichard, P., 1993, The anaerobic ribonucleotide reductase from *Escherichia coli*. *J. Biol. Chem.* **268**:8383–8386.

Reitzer, R., Gruber, K., Jogl, G., Wagner, U. G., Bothe, H., Buckel, W., and Kratky, C. 1999, Structure of coenzyme B₁₂ dependent enzyme glutamate mutase from *Clostridium cochlearium*, *Structure* **7**:891–902.

Retey, J., 1990, Enzymatic-reaction selectivity by negative catalysis or how do enzymes deal with highly reactive intermediates. *Angew. Chem. Intl. Ed. Engl.* **29**:355–361.

Rickes, E. L., Brink, N. G., Koniuszy, F. R., Wood, T. R., and Folkers, K., 1948, Crystalline vitamin B₁₂. *Science* **107**:396–397.

Rosenberg, L. E., and Fenton, W. A., 1989, Disorders of propionate and methylmalonate metabolism. In *The Metabolic Basis of Inherited Disease* 6th edit. (C. R. Scriver, A. L. Beaudet, W. S. Sly, and D. Valle, eds.), McGraw Hill, New York, pp. 822–844.

Rossmann, M. G., Moras, D., and Olsen, K. W., 1974, Chemical and biological evolution of a nucleotide-binding protein. *Nature* **250**:194–199.

Roymoulik, I., and Marsh, E. N. G., 1999, A new adenosylcobalamin-dependent rearrangement catalyzed by glutamate mutase. *Submitted for publication.*

Roymoulik, I., Chen, H.-P., and Marsh, E. N. G., 1999, The reaction of the substrate analog 2-ketoglutarate with adenosylcobalmin-dependent glutamate mutase, *J. Biol. Chem.* **274**:11619–11622.

Sauer, K., Harms, U., and Thauer, R. K., 1997, Methanol:coenzyme M methyltransferase from *Methanosarcina barkeri*. Purification, properties and encoding genes of the corrinoid protein MT1, *Eur. J. Biochem.* **243**:670–677.

Scheuring, E., Padmakumar, R., Banerjee, R., and Chance, M. R., 1997, Extended X-ray absorption fine structure analysis of coenzyme $B_{12}$ bound to methylmalonyl-coenzyme A mutase using global mapping techniques, *J. Am. Chem. Soc.* **119**:12192–12200.

Shibata, N., Masuda, J., Tobimatsu, T., Toraya, T., Suto, K., Morimoto, Y., and Yasuoka, N., 1999, A new mode of $B_{12}$ binding and the direct participation of a potassium ion in enzyme catalysis: X-ray structure of diol dehydrase. *Structure* **7**:997–1008.

Sirovatka, J. M., and Finke, R. G., 1997, Coenzyme $B_{12}$ chemical precedent studies: Probing the role of the imidazole base-on motif found in $B_{12}$-dependent methylmalonyl-CoA mutase, *J. Am. Chem. Soc.* **119**:3057–3067.

Steiner, U. E., and Ulrich, T., 1989, Magnetic-field effects in chemical kinetics and related phenomena. *Chem. Rev.* **89**:51–147.

Stubbe, J., and RiggsGelasco, P., 1998, Harnessing free radicals: Formation and function of the tyrosyl radical in ribonucleotide reductase, *Trends Biochem. Sci.* **23**(11):438–443.

Stupperich, E., 1993, Recent advances in elucidation of biological corrinoid functions. *FEMS Microbiol. Rev.* **12**:349–366.

Stupperich, E., Eisinger, H. J., and Albracht, S. P. J., 1990, Evidence for a super-reduced cobamide as the major corrinoid fraction in vivo and a histidine residue as a cobalt ligand of the *para*-cresolyl cobamide in the acetogenic bacterium *Sporomusa-ovata, Eur. J. Biochem.* **193**:105–109.

Switzer, R. L., 1982, Glutamate mutase. In $B_{12}$ (D. Dolphin, ed.), Vol. 2, John Wiley and Sons, New York, pp. 289–305.

Tamao, Y., and Blakely, R. L., 1973, Direct spectroscopic observation of an intermediate formed from deoxyadenosylcobalamin in ribonucleotide reduction. *Biochemistry* **12**:24–34.

Tan, S. L., Kopczynski, M. G., Bachovchin, W. W., Orme-Johnson, W. H., and Babior, B. M., 1986, Electron spin-echo studies of the composition of the paramagnetic intermediate formed during the deamination of propanolamine by ethanolamine ammonia-lyase, and AdoCbl-dependent enzyme, *J. Biol. Chem.* **261**:3483–3485.

Thoma, N. H., Meier, T. W., Evans, P. R., and Leadlay, P. F., 1998, Stabilization of radical intermediates by an active-site tyrosine residue in methylmalonyl-CoA mutase. *Biochemistry* **37**:14386–14393.

Tobimatsu, T., Sakai, T., Hashida, Y., Mizoguchi, N., Miyoshi, S., and Toraya, T., 1997, Heterologous expression, purification, and properties of diol dehydratase, an adenosylcobalamin-dependent enzyme of *Klebsiella oxytoca, Arch. Biochem. Biophys.* **347**:132–140.

Tobimatsu, T., Azuma, M., Hayashi, S., Nishimoto, K., and Toraya, T., 1998, Molecular cloning, sequencing and characterization of the genes for adenosylcobalamin-dependent diol dehydratase of *Klebsiella pneumoniae. Biosci. Biotechnol. Biochem.* **62**(9):1774–1777.

Tollinger, M., Konrat, R., Hilbert, B. H., Marsh, E. N. G., and Kräutler, B., 1998, How a protein prepares for $B_{12}$ binding: structure and dynamics of the $B_{12}$-binding subunit of glutamate mutase from *Clostridium tetanomorphum*. *Structure* **6**:1021–1033.

Toraya, T., 1998, Recent structure-function studies of $B_{12}$ coenzymes in diol dehydrase. In *Vitamin $B_{12}$ and $B_{12}$-proteins* (B. Krautler, D. Arigoni, and B. T. Golding, eds.), Wiley-VCH, Weinheim, pp. 303–320.

Toraya, T., 1994, Diol dehydrase and glycerol dehydrase, coenzyme $B_{12}$-dependent isozymes, In *Metal ions in biological systems* (H. Sigel and A. Sigel, eds.), Vol. 30, Marcel Dekker, New York, pp. 217–254.

Toraya, T., Shirakashi, T., Kosuga, T., and Fukui, S., 1976, Substrate specificity of coenzyme $B_{12}$-dependent diol dehydrase: glycerol as both a good substrate and a potent inactivator. *Biochem. Biophys. Res. Commun.* **69**:475–480.

Toraya, T., Miyoshi, S., Mori, M., and Wada, K., 1994, The synthesis of a pyridyl analog of adenosylcobalamin and its coenzymic function in the diol dehydratase reaction, *Biochim. Biophys. Acta.* **1204**:169–174.

Valinsky, J. E., Abeles, R. H., and Fee, J. A., 1974, Electron spin resonance studies on diol dehydrase. III. Rapid kinetic studies on the rate of formation of radicals in the reaction with propanediol, *J. Am. Chem. Soc.* **96**:4709–4710.

West, R., 1948, Activity of vitamin $B_{12}$ in Addisonian pernicious anaemia. *Science* **107**:398.

Wirt, M. D., Sagi, I., Chen, E., Frisbie, S. M., Lee, R., and Chance, M. R., 1991, Geometric conformations of intermediates of $B_{12}$ catalysis by X-ray edge spectroscopy: Co(I) $B_{12}$, Co(II) $B_{12}$, and base-off adenosylcobalamin, *J. Am. Chem. Soc.* **113**:5299–5304.

Wollowitz, S., and Halpern, J., 1984, Free-radical rearrangement involving the 1,2-migration of a thioester group—Model for the coenzyme-B12 dependent methylmalonyl-CoA mutase reaction, *J. Am. Chem. Soc.* **106**:8319–8321.

Wollowitz, S., and Halpern, J., 1988, 1,2-Migrations in free-radicals related to coenzyme-$B_{12}$ dependent rearrangements. *J. Am. Chem. Soc.* **110**:3112–3120.

Yamanishi, M., Yamada, S., Ishida, A., Yamauchi, J., and Toraya, T., 1998a, EPR spectroscopic evidence for the mechanism-based inactivation of adenosylcobalamin-dependent diol dehydratase by coenzyme analogs, *J. Biochem. (Tokyo)* **124**:598–601.

Yamanishi, M., Yamada, S., Muguruma, H., Murakami, Y., Tobimatsu, T., and Toraya, T., 1998b, Evidence for axial coordination of 5,6-dimethylbenzimidazole to the cobalt atom of adenosylcobalamin bound to diol dehydratase, *Biochemistry* **37**:4799–4803.

Zagalak, B., Frey, P. A., Karabatsos, G. L., and Abeles, R. H., 1966, The stereochemistry of the conversion of D- and L-1,2-propanediols to propionaldehyde, *J. Biol. Chem.* **241**:3028–3035.

Zelder, O., Beatrix, B., and Buckel, W., 1994, Cloning, sequencing and expression in *Escherichia coli* of the gene encoding component S of the coenzyme $B_{12}$-dependent glutamate mutase from *Clostridium cochlearium*. *FEMS Microbiol. Lett.* **118**:15–22.

Zelder, O., Beatrix, B., Kroll, F., and Buckel, W., 1995, Coordination of a histidine residue of the protein-component S to the cobalt atom in coenzyme $B_{12}$-dependent glutamate mutase from *Clostridium cochlearium*. *FEBS Lett.* **369**:252–254.

Zhao, Y., Such, P., and Rétey, J., 1992, Radical intermediates in the coenzyme $B_{12}$ dependent methylmalonyl-CoA mutase reaction shown by EPR spectroscopy, *Angew. Chem. Int. Ed. Engl.* **31**:215–216.

Zhao, Y., Abend, A., Kunz, M., Such, P., and Retey, J., 1994, Electron paramagnetic resonance studies of the methylmalonyl-CoA mutase reaction. Evidence for radical intermediates using natural and artificial substrates as well as the competitive inhibitor 3-carboxypropyl-CoA, *Eur. J. Biochem.* **225**:891–896.

*Chapter 12*

# Ribonucleotide Reductase

## A Virtual Playground for Electron Transfer Reactions

Margareta Sahlin and Britt-Marie Sjöberg

## 1. INTRODUCTION

Beyond doubt, ribonucleotide reductase (RNR) is a very complicated—and fascinating—enzyme! It catalyses a demanding chemical reaction, the site-specific reduction of the sugar moiety of a ribonucleotide to its corresponding deoxyribonucleotide. Biosynthetically as well as chemically, this reaction requires free radical chemistry. As radicals in general are extremely reactive, it is essential that the enzyme has evolved to master the power of the radicals. Interestingly, and as we will see below, the currently known different classes of RNRs have explored fundamentally different ways of achieving this control.

---

Abbreviations: AdoCbl, deoxyadenosylcobalamin; AdoMet, S-adenosyl methionine; dopa, 3,4-dihydroxyphenylalanine; ENDOR, electron nuclear double resonance; EPR, electron paramagnetic resonance; NMR, nuclear magnetic resonance; RNR, ribonucleotide reductase; RTP, radical transfer pathway.

---

**MARGARETA SAHLIN and BRITT-MARIE SJÖBERG**     Department of Molecular Biology, Stockholm University, SE-10691 Stockholm, Sweden.
*Subcellular Biochemistry, Volume 35: Enzyme-Catalyzed Electron and Radical Transfer*, edited by Holzenburg and Scrutton. Kluwer Academic / Plenum Publishers, New York, 2000.

Ribonucleotide reductase plays a critical role in the life cycle of all living organisms. By catalysing the conversion of ribonucleotides to deoxyribonucleotides, it holds a unique position at the biological crossroads between RNA synthesis and DNA synthesis. Its control of DNA synthesis and cell proliferation is mediated both by providing all precursors for replication and by keeping a balanced supply between them. Throughout the years, the metabolic key-role of RNR has been explored successfully in a variety of antiviral and antiproliferative therapies.

In addition, RNR ought to tickle the mind of anyone interested in evolution. It is now generally accepted that life on Earth evolved as an RNA world that later developed into the DNA world that modern life is based on. Ribonucleotide reductase is an absolute requirement for such a transition to occur. Currently, three different classes of RNR are known—classes I, II and III. Results of recent years imply that the catalytic core of all RNRs has a common origin, yet some RNR components seem evolutionary related to widely different metabolic enzymes, e.g. methane monooxygenase, fatty acid desaturase, pyruvate formate-lyase, and biotin synthetase. In the footsteps of the Genome Projects we have learnt that many prokaryotes, especially those that experience a variety of different growth conditions, may encode genes for several different RNRs.

This review will primarily focus on electron transfer reactions in class I RNR, the RNR found in essentially all eukaryotes, including humans and mammals, and in several prokaryotes. As indicated in the title, RNR is a virtual playground for electron transfer reactions, and in class I RNR such transfer reactions occur both during enzyme activation and during catalysis. We will discuss these two levels extensively, and towards the end make relevant comparisons to class II and III RNRs, found exclusively in prokaryotes (except for a class II RNR found in an early diverging eukaryote). Other aspects of RNRs have been covered in several recent review articles (Eklund and Fontecave, 1999; Jordan and Reichard, 1998; Stubbe, 1998; Sjöberg, 1997; Gräslund and Sahlin, 1996; Nocentini, 1996).

## 2.  THREE DIFFERENT RIBONUCLEOTIDE REDUCTASE CLASSES

The common denominators of RNRs are radical chemistry involving redox active cysteines (see sections 3.1 and 3.3) and with few exceptions an allosteric control mediated by deoxyribonucleoside triphosphates (dNTPs) and ATP (Reichard, 1997). Yet, several other characteristics distinguish the three RNR classes (Table 1) from each other:

## Table 1
### Characteristics of RNR classes

| Type of class | Polypeptide composition | Gene nomenclature | Stable radical/ cofactor | Metal site/ cofactor | Substrate | Reductant | $k_{cat}$ $(s^{-1})^a$ |
|---|---|---|---|---|---|---|---|
| I | $\alpha_2\beta_2$ | nrdA/nrdE nrdB/nrdF | Tyr$\bullet^b$ | Diiron-oxo$^b$ | NDP | Thioredoxin Glutaredoxin NrdH-redoxin | 4–11 |
| II | $\alpha$ or $\alpha_2$ | nrdJ | AdoCbl | Cobalamin | NDP/NTP | Thioredoxin | 2 |
| III | $\alpha_2(\beta_2)^c$ | nrdD nrdG | Gly$\bullet^c$ | Iron-sulphur$^d$ AdoMet$^d$ | NTP | Formate | 3 |

$^a$ $k_{cat}$ values are from (Andersson, Sahlin and Sjöberg, manuscript in preparation (Licht et al., 1999; Sjöberg, 1997).
$^b$ The tyrosyl radical and the diiron-oxo centre is confined to the $\beta_2$ (or NrdB/NrdF) component.
$^c$ The glycyl radical is confined to the $\alpha_2$ (or NrdD) component.
$^d$ Recent studies indicate that $\beta_2$ (or NrdG) is an activase needed for introduction of the glycyl radical in $\alpha_2$ (or NrdD), and plausibly not needed during catalysis (Tamarit et al., 1999). The Fe$_4$S$_4$ cluster is confined to $\beta_2$ and AdoMet plausibly binds close by (Ollagnier et al., 1999; Ollagnier et al., 1997).

(i) In its simplest form, e.g. in *Lactobacillus leichmannii* class II RNR, the holoenzyme can comprise only one polypeptide chain, denoted $\alpha$. Class I RNRs are tetrameric enzymes comprising an $\alpha_2$ and a $\beta_2$ homodimer, where $\alpha_2$ carries the active site and $\beta_2$ a stable radical essential for catalysis. Class III RNRs have also been described to consist of two different homodimers, $\alpha_2$ and $\beta_2$, where $\alpha_2$ carries both the active site and a stable radical. The $\beta_2$ homodimers of class I and class III RNRs are structurally different and fulfil completely different functions. The $\beta_2$ component of class I RNR (denoted R2 below) is an intrinsic part of the holoenzyme required throughout the reaction mechanism. The $\beta_2$ component of class III RNR functions to activate the $\alpha_2$ component, and recent results suggest that the class III RNR proper may be an $\alpha_2$ enzyme (Tamarit *et al.*, 1999).

(ii) In class I RNRs the radical chain initiator is a stable tyrosyl radical localised to the R2 component. In class II RNRs the radical chain is initiated by cleavage of the cofactor AdoCbl, and in class III RNRs the initiator is a stable glycyl radical localised to the $\alpha_2$ component.

(iii) Activation of class I RNRs requires oxygen, whereas class II RNRs are indifferent to oxygen, and class III RNRs are irreversibly inactivated by oxygen.

(iv) The phosphorylation level of the ribonucleotide substrates differs, and is either diphosphate or triphosphate as shown in the overall equations below:

Class I:   $rNDP + RNR\text{-}(SH)_2 \rightarrow dNDP + RNR\text{-}S_2 + H_2O$        (1a)

Class II:   $rND(T)P + RNR\text{-}(SH)_2 \rightarrow dND(T)P + RNR\text{-}S_2 + H_2O$       (1b)

Class III:  $rNTP + RNR + HCOOH \rightarrow dNTP + RNR + CO_2 + H_2O$    (1c)

(v) The physiological reducing system(s) differs for the different classes, as specified in Table 1. Notably, class I RNRs can be further subdivided into class Ia (NrdA, NrdB) and class Ib (NrdE, NrdF) based on sequence similarities and physiological reduction systems, where class Ia utilises the thioredoxin and the glutaredoxin systems (Holmgren and Björnstedt, 1995; Holmgren and Åslund, 1995), and class Ib utilises the NrdH-redoxin system (Jordan *et al.*, 1997).

Amazingly, and despite these seemingly major differences, the RNR $\alpha$ polypeptides plausibly have a common evolutionary origin with a preserved structural core comprised of two antiparallel five-stranded $\beta/\alpha$ halfbarrels (Figure 1). The resulting ten-stranded $\beta/\alpha$ barrel is wide enough to accom-

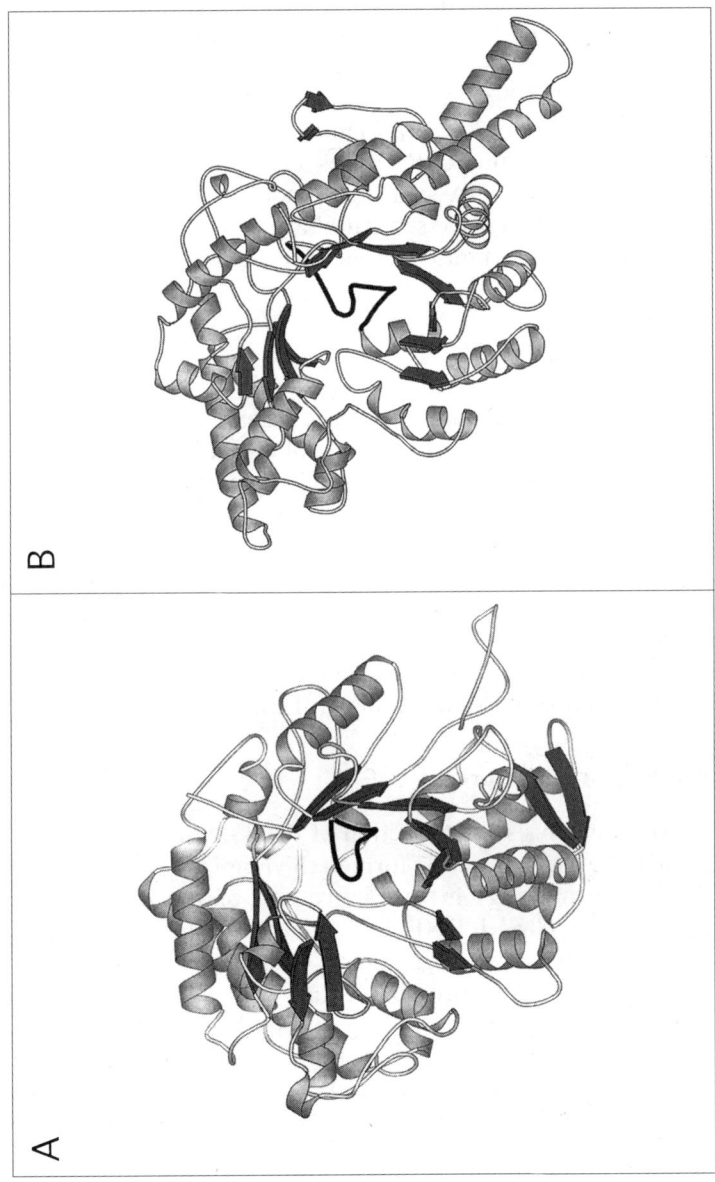

**FIGURE 1.** Three-dimensional structure of catalytic core regions in *Escherichia coli* class I (A) and bacteriophage T4 class III (B) RNRs (Logan *et al.*, 1999; Uhlin and Eklund, 1994). The ten-stranded β/α barrel comprising the active site region is shown. The finger-loop, which contains the active site cysteine, is shown as a solid line.

modate a "finger"-loop, with a mechanistically essential cysteine residue at its tip. The reaction mechanism, based on concerted radical chemistry, is initiated by formation of a transient cysteinyl radical at the finger-loop. The intermediate steps of the reaction mechanism are similar for class I and II, but slightly different for class III RNRs (cf. (Eklund and Fontecave, 1999; Stubbe and van der Donk, 1998). As indicated above, however, the reaction path leading to formation of the transient cysteinyl radical is entirely different for the three different RNR classes. Finally, the balanced supply of deoxyribonucleotides for DNA synthesis is controlled via an elaborate allosteric regulation, with extensive structural and mechanistic similarities, but with physiologically important differences (Reichard, 1997).

## 3.   REACTION MECHANISM

The proposed reaction mechanism is based on experiments with RNRs and model compounds, as well as on theoretical considerations. The RNR studies have involved specific isotope labelling experiments, x-ray crystal-lography, EPR, and studies with substrate analogues and active site mutant enzymes. Over the years several independent research groups have contributed important details to the consensus mechanism for class I RNRs (Lawrence *et al.*, 1999; Persson *et al.*, 1998; Eriksson *et al.*, 1997; Mao *et al.*, 1992a; Åberg *et al.*, 1989; Larsson and Sjöberg, 1986; Stubbe and Ackles, 1980; Thelander and Larsson, 1976; Thelander, 1974; Ehrenberg and Reichard, 1972; Durham *et al.*, 1967) shown in Figure 2.

### 3.1.   Radical Chemistry At Work

The first step in catalysis is abstraction of the 3'-hydrogen of the sub-strate ribose moiety by the transiently formed thiyl radical. In class I RNRs, Cys439 (in *E. coli* numbering) in the active site of the $\alpha_2$ component (denoted protein R1 in class I RNRs) is the site of this transient thiyl radical. It is generated by long range radical transfer between the stable tyrosyl radical at position 122 in protein R2 and Cys439 in R1 (see section 4). The 3'-hydrogen abstraction step was proposed by Stubbe almost two decades ago, and subsequently was corroborated by use of specifically tritiated substrates (Stubbe and Riggs-Gelasco, 1998; Stubbe and Ackles, 1980). The next step, protonation and abstraction of the 2'-hydroxyl group is facilitated by hydrogen bonds between the 2'- and 3'-hydroxyls of the substrate and the conserved residues Asn437 and Glu441 (in *E. coli* numbering) in the active site. On theoretical grounds, it has been suggested that the resulting water molecule may remain in the active site and

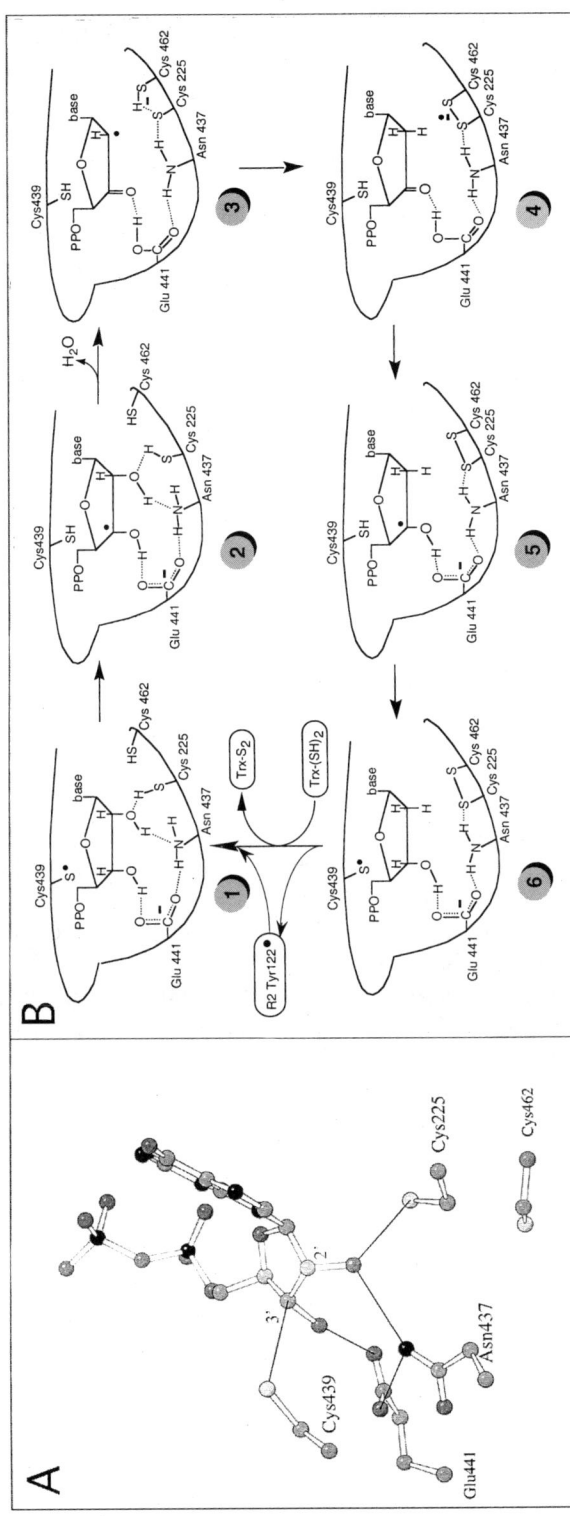

**FIGURE 2.** Three-dimensional structure of conserved residues and the substrate GDP in the active site of *E. coli* protein R1 (A) (Eriksson *et al.*, 1997), and proposed reaction mechanism for RNR (B). A turnover is initiated by long-range RTP between Tyr122• in R2 and Cys439 in R1, and completed by a reversal of this RTP (thin arrows). Oxidised R1 is reduced by the thioredoxin (Trx), or glutaredoxin system.

facilitate progression of later steps (Siegbahn, 1998). The actual reduction of the 2'-position is achieved by stepwise oxidation of a dithiol pair (Cys225 and Cys462 in *E. coli* numbering) and results in a disulphide bridge between the two cysteines. The theoretical studies indicate that these reductive steps require H-bonded participation of the conserved Glu441 and Asn437, and a water molecule (Siegbahn, 1998). Formation of the deoxyribonucleotide product and completion of the reaction involves reintroduction of the 3'-hydrogen and reformation of the transient thiyl radical at Cys439.

At the end of a catalytic cycle, Cys439• in R1 acquires a hydrogen atom via the long-range radical transfer pathway (RTP) and the radical is transferred to the stable position at Tyr122 in R2. This makes protein R2 ready for a new catalytic cycle. The disulphide in the active site region of R1 is reduced by the physiological glutaredoxin system or thioredoxin system (cf. Table 1 and Figure 2B) prior to the next catalytic turnover. *In vitro* this reduction can be performed by e.g. dithiothreitol. If reduction of oxidised Cys225-Cys462 in R1 is slower than the reduction of a ribonucleotide, a high $k_{cat}$ could still be sustained if a dissociated active R2 could rapidly associate with another reduced R1. Such a mechanism would allow an efficient substrate reduction even in situations when the *in vivo* concentration of R2 is lower than that of R1.

## 3.2. Substrate Analogues

A series of substrate analogues have been instrumental in elucidating the reaction mechanism of RNRs. All have in common a substitution at the 2'-position of the ribose moiety of the substrate nucleotide, i.e. the position to be reduced in the enzymatic reaction. Examples of 2'-substitutions are azide, chloride and other halide, thiol, and fluorinated alkane/ene groups (Stubbe and van der Donk, 1995). These analogues function as $k_{cat}$ or suicide inhibitors of the enzyme, i.e. they are chemically unreactive in the absence of a functional RNR. By utilising their binding specificity to RNR they are activated by the enzyme and inhibit it irreversibly during catalysis. The changed leaving group in the 2'-analogues prevents the chemical reaction from completing a full turnover cycle. The radical transfer only works in one direction (i.e. the first few steps in Figure 2B), and initially generates a transient radical in the substrate analogue. Early studies with 2'-azido-2'-deoxy-CDP demonstrated a kinetic coupling between loss of the tyrosyl radical in R2 and appearance of a new radical derived from the substrate analogue (Sjöberg *et al.*, 1983), later shown to be a covalent attachment of one azide nitrogen to an active site cysteine, presumably Cys225 (van der Donk *et al.*, 1995). In this respect a radical transfer between R1 and R2 in

*E. coli* class I RNR was evident already in 1983, long before the three-dimensional structures were known or protein engineering studies had been adopted. A common net result of incubations with 2′-substituted substrate analogues is inactivation of the R2 component by loss of the tyrosyl radical, and formation of a transient radical localised to the active site of the enzyme which eventually leads to inactivation of protein R1 by covalent modification. This effective dual inhibition of RNR by some of the 2′-substituted substrate analogues is currently explored in antiproliferative treatment in clinical trials (Nocentini, 1996).

### 3.3. Studies in Active Site Mutant Enzymes

Early protein chemical studies, followed by site-directed mutagenesis experiments in *E. coli*, identified five catalytically important cysteine residues in R1 (Mao *et al.*, 1992a; Åberg *et al.*, 1989; Lin *et al.*, 1987) (cf. Figure 2). These cysteines are:

(i) Cys439, the active site cysteine that transiently forms a thiyl radical during catalysis
(ii) Cys225 and Cys462, the redox active cysteine pair that is oxidised to a disulphide during catalysis
(iii) Cys754 and Cys759, a C-terminal redox active cysteine pair that presumably shuffles reducing equivalents between the glutaredoxin/thioredoxin systems and the Cys225-Cys462 disulphide.

In cocrystals of protein R1 with substrate (cf. Figure 2A) it is evident that the Cys225-Cys462 pair is on one side of the ribose moiety (i.e. the side where the 2′- and 3′-OH groups protrude) and that Cys439 is on the other side of the ribose ring (Eriksson *et al.*, 1997). The C3′ and C2′ positions of the sugar are at rather similar distances (3.4 and 4.1 Å) from the sulphur of Cys439 (Eriksson *et al.*, 1997). Evidently, additional features of the active site, e.g. the 3′-OH group H-bonding to a carboxylate, contribute to the selection of the 3′-position in the hydrogen abstraction step. Cys225 is at hydrogen-bonding distance from the ribose 2′-OH (Eriksson *et al.*, 1997). In the reduced form of R1, Cys462 is too far away from the active site to contribute to substrate binding. However, in oxidised R1 both Cys225 and Cys462 have moved and partly fill the area previously occupied by the substrate. The C-terminal part of R1 is not deducible in the crystal structure; presumably it needs to be highly flexible during turnover.

As expected, mutations at Cys439 (C439A, C439S) are catalytically inert (Mao *et al.*, 1992b; Åberg *et al.*, 1989). Similar results were obtained with mutations affecting the cysteine pair (C225A, C462A, C462S) (Mao *et al.*, 1992a; Åberg *et al.*, 1989). A very interesting and highly enlightening result was obtained with the mutant enzyme C225S, which commits mechanism-based suicide (van der Donk *et al.*, 1996). Apparently, the reaction mechanism is initiated and proceeds to step 3 (Figure 2B), but the reduction *per se* cannot be promoted by the Ser225 side chain. Instead, the α-polypeptide is truncated site-specifically at the 225 position, the stable tyrosyl radical character at Tyr122 in R2 is not reformed, and the prolonged radical character of the substrate leads to decomposition of the nucleotide. This virtual massacre that leads to irreversible inactivations of R1, R2, and the substrate, is reminiscent of the net result obtained with the $k_{cat}$ inhibitors mentioned above, and strongly supports the first few steps of the proposed reaction mechanism (Figure 2B).

Recent results obtained with site-directed mutagenesis of Glu441 have provided additional insight and support for the mechanism outlined in Figure 2 (Lawrence *et al.*, 1999; Persson *et al.*, 1999; Persson *et al.*, 1998; Persson *et al.*, 1997). Glu441 has a crucial position in the active site, providing hydrogen bonds to the 3'-OH and the nearby conserved Asn437, which in its turn has a hydrogen bond to the 2'-OH. Glu441 is believed to participate as a general base, and as part of an electron transfer pathway used during the reduction step. Characterisation of E441A and E441D clearly showed that a carboxylate function is needed in this position. Most intriguing results were obtained in studies of the E441Q mutant, which as C225S, turns out to be a suicidal enzyme. At the expense of the tyrosyl radical, incubation of E441Q with native R2 and substrate leads to formation of a series of consecutive transient radical intermediates, and eventually decomposition of the substrate nucleotide (Figure 3). However, different from the C225S reaction, the E441Q reaction does not lead to any truncation of the polypeptide chain (Persson *et al.*, 1998). A combination of specific isotope-labelling experiments and advanced spectroscopic techniques allowed assignment of the transient radicals (Lawrence *et al.*, 1999; Persson *et al.*, 1999; Persson *et al.*, 1998; Persson *et al.*, 1997) and formulation of the reaction sequence in Figure 3. Decay of Tyr122• in R2 is followed by formation of another tyrosyl radical in R2, presumably at Tyr356 (Persson *et al.*, 1999; Persson *et al.*, 1998). Decay of this transient tyrosyl radical in R2, is followed by formation of a disulphide anion radical in R1, presumably at the Cys225-Cys462 pair. Decay of the disulphide anion radical in R1 leads to formation of a carbon-centred nucleotide radical and eventually to decomposition of the nucleotide and formation of a furanone-derived covalent adduct to R1. Plausibly, this highly informative molecular damage has identified at least two of the transient radical intermediates in the proposed native reaction mechanism (Figure 2).

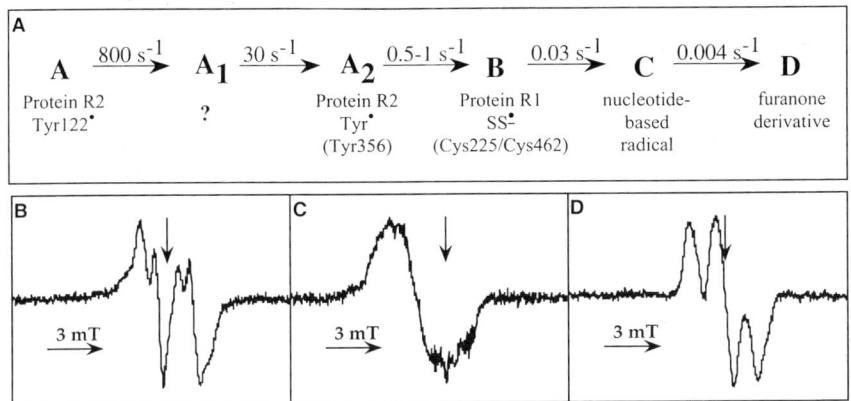

**FIGURE 3.** Consecutive radical intermediates observed in the suicidal reaction between *E. coli* E441Q, wild type R2 and substrate (A), including EPR spectra for the transient tyrosyl radical in R2 presumably at Tyr356 (B), the disulphide anion radical at Cys225–Cys462 in R1 (C), and the nucleotide-derived radical (D). The arrows in B-D indicate g = 2.005.

## 4. CLASS I RIBONUCLEOTIDE REDUCTASE—THE RADICAL TRANSFER PATHWAY

The demanded electron transfer in *E. coli* RNR could present a real enigma! As estimated from the docked complex of the two separately solved three-dimensional structures of R1 and R2 (Figure 4), the site of the stable tyrosyl radical in protein R2 is at a distance of approximately 30–40 Å from the active site in protein R1 (Eklund *et al.*, 1997; Uhlin and Eklund, 1994). According to theoretical as well as experimental considerations of electron transfer reactions in proteins (Gray and Winkler, 1996; Beratan *et al.*, 1992; Moser *et al.*, 1992; Marcus and Sutin, 1985), the electron transfer rate over such a distance would be on the order of $1$–$0.001\,s^{-1}$. As shown in Table 1, $k_{cat}$ for *E. coli* RNR is $4$–$11\,s^{-1}$. This overall rate comprises several steps: an initial electron transfer between Tyr122 in R2 and Cys439 in R1, the chemical steps at the active site, and finally the reversal (Cys• → Tyr•) of the initial electron transfer.

Fortunately, the three-dimensional structures of R1 and R2, and the docked holoenzyme complex, suggest a plausible explanation to the problem (Figure 4). Cys439 in R1 is in contact with the surface of the protein via two hydrogen bonded side chains. Likewise, Fe1, the ferric ion closest to Tyr122, is in contact with the surface of R2 via three hydrogen bonded side chains. In the docked complex of R1 and R2, these two hydrogen-bonded pathways meet at the R1–R2 interface. All amino acid residues involved in this proposed electron transfer pathway are conserved

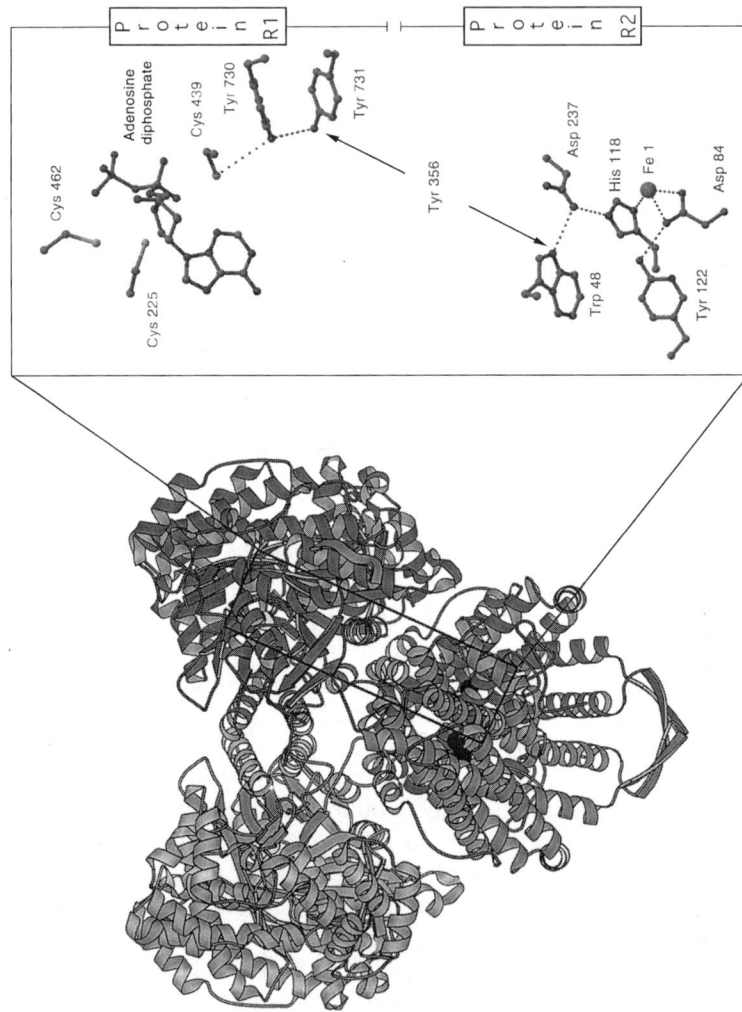

**FIGURE 4.** The class I RNR holoenzyme complex model built from the separately solved three-dimensional structures of *E. coli* protein R1 (Uhlin and Eklund, 1994) and protein R2 (Nordlund and Eklund, 1993; Nordlund *et al.*, 1990). The path of the radical transfer is indicated, and a blow-up of the participating residues and a substrate nucleotide is shown in the box.

throughout the class I RNR family and are hydrogen-bonded to each other. It was therefore suggested that the electron transfer in RNR is coupled with proton transfer, i.e. radical transfer. Such a radical transfer, mediated via change of H-bond donors/acceptors along the hydrogen-bonded array, would avoid the energetically demanding problem with charge separation. It would also offer a plausible explanation for the return radical transfer, which can occur by a simple reversal of the H-bond acceptor/donor array (Figure 4 and section 4.3). Below, we will summarise molecular engineering experiments performed to probe the importance of the RTP in *E. coli* and mouse class I RNRs (cf. Table 2).

**Table 2**
**Characteristics of mutations affecting the proposed RTP**

| Organism and RNR component | Mutation | Comment | Ref. |
|---|---|---|---|
| *E. coli* R1 | C439A/S | Both catalytically inert | a |
| | Y730F | Catalytically inert, π–π interaction with Tyr731 | b |
| | Y731F | Catalytically inert, π–π interaction with Tyr730 | b |
| *E. coli* R2 | Y122F | Catalytically inert, native iron site, no radical | c |
| | D84A/H/E | Ala84 catalytically inert, changed iron site, transient Tyr• | d, e |
| | | His84 changed iron site, no Tyr• | |
| | | Glu84 transient peroxo, native iron site, forms Tyr• | |
| | H118A | Catalytically inert, changed iron site, transient Tyr• | d |
| | D237A/N/E | Ala237 unstable protein | f |
| | | Asn237 catalytically inert, native iron site, transient Tyr• | |
| | | Glu237 6–7% activity, native iron site, Tyr•, H-bonded RTP | |
| | W48Y | Tyr48 catalytically inert, changed iron site, Tyr• | g |
| | Y356A/F/W | All catalytically inert, native iron site, Tyr•, R1 complex | h |
| Mouse R2 | Y177F/C/W | All catalytically inert, changed iron site, no Tyr• (Phe177/Cys177), Tyr• (Trp177) | i |
| | D266A | Catalytically inert, native iron site, Tyr• | j |
| | W103F/Y | Both catalytically inert, native iron site, no Tyr• (Phe103), Tyr• (Trp103) | j |
| | Y370F/W | Native iron site, Tyr•, R1 complex, catalytically inert (Phe370), 1.7% activity (Trp370) | k |

References: (a) Åberg *et al.*, 1989; Mao *et al.*, 1992a; (b) Ekberg *et al.*, 1996; (c) Larsson and Sjöberg, 1986; (d) Persson *et al.*, 1996; (e) Bollinger *et al.*, 1998; (f) Ekberg *et al.*, 1998; (g) Huque *et al.*, unpublished; (h) Climent *et al.*, 1992; Ekberg, Slaby and Sjöberg, unpublished; Tong *et al.*, 1998; (i) Pötsch *et al.*, 1999; (j) Rova *et al.*, 1995; (k) Rova *et al.*, 1999.

## 4.1. Protein R1

The H-bonded radical transfer array in protein R1 involves residues Cys439, Tyr730 and Tyr731. The two tyrosine rings are arranged in π–π stacking and H-bonded to each other; Tyr730 is also H-bonded to the sulphur of Cys439. This arrangement connects the active site with the surface of R1, close to the proposed interaction area with R2 (Figure 4). Mutations in Cys439, C439A/S, are catalytically inert, as has been discussed above (section 3.3). Separate engineering of Tyr730 and Tyr731 to phenylalanines provided the first solid evidence for the importance of the hydrogen-bonded pathway (Ekberg *et al.*, 1996). Crystallographic studies of Y730F and Y731F showed that the geometry of the side chains was preserved in the mutant proteins, still allowing for π–π stacking of the aromatic rings. As both these mutant proteins were catalytically inert, despite unperturbed interactions with active R2, substrates and effector nucleotides, we could conclude that radical transfer in R1 required an intact hydrogen-bonded pathway.

## 4.2. Protein R2

### 4.2.1. The Tyrosyl Radical and the Iron Centre

At the heart of the RTP in R2 is of course the radical at Tyr122. The unexpected finding that the enzyme activity of RNR was linked to the presence of a stable radical in protein R2 was reported by Ehrenberg & Reichard already in 1972 (Ehrenberg and Reichard, 1972). Identification of the radical to a tyrosine side chain (Sjöberg *et al.*, 1978) and its localisation to Tyr122 in R2 (Larsson and Sjöberg, 1986) required cloning of the gene in efficient expression derivatives (Thelander *et al.*, 1978; Eriksson *et al.*, 1977). The tyrosyl radical is an oxidised phenoxy radical with fixed ring geometry (see section 6.1 below). The spin density is "odd-alternate", residing mainly on C1, C3, C5 and the phenolic oxygen (cf. inset of Figure 7B). Since the three-dimensional structure of active R2 is unknown, the metR2 (with a diferric-oxo centre, but lacking a radical at Tyr122) structure has been used to describe the immediate surroundings of the tyrosine side chain. The phenolic oxygen of Tyr122 is at 5.3 Å distance from the closest iron and at 3.2–3.5 Å distance from one carboxylate oxygen of the bidentate Fe ligand Asp84 (Nordlund and Eklund, 1993; Nordlund *et al.*, 1990); Figure 5a). No other contacts to the phenolic oxygen are obvious in the 2.2 Å resolution map of metR2.

The diiron centre is enclosed by a four helix bundle and ligated to the protein matrix via six side chains (Figure 5a), two histidines (His118,

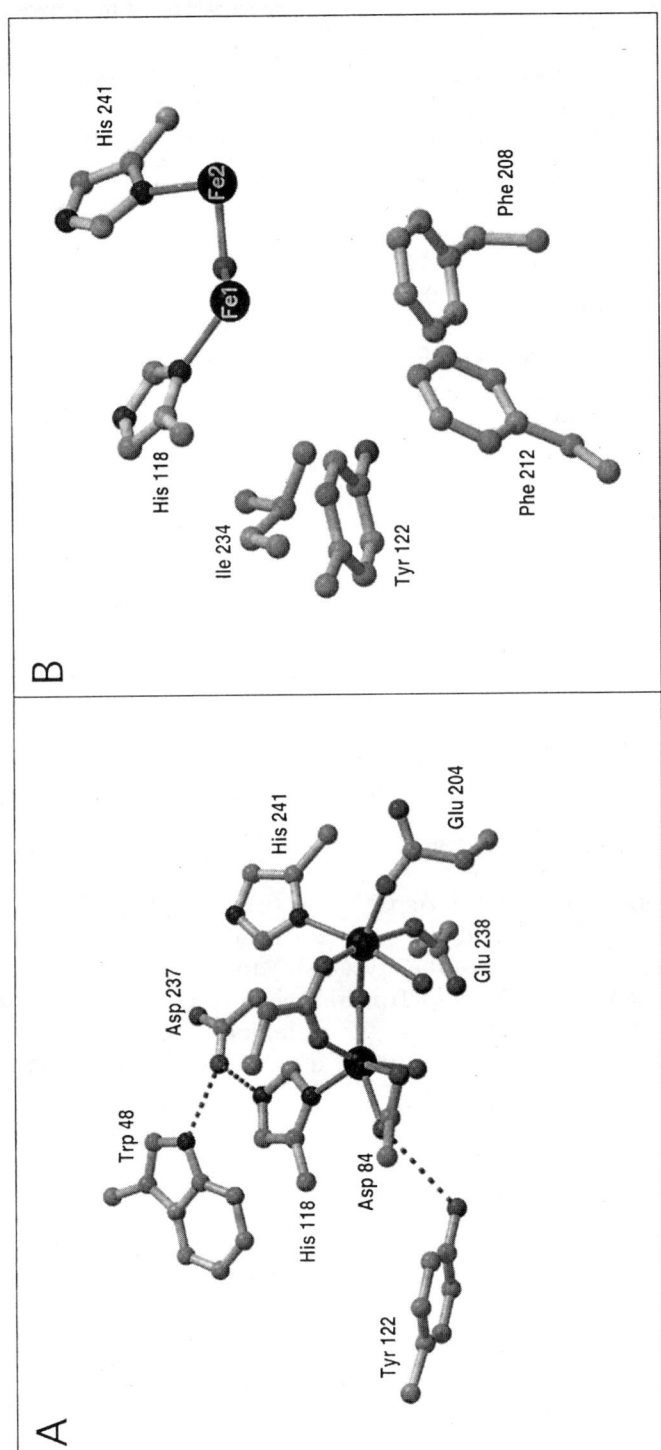

**FIGURE 5.** Three-dimensional structure of the radical-diiron site in protein R2. The iron centre, the RTP, and Tyr122 in met R2 (A). The hydrophobic pocket close to Tyr122 (B). For clarity, some iron ligands have been omitted from B.

His241) and four carboxylates (Asp84, Glu115, Glu204, Glu238). Glu115 bridges the two iron ions, whereas all other protein ligands are terminal. Apart from Asp84, which is bidentate in the metR2 structure, all other protein ligands are monodentate. Additionally, the diiron site of metR2 contains three low molecular weight ligands, one water ligand per Fe, and a bridging μ-oxo ligand, making each Fe ion six-coordinated (Nordlund and Eklund, 1993; Nordlund et al., 1990).

The driving force for the enzymatic reaction, as well as the stability of the tyrosyl radical, should in part be determined by the redox properties of the iron-radical site. However, there is no logic to the directionality of the initial electron transfer reaction in the experimentally estimated reduction potential of $1.00 \pm 0.10$ V for Tyr122• in R2 (Silva et al., 1995) (cf. $E_0 = 0.94$ V for a TyrO•/TyrOH couple) and the expected redox potential of 1.33 V for a RS•/RSH couple (Surdhar and Armstrong, 1987). One or both of these potentials are plausibly perturbed in the ternary holoenzyme complex during catalysis. Interestingly, advanced electrochemical titrations showed that the midpoint potential of the diiron site (−115 mV in metR2) changed to −228 mV upon binding of protein R1 (Silva et al., 1995). The estimated midpoint potential for mouse metR2 is ca. 100 mV (Davydov et al., 1996a; Atta et al., 1994), i.e. considerably more positive than E. coli metR2. The differences in reduction potentials give the diiron sites slightly different properties, such as the ability of mouse R2 to stabilise a mixed-valent iron site, which is not easily trapped in E. coli R2.

### 4.2.2.   A hydrogen-bonded Triad

A H-bonded triad His118-Asp237-Trp48 (Figures 4 and 5A) between the diiron centre of R2 and the surface of the protein was recognised already in the initial report of the crystal structure of R2 (Nordlund et al., 1990). It mimicked a His-Asp-Trp triad in cytochrome c peroxidase (Edwards et al., 1988) that connects the metal ion of the heme group with a tryptophan residue known to harbour a transient radical (Huyett et al., 1995).

The importance of the H-bonded pathway in R2 has since been probed by molecular engineering experiments in mouse R2 as well as in E. coli R2. In mouse R2 the triad consists of residues His173-Asp266-Trp103 (Rova et al., 1995). Of three substitutions, only W103Y and D266A, but not W103F, were capable of stabilising a tyrosyl radical. However, the two former constructs were catalytically inert, corroborating the importance of these two side chains in ribonucleotide reduction. A molecular engineering study at Asp237 in E. coli R2 focused on the importance of its hydrogen bonds (Ekberg et al., 1998). The mutant D237N was catalytically inert, and D237A turned out to be a highly unstable protein, thereby underpinning the importance of this side chain as a connecting link between the iron ligand His118 and the surface residue Trp48. Only D237E showed some enzyme activity (ca. 6–7% compared to wild type). The crystal structure of the D237E protein identified a preserved H-bonded triad. A slightly longer hydrogen bond between the carboxylate and Trp48 (3.3 Å in the mutant protein as compared to 2.9 Å in wild type R2) most likely relates to the impaired $k_{cat}$ of the mutant.

### 4.2.3. The Flexible C-terminal Domain

The C-terminal end of protein R2 is highly flexible, and its three-dimensional structure cannot be deduced from the electron density maps of *E. coli* or mouse R2. Yet, several independent studies have contributed important knowledge about the function and overall structure of the C-terminus. Two seminal studies published in 1986 (Cohen *et al.*, 1986; Dutia *et al.*, 1986) showed that peptides mimicking the C-terminal part of R2 from herpes simplex virus were potent inhibitors of herpes simplex virus RNR, presumably by inhibiting holoenzyme formation. Similar kinetically supported studies with peptide mimics of *E. coli* R2 showed that they were competitive inhibitors of R2 for R1 interaction (Climent *et al.*, 1991). Based on this principle, C-terminal biomimetics have been explored as lead compounds in antiviral therapy (Nocentini, 1996). NMR studies on R2 proteins from mouse, herpes simplex virus and *E. coli* have given limited structural information (Lycksell and Sahlin, 1995; Laplante *et al.*, 1994; Lycksell *et al.*, 1994). The NMR data confirmed the extraordinary high mobility of the C-terminal part of R2 proteins, which becomes immobilised in the R1:R2 complexes. This C-terminal part of R2 proteins, comprising 7–30 residues, may be considered as a separate domain of the protein designated to interact with R1. The current three-dimensional structure of R1 is obtained from crystals grown in presence of a 20-mer polypeptide corresponding to the C-terminal domain of R2. The 14–18 most C-terminal residues of the peptide are visible and are bound to R1 in a shallow groove between a barrel helix and another R1 domain (Uhlin and Eklund, 1994). Collectively, these studies show that the C-terminal end of a R2 protein forms a highly flexible domain, presumably because its function is to fit snugly into its species specific R1 structure upon complex formation, whereby the final structure of the C-terminus is achieved.

Of importance to the RTP, is the occurrence of the invariant residue, Tyr356, in the flexible C-terminal domain of R2. We have suggested that Tyr356 connects the H-bonded arrays of R1 and R2, by making H-bonds directly or indirectly to Trp48 in R2 and to Tyr731 in R1 (Sjöberg, 1997;

Climent *et al.*, 1992). Molecular engineering studies in *E. coli* R2 showed
that Y356A (Climent *et al.*, 1992), Y356W (Ekberg, Slaby & Sjöberg, unpublished), and Y356F (Tong *et al.*, 1998) proteins were catalytically inert and
incapable of radical transfer during assay conditions. A recent engineering
of the corresponding Tyr370 in mouse R2 shows that, whereas Y370F has
the same phenotype as the bacterial mutants, some residual activity (1.7%
of the wild type activity) remains in the Y370W mutant protein. However,
Y370W also loses substantial amounts of its radical during the course of the
assay. As a tryptophan side chain has hydrogen bonding capacity, it was
speculated that Y370W may preserve a RTP between R1 and R2. Together,
these studies corroborate the importance of the conserved C-terminal
tyrosine in the catalytic RTP.

### 4.2.4.    Is the Tyrosyl Radical Hydrogen-bonded to the Radical Transfer Pathway?

Hydrogen bonding of the phenolic oxygen of the tyrosyl radical in part
determines the properties of the radical-iron site in mouse R2 and herpes
simplex virus R2 (van Dam *et al.*, 1998). Presumably, the hydrogen bond
involves a water ligand of Fe1 in these R2 proteins. In contrast, the phenolic oxygen of the tyrosyl radical in *E. coli* class I RNR has not been observed
to be hydrogen bonded, neither in the isolated R2 protein (van Dam *et al.*,
1998), nor in a catalytically competent R1:R2 holoenzyme complex
(Schmidt, Sjöberg & Gräslund, unpublished). Also the tyrosyl radicals in
*Salmonella typhimurium* and *Mycobacterium tuberculosis* class Ib RNRs
lack detectable hydrogen bonds (Liu *et al.*, 1998; van Dam *et al.*, 1998). It
seems, however, reasonable that the tyrosyl radical of any R2 protein is
close enough to a hydrogen-bonding partner that a conformational change
in the ternary holoenzyme complex can promote formation of a hydrogen
bond during catalysis. Even given this assumption, it is hard to offer a
detailed description of how the catalytic RTP proceeds in the immediate
vicinity of the iron centre (cf. section 4.3).

### 4.3.    Theoretical Considerations on Radical Transfer and Protein Dynamics

Protein crystal structures may entice the illusion that the individual
atoms in a protein are at fixed positions. However, substantial movements
do occur. For instance, the C-terminal domain of protein R2 is too mobile
to be resolved in the crystal structure. In solution even larger flexibility is

allowed and is required e.g. to enable docking of proteins to each other. In the case of RNR fine tuning of the RTP is needed after binding of substrate (and effector), such that the hydrogen atom can move from Cys439 in R1 to Tyr122 in R2 and back. Ehrenberg recently used protein dynamics to describe the radical transfer in RNR (Ehrenberg, 1998). As a result of protein dynamics hydrogen bonded side chains will reorient relative to each other and hydrogen bonds will be broken and formed. If at one end of a hydrogen bonded chain there is an electron hole (e.g. a neutral oxidised tyrosyl radical) an electron and a proton can be gained by changing the "ownership" of the hydrogen bonded proton and radicalising the hydrogen-bonded side chain. Given that all orbitals overlap, this transfer of the electron hole, i.e. transfer of an electron and a proton, will propagate to the end of the chain. In RNR the end of the chain is Cys439, but as it is essential that the enzyme does not commit suicide by exposing the Cys439• to solvent, the array of bonds in RNR may not be precisely arranged unless substrate is bound. However, in presence of substrate orbital overlap will eventually lead to 3'-hydrogen abstraction. After reduction of the substrate, Cys439• is reformed and as the hydrogen bonds have been reversed the hydrogen atom can now be abstracted from Tyr122, and the stable Tyr122• is formed. The dynamics picture is enlightening since it predicts that only a very small population of the holoenzyme complexes at any given time has the correct orbital overlaps to promote radical transfer. The predicted short lifetime of amino acid radicals along the hydrogen bonded network is in agreement with the failure to trap any intermediates in RNR during catalysis and the fact that no decrease in the tyrosyl radical content is observed during turnover.

Another more elaborate study which also emphasises the importance of side chain motions is based on quantum chemical models of hydrogen transfer between amino acids in the presence of radicals (Siegbahn, 1998; Siegbahn et al., 1998; Siegbahn et al., 1997). This study introduced the concept of a neutral hydrogen atom passing along the RTP in RNR to avoid charge separation, which would be energetically costly inside a protein with a low dielectric constant. Siegbahn points out that a single concerted motion of hydrogen atoms along the RTP is not necessary and that the probability of finding the radical at a particular amino acid at a given time is governed by the coupling of the kinetics of the individual reactions (Siegbahn, 1998). The direction of the transfer will depend on the bond strengths of the first and the last member of the chain (but not on the strength of the intervening hydrogen bonds), where the hydrogen donor should have a weaker bond than the acceptor. Whether the hydrogen atom transfer theory holds for the RTP in RNR still remains to be experimentally proven.

## 5. GENERATION OF THE STABLE TYROSYL RADICAL IN PROTEIN R2

### 5.1. Radical Generation Involves the Radical Transfer Pathway

The generation of the catalytically essential tyrosyl radical-diferric iron centre has been studied extensively with numerous time-resolved techniques on R2 proteins, mostly from *E. coli* and from mouse. An initial demonstration by Atkin *et al.* that addition of ferrous ions to apoR2 spontaneously generated the radical-iron cofactor in active R2 (Atkin *et al.*, 1973) and subsequent demonstration by Petersson *et al.* that generation of active R2 required molecular oxygen (Petersson *et al.*, 1980), enabled formulation of the overall reaction:

$$\boxed{TyrOH} + 2Fe^{2+} + O_2 + e^- + H^+ \rightarrow \boxed{TyrO\bullet \;/\; Fe(III)\text{-}O^{2-}\text{-}Fe(III)} \; + H_2O \qquad (2)$$

apoR2                                              active R2

The oxygen in the μ-oxo bridge comes from dioxygen (Burdi *et al.*, 1998; Ling *et al.*, 1994). Recently, it has also been shown that the other dioxygen atom ends up as an exchangeable hydroxyl/water ligand of the diiron site (Burdi *et al.*, 1998). It is obvious that three of the four electrons and one of the protons needed to formally reduce dioxygen to water are supplied by the forming radical-diferric centre inside the protein. However, the fourth electron and the last proton have to be supplied from an external source. All experimental data suggest that they enter via the RTP in protein R2, i.e. via residues His118, Asp237, Trp48 and possibly Tyr356 (Rova *et al.*, 1999; Bollinger *et al.*, 1998; Ekberg *et al.*, 1998; Parkin *et al.*, 1998; Schmidt *et al.*, 1998; Persson *et al.*, 1996), see Figure 4. *In vitro* the ultimate "fourth electron" donor can be a reductant like ferrous ion or ascorbic acid (Bollinger *et al.*, 1994a; Bollinger *et al.*, 1994b; Ochiai *et al.*, 1990) but the *in vivo* source is not known.

Through the work of several research groups during the last decade the mechanism has become increasingly more detailed (Andersson *et al.*, 1999; Bollinger *et al.*, 1998; Schmidt *et al.*, 1998; Sturgeon *et al.*, 1996; Bollinger *et al.*, 1991) and the generation of active R2 from apoR2 can currently be divided into the following steps (Figure 6):

(i)   binding of two $Fe^{2+}$ to apoR2 to give a diferrous species (reduced R2)
(ii)  binding of dioxygen to the diferrous site
(iii) two-electron reduction of dioxygen to give a diferric peroxo species
(iv)  uptake of an electron and a proton to give a μ-oxo bridged ferryl species, denoted species X, which formally has been described as

Fe(III)/Fe(IV) but possibly has its unpaired spin partly delocalised over the ligands of the iron site

(v) transfer of an electron and a proton from Tyr122 to the iron site to give active R2 with the radical/diferric-oxo site.

The crystal structures have been determined for apoR2, reduced R2, and metR2 of *E. coli* (Logan *et al.*, 1996; Åberg *et al.*, 1993a; Nordlund *et al.*, 1990), but as pointed out earlier the radical containing structure has not been determined. The three-dimensional structures are also known for a number of mutant proteins affecting side chains in the first and second sphere of the iron-radical centre (Figure 5). Strikingly, the iron ligands display very different flexibility. Functionally important changes in co-ordination are observed for Glu238 and Asp84 in the different redox states (cf. Figure 6). This type of flexibility, denoted carboxylate shift, was first described for model compounds (Rardin *et al.*, 1991) and has also been seen in methane monooxygenase (Rosenzweig *et al.*, 1995).

A recently solved structure of the reduced form of *E. coli* F208A/Y122F in complex with azide suggested how dioxygen may bind to and activate reduced R2 (Andersson *et al.*, 1999). The structure revealed a new coordination where Glu238 uses one of its carboxylate oxygens to bridge the two iron atoms, whereas the second oxygen is coordinated only to Fe2. The azide molecule is bound end on to Fe2. Assuming that dioxygen binds as the azide molecule does, Andersson *et al.* presented the model for radical formation depicted in Figure 6. Especially attractive is that very small rearrangements of the diiron site occur once the dioxygen is bound. The first reduction step giving the peroxo species makes both iron atoms six coordinated and they then remain six coordinate through the reaction. Formation of X from the peroxide state may involve two steps. Possibly, the peroxo form is cleaved in a process giving a diferryl-oxo species, which then relaxes after uptake of the electron and the proton via the RTP chain. Resonance forms of X can direct the final step, which oxidises Tyr122. In the final step carboxylate shifts of the ligands Glu238 and Asp84 occur and coordination of a water molecule to Fe2 gives active R2.

The activation sequence in Figure 6, with two bridging carboxylates in X, is consistent with the experimentally observed short iron-iron distance of 2.49 Å for X in *E. coli* R2 (Riggs-Gelasco *et al.*, 1998). A recent study on reconstitution of mouse R2 came to the unexpected finding that X was not an observable precursor of the mouse tyrosyl radical; rather the small amounts of X that were observed resulted from a side reaction in protomers only giving metR2 (Schmidt *et al.*, 1998). Although the same pathway and intermediates may exist for the reconstitution of *E. coli* and mouse R2, the kinetic differences between the two systems may be considerable and not

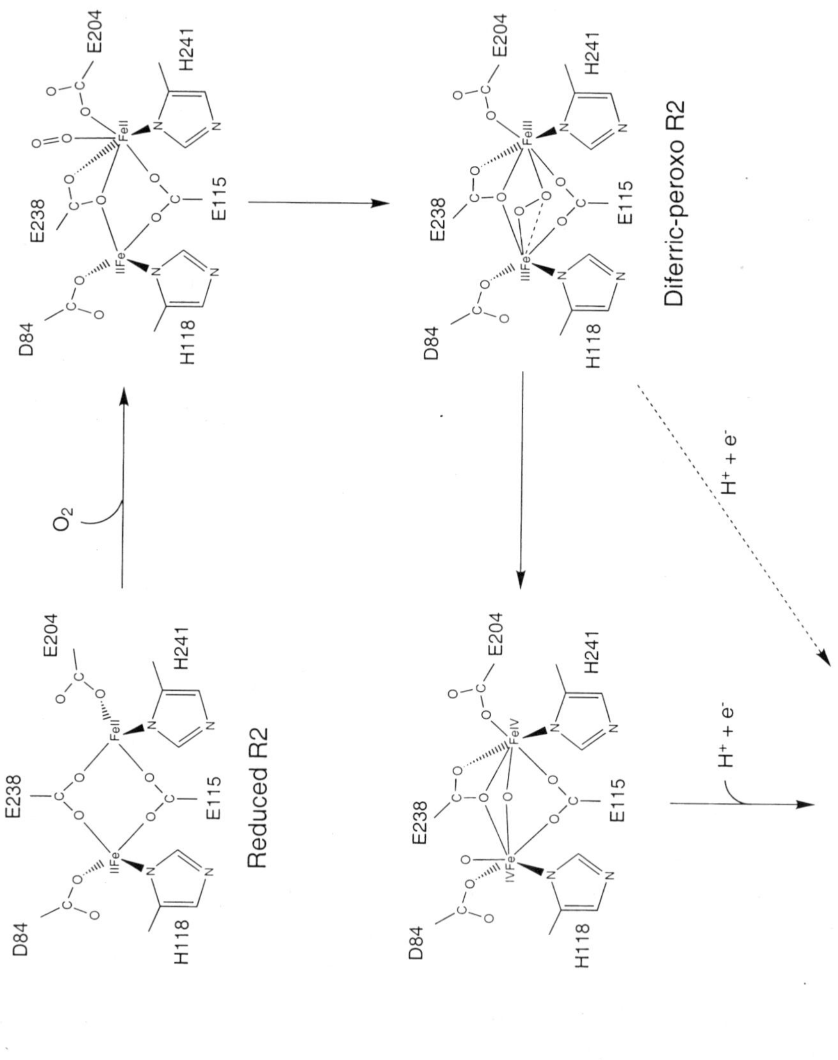
Reduced R2
Diferric-peroxo R2
O2
H+ + e-
H+ + e-
D84
E238
E204
E115
H118
H241

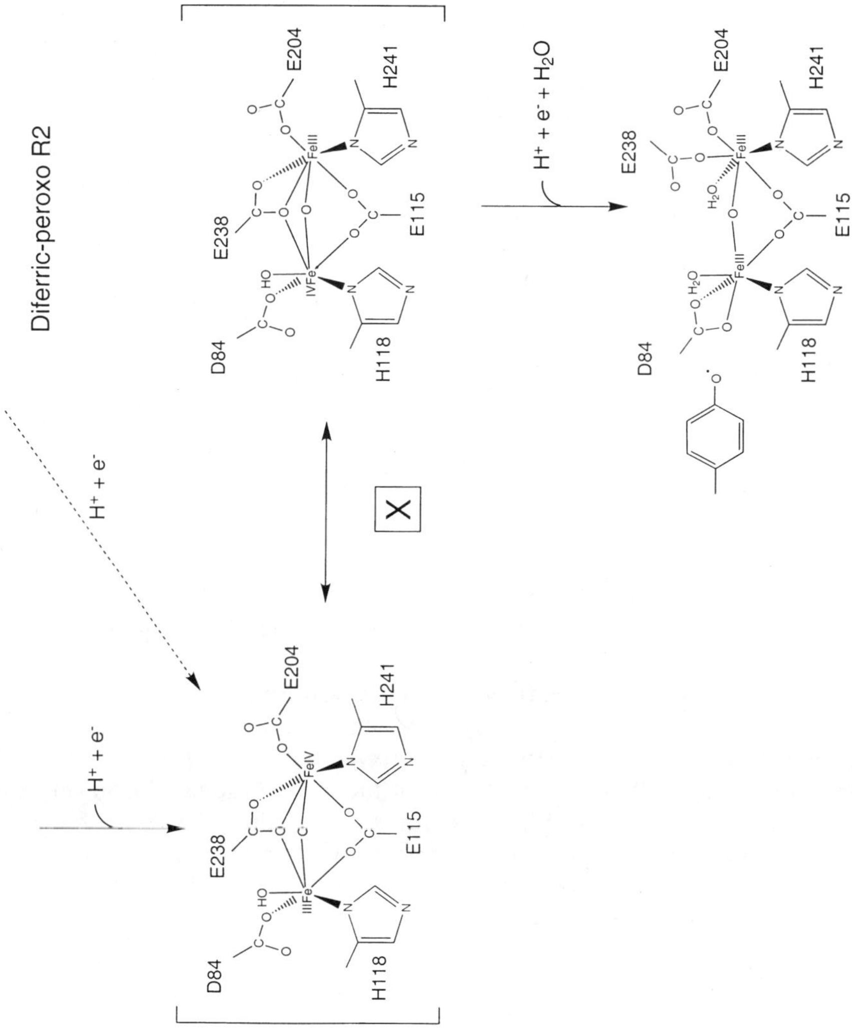

**FIGURE 6.** Proposed formation of the tyrosyl radical in class I RNR (from Andersson *et al.*, 1999).

allow significant accumulation of X in the mouse R2. A structural difference (Kauppi et al., 1996), reflected in e.g. a shorter distance between the iron site and the tyrosine residue in mouse R2, as compared to that in E. coli R2, was suggested as a reason for the rapid Tyr• formation from X (Schmidt et al., 1998).

Several of the intermediates proposed in Figure 6 are too short-lived to be observed. However, protein engineering has been a successful approach to increase the probability of trapping such intermediates. The first choice for site-directed mutagenesis was the Tyr122 side chain. Formation of the diferric-oxo site is considerably slowed down in the mutant Y122F compared to the wild type since the electron and proton supplied by Tyr122 is no longer available (Equation 2). Another advantage is that EPR visible intermediates will no longer be obscured by the appearance of a forming Tyr122•. Studies on Y122F greatly facilitated the assignment of species X (Figure 6) (Sturgeon et al., 1996; Bollinger et al., 1994b; Ravi et al., 1994; Bollinger et al., 1991) and analogous studies enabled identification of a Tyr356• intermediate, implicating that the RTP chain was used during the reactivation process (Sahlin et al., 1995). Transient formation of a Trp48• was suggested from observations of a peroxide adduct to Trp48 in the Y122F protein, implicating that the Trp48 radical had reacted with a second dioxygen molecule (Sahlin, unpublished data). Kinetic evidence with mouse R2 mutants W103Y and D266A (corresponding to Trp48 and Asp237 in E. coli R2) also implicated involvement of the RTP. These mutants were 20-fold slower in generating the tyrosyl radical and furthermore no radical was formed when the mutant W103F was reconstituted (Schmidt et al., 1998).

The peroxo species has only been observed in the E. coli R2 mutant D84E (Bollinger et al., 1998), but is a well-known intermediate in the analogous catalytic cycle of methane monooxygenase (Liu et al., 1995). The peroxo species in D84E is a precursor of the tyrosyl radical-diiron centre, however it was not possible to determine whether the RTP mutant stabilised the intermediate enough to be observed, or if the mutation caused a new intermediate to follow a reaction sequence more similar to that of methane monooxygenase.

## 5.2. Interactions Between Metal Sites

Due to a number of puzzling observations one might ask whether the iron sites in the two protomers of R2 are identical. The purified protein from E. coli normally contains 1–1.2 Tyr• and ca 3 Fe per R2 as opposed to the expected integer values of 2 and 4, respectively. The crystal structure indicates at least 70–80% occupancy at the iron sites. In a recent publication Miller *et al.* proposed that the iron sites are different, as radical-associated diferric sites responded differently to reduction than did the non-radical associated sites (Miller *et al.*, 1999). Reduction by γ-irradiation induced electrons has previously identified a difference in iron centres in active R2 depending on whether there is a neighbouring tyrosyl radical or not (Davydov *et al.*, 1996b). These and other observations have evoked hypotheses about different functions for the iron sites in which one iron site could work as a chaperone for another iron site by providing reducing equivalents via inter- or intramolecular electron (and proton) transfer (Miller *et al.*, 1999; Elgren *et al.*, 1991). A related situation has been reported for yeast RNR. *Saccharomyces cerevisiae* encodes two R2 genes, *rnr2* and *rnr4*, both of which are expressed. Studies from several different laboratories indicate that complexes between the two gene products exist *in vivo* and that an enzymatically active yeast R2 component is only achieved in the presence of both gene products (Nguyen *et al.*, 1999; Huang and Elledge, 1997; Wang *et al.*, 1997).

### 5.3. Non-Native Radicals and Secondary Radical Transfer Pathways Observed in Mutant R2 Proteins

The violent reaction that goes on in the interior of the protein when dioxygen is reduced opens for other ways to import the electron and proton when the site is deprived of the tyrosine residue at position 122. This is revealed by the observation of long-lived radicals occurring on other hydrogen bonded networks connecting the iron with oxidisable amino acid residues closer to the surface. Identified hydrogen bonded networks connected to Fe2 are Trp111-Glu204-Fe2 and Trp107-$H_2$O-His241-Fe2 where Trp111 and Trp107 are ca 4 and 8 Å respectively from Fe2. Radical intermediates were trapped at both these tryptophans and identified as neutral oxidised radicals (Katterle *et al.*, 1997; Lendzian *et al.*, 1996; Sahlin *et al.*, 1995). In mouse R2 where corresponding tryptophan residues are absent, mutants Y177F or Y177C at the native radical site do not give rise to any observable transient side chain radicals, but form a diferric centre when reconstituted (Pötsch *et al.*, 1999).

Can other oxidisable amino acids replace the tyrosine in stabilising a radical and initiating catalysis? In order to answer this question the mouse R2 Tyr177 was engineered to tryptophan and to cysteine, in addition to the phenylalanine control (Pötsch *et al.*, 1999). All constructs were enzymatically inert. Only Y177W formed a transient neutral tryptophan radical (Figure 7B), from light absorption spectra, EPR and ENDOR analyses concluded to be at position 177. A species judged to be similar to X was observed in the time window 0–15 s with the mutants Y177C and Y177F.

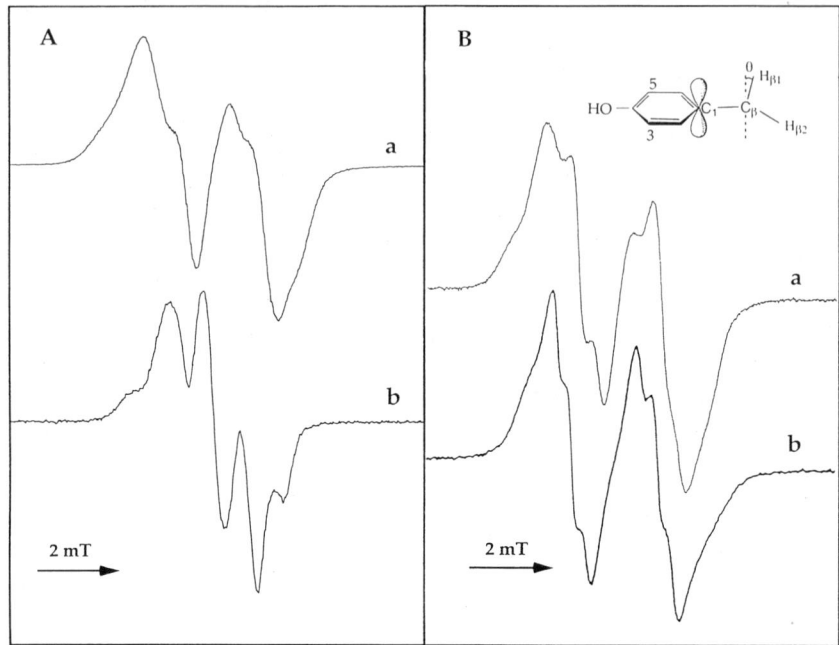

**FIGURE 7.** Typical EPR spectra of stable RNR radicals. A. The Tyr122 radical in *E. coli* class Ia RNR and the Tyr105 radical in *S. typhimurium* class Ib RNR. B. The Tyr177 in mouse RNR and the Trp177 radical in mutant mouse RNR. *Inset:* illustration of the coupling of β-protons to the unpaired electron at C1 of a tyrosyl radical.

This is a considerably longer lifetime than for X in mouse and *E. coli* wild type proteins. Possibly, an improper positioning of Cys177 and Phe177 vis-à-vis the iron site and the lack in mouse R2 of hydrogen bonded oxidisable side chains connected to Fe2 make it difficult for the highly reactive X to acquire the last reducing equivalent. Further studies are needed to determine the precise chemistry leading to the diiron centre.

The inability of Trp177• to initiate RNR catalysis in the mouse holoenzyme can be due to a change in the redox potential and/or structural changes perturbing the RTP chain. From the badly resolved UV-vis spectrum of the mutant protein it seems likely that the interaction with the diiron centre is changed in the mutant protein (Pötsch *et al.*, 1999). Efforts to engineer the Y122C/W mutants in *E. coli* R2 failed due to instability of the mutated proteins (Pötsch, Sjöberg and Gräslund, unpublished data).

## 5.4.  Unexpected Hydroxylation Reactions

The introduction of designed mutations sometimes gives unexpected reactions. In two mutant R2 proteins the iron-oxygen reaction results in hydroxylation of the side chain at position 208 close to the diiron site. The first case is observed in the mutant F208Y, where Tyr208 is a ligand to Fe1 in the reduced protein (Logan et al., 1998). Incubation of the apo-form with ferrous ions and dioxygen gives a 3,4-dihydroxyphenylalanine (dopa208), which is bidently ligated to Fe1 (Logan et al., 1998; Åberg et al., 1993b). In the presence of excess external reductant, the generation of Tyr122• competes with the hydroxylation of Tyr208, but in the absence of an external reductant or if the external RTP is interrupted as in the double mutant W48F/F208Y only dopa208 and no Tyr122• is formed (Parkin et al., 1998). The second case of hydroxylation, now at Phe208, is observed in the double mutant Y122F/E238A (Logan et al., 1998). In this case the mutant protein has been deprived both of a flexible iron ligand normally involved in carboxylate shifts during the iron-oxygen reaction (cf. Figure 6), and of the electron and proton donating Tyr122. The crystal structure of the mutant protein shows that the hydroxylation occurs at the 3/5 position of Phe208 (Logan et al., 1998). These examples emphasise the importance of a very precise arrangement of the iron-radical site as well as of an intact hydrogen bonded triad between the iron site and the surface for proper dioxygen cleavage and tyrosyl radical formation to occur.

## 6.  STABILITY OF THE TYROSYL RADICAL

It is still not fully understood why the tyrosyl radical in protein R2 is so stable. In E. coli R2 stored at 4°C the radical remains for weeks, and stored at −80°C for years (Sahlin et al., 1989). Even though the lifetime of the radical varies considerably in preparations of R2 from different species, especially when handled above 0°C, the radical in these R2 proteins generally withstands low temperature storage well.

## 6.1.  Different Tyrosyl Radical Conformers

The packing of the polypeptide(s) in a protein to give its tertiary (and quaternary) structure imposes restrictions as to how the side chains are allowed to orient and move. The geometry of a radical-containing side chain relative to its Cα-Cβ bond can be studied by EPR and estimated from the relative magnitudes of the two β proton hyperfine coupling constants. If the side chain were free to rotate the two protons would be equivalent (Borg and Elmore, 1967). When the side chain is locked in a particular conformation by the protein, one or both protons may have orbital overlap with the orbital of the unpaired electron (cf. Figure 7B inset). The magnitude of the coupling is strongly geometry dependent. An empirical equation allows the calculation of the dihedral angle θ, between the radical z orbital axis and the projected Cβ–Hβ-bond (Stone and Maki, 1962). In the case of tyrosyl radicals their EPR spectra are reflections of how the aromatic ring is twisted compared to the β-protons, i.e. how the orbitals of the β-protons overlap with unpaired spin residing at C1 of the tyrosyl ring (spin density 0.4). As the spin density distribution in tyrosyl radicals usually are of the "odd-alternate" type, considerable spin density also occur at C3, C5 and the phenolic oxygen (0.25, 0.25, and 0.29 respectively in *E. coli* (Hoganson *et al.*, 1996).

In biological systems, tyrosyl radicals appear to have only two preferred geometries, as reflected in their EPR spectra. One type is found in *E. coli* protein R2 (Bender *et al.*, 1989), where a θ of ca. 30° gives a large enough coupling to one of the β-protons to give a doublet character of the EPR spectrum (Figure 7A). All other observed class Ia RNR tyrosyl radicals have this geometry. The other type is e.g. found in *S. typhimurium* class Ib RNR (Jordan *et al.*, 1994), and in the photosystem II tyrosyl radical (Warncke *et al.*, 1994), where a θ of ca. 70° gives a coupling that is only about half of the coupling in *E. coli* R2 (Figure 7A). Class Ib RNRs from *Corynebacterium ammoniagenes* (Huque, Sahlin and Sjöberg, unpublished) and *M. tuberculosis* (Liu *et al.*, 1998) also have this type of geometry. Additional hyperfine features in the EPR spectra come from couplings to the C3 and C5 protons, whereas the coupling of the second β-proton usually is too small to be resolved in either type of geometry.

A recent theoretical calculations study offered an explanation to why there are only two observed conformers of the tyrosyl radicals (Himo *et al.*, 1997). For an isolated *p*-ethylphenoxyl radical two different conformers are energetically favourable, one at a global minimum corresponding to the photosystem II radical and one at a local minimum corresponding to the conformation of *E. coli* R2. Although the energy barrier for the local minimum is low enough (0.2 kcal/mol) to be easily overcome by thermal motion, it is striking that the studied class Ia tyrosyl radicals falls within a calculated dihedral angle of 20–40°. The protein will of course constrain and slightly change the geometry of the tyrosine side chain. The class Ib radicals have very similar ring geometry and e.g. the EPR spectra of the *S. typhimurium* and *C. ammoniagenes* radicals are superimposable. The class Ia radicals, on the other hand, seem more sensitive to the protein environment and hence every species has its own EPR fingerprint. For instance, the tyrosyl radical in bacteriophage T4 R2 is twisted so that also the second β-proton is resolved (Sahlin *et al.*, 1982). However, the fact that the same EPR spectrum is always observed for a given species, indicates that the side chain is kept in a fixed position by the surrounding environment of the protein even at room temperature.

## 6.2.  The Tyrosyl Radical Environment

With the advent of the *E. coli* R2 crystal structure, it became obvious that the radical was sheltered inside the protein matrix, and protected from the solvent by an adjacent hydrophobic pocket (Nordlund *et al.*, 1990). Three residues of the hydrophobic pocket (Figure 5b) are conserved throughout the class I RNRs (Phe208, Phe212 and Ile234). The influence of these hydrophobic side chains on radical stability was tested by site-directed mutagenesis (F208Y, F212Y/W, I234N). All mutants can form a Tyr122•, but the life time of the radical is significantly reduced in either case, indicating that the major function of the hydrophobic pocket is to stabilise the tyrosyl radical (Ormö *et al.*, 1995).

In a related study each iron ligand (cf. Figure 5a) was changed to an alanine residue (Persson *et al.*, 1996). Even though these mutant R2 proteins could form a dinuclear iron site (except mutations involving Glu115 which bridges the iron atoms in all oxidations states), a tyrosyl radical was only observed in D84A, H118A, and E204A. In addition, these radicals were formed in low yield and were highly unstable. Interestingly, only the mutant E204A affecting a Fe2 ligand displayed some enzyme activity, whereas the other two mutants, affecting Fe1 ligands proposed to participate in the RTP, were inactive. This study shows that the properties of the diiron site influence the stability of the tyrosyl radical.

The mouse R2 radical is considerably less stable than the *E. coli* R2 radical (Nyholm *et al.*, 1993). Strikingly, the iron ions are not very tightly bound to the mouse protein, and purified mouse R2 is obtained with only 0.1 Fe/R2 and has to be reconstituted before use (Mann *et al.*, 1991). This is probably an effect of the more open structure of the iron-radical site as seen from the crystal structure (Kauppi *et al.*, 1996). Possibly, the relatively high midpoint potential of ca. 100 mV for the mouse diiron site also affects the radical stability.

## 6.3.  Radical Stability during Catalysis

Radical containing enzymes are often probed with radical scavengers. One such scavenger, used extensively in cancer therapy, is hydroxyurea (Nocentini, 1996). In mouse R2 both the tyrosyl radical and the iron site are reduced by hydroxyurea (Nyholm *et al.*, 1993). In *E. coli* R2 only the tyrosyl radical is reduced by hydroxyurea, but not the diferric site (Karlsson *et al.*, 1992; Atkin *et al.*, 1973). Interestingly, it has been shown that *E. coli* RNR holoenzyme complex exhibits an increased sensitivity to hydroxyurea when the RTP chain is properly arranged for radical shuttling, i.e. in the presence of substrate and allosteric effector (Karlsson *et al.*, 1992). In retrospect, this result may relate to the theoretical considerations on RTP discussed in section 4.3.

## 7.  RADICAL TRANSFER REACTIONS IN CLASS II AND III RIBONUCLEOTIDE REDUCTASES

Are there long-range RTPs in class II and III RNRs? No, in neither of these two classes is the radical chain initiator site separated very far from the active site and the cysteine residue that transiently forms a thiyl radical.

### 7.1.  Class II RNRs

As mentioned in the introduction, class II RNRs do not harbour a stable protein radical. Instead, homolytic cleavage of the cofactor AdoCbl (Table 1) is utilised as a radical chain initiator during catalytic turnover. Unfortunately, the three-dimensional structure of a class II RNR has not been solved yet, but since several class II RNRs have been cloned and efficiently overexpressed lately, we may not be on tenterhooks much longer. Meanwhile, EPR analysis on *L. leichmannii* RNR identified a magnetic interaction between the Co(II) nucleus, resulting from the AdoCbl homolysis, and a thiyl radical in the enzyme, presumably at Cys408 (Licht *et al.*, 1996). An upper limit of 8 Å between the two paramagnetic centres was estimated (Gerfen *et al.*, 1996). Formation of the thiyl radical is plausibly concerted with homolysis of AdoCbl (Figure 8A), putting further constraints on the distance between the cofactor and the active site cysteine in *L. leichmannii* RNR. Collectively, these results imply that there are no long-range radical transfer reactions occurring in the class II enzymes.

### 7.2.  Class III RNRs

The three-dimensional structure of the class III RNR from bacteriophage T4 was recently solved to high resolution for the mutant enzyme G580A, which has an alanine residue at the position of the stable glycyl radical in the native enzyme (Logan *et al.*, 1999). This prevents oxygen-dependent irreversible truncation of the polypeptide chain (Young *et al.*, 1996). The three-dimensional structure of the mutant enzyme showed that

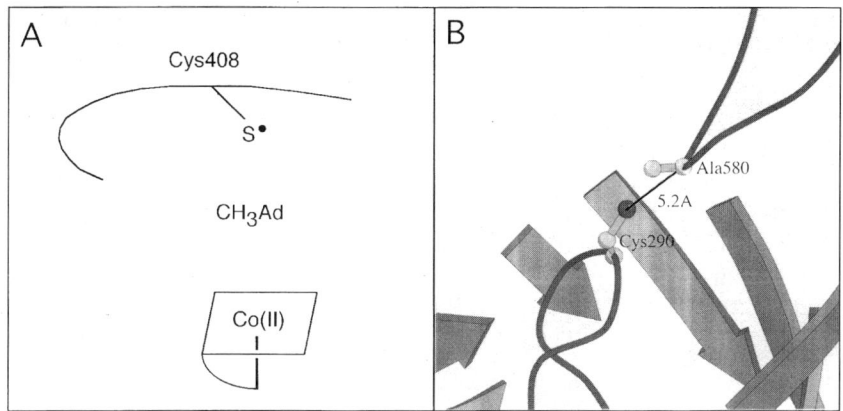

**FIGURE 8.** Approximate arrangements of radical initiator and transient thiyl radical in class II and III RNR. A. The proposed members in a concerted pathway for generation of a thiyl radical in class II RNR from L. Leichmannii, where AdoCbl is homolytically cleaved for generation of the thiyl radical, which shows magnetic interaction with Co(II), the deoxyadenosine moiety remains in the active site during catalysis (from Stubbe and van der Donk, 1998). B. Three-dimensional structure of the active site region of class III RNR from bacteriophage T4 (from Logan *et al.*, 1999). Arrows denote β-strands, and thick grey lines denote loops in the backbone structure. The distance between the sulphur of Cys290 and the Cα-position of Ala580 is shown.

Cα at position 580 is 5.2 Å from the sulphur in Cys290, the cysteine proposed to harbour the 3′-H-abstracting thiyl radical in this enzyme (Figure 8B). There is neither a nearby side chain or backbone group, nor any unaccountable electron density occupying the space between position 580 and Cys290. It was therefore suggested that the initial step in catalysis involves a direct H atom abstraction from Cys290 mediated by the glycyl radical at position 580 (Logan *et al.*, 1999). In essence this interpretation implies that there is no factual RTP involved during catalysis by class III RNRs.

Is there a RTP involved during generation of the glycyl radical? The current three-dimensional structure of the T4 class III enzyme does not permit any structural conclusions, as the activating NrdG component is not visible in the current electron density maps. However, position 580 is close to the surface of the NrdD protein. In addition, the interaction between AdoMet and the NrdG component, has only been demonstrated indirectly, even though cleavage of AdoMet to the deoxyadenosyl radical species is mediated by the $Fe_4S_4$ cluster of NrdG (Ollagnier *et al.*, 1999; Ollagnier *et al.*, 1997). We therefore suggest that glycyl radical generation may plausibly occur in a concerted manner involving the iron-sulphur cluster of

NrdG, AdoMet and Gly580 of NrdD, and that also the RTP for activation of class III RNR is short.

ACKNOWLEDGMENTS. We thank Anders Ehrenberg, Astrid Gräslund and Margareta Karlsson for helpful discussions and criticism, and Martin Andersson and Derek Logan for help with the structural illustrations. Support for research performed in the authors' laboratory comes from the Swedish Cancer Foundation, the Swedish Natural Science Research Council, the Foundation for Strategic Research, the Training and Mobility Program (EU), Carl Trygger Foundation, and Magn. Bergvall Foundation.

## 8. REFERENCES

Andersson, M. E., Högbom, M., Rinaldo-Matthis, A., Andersson, K. K., Sjöberg, B.-M., and Nordlund, P., 1999, The crystal structure of an azide complex of the diferrous R2 subunit of ribonucleotide reductase displays a novel carboxylate shift with important mechanistic implications for diiron-catalyzed oxygen activation. *J. Amer. Chem. Soc.* **121**: 2346–2352.

Atkin, C. L., Thelander, L., Reichard, P., and Lang, G., 1973, Iron and free radical in ribonucleotide reductase. Exchange of iron and Mössbauer spectroscopy of the protein B2 subunit of the *Escherichia coli* enzyme. *J. Biol. Chem.* **248**:7464–7472.

Atta, M., Andersson, K. K., Ingemarson, R., Thelander, L., and Gräslund, A., 1994, EPR studies of mixed-valent [$Fe^{II}Fe^{III}$] clusters formed in the R2 subunit of ribonucleotide reductase from mouse or herpes simplex virus: Mild chemical reduction of the diferric centers. *J. Am. Chem. Soc.* **116**:6429–6430.

Bender, C. J., Sahlin, M., Babcock, G. T., Barry, B. A., Chandrashekar, T. K., Salowe, S. P., Stubbe, J., Lindström, B., Petersson, L., Ehrenberg, A., and Sjöberg, B.-M., 1989, An ENDOR study of the tyrosyl free radical in ribonucleotide reductase from *Escherichia coli*. *J. Am. Chem. Soc.* **111**:8076–8083.

Beratan, D. N., Nelson Onuchic, J., Winkler, J. R., and Gray, H. B., 1992, Electron-tunneling pathways in proteins. *Science* **258**:1740–1741.

Bollinger, J. M., Krebs, C., Vicol, A., Chen, S. X., Ley, B. A., Edmondson, D. E., and Huynh, B. H., 1998, Engineering the diiron site of *Escherichia coli* ribonucleotide reductase protein R2 to accumulate an intermediate similar to H-peroxo, the putative peroxodiiron(III) complex from the methane monooxygenase catalytic cycle. *J. Am. Chem. Soc.* **120**:1094–1095.

Bollinger, J. M., Tong, W. H., Ravi, N., Huynh, B. H., Edmondson, D. E., and Stubbe, J., 1994a, Mechanism of assembly of the tyrosyl radical-diiron(III) cofactor of *E. coli* ribonucleotide reductase: II. Kinetics of the excess $Fe^{2+}$ reaction by optical, EPR, and Mössbauer spectroscopies. *J. Am. Chem. Soc.* **116**:8015–8023.

Bollinger, J. M., Tong, W. H., Ravi, N., Huynh, B. H., Edmondson, D. E., and Stubbe, J., 1994b, Mechanism of assembly of the tyrosyl radical-diiron(III) cofactor of *E. coli* ribonucleotide reductase: III. Kinetics of the limiting $Fe^{2+}$ reaction by optical, EPR, and Mössbauer spectroscopies. *J. Am. Chem. Soc.* **116**:8024–8032.

Bollinger, J. M. J., Edmondson, D. E., Huynh, B. H., Filley, J., Norton, J. R., and Stubbe, J., 1991, Mechanism of assembly of the tyrosyl radical-dinuclear iron cluster cofactor of ribonucleotide reductase. *Science* **253**:292–298.

Borg, D. C., and Elmore, J. J. J., 1967, Evidence for restricted molecular conformation and for hindered rotation of side chain groups from EPR of labile free radicals. *In* "Magnetic Resonance in Biological Systems" (A. Ehrenberg, B. G. Malmström, and T. Vänngård, eds.), pp. 341–350. Symposium Publications Division, Pergamon Press.

Burdi, D., Willems, J. P., RiggsGelasco, P., Antholine, W. E., Stubbe, J., and Hoffman, B. M., 1998, The core structure of X generated in the assembly of the diiron cluster of ribonucleotide reductase: $^{17}O_2$ and $H_2^{17}O$ ENDOR. *J. Amer. Chem. Soc.* **120**:12910–12919.

Climent, I., Sjöberg, B.-M., and Huang, C. Y., 1991, Carboxyl-terminal peptides as probes for *Escherichia coli* ribonucleotide reductase subunit interaction: Kinetic analysis of inhibition studies. *Biochemistry* **30**:5164–5171.

Climent, I., Sjöberg, B.-M., and Huang, C. Y., 1992, Site-directed mutagenesis and deletion of the carboxyl terminus of *Escherichia coli* ribonucleotide reductase protein R2. Effects on catalytic activity and subunit interaction. *Biochemistry* **31**:4801–4807.

Cohen, E. A., Gaudreau, P., Brazeau, P., and Langelier, Y., 1986, Specific inhibition of herpesvirus ribonucleotide reductase by a nonapeptide derived from the carboxy terminus of subunit 2. *Nature* **321**:441–443.

Davydov, A., Schmidt, P. P., and Gräslund, A., 1996a, Reversible red-ox reactions of the diiron site in the mouse ribonucleotide reductase R2 protein. *Biochem. Biophys. Res. Commun.* **219**:213–218.

Davydov, R., Sahlin, M., Kuprin, S., Gräslund, A., and Ehrenberg, A., 1996b, Effect of the tyrosyl radical on the reduction and structure of the *Escherichia coli* ribonucleotide reductase protein R2 diferric site as probed by EPR on the mixed-valent state. *Biochemistry* **35**:5571–5576.

Durham, L. J., Larsson, A., and Reichard, P., 1967, Enzymatic synthesis of deoxyribonucleotides: II. The mechanism of hydrogen transfer of the ribonucleoside diphosphate reductase system from *Escherichia coli* studied with nuclear magnetic resonance. *European J. Biochem.* **1**:92–95.

Dutia, B. M., Frame, M. C., Subak-Sharpe, J. H., Clark, W. N., and Marsden, H. S., 1986, Specific inhibition of herpesvirus ribonucleotide reductase by synthetic peptides. *Nature* **321**:439–441.

Edwards, S. L., Kraut, J., and Poulos, T. L., 1988, Crystal structure of nitric oxide inhibited cytochrome *c* peroxidase. *Biochemistry* **27**:8074–8081.

Ehrenberg, A., 1998, Protein dynamics, free radical transfer and reaction cycle of ribonucleotide reductase. *In* "Biological Physics. Third International Symposium" (H. Frauenfelder, G. Hummer, and R. Garcia, eds.), pp. 163–174. American Institute of Physics, Santa Fé, NM, USA.

Ehrenberg, A., and Reichard, P., 1972, Electron spin resonance of the iron-containing protein B2 from ribonucleotide reductase. *J. Biol. Chem.* **247**:3485–3488.

Ekberg, M., Pötsch, S., Sandin, E., Thunnissen, M., Nordlund, P., Sahlin, M., and Sjöberg, B.-M., 1998, Preserved catalytic activity in an engineered ribonucleotide reductase R2 protein with a nonphysiological radical transfer pathway—The importance of hydrogen bond connections between the participating residues. *J. Biol. Chem.* **273**:21003–21008.

Ekberg, M., Sahlin, M., Eriksson, M., and Sjöberg, B.-M., 1996, Two conserved tyrosine residues in protein R1 participate in an intermolecular electron transfer in ribonucleotide reductase. *J. Biol. Chem.* **271**:20655–20659.

Eklund, H., Eriksson, M., Uhlin, U., Nordlund, P., and Logan, D., 1997, Ribonucleotide reductase—Structural studies of a radical enzyme. *Biol. Chem.* **378**:821–825.

Eklund, H., and Fontecave, M., 1999, Glycyl radical enzymes: a conservative structural basis for radicals. *Structure* **7**:R257–R262.

Elgren, T. E., Lynch, J. B., Juarez-Garcia, C., Münck, E., Sjöberg, B.-M., and Que, L. J., 1991, Electron transfer associated with oxygen activation in the B2 protein of ribonucleotide reductase from *Escherichia coli. J. Biol. Chem.* **266**:19265–19268.

Eriksson, M., Uhlin, U., Ramaswamy, S., Ekberg, M., Regnström, K., Sjöberg, B.-M., and Eklund, H., 1997, Binding of allosteric effectors to ribonucleotide reductase protein R1: reduction of active-site cysteines promotes substrate binding. *Structure* **5**:1077–1092.

Eriksson, S., Sjöberg, B.-M., and Hahne, S., 1977, Ribonucleoside diphosphate reductase from *Escherichia coli*. An immunological assay and a novel purification from an overproducing strain lysogenic for phage λdnrd. *J. Biol. Chem.* **252**:6132–6138.

Gerfen, G. J., Licht, S., Willems, J. P., Hoffman, B. M., and Stubbe, J., 1996, Electron paramagnetic resonance investigations of a kinetically competent intermediate formed in ribonucleotide reduction: Evidence for a thiyl radical-Cob(II) alamin interaction. *J. Am. Chem. Soc.* **118**:8192–8197.

Gray, H. B., and Winkler, J. R., 1996, Electron transfer in proteins. *Annu. Rev. Biochem.* **65**:537–561.

Gräslund, A., and Sahlin, M., 1996, Electron paramagnetic resonance and nuclear magnetic resonance studies of class I ribonucleotide reductase. *Annu. Rev. Biophys. Biomol. Struc.* **25**:259–286.

Himo, F., Gräslund, A., and Eriksson, L. A., 1997, Density functional calculations on model tyrosyl radicals. *Biophys. J.* **72**:1556–1567.

Hoganson, C. W., Sahlin, M., Sjöberg, B.-M., and Babcock, G. T., 1996, Electron magnetic resonance of the tyrosyl radical in ribonucleotide reductase from *Escherichia coli. J. Am. Chem. Soc.* **118**:4672–4679.

Holmgren, A., and Björnstedt, M., 1995, Thioredoxin and thioredoxin reductase. *In* "Biothiols, Pt B" (L. Packer, ed.), vol. 252, pp. 199–208. Academic Press Inc, 525 B Street, Suite 1900, San Diego, CA 92101-4495.

Holmgren, A., and Åslund, F., 1995, Glutaredoxin. *In* "Biothiols, Pt B" (L. Packer, ed.), vol. 252, pp. 283–292. Academic Press Inc, 525 B Street, Suite 1900, San Diego, CA 92101-4495.

Huang, M. X., and Elledge, S. J., 1997, Identification of RNR4, encoding a second essential small subunit of ribonucleotide reductase in *Saccharomyces cerevisiae. Mol. Cell. Biol.* **17**:6105–6113.

Huyett, J. E., Doan, P. E., Gurbiel, R., Houseman, A. L. P., Sivaraja, M., Goodin, D. B., and Hoffman, B. M., 1995, Compound ES of cytochrome *c* peroxidase contains a Trp π-cation radical: Characterization by CW and pulsed Q-band ENDOR spectroscopy. *J. Am. Chem. Soc.* **117**:9033–9041.

Jordan, A., Pontis, E., Atta, M., Krook, M., Gibert, I., Barbé, J., and Reichard, P., 1994, A second class I ribonucleotide reductase in *Enterobacteriaceae:* Characterization of the *Salmonella typhimurium* enzyme. *Proc. Natl. Acad. Sci. USA* **91**:12892–12896.

Jordan, A., and Reichard, P., 1998, Ribonucleotide reductases. *Annu. Rev. Biochem.* **67**:71–98.

Jordan, A., Åslund, F., Pontis, E., Reichard, P., and Holmgren, A., 1997, Characterization of *Escherichia coli* NrdH. A glutaredoxin-like protein with a thioredoxin-like activity profile. *J. Biol. Chem.* **272**:18044–18050.

Karlsson, M., Sahlin, M., and Sjöberg, B.-M., 1992, *Escherichia coli* ribonucleotide reductase. Radical susceptibility to hydroxyurea is dependent on the regulatory state of the enzyme. *J. Biol. Chem.* **267**:12622–12626.

Katterle, B., Sahlin, M., Schmidt, P. P., Pötsch, S., Logan, D. T., Gräslund, A., and Sjöberg, B.-M., 1997, Kinetics of transient radicals in *Escherichia coli* ribonucleotide reductase— Formation of a new tyrosyl radical in mutant protein R2. *J. Biol. Chem.* **272**:10414–10421.

Kauppi, B., Nielsen, B. A., Ramaswamy, S., Kjøller Larsen, I., Thelander, M., Thelander, L., and Eklund, H., 1996, The three-dimensional structure of mammalian ribonucleotide reductase protein R2 reveals a more-accessible iron-radical site than *Escherichia coli* R2. *J. Mol. Biol.* **262**:706–720.

Laplante, S. R., Aubry, N., Liuzzi, M., Thelander, L., Ingemarson, R., and Moss, N., 1994, The critical C-terminus of the small subunit of herpes simplex virus ribonucleotide reductase is mobile and conformationally similar to C-terminal peptides. *Int. J. Pept. Protein Res.* **44**:549–555.

Larsson, A., and Sjöberg, B.-M., 1986, Identification of the stable free radical tyrosine residue in ribonucleotide reductase. *EMBO J.* **5**:2037–2040.

Lawrence, C. C., Bennati, M., Obias, H. V., Bar, G., Griffin, R. G., and Stubbe, J., 1999, High-field EPR detection of a disulfide radical anion in the reduction of cytidine 5'-diphosphate by the E441Q R1 mutant of *Escherichia coli* ribonucleotide reductase. *Proc. Natl. Acad. Sci. U S A* **96**:8979–8984.

Lendzian, F., Sahlin, M., Macmillan, F., Bittl, R., Fiege, R., Pötsch, S., Sjöberg, B.-M., Gräslund, A., Lubitz, W., and Lassmann, G., 1996, Electronic structure of neutral tryptophan radicals in ribonucleotide reductase studied by EPR and ENDOR spectroscopy. *J. Am. Chem. Soc.* **118**:8111–8120.

Licht, S., Gerfen, G. J., and Stubbe, J. A., 1996, Thiyl radicals in ribonucleotide reductases. *Science* **271**:477–481.

Licht, S. S., Lawrence, C. C., and Stubbe, J., 1999, Class II ribonucleotide reductases catalyze carbon-cobalt bond reformation on every turnover. *J. Amer. Chem. Soc.* **121**:7463–7468.

Lin, A. N., Ashley, G. W., and Stubbe, J., 1987, Location of the redox-active thiols of ribonucleotide reductase: sequence similarity between the *Escherichia coli* and *Lactobacillus leichmannii* enzymes. *Biochemistry* **26**:6905–6909.

Ling, J. S., Sahlin, M., Sjöberg, B.-M., Loehr, T. M., and Sanders-Loehr, J., 1994, Dioxygen is the source of the μ-oxo bridge in iron ribonucleotide reductase. *J. Biol. Chem.* **269**:5595–5601.

Liu, A., Pötsch, S., Davydov, A., Barra, A. L., Rubin, H., and Gräslund, A., 1998, The tyrosyl free radical of recombinant ribonucleotide reductase from *Mycobacterium tuberculosis* is located in a rigid hydrophobic pocket. *Biochemistry* **37**:16369–16377.

Liu, K. E., Valentine, A. M., Qiu, D., Edmondson, D. E., Appelman, E. H., Spiro, T. G., and Lippard, S. J., 1995, Characterization of a diiron(III) peroxo intermediate in the reaction cycle of methane monooxygenase hydroxylase from *Methylococcus capsulatus* (Bath). *J. Am. Chem. Soc.* **117**:4997–4998.

Logan, D. T., Andersson, J., Sjöberg, B.-M., and Nordlund, P., 1999, A glycyl radical site in the crystal structure of a class III ribonucleotide reductase. *Science* **283**:1499–1504.

Logan, D. T., deMaré, F., Persson, B. O., Slaby, A., Sjöberg, B.-M., and Nordlund, P., 1998, Crystal structures of two self-hydroxylating ribonucleotide reductase protein R2 mutants: Structural basis for the oxygen-insertion step of hydroxylation reactions catalyzed by diiron proteins. *Biochemistry* **37**:10798–10807.

Logan, D. T., Su, X. D., Åberg, A., Regnström, K., Hajdu, J., Eklund, H., and Nordlund, P., 1996, Crystal structure of reduced protein R2 of ribonucleotide reductase: The structural basis for oxygen activation at a dinuclear iron site. *Structure* **4**:1053–1064.

Lycksell, P.-O., and Sahlin, M., 1995, Demonstration of segmental mobility in the functionally essential carboxyl terminal part of ribonucleotide reductase protein R2 from *Escherichia coli*. *FEBS Lett.* **368**:441–444.

Lycksell, P. O., Ingemarson, R., Davis, R., Gräslund, A., and Thelander, L., 1994, $^1$H NMR studies of mouse ribonucleotide reductase—The R2 protein carboxyl-terminal tail, essential for subunit interaction, is highly flexible but becomes rigid in the presence of protein R1. *Biochemistry* **33**:2838–2842.

Mann, G. J., Gräslund, A., Ochiai, E., Ingemarson, R., and Thelander, L., 1991, Purification and characterization of recombinant mouse and herpes simplex virus ribonucleotide reductase R2 subunit. *Biochemistry* **30**:1939–1947.

Mao, S. S., Holler, T. P., Yu, G. X., Bollinger, J. M., Booker, S., Johnston, M. I., and Stubbe, J., 1992a, A model for the role of multiple cysteine residues involved in ribonucleotide reduction—Amazing and still confusing. *Biochemistry* **31**:9733–9743.

Mao, S. S., Yu, G. X., Chalfoun, D., and Stubbe, J., 1992b, Characterization of C439SR1, a mutant of *Escherichia coli* ribonucleotide diphosphate reductase: Evidence that C439 is a residue essential for nucleotide reduction and C439SR1 is a protein possessing novel thioredoxin-like activity. *Biochemistry* **31**:9752–9759.

Marcus, R. A., and Sutin, N., 1985, Electron transfers in chemistry and biology. *Biochim. Biophys. Acta.* **811**:265–322.

Miller, M. A., Gobena, F. T., Kauffmann, K., Munck, E., Que, L., and Stankovich, M. T., 1999, Differing roles for the diiron clusters of ribonucleotide reductase from aerobically grown *Escherichia coli* in the generation of the Y122 radical. *J. Amer. Chem. Soc.* **121**:1096–1097.

Moser, C. C., Keske, J. M., Warncke, K., Farid, R. S., and Dutton, P. L., 1992, Nature of biological electron transfer. *Nature* **355**:796–802.

Nguyen, H. H., Ge, J., Perlstein, D. L., and Stubbe, J., 1999, Purification of ribonucleotide reductase subunits Y1, Y2, Y3, and Y4 from yeast: Y4 plays a key role in diiron cluster assembly. *Proc. Natl. Acad. Sci. U S A* **96**:12339–12344.

Nocentini, G., 1996, Ribonucleotide reductase inhibitors: new strategies for cancer chemotherapy. *Crit. Rev. Oncol. Hematol.* **22**:89–126.

Nordlund, P., and Eklund, H., 1993, Structure and function of the *Escherichia coli* ribonucleotide reductase protein R2. *J. Mol. Biol.* **232**:123–164.

Nordlund, P., Sjöberg, B.-M., and Eklund, H., 1990, Three-dimensional structure of the free radical protein of ribonucleotide reductase. *Nature* **345**:593–598.

Nyholm, S., Thelander, L., and Gräslund, A., 1993, Reduction and loss of the iron center in the reaction of the small subunit of mouse ribonucleotide reductase with hydroxyurea. *Biochemistry* **32**:11569–11574.

Ochiai, E., Mann, G. J., Gräslund, A., and Thelander, L., 1990, Tyrosyl free radical formation in the small subunit of mouse ribonucleotide reductase. *J. Biol. Chem.* **265**:15758–15761.

Ollagnier, S., Meier, C., Mulliez, E., Gaillard, J., Schuenemann, V., Trautwein, A., Mattioli, T., Lutz, M., and Fontecave, M., 1999, Assembly of 2Fe-2S and 4Fe-4S clusters in the anaerobic ribonucleotide reductase from *Escherichia coli*. *J. Amer. Chem. Soc.* **121**:6344–6350.

Ollagnier, S., Mulliez, E., Schmidt, P. P., Eliasson, R., Gaillard, J., Deronzier, C., Bergman, T., Gräslund, A., Reichard, P., and Fontecave, M., 1997, Activation of the anaerobic ribonucleotide reductase from *Escherichia coli*—The essential role of the iron-sulfur center for S-adenosylmethionine reduction. *J. Biol. Chem.* **272**:24216–24223.

Ormö, M., Regnström, K., Wang, Z. G., Que, L., Sahlin, M., and Sjöberg, B.-M., 1995, Residues important for radical stability in ribonucleotide reductase from *Escherichia coli*. *J. Biol. Chem.* **270**:6570–6576.

Parkin, S. E., Chen, S. X., Ley, B. A., Mangravite, L., Edmondson, D. E., Huynh, B. H., and Bollinger, J. M., 1998, Electron injection through a specific pathway determines the outcome of oxygen activation at the diiron cluster in the F208Y mutant of *Escherichia coli* ribonucleotide reductase protein R2. *Biochemistry* **37**:1124–1130.

Persson, A. L., Eriksson, M., Katterle, B., Pötsch, S., Sahlin, M., and Sjöberg, B.-M., 1997, A new mechanism-based radical intermediate in a mutant R1 protein affecting the catalytically essential Glu(441) in *Escherichia coli* ribonucleotide reductase. *J. Biol. Chem.* **272**:31533–31541.

Persson, A. L., Sahlin, M., and Sjöberg, B.-M., 1998, Cysteinyl and substrate radical formation in active site mutant E441Q of *Escherichia coli* class I ribonucleotide reductase. *J. Biol. Chem.* **273**:31016–31020.

Persson, A. L., Salin, M., and Sjöberg, B.-M., 1999, Transient free radicals in the reaction with mutant E441Q R1, wild-type R2 and CDP of *Escherichia coli* class Ia ribonucleotide reductase. Manuscript in preparation.

Persson, B. O., Karlsson, M., Climent, I., Ling, J. S., Sanders Loehr, J., Sahlin, M., and Sjöberg, B.-M., 1996, Iron ligand mutants in protein R2 of *Escherichia coli* ribonucleotide reductase—Retention of diiron site, tyrosyl radical and enzymatic activity in mutant proteins lacking an iron-binding side chain. *J. Biol. Inorg. Chem.* **1**:247–256.

Petersson, L., Gräslund, A., Ehrenberg, A., Sjöberg, B.-M., and Reichard, P., 1980, The iron center in ribonucleotide reductase from *Escherichia coli*. *J. Biol. Chem.* **255**:6706–6712.

Pötsch, S., Lendzian, F., Ingemarson, R., Hornberg, A., Thelander, L., Lubitz, W., Lassmann, G., and Gräslund, A., 1999, The iron-oxygen reconstitution reaction in protein R2-tyr-177 mutants of mouse ribonucleotide reductase—EPR and electron nuclear double resonance studies on a new transient tryptophan radical. *J. Biol. Chem.* **274**:17696–17704.

Rardin, R. L., Tolman, W. B., and Lippard, S. J., 1991, Monodentate carboxylate complexes and the carboxylate shift: Implications for polymetalloprotein structure and function. *New. J. Chem.* **15**:417–430.

Ravi, N., Bollinger, J. M., Huynh, B. H., Edmondson, D. E., and Stubbe, J., 1994, Mechanism of assembly of the tyrosyl radical-diiron(III) cofactor of *E. coli* ribonucleotide reductase: I. Mössbauer characterization of the diferric radical precursor. *J. Am. Chem. Soc.* **116**:8007–8014.

Reichard, P., 1997, The evolution of ribonucleotide reduction. *Trends. Biochem. Sci.* **22**:81–85.

Riggs-Gelasco, P. J., Shu, L. J., Chen, S. X., Burdi, D., Huynh, B. H., Que, L., and Stubbe, J., 1998, EXAFS characterization of the intermediate X generated during the assembly of the *Escherichia coli* ribonucleotide reductase R2 diferric tyrosyl radical cofactor. *J. Am. Chem. Soc.* **120**:849–860.

Rosenzweig, A. C., Nordlund, P., Takahara, P. M., Frederick, C. A., and Lippard, S. J., 1995, Geometry of the soluble methane monooxygenase catalytic diiron center in two oxidation states. *Chem. Biol.* **2**:409–418.

Rova, U., Adrait, A., Pötsch, S., Gräslund, A., and Thelander, L., 1999, Evidence by mutagenesis that Tyr(370) of the mouse ribonucleotide reductase R2 protein is the connecting link in the intersubunit radical transfer pathway. *J. Biol. Chem.* **274**:23746–23751.

Rova, U., Goodtzova, K., Ingemarson, R., Behravan, G., Gräslund, A., and Thelander, L., 1995, Evidence by site-directed mutagenesis supports long-range electron transfer in mouse ribonucleotide reductase. *Biochemistry* **34**:4267–4275.

Sahlin, M., Gräslund, A., Ehrenberg, A., and Sjöberg, B.-M., 1982, Structure of the tyrosyl radical in bacteriophage T4-induced ribonucleotide reductase. *J. Biol. Chem.* **257**:366–369.

Sahlin, M., Gräslund, A., Petersson, L., Ehrenberg, A., and Sjöberg, B.-M., 1989, Reduced forms of the iron-containing small subunit of ribonucleotide reductase from *Escherichia coli*. *Biochemistry* **28**:2618–2625.

Sahlin, M., Lassmann, G., Pötsch, S., Sjöberg, B.-M., and Gräslund, A., 1995, Transient free radicals in iron/oxygen reconstitution of mutant protein R2 Y122F—Possible participants in electron transfer chains in ribonucleotide reductase. *J. Biol. Chem.* **270**:12361–12372.

Schmidt, P. P., Rova, U., Katterle, B., Thelander, L., and Gräslund, A., 1998, Kinetic evidence that a radical transfer pathway in protein R2 of mouse ribonucleotide reductase is involved in generation of the tyrosyl free radical. *J. Biol. Chem.* **273**:21463–21472.

Siegbahn, P. E. M., 1998, Theoretical study of the substrate mechanism of ribonucleotide reductase. *J. Am. Chem. Soc.* **120**:8417–8429.

Siegbahn, P. E. M., A., B. M. R., and M., P., 1998, A comparison of electron transfer in ribonucleotide reductase and the bacterial photosynthetic reaction center. *Chem. Phys. Lett.*
**292**:421–430.

Siegbahn, P. E. M., Blomberg, M. R. A., and Crabtree, R. H., 1997, Hydrogen transfer in the
presence of amino acid radicals. *Theor. Chem. Acc.* **97**:289–300.

Silva, K. E., Elgren, T. E., Que, L., and Stankovich, M. T., 1995, Electron transfer properties of
the R2 protein of ribonucleotide reductase from *Escherichia coli*. *Biochemistry*
**34**:14093–14103.

Sjöberg, B.-M., 1997, Ribonucleotide reductases—A group of enzymes with different metal-
losites and similar reaction mechanism. *Structure & Bonding* **88**:139–173.

Sjöberg, B.-M., Gräslund, A., and Eckstein, F., 1983, A substrate radical intermediate in
the reaction between ribonucleotide reductase from *Escherichia coli* and 2'-azido-2'-
deoxynucleoside diphosphates. *J. Biol. Chem.* **258**:8060–8067.

Sjöberg, B.-M., Reichard, P., Gräslund, A., and Ehrenberg, A., 1978, The tyrosine free radical
in ribonucleotide reductase from *Escherichia coli*. *J. Biol. Chem.* **253**:6863–6865.

Stone, E. W., and Maki, A. H., 1962, Hindered internal rotation and ESR spectroscopy. *J. Chem.
Phys.* **37**:1326–1333.

Stubbe, J., 1998, Ribonucleotide reductases in the twenty-first century. *Proc. Natl. Acad. Sci.
USA* **95**:2723–2724.

Stubbe, J., and Ackles, D., 1980, On the mechanism of ribonucleoside diphosphate reductase
from *Escherichia coli*. Evidence for 3'-C-H bond cleavage. *J. Biol. Chem.* **255**:8027–8030.

Stubbe, J., and Riggs-Gelasco, P., 1998, Harnessing free radicals: formation and function of the
tyrosyl radical in ribonucleotide reductase. *Trends. Biochem. Sci.* **23**:438–443.

Stubbe, J. A., and van der Donk, W. A., 1995, Ribonucleotide reductases: Radical enzymes with
suicidal tendencies. *Chem. Biol.* **2**:793–801.

Stubbe, J. A., and van der Donk, W. A., 1998, Protein radicals in enzyme catalysis. *Chem. Rev.*
**98**:705–762.

Sturgeon, B. E., Burdi, D., Chen, S. X., Huynh, B. H., Edmondson, D. E., Stubbe, J., and
Hoffman, B. M., 1996, Reconsideration of X, the diiron intermediate formed during
cofactor assembly in *E. coli* ribonucleotide reductase. *J. Am. Chem. Soc.* **118**:7551–
7557.

Surdhar, P. S., and Armstrong, D. A., 1987, Reduction potentials and exchange-reactions of
thiyl radicals and disulfide anion radicals. *J. Phys. Chem.* **91**:6532–6537.

Tamarit, J., Mulliez, E., Meier, C., Trautwein, A., and Fontecave, M., 1999, The anaerobic ribonucleotide reductase from *Escherichia coli*. The small protein is an activating enzyme containing a [4Fe-4S]$^{2+}$ center. *J. Biol. Chem.* **274**:31291–31296.

Thelander, L., 1974, Reaction mechanism of ribonucleoside diphosphate reductase from
*Escherichia coli*. Oxidation-reduction-active disulfides in the B1 subunit. *J. Biol. Chem.*
**249**:4858–4862.

Thelander, L., and Larsson, B., 1976, Active site of ribonucleoside diphosphate reductase from
*Escherichia coli*. Inactivation of the enzyme by 2'-substituted ribonucleoside diphosphates. *J. Biol. Chem.* **251**:1398–1405.

Thelander, L., Sjöberg, B. M., and Eriksson, S., 1978, Ribonucleoside diphosphate reductase
(*Escherichia coli*). *Methods Enzymol.* **51**:227–237.

Tong, W., Burdi, D., Riggs-Gelasco, P., Chen, S., Edmondson, D., Huynh, B. H., Stubbe, J., Han,
S., Arvai, A., and Tainer, J., 1998, Characterization of Y122F R2 of *Escherichia coli* ribonucleotide reductase by time-resolved physical biochemical methods and X-ray crystallography. *Biochemistry* **37**:5840–5848.

Uhlin, U., and Eklund, H., 1994, Structure of ribonucleotide reductase protein R1. *Nature*
**370**:533–539.

van Dam, P. J., Willems, J. P., Schmidt, P. P., Pötsch, S., Barra, A. L., Hagen, W. R., Hoffman, B. M., Andersson, K. K., and Gräslund, A., 1998, High-frequency EPR and pulsed Q-Band ENDOR studies on the origin of the hydrogen bond in tyrosyl radicals of ribonucleotide reductase R2 proteins from mouse and herpes simplex virus type 1. *J. Am. Chem. Soc.* **120**:5080–5085.

van der Donk, W. A., Stubbe, J., Gerfen, G. J., Bellew, B. F., and Griffin, R. G., 1995, EPR investigations of the inactivation of *E. coli* ribonucleotide reductase with 2′-azido-2′-deoxyuridine 5′-diphosphate: Evidence for the involvement of the thiyl radical of C225-R1. *J. Am. Chem. Soc.* **117**:8908–8916.

van der Donk, W. A., Zeng, C. H., Biemann, K., Stubbe, J., Hanlon, A., and Kyte, J., 1996, Identification of an active site residue of the R1 subunit of ribonucleotide reductase from *Escherichia coli*: Characterization of substrate-induced polypeptide cleavage by C225SR1. *Biochemistry* **35**:10058–10067.

Wang, P. J., Chabes, A., Casagrande, R., Tian, X. C., Thelander, L., and Huffaker, T. C., 1997, Rnr4p, a novel ribonucleotide reductase small-subunit protein. *Mol. Cell Biol.* **17**:6114–6121.

Warncke, K., Babcock, G. T., and McCracken, J., 1994, Structure of the $Y_D$ tyrosine radical in photosystem II as revealed by $^2H$ electron spin echo envelope modulation (ESEEM) spectroscopic analysis of hydrogen hyperfine interactions. *J. Am. Chem. Soc.* **116**:7332–7340.

Young, P., Andersson, J., Sahlin, M., and Sjöberg, B.-M., 1996, Bacteriophage T4 anaerobic ribonucleotide reductase contains a stable glycyl radical at position 580. *J. Biol. Chem.* **271**:20770–20775.

Åberg, A., Hahne, S., Karlsson, M., Larsson, A., Ormö, M., Åhgren, A., and Sjöberg, B.-M., 1989, Evidence for two different classes of redox-active cysteines in ribonucleotide reductase of *Escherichia coli*. *J. Biol. Chem.* **264**:12249–12252.

Åberg, A., Nordlund, P., and Eklund, H., 1993a, Unusual clustering of carboxyl side chains in the core of iron-free ribonucleotide reductase. *Nature* **361**:276–278.

Åberg, A., Ormö, M., Nordlund, P., and Sjöberg, B.-M., 1993b, Autocatalytic generation of dopa in the engineered protein R2 F208Y from *Escherichia coli* ribonucleotide reductase and crystal structure of the dopa-208 protein. *Biochemistry* **32**:9845–9850.

*Chapter 13*

# Molybdenum Enzymes

Russ Hille

## 1. INTRODUCTION

### 1.1. Molybdenum Enzymes and the Reactions They Catalyze

Molybdenum is found in the active site of nitrogenase as part of a multinuclear cluster with iron (the so-called M-cluster), but in all other enzymes it comprises a mononuclear center (Hille, 1996; Kisker *et al.*, 1997a). The distinguishing characteristic of this latter group of enzymes is that, while other redox-active centers are often present, the molybdenum center itself possesses only a single equivalent of the transition metal. This large and diverse group of enzymes is the subject of the present account.

As a rule, the mononuclear molybdenum enzymes catalyze either the hydroxylation of aldehydes and heterocycles or oxygen atom transfer to or from a suitable acceptor or donor (typically at nitrogen or sulfur, but in the case of CO dehydrogenase a carbon atom). Having said this, several enzymes are known that catalyze other types of reaction, including simple oxidation-reduction and hydroxyl group transfer. In general, the molybdenum center functions as a reversible two-electron acceptor, cycling between the Mo(VI) and Mo(IV) oxidation states in carrying out this chemistry.

**RUSS HILLE**    Department of Medical Biochemistry, The Ohio State University, 333 Hamilton Hall, 1645 Neil Avenue, Columbus, OH 43210-1218 USA.

*Subcellular Biochemistry, Volume 35: Enzyme-Catalyzed Electron and Radical Transfer*, edited by Holzenburg and Scrutton. Kluwer Academic / Plenum Publishers, New York, 2000.

Examples of reactions catalyzed by molybdenum enzymes are shown in Figure 1.

## 1.2 Sequence Homologies and Classification of the Molybdenum Enzymes

From a comparison of amino acid sequences that have been obtained over the past several years, it has become apparent that the molybdenum enzymes other than nitrogenase consist of three principal families (Hille, 1996). The first of these is the molybdenum hydroxylases, enzymes from organisms ranging from thermophilic archaea to humans that catalyze the hydroxylation of a variety of aromatic heterocycles (including purines, pteridines and related compounds), or the hydroxylation of aldehydes (to form biogenic carboxylic acids such as retinoic acid, abscissic acid and indole-3-acetic acid) to the corresponding carboxylic acid. The molybdenum-containing CO dehydrogenase from *Oligotropha carboxidovorans* (not to be confused with the nickel-containing enzyme of the same name from sources such as *Clostridium thermoaceticum*) is also a member of this family, although the reaction catalyzed is not, strictly speaking, a hydroxylation (Meyer *et al.*, 1993). At present approximately 30 molybdenum hydroxylases have been identified biochemically or inferred from gene sequences; of these, the best characterized from a mechanistic standpoint is xanthine oxidase from bovine milk (see below). The crystal structures of two members of this family, the aldehyde oxidoreductase from *Desulfovibrio gigas* (Huber *et al.*, 1996; Romão *et al.*, 1995) and the CO dehydrogenase from *Oligotropha carboxidovorans* (Dobbek *et al.*, 1999), have been reported.

The second family of molybdenum enzymes consists of vertebrate sulfite oxidases and the assimilatory nitrate reductases from higher plants and fungi (enzymes that are involved in the reductive assimilation of nitrate to ammonia for utilization in protein synthesis, *etc.*). Here, these enzyme will be referred to as the eukaryotic oxotransferases, as both types of enzyme catalyze reactions that involve oxygen atom transfer either to or from a substrate lone pair of electrons (to the sulfur atom of sulfite to form sulfate, from the nitrogen atom of nitrate to yield nitrite). Both sulfite oxidase and nitrate reductase have been extensively studied using a wide range of methods, including sophisticated spectroscopic techniques such as x-ray absorption and electron paramagnetic resonance spectroscopy. The crystal structure of sulfite oxidase from chicken liver has been reported (Kisker *et al.*, 1997b).

The final family of molybdenum enzymes consists of enzymes from bacterial and archaeal sources that also catalyze oxygen atom transfer reactions for the most part, although formate dehydrogenase and polysulfide

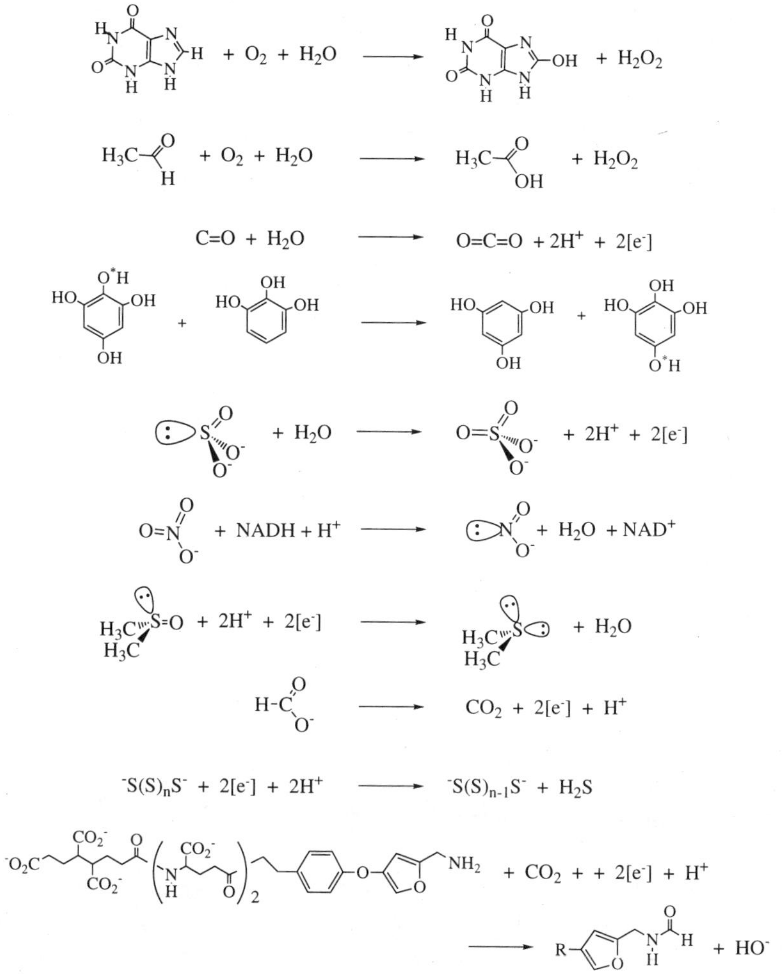

**FIGURE 1.** Examples of reactions catalyzed by molybdenum containing enzymes. From *top* to *bottom*, hydroxylation of xanthine, hydroxylation of acetaldehyde, dehydrogenation of carbon monoxide, transhydroxylation of pyrogallol, oxidation of sulfite, reduction of nitrate, reduction of dimethylsulfoxide, oxidation of formate, reduction of polysulfide and formation of formylmethanofuran.

reductase (to name just two exceptions) catalyze other types of reaction. This family is considerably more diverse than the other two, and includes enzymes such as the dimethylsulfoxide (DMSO) reductases, dissimilatory nitrate reductases (enzymes that serve as terminal oxidases in respiratory electron transfer chains rather than in the assimilation of nitrate) and formate dehydrogenase. Of these, the best characterized are the DMSO reductases from *Escherichia coli* and *Rhodobacter* species (both *sphaeroides* and *capsulatus*). Crystal structures have been reported for both of the *Rhodobacter* enzymes (Schindelin *et al.*, 1996; Schneider *et al.*, 1996; McAlpine *et al.*, 1997; McAlpine *et al.*, 1998), as have structures for formate dehydrogenase from *E. coli* (Boyington *et al.*, 1997), trimethylamine-N-oxide from *Shewanella massilia* (Czjzek *et al.*, 1998) and the dissimilatory nitrate reductase (the Nap gene product) from *Desulfovibrio desulfuricans* (Dias *et al.*, 1999).

The active sites of these families are each distinct, as indicated in Figure 2. In the case of the molybdenum hydroxylases, the molybdenum center (in its oxidized Mo(VI) form) consists of LMoOS(OH), with L representing an unusual pterin cofactor that is common to all the enzymes; this cofactor is coordinated to the metal via an enedithiolate side chain as shown. In the case of the *D. gigas* aldehyde oxidoreductase, the molybdenum center possesses a distorted square pyramidal coordination geometry, with the Mo=S (terminal sulfido) occupying the apical position (Romão *et al.*, 1995). As will be discussed further below, it is the Mo—OH ligand that is transferred to substrate in the course of hydroxylation, regenerated with hydroxide from solvent at the end of each catalytic sequence. The terminal sulfido group of the active site can be removed by reaction with cyanide to yield thiocyanate and a "desulfo" form of the molybdenum center (Massey and Edmondson,

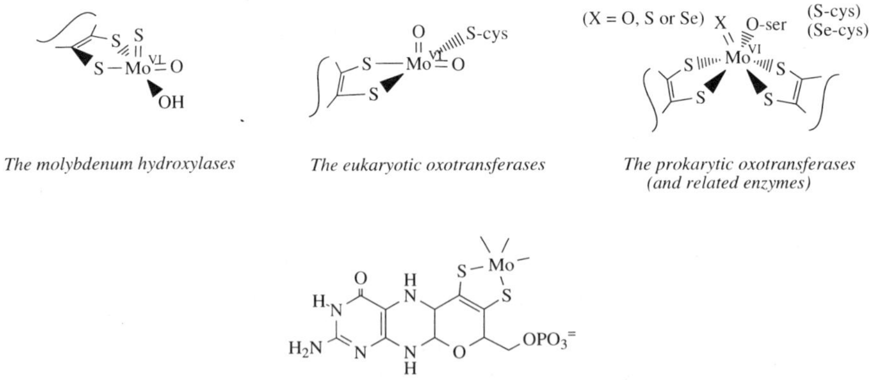

*The molybdenum hydroxylases*          *The eukaryotic oxotransferases*          *The prokarytic oxotransferases*
                                                                              *(and related enzymes)*

**FIGURE 2.** The active site structures for the three families of mononuclear molybdenum enzymes. From *left* to *right*, the molybdenum hydroxylases (including by aldehyde oxido-reductase and xanthine oxidase), the eukaryotic oxotransferases (consisting of sulfite oxidase and the assimilatory nitrate reductases of higher plants), and the prokaryotic oxotransferases and related enzymes (including the DMSO reductases, dissimilatory nitrate reductases and formate dehydrogenase). The pterin cofactor common to all these enzymes is also shown, *bottom*, indicating coordination to the molybdenum via the enedithiolate portion of its pyran ring.

1970) that possesses a $MoO_2$ core and is catalytically inactive. For the eukaryotic oxotransferases, the molybdenum center core is best described as having a $LMoO_2(S—Cys)$ core, with L again representing the pterin cofactor. As in the case of the molybdenum hydroxylases, the coordination geometry is distorted square pyramidal, with one of the two Mo=O (terminal oxo) groups occupying the apical position (Kisker et al., 1997b). On the basis of the crystal structure for the chicken sulfite oxidase, it is most likely that the equatorial Mo=O group is the source of the oxygen atom transferred to sulfite in the course of catalysis, as this is positioned best for interaction with substrate (but see below). The prokaryotic oxotransferases and related enzymes are unique in that they possess two equivalents of the pterin cofactor coordinated to the molybdenum rather than just one. The remainder of the coordination sphere usually consists of a terminal Mo=O oxo group (although a Mo=S may be present in the formate dehydrogenase from *Methanobacterium thermoautotrophicum*, on the basis of its inhibition by cyanide; Barber et al., 1986) and a ligand contributed by the polypeptide. In the DMSO reductases this peptide ligand is a serine residue (Schindelin et al., 1995), while in the dissimilatory nitrate reductases it is a cysteine (Dias et al., 1999); remarkably, in the formate dehydrogenase from *E. coli* this ligand is a selenocysteine (Axley et al., 1990; Gladyshev et al., 1994; Boyington et al., 1997). In all these enzymes, the coordination geometry for the oxidized center is best described as distorted trigonal prismatic, as shown in Figure 2 (but see further below). With the *R. sphaeroides* DMSO reductase, mutation of the molybdenum-coordinating serine to a cysteine residue results in only a 60% reduction in reactivity toward DMSO but, surprisingly, a four-fold increase in reactivity toward adenosine N-oxide (Hilton et al., 1999).

The pterin cofactor common to all these enzymes possesses the structure shown in Figure 2. As can be seen, in addition to the pteridine component there is a pyran ring with an enedithiolate (or dithiolene) group and a phosphorylated side chain; the pyranopterin cofactor (frequently referred to as "molybdopterin", although the identical cofactor is also found in tungsten-containing enzymes, see below) coordinates to the metal via the enedithiolate. In enzymes from eukaryotic sources the cofactor has the structure shown, but in prokaryotic enzymes the cofactor is elaborated as a "dinucleotide" with guanine, cytosine, adenine or (less frequently) hypoxanthine. In the case of the prokaryotic oxotransferases, the two molecules of the pterin cofactor are not equivalent. That designated P in the crystal structure of the *R. sphaeroides* DMSO reductase (Schindelin et al., 1996) has somewhat shorter Mo—S distances, and on the basis of resonance Raman work (Garton et al., 1997) appears best described as having considerable π-delocalization over the four atoms of the enedithiolate unit. The

Q pterin, on the other hand, has longer Mo—S distances (in one structure of *the R. capsulatus* enzyme it is fully dissociated from the metal, being replaced by a pair of terminal oxo groups) and has an electronic structure that reflects discrete dithiolate character. The pterin cofactor found in the other two families of enzyme are thought to be of the $\pi$-delocalized variety. Not surprisingly given its structure, biosynthesis of the pterin cofactor is quite involved, with the involvement of no fewer than twelve gene products in its maturation. Cofactor biosynthesis in *Escherichia coli, Drosophila melanogaster, Arabidopsis melanogaster* and *Nicotiana tabacum* has been studied extensively, and detailed reviews are available (Rajagopalan, 1992; Rajagopalan and Johnson, 1992; Mendel and Schwartz, 1999).

The identical pterin cofactor is also found in tungsten-containing enzymes (Johnson *et al.*, 1996). Tungsten lies immediately below molybdenum in the periodic table, and is the only element of the sixth row of the periodic table (as molybdenum is of the fifth) to have a clearly established biological function. Tungsten is found in the active sites of archaeal enzymes such as aldehyde:ferredoxin oxidoreductase from *Pyrococcus furiosus*. Presently, it is thought that tungsten-containing enzymes can be grouped into two families on the basis of amino acid sequence homologies, one of which is closely related to the prokaryotic oxotransferase family of molybdenum enzymes (Johnson *et al.*, 1996). Indeed at least one enzyme (the formylmethanofuran dehydrogenase from *Methanobacterium wolfei*; Schmitz *et al.*, 1992) can be found in an active form with either molybdenum or tungsten, depending on the growth conditions. Tungsten can also be incorporated into other molybdenum enzymes from microorganisms grown on tungsten-rich medium (or from animals fed high-tungsten diets), but these enzyme forms are typically inactive. It has been suggested that molybdenum enzymes are descended from the tungsten-containing enzymes (Johnson *et al.*, 1996). The strictly anaerobic and typically quite thermophilic organisms found to possess tungsten enzymes grow in environments that presumably prevailed in the early Earth, conditions under which tungsten would have been readily available to the early biosphere. These organisms also appear to possess a quite low internal oxidation-reduction poise, which would also favor tungsten given the considerably lower reduction potentials exhibited by tungsten compounds relative to the corresponding molybdenum species. As the atmosphere of the Earth became more aerobic, so the argument goes, molybdenum (in the form of very water-soluble molybdate salts) became more bio-available at the expense of tungsten. The aerobic organisms that arose would also have been able to make better use of higher-potential molybdenum centers. While this hypothesis remains unproved, it provides an attractive rationale for the close relationship that exists between at least certain molybdenum and tungsten enzymes.

## 1.3. Structural and Catalytic Variations Within the Three Families of Molybdenum Enzymes

Although the organization of molybdenum enzymes into the three families given above has become increasingly justified over the past several years as more amino acid sequences and crystal structures have become available, it is important to recognize that there are significant variations among members of a given family with respect to both structure and reaction catalyzed. Among the molybdenum hydroxylases, for example, the vast majority of enzymes are dimers with an overall structure that proceeds from the N-terminus to trace out discrete domains for a pair of $Fe_2S_2$ (spinach ferredoxin-like) centers, a third domain that binds FAD and finally a pair of large domains that together constitute the molybdenum-binding portion of the protein. Some enzymes from this family (the *D. gigas* aldehyde oxidoreductase being a notable example) lack the FAD-binding domain, however, and others have separate subunits for the iron-sulfur-, flavin- and molybdenum-binding segments rather than a single polypeptide, being organized as $\alpha_2\beta_2\gamma_2$ hexamers (Hille, 1996). Functionally, the vast majority of the known enzymes of this family catalyze hydroxylation of compounds such as purines and aldehydes, but there are some interesting exceptions. The CO dehydrogenase from *Oligotropha carboxidovorans* has already been mentioned, and a molybdenum-requiring nitrogenase from the aerobe *Streptomyces thermoautotrophicus* has been reported that may well be a member of the molybdenum hydroxylase (or another) family of mononuclear molybdenum enzymes (Ribbe *et al.*, 1997; the conventional nitrogenase is from obligate anaerobes, and the enzyme is extremely oxygen-sensitive). There is also an acetylene hydratase in this family (Rosner *et al.*, 1997). Finally, pyrogallol transhydroxylase from *Pelobacter acidigallici* catalyzes a hydroxyl group transfer reaction yielding phloroclucinol and 1,2,3,5 tetrahydroxybenzene (Figure 1) in a reaction that possibly involves consecutive dehydroxylation-hydroxylation steps (Reichenbecher *et al.*, 1996; Hille *et al.*, 1999).

For the eukaryotic oxotransferases, sulfite oxidase is organized with an N-terminal heme domain, a central molybdenum-binding domain and a C-terminal domain responsible for the subunit contacts in the dimeric protein (Kisker *et al.*, 1997b). In nitrate reductase, the molybdenum-binding portion of the protein is at the N-terminus, with succeeding domains possessing heme and FAD (Kubo *et al.*, 1988). The heme domains of both sulfite oxidase and nitrate reductase possess significant sequence homology to cytochrome $b_5$ and their molybdenum domains possess some degree of sequence similarity, but this is not as pronounced as the homologies seen among the molybdenum hydroxylases (Wootton *et al.*, 1991). The reactions catalyzed by sulfite oxidase and nitrate reductase are in essence the reverse of one another, with oxygen atom transfer to a lone pair of sulfite, and oxygen atom transfer from nitrate. This situation has been justified on thermodynamic grounds from a comparison of the relative stabilities of the S—O and N—O bonds in the sulfite:sulfate and nitrite:nitrate couples (Holm, 1987; Holm and Donohue, 1993).

As indicated above, the prokaryotic oxotransferases constitute the most diverse family of molybdenum enzymes, and for reasons in addition to the above-mentioned variations in serine, cysteine and selenocysteine coordination to the metal. Several of these enzymes, including the *Rhodobacter* DMSO reductases (Satoh and Kurihara, 1987; McEwan *et al.*, 1991) and biotin-S-oxide reductase from *E. coli* (Pierson and Campbell, 1990)), are unusual in that they possess no redox-active sites other than the molybdenum center. These enzymes are typically found in the periplasmic space of the host organism, and appear to function in the dissipation of excess reducing equivalents (generated under certain growth conditions) without contributing to the transmembrane electrochemical potential. The DMSO reductase from *E. coli*, on the other hand, is an integral membrane protein that does contribute to the transmembrane potential (Weiner *et al.*, 1988; Weiner *et al.*, 1992). It is comprised of three subunits (the products of the *dmsABC* operon; Bilous and Weiner, 1988): one possessing the molybdenum center, a second with four $Fe_4S_4$ clusters, and a third that serves as a transmembrane anchor. Similarly, the three distinct nitrate reductases of *E. coli* (which are expressed under differing growth conditions) are integral membrane proteins with multiple iron-sulfur centers—or in the case of the Nap gene product, hemes (Blasco *et al.*, 1989; Blasco *et al.*, 1990; Berks *et al.*, 1994; Choe and Reznikoff, 1993; Thomas *et al.*, 1999). The molybdenum- and iron-sulfur containing subunits of these enzymes project into the cytoplasm and can be isolated in a soluble form (Buc *et al.*, 1995). The transmembrane anchor of the two major nitrate reductases is distinct from that of the DMSO reductase in possessing a b-type cytochrome. Other enzymes of this family are less well-characterized, but it may be expected that additional variations in subunit makeup, stoichiometry and cofactor constitution will be identified.

As for the reactions catalyzed by members of this third family of molybdenum enzymes, there are several variations from the principal theme of oxygen atom transfer. Formate dehydrogenase from *E. coli* catalyzes the oxidation of formate to $CO_2$, a reaction that isotope studies have shown does not pass through a bicarbonate intermediate (Khangulov *et al.*, 1998). Instead, it appears likely that $CO_2$ is formed by direct hydride transfer from substrate to the molybdenum center. Polysulfide reductase is another molybdenum enzyme that catalyzes a non-canonical reaction: the reductive cleavage of sulfide from substrate polysulfide: $^-S(S)_nS^-$ + $2[e^-]$ + $H^+ \rightarrow {}^-S(S)_{n-1}S^-$ + $H_2S$. In species such as *Wolinella succinogenes*, the reducing equivalents used in this reaction are obtained from the formate dehydrogenase of the organism (Jankielewicz *et al.*, 1995). A final reaction is that catalyzed by formylmethanofuran dehydrogenase from various methanogenic organisms, including *Methanobacterium thermoautotrophicum*, which involves the reductive formylation of the terminal amine group of methanofuran using $CO_2$ as carbon source (this is the first step in methanogenesis in these organisms): $R$—$NH_2$ + $CO_2$ + $2[e^-]$ + $H+ \rightarrow$ R—NHCHO + $HO^-$ (Thauer, 1998).

With this background information, the three families of molybdenum enzymes will be considered in turn, focusing in each case first on structural aspects followed by a discussion of reaction mechanism.

## 2. THE MOLYBDENUM HYDROXYLASES

### 2.1. Structural Studies

As indicated above, the two members of the molybdenum hydroxylase family for which a crystal structure has been reported are the aldehyde oxidoreductase from *D. gigas* (Huber *et al.*, 1996; Romão *et al.*, 1995) and the CO dehydrogenase from *O. carboxidovorans* (Dobbek *et al.*, 1999). The first of these enzymes is an $\alpha_2$ dimer of molecular weight 194,000 Da with each subunit made up of four domains, as shown in Figure 3. Two domains at the N-terminus each possess a $Fe_2S_2$ center, which are followed by a long meander then a pair of large domains with the molybdenum center (possessing the cytosine dinucleotide form of the cofactor) at their interface. The first of the $Fe_2S_2$ domains possesses a polypeptide fold similar to that of spinach ferredoxin and consists primarily of β-sheet, but the second has a unique fold made up of a pair of long α-helices flanked by a pair of two shorter helices; the domain possesses a pseudo two-fold axis of symmetry that bisects the metal center, which lies at one end of the two longer helices. The two quite elongated molybdenum-binding domains lie across one another at approximately 90°, and each consists of a pair of subdomains having a complex αβ structure. The molybdenum center lies at the interface between these two domains, but most of the protein contacts with the cofactor (present as the cytosine dinucleotide) are in the C-terminal domain. The polypeptide chain traces out each of the four domains of aldehyde oxidoreductase in succession from the N-terminus and each is encoded by a single, contiguous segment of the gene for the protein (by contrast, there are multiple passes between the two subdomains of each of

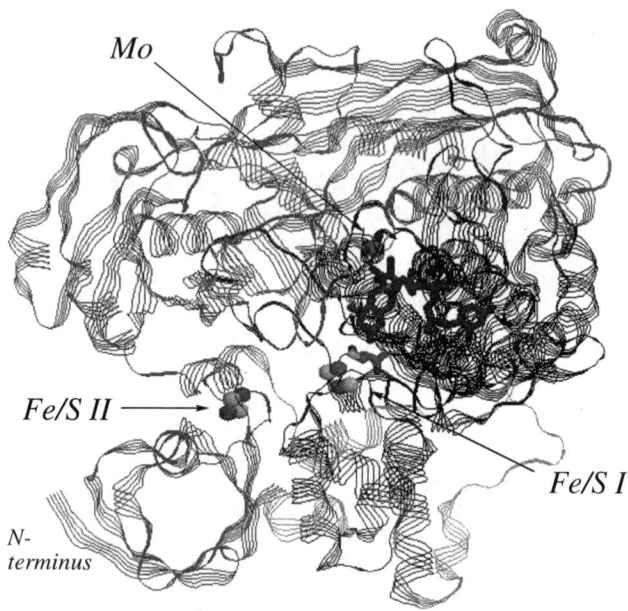

**FIGURE 3.** The overall protein fold of the *D. gigas* aldehyde oxidoreductase. The two $Fe_2S_2$ centers and the molybdenum center are indicated in wireframe/spacefilling representation. The two $Fe_2S_2$ domains are at bottom left and bottom center, respectively (structure rendered from the data of Romão *et al.*, 1995).

the molybdenum-binding domains). The implication is that the enzyme was built up over the course of evolution by gene duplication/fusion events from smaller, structurally autonomous elements possessing the redox-active centers.

The molybdenum center of aldehyde oxidoreductase has the overall structure shown in Figure 1 above, with the terminal sulfido group occupying the apical position of a distorted square pyramidal coordination sphere. Access to the molybdenum center is via a 15 Å-long solvent channel, which opens onto the active site adjacent to the Mo—OH group. (Although initially assigned as a metal-coordinated water on the basis of a long Mo—O distance of 2.2 Å, this distance may represent an overestimate due to truncation artifacts necessarily introduced in the data analysis (Schindelin *et al.*, 1997). For the purposes of the present discussion, therefore, this ligand will be rendered as hydroxide rather than water.) As suggested on the basis of the crystal structure, and subsequently demonstrated by ENDOR and EPR analysis of the related enzyme xanthine oxidase (see below), it is this oxygen that is transferred to substrate in the course of hydroxylation (ultimately regenerated by hydroxide from solvent).

The redox-active centers in aldehyde oxidoreductase are not particularly close to one another. The molybdenum is 14.5 Å from the second $Fe_2S_2$ center and 27 Å from the first; the two $Fe_2S_2$ centers are 12 Å from one another, and the angle between the three centers is approximately 150°. Interestingly, the distal amino group of the pterin cofactor is hydrogen-bonded to a cysteine residue coordinating one of the irons of the closer iron-sulfur cluster, and it is evident that the pterin cofactor plays a role in mediating electron transfer between the two centers (Huber et al., 1996). Indeed, a specific mechanism by which this occurs has recently been suggested, with electron transfer being mediated by σ rather than π interactions (Inscore et al., 1999). The central aspect of this proposal is based on the expectation that the Mo=O group of the molybdenum coordination sphere will define the z axis of the ligand field about the molybdenum, with the electrons of Mo(V) and (IV) found in the $d_{xy}$ orbital. This orbital must interact with ligands in the xy plane of the metal coordination sphere (such as the enedithiolate of the pterin cofactor) in a σ rather than π fashion. (In addition to Inscore et al. (1999), see Carducci et al. (1994), Nipales and Westmoreland (1997), Swann and Westmoreland (1997) and Balagopalakrishna et al. (1996) for a fuller description of the electronic structures of molybdenum complexes relevant to the active sites of the molybdenum enzymes.)

The two $Fe_2S_2$ centers of the molybdenum hydroxylases have long been known to be distinct with readily discernible EPR properties. One has more or less typical features for a spinach ferredoxin-type center and the other with considerably broader features, designated Fe/S I and II, respectively (Palmer and Massey, 1969). In the case of xanthine oxidase and other enzymes, a relatively strong magnetic interaction has been observed between the molybdenum center and the iron-sulfur center designated Fe/S I (Lowe and Bray, 1978), and on the basis of the crystal structure this is most likely the cluster that lies closer (within 15 Å) to the molybdenum atom. This assignment has recently been confirmed in site-directed mutagenesis studies of xanthine oxidase and spectroscopic studies of CO dehydrogenase, indicating that physiological electron transfer occurs in the following direction: Mo → Fe/S I → Fe/S II. It is somewhat surprising that the iron-sulfur center with the more typical features is found in the domain with the more unusual fold (and that having the unique EPR features is found in the domain having the standard ferredoxin fold). It is possible that the unusual EPR features of Fe/S II are attributable to an unusual degree of solvent accessibility for the cluster (Romão et al., 1995).

Considering now CO dehydrogenase, the enzyme is made up of three subunits designated Small (S), Medium ((M) and Large (L), having molecular masses of 18, 30 and 89 kDa to give an overall molecular mass for the hexamer of 274 kDa (Schübel *et al.*, 1996). The S and L subunits of this enzyme contain the $Fe_2S_2$ and molybdenum centers of the protein, respectively, and possess a high degree of structural homology to the corresponding portions of the *D. gigas* aldehyde oxidoreductase. The active site also has a conserved glutamate residue that in the case of aldehyde oxidoreductase has been proposed to act as an active site base (Dobbek *et al.*, 1999), and a glutamine (rather than a histidine, as is found in the case of the *D. gigas* enzyme) that is hydrogen-bonded to one of two $Mo{=}O$ groups of the metal. CO dehydrogenase is unusual among the molybdenum hydroxylases in not having a cyanolyzable $Mo{=}S$ group, although this is not necessarily surprising, given the quite distinct reaction catalyzed by this enzyme. It has been known for some time, however, that the enzyme possesses a cyanolyzable selenium rather than sulfur (Meyer and Rajagopalan, 1984), and in the crystal structure of the enzyme this catalytically essential selenium is found as selanylcysteine (*i.e.*, Cys—S—Se$^-$) at position 388. The selenium atom itself is located approximately 3.7 Å from the molybdenum, near the Mo—OH group (Dobbek *et al.*, 1999), and a possible mechanistic role of this selanylcysteine residue has been suggested (see below).

The flavin-binding subunit M of CO dehydrogenase, which lacks a corresponding structural element in the aldehyde oxidoreductase, consists of three distinct domains (Dobbek *et al.*, 1999). The N-terminal and central domains are structurally quite closely related to the corresponding portions of vanillyl alcohol oxidase of *Penicillium simplicissimum* and UDP-acetylenolpyruvylglucosamine reductase (MurB) of *E. coli*, and together these proteins make up a new structural family of flavin-containing enzymes. Each of these proteins also possesses a third domain, but these are not closely related structurally and presumably are important in the specific physiological roles played by each of these enzymes. The N-terminal domain of the CO dehydrogenase subunit M consists of a three-stranded parallel β-sheet flanked by two α-helices. The middle domain has a 5-stranded antiparallel β-sheet with several small α-helical insertions that contact the adenosine portion of FAD. The C-terminal domain has a three-stranded antiparallel β-sheet associated with a bundle of three α-helices. The N-terminal and middle domains have characteristic FAD-binding sequences, AGGHS and TIGG, respectively, that are fairly well conserved among the molybdenum hydroxylases. It is thus likely that the flavin domains of other enzymes of this family will resemble that seen in CO dehydrogenase. Subunit M of CO dehydrogenase is oriented with respect to the remainder of the protein in the crystal structure in such a way that the dimethylbenzene ring of the flavin is directed toward the N-terminal $Fe_2S_2$ center of Subunit S (Fe/S II). Thus the four redox-active centers are laid out in such a way that the sequence of electron transfer within the enzyme is: Mo → Fe/S I → Fe/S II → FAD. This sequence is consistent with the results of previous pulse radiolysis studies of xanthine oxidase, another molybdenum hydroxylase, in which it has been found that electron transfer from the molybdenum center to the FAD of the enzyme involved transient reduction of the Fe/S centers at a rate of approximately $8500\,s^{-1}$ (Hille and Anderson, 1991); electron transfer on to the flavin occurred much more slowly, with a rate constant of $\sim300\,s^{-1}$, depending on the pH (Hille and Massey, 1986).

From the standpoint of human physiology and pathology, xanthine oxidase has long been recognized to occur in two forms, a proper oxidase utilizing only $O_2$ as oxidizing substrate, and a dehydrogenase form that utilizes $NAD^+$. The naturally occurring form in vertebrate liver is thought to be the dehydrogenase, and the enzyme can be isolated as such when the purification is carried out rapidly or in the presence of thiol reagents (Rajagopalan and Handler, 1967)—the same is true for the bovine milk protein (Hunt and Massey, 1992). The dehydrogenase form of the protein can be reversibly converted to the oxidase by thiol oxidants such as 4,4'-dithiopyridine, or irreversibly by limited proteolysis (Amaya et al., 1990). These procedures presumably alter the structure of the flavin domain of the enzyme, where the reaction with $O_2$ or $NAD^+$ occurs, in such a way that an $NAD^+$ binding site is lost. By contrast, the enzyme from chicken liver or *Drosophila* cannot be converted to an oxidase form (Nishino et al., 1989). The ability to utilize $NAD^+$ as oxidizing substrate distinguishes the xanthine-utilizing enzymes (sometimes referred to as oxidoreductases), reflecting the fact that the oxidase and dehydrogenase forms are products of the same gene) from the aldehyde oxidizing members of the molybdenum hydroxylases. These latter enzymes are always found as oxidases, and indeed contain a distinct consensus sequence in their flavin domains (KCPxAD) that differs from the GYRKTL consensus over the same sequence for the $NAD^+$-utilizing enzymes (Li Calzi et al., 1995; Turner et al., 1995)). The tyrosine residue in the dehydrogenase sequence can be modified with 5'-p-fluorosulfonylbenzoyladenosine, thereby abolishing reactivity with $NAD^+$ and confirming this as part of the nucleotide binding site (Nishino and Nishino, 1987; Nishino and Nishino, 1989).

Conversion of xanthine dehydrogenase to an oxidase form has been suggested to be important in ischemia/reperfusion injury, and its role has been the subject of much speculation and investigation (for a review see Nishino, 1994). Conversion of the dehydrogenase to the oxidase during ischemia has been proposed to leave the vascular endothelium predisposed to generate deleterious radical species upon recovery of circulation to the damaged tissue (McCord et al., 1985). Upon reperfusion, the newly formed oxidase can oxidize the substantial amounts of hypoxanthine that accumulate during ischemia, generating superoxide, peroxide and possibly hydroxyl radical (through Fenton-like chemistry). These reactive oxygen species are thought to then damage cell membranes and other components. Controversy remains, however, as to whether the anticipated protection against reperfusion damage upon administration of the xanthine oxidase inhibitor allopurinol is in fact observed, and whether any legitimate effect is due to enzyme inhibition or to scavenging of radical species by the drug directly. Further, it has not been generally recognized that the dehydrogenase itself reacts readily with $O_2$ generates considerable amounts of superoxide (Nishino et al., 1989), although under normal conditions $NAD^+$ would be expected to effectively compete with molecular oxygen for reduced enzyme. While suggestive evidence exists for the involvement of a dehydrogenase/oxidase conversion in ischemia/reperfusion damage, it has not yet been demonstrated unequivocally.

## 2.2.  Mechanistic Aspects

It has been known for some time that the molybdenum hydroxylases carry out this ubiquitous metabolic reaction in a fundamentally different way than do other biological systems possessing flavin, pterin, heme or non-heme iron in their active sites (Hille, 1994). Solvent rather than dioxygen is the ultimate source of oxygen incorporated into product (Murray et al., 1966), and reducing equivalents are generated rather than consumed in the course of the reaction. In the case of xanthine oxidase it is also well-established that a catalytically labile oxygen site on the enzyme is the proximal donor of oxygen to substrate, regenerated in the course of each catalytic cycle with oxygen derived from water (Hille and Sprecher, 1987). This is presumably also the case for other molybdenum hydroxylases. Given that xanthine oxidase is by far the best understood member of the molybdenum hydroxylases, mechanistic aspects of this family will be discussed in the context of this enzyme.

Xanthine oxidase is an $\alpha_2$ dimer (molecular mass ~280,000 Da) possessing the typical set of redox-active centers for the molybdenum hydroxylases: two $Fe_2S_2$ iron-sulfur centers, FAD and the molybdenum center. Analysis of the gene sequence indicates these centers are laid out as in the D. gigas aldehyde oxidoreductase, with an FAD-binding domain homologous to the M subunit of the O. carboxidovorans CO dehydrogenase inserted between the second $Fe_2S_2$ domain and the molybdenum-binding portion of the protein. Although no x-ray crystal structure for xanthine oxidase from any source is available, a considerable amount of work using x-ray absorption spectroscopy (XAS) has established the immediate coordination environment of the molybdenum (Bordas *et al.*, 1980; Cramer *et al.*, 1981; Cramer and Hille, 1985; Hille *et al.*, 1989; Turner *et al.*, 1989). The results of these studies are consistent with the crystal structure of the *D. gigas* enzyme and a structure for the oxidized enzyme as shown in Figure 1. Furthermore, XAS studies have shown that reduction of the oxidized enzyme from the Mo(VI) to the Mo(IV) oxidation state results in protonation of the Mo$^{VI}$=S unit to give Mo$^{IV}$—SH (Cramer *et al.*, 1981). This protonation is quite typical of even the simplest inorganic complexes of molybdenum (Stiefel, 1977a; Stiefel, 1977b) and accounts for the long-known sensitivity of the reduced enzyme (but not the oxidized) to thiol reagents such as arsenite (Coughlan *et al.*, 1969). EPR studies of the several Mo(V) species seen with xanthine oxidase have demonstrated that the one designated "very rapid" (on the basis of the millisecond time scale of its formation and decay in the course of the reaction of enzyme with xanthine) arises from a true catalytic intermediate that has product coordinated to the molybdenum center (Bray and Vånngård, 1969; Turner *et al.*, 1978; Bray, 1988). The same EPR signal is seen with 2-hydroxy-6-methylpurine, but in the case of this "poor" substrate formation and decay of the signal-giving species is on the second rather than millisecond time scale (Bray and George, 1985). Experiments with this slow substrate have shown that at pH 10 the EPR-active Mo(V) species is an obligatory intermediate and is preceded in the catalytic sequence by a Mo(IV) species that can be detected by UV-visible spectroscopy (McWhirter and Hille, 1991). These results can be understood in the context of the kinetic mechanism shown in Figure 4 (*top*), in which substrate binds to the enzyme, is converted directly to product complexed to a now fully reduced Mo(IV) center, which is then reoxidized in sequential steps to give first the EPR-active Mo(V) species and finally (with product release) the initial Mo(VI) state. At lower pH or with other substrates (including xanthine) product dissociation prior to reoxidation of the Mo(IV) intermediate occurs to an appreciable extent, however, and the signal-giving species is thereby circumvented. By following product formation discretely in a rapid quench experiment, it has been shown that formation of the C—O bond of product has formed in the first step of the reaction (*i.e.*, formation of the Mo(IV)-P complex; Mcwhirter and Hille, 1991). It is significant that formation of the EPR-active Mo(V) species is via this Mo(IV) precursor: from the standpoint of the molybdenum center, formation of the "very rapid" species is an oxidative rather than reductive event, involving electron transfer to the other redox-active

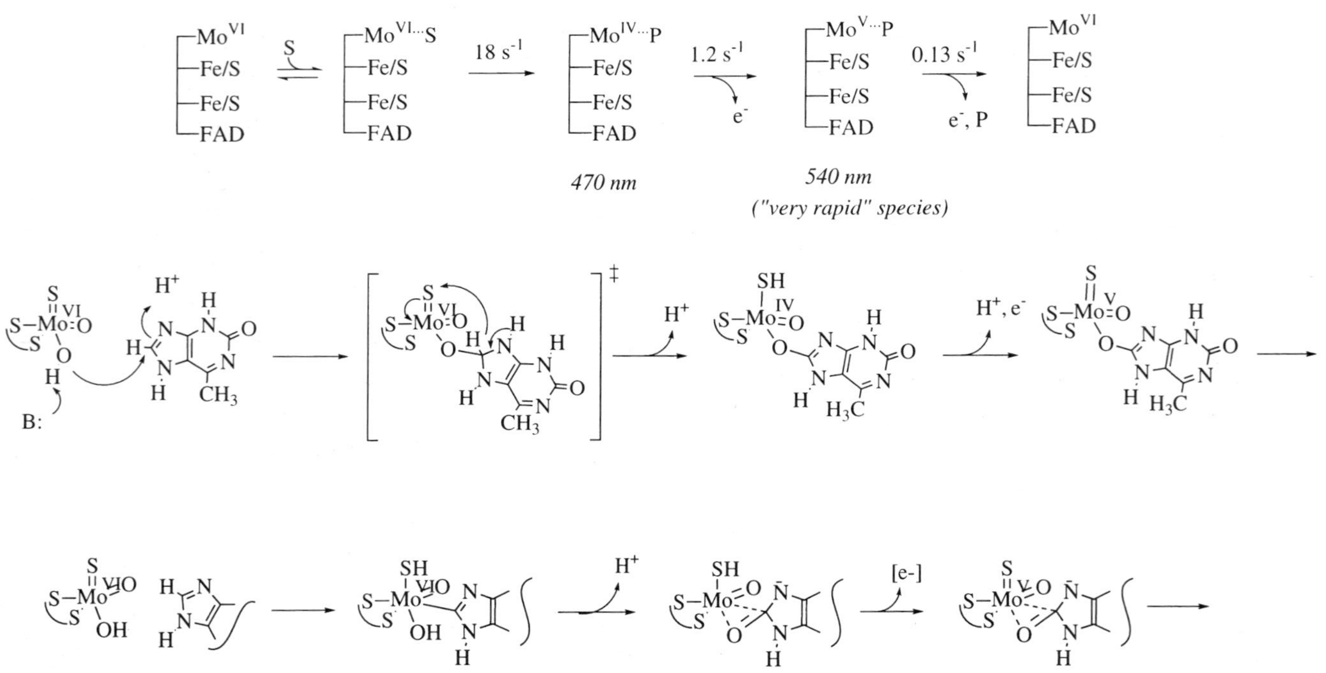

**FIGURE 4.** The reaction mechanism of xanthine oxidase. *Top,* a kinetic mechanism for the enzyme based on its reaction with 2-hydroxy-6-methylpurine. *Middle,* a chemical mechanism (through to formation of the species yielding the "very rapid" EPR signal) in which catalysis is initiated by nucleophilic attack of the active site Mo—OH on substrate (Greenwood *et al.,* 1993; Huber *et al.,* 1996; Xia *et al.* 1999). *Bottom,* a chemical mechanism (again through formation of the "very rapid" species) in which catalysis is initiated by insertion of the C8—H bond across the Mo≡S of the molybdenum center to give a species possessing a covalent Mo—C bond (Howes *et al.,* 1996).

centers of the enzyme. Although it is not at present possible to discount a mechanism in which the initial Mo(IV) intermediate is in fact formed in rapid, sequential one-electron steps, the present evidence indicates that the EPR-detectable Mo(V) species itself is not formed via radical chemistry, as has been proposed previously (Symons et al., 1989).

It is now generally recognized that the Mo—OH group of the active site in xanthine oxidase represents the catalytically labile oxygen that has been known for some time. Two different types of experiment have led to the conclusion that it is this group rather than the Mo=O that is transferred to substrate in the course of the reaction. In the first study (Howes et al., 1996), electron-nuclear double resonance (ENDOR) spectroscopy was used to examine coupling of 17—O in the "very rapid" EPR signal generated with enzyme equilibrated in [17—O]—$H_2O$. A strongly coupled oxygen was observed ($A_{av} = 34.2\,MHz$) that was ascribed to the catalytically labile oxygen now transferred to bound product. However, no evidence was obtained for a second, weakly coupled oxygen (as would be expected were the Mo=O group expected in the signal-giving species to have been labeled with $^{17}O$). It was thus concluded that water or hydroxide "in the vicinity of the molybdenum and quite possibly ligated to it" represents the catalytically labile oxygen. In the second study (Xia et al., 1999), a protocol was developed whereby 17—O could be incorporated into the enzyme molybdenum center in the course of a single turnover, and the strength of the magnetic coupling to Mo(V) ascertained. On the basis of work with Mo(V) model compounds (Greenwood et al., 1993), a Mo=O group would be expected to be weakly coupled, while a Mo—OH group would be more strongly so. The results of the experiment clearly demonstrated strong, anisotropic coupling, which on the basis of the model compound work provide clear evidence that 17—O has been incorporated into the Mo—OH and not the Mo=O of the molybdenum center in the course of a single turnover, and that this group must therefore represent the catalytically labile site.

Although the nature of the catalytically labile oxygen is now generally agreed upon, alternate mechanisms whereby it is transferred to substrate have been proposed. In the first, the Mo—OH group undergoes nucleophilic attack on substrate (Greenwood et al., 1993; Huber et al., 1996; Xia et al., 1999), which then transfers hydride from C—8 to the Mo(VI)=S to give Mo(IV)—SH and product bound to the (reduced) molybdenum as Mo—OR (Figure 4, middle). In the second mechanism (Howes et al., 1996), a somewhat more complicated reaction is proposed wherein the $C_8$—H of substrate adds across the Mo=S bond to yield a species that is subsequently attacked by a "buried" water (possibly the Mo—OH) and rearranges to give a Mo(IV)•P complex in which the $C_8$=O of product is bound side-on in a so-called $\eta^2$ bond with significant Mo—C character to the interaction (Figure 4, *bottom*). This mechanism has been proposed on the basis of 13—C ENDOR studies of the "very rapid" species, in which a Mo...C distance no greater than 2.4 Å has been inferred from the anisotropy of the hyperfine interaction (Howes *et al.*, 1996). A number of difficulties attend distance estimations of this type, however, including problems associated with the general practice of approximating the two interacting magnetic moments (in this case, those of the unpaired electron and the I=1/2 nucleus of 13—C) as point-dipoles. In particular, the unpaired electron in the system is significantly delocalized, even if fully confined to the $d_{xy}$ orbital of molybdenum, compromising the accuracy of the determination. Indeed, other workers using both ENDOR and pulsed EPR methods have obtained a distance estimate of 2.8–3.0 Å (Grant, C., Britt, D., Choi, E.-Y., and Hille, R., unpublished), more consistent with a simple bonding interaction such as would be encountered in a mechanism involving simple nucleophilic attack and hydride transfer. Recent theoretical studies of the hydroxylation of formamide by xanthine oxidase also provide clear evidence in support of a mechanism initiated by nucleophilic attack, with concomitant hydride transfer in the course of the reaction. Particularly with regard to hydride transfer *per se*, the calculations indicate that substantial negative charge accumulates on the hydrogen in the course of transfer (Ilich and Hille, 1999). Significantly, a species possessing a Mo—C bond could be observed in these calculations, but lay a minimum of 70 kcal/mol higher in energy than intermediates along a reaction coordinate involving nucleophilic attack. Although at present most workers in the field appear to favor a reaction mechanism initiated by nucleophilic attack, additional work is clearly required to resolve this central mechanistic issue.

Along different lines, a comprehensive temperature dependence study of the reaction of xanthine oxidase with xanthine has recently been done (Mondal and Mitra, 1994). Using a combination of steady-state and rapid reaction kinetics, the microscopic rate constants for each step in the conversion of substrate to product have been resolved, permitting several key observations to be made. First, binding of substrate is entropically driven, likely due to desolvation of the aromatic substrate with its (presumably) hydrophobic binding site. Second, the binding process involves multiple kinetic steps, a result that is consistent with earlier kinetic studies of product binding to reduced enzyme (Hille and Stewart, 1984) and substrate analog binding to oxidized enzyme (Kim and Hille, 1994). Several apparent substrate docking sites have been identified in the *D. gigas* aldehyde oxidoreductase along the tunnel connecting the active site to bulk solvent (Romão *et al.*, 1995), and on the basis of the available kinetic evidence it is likely that at least one such docking site is also present in xanthine oxidase. Finally, the work of Mondal and Mitra (1994) has shown that there is a change in rate-limiting step in the reaction of xanthine oxidase as the temperature is changed: at lower temperatures final dissociation of product dissociation from enzyme is rate-limiting, while at higher temperatures an internal step in the mechanism becomes rate-limiting. This result must be interpreted in the context of isotope effect studies comparing the kinetics of enzyme reaction with [8—$^1$H]xanthine and [8—$^2$H]-xanthine (D'Ardenne and Edmondson, 1990). Specifically, it has been shown that while the intrinsic deuterium isotope effect on the C—H bond-breaking step of the reaction is 7.4 for the bovine milk xanthine oxidase (and 4.2 for the chicken liver xanthine dehydrogenase), the isotope effect on steady-state $k_{cat}$ is much more modest (~1.1). On the basis of these results, it has been estimated that the chemical step involving C—H bond cleavage must be faster than the rate-limiting step of the reaction by a factor of 10 (for the dehydrogenase) to 75 (for the oxidase). Separate rapid quench experiments have shown that the product uric acid has formed quantitatively within 10 ms in the reaction of enzyme with xanthine (D'Ardenne and Edmondson, 1990), indicating that the chemical step of the reaction is indeed very fast. In light of this work, the second rate-limiting step established in the above temperature dependence work cannot be the chemical step of the reaction, but possibly involves product going from the active site to a nearby docking site prior to dissociation from the enzyme.

In addition to hydroxylating xanthine and a wide range of purines, pteridines and similar compounds, xanthine oxidase is able to oxidize aromatic and aliphatic aldehydes to the corresponding carboxylic acids. Clearly then, there is a close functional as well as structural relationship between xanthine oxidase and the eukaryotic aldehyde oxidases and prokaryotic oxidoreductases. With xanthine oxidase, the pH dependence of the kinetic parameter $k_{red}/K_d$ (obtained from the substrate concentration dependence of the rate constant for enzyme reduction in pre-steady-state experiments and reflecting the reaction of free substrate with free enzyme through the first irreversible step of the reaction—in this case cleavage of the C—H bond of substrate) is the same for both xanthine (Kim et al., 1996) and 2,5-dihydroxybenzaldehyde (Xia et al., 1999). In both cases the data provide clear evidence for enzyme acting on neutral rather than deprotonated substrate as well as the presence of an active site base having a p$K_a$ of 6.4, suggesting that the enzyme hydroxylates these two very different classes of compounds in fundamentally the same way. A highly conserved active site glutamate residue has been identified in the molybdenum hydroxylases, and it has been suggested that this residue represents the catalytically essential

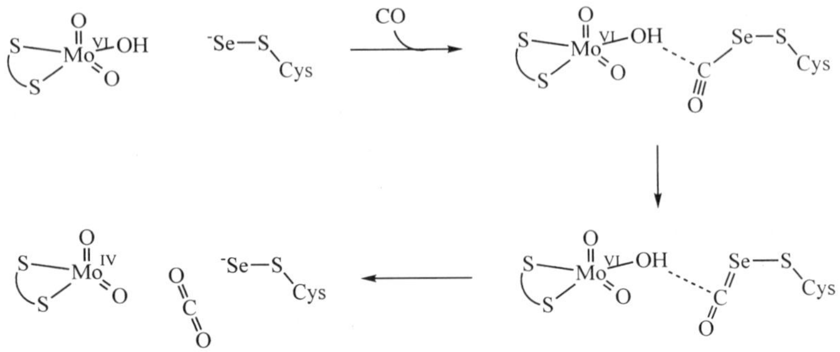

**FIGURE 5.** A proposed reaction mechanism for CO dehydrogenase, involving the active site Cys—S—Se⁻ group (after Dobbek *et al.*, 1999).

base (Huber *et al.*, 1996; Xia *et al.*, 1999). In a mechanism initiated by nucleophilic attack on substrate, this glutamate residue serves to deprotonate the Mo—OH group so that it becomes a more effective nucleophile (Figure 4, *middle*).

Carbon monoxide dehydrogenase stands out among members of the molybdenum hydroxylase family in catalyzing an atypical reaction. Although clearly a member of this group of enzymes on the basis of sequence and structural homology, the reaction involves (at face value) a simpler oxygen atom transfer to a lone-pair of CO, analogous to the reactions catalyzed by the other two families of molybdenum enzymes. A unique feature of CO dehydrogenase is the presence of a cyanide-sensitive selenium ion, found in the crystal structure of the enzyme as a Cys—S—Se⁻ selanylcysteine residue (Dobbek *et al.*, 1999). This residue is in the immediate vicinity of the molybdenum center, and a reaction mechanism has been proposed in which it plays a central catalytic role (Dobbek *et al.*, 1999). As shown in Figure 5, it has been suggested that catalysis is initiated by formation of a transient O=C=Se—S—cys species, which breaks down to give $CO_2$, the oxygen atom coming from the Mo—OH group of the molybdenum coordination sphere (in loose analogy to the case with the *bona fide* molybdenum hydroxylases). This reaction bears a strong resemblance to the mechanism for the selenium-catalyzed synthesis of carbonates from CO. Although the point in the course of the reaction in which the molybdenum becomes reduced (as well as the point at which the Mo—OH group becomes regenerated) remains to be established, a strong test of the most basic aspects of the proposed mechanism would be demonstration of the ability of SeCO to reactivate the deseleno form of the enzyme generated upon reaction with cyanide.

## 3.  THE EUKARYOTIC OXOTRANSFERASES

### 3.1.  Structural Studies

The X-ray crystal structure of sulfite oxidase from chicken liver at a resolution of 1.9 Å has recently been reported by Kisker *et al.* (1997b). The $\alpha_2$ dimeric protein, approximately 110 kDa in mass, consists of an N-terminal heme domain and two larger domains that constitute the molybdenum-binding and subunit interface portions of the protein, respectively (Figure 6). As indicated above, the active site molybdenum center has a square pyramidal coordination geometry, but in the crystal structure the equatorial oxygen was found at a distance of 2.3 Å from the molybdenum, consistent with water or possibly hydroxide, but too long to represent

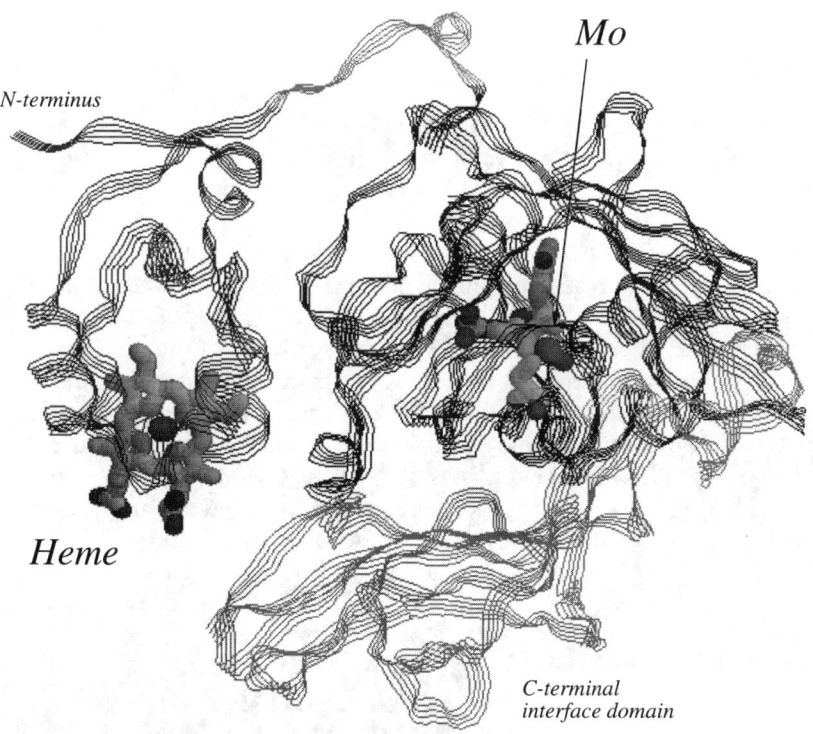

*Mo*

*N-terminus*

*Heme*

*C-terminal
interface domain*

**FIGURE 6.** The structure of chicken liver sulfite oxidase (from Kisker *et al.*, 1997). The heme domain is shown at left, the molybdenum-binding portion of the protein at right. The redox-active centers are rendered in wireframe representation.

the second Mo=O expected on the basis of earlier XAS studies (George *et al.*, 1989; George *et al.*, 1996). The results suggest strongly that the crystalline enzyme had become reduced in the course of sample preparation or data acquisition. Solvent access to the deeply buried active site is via a channel that opens onto the active site adjacent to the equatorial Mo—O group. In the crystal structure, a sulfate ion (from the crystallization mother liquor) is found at a distance of 2.4 Å from the equatorial Mo—O, occupying the presumed substrate binding site. As expected for binding an anionic substrate, this binding binding site is very positively charged and consists of Arg 138, Arg 190, Arg 450, Trp 204 and Tyr 322.

As in the case of aldehyde oxidoreductase and CO dehydrogenase, the disposition of the two subunits within the dimer indicates that their redox-active centers function independently of one another, with no readily evident pathway for electron transfer between subunits found in the structure. The two heme domains of the sulfite oxidase dimer are connected to their respective molybdenum domains via a short polypeptide sequence that is disordered in the crystal structure, and may represent a tether providing considerable mobility for the domain. Indeed, the two heme domains do not occupy identical positions in the dimer, suggesting that they may indeed be quite mobile. Interestingly, in both orientations the heme group is far removed from the molybdenum center, with a metal-metal distance of ~32 Å, again suggesting that domain motions are important in bringing the two redox-active centers sufficiently close together as to permit efficient electron transfer between them (electron transfer is known to be fast in sulfite oxidase, with rate constants as high as $1000 \, s^{-1}$—Sullivan *et al.*, 1992; Sullivan *et al.*, 1993).

A number of genetic deficiencies in biosynthesis of the pterin cofactor in have been identified in humans, and the clinical symptoms manifested by these patients appear to be more closely associated with loss of sulfite oxidase rather than xanthine or aldehyde oxidase activity (Johnson and Wadman, 1989; Garrett *et al.*, 1998). The most serious of these symptoms are neurological in origin (particularly involving dysfunction of the central nervous system and including frequently severe mental retardation), and it appears that loss of sulfite oxidase activity, resulting in the deleterious accumulation of sufficient concentrations of sulfite (a good nucleophile capable of reacting indiscriminately with a wide range of cellular components), is the principal reason for the observed symptoms. Particularly given the prevalence of sulfatides and related sulfoglycolipids in the myelin sheath and other membranes of the central nervous system (Vos *et al.*, 1994), the neurological effects of compromised sulfite oxidase function are not surprising. This does illustrate the point, however, that sulfite oxidase deficiency likely represents a genetic lesion in lipid rather than protein metabolism. Several point mutations in the structural gene for human sulfite oxidase have been identified in patients manifesting specific sulfite oxidase deficiency (*i.e.*, a loss of activity for this enzyme specifically, rather than for all enzymes requiring the pterin cofactor; Kisker *et al.*, 1997b). Although one of these maps to the active site (an Arg160Gln mutation at the substrate binding site of the human protein, equivalent to Arg 138 in the chicken enzyme), others map to the interface domain of the protein and appear to destabilize the protein fold and/or stability of the dimer (Kisker *et al.*, 1997b).

Although no crystal structure exists for full-length nitrate reductase, it is possible to draw some structural conclusions in light of the structure for sulfite oxidase. Nitrate reductase is organized in the reverse fashion of sulfite oxidase, with a large molybdenum-binding domain at the N-terminus, followed by a central heme domain and a C-terminal domain containing FAD; the dimeric protein has a molecular mass of approximately 200 kDa. Effective expression systems have been developed for the heme domain from *Chlorella vulgaris* (Cannons *et al.*, 1991), flavin domain from corn (Hyde and Campbell, 1990; a fragment with NADH:cytochrome $b$ reductase activity) and a combined heme-flavin protein also from corn (Campbell, 1992; a fragment with NADH:cytochrome $c$ reductase activity); very recently the holoenzyme from *Arabidopsis thaliana* has been expressed in the yeast *Pichia pastoris* (Su *et al.*, 1999). The crystal structure of the flavin fragment has been reported (Lu *et al.*, 1994), and on the basis of the strong homology to the crystallographically known bovine cytochrome $b_5$, a model for the latter constructed in which one of the heme propionate groups is found in close proximity to the flavin C—8 methyl group (Lu *et al.*, 1995). The reaction of each of these fragments with the physiological reductant NADH has been studied, demonstrating that in the course of the reaction a transient charge-transfer complex between the reduced flavin and $NAD^+$ is formed (Ratnam *et al.*, 1992; Ratnam *et al.*, 1997). Significantly, electron transfer from the flavin to the heme of the cytochrome c reductase fragment does not occur until this charge-transfer intermediate decays, presumably by dissociation of the $NAD^+$ (Ratnam *et al.*, 1997). A physical structure for the holoenzyme indicating the dispositions of the several domains with respect to one another is obviously highly desirable, and in the absence of a crystal structure it is tempting to extend the modeling work by incorporating the molybdenum domain of sulfite oxidase. This exercise may well not prove meaningful, however, given that the heme domain of sulfite oxidase lies N-terminal to the molybdenum domain while in nitrate reductase it lies C-terminal. There may thus be major differences in the conectivity of the domains in the two proteins.

Nitrate reductase is unique among known molybdenum enzymes in being tightly regulated by post-translational modification, a consequence of its physiological role in catalyzing the first and rate-limiting step of nitrogen assimilation in photosynthetic organisms (Kaiser and Huber, 1994; Campbell, 1996; Mendel and Schwartz, 1999). The enzyme is inactivated at night apparently so that substantial concentrations of product nitrite will not accumulate when photosynthetically generated reducing equivalents are not available to take the reaction on to ammonia (a reaction catalyzed by nitrite reductase). Nitrate reductase is inactivated in a process that involves initial phosphorylation at a specific serine residue in the linker region between the molybdenum- and heme-binding domains of the protein (Ser 534 in the enzyme from *Arabidopsis thaliana*; Douglas *et al.*, 1995); this reaction is catalyzed by a specific nitrate reductase kinase of calmodulin-dependent kinase family that has recently been purified and characterized (Douglas *et al.*, 1998). Phosphorylation creates a recognition site for a second protein, termed GF14ω in the case of *Arabidopsis* (Athwal *et al.*, 1998), a member of the 14-3-3 class of proteins that are involved in a wide range of cell signaling pathways. It is binding of this second protein that is responsible for enzyme inhibition, possibly by preventing effective intra-domain electron transfer rather than directly inhibiting nitrate reduction at the molybdenum center. This would be a particularly attractive mechanism of enzyme inhibition were mobility between the molybdenum and heme domains of nitrate reductase to prove as important as appears to be the case in sulfite oxidase. Interaction of the 14-3-3 protein with phosphorylated nitrate reductase is modulated by a number of physiological factors (including phosphate, chloride and AMP) that stimulate activity, presumably by weakening the inhibitory binding to nitrate reductase (although stimulation of a specific phosphatase activity must also be considered).

### 3.2. Mechanistic Aspects

From a mechanistic standpoint, the catalytic sequences of sulfite oxidase and nitrate reductase are the most relevant to the considerable work that has been done with inorganic complexes of molybdenum (Berg and Holm, 1985; Gheller *et al.*, 1992; Xiao *et al.*, 1992; Schultz *et al.*, 1993). These model systems are able to cycle effectively between dioxo Mo(VI) and monooxo Mo(IV) species in the presence of a suitable oxygen atom acceptor and donor, as indicated in Figure 7. Several of these systems are quite robust and can cycle catalytically hundreds to thousands of times without evident degradation. The catalytic lability of the Mo=O bond in these systems is surprising, given the high enthalpic cost of cleaving the

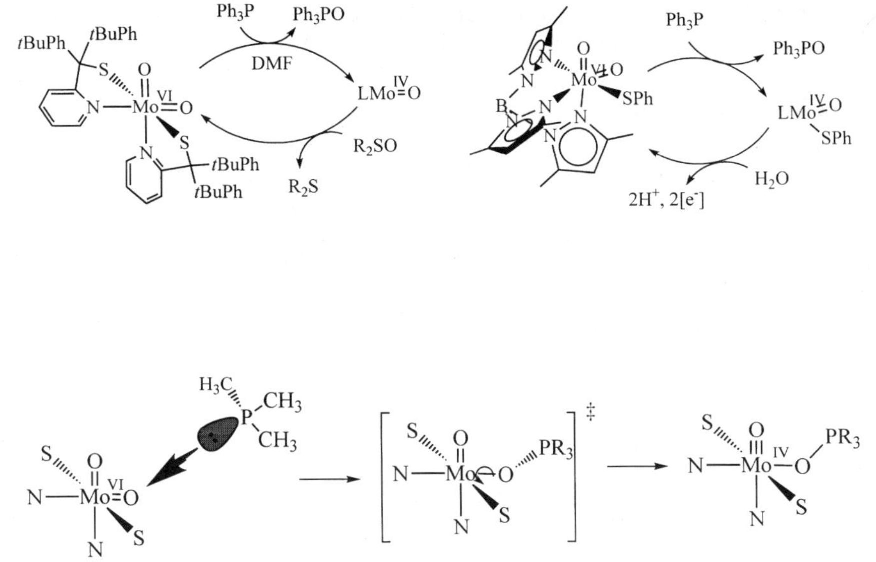

**FIGURE 7.** The catalytic properties of dioxo Mo(VI) compounds relevant to the eukaryotic oxotransferases.

bond (40 kcal/mol or more). It has been shown, however, that a significant portion of this energy is recovered by a strengthening of the remaining Mo=O bond in the monooxo Mo(IV) species, a phenomenon referred to as the "spectator oxo" effect (Rappé and Goddard, 1980; Rappé and Goddard, 1982). This arises because in the dioxo Mo(VI) species, the two oxo groups compete with one another for the molybdenum d orbitals into which they donate electron density in the metal-ligand bonding interaction: loss of one oxo group on going to Mo(IV) permits the remaining oxo to bind the metal more tightly.

The reaction of the dioxo Mo(VI) species with an oxo acceptor such as a phosphine has been shown to proceed with a large and negative entropy of activation ($\sim$ −30 eu), which has been interpreted as reflecting a highly ordered, associative transition state (Caradonna et al., 1988). In an elegant demonstration of the power of computational approaches to systems such as these, Pietsch and Hall (1996) have analyzed a model for this type of reaction in considerable detail. Using a $MoO_2$ system whose metal coordination sphere includes a pair of ammonia and thiolate ligands (in an octahedral coordination geometry analogous to that seen for the tris-pyrazoloborate system in Figure 6, *bottom*), oxygen atom transfer from the dioxo Mo(VI) parent to trimethylphosphine has been examined using perturbation theory (using a Møller-Plesset approach at the MP3 level). In order to optimize orbital overlap for efficient O—P bond formation (and also to populate the Mo $d_{xy}$ orbital, thereby formally reducing the metal to the Mo(IV) level), the attacking phosphine lone pair approaches the metal complex normal to the $MoO_2$ plane at an angle of ~130° to the Mo—O bond. To facilitate release of the phosphine oxide subsequent to formation of the O—P bond and reduction of the molybdenum, the system must pass through an intermediate in which the Mo—O bond is rotated by 90°, bringing the Mo—O—P unit coplanar with the remaining Mo=O group. This rotation weakens the Mo—$OPR_3$ bond prior to dissociation of the product phosphine oxide, and at the same time strengthens the remaining ("spectator") metal oxo bond. The spectator oxygen effect in this system thus plays a significant role in lowering the activation barrier for the reaction and stabilizing this intermediate. Displacement of the phosphine oxide by water (in a reaction analogous to that seen in the enzyme reaction) also occurs via an associative mechanism in which water binds to the molybdenum in the xy plane (the z axis being defined by the Mo(IV)=O bond), transiently expanding the coordination number of the metal. The free energies of activation for the reaction of the Mo(VI)$O_2$ complex with phosphine and subsequent displacement of the phosphine oxide from the Mo(IV)O complex by water are found to be relatively modest (15–20 kcal/mol), consistent with the established catalytic power of these $MoO_2$ complexes.

Chemistry directly analogous to that described above appears to pay out in the reaction of sulfite oxidase with sulfite (although it remains to be experimentally established that sulfite oxidase indeed possesses a catalytically labile oxygen and indeed operates via dioxoMo(VI)/monooxoMo(IV) catalytic sequence, it is extremely likely that this will prove to be the case). In particular, a comparison of the reaction of sulfite and dimethylsulfite with the enzyme has demonstrated the key role played by the substrate lone-pair (Brody and Hille, 1995). From the substrate concentration dependence of the rate of enzyme reduction in pre-steady-state experiments, it has been shown that while methylation of the two oxyanion groups of sulfite decreases the affinity of enzyme for substrate by a factor of 300 ($K_d$ increases from $33 \mu M$ to $11 mM$ at pH 8.0), the rate of breakdown of the Michaelis complex, once formed at high substrate concentrations, is essentially unchanged: $170 s^{-1}$ for dimethylsulfite as compared to $194 s^{-1}$ for sulfite. The results indicate that the negatively charged oxygens of substrate are important for substrate binding, but not for the subsequent chemical transformation of bound substrate to product. Interestingly, human sulfite oxidase has recently been cloned and expressed in *E. coli*, and a mutant in which Arg 160 (equivalent to Arg 138 in the chicken protein) has been isolated and characterized (Garrett *et al.*, 1998). The mutant protein, corresponding to a point mutation previously identified clinically, exhibits a 1000-fold lower steady-state value for $k_{cat}/K_m$, comparable to the extent to which catalysis is compromised in native enzyme when dimethylsulfite is used as substrate. This suggests that of the two substrate-binding arginine residues, Arg 160 is the more important, a question that can be tested by further site-directed mutagenesis studies. In light of this work and the crystal structure of sulfite oxidase (in which a bound sulfate ion is found at the presumed substrate binding site), a detailed physical picture of the Michaelis complex emerges, as shown in Figure 8. In particular, as pointed out in the report of the crystal structure (Kisker *et al.*, 1997b), it is the equatorial rather than axial oxo group of the molybdenum center of oxidized enzyme that is appropriately positioned for transfer to substrate. On the basis of the kinetic work with the enzyme and in light of the model compound and theoretical work described above, the key element of catalysis in sulfite oxidase can be envisaged as nucleophilic attack of the (dimethyl)sulfite lone pair on the equatorial Mo=O group of the molybdenum center (Figure 8). This is obligatory two-electron chemistry, and as in the case of the molybdenum hydroxylases there is no evidence at present to support a mechanism in which enzyme reduction occurs by sequential one-electron steps.

The pH dependence of both the oxidative and reductive half-reactions of sulfite oxidase has recently been examined (Brody and Hille, 1999). From a comparison of the pH dependence of $k_{cat}$ with the limiting rate constants for the two half-reactions, $k_{red}$ and $k_{ox}$, respectively, it is evident that $k_{red}$ is principally rate-limiting above pH 7, but at lower pH $k_{ox}$ becomes increasingly important. $k_{red}$ is essentially pH-independent, consistent with a reaction mechanism in which nucleophilic attack by the substrate lone pair on a Mo=O group initiates the catalytic sequence. The pH dependence of $k_{red}/K_d^{sulfite}$ indicates an active site group having a p$K_a$ of ~9.3 must be deprotonated for reaction of oxidized enzyme with free sulfite, possibly Tyr 322 which from the crystal structure of the enzyme constitutes part of the

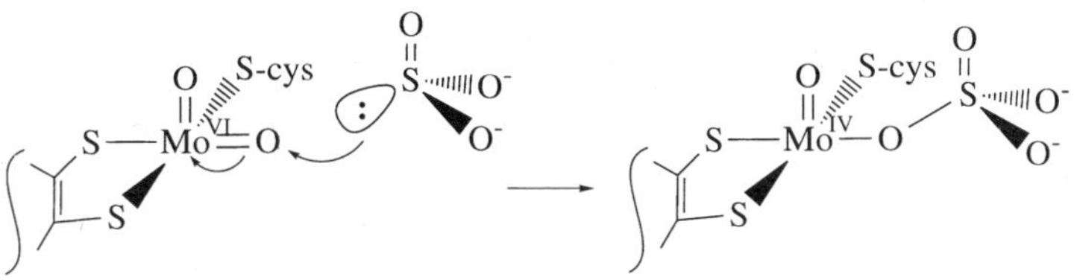

**FIGURE 8.** A representation of the complex of sulfite oxidase with sulfite, with the orientation of the substrate lone pair with respect to the molybdenum center indicated. The reaction likely proceeds via nucleophilic attack on an equatorial Mo=O, as shown.

substrate binding site. From the reaction of reduced sulfite oxidase with cytochrome c (its physiological oxidant), the pH profile for $k_{ox}/K_d^{cytc}$ indicates that two groups with $pK_a$ ~8 are involved in the reaction of free reduced enzyme with cytochrome $c$, one of which must be deprotonated and the other protonated, consistent with the known electrostatic nature of the interaction of cytochrome $c$ with its physiological partners.

The reaction of nitrate reductase with nitrite is thought to proceed in the reverse of that seen for the reaction of sulfite oxidase with sulfite, with nitrate binding to a reduced monooxo Mo(IV) molybdenum center (generated by reaction of enzyme with NADH), presumably displacing a water or hydroxide ligand. Following an electronic rearrangement that corresponds to the microscopic reverse of that described above for sulfite oxidase and the model compounds, product nitrite dissociates leaving an oxidized dioxo Mo(VI) center. Again, the fact that one enzyme oxidizes sulfite and the other reduces nitrate is principally attributable to the stabilities of S=O and N=O bonds relative to the Mo=O bonds of the molybdenum center (Holm, 1987; Holm and Donohue, 1993). Having said this, nitrate will not reoxidize reduced sulfite oxidase and sulfite will not reduce nitrate reductase. Thus, there is an essential element of substrate specificity in both cases and it has not proven possible to demonstrate sulfite:nitrate oxidoreductase activity with either enzyme (although this is an attractive issue to address by site-directed mutagenesis).

## 4.  THE PROKARYOTIC OXOTRANSFERASES

### 4.1.  Structural Studies

Crystal structures have been reported for five members of this family of molybdenum enzymes: DMSO reductase from *Rhodobacter sphaeroides* (Schindelin *et al.*, 1996) and *R. capsulatus* (Schneider *et al.*, 1996; McAlpine *et al.*, 1997; McAlpine *et al.*, 1998), formate dehydrogenase from *Escherichia coli* (Boyington *et al.*, 1997), trimethylamine-N-oxide reductase from *Shewanella massilia* (Czjzek *et al.*, 1998), and the dissimilatory (Nap) nitrate reductase from *Desulfovibrio desulfuricans* (Dias *et al.*, 1999). These proteins all possess related polypeptide folds, and are organized into four domains. The structure for the DMSO reductase from *R. sphaeroides* is shown in Figure 9. Each of the four domains of the protein has a compound α/β architecture, with Domains II and III responsible for most of the interactions with the two equivalents of the pterin cofactor. In the *E. coli* formate dehydrogenase (Boyington *et al.*, 1997) and nitrate reductase from *D. desulfuricans* (Dias *et al.*, 1999) Domain I possesses a 4Fe/4S iron-sulfur cluster

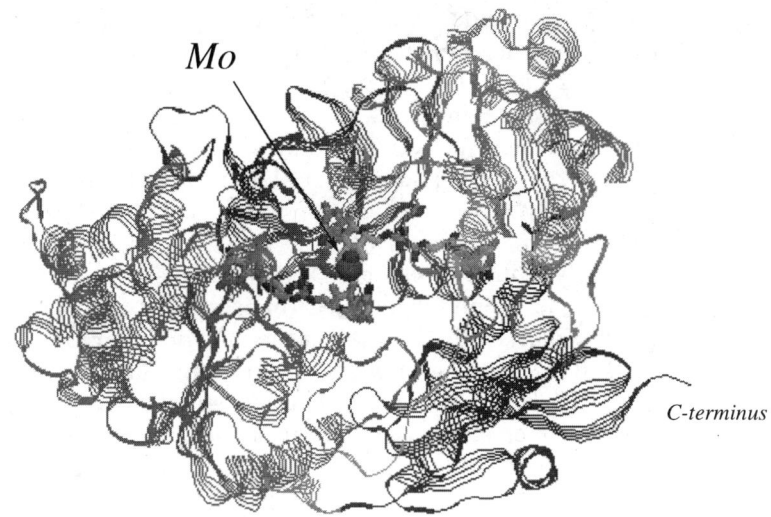

**FIGURE 9.** The structure of DMSO reductase from *Rhodobacter sphaeroides* (Schindelin *et al.*, 1996). The four domains referred to in the text are indicated. The molybdenum center is rendered in wireframe representation.

that occupies a position in the protein approximately 12–14 Å from the molybdenum, with the P pterin intervening. In formate dehydrogenase a selenocysteine constitutes the sixth ligand to the metal, and in the dissimilatory nitrate reductase it is a cysteine, as opposed to the serine residue found in the *Rhodobacter* DMSO reductases.

The general agreement concerning the overall polypeptide fold of this family of molybdenum enzymes notwithstanding, very considerable differences are found at their molybdenum centers. Even among the three structures reported for the *Rhodobacter* DMSO reductases, very different structures have been reported for the active site, even though the protein folds are essentially identical. In the (oxidized) *R. sphaeroides* protein (Schindelin *et al.*, 1996), the overall coordination geometry of the molybdenum center is trigonal prismatic, with two equivalents of the pterin cofactor, a terminal Mo=O group and the serine residue making up the molybdenum coordination sphere, as indicated in Figure 1. The active site of oxidized formate dehydrogenase possesses the same trigonal prismatic coordination geometry, but has a much longer Mo—O ligand, best formulated as hydroxide or water rather than a terminal oxo (Boyington *et al.*, 1997). In the first of the two structures reported for the *R. capsulatus* protein, however, one of the two pterin cofactors (that designated Q) has fully dissociated from the molybdenum to be replaced by a pair of terminal oxo groups (Schneider *et al.*, 1996). A very similar structure is reported for the active site of trimethylamine-N-oxide reductase from *S. massilia* (Czjzek *et al.*, 1998). In the second *R. capsulatus* structure (McAlpine *et al.*, 1997; McAlpine *et al.*, 1998), the two pterin cofactors are bound essentially equivalently, but the coordination sphere contains a second long Mo—O(H) group in addition to a short terminal oxo and serine ligand.

Amazingly, the pterin cofactors occupy virtually identical positions with respect to the polypeptide in all three DMSO reductase structures, and the root-mean-square deviation for the $C_\alpha$ carbons is small (at ~0.5 Å); the structures differ principally in the position of the molybdenum ion relative to the dithiolene sulfurs, and this changes by no more than 1 Å in going from one structure to another. These differences in structure are not likely to be due to "incorrect" crystallographic analysis—each active site structure represents a legitimate interpretation of the experimentally determined electron density—but rather to actual variability in the active site from one preparation of enzyme to another. Indeed, an almost bewildering variety of forms of the *Rhodobacter* DMSO reductase have been identified by EPR spectroscopy (Bennett *et al.*, 1998). At one level, the present uncertainties concerning active site structure play to an inherent weakness of protein crystallography, particularly with regard to metalloproteins: in the absence of a defined structural model to fit into a given portion of the electron density it is frequently difficult to infer structure (even structures involving strongly scattering atoms such as transition metal clusters) on the basis of electron density maps in the 1.5–2.5 Å resolution range. Indeed, artifacts are inevitably generated in the mathematical manipulation of the data (truncation of an infinite Fourier series is required to generate an electron density map from the primary diffraction data), a problem that can be compounded in the immediate vicinity of a strongly scattering atom, introducing considerable error in the estimation of interatomic distances. These effects have recently been discussed specifically in the context of the structures of interest here (Schindelin *et al.*, 1997). The principal issue presently before workers in the field is correlating one crystal structure or another with the several spectroscopically distinct forms of the enzyme that have been described (Bennett *et al.*, 1998; Adams *et al.*, 1999). At present, it appears likely that the most catalytically relevant structure for the oxidized molybdenum centers of these enzymes is six coordinate with a terminal Mo=O, a single protein ligand and two bidentate pterin cofactors bound more or less equivalently to the metal (at least from the standpoint of the physical structure, they remain distinct in their electronic structures as discussed above).

Both X-ray crystallographic (Schindelin *et al.*, 1995) and XAS (George *et al.*, 1998) analysis of reduced DMSO reductase indicate that the Mo(IV) center has lost the short Mo=O ligand seen in the oxidized enzyme, and its active site appears to be best described as having a distorted square-pyramidal coordination geometry. The pterin enedithiolates constitute the equatorial plane and the serine residue occupies the apical position (although one of the sulfurs of the Q pterin appears to have pulled away from the molybdenum to a significant degree in the crystal structure of the reduced enzyme, with Mo—S distance of 3.7 Å, this is less evident in the XAS of the reduced enzyme; George *et al.*, 1996; George *et al.*, 1999). A similar change in coordination geometry from trigonal prismatic to square pyramidal, with loss of the oxygen ligand, occurs upon reduction of formate dehydrogenase (Boyington *et al.*, 1997).

### 4.2. Mechanistic Aspects

It has been known for some time that DMSO reductase possesses a catalytically labile oxygen (Schultz *et al.*, 1995). Reduced enzyme can be reoxidized with 18—O labeled DMSO, and the enzyme separated from the reaction mix by size-exclusion chromatography (in unlabeled water). Subsequent reaction of the enzyme with a water-soluble phosphine yields [18—O]-labeled phosphine oxide, indicating that the enzyme is able to function quantitatively as an oxo transferase. These studies have subsequently extended using resonance Raman spectroscopy to demonstrate that a Mo=O site on the reoxidized enzyme is labeled on reaction with labeled DMSO (Garton *et al.*, 1997). In agreement with the XAS studies, the resonance Raman work has also been interpreted as indicating that the oxidized enzyme possesses a single Mo=O, and that the reduced enzyme lacks a Mo=O group (although a dioxo/monooxo catalytic cycle could not be rigorously discounted; Garton *et al.*, 1997).

With the above caveats in mind concerning uncertainties in the structure of the active site of DMSO reductase and related enzymes, some degree of circumspection must be attend any discussion of reaction mechanism. Assuming a monooxo Mo(VI) center (with approximately trigonal prismatic coordination geometry) in the oxidized enzyme, however, and a des-oxo Mo(IV) center (with square pyramidal coordination geometry) in the reduced, a reaction mechanism can be proposed in which the oxidized enzyme takes electrons from its physiological oxidant (a periplasmic, penta-heme c-type cytochrome encoded by the *dorC* gene, Shaw *et al.*, 1999), with the Mo=O group protonating to give water and dissociating upon full reduction to the Mo(IV) level. Reoxidation by DMSO then involves binding of substrate to the molybdenum via its S=O followed by chemistry that gives rise to dimethylsulfide and oxidized enzyme, then finally product dissociation. Such a mechanism is based on that originally proposed by Schindelin *et al.*, and is shown in Figure 10. It has recently been shown that addition of dimethylsulfide to oxidized enzyme gives rise to a pronounced spectral change, yielding enzyme with a distinct red color (McAlpine *et al.*, 1998). This red species has also been shown to form in the reaction of reduced enzyme with DMSO (Adams *et al.*, 1999), and appears to represent a kinetically viable $E_{red}\bullet DMSO$ Michaelis complex, which rapidly achieves equilibrium with the corresponding $E_{ox}\bullet DMS$ species in the enzyme active site. Consistent with this, recent kinetic work with DMSO reductase has demonstrated that the enzyme is kinetically competent to oxidized DMS as well as reduce DMSO (Adams *et al.*, 1999). Interestingly, the former reaction is sensitive to extended preincubation of enzyme with DMS under aerobic (but not anaerobic), but the latter reaction is not. The spectrum of the DMS-modified enzyme resembles that exhibited by recombinant as-isolated enzyme heterologously expressed in *E. coli*, which can be converted to native enzyme by reduction and reoxidation (Hilton *et al.*, 1999). It is thus becoming increasingly clear that DMSO reductase can be obtained in a variety of forms, many of which can be reduced and reoxidized to obtain an apparently homogeneous enzyme preparation.

The reaction mechanism given in Figure 10 is consistent with the bulk of the presently available information, but raises the interesting question as to how, in the absence of a second (spectator) oxo group, the Mo=O bond of the oxidized enzyme is made labile. Indeed one possibility is that the oxidized site does possess a spectator oxo (McAlpine *et al.*, 1997; McAlpine *et al.*, 1997), although this presently appears to be a minority opinion in the field. In the context of the mechanism given in Figure 10, the Mo=O bond of the molybdenum center must be sufficiently labile that

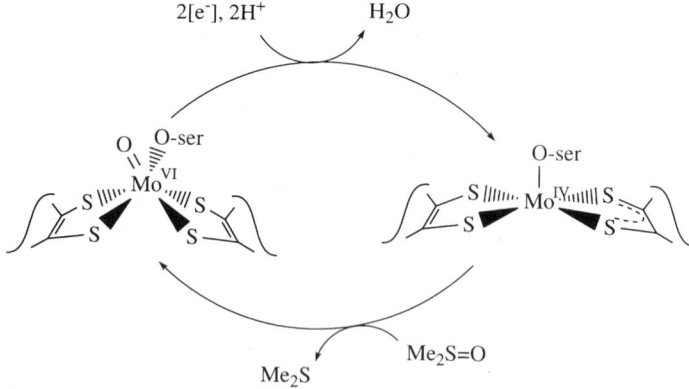

**FIGURE 10.** A reaction mechanism for DMSO reductase, based on that proposed by Schindelin *et al.* (1996).

upon reduction of the metal it protonates and spontaneously dissociates from the metal. One obvious possibility is that the bis-dithiolene unit contributed by the two equivalents of the pterin cofactor bring this about. In particular, given the substantive change in coordination geometry that accompanies reduction of the enzyme, it is possible that a portion of the enthalpic cost of loss of the Mo=O of oxidized enzyme is recovered in a more stable interaction between the metal and dithiolenes in the square pyramidal coordination geometry of the reduced enzyme. In fact, the XAS evidence indicates that the mean Mo—S distance decreases significantly, from 2.44 to 2.33 Å, reflecting a substantial strengthening of the metal-ligand interactions and consistent the idea that the bis-dithiolene unit plays a major role in stabilizing a des-oxo structure in the reduced enzyme. It remains to be seen, however, whether this is indeed the case, and if so how the electronic structure of the cluster brings this about.

## 5. CONCLUDING REMARKS

It is evident from the above that considerable progress has been made over the past several years in understanding the physical structures of a variety of molybdenum enzymes, and the mechanisms by which they catalyze their respective reactions. Several interesting new molybdenum-containing enzymes have been identified, and it can only be expected that more will be become known as genomic analyses of a range of organisms increasingly become available. Although not dealt with extensively here, rapid progress is also being made in understanding the biosynthesis of the pterin cofactor found in molybdenum (and tungsten) enzymes, with the roles of specific gene products in the pathway becoming established in a variety of species.

From a mechanistic standpoint, progress with regard to the mechanism of action of members of each of the three families of mononuclear molybdenum enzymes has made it possible to frame precisely the principal outstanding questions concerning the process by which substrate is converted to product (*e.g.*, whether or not a transient Mo—C bond is seen in the molybdenum hydroxylases; whether sulfite oxidase possesses a catalytically labile oxo group; establishing the precise structural context in which DMSO reductase carries out its reaction). Particularly for enzymes catalyzing non-canonical reactions (*e.g.*, CO dehydrogenase, formate dehydrogenase, polysulfide reductase and the $O_2$-insensitive nitrogenase), it can be hoped that future work will establish the common mechanistic themes as well as the distinguishing features of these reactions. The important insights into reaction mechanism provided by computational approaches can be expected to further our understanding not only of intermediates in a given reaction mechanism but also the structures of the transition states that separate them. Given the central role of transition state stabilization in theories of enzyme catalysis, a deeper understanding of transition state structure is essential to our emerging picture of how these enzymes function. A key ingredient in a comprehensive picture of enzyme function must also incorporate an understanding of the electronic structure of the active site, and spectroscopic techniques such as resonance Raman and magnetic circular dichroism can be expected to complement the x-ray crystallographic and XAS techniques that have elucidated their physical structures.

Finally, the recent development of suitable expression systems for recombinant molybdenum enzymes makes possible the use of site-directed mutagenesis to incisively probe the roles of specific active site residues in catalysis. Although these systems have been somewhat slow in development, presumably due to the intricacies of assembling and inserting the molybdenum center into the overexpressed apoenzyme, at least two such systems have now been developed and more can be expected in the immediate future. Future work can be expected to critically evaluate the involvement of specific amino acid residues in each step of a given catalytic sequence.

ACKNOWLEDGMENTS. The author wishes to thank Dr. Ortwin Meyer for his generous communication of data prior to publication. Work in the author's laboratory is supported by grants from the National Institutes of Health (GM 59953 and GM 58481).

## 6. REFERENCES

Adams, B., Smith, A. T., Bailey, S., McEwan, A. G., and Bray, R. C., 1999, Reactions of dimethylsulfoxide reductase from *Rhodobacter capsulatus* with dimethylsulfide and with dimethylsulfoxide: complexities revealed by conventional and stopped-flow spectrophotometry, *Biochemistry* **38**:8501–8511.

Amaya, Y., Yamazaki, X., Sato, M., Noda, K., Nishino, T., and Nishino, T., 1990, Proteolytic conversion of xanthine dehydrogenase from the NAD-dependent type to the $O_2$-dependent type, *J. Biol. Chem.* **265**:14170–14175.

Athwal, G. S., Huber, J. L., and Huber, S. C., 1998, Phosphorylated nitrate reductase and 14-3-3 proteins, *Plant Physiol.* **118**:1041–1048.

Axley, M. J., Grahame, D. A., and Stadtman, T. C., 1990, *Ecsherichia coli* formate-hydrogen lyase. Purification and properties of the selenium-dependent formate dehydrogenase complex, *J. Biol. Chem.* **265**:18213–18218.

Balagopalakrishna, C., Kimbrough, J. T., and Westmoreland, T. D., 1996, Electronic structural contributions to g values and molybdenum hyperfine coupling constants in oxyhalide anions of molybdenum(V), *Inorg. Chem.* **35**:7758–7768.

Barber, M. J., May, H. D., and Ferry, J. G., 1986, Inactivation of formate dehydrogenase from *Methanobacterium formicicum* by cyanide, *Biochemistry* **25**:8150–8155.

Bennett, B., Benson, N., McEwan, A. G., and Bray, R. C., 1994, Multiple states of the molybdenum centre of dimethylsulfoxide reductase from *Rhodobacter capsulatus* revealed by EPR spectroscopy, *Eur. J. Biochem.* **225**:321–331.

Berg, J. M., and Holm, R. H., 1985, A model for the active sites of oxo-transfer molybdoenzymes: reactivity, kinetics and catalysis, *J. Am. Chem. Soc.* **107**:925–932.

Berks, B. C., Richardson, D. J., Robinson, C., Reilly, A., Aplin, R. T., and Ferguson, S. J., 1994, Purification and characterization of the periplasmic nitrate reductase from *Thiosphaera pantetropha.*, *Eur. J. Biochem.* **220**:117–124.

Bilous, P. T., and Weiner, J. H., 1988, Molecular cloning and expression of the *Escherichia coli* dimethyl sulfoxide reductase operon, *J. Bacteriol.* **170**:1511–1518.

Blasco, F., Iobbi, C., Giordano, G., Chippaux, M., and Bonnefoy, V., 1989, Nitrate reductase from *Escherichia coli*: completion of the nucleotide sequence of the *nar* operon and reassessment of the role of the α and β subunits in iron binding and electron transfer, *Mol. Gen. Genet.* **218**:249–256.

Blasco, F., Iobbi, C., Ratouchniak, J., Bonnefoy, V., and Chippaux, M., 1990, Nitrate reductases of *Escherichia coli*: sequence of the second nitrate reductase and comparison with that encoded by the *narGHJI* operon, *Mol. Gen. Genet.* **222**:104–111.

Bordas, J., Bray, R. C., Garner, C. D., Gutteridge, S., and Hasnain, S. S., 1980, X-ray absorption spectroscopy of xanthine oxidase: the molybdenum centres of the functional and the desulfo forms, *Biochem. J.* **191**:499–508.

Boyington, J. C., Gladyshev, V. N., Khangulov, S. V., Stadtman, T. C., and Sun, P. D., 1997, Crystal structure of formate dehydrogenase H: catalysis involving Mo, molybdopterin, selenocysteine and an $Fe_4S_4$ cluster, *Science* **275**:1305–1308.

Bray, R. C., 1988, The inorganic biochemistry of molybdoenzymes, *Quart. Rev. Biophys.* **21**:299–329.

Bray, R. C., and George, G. N., 1985, Electron paramagnetic resonance studies using pre-steady-state kinetics and substitution with stable isotopes on the mechanism of action of molybdoenzymes, *Biochem. Soc. Trans.* **13**:561–567.

Bray, R. C., and Vånngård, T., 1969, 'Rapidly appearing' molybdenum electron paramagnetic resonance signals from reduced xanthine oxidase, *Biochem. J.* **114**:725–734.

Brody, M. S., and Hille, R., 1995, The reaction of chicken liver sulfite oxidase with dimethylsulfite, *Biochim. Biophys. Acta.* **1253**:133–135.

Brody, M. S., and Hille, R., 1999, The kinetic behavior of chicken liver sulfite oxidase. *Biochemistry* **38**:6668–6677.

Buc, J., Santini, C.-L., Blasco, F., Giordani, R., Cárdenas, M. L., Chippaux, M., Cornish-Bowden, A., and Giordano, G., 1995, Kinetic studies of a soluble αβ complex of nitrate reductase A from *Escherichia coli*. Use of various αβ mutants with altered β subunits, *Eur. J. Biochem.* **234**:766–772.

Campbell, W. H., 1992, Expression in *Escherichia coli* of cytochrome c reductase activity from a maize NADH:nitrate reductase complementary DNA, *Plant Physiol.* **99**:693–699.

Campbell, W. H., 1996, Nitrate reductase biochemistry comes of age, *Plant Physiol.* **111**:355–361.

Cannons, A. C., Iida, N., and Solomonson, L. P., 1991, Expression of a cDNA clone encoding the haem-binding domain of *Chlorella* nitrate reductase, *Biochem. J.* **278**:203–209.

Caradonna, J. P., Reddy, P. R., and Holm, R. H., 1988, Kinetics, mechanisms, and catalysis of oxygen atom transfer reactions of *S* Oxide and pyridine *N* Oxide substrates with molybdenum (IV,VI) complexes: relevance to molybdoenzymes, *J. Am. Chem. Soc.* **110**:2139–2144.

Choe, M., and Reznikoff, W. S., 1993, Identification of the regulatory sequence of anaerobically expressed locus *aeg-46.5*, *J. Bacteriol.* **175**:1165–1172.

Coughlan, M. P., Rajagopalan, K. V., and Handler, P., 1969, The role of molybdenum in xanthine oxidase and related enzymes. Reactivity with cyanide, arsenite, and methanol, *J. Biol. Chem.* **244**:2658–2663.

Cramer, S. P., Wahl, R., and Rajagopalan, K. V., 1981, Molybdenum sites of sulfite oxidase and xanthine dehydrogenase, *J. Am. Chem. Soc.* **103**:7721–7727.

Cramer, S. P., and Hille, R., 1985, Arsenite-inhibited xanthine oxidase-determination of the Mo—S—As geometry by EXAFS, *J. Am. Chem. Soc.* **107**:8164–8169.

Czjzek, M., Dos Santos, J.-P., Pommier, J., Giordano, G., Méjean, V., and Haser, R., 1998, Crystal structure of oxidized trimethylamine-N-oxide reductase from *Shewanella massilia* at 2.5 Å resolution, *J. Mol. Biol.* **284**:435–447.

D'Ardenne, S. C., and Edmondson, D. E., 1990, Kinetic isotope effect studies on milk xanthine oxidase and on chicken liver xanthine dehydrogenase, *Biochemistry* **29**:9046–9052.

Dias, J. M., Than, M. E., Humm, A., Huber, R., Bourenkov, G. P., Bartunik, H. D., Bursakov, S., Calvete, J., Caldiera, J., Carniero, C., Moura, J. J. G., Moura, I., and Romão, M. J., 1999, Crystal structure of the first dissimilatory nitrate reductase at 1.9 Å solved by MAD methods, *Structure Fold Res.* **7**:65–79.

Dobbek, H., Gremer, L., Meyer, O., and Huber, R., 1999, Crystal structure and mechanism of CO dehydrogenase, a molybdo iron-sulfur flavoprotein containing S-selanylcysteine, *Proc. Natl. Acad. Sci. (USA)* **96**:8884–8889.

Douglas, P., Morrice, N., and MacKintosh, C., 1995, Identification of a regulatory phosphorylation site in the hinge 1 region of nitrate reductase from spinach (*Spinacea oleracea*) leaves, *FEBS Lett.* **177**:113–117.

Douglas, P., Moorhead, G., Hong, Y., Morrice, N., and MacKintosh, C., 1998, Purification of a nitrate reductase kinase from *Spinaceaa oleracea* leaves, and its identification as a calmodulin-domain protein kinase, *Planta* **206**:435–442.

Garrett, R. M., Johnson, J. M., Graf, T. N., Feigenbaum, A., and Rajagopalan, K. V., 1998, Human sulfite oxidase R160Q: identification of the mutation in a sulfite oxidase-deficient patient and expression and characterization of the mutant enzyme, *Proc. Natl. Acad. Sci. (USA)* **95**:6394–6398.

Garton, S. D., Hilton, J., Oku, H., Crouse, B. R., Rajagopalan, K. V., and Johnson, M. K., 1997, Active site structures and catalytic mechanism of *Rhodobacter sphaeroides* dimethyl sulfoxide reductase as revealed by resonance Raman spectroscopy, *J. Am. Chem. Soc.* **119**:12906–12916.

George, G. N., Kipke, C. A., Prince, R. C., Sunde, R. A., Enemark, J. H., and Cramer, S. P., 1989, Structure of the active site of sulfite oxidase. X-ray absorption spectroscopy of the Mo(IV), Mo(V) and Mo(VI) oxidation states, *Biochemistry* **28**:5075–5080.

George, G. N., Garrett, R. M., Prince, R. C., and Rajagopalan, K. V., 1996, The molybdenum center of sulfite oxidase: a comparison of wild-type and and the cysteine 207 to serine mutant using x-ray absorption spectroscopy, *J. Am. Chem. Soc.* **118**:8588–8592.

George, G. N., Hilton, J., and Rajagopalan, K. V., 1996, X-ray absorption spectroscopy of dimethyl-sulfoxide reductase from *Rhodobacter sphaeroides*, *J. Am. Chem. Soc.* **118**:1113–1117.

George, G. N., Hilton, J., Temple, C., and Rajagopalan, K. V., 1999, Structure of the molybdenum site of dimethylsulfoxide reductase, *J. Am. Chem. Soc.* **121**:1256–1266.

Gheller, S. F., Schultz, B. E., Scott, M. J., and Holm, R. H., 1992, A broad-substrate analogue reaction system of the molybdenum oxotransferases, *J. Am. Chem. Soc.* **114**:6934–6935.

Gladyshev, V. N., Khangulov, S. V., Axley, M. J., and Stadtman, T. C., 1994, Coordination of selenium to molybdenum in formate dehydrogenase H from *Escherichia coli*, *Proc. Natl. Acad. Sci. (USA)* **91**:7708–7711.

Greenwood, R. J., Wilson, G. L., Pilbrow, J. R., and Wedd, A. G., 1993, Molybdenum(V) sites in xanthine oxidase and relevant analog complexes: comparison of oxygen-17 hyperfine coupling, *J. Am. Chem. Soc.* **115**:5385–5392.

Hille, R., 1994, The reaction mechanism of oxomolybdenum enzymes, *Biochim. Biophys. Acta* **1184**:143–169.

Hille, R., 1996, The mononuclear molybdenum enzymes, *Chem. Rev.* **96**:2757–2816.

Hille, R., and Anderson, R. F., 1991, Electron Transfer in Milk Xanthine Oxidase as Studied by Pulse Radiolysis, *J. Biol. Chem.* **266**:5608–5615.

Hille, R., and Massey, V., 1986, The equilibration of reducing equivalents within milk xanthine oxidase, *J. Biol. Chem.* **261**:1241–1247.

Hille, R., and Sprecher, H., 1987, On the mechanism of action of xanthine oxidase, *J. Biol. Chem.* **262**:10914–10917.

Hille, R., and Stewart, R. C., 1984, The interaction of xanthine oxidase with 8-bromoxanthine, *J. Biol. Chem.* **259**:1570–1576.

Hille, R., George, G. N., Eidsness, M. K., and Cramer, S. P., 1989, EXAFS of xanthine oxidase complexes with alloxanthine, violapterin, and 6-pteridylaldehyde, *Inorg. Chem.* **28**:4018–4022.

Hille, R., Rétey, Bartlewski-Hof, U., Reichenbacher, W., and Schink, B., 1999, Mechanistic aspects of molybdenum-containing enzymes, *FEMS Microbiol. Rev.* **22**:489–501.

Hilton, J. C., Temple, C. A., and Rajagopalan, K. V., 1999, Re-design of *Rhodobacter sphaeroides* dimethyl sulfoxide reductase. Enhancement of adenosine N1-oxide reductase activity, *J. Biol. Chem.* **274**:8428–8436.

Holm, R. H., 1987, Metal-centered oxygen atom transfer reactions, *Chem. Rev.* **87**:1401–1449.

Holm, R. H., and Donahue, J. P., 1993, A thermodynamic scale for oxygen atom transfer reactions, *Polyhedron* **12**:571–589.

Howes, B. D., Bray, R. C., Richards, R. L., Turner, N. A., Bennett, B., and Lowe, D. J., 1996, Evidence favoring molybdenum-carbon bond formation in xanthine oxidase action: [17]O- and [13]C-ENDOR and kinetic studies, *Biochemistry* **35**:1432–1443.

Huber, R., Hof, P., Duarte, R. O., Moura, J. J. G., Moura, I., LeGall, J., Hille, R., Archer, M., and Romão, M., 1996, A structure-based catalytic mechanism for the xanthine oxidase family of molybdenum enzymes, *Proc. Natl. Acad. Sci. (USA)* **93**:8846–8851.

Hunt, J., and Massey, V., 1992, Purification and properties of milk xanthine dehydrogenase, *J. Biol. Chem.* **267**:21479–21485.

Hyde, G. E., and Campbell, W. H., 1990, High-level expression in *Escherichia coli* of the catalytically active flavin domain of corn leaf NADH:nitrate reductase and its comparison to the human NADH:cytochrome b5 reductase, *Biochem. Biophys. Res. Commun.* **168**:1285–1291.

Ilich, P., and Hille, R., 1999, Mechanism of formamide hydroxylation catalyzed by a molybdenum-dithiolene complex: a model of xanthine oxidase reactivity, *J. Phys. Chem. B.* **103**:5406–5412.

Inscore, F. E., McNaughton, R., Westcott, B. L., Helton, M. E., Jones, R., Dhawan, I., Enemark, J. H., and Kirk, K. L., 1999, Spectroscopic evidence for a unique bonding interaction in oxo-molybdenum dithiolate complexes: implications for σ electron transfer pathways in the pyranopterin dithiolate centers of enzymes, *Inorg. Chem.* **38**:1401–1410.

Jankielewicz, A., Klimmeck, O., and Kröger, A., 1995, The electron transfer from hydrogenase and formate dehydrogenase to polysulfide reductase in the membrane of *Wolinella succinogenes*, *Biochim. Biophys. Acta* **1231**:157–162.

John, J. L., and Wadman, S. K., 1989, Molybdenum cofactor deficiency, in: *The Metabolic Basis of Inherited Disease*, 6th edition (C. R. Scriver, A. L. Beaudet, W. S. Sly, and D. L. Valle, eds.), McGraw-Hill, New York, pp. 1463–1475.

Johnson, K. K., Rees, D. C., and Adams, M. W. W., 1996, Tungstoenzymes, *Chem. Rev.* **96**:2817–2839.

Kaiser, W. M., and Huber, S. C., 1994, Posttranslational regulation of nitrate reductase in higher plants, *Plant Physiol.* **106**:817–821.

Khangulov, S. V., Gladyshev, V. N., Dismukes, G. C., and Stadtman, T. C., 1998, Selenium-containing formate dehydrogenase H from *Escherichia coli*: a molybdopterin enzyme that catalyzes formate oxidation without oxygen transfer, *Biochemistry* **37**:3518–3528.

Kim, J. H., and Hille, R., 1994, Studies of substrate binding to xanthine oxidase by using a spin-labeled analog, *J. Inorg. Biochem.* **55**:295–303.

Kim, J. H., Ryan, M. G., Knaut, H., and Hille, R., 1996, The reductive half-reaction of xanthine oxidase: a pH dependence and solvent kinetic isotope effect study, *J. Biol. Chem.* **271**:6771–6780.

Kisker, C., Schindelin, H., and Rees, D. C., 1997a, Molybdenum-cofactor-containing enzymes: structure and mechanism, *Ann. Rev. Biochem.* **66**:233–268.

Kisker, C., Schindelin, H., Pacheco, A., Wehbi, W. A., Garrett, R. M., Rajagopalan, K. V., Enemark, J. H., and Rees, D. C., 1997b, Molecular basis of sulfite oxidase deficiency from the structure of sulfite oxidase, *Cell* **91**:973–983.

Kubo, Y., Ogura, N., and Nakagawa, H., 1988, Limited proteolysis of the nitrate reducase from spinach leaves, *J. Biol. Chem.* **263**:19684–19689.

Li Calzi, Raviolo, C., Ghibaudi, E., De Gioia, Salmona, M., Cazzaniga, G., Kurosaki, M., Terao, M., and Garattini, E., 1995, Purification, cDNA cloning and tissue distribution of bovine liver aldehyde oxidase, *J. Biol. Chem.* **270**:31037–31045.

Lowe, D. J., and Bray, R. C., 1978, Magnetic coupling of the molybdenum and iron-sulfur centres in xanthine oxidase and xanthine dehydrogenase, *Biochem. J.* **169**:471–479.

Lu, G., Campbell, W. H., Schneider, G., and Lindqvist, Y., 1994, Crystal structure of the FAD-containing fragment of corn nitrate reductase at 2.5 A resolution: relationship to other flavoprotein reductases, *Structure* **2**:809–821.

Lu, G., Lindqvist, Y., Schneider, G., Dwivedi, U., and Campbell, W. H., 1995, Structural studies on corn nitrate reductase: refined structure of the cytochrome b reductase fragment at 2.5 A, its ADP complex and an active-site mutant and modeling of the cytochrome b domain, *J. Mol. Biol.* **248**:931–948.

Massey, V., and Edmondson, D., 1970, On the mechanism of inactivation of xanthine oxidase by cyanide, *J. Biol. Chem.* **245**:6595–6598.

McAlpine, A. S., McEwan, A. G., and Bailey, S., 1998, The high resolution crystal structure of DMSO reductase in complex with DMSO, *J. Mol. Biol.* **275**:613–623.

McAlpine, A. S., McEwan, A. G., Shaw, A. L., and Bailey, S., 1997, Molybdenum active centre of DMSO reductase from *Rhodobacter capsulatus*: crystal structure of the oxidized enzyme at 1.82 Å resolution and the dithionite reduced enzyme at 2.8 Å resolution, *J. Biol. Inorg. Chem.* **2**:690–701.

McCord, J. M., Roy, R. S., and Schaffer, S. W., 1985, Free radicals and myocardial ischemia. The role of xanthine oxidase, *Adv. Myocardiol.* **5**:183–189.

McEwan, A. G., Ferguson, S. J., and Jackson, J. B., 1991, Purification and properties of dimethyl-sulfoxide reductase from *Rhodobacter capsulatus*, *Biochem. J.* **274**:305–307.

McWhirter, R. B., and Hille, R., 1991, The reductive half-reaction of xanthine oxidase. Spectral intermediates in the hydroxylation of 2-hydroxy-6-methylpurine, *J. Biol. Chem.* **266**:23724–23731.

Mendel, R. R., and Schwartz, G., 1999, Molybdoenzymes and molybdenum cofactor in plants, *Crit. Rev. Plant Sci.* **18**:33–69.

Meyer, O., and Rajagopalan, K. V., 1984, Selenite binding to carbon monoxide oxidase from *Pseudomonas carboxidovorans*. Selenium binds covalently to the protein and activates specifically the CO → methylene blue reaction, *J. Biol. Chem.* **259**:5612–5617.

Meyer, O., Frunzke, K., and Mördorf, G., 1993, Biochemistry of the aerobic utilization of carbon monoxide, in: *Microbial growth on C1 compounds* (J. C. Murrell and D. P. Kelly, eds.), Intercept Press, Andover, pp. 433–459.

Mondal, M. S., and Mitra, S., 1994, Kinetics and thermodynamics of the molecular mechanism of the reductive half-reaction of xanthine oxidase, *Biochemistry* **33**:10305–10312.

Murray, K. N., Watson, J. G., and Chaykin, S., 1966, Catalysis of the direct transfer of oxygen from nicotinamide N-oxide to xanthine by xanthine oxidase, *J. Biol. Chem.* **241**:4798–4801.

Nipales, N. S., and Westmoreland, T. D., 1997, Correlation of EPR parameters with electronic structure in the homologous series of low-symmetry complexes Tp*MoOX$_2$ (Tp*=hydrotris(3,5-dmethylpyrazol-1-yl)borate; X=F, Cl, Br), *Inorg. Chem.* **36**:756–757.

Nishino, T., 1994, The conversion of xanthine dehydrogenase to xanthine oxidase and the role of the enzyme in reperfusion injury, *J. Biochem.* **116**:1–6.

Nishino, T., and Nishino, T., 1987, Evidence for a tyrosine residue in the nicotinamide adenine dinucleotide binding site of chicken liver xanthine dehydrogenase, *Biochemistry* **26**:3068–3072.

Nishino, T., and Nishino, T., 1989, The nicotinamide adenine dinucleotide-binding site of chicken liver xanthine dehydrogenase, *J. Biol. Chem.* **264**:5468–5473.

Nishino, T., Nishino, T., Schopfer, L. M., and Massey, V., 1989, The reactivity of chicken liver xanthine dehydrogenase with molecular oxygen, *J. Biol. Chem.* **264**:2518–2527.

Palmer, G., and Massey, V., 1969, Spectroscopic studies of xanthine oxidase, *J. Biol. Chem.* **244**:2614–2622.

Pierson, D. E., and Campbell, A., 1990, Cloning and nucleotide sequence of *bisC*, the structural gene for biotin sulfoxide reductase in *Escherichia coli*, *J. Bacteriol.* **172**:2194–2198.

Pietsch, M. A., and Hall, M. B., 1996, Theoretical studies on models for the oxo-transfer reaction of dioxomolybdenum enzymes, *Inorg. Chem.* **35**:1273–1278.

Rajagopalan, K. V., 1992, Novel aspects of the biochemistry of the molybdenum cofactor, *Adv. Enzymol.* **64**:215–290.

Rajagopalan, K. V., and Handler, P., 1967, Purification and properties of chicken liver xanthine dehydrogenase, *J. Biol. Chem.* **242**:4097–4107.

Rajagopalan, K. V., and Johnson, J. L., 1992, The pterin molybdenum cofactors, *J. Biol. Chem.* **267**:10199–10202.

Rappé, A. K., and Goddard, W. A., III, 1980, Bivalent spectator oxo bonds in metathesis and epoxidation of alkenes, *Nature* **285**:311–314.

Rappé, A. K., and Goddard, W. A., III, 1982, Olefin metathesis. A mechanistic study of high-valent group 6 catalysts, *J. Am. Chem. Soc.* **104**:3287–3289.

Ratnam, K., Shiraishi, N., Campbell, W. H., and Hille, R., 1995, Spectroscopic and kinetic characterization of the recombinant wild-type and C242S mutant of the cytochrome *b* reductase fragment of nitrate reductase, *J. Biol. Chem.* **270**:24067–24072.

Ratnam, K., Shiraishi, N., Campbell, W. H., and Hille, R., 1997, Spectroscopic and kinetic characterization of the heme- and flavin-containing cytochrome *c* reductase fragment of nitrate reductase, *J. Biol. Chem.* **272**:2122–2128.

Reichenbecher, W., Rüdiger, A., Kroneck, P. M. H., and Schink, B., 1996, One molecule of molybdopterin guanine dinucleotide is associated with each subunit of the heterodimeric Mo—Fe—S protein transhydroxylase of *Pelobacter acidigallici* as determined by SDS/PAGE electrophoresis and mass spectrometry, *Eur. J. Biochem.* **237**:413–419.

Ribbe, M., Gadkari, D., and Meyer, O., 1997, $N_2$ fixation by *Streptomyces thermoautotrophicus* involves a molybdenum-dinitrogenase and a manganese-superoxide oxidoreductase that couple $N_2$ reduction to the oxidation of superoxide produced from $O_2$ by a molybdenum—CO dehydrogenase, *J. Biol. Chem.* **272**:26627–26633.

Romão, M. J., Archer, M., Moura, I., Moura, J. J. G., LeGall, J., Engh, R., Schneider, M., Hof, P., and Huber, R., 1995, Crystal structure of the xanthine oxidase-related aldehyde oxidoreductase from *D. gigas*, *Science* **270**:1170–1176.

Rosner, B. M., Rainey, F. A., Kroppenstedt, R. M., and Schink, B., 1997, Acetylene degradation by new isolates of aerobic bacteria and comparison of acetylene hydratase enzymes, *FEMS Microbiol. Lett.* **148**:175–180.

Satoh, T., and Kurihara, F. N., 1987, Purification and properties of dimethylsulfoxide reductase containing a molybdenum cofactor from a denitrifier, *Rhodopseudomonas sphaeroides* f.s. *denitrificans*, *J. Biochem.* **102**:191–197.

Schindelin, H., Kisker, C., Hilton, J., Rajagopalan, K. V., and Rees, D. C., 1996, Crystal structure of DMSO reductase: redox-linked changes in molybdopterin coordination, *Science* **272**:1615–1621.

Schindelin, H., Kisker, C., and Rees, D. C., 1997, The molybdenum cofactor: a crystallographic perspective, *J. Biol. Inorg. Chem.* **2**:773–781.

Schmitz, R. A., Albracht, S. P. J., and Thauer, R. K., 1992, A molybdenum and a tungsten isoenzyme of formylmethanofuran dehydrogenase in the thermophilic archaeon *Methanobacterium wolfei*, *Eur. J. Biochem.* **209**:1013–1018.

Schneider, F., Loewe, J., Huber, R., Schindelin, H., and Kisker, C., 1996, Crustal structure of dimethyl sulfoxide reductase from *Rhodobacter capsulatus* at 1.88 Å resolution, *J. Mol. Biol.* **263**:53–63.

Schübel, U., Kraut, M., Mördorf, G., and Meyer, O., 1996, Molecular characterization of the gene cluster *coxMSL* encoding the molybdenum-containing carbon monoxide dehydrogenase of *Oligotropha carboxidovorans*, *J. Bacteriol.* **177**:2197–2203.

Schultz, B. E., Gheller, S. F., Muetterties, M. C., Scott, M. J., and Holm, R. H., 1993, Molybdenum-mediated oxygen atom transfer: an improved analogue reaction system of the molybdenum oxotransferases, *J. Am. Chem. Soc.* **115**:2714–2722.

Schultz, B. E., Hille, R., and Holm, R. H., 1995, Direct oxygen atom transfer in the mechanism of action of *Rhodobacter sphaeroides* dimethylsulfoxide reductase, *J. Am. Chem. Soc.* **117**:827–828.

Shaw, A. L., Hochkoeppler, A., Bonora, P., Zannoni, D., Hanson, G., and McEwan, A. G., 1999, Characterization of DorC from *Rhodobacter capsulatus*, a c-type cytochrome involved in electron transfer to dimethylsulfoxide reductase, *J. Biol. Chem.* **274**:9911–9914.

Stiefel, E. I., 1977a, Proposed molecular mechanism of the action of molybdenum in enzymes: coupled proton and electron transfer, *Proc. Natl. Acad. Sci.* **70**:988–992.

Stiefel, E. I., 1977b, The coordination and bioinorganic chemistry of molybdenum, *Prog. Inorg. Chem.* **21**:1–221.

Su, W., Mertens, J. A., Kanamaru, K., Campbell, W. H., and Crawford, N. M., 1999, Analysis of wild-type and mutant plant nitrate reductase expressed in the methylotrophic yeast *Pichia pastoris*, *Plant Physiol.* **115**:1135–1143.

Sullivan, E. P., Jr., Hazzard, J. T., Tollin, G., and Enemark, J. H., 1992, Inhibition of intramolecular electron transfer in sulfite oxidase by anion binding, *J. Am. Chem. Soc.* **114**:9662–9663.

Sullivan, E. P., Jr., Hazzard, J. T., Tollin, G., and Enemark, J. H., 1993, Electron transfer in sulfite oxidase: effects of pH and anions on transient kinetics, *Biochemistry* **32**:12465–12470.

Swann, J., and Westmoreland, T. D., 1997, Density functional calculations of *g* values and molybdenum hyperfine coupling constants for a series of molybdenum(V) oxyhalide anions, *Inorg. Chem.* **36**:5348–5357.

Symons, M. C. R., Taiwo, F. A., and Peterson, R. L., 1989, Electron addition to xanthine oxidase, *J. Chem. Soc. Faraday Trans.* **85**:4063–4074.

Thauer, R. K., 1998, Biochemistry of methanogenesis, *Microbiology* **144**:2377–2406.

Thomas, G., Potter, L., and Cole, J. A., 1999, The periplasmic nitrate reductase from *Escherichia coli*: a heterodimeric molybdoprotein with a double-arginine signal sequence and an unusual leader peptide cleavage site, *FEMS Microbiol. Lett.* **174**:167–171.

Tanner, S. J., Bray, R. C., and Bergmann, F., 1978, $^{13}$C hyperfine splitting of some molybdenum electron paramagnetic resonance signals of xanthine oxidase, *Biochem. Soc. Trans.* **6**:1328–1330.

Turner, N. A., Bray, R. C., and Diakun, G. P., 1989, Information from e.x.a.f.s. spectroscopy on the structures of different forms of molybdenum in xanthine oxidase and the catalytic mechanism of the enzyme, *Biochem. J.* **260**:563–571.

Turner, N. A., Doyle, W. A., Ventom, A. M., and Bray, R. C., 1995, Properties of liver aldehyde oxidase and the relationship of the enzyme to xanthine oxidase and dehydrogenase, *Eur. J. Biochem.* **232**:646–657.

Vos, J. P., Lopes-Cardozo, M., and Gadella, B. M., 1994, Metabolic and functional aspects of sulfogalactolipids, *Biochim. Biophys. Acta* **1211**:125–149.

Weiner, J. H., MacIsaac, D. P., Bishop, R. E., and Bilous, P. T., 1988, Purification and properties of *Escherichia coli* dimethylsulfoxide reductase, an iron-sulfur molybdoenzyme with broad substrate specificity, *J. Bacteriol.* **270**:2505–2520.

Weiner, J. H., Rothery, R. A., Sambasivarao, D., and Trieber, C. A., 1992, Molecular analysis of dimethylsulfoxide reductase: a complex iron-sulfur molybdoenzyme of *Escherichia coli*, *Biochim. Biophys. Acta* **1102**:1–12.

Wootton, J. C., Nicolson, R. E., Cock, J. M., Walters, J. W., Burke, J. F., Doyle, W. A., and Bray, R. C., 1991, Enzymes depending on the pterin molybdenum cofactor: sequence families, spectral properties and possible cofactor binding domains, *Biochim. Biophys. Acta* **1057**:157–185.

Xia, M., Dempski, R., and Hille, R., 1999, The reductive half-reaction of xanthine oxidase. Reaction with aldehyde substrates and identification of the catalytically labile oxygen, *J. Biol. Chem.* **274**:3323–3330.

Xiao, Z., Young, C. G., Enemark, J. H., and Wedd, A. G., 1992, A single model displaying all the important centers and processes involved in catalysis by molybdoenzymes containing $[Mo^{VI}O_2]^{2+}$ active sites, *J. Am. Chem. Soc.* **114**:9194–9195.

*Chapter 14*

# Nickel Containing CO Dehydrogenases and Hydrogenases

Stephen W. Ragsdale

## 1. BACKGROUND

### 1.1. Importance of CO and $H_2$ Metabolism for Anaerobic Microbes

### 1.1.1. Why CO?

Why do some microbes have an enzyme that metabolizes carbon monoxide as the cornerstone of their autotrophic metabolism? CO is a very important substrate for anaerobic microbes that fix $CO_2$ by the Wood Ljungdahl pathway. The atmospheric concentrations of CO range from about 0.1 ppm in rural to 200 ppm in urban settings (Meyer, 1985). How then can acetogenic and methanogenic microbes base their autotrophic metabolism on the use of such trifling amounts of CO? In fact, they and other anaerobes make their own CO. Anaerobic microbes produce the extremely toxic (at least for aerobic organisms) gas, carbon monoxide, from $CO_2$ or pyruvate as an

**STEPHEN W. RAGSDALE**    Department of Biochemistry, The Beadle Center, University of Nebraska P.O. Box 880664, Lincoln, NE 68588-0664, USA

*Subcellular Biochemistry, Volume 35: Enzyme-Catalyzed Electron and Radical Transfer*, edited by Holzenburg and Scrutton. Kluwer Academic / Plenum Publishers, New York, 2000.

intermediate in the synthesis of acetyl-CoA (Menon and Ragsdale, 1996). Even humans appear to produce CO as a neurotransmitter (Verma *et al.*, 1993).

Why is CO used for these roles when formic acid is at the same redox state as CO and is much less toxic? Barring one problem of toxicity for aerobic organisms because of its high affinity for cytochrome oxidase and hemoglobin, CO is an ideal metabolite. It can serve as a carbon and electron source. As an electron donor, it is more potent than NADH. The $CO_2/CO$ half cell reaction has a midpoint reduction potential of $-520\,mV$. These, however, are also properties of formate. CO binds strongly to the low valent states of metals. This is a two edged sword since it makes CO a strong inhibitor of hemeproteins. As a carbon source, CO is already dehydrated and at the oxidation state of the carbonyl group of acetyl-CoA, which is an important building block for cellular anabolism and an important source of ATP, derived through substrate-level phosphorylation.

It is likely that the use of CO remains from the early atmospheric conditions when life first evolved around 4 billion years ago. This follows from the hypothesis that the first organisms were autotrophic (Huber and Wachtershauser, 1997; Russell *et al.*, 1998). Volcanic gases can contain as high as 1% CO. Early life forms evolving in volcanic sites or hydrothermal vents could have used CO as their carbon and energy source. If this scenario is correct, CO metabolism today can be viewed as the extant survivor of early metabolic processes (Huber and Wachtershauser, 1997). The ability to metabolize CO is still important today since about $10^8$ tons of CO are removed from the lower atmosphere of the earth by bacterial oxidation every year (Bartholomew and Alexander, 1979). This helps to maintain CO below toxic levels, except in extreme cases.

If these organisms are generating CO during growth, why aren't they considered environmental biohazards? On the contrary, we harbor acetogens and methanogens in our gastrointestinal tract (Wolin and Miller, 1994; Dore *et al.*, 1995). Why aren't they toxic? Cultures of *C. thermoaceticum* growing on glucose with a 100% $CO_2$ gas phase produce about 50 ppb ($1\,\mu M$) CO (Diekert *et al.*, 1984). The amount of CO produced would be strongly dependent upon the concentration of $CO_2$ present in the medium. There is strong evidence that the CO produced during acetyl-CoA biosynthesis is sequestered in a molecular channel, which would serve two purposes. It would keep the CO concentration below toxic levels for the host organisms and would allow organisms to retain this valuable carbon and energy source without having it escape into the environment. The channeling of CO and $CO_2$ is discussed in more detail below.

## 1.1.2.   Where Does the CO Come from and Where Does It Go?

As mentioned above, anaerobic microbes that use the Wood Ljungdahl pathway catalyze the reduction of $CO_2$ to CO with CO dehydrogenase. $CO_2$ can come from the growth medium, from pyruvate, which is generated from sugars by the Embden-Meyerhof pathway, or from the carboxyl group of benzoic acid (Hsu *et al.*, 1990; Hsu *et al.*, 1990) and its derivatives. The coupling of pyruvate oxidation to acetyl-CoA formation has been well studied in *Clostridium thermoaceticum* (Menon and Ragsdale, 1996; Menon and Ragsdale, 1997). Pyruvate ferredoxin oxidoreductase catalyzes the conversion of pyruvate to acetyl-CoA and $CO_2$ (Eq. 1). Then, CO dehydrogenase reduces $CO_2$ to CO (Eq. 2). Finally, acetyl-CoA synthase, which is the other subunit of the bifunctional CO dehydrogenase/acetyl-CoA synthase, catalyzes the condensation of CO with a methyl group and coenzyme A to form acetyl-CoA (Eq. 3). Acetyl-CoA is then used in ATP generation or in cellular biosynthesis (Eq. 4–5). Thus, the bifunctional enzyme, CO dehydrogenase/acetyl-CoA synthase, plays an essential role in energy metabolism and cellular biosynthesis.

$$\text{Pyruvate} + \text{CoA} \rightarrow \text{acetyl-CoA} + CO_2 + 2\ \text{electrons} \qquad (1)$$

$$2\ \text{electrons} + 2H^+ + CO_2 \rightarrow CO + H_2O \qquad (2)$$

$$CO + \text{``}CH_3+\text{''} + \text{CoA} \rightarrow \text{acetyl-CoA} \qquad (3)$$

$$\text{Acetyl-CoA} + \text{ADP} \rightarrow \text{ATP} + \text{Acetate} \qquad (4)$$

$$\text{Acetyl-CoA} \rightarrow \text{Cell Material} \qquad (5)$$

## 1.1.3.   Why $H_2$?

As Oparin suggested early in this century, $H_2$ metabolism is considered to have arisen in the early stages of life's evolution. Also, like CO, molecular hydrogen is an excellent electron donor, with a standard reduction potential significantly more negative than that of the NAD(P)/NAD(P)H couple. An important biofuel throughout evolutionary time, $H_2$ is receiving attention as the fuel of the future because its combustion generates no waste (only $H_2O$), it can be transported in pipelines, and can be produced relatively cheaply from many sources.

### 1.1.4.  Where Does the $H_2$ Come from and Where Does It Go?

When the sun ignited nearly 5000 million years ago, it's cosmic storm propelled gases to the limits of the solar system. $H_2$, which is the lightest of the gases, became concentrated around the coldest and largest planets with the strongest gravitational force. In the current earth's atmosphere, molecular hydrogen is present at a concentration of about 0.55 ppmv and is produced by natural and industrial processes. However, soil air contains only about 10–30 ppbv $H_2$, which is too low for uptake by known microbes. This presents a conundrum since $H_2$ is consumed by soil microbes at a rate of $10^{14}$ g/year. Who are the $H_2$ utilizers? Apparently, the $H_2$-utilizing bacteria live in close association with microbes that produce $H_2$ during fermentation of organic material (for review see (Conrad, 1995)). In the early 1920's, the hydrogen-oxidizing bacteria were classed in one genus, called hydrogenomonas, however, it quickly became clear that this characteristic was shared by many types of organisms.

The biological production of molecular hydrogen presents less of a supply-and-demand problem since protons are abundant. The direct electron donor for proton reduction is often ferredoxin or flavodoxin, although there are examples of the direct and efficient coupling of redox enzymes like CO dehydrogenase to hydrogenase in the absence of a mediator (Ragsdale and Ljungdahl, 1984). Some hydrogenases use NADH as the direct electron donor (Eitinger and Friedrich, 1997).

### 1.1.5.  Supercellular Biochemistry: Interspecies $H_2$ Transfer

The ability to produce or use $H_2$ during growth is an important biochemical and ecological activity for anaerobic microbes (Adams, 1990). As pointed out above, the levels of $H_2$ in the soil are very low. Competition for these essential reducing equivalents can be severe, for example, between methanogens and sulfate reducers (Odom and Peck, 1984). At high sulfate concentrations, the sulfate reducer is the champion; at low sulfate and high $CO_2$ levels, the methanogen can outcompete. A strategy for efficient utilization of $H_2$ used by many microbes is syntrophic growth. For example, fatty and aromatic acids are degraded by a fermentative bacterium in syntrophic association with a hydrogen-using microbe. The reason for this association is that degradation of fatty and aromatic acids is thermodynamically unfavorable unless the hydrogen concentration is kept very low by the hydrogen-using bacterium (Eq. 6) (Jackson et al., 1999). By this interspecies $H_2$ transfer process, the fermenter provides the reductant for the chemolithotroph as it excretes the $H_2$ that inhibits its growth. Expression of genes involved in $H_2$ utilization and uptake can be regulated by $H_2$

levels (Reeve *et al.*, 1997). This leads to inhibition of methanogenesis by low levels of $H_2$.

$$CH_3CH_2COO^- + 2H_2O \rightarrow CH_3COO^- + 3H_2\Delta G^{o'} = +76\,kJ/mol \quad (6)$$

## 1.2.  Introduction to the Enzymes

### 1.2.1.  CO Dehydrogenase and Acetyl-CoA Synthase

There are three classes of CO dehydrogenase: the Ni—CO dehydrogenase, the Ni—CO dehydrogenase/acetyl-CoA synthase, and the Mo—CO dehydrogenase. I will focus only on the first two enzyme classes in this review.

Why is CO dehydrogenase important? CO dehydrogenase plays an essential role in CO oxidation by many microbes. CO is oxidized to $CO_2$, which then is fixed into cellular carbon by one of the reductive $CO_2$ fixation pathways, like the Calvin-Benson-Bassham Cycle, the reverse TCA cycle, or the Wood-Ljungdahl pathway. For example, *R. rubrum* senses the presence of CO by a heme sensor protein (Aono *et al.*, 1996; Shelver *et al.*, 1997) that activates the expression of a battery of structural, metal incorporation, and maturation genes (Fox *et al.*, 1996). In microbes that use the Wood-Ljungdahl pathway, there is an unusual twist in which $CO_2$ is reduced to CO, which is the precursor of the carbonyl group of acetyl CoA. Here, CO dehydrogenase, which catalyzes $CO_2$ reduction, is one subunit of a bifunctional protein that also contains the acetyl-CoA synthase subunit, which is responsible for the final steps in the Wood-Ljungdahl pathway (Figure 1).

### 1.2.2.  Hydrogenase

Marjory Stephenson and Leonard Stickland named the enzyme in 1931 (Stephenson and Stickland, 1931). There are five types of hydrogenase. Three of these are Ni proteins and one contains iron as the only transition metal. These proteins catalyze the reversible oxidation of dihydrogen gas (Eq. 7). A fifth class of hydrogenase lacks metals and catalyzes Eq. 8, but not Eq. 7. The Ni enzymes are classed into the NiFe, the NiFeSe, and the hydrogen sensors. All of these enzymes are bidirectional, although different enzymes exhibit a catalytic bias toward making or oxidizing $H_2$.

The hydrogenase reaction is very important in the energy metabolism of organisms. Besides serving as a reductant for energy generation, the hydrogenase reaction allows organisms to siphon off excess reducing equivalents by transferring electrons to protons. Some organisms even localize

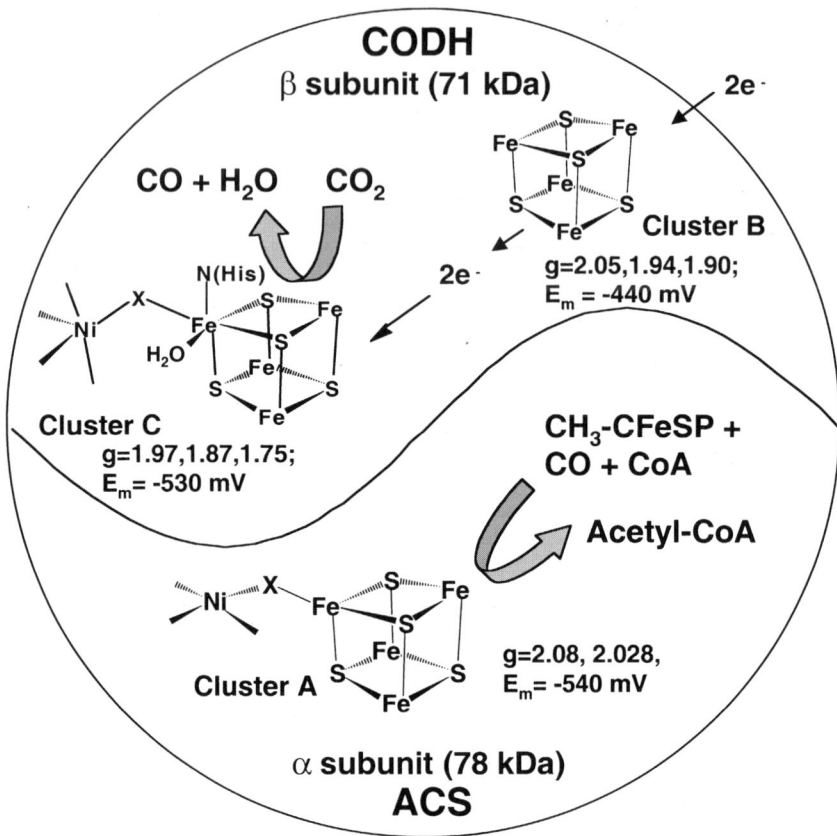

**FIGURE 1.** Cartoon of the two active sites of the bifunctional CO dehydrogenase/acetyl-CoA synthase.

distinct hydrogenases in different cellular compartments and couple electron transport to proton translocation. A related example of this bioenergetic mechanism, termed hydrogen cycling, involves the periplasmic oxidation of formate to $CO_2$ and protons coupled to the cytoplasmic reduction of fumarate to succinate (Kröger, 1978; Odom and Peck, 1984). Only electrons are transferred across the membrane and the protons remain in the periplasm, generating a proton gradient across the cytoplasmic membrane.

$$H_2 \leftrightarrow 2H^+ + 2\ e^- \quad Eo = -414\,mV \text{ at pH 7 and 1 bar of } H_2. \quad (7)$$

$$\text{Methenyltetrahydromethanopterin} + H_2 \\ \leftrightarrow \text{methylenetetrahydromethanopterin} + H^+ \quad (8)$$

## 2. CO OXIDATION AND $CO_2$ REDUCTION BY CO DEHYDROGENASE

### 2.1. The Catalytic Redox Machine: A Ni—Fe—S Cluster

The CO dehydrogenase reaction has been extensively studied in *Rhodospirillum rubrum* and *C. thermoaceticum*. For studying this reaction, *R. rubrum* offers the advantage that it lacks Cluster A and the entire acetyl-CoA synthase subunit, which also contains Ni. In addition, a Ni-deficient protein containing all Fe components of the holoenzyme can be isolated. Furthermore, the crystal structure of this protein will be known in the next few months.

Two clusters in CO dehydrogenase are required for oxidation of CO or reduction of $CO_2$ (Figure 1). The catalytic site is a nickel iron sulfur cluster called Cluster C. The two electrons involved in this redox reaction are transferred to or from a ferredoxin-like [4Fe—4S] cluster called Cluster B. Cluster C is a NiFeS center; whereas, Cluster B is most certainly a typical $[4Fe—4S]^{2+/1+}$ cluster (Ragsdale *et al.*, 1982; Lindahl *et al.*, 1990; Lindahl *et al.*, 1990).

The C-cluster of carbon monoxide dehydrogenase is the active site for the oxidation of CO to $CO_2$. This conclusion is based on rapid kinetic studies in which changes in the spectra of Cluster C undergo changes at rates commensurate with the rate of CO oxidation (Kumar *et al.*, 1993). In addition, cyanide, which is a relatively specific inhibitor of CO oxidation, binds specifically to Cluster C (Anderson *et al.*, 1993).

What is the structure of Cluster C? Until recently, spectroscopic studies have been interpreted to indicate a high spin (S = 1) Ni-II site coupled through a bridge to a $[4Fe—4S]^{2+/1+}$ cluster (Hu *et al.*, 1996) (Figure 2). Apparently, one of the irons in this cluster is pentacoordinate and is considered to be the CN binding site as well as the bridge to the Ni site (DeRose *et al.*, 1998). Based on ENDOR studies, it was suggested that a water or hydroxide group, which donates an oxygen to CO during catalysis, also binds to this special iron site (DeRose *et al.*, 1998). Recently, it was suggested that Cluster C might contain a binuclear NiFe cluster bridged to the cubane $[4Fe—4S]^{2+/1+}$ cluster (Heo *et al.*, 1999; Staples *et al.*, 1999). This is based in part on the observation that a variant of the *R. rubrum* CO dehydrogenase (C531A) exhibits an EPR signal similar to the Ni—C signal of the binuclear NiFe center of hydrogenase. This signal lacks [61]Ni hyperfine interactions indicating that the electron density resides on some other atom. In this mutant, the bridge, which is proposed to be C531, between Ni and the 4Fe cluster apparently is broken. Furthermore, the metal content consistently shows 9Fe per monomeric enzyme, implying the presence of 4 irons in Cluster B and 5 in Cluster C.

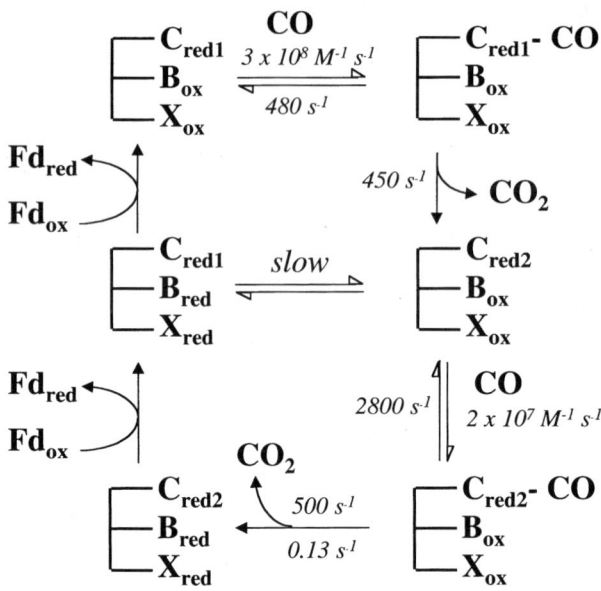

**FIGURE 2.** Proposed catalytic mechanism of CO oxidation. Results indicate that, at high CO concentrations, two mol of CO bind to generate a four-electron reduced enzyme during each catalytic cycle. At low CO concentrations, the upper cycle is the predominant reaction, since electron transfer to Cluster B would then occur faster than binding of CO to the cred2 form of the enzyme. Modified from (Seravalli *et al.*, 1997).

What is the mechanism by which an enzyme like the *C. thermoaceticum* CO dehydrogenase can catalyze the oxidation of CO to $CO_2$ with a turnover number of over $3200 s^{-1}$ (Seravalli *et al.*, 1995)? Furthermore, it can catalyze the reduction of $CO_2$ to CO with a turnover number of $10 s^{-1}$ (Kumar *et al.*, 1994) without imposing an overpotential. In the current schemes for industrial activation of $CO_2$, an overpotential of about a volt is applied to the reaction. Redox changes at Clusters B and C are fundamental to the catalytic mechanism, which is summarized in Figure 2. The catalytic cycle involves two half reactions. The first is reduction of CO dehydrogenase (Clusters C and B) and the second one is electron transfer from Cluster B to ferredoxin. An unexpected feature of the reductive half reaction is that CO dehydrogenase can undergo a two-electron reduction to release one equivalent of $CO_2$ and then bind and oxidize another CO molecule before being oxidized by an electron carrier (Seravalli *et al.*, 1995). Thus, the enzyme has the capability to undergo four-electron reduction during a single catalytic cycle. The binding of CO to Cluster C occurs at near diffusion controlled rates. The oxidation of CO to $CO_2$ also is extremely fast. At high concentrations of CO, the rate limiting step is oxidation of the reduced enzyme by the electron acceptor; whereas, at low CO concentrations, it is the intramolecular electron transfer from Cluster C to Cluster B (Kumar et al., 1993; Seravalli et al., 1997). The intramolecular electron transfer reaction is also rate limiting when CO dehydrogenase is rapidly mixed with CO during single turnover kinetics in the absence of electron carriers.

## 2.2. The Intramolecular Wire

Electrons zip along an intramolecular wire between the catalytic Cluster C and Cluster B at a rate of $3200\,s^{-1}$ at 55°C under optimal conditions. The mechanism of electron transfer from CO to Cluster C to Cluster B is relatively well understood, since it has been studied by a variety of rapid kinetics and electrochemical methods.

Cluster C has four redox states: $C_{ox}$, $C_{red1}$, $C_{dia}$, and $C_{red2}$. The midpoint potential of the $C_{ox}/C_{red1}$ couple is approximately $-200\,mV$, whereas that for the formation of $C_{red2}$ is $\sim-530\,mV$ (Lindahl et al., 1990). $C_{red1}$ appears to be the state that reacts with CO. Transfer of two electrons to Cluster C forms $CO_2$ and the $C_{red2}$ state. Then, apparently another CO can bind to the red2 state of the C cluster (Seravalli et al., 1997). The subsequent oxidation of CO is coupled to the two-electron reduction of Cluster B and another redox center in the protein, which has been called "X". The reduction of X is required to balance the electron transfer reaction and to explain the kinetics and the potentiometric titrations of CO dehydrogenase with CO and other reductants. One possibility is the "X" is the Fe center in the binuclear cluster recently proposed by Ludden (Staples et al., 1999). Thus, at the end of this part of the catalytic cycle, CO dehydrogenase has undergone a four-electron reduction. Closure of the catalytic cycle involves sautering the intramolecular wire to an intermolecular wire.

## 2.3. The Intermolecular Wires: How Electrons Enter and Exit CO Dehydrogenase

Reduced CO dehydrogenase is like a junction box that can couple to many wires. When CO dehydrogenase was first purified, it was surprising how the enzyme could efficiently transfer electrons to rubredoxin, methylene blue, a 4Fe and 8Fe ferredoxin, benzyl viologen, flavodoxin, and methyl viologen, with the rates decreasing in that order (Ragsdale et al., 1983). As an electron acceptor, CO dehydrogenase efficiently accepts electrons directly from pyruvate ferredoxin oxidoreductase ($k_{cat}/K_m = 7 \times 10^7\,M^{-1}s^{-1}$, similar to the specificity for ferredoxin) (Menon and Ragsdale, 1996). It also donates electrons to an Fe-only hydrogenase (Ragsdale and Ljungdahl, 1984) and the corrinoid iron-sulfur protein (Ragsdale et al., 1987). It also reduces very low potential acceptors like triquat and very high potential acceptors like cytochrome c. Our attempts to study proton uptake or release by CO dehydrogenase has been plagued by the ability of CO dehydrogenase to reduce nearly every pH indicator that we have tried to use (unpublished information). However, one electron acceptor that can not couple to CO dehydrogenase is NAD(P).

## 2.4. By Channel or By Sea? How Carbon Enters and Exits CO Dehydrogenase

As described in section 1.1.ii, when microbes grow on glucose, they produce pyruvate by the Embden Meyerhof pathway. Then, pyruvate:ferredoxin oxidoreductase converts pyruvate to acetyl-CoA and $CO_2$; CO dehydrogenase reduces $CO_2$ to CO; and acetyl-CoA synthase condenses CO with CoA, and the methyl group to form acetyl-CoA (Menon and Ragsdale, 1996). In 1973, Marv Schulman et al. published an intriguing paper (Schulman et al., 1973). One can interpret their results to indicate that the carboxyl group of pyruvate is incorporated into the carbonyl group of acetyl-CoA without equilibrating with $CO_2$ in solution. This result is consistent with the hypothesis that there is a long macromolecular channel connecting PFOR to CO dehydrogenase and CO dehydrogenase to acetyl-CoA synthase. First, $CO_2$ would be channeled from the thiamine pyrophosphate binding site on PFOR to Cluster C on CO dehydrogenase, without $CO_2$ entering free solution. Channels have been located in the PFOR structure that lead from the surface to the thiamine pyrophosphate binding site, which is buried $30 \text{ Å}$ from the surface (Chabriere et al., 1999). If there is indeed a $CO_2$ channel between PFOR and CO dehydrogenase, the pathway of carbon flow in the acetyl-CoA pathway becomes extremely interesting because there is strong evidence for a channel that couples the active sites of CO oxidation and acetyl-CoA synthase.

## 2.5. Coupling $CO_2$ Reduction and Acetyl-CoA Synthesis

Recent results strongly indicate that CO derived from $CO_2$ migrates from Cluster C in the small CO dehydrogenase subunit through a channel in the interior of the protein to the large acetyl-CoA synthase subunit (Seravalli and Ragsdale, 2000; Maynard and Lindahl, 1999). Here it reacts with Cluster A to form a metal-carbonyl species that condenses with the methyl group and CoA to form acetyl-CoA. The results are based on studies of an isotope exchange reaction between labeled $CO_2$ and the carbonyl group of acetyl-CoA with the CO dehydrogenase/acetyl-CoA synthase from *Clostridium thermoaceticum*. When solution CO is provided at saturating levels, only $CO_2$-derived CO is incorporated into the carbonyl group of acetyl-CoA. Furthermore, hemoglobin, which tightly bind CO, only partially inhibit the synthesis of acetyl-CoA from $CH_3$-$H_4$folate, CoA and pyruvate (Menon and Ragsdale, 1996) or $CO_2$ (Maynard and Lindahl, 1999). Hemoglobin and myoglobin also only partially inhibit the exchange reaction between $CO_2$ and acetyl-CoA.

These results provide strong evidence for the existence of a CO channel between Cluster C in the CO dehydrogenase subunit and Cluster A in the acetyl-CoA synthase subunit. Such a channel would tightly couple CO production and utilization and help explain why high levels of this toxic gas do not escape into the environment. Instead, microbes sequester the CO as an energy rich carbon source. The channel would also protect aerobic organisms that harbor the microbes and help keep CO below toxic levels.

## 2.6. Acetyl-CoA Synthase: Another Catalytic NiFeS Cluster

Cluster A of the acetyl-CoA synthase subunit of the CO dehydrogenase/acetyl-CoA synthase complex will only briefly be discussed here. Several rather recent reviews are available (Ragsdale *et al.*, 1996; Ragsdale and Riordan, 1996; Ragsdale, 1997; Ragsdale *et al.*, 1998). Cluster A is the site of assembly of the methyl, carbonyl, and SCoA moieties of acetyl-CoA. There is unambiguous evidence that a Ni center is bonded to an FeS cluster and that when CO binds, it forms an adduct with this NiFeS cluster (see Figure 1). This is based on the observation of an EPR signal at g = 2.08, 2.02, thus a net $S = 1/2$ species, that exhibits $^{61}Ni$, $^{13}C$, and $^{57}Fe$ hyperfine interactions (Ragsdale *et al.*, 1985; Fan *et al.*, 1991). The Ni site apparently adopts a distorted square-planar geometry with two nitrogen or oxygen and two sulfur ligands (Russell *et al.*, 1998). The Fe site apparently is a [4Fe—4S]$^{2+/1+}$ cluster (Lindahl *et al.*, 1990; Lindahl *et al.*, 1990; Fan *et al.*, 1991). The bridge between Ni and Fe is unknown.

The chemistry of acetyl-CoA synthesis is thought to resemble the Monsanto process for acetate synthesis in that a metal center binds a methyl group and CO and the CO undergoes a carbonyl insertion into the methyl-metal bond. Elimination of the acetyl group is catalyzed by a strong nucleophile, iodide in the industrial process and CoA in the biochemical one. Currently, there are two views of the catalytic mechanism.

In our model of the acetyl-CoA synthase catalytic mechanism (Figure 3), CO binds first to Cluster A (actually, the order of substrate binding is uncertain) to form a paramagnetic adduct. Fourier transform infrared

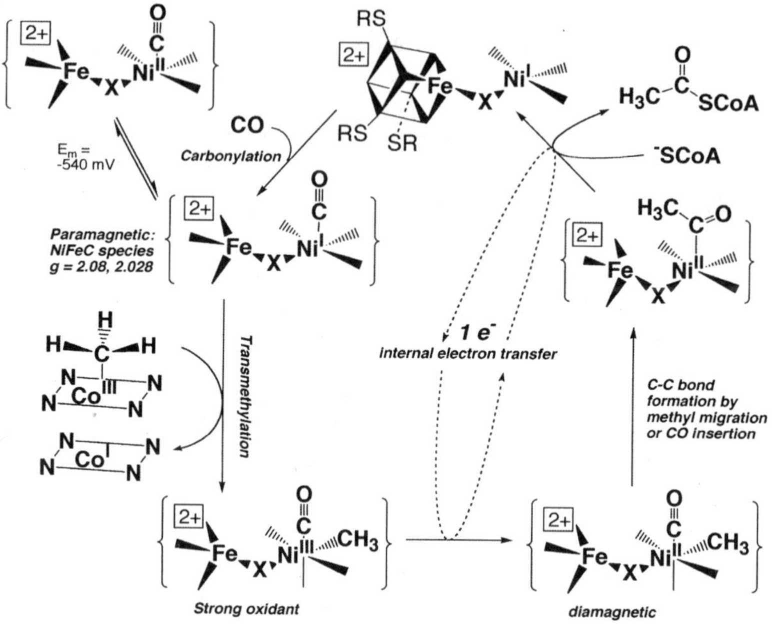

**FIGURE 3.** Proposed acetyl-CoA synthase catalytic mechanism. An internal electron transfer reaction between the CO dehydrogenase and acetyl-CoA synthase subunits is proposed to be an important part of the catalytic cycle.

(FTIR) studies show that CO is terminally bound (Kumar and Ragsdale, 1992) to a metal center in Cluster A. Although the catalytic relevance of this adduct has been questioned (Grahame and Demoll, 1995; Barondeau and Lindahl, 1997), many studies (including demonstration of catalytic competence) support its identity as the precursor of the carbonyl group of acetyl-CoA (Ragsdale *et al.*, 1985; Shanmugasundaram *et al.*, 1988; Kumar *et al.*, 1993; Menon and Ragsdale, 1996).

Next, methyl group transfer from the methylated corrinoid iron-sulfur protein to CO dehydrogenase/acetyl-CoA synthase occurs by the nucleophilic attack of Cluster A on the methyl group of methyl-corrinoid iron-sulfur protein (Menon and Ragsdale, 1998; Menon and Ragsdale, 1999). The final steps in the mechanism include carbon-carbon bond formation by condensation of the methyl and carbonyl groups to form an acetyl-metal species. Then, CoA cleaves the acetyl group with C—S bond formation. As shown in the Figure, an intramolecular electron transfer step is included. This is based on the highly oxidizing character of methyl-Ni(III) (Thauer, 1998) and on the requirement in the CO/acetyl-CoA exchange reaction for an electron carrier (Ragsdale and Wood, 1985). A catalytic cycle involving only diamagnetic intermediates also has been proposed (Barondeau and Lindahl, 1997).

## 3.  $H_2$ OXIDATION AND PROTON REDUCTION BY HYDROGENASE

### 3.1.  The Catalytic Redox Machine: Ni—Fe—S and Fe—FeS Clusters

The oxidation of $H_2$ is catalyzed by the Ni and the Fe-only metalloenzymes described above. These are extremely active catalysts, rapidly interconverting protons and $H_2$ without the need for an overpotential. When the NiFe hydrogenase was studied by voltammetric methods, the rate of $H_2$ oxidation was diffusion controlled ($k_{cat}/K_m$ of $10^8$–$10^9 M^{-1} s^{-1}$), with a turnover number reaching $9000 s^{-1}$ at $30°C$ (Pershad *et al.*, 1999). This is remarkably rapid given that the strength of the H—H bond is 103 kcal/mol and the active site is 30 Å from the surface. As was first demonstrated by studying an H—D exchange reaction using $D_2$ and $H_2O$ (Yagi, 1959), these enzymes all cleave $H_2$ heterolytically to generate a hydride and a proton.

An important review of metal dihydrogen chemistry is an essential primer to the understanding of the hydrogenase catalytic mechanism (Goldman and Mascharak, 1995). Here it is pointed out how the coordination of $H_2$ to a transition metal significantly enhances the potential for heterolysis, dropping the pKa of $H_2$ by 15 pH units. There are many examples of base-assisted generation of metal hydrides from $H_2$ and of protonation of metal hydrides to form $H_2$. Thus, the enzyme needs a basic amino acid residue in the active site to promote heterolytic cleavage of $H_2$. It also is pointed out that, although metal hydrides are relatively stable, metal-dihydrogen species are not very likely to be present in stable states of the enzyme, like the Ni—C state. The importance of sulfur coordination in the NiFe center has been pointed out (Kruger and Holm, 1990; Farmer *et al.*, 1993; Goldman and Mascharak, 1995). Sulfur or, in the NiSe enzymes, selenium coordination is proposed to allow the reduced Ni site to bind CO and hydride (Goldman and Mascharak, 1995).

### 3.1.1.  The NiFe Hydrogenase

The first example of a naturally occuring center containing nickel and iron was in CO dehydrogenase/acetyl-CoA synthase (Ragsdale *et al.*, 1985). Ten years later, it was found that the NiFe hydrogenase also contains a bimetallic NiFe center coordinated to the protein through four cysteinyl ligands. In the past four years, several X-ray structures of the NiFe hydrogenase have been solved. The first was the 2.85 Å structure of the *Desulfovibrio gigas* hydrogenase (Volbeda *et al.*, 1995; Volbeda *et al.*, 1996). A 1.8 Å resolution X-ray structure of the hydrogenase from *Desulfovibrio vulgaris* Miyazaki has been solved (Higuchi *et al.*, 1997) and recently the structure of the reduced enzyme was determined at 1.4 A resolution (Higuchi *et al.*, 1999). There is general agreement about the basic structure of the active site (Figure 4), which includes a binuclear NiFe center with the two metals separated by 2.9 Å and bridged by the thiolates from two cysteine residues. The Ni site contains two other cysteine ligands, whereas the iron contains three diatomic ligands (two cyanides and one carbon monoxide). One of the terminal cysteines is replaced by selenocysteine in the NiFeSe protein (Sorgenfrei *et al.*, 1993). The existence of the diatomic ligands was also revealed by infrared spectroscopic studies (Bagley *et al.*, 1995; deLacey *et al.*, 1997; Happe *et al.*, 1997). Controversy remains about the nature of a bridging ligand between Ni and Fe. It is proposed that there is a bridging oxo (Volbeda *et al.*, 1995) or a sulfide that is released upon reduction with $H_2$ (Higuchi *et al.*, 1997; Higuchi *et al.*, 1999). A 2.15 Å resolution structure of the periplasmic [NiFeSe] hydrogenase from *Desulfomicrobium baculatum* has been determined in its reduced, active form (Garcin *et al.*, 1999). In the reduced enzyme, the nickel-iron distance is significantly (0.4 Å) shorter than in the oxidized enzyme.

Although it is clear that there are several redox states of the NiFe center, there has been much controversy about assignment of the oxidation states of nickel. The nomenclature implies changes in the Ni redox state, which is controversial. There are two catalytically inactive oxidized states, called Ni—A and Ni—B (Roberts and Lindahl, 1994). Upon reduction, an EPR-silent state, called Ni—SI, is formed. Further reduction with hydrogen generates the Ni—C state, which is partially reduced with a bound H species (Albracht, 1994). Further reduction generates the Ni-R form, which is thought to be one electron more reduced than Ni—C. The iron remains low spin in all redox states of the enzyme because of the strong-field diatomic ligands bound to it. The low spin character is viewed to be important to partly deshield the metal's positive nucleus, which promotes electrostatic attraction for $H_2$ (Pavlov *et al.*, 1998).

Oxidation of $H_2$ should generate at least a transient reduction of the active site NiFe center. Where do the electrons go? There is evidence that Fe is diamagnetic in all states of the enzyme (Dole *et al.*, 1997). Then, does Ni undergo the redox changes? Proposals range from a Ni(0) to Ni(IV) redox interconversion to no oxidation state changes in Ni at all. X-ray absorption experiments show that the electron density at Ni does not change as a function of redox state (Bagyinka *et al.*, 1993). These results

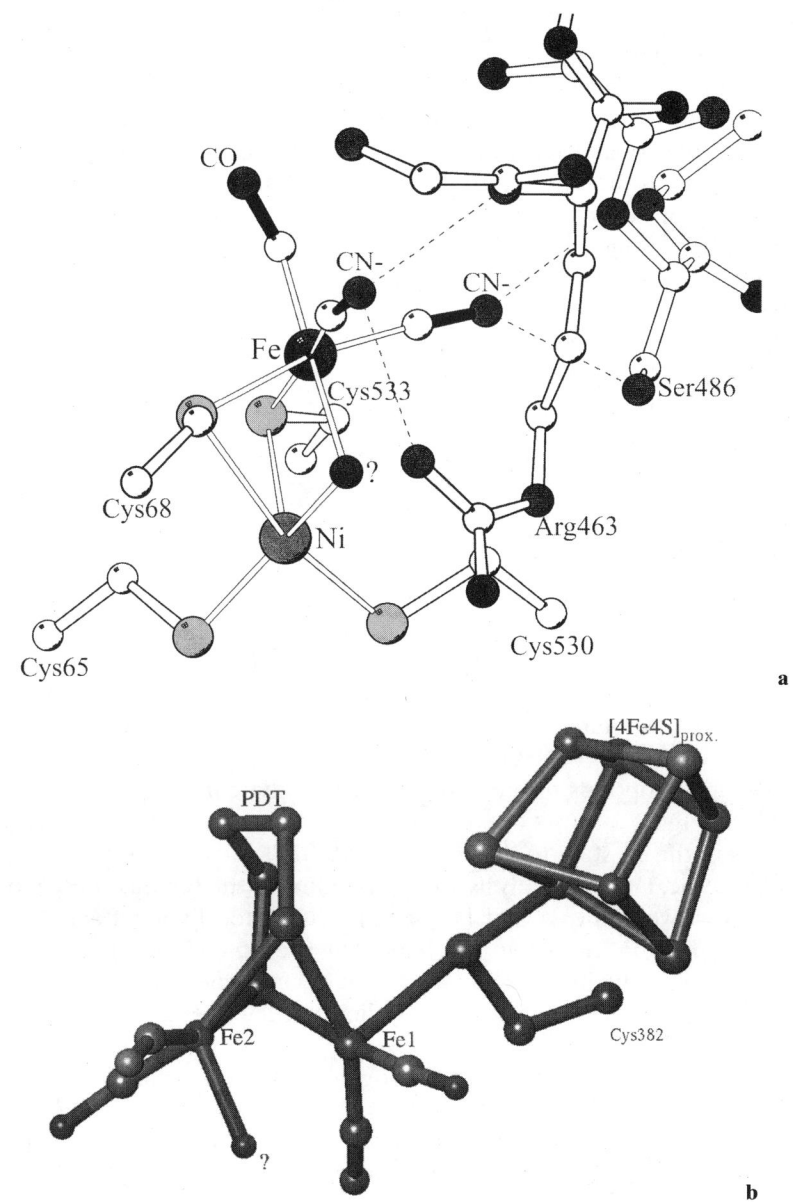

**FIGURE 4.** The active site of hydrogenase. a. Structure of the [NiFe] cluster. b. Structure of the H cluster of the Fe-only hydrogenase. From (Nicolet *et al.*, 1999).

strongly indicate that Ni does not change redox state during catalysis since one would expect a ~2 eV shift in the edge energy associated with each change in oxidation state. One possible problem with the studies so far described is that spectroscopic states of the enzyme are being assigned to catalytic intermediates. However, the spectroscopically characterized states have been observed after incubation of hydrogenase over long periods of time (minutes to hours) with oxidants, reductants, or other reagents. Catalytic turnover occurs in the msec time scale. The first rapid mixing experiment with hydrogenase has just been reported (Happe *et al.*, 1999); this approach is expected to help in relating states of the enzyme to catalytic intermediates.

There also is a related controversy about where $H_2$ or the active hydride binds to the enzyme. It has been proposed that Ni is the binding site for the active hydrogen species. This is based on several results. Photolysis of Ni—C is six-fold slower in $D_2O$ than $H_2O$ (Van der Zwaan *et al.*, 1985). A solvent-exchangeable proton with a 17 MHz coupling constant is observed by ENDOR spectroscopy of the Ni—C state and photolysis led to loss of this ENDOR resonance (Whitehead *et al.*, 1993). These results are consistent with a Ni—H interaction; however, they are also consistent with any model containing a bond between "hydride" or "$H_2$" and the "spin-system" which includes Ni, Fe, and the ligands. Maroney suggests that the structure of the active Ni—C state could be a Fe—H complex or a protonated ligand thiolate (Maroney *et al.*, 1998). He also points out that a metal-hydride should have a much stronger hyperfine interaction than that observed in the ENDOR experiment.

Several proposals of the hydrogenase catalytic mechanism have been described. Figure 5 shows the conversion of the inactive NiA state in which the Ni is in the 3+ state to the Ni—SI2 state that is capable of entering the catalytic cycle. During catalysis, the Fe remains in the ferrous state and the Ni cycles between the 2+ and 1+ (or 3+) state. A terminal cysteine residue is proposed to be the active site base that promotes heterolysis of $H_2$ to generate a bridging hydride and a protonated cysteine ligand. Some mechanisms have included a sulfur cation radical at the cysteine residue that acts as the active site base.

The mechanism of the NiFe hydrogenase has been treated by calculational methods, with some interesting conclusions (Pavlov *et al.*, 1998). Scheme 1 of this reference proposes a catalytic cycle based on these results. It was proposed that Fe binds $H_2$ and that a low spin Fe is essential for heterolytic cleavage of the H—H bond. The next step is proposed to be hydride transfer to Fe and proton transfer to a ligated cysteine thiolate, which leads to decoordination of the cysteine and concurrent bridging of the N of CN

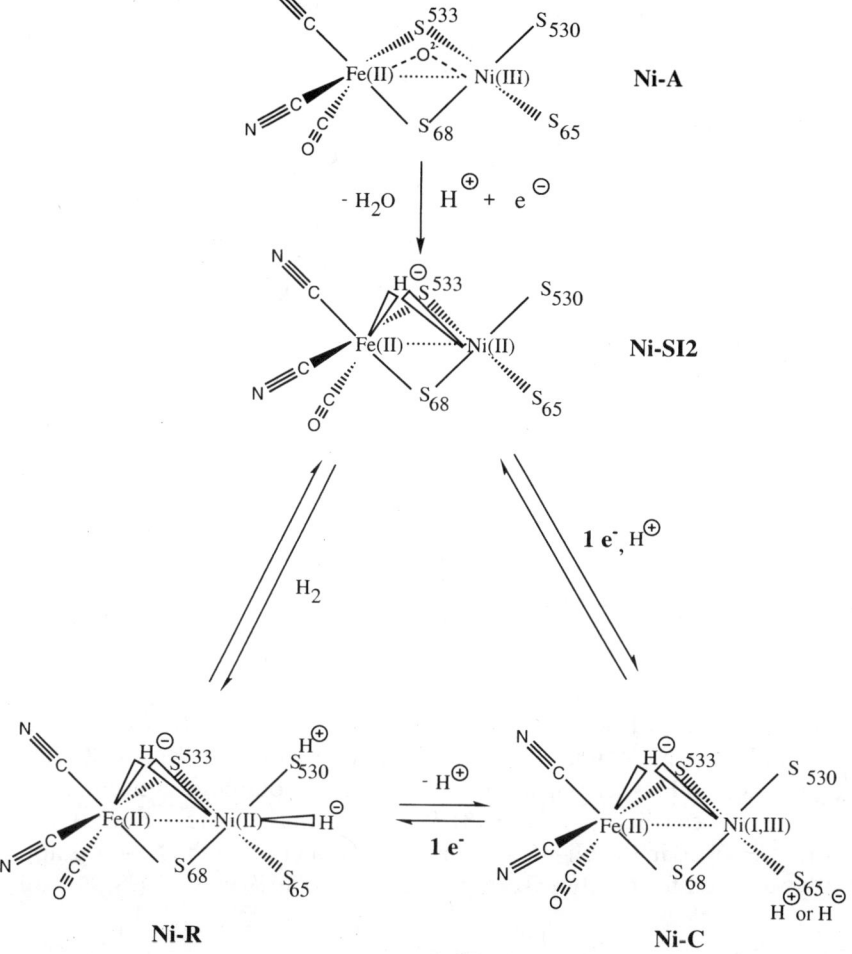

**FIGURE 5.** Proposed catalytic mechanism of hydrogenase catalysis. See text for description. From (Amara *et al.*, 1999).

to Ni. Hydride transfer to Ni is then thought to be important in the subsequent steps of proton transfer out of the active site. The major activation barrier (10 kcal/mol) in the reaction is at the H—H activation step. In the transition state (Figure 6), the proton is equally bonded between the hydride and the sulfur of the bridging cysteine, which is in the process of dissociating from the Ni as the N of the CN begins to form a bridge.

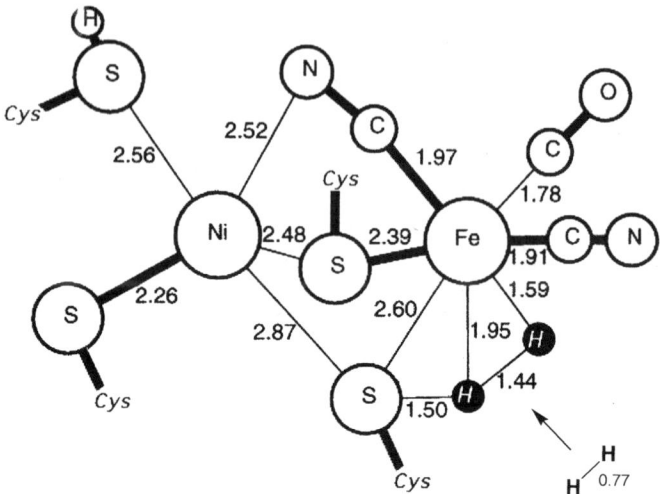

**FIGURE 6.** Calculated hydrogenase transition state. From (Pavlov *et al.*, 1998).

### 3.1.2. The NiFe Regulatory Hydrogenase

A new type of oxygen insensitive NiFe hydrogenase has been recently discovered that has very low catalytic activity, but functions as a hydrogen sensor (Lenz and Friedrich, 1998). This protein has spectroscopic properties similar to the standard NiFe hydrogenase with 2 CN and 1 CO bound to the low spin Fe site, indicating high structural homology between the two enzyme classes (Pierik *et al.*, 1998). Why isn't this enzyme active? It apparently is blocked in the Ni—C state. Further reduction to the Ni—R is apparently required for catalytic turnover (above) (Pierik *et al.*, 1998).

### 3.1.3. The Fe-Only Hydrogenase

Many microbes contain a hydrogenase that lacks nickel. The best studied of these proteins is the hydrogenase from *C. pasteurianum* (Adams, 1990). The active site of this protein has been called the H cluster. Two structures of this protein recently appeared, a 1.6 Å structure of the heterodimeric *Desulfovibrio desulfuricans* enzyme (Nicolet *et al.*, 1999) and the 1.8 Å structure of the Cpl enzyme from *Clostridium pasteurianum* (Peters *et al.*, 1998). The H cluster at the active site contains 6Fe as proposed earlier (Adams *et al.*, 1989). There are two subcomponents: a typical [4Fe—4S] cubane bridged by a cysteine residue to an active site binuclear Fe center (Figure 4b). The unusual nature of the iron site in the NiFe hydrogenase is conserved in the Fe-only hydrogenase—it also contains CO and CN ligands (Peters *et al.*, 1998; Pierik *et al.*, 1998; Nicolet *et al.*, 1999). Apparently a low-spin iron is essential to the ability to reversibly reduce protons to $H_2$. Five diatomics were assigned in the *C. pasteurianum* structure (Peters *et al.*, 1998) and 3 (2 cyanide and one CO) in the *D. desulfuricans* structure (Nicolet *et al.*, 1999) and an IR study (Pierik *et al.*, 1998). An unusual bridging ligand, 1, 3-propanedithiol, was observed in the structure of the *D. desulfuricans* enzyme (Nicolet *et al.*, 1999). Because of the tenuous connection to the protein of the iron that binds CO and CN, it was suggested that this center might have been "imported from the inorganic world as an already functional unit" (Nicolet *et al.*, 1999).

### 3.1.4. Models of Hydrogenase

Both irons in the binuclear 2Fe site and the iron in the binuclear NiFe site are hexacoordinate. A pentacoordinate (Hsu *et al.*, 1997) and a hexacoordinate (Lai *et al.*, 1998) low-spin iron model complex have been found to elicit IR spectra that are very similar to that of the NiFe and Fe hydrogenases (Lai *et al.*, 1998). Mossbauer spectroscopy showed conclusively that the model compound and the Fe site at the enzyme's active site (since the properties of the model are so similar to the enzyme) are low spin (Hsu *et al.*, 1997). The hexacoordinate complex was extensively studied by IR and electrochemistry and some interesting conclusions were reached regarding the chemistry of biological hydrogen activation. The inactive states of the enzyme (NiA, NiB, and a silent form) are proposed to be hexacoordinate; however, the active states appear to have an altered coordination number. Considering an Fe-based catalytic site, the loss of a ligand could promote the $\eta^2$ bonding of $H_2$ to the vacant coordination site. It was suggested that the reduction of Ni (II) to Ni(I) could promote loss of the water ligand and promote formation of the pentacoordinate Fe site. On the other hand, if the enzyme uses a Ni-based catalytic mechanism, the $Fe(CO)(CN)_2$ unit might be present to tune the redox potential of the Ni center to the range required for biological catalysis. A square planar selenolate-rich Ni(II)—CO complex has recently been studied (Liaw *et al.*, 1997). Furthermore, Ni(I)-hydride complexes containing thiolate rich coordination spheres have been prepared that have similar EPR spectra to the active Ni—C state in hydrogenase (see (Goldman and Mascharak, 1995) for review).

### 3.1.5. The Metal-Free Hydrogenase

In superacids, electrophilic carbenium ions can react with $H_2$ to form a two-electron three atom pentavalent carbonium ion, in essence

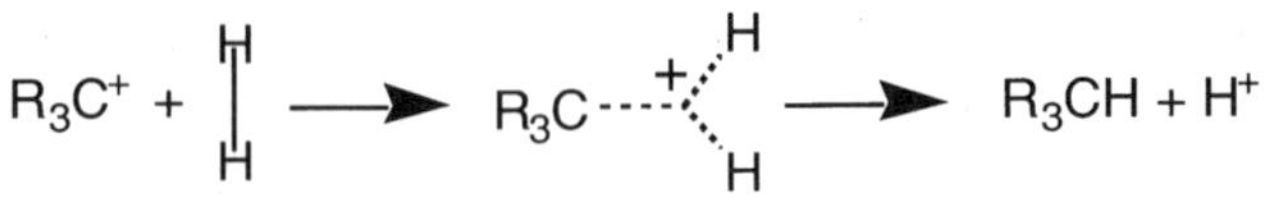

**FIGURE 7.** Cleavage of $H_2$ by a carbenium ion in superacid solution. The reaction involves a pentavalent carbonium ion, formed by transferring a hydride equivalent to the carbenium ion (Olah *et al.*, 1995).

transferring a hydride equivalent to the carbenium ion (Olah *et al.*, 1995) (Figure 7). The reaction of the nonmetal hydrogenase, which generates $H_2$ from methylenetetrahydromethanopterin (Equation 8, above), has been proposed to occur through this type of mechanism (Geierstanger *et al.*, 1998). Instead of having a metal ion(s) to activate $H_2$, which is the step with the highest activation barrier, this enzyme appears to generate a carbenium ion intermediate at C14a of the substrate that can react directly with $H_2$. The net result is the transfer of a hydride from $H_2$ to methenyltetra-hydromethanopterin to yield methylenetetrahomethanopterin and a proton.

### 3.2. Proton Transfer Pathway

The initial proton release associated with $H_2$ cleavage is promoted by a base. For hydrogen evolution, net proton uptake from the medium is necessary. Conversely, for $H_2$ oxidation protons are transferred from the active site to solution. The transfer of protons within a protein is considered to involve small (<1 Å) movements of the amino acids that participate in the pathway (Williams, 1995). The proton transfer would involve a rotation of each individual donor and acceptor. Analyses of the crystal structures have suggested proton transfer pathways for the NiFe and Fe-only hydrogenases.

For the Fe-only *D. desulfuricans* hydrogenase, it is proposed that, when $H_2$ is cleaved, the proton is transferred to the $CN^-$ ligand of Fe (Nicolet *et al.*, 1999). A proposed pathway includes proton transfer from cyanide sequentially to a lysine residue (K237), a glutamate residue (E240), three water molecules, and a surface glutamate residue (E245) (Figure 8). In the *C. pasteurianum* Fe-only hydrogenase, the proposed proton transfer pathway involves the cluster-bound water (above), a cysteine ligand (C299), 2 glutamates, 1 serine, and a water molecule; however, the actual residues were not designated (Peters *et al.*, 1998). Peters *et al.* suggest that proton reduction could involve displacement of the terminal water ligand bound to one of the irons in the binuclear cluster (Fe2) upon reduction, followed by formation of an Fe-hydride intermediate. Cys299 could act as a proton

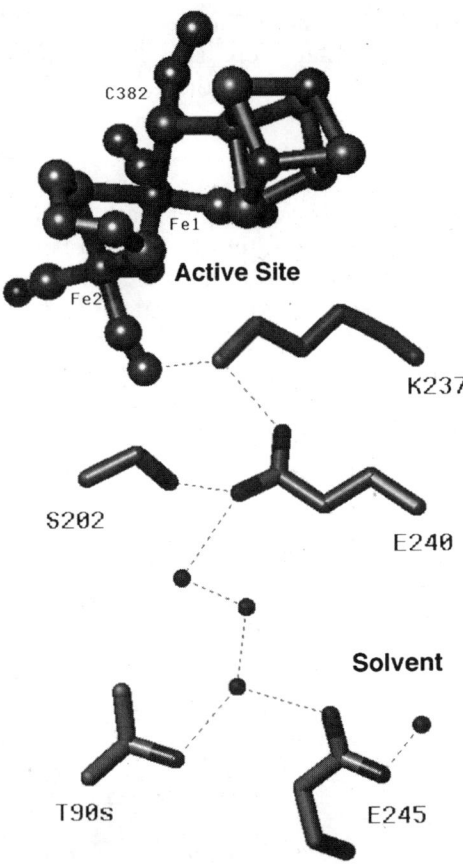

**FIGURE 8.** Proposed proton transfer pathway in the Fe-only hydrogenase. From (Nicolet *et al.*, 1999).

donor for the formation of dihydrogen, or, in the reverse reaction, as the proton acceptor from the water ligand.

The active site of the NiFe enzyme is 30 Å from the surface. Volbeda *et al.* proposed a proton transfer chain beginning at the NiFe cluster to His72 to His 536 through two water molecules finally to Glu46 at the surface. All of these residues are highly conserved among the NiFe hydrogenases. Two other histidine residues were also considered to have a possible role in proton transfer.

As important as how protons enter and exit hydrogenase is how $H_2$ migrates between the surface and the active site. A cavity map was constructed by calculational approaches and by solving the structure of the Xe derivative of the NiFe hydrogenase (Montet *et al.*, 1997; Montet *et al.*, 1998). Both approaches suggest that there is a hydrophobic channel for $H_2$ between the surface and the Ni ion in the active site.

## 3.3.   The Intramolecular Wire

The intramolecular electron transfer pathway in the hydrogenases involves a series of FeS clusters. Although the two Fe-hydrogenases whose structures are known are different in their cluster composition, the Fe—S clusters in each cases are located at about 10 Å from each other. For example, in the Fe-only hydrogenase from *C. pasteurianum* (Figure 9) in the direction of $H_2$ oxidation, one can imagine electron transfer from the H Cluster to FS4A to FE4B. Then, FE4B is at a junction point and could donate electrons to FS4C or FS2, depending upon which external electron

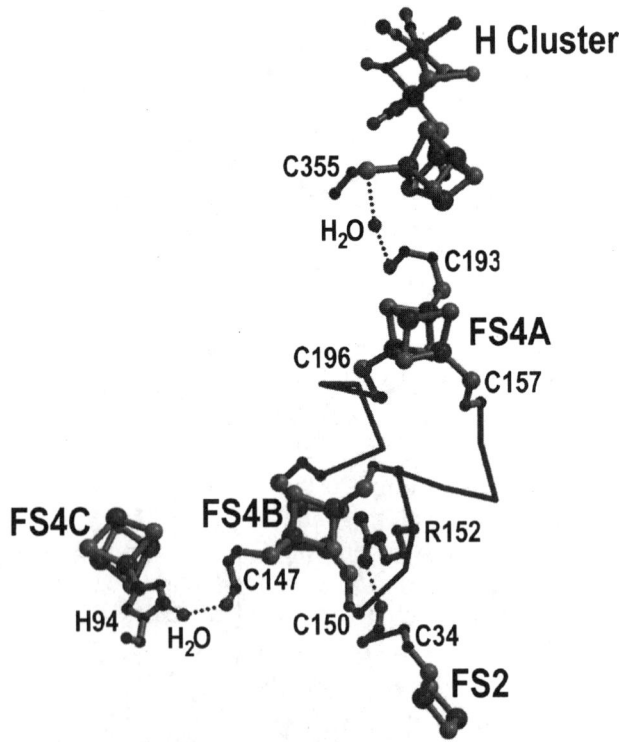

**FIGURE 9.** The proposed electron transfer pathway for the *C. pasteurianum* Fe-only hydrogenase. Intributed by (Peters *et al.*, 1998).

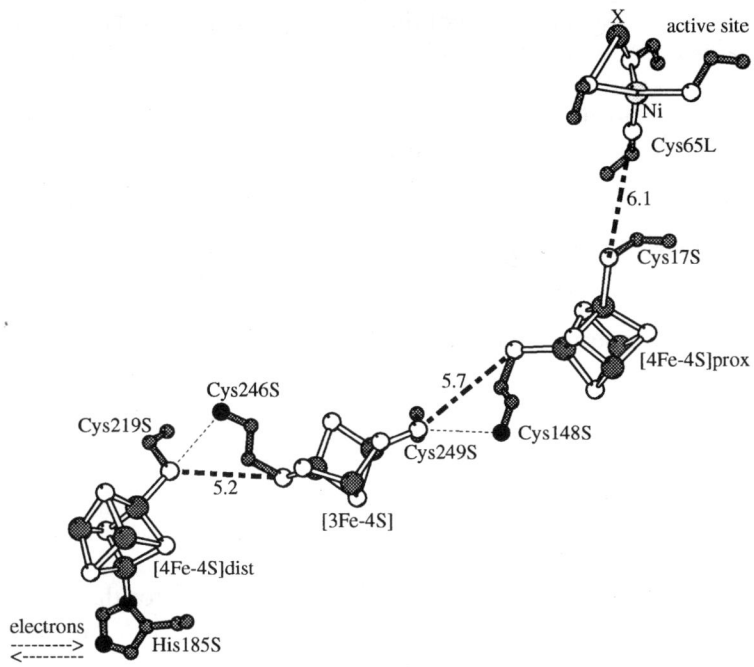

**FIGURE 10.** The proposed electron transfer pathway from the proximal to the distal FeS cluster in the *D. gigas* NiFe hydrogenase. From Figure 9 of (Fontecilla-Camps and Ragsdale, 1999).

donor is present. (Peters *et al.*, 1998). The electron transfer pathway in the *D. desulfuricans* protein would involve electron transfer from the H cluster sequentially to two [4Fe4S] clusters (Nicolet *et al.*, 1999). This protein lacks the terminal 2Fe cluster.

For the NiFe hydrogenase, electrons are transferred from the NiFe cluster in the large subunit to the small subunit, which contains three FeS clusters. The FeS clusters of the *D. gigas* hydrogenase are arranged such that a 3Fe—4S cluster is equidistant from the proximal and the distal (surface) 4Fe4S clusters (Figure 10). This is an unusual arrangement because the redox potential of the 3Fe cluster is 300 mV more positive than that of either of the 4Fe clusters. Can this cluster play a role in electron transfer? The catalytic electron transfer reaction must take place at a redox potential nearly 200 mV more negative than the midpoint potential of this cluster (Pershad *et al.*, 1999). Replacement of this 3Fe cluster by a 4Fe cluster in the *Desulfovibrio fructosovorans* NiFe hydrogenase causes only minor changes in the rates of $H_2$ oxidation or proton reduction (Rousset *et al.*,

1998). Another distinct feature of the NiFe hydrogenase is the ligation of a surface histidine residue to the distal cluster.

The proposed electron transfer pathway from the proximal to the distal FeS cluster is shown in Figure 10. It involves a series of hydrogen bonds over the 16 Å distance between these two 4Fe clusters. The thin dotted lines indicate the most likely pathway; whereas, the thick lines include two "through-space jumps", which are considered to incur a penalty in terms of electron transfer rate. The pathway is predominantly through the Fe—S clusters, the cysteine ligands, and two histidines. The final site at which electrons are transferred to the electron donor or acceptor appears to be histidine 185, the ligand to the distal cluster.

## 3.4. The Intermolecular Wires: How Electrons Enter and Exit Hydrogenase

Several terminal electron acceptors are coupled to $H_2$ oxidation. These include $O_2$, nitrate, sulfate, carbon dioxide, fumarate, and halogenated organics. Electron donors that couple to proton reduction include pyruvate and carbon monoxide. The electron donors or acceptors for the *D. desulfuricans* Fe-only hydrogenase is cytochrome c3 or c6 (Guerlesquin *et al.*, 1994; Verhagen *et al.*, 1994).

In the NiFe hydrogenase, a "crown" of acidic residues surround the solvent-exposed histidine ligand to the distal cluster (Volbeda *et al.*, 1995). It was suggested that this patch could serve as a site for recognition by the redox partner, which is a monoheme or polyheme cytochrome.

## 3.5. Tunnel-diodes and Catalytic Bias

Is there a bias for an enzyme to preferentially catalyze a reaction in a particular direction? Such a situation can be compared with a tunnel diode, which exhibits negative resistance over a certain range of potential bias. One possibility is that the redox enzyme catalyzes the oxidative reaction when the best electron acceptor is present and the reductive reaction when coupled to the strongest electron donor. The other possibility is that the enzyme itself has evolved to be a better catalyst in one direction than the other. For example, a hydrogenase from an organism that needs to take up $H_2$ to fuel its metabolic reactions might be better at oxidation of $H_2$ than at reduction of protons. It is thought that generally the NiFe hydrogenases are poised towards $H_2$ oxidation; whereas, the Fe protein prefers proton reduction.

Voltammetric methods in which the enzyme is adsorbed to an electrode are considered to measure catalytic bias. In these "direct" (unmediated by redox dyes) electrochemical experiments, one can decouple redox catalysis (for example, $H^+$ reduction) from the oxidation of the electron donor. The *Chromatium vinosum* NiFe hydrogenase is thought to be poised to preferentially catalyze $H_2$ oxidation. In agreement with this concept, over 3 pH unit range, the predominant catalytic wave is from the oxidation reaction (Pershad *et al.*, 1999). On the other hand, for the Fe-hydrogenase from *Megasphaera elsdenii*, the reductive reaction generates the predominant catalytic wave, strongly suggesting that this protein preferentially catalyzes proton reduction (Butt *et al.*, 1997). Succinate dehydrogenase apparently is biased toward fumarate reduction; whereas, fumarate reductase does not exhibit a strong tunnel diode effect (Hirst *et al.*, 1996). On the other hand, succinate dehydrogenase can replace a fumarate reductase mutant (Maklashina *et al.*, 1998) and vice versa (Guest, 1981).

How could this catalytic bias be controlled? One possibility is that the proton transfer pathway could contribute to specificity (Peters *et al.*, 1998). Another possibility is that differences in midpoint potential of the FeS clusters (or other redox sites) that constitute the intramolecular wire could be tuned to facilitate one of the two directions of the reaction. For example, these redox sites could best match the midpoint potentials of a particular oxidized or reduced electron carrier (Holm and Sander, 1999). Apparently, a conformational change in succinate dehydrogenase, coupled to the reduction of FAD, is responsible for its catalytic bias for fumarate reduction (Hirst *et al.*, 1996).

## 4. SUMMARY

The two redox catalysts described here can generate very low potential electrons in one direction and perform chemically difficult reductions in the other. The chemical transformations occur at unusual metal clusters. Spectroscopic, crystallographic, and kinetic analyses are converging on answers to how the metals in these clusters are arranged and how they are involved in the chemical and redox steps. The first structure of CO dehydrogenase, which will appear in the next year, will help define a firm chemical basis for future mechanistic studies. In the immediate future, we hope to learn whether the hydride intermediate in hydrogenase or the carbonyl intermediate in CO dehydrogenase bind to the Ni or Fe subsites in these heterometallic clusters. Or perhaps could they be bridged to two metals? Inter- and intramolecular wires have been proposed that connect the catalytic redox machine to proximal redox centers leading eventually to the ultimate redox partners. Elucidating the pathways of electron flow is a priority for the future. There is evidence for molecular channels delivering substrates to the active sites of these enzymes. In the next few years, these channels will be better defined. The products of $CO_2$ and proton reduction are passed to the active sites of other enzymes and, in the case of $H_2$, even passed from one organism to another. In the future, the mechanism of gas transfer will be uncovered. General principles of how these redox reactions are catalyzed are becoming lucid as the reactions are modeled theoretically and experimentally. Proton and $CO_2$ reduction and the generation of C-C bonds from simple precursors are important reactions in industry. $H_2$ could be the clean fuel of the future. Hopefully, the knowledge gained from studies of hydrogenase, CO dehydrogenase, and acetyl-CoA synthase can be used to improve life on earth.

ACKNOWLEDGMENTS. I wish to thank Juan Fontecilla-Camps, Anne Volbeda, Patricia Amara, Yvain Nicolet, and John Peters for supplying figures for this review.

## 5.  REFERENCES

Adams, M. W., 1990, The structure and mechanism of iron-hydrogenases, *Biochim. Biophys. Acta.* **1020**(2):115–145.

Adams, M. W., Eccleston, E., and Howard, J. B., 1989, Iron-sulfur clusters of hydrogenase I and hydrogenase II of Clostridium pasteurianum, *Proc. Natl. Acad. Sci. USA* **86**(13): 4932–4936.

Albracht, S. P. J., 1994, Nickel hydrogenases: In search of the active site, *Biochim. Biophys. Acta.* **1188**(3):167–204.

Amara, P., Volbeda, A., Fontecilla-Camps, J. C., and Field, M. J., 1999, A Hybrid Density Functional Theory/Molecular Mechanics Study of Nickel-Iron Hydrogenase: Investigation of the Active Site Redox States, *J. Am. Chem. Soc.* **121**(18):4468–4477.

Anderson, M. E., DeRose, V. J., Hoffman, B. M., and Lindahl, P. A., 1993, Identification of a cyanide binding site in CO dehydrogenase from *Clostridium thermoaceticum* using EPR and ENDOR spectroscopies, *J. Am. Chem. Soc.* **115**:12204–12205.

Aono, S., Nakajima, H., Saito, K., and Okada, M., 1996, A novel heme protein that acts as a carbon monoxide-dependent transcriptional activator in Rhodospirillum rubrum, *Biochem. Biophys. Res. Commun.* **228**(3):752–756.

Bagley, K. A., Duin, E. C., Roseboom, W., Albracht, S. P. J., and Woodruff, W. H., 1995, Infrared-detectable groups sense changes in charge density on the nickel center in hydrogenase from *Chromatium vinosum, Biochem.* **34**(16):5527–5535.

Bagyinka, C., Whitehead, J. P., and Maroney, M. J., 1993, An X-ray absorption spectroscopic study of nickel redox chemistry in hydrogenase, *J. Am. Chem. Soc.* **115**:3576–3585.

Barondeau, D. P., and Lindahl, P. A., 1997, Methylation of carbon monoxide dehydrogenase from Clostridium thermoaceticum and mechanism of acetyl coenzyme A synthesis, *J. Am. Chem. Soc.* **119**(17):3959–3970.

Bartholomew, G. W., and Alexander, M., 1979, *Appl. Environ. Microbiol.* **37**:932–937.

Butt, J. N., Filipiak, M., and Hagen, W. R., 1997, Direct electrochemistry of Megasphaera elsdenii iron hydrogenase—Definition of the enzyme's catalytic operating potential and quantitation of the catalytic behaviour over a continuous potential range, *Eur. J. Biochem.* **245**(1):116–122.

Chabriere, E., Charon, M.-H., Volbeda, A., Pieulle, L., Hatchikian, E. C., and Fontecilla-Camps, J.-C., 1999, Crystal structures of the key anaerobic enzyme pyruvate:ferredoxin oxido-reductase, free and in complex with pyruvate, *Nat. Struct. Biol.* **6**(2):182–190.

Conrad, R., Ed. 1995, *Soil microbial processes involved in production and consumption of atmospheric trace gases.* Advances in microbial ecology. New York, Plenum Press.

deLacey, A. L., Hatchikian, E. C., Volbeda, A., Frey, M., FontecillaCamps, J. C., and Fernandez, V. M., 1997, Infrared spectroelectrochemical characterization of the [NiFe] hydrogenase of Desulfovibrio gigas, *J. Am. Chem. Soc.* **119**(31):7181–7189.

DeRose, V. J., Telser, J., Anderson, M. E., Lindahl, P. A., and Hoffman, B. M., 1998, A multinuclear ENDOR study of the C-cluster in CO dehydrogenase from Clostridium thermoaceticum: Evidence for HxO and histidine coordination to the [Fe4S4] center, *J. Am. Chem. Soc.* **120**(34):8767–8776.

Diekert, G., Hansch, M., and Conrad, R., 1984, Acetate synthesis from 2 $CO_2$ in acetogenic bacteria: is carbon monoxide an intermediate?, *Arch. Microbiol.* **138**:224–228.

Dole, F., Fournel, A., Magro, V., Hatchikian, E. C., Bertrand, P., and Guigliarelli, B., 1997, Nature and electronic structure of the Ni—X dinuclear center of Desulfovibrio gigas hydrogenase. Implications for the enzymatic mechanism, *Biochem.* **36**(25):7847–7854.

Dore, J., Morvan, B., Rieu-Lesme, F., Goderel, I., Gouet, P., and Pochart, P., 1995, Most probable number enumeration of H2-utilizing acetogenic bacteria from the digestive tract of animals and man, *FEMS Microbiol. Lett.* **130**(1):7–12.

Eitinger, T., and Friedrich, B., 1997, Microbial nickel transport and incorporation into hydrogenases. *Transition Metals in Microbial Metabolism.* G. Winkelmann and C. Carrano. London, Harwood Academic Publishers: 235–256.

Fan, C., Gorst, C. M., Ragsdale, S. W., and Hoffman, B. M., 1991, Characterization of the Ni—Fe—C complex formed by reaction of carbon monoxide with the carbon monoxide dehydrogenase from *Clostridium thermoaceticum* by Q-band ENDOR, *Biochem.* **30**:431–435.

Farmer, P. J., Reibenspies, J. H., Lindahl, P. A., and Darensbourg, M. Y., 1993, *J. Am. Chem. Soc.* **115**:4665–4574.

Fontecilla-Camps, J.-C., and Ragsdale, S. W., 1999, Nickel-iron-sulfur active sites: hydrogenase and CO dehydrogenase. *Advances in Inorganic Chemistry.* A. G. Sykes and R. Cammack. San Diego, Academic Press, *Inc.* **47**:283–333.

Fox, J. D., He, Y. P., Shelver, D., Roberts, G. P., and Ludden, P. W., 1996, Characterization of the region encoding the CO-induced hydrogenase of Rhodospirillum rubrum, *J. Bacteriol.* **178**(21):6200–6208.

Garcin, E., Vernede, X., Hatchikian, E., Volbeda, A., Frey, M., and Fontecilla-Camps, J., 1999, The crystal structure of a reduced [NiFeSe] hydrogenase provides an image of the activated catalytic center, *Structure* **7**(5):557–566.

Geierstanger, B. H., Prasch, T., Griesinger, C., Hartmann, G., Buurman, G., and Thauer, R. K., 1998, Catalytic mechanism of the metal-free hydrogenase from methanogenic archaea: Reversed stereospecificity of the catalytic and noncatalytic reaction, *Angew. Chem. Int. Ed.* **37**(23):3300–3303.

Goldman, C. M., and Mascharak, P. K., 1995, Reactions of H2 with the Nickel Site(s) of the [FeNi] and [FeNiSe] Hydrogenases: What do the Model Complexes Suggest?, *Comments Inorg. Chem.* **18**:1–25.

Grahame, D. A., and Demoll, E., 1995, Substrate and accessory protein requirements and thermodynamics of acetyl-CoA synthesis and cleavage in *Methanosarcina barkeri*, *Biochem.* **34**(14):4617–4624.

Guerlesquin, F., Dolla, A., and Bruschi, M., 1994, Involvement of electrostatic interactions in cytochrome c complex formations, *Biochimie* **76**(6):515–23.

Guest, J. R., 1981, Partial replacement of succinate dehydrogenase function by phage- and plasmid-specified fumarate reductase in *Escherichia coli*, *J. Gen. Microbiol.* **122**:171–179.

Happe, R. P., Roseboom, W., and Albracht, S. P., 1999, Pre-steady-state kinetics of the reactions of [NiFe]-hydrogenase from Chromatium vinosum with H2 and CO, *Eur. J. Biochem.* **259**(3):602–8.

Happe, R. P., Roseboom, W., Pierik, A. J., Albracht, S. P. J., and Bagley, K. A., 1997, Biological activation of hydrogen, *Nature* **385**(6612):126–126.

Heo, J., Staples, C. R., and Ludden, P. W., 1999, *Rhodospirillum rubrum* CO dehydrogenase. Part 2. Spectroscopic investigation and assignment of spin-spin coupling signals, *J. Am. Chem. Soc.*: in press.

Higuchi, Y., Ogata, H., Miki, K., Yasuoka, N., and Yagi, T., 1999, Removal of the bridging ligand atom at the Ni—Fe active site of [NiFe] hydrogenase upon reduction with H2, as revealed by X-ray structure analysis at 1.4 A resolution [In Process Citation], *Structure Fold. Des.* **7**(5):549–56.

Higuchi, Y., Yagi, T., and Yasuoka, N., 1997, Unusual ligand structure in Ni-Fe active center and an additional Mg site in hydrogenase revealed by high resolution X-ray structure analysis, *Structure* **5**:1671–1680.

Hirst, J., Sucheta, A., Ackrell, B. A. C., and Armstrong, F. A., 1996, Electrocatalytic Voltammetry of Succinate Dehydrogenase: Direct Quantification of the Catalytic Properties of a Complex Electron-Transport Enzyme, *J. Am. Chem. Soc.* **118**(21):5031–5038.

Holm, L., and Sander, C., 1999, Protein folds and families: sequence and structure alignments, *Nucleic Acids Res.* **27**(1):244–247.

Hsu, H. F., Koch, S. A., Popescu, C. V., and Munck, E., 1997, Chemistry of iron thiolate complexes with CN- and CO. Models for the [Fe(CO)(CN)(2)] structural unit in Ni—Fe hydrogenase enzymes, *J. Am. Chem. Soc.* **119**(35):8371–8372.

Hsu, T., Daniel, S. L., Lux, M. F., and Drake, H. L., 1990, Biotransformations of carboxylated aromatic compounds by the acetogen Clostridium thermoaceticum: generation of growth-supportive CO2 equivalents under CO2-limited conditions, *J. Bacteriol.* **172**(1):212–217.

Hsu, T., Lux, M. F., and Drake, H. L., 1990, Expression of an aromatic-dependent decarboxylase which provides growth-essential $CO_2$ equivalents for the acetogenic (Wood) pathway of *Clostridium thermoaceticum*, *J. Bacteriol.* **172**:5901–5907.

Hu, Z. G., Spangler, N. J., Anderson, M. E., Xia, J. Q., Ludden, P. W., Lindahl, P. A., and Münck, E., 1996, Nature of the C-cluster in Ni-containing carbon monoxide dehydrogenases, *J. Am. Chem. Soc.* **118**(4):830–845.

Huber, C., and Wachtershauser, G., 1997, Activated acetic acid by carbon fixation on (Fe,Ni)S under primordial conditions [see comments], *Science* **276**(5310): 245–7.

Jackson, B. E., Bhupathiraju, V. K., Tanner, R. S., Woese, C. R., and McInerney, M. J., 1999, Syntrophus aciditrophicus sp. nov., A new anaerobic bacterium that degrades fatty acids and benzoate in syntrophic association with hydrogen-using microorganisms [In Process Citation], *Arch. Microbiol.* **171**(2):107–14.

Kröger, A., 1978, *Biochim. Biophys. Acta.* **505**:129–145.

Kruger, H.-J., and Holm, R. H., 1990, *J. Am. Chem. Soc.* **112**:2955–2963.

Kumar, M., Lu, W.-P., Liu, L., and Ragsdale, S. W., 1993, Kinetic evidence that CO dehydrogenase catalyzes the oxidation of CO and the synthesis of acetyl-CoA at separate metal centers, *J. Am. Chem. Soc.* **115**:11646–11647.

Kumar, M., Lu, W.-P., and Ragsdale, S. W., 1994, Binding of carbon disulfide to the site of acetyl-CoA synthesis by the nickel-iron-sulfur protein, CO dehydrogenase, from *Clostridium thermoaceticum*, *Biochem.* **33**:9769–9777.

Kumar, M., and Ragsdale, S. W., 1992, Characterization of the CO binding site of carbon monoxide dehydrogenase from *Clostridium thermoaceticum* by infrared spectroscopy, *J. Am. Chem. Soc.* **114**:8713–8715.

Lai, C. H., Lee, W. Z., Miller, M. L., Reibenspies, J. H., Darensbourg, D. J., and Darensbourg, M. Y., 1998, Responses of the Fe(CN)(2)(CO) unit to electronic changes as related to its role in [NiFe]hydrogenase, *J. Am. Chem. Soc.* **120**(39):10103–10114.

Lenz, O., and Friedrich, B., 1998, A novel multicomponent regulatory system mediates H-2 sensing in Alcaligenes eutrophus, *Proc. Natl. Acad. Sci. USA* **95**(21):12474–12479.

Liaw, W. F., Horng, Y. C., Ou, D. S., Ching, C. Y., Lee, G. H., and Peng, S. M., 1997, Distorted square planar Ni(II)-chalcogenolate carbonyl complexes [Ni(CO)(SPh)(n)(SePh)(3-n)]-(-) (n = 0, 1, 2): Relevance to the nickel site in CO dehydrogenases and [NiFeSe] hydrogenase, *J. Am. Chem. Soc.* **119**(39):9299–9300.

Lindahl, P. A., Münck, E., and Ragsdale, S. W., 1990, CO dehydrogenase from *Clostridium thermoaceticum*: EPR and electrochemical studies in CO₂ and argon atmospheres, *J. Biol. Chem.* **265**:3873–3879.

Lindahl, P. A., Ragsdale, S. W., and Münck, E., 1990, Mössbauer studies of CO dehydrogenase from *Clostridium thermoaceticum*, *J. Biol. Chem.* **265**:3880–3888.

Maklashina, E., Berthold, D. A., and Cecchini, G., 1998, Anaerobic expression of Escherichia coli succinate dehydrogenase: functional replacement of fumarate reductase in the respiratory chain during anaerobic growth, *J. Bacteriol.* **180**(22):5989–5996.

Maroney, M. J., Davidson, G., Allan, C. B., and Figlar, J., 1998, The structure and function of nickel sites in metalloproteins, *Structure and Bonding* **92**:1–65.

Maynard, E. L., and Lindahl, P. A., 1999, Kinetic mechanism of acetyl-CoA synthesis catalyzed by CO dehydrogenase/acetyl-CoA synthase: preliminary evidence for a molecular tunnel, *J. Biol. Inorg. Chem.* **74**:227.

Menon, S., and Ragsdale, S. W., 1996, Evidence that carbon monoxide is an obligatory intermediate in anaerobic acetyl-CoA synthesis, *Biochem.* **35**(37):12119–12125.

Menon, S., and Ragsdale, S. W., 1997, Mechanism of the *Clostridium thermoaceticum* pyruvate:ferredoxin oxidoreductase: Evidence for the common catalytic intermediacy of the hydroxyethylthiamine pyropyrophosphate radical, *Biochem.* **36**:8484–8494.

Menon, S., and Ragsdale, S. W., 1998, Role of the [4Fe—4S] cluster in reductive activation of the cobalt center of the corrinoid iron-sulfur protein from *Clostridium thermoaceticum* during acetyl-CoA synthesis, *Biochem.* **37**(16):5689–5698.

Menon, S., and Ragsdale, S. W., 1999, The role of an iron-sulfur cluster in an enzymatic methylation reaction: methylation of CO dehydrogenase/acetyl-CoA synthase by the methylated corrinoid iron-sulfur protein, *J. Biol. Chem.* **274**(17):11513–11518.

Meyer, O., 1985, *Microbial gas metabolism, mechanistic, metabolic, and biotechnological aspects*. P. R. K. and D. C. S. London, Academic press: 131–151.

Montet, Y., Amara, P., Volbeda, A., Vernede, X., Hatchikian, E. C., Field, M. J., Frey, M., and Fontecilla-Camps, J. C., 1997, Gas access to the active site of Ni—Fe hydrogenases probed by X-ray crystallography and molecular dynamics [letter], *Nat. Struct. Biol.* **4**(7):523–526.

Montet, Y., Garcin, E., Volbeda, A., Hatchikian, E. C., Frey, M., and Fontecilla-Camps, J. C., 1998, Structural basis for the catalytic mechanism of NiFe hydrogenase, *Pure and Applied Chemistry* **70**(1):25–31.

Nicolet, Y., Piras, C., Legrand, P., Hatchikian, C. E., and Fontecilla-Camps, J. C., 1999, Desulfovibrio desulfuricans iron hydrogenase: the structure shows unusual coordination to an active site Fe binuclear center, *Structure* **7**(1):13–23.

Odom, J. M., and Peck, H. D., Jr., 1984, Hydrogenase, electron-transfer proteins, and energy coupling in the sulfate-reducing bacteria Desulfovibrio, *Annu. Rev. Microbiol.* **38**:551–592.

Olah, G. A., Hartz, N., Rasul, G., and Prokash, G. K. S., 1995, Electrophilic substitution of methane revisited, *J. Am. Chem. Soc.* **117**:1336–1343.

Pavlov, M., Siegbahn, P. E. M., Blomberg, M. R. A., and Crabtree, R. H., 1998, Mechanism of H—H activation by nickel-iron hydrogenase, *J. Am. Chem. Soc.* **120**(3):548–555.

Pershad, H. R., Duff, J. L., Heering, H. A., Duin, E. C., Albracht, S. P., and Armstrong, F. A., 1999, Catalytic electron transport in chromatium vinosum [NiFe]-hydrogenase: application of voltammetry in detecting redox-active centers and establishing that hydrogen oxidation is very fast even at potentials close to the reversible H(+)/H(2) value [In Process Citation], *Biochem.* **38**(28):8992–8999.

Peters, J. W., Lanzilotta, W. N., Lemon, B. J., and Seefeldt, L. C., 1998, X-ray crystal structure of the Fe-only hydrogenase (Cpl) from Clostridium pasteurianum to 1.8 angstrom resolution, *Science* **282**(5395):1853–1858.

Pierik, A. J., Hulstein, M., Hagen, W. R., and Albracht, S. P., 1998, A low-spin iron with CN and CO as intrinsic ligands forms the core of the active site in [Fe]-hydrogenases, *Eur. J. Biochem.* **258**(2):572–578.

Pierik, A. J., Schmelz, M., Lenz, O., Friedrich, B., and Albracht, S. P. J., 1998, Characterization of the active site of a hydrogen sensor from Alcaligenes eutrophus, *FEBS Lett.* **438**(3):231–235.

Ragsdale, S. W., 1997, The Eastern and Western branches of the Wood/Ljungdahl pathway: how the East and West were won, *BioFactors* **9**:1–9.

Ragsdale, S. W., Clark, J. E., Ljungdahl, L. G., Lundie, L. L., and Drake, H. L., 1983, Properties of purified carbon monoxide dehydrogenase from *Clostridium thermoaceticum* a nickel, iron-sulfur protein, *J. Biol. Chem.* **258**:2364–2369.

Ragsdale, S. W., Kumar, M., Seravalli, J., Qiu, D., and Spiro, T. G., 1996, Anaerobic carbon monoxide dehydrogenase. *Microbial Growth on C1 compounds*. M. E. Lidstrom and F. R. Tabita. Dordecht, Kluwer Publications: 191–196.

Ragsdale, S. W., Kumar, M., Zhao, S., Menon, S., Seravalli, J., and Doukov, T., 1998, Discovery Of A Bio-Organometallic Reaction Sequence Involving Vitamin $B_{12}$ And Nickel/Iron-Sulfur Clusters. *Vitamin $B_{12}$ and $B_{12}$-Proteins*. B. Krautler. Weinheim, Germany, Wiley-VCH: 167–177.

Ragsdale, S. W., Lindahl, P. A., and Münck, E., 1987, Mössbauer, EPR, and optical studies of the corrinoid/Fe—S protein involved in the synthesis of acetyl-CoA by *Clostridium thermoaceticum*, *J. Biol. Chem.* **262**:14289–14297.

Ragsdale, S. W., and Ljungdahl, L. G., 1984, Hydrogenase from *Acetobacterium woodii*, *Arch. Microbiol.* **139**:361–365.

Ragsdale, S. W., Ljungdahl, L. G., and DerVartanian, D. V., 1982, EPR evidence for nickel substrate interaction in carbon monoxide dehydrogenase from *Clostridium thermoaceticum*, *Biochem. Biophys. Res. Commun.* **108**:658–663.

Ragsdale, S. W., and Riordan, C. G., 1996, The Role Of Nickel In Acetyl-CoA Synthesis By The Bifunctional Enzyme CO Dehydrogenase/Acetyl-CoA Synthase: Enzymology And Model Chemistry, *J. Bioinorganic Chemistry* **1**:489–493.

Ragsdale, S. W., and Wood, H. G., 1985, Acetate biosynthesis by acetogenic bacteria: evidence that carbon monoxide dehydrogenase is the condensing enzyme that catalyzes the final steps of the synthesis, *J. Biol. Chem.* **260**:3970–3977.

Ragsdale, S. W., Wood, H. G., and Antholine, W. E., 1985, Evidence that an iron-nickel-carbon complex is formed by reaction of CO with the CO dehydrogenase from *Clostridium thermoaceticum*, *Proc. Natl. Acad. Sci. USA* **82**:6811–6814.

Reeve, J. N., Nolling, J., Morgan, R. M., and Smith, D. R., 1997, Methanogenesis: Genes, genomes, and who's on first?, *J. Bacteriol.* **179**(19):5975–5986.

Roberts, L. M., and Lindahl, P. A., 1994, Analysis of oxidative titrations of *Desulfovibrio gigas* hydrogenase; Implications for the catalytic mechanism, *Biochem.* **33**:14339–14350.

Rousset, M., Montet, Y., Guigliarelli, B., Forget, N., Asso, M., Bertrand, P., FontecillaCamps, J. C., and Hatchikian, E. C., 1998, [3Fe—4S] to [4Fe—4S] cluster conversion in Desulfovibrio fructosovorans [NiFe] hydrogenase by site-directed mutagenesis, *Proc. Natl. Acad. Sci. USA* **95**(20):11625–11630.

Russell, M. J., Daia, D. E., and Hall, A. J., 1998, The emergence of life from FeS bubbles at alkaline hot springs in an acid ocean. *Thermophiles: The keys to molecular evolution and the origin of life?* M. W. W. Adams, L. G. Ljungdahl, and J. Wiegel. Washington, D.C., Taylor and Francis.

Russell, W. K., Stalhandske, C. M. V., Xia, J. Q., Scott, R. A., and Lindahl, P. A., 1998, Spectroscopic, redox, and structural characterization of the Ni-labile and nonlabile forms of the acetyl-CoA synthase active. Site of carbon monoxide dehydrogenase, *J. Am. Chem. Soc.* **120**(30):7502–7510.

Schulman, M., Ghambeer, R. K., Ljungdahl, L. G., and Wood, H. G., 1973, Total synthesis of acetate from $CO_2$. VII. Evidence with *Clostridium thermoaceticum* that the carboxyl of acetate is derived from the carboxyl of pyruvate by transcarboxylation and not by fixation of $CO_2$, *J. Biol. Chem.* **248**:6255–6261.

Seravalli, J., Kumar, M., Lu, W.-P., and Ragsdale, S. W., 1997, Mechanism of carbon monoxide oxidation by the carbon monoxide dehydrogenase/acetyl-CoA synthase from *Clostridium thermoaceticum*: Kinetic characterization of the intermediates, *Biochem.* **36**:11241–11251.

Seravalli, J., Kumar, M., Lu, W. P., and Ragsdale, S. W., 1995, Mechanism of CO oxidation by carbon monoxide dehydrogenase from Clostridium thermoaceticum and its inhibition by anions, *Biochem.* **34**(24):7879–7888.

Seravalli, J., and Ragsdale, S. W., 2000, Channeling of Carbon Monoxide During Anaerobic Carbon Dioxide Fixation, *Biochemistry* **39**:1274–1277.

Shanmugasundaram, T., Ragsdale, S. W., and Wood, H. G., 1988, Role of carbon monoxide dehydrogenase in acetate synthesis by the acetogenic bacterium, *Acetobacterium woodii*, *BioFactors* **1**:147–152.

Shelver, D., Kerby, R. L., He, Y. P., and Roberts, G. P., 1997, CooA, a CO-sensing transcription factor from Rhodospirillum rubrum, is a CO-binding heme protein, *Proc. Natl. Acad. Sci. USA* **94**(21):11216–11220.

Sorgenfrei, O., Klein, A., and Albracht, S. P. J., 1993, *FEBS Lett.* **332**:291–297.

Staples, C. R., Heo, J., Spangler, N. J., Kerby, R. L., Roberts, G. P., and Ludden, P. W., 1999, *Rhodospirillum rubrum* CO dehydrogenase. Part 1. Spectroscopic studies of CODH variant C531A indicate the presence of a binuclear [FeNi] Cluster, *J. Am. Chem. Soc.*: in press.

Stephenson, M., and Stickland, L. H., 1931, XXVII. Hydrogenase: a bacterial enzyme activating molecular hydrogen. I. The properties of the enzyme, *Biochem. J.* **25**(205–214).

Thauer, R. K., 1998, Biochemistry of methanogenesis: a tribute to Marjory Stephenson, *Microbiology Uk* **144**:2377–2406.

Van der Zwaan, J. W., Albracht, S. P. J., Fontijn, R. D., and Slater, E. C., 1985, Monovalent nickel in hydrogenase from Chromatium vinosum. Light sensitivity and evidence for direct interaction with hydrogen, *FEBS Lett.* **179**:271–277.

Verhagen, M. F., Wolbert, R. B., and Hagen, W. R., 1994, Cytochrome c553 from Desulfovibrio vulgaris (Hildenborough). Electrochemical properties and electron transfer with hydrogenase, *Eur. J. Biochem.* **221**(2):821–829.

Verma, A., Hirsch, D. J., Glatt, C. E., Ronnett, G. V., and Snyder, S. H., 1993, Carbon monoxide: a putative neural messenger, *Science* **259**:381–384.

Volbeda, A., Charon, M. H., Piras, C., Hatchikian, E. C., Frey, M., and Fontecilla-Camps, J. C., 1995, Crystal structure of the nickel-iron hydrogenase from *Desulfovibrio gigas*, *Nature* **373**(6515):580–587.

Volbeda, A., Garcia, E., Piras, C., deLacey, A. L., Fernandez, V. M., Hatchikian, E. C., Frey, M., and FontecillaCamps, J. C., 1996, Structure of the [NiFe] hydrogenase active site: Evidence for biologically uncommon Fe ligands, *J. Am. Chem. Soc.* **118**(51):12989–12996.

Whitehead, J. P., Gurbiel, R. J., Bagyinka, C., Hoffman, B. M., and Maroney, M. J., 1993, *J. Am. Chem. Soc.* **115**:5629–5635.

Williams, R. J. P., 1995, Purpose of proton pathways, *Nature* **376**:643.

Wolin, M. J., and Miller, T. L., 1994, Acetogenesis from $CO_2$ in the human colonic ecosystem. *Acetogenesis*. H. L. Drake. New York, Chapman and Hall: 365–385.

Yagi, T., 1959, Enzymic oxidation of carbon monoxide, *Biochim. Biophys. Acta.* **30**:194–195.

*Chapter 15*

# Cytochrome $cd_1$ Nitrite Reductase Structure Raises Interesting Mechanistic Questions

Stuart J. Ferguson and Vilmos Fülöp

## 1. INTRODUCTION

There are several types of nitrite reduction reaction in biology and the use of the common name "nitrite reductase" for the enzymes catalysing these reactions causes endless confusion. Thus we begin this article by outlining the different types and functions of the nitrite reductases before focusing on the structure/function relationships for one type of enzyme, cytochrome $cd_1$ nitrite reductase.

In a plant, nitrite reductase catalyses reduction of its substrate to ammonia, which is subsequently incorporated into cell material. This type of nitrite reductase is classified as an assimilatory enzyme. The plant enzyme is found in the chloroplasts and acquires reductant from ferredoxin which is in turn reduced by the action of the photosystems that derive electrons

**STUART J. FERGUSON**   Department of Biochemistry and Oxford Centre for Molecular Sciences, University of Oxford, South Parks Road, Oxford OX1 3QU, U.K.   **VILMOS FÜLÖP**   Department of Biological Sciences, University of Warwick, Gibbet Hill Road, Coventry CV4 7AL, U.K.

*Subcellular Biochemistry, Volume 35: Enzyme-Catalyzed Electron and Radical Transfer*, edited by Holzenburg and Scrutton. Kluwer Academic / Plenum Publishers, New York, 2000.

from water. This plant nitrite reductase contains siroheme at the active site (Cole, 1988).

Many species of bacteria also have an assimilatory nitrite reductase which is located in the cytoplasm. There is relatively little known about such enzymes but the electron donor is throught to be NADPH and the active site again has siroheme (Cole, 1988). The assimilatory nitrite reductases of both plants and bacteria use nitrite that is provided as the product of the assimilatory nitrate reductases. Nitrate is a very common natural N source for plant and bacterial growth.

In various species of bacteria several different types of non-assimilatory nitrite reductases are found. *Escherichia coli* has a cytoplasmic NAD(P)H-dependent enzyme whose role seems to be detoxification of nitrite. This type of enzyme, coded for by the *nirB* gene, also contains siroheme as the redox active catalytic center (Cole, 1988). Additionally in *E. coli*, and expressed under different conditions to the cytoplasmic enzyme, is a periplasmic nitrite reductase that catalyses formation of ammonia from nitrite (Cole, 1988). This enzyme has five *c*-type (Figure 1) hemes per polypeptide chain; one of these hemes, the catalytic site, has the unique CXXCK sequence as its attachment site (Einsle *et al.*, 1999). Electrons reach this type of nitrite reductase, which is fairly widely distributed amongst the microbial world, from the cytoplasmic membrane electron transfer chain. The exact electron donor partner from such chains for this type of nitrite reductase is unknown (Berks *et al.*, 1995).

A further type of bacterial nitrite reductase, also located in the periplasm, which is separated from the external medium by the cell wall and from the cytoplasm by the cytoplasmic membrane, produces nitric oxide as reaction product and thus participates in the overall respiratory process of denitrification (Averill, 1996; Berks *et al.*, 1995; Richardson and Watmough, 1999; Zumft, 1997). There are two distinct types of enzyme catalysing this reaction, both of which receive electrons from the underlying electron transport system in the cytoplasmic membrane. One of the nitrite reductases of denitrification is a trimeric copper enzyme containing six copper atoms in total. Three of these coppers can be classified as type 1 copper whilst the remaining three are at active sites and can be considered as type 2 copper. Crystal structures and mechanistic proposals are available (Dodd *et al.*, 1997 and 1998; Murphy *et al.*, 1997). The second type of nitrite reductase involved in denitrification is known as cytochrome $cd_1$. This name reflects the presence in the enzyme of both a *c*-type cytochrome center and a $d_1$ heme center. The former center contains a protoheme IX (or heme *b*) molecule that is covalently bound to the protein by the addition of two cysteine thiol groups to the two vinyl groups of the heme (Figure 1). Strictly speaking, there is no such species as *c*-type heme, but hereafter we will use

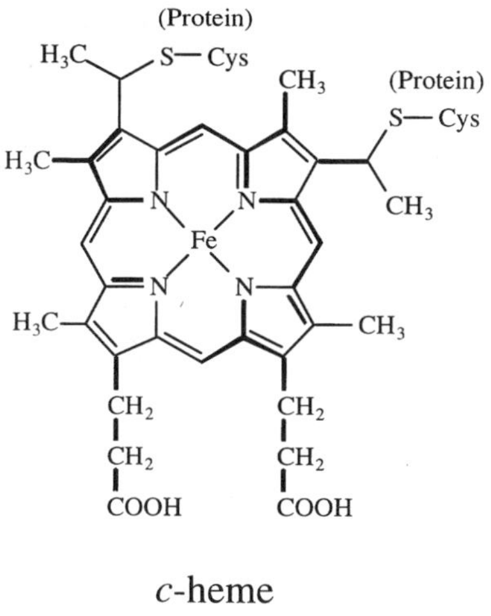

FIGURE 1. The heme group in a $c$-type cytochrome showing the covalent attachment.

$c$-heme

the term to mean the $c$-type cytochrome center in the protein. The $d_1$ heme center comprises the non-covalently bound heme $d_1$ (Figure 2) which is unique to this type of enzyme. It is clear from Figure 2 that this heme is tailored (note the two carbonyl groups) in very particular ways relative to the protoheme IX. The possible reasons for this tailoring, which requires a special biosynthetic effort on the part of the cells catalysing this reaction (Zumft, 1997), is addressed later. Except where comparison with the copper protein is appropriate, this article is concerned with the structure/function relationships of the $cd_1$ type of enzyme.

Cytochrome $cd_1$ needs three things to reduce nitrite to nitric oxide, substrates nitrite and protons plus electrons. The first two are simply supplied from the environment but where do the electrons come from? Cytochrome $cd_1$ derives its electrons from the electron transport system in the cytoplasmic membranes of bacteria. Thus, for example, electrons can originate from NADH and pass via NADH dehydrogenase, ubiquinone/ubiquinol and the cytochrome $bc_1$ complex to nitrite reductase (Berks *et al.*, 1995; Zumft, 1997). It is known that the cytochrome $bc_1$ complex is used because electron transfer from physiological donors, e.g. NADH, to nitrite reductase is blocked by specific inhibitors, e.g. myxothiazol, of this complex. An important issue arises when we consider how electrons are transferred from the cytochrome $bc_1$ complex to cytochrome $cd_1$, which, recall, is a water-soluble protein located in the periplasm. The structure of the cytochrome $bc_1$ complex (determined for the mitochondrial protein but we can assume that the bacterial counterparts are similarly organised) shows that the

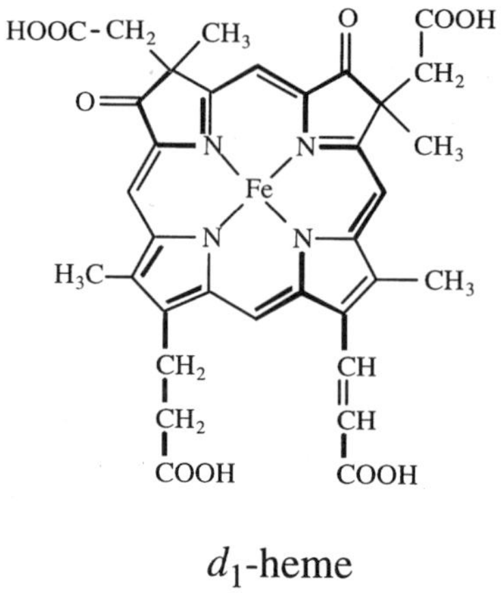

$d_1$-heme

**FIGURE 2.** The $d_1$ heme group.

cytochrome $c_1$ will project some way into the periplasm, thus raising the possibility that direct electron transfer from cytochrome $c_1$ to nitrite reductase can occur. It is more usual to think that low molecular weight mono heme $c$-type cytochromes or cupredoxins are involved. We will return to this question later.

## 2. STRUCTURE OF *PARACOCCUS PANTOTROPHUS* CYTOCHROME $cd_1$

If we depart from the premise that nitrite reduction requires a protein that can acquire electrons, bind nitrite and protons to yield nitric oxide and water as products, we might imagine that a monomeric protein containing a catalytic center with a suitable metal binding site for the reaction would suffice. However, the structure of *Paracoccus pantotrophus* (fomerly *Thiospharea pantotropha*, Rainey *et al.*, 1999) cytochrome $cd_1$, the first of this type to have its crystal structure solved (Fülöp *et al.*, 1995), shows that the enzyme is not so simple. It is a homodimer and each monomer contains both a $c$-type cytochrome center and a $d_1$ heme center (Figure 3). The $d_1$ heme (Figure 2) is unique to this class of enzyme and on that basis alone might be expected to be the catalytic site. $c$-type cytochrome centers, which are defined by the covalent attachment of the heme are usually, but not always, involved in electron transfer. Thus if nothing functional was known about the enzyme, which was in fact close to the case for the enzyme from *P. pantotrophus* when the structure was obtained, one might reasonably surmise that the $c$-type cytochrome center was the point of entry of

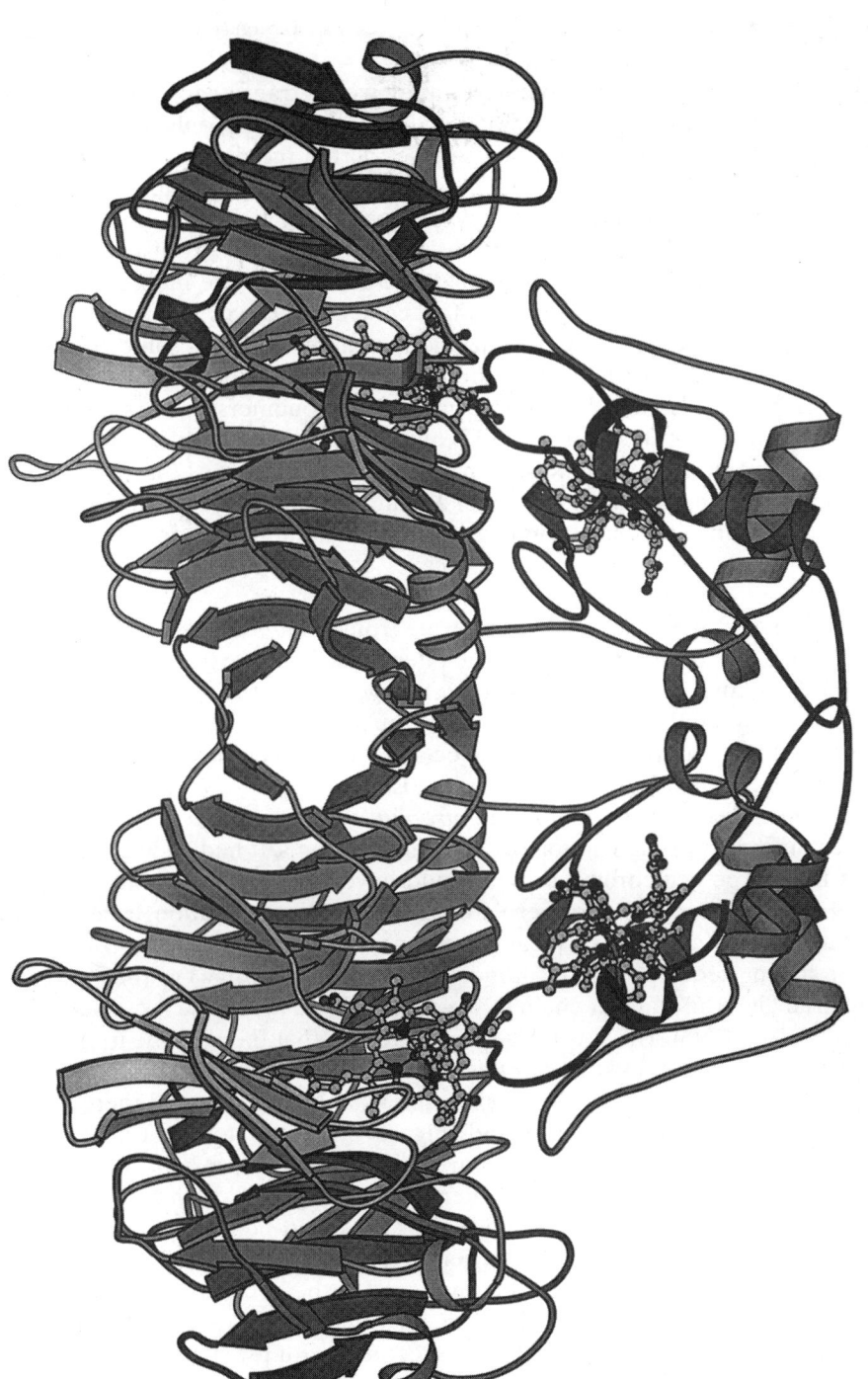

**FIGURE 3.** The X-ray crystal structure of the oxidised state of cytochrome $cd_1$ nitrite reductase from *P. pantotrophus*. (Drawn from PDB entry 1qks)

electrons into the enzyme from which they would transfer to the $d_1$ heme. This transfer is expected, but not strictly proven, to be only within one monomer because the interheme distances across the dimer interface are of the order of 30 Å or more. Such a distance is generally thought to be incompatible with biological activity because predicted rates of electron transfer over such distances are of the order of one or two events per hour or more, scarcely compatible with enzyme turnover on the millisecond or second timescale (Page *et al.*, 1999). This conclusion leaves unanswered, of course, the question of why the enzyme is a dimer. There is some crystallographic evidence that the two monomers are not independent of one another because within the dimer each of the monomers, called A and B (Fülöp *et al.*, 1995; Baker *et al.*, 1997) always has a slightly different structure. For example in the oxidised enzyme Nδ1 of histidine 17 is hydrogen bonded to the main chain carbonyl of alanine 101 and two a water whereas in the B subunit the corresponding histidine residue makes only a hydrogen bond only to a water molecule (Baker *et al.*, 1997). More differences between the two monomers will be described later.

The structure of the oxidised enzyme from *P. pantotrophus* shows that within one monomer the edge to edge inter-heme distance is circa 12 Å (the Fe—Fe distance is 20 Å). This distance suggests that the rate of electron transfer between the hemes should be on the microsecond timescale, unless electron transport is obligatorily associated with other slower events, for example rearrangement of chemical bonds.

The oxidised structure of *P. pantotrophus* cytochrome $cd_1$ provided surprises in respect of the ligation of the heme Fe atoms. It had been expected that the $c$-type cytochrome center would have His/Met coordination, but His/His is observed. The former is the more usual coordination, especially at the high potential end (meaning components of the electron transfer chain having redox potentials of more than approx. 200 mV) of the typical bacterial electron transfer chain to which the nitrite reductase is connected (Berks *et al.*, 1995). The second curious feature is that the $d_1$ heme iron was also six coordinate; thus the enzyme did not offer a substrate binding site at either heme. In addition to an expected axial histidine ligand there was an axial tyrosine (residue 25) ligand (Figure 4a). Each monomer is organised into two domains, with the $c$ heme being in a mainly α-helical domain and the $d_1$ heme being in a β-propeller domain (Figure 3). A notable point is that the tyrosine 25 ligand is provided by the α-helical cytochrome $c$ domain. Positioned above the $d_1$ heme ring are two histidine residues (Figure 4) which could readily be envisaged to be proton donors to one oxygen of nitrite, thus generating water as one of the reaction products (Fülöp *et al.*, 1995). This would be in agreement with previous mechanistic proposals (Averill, 1996).

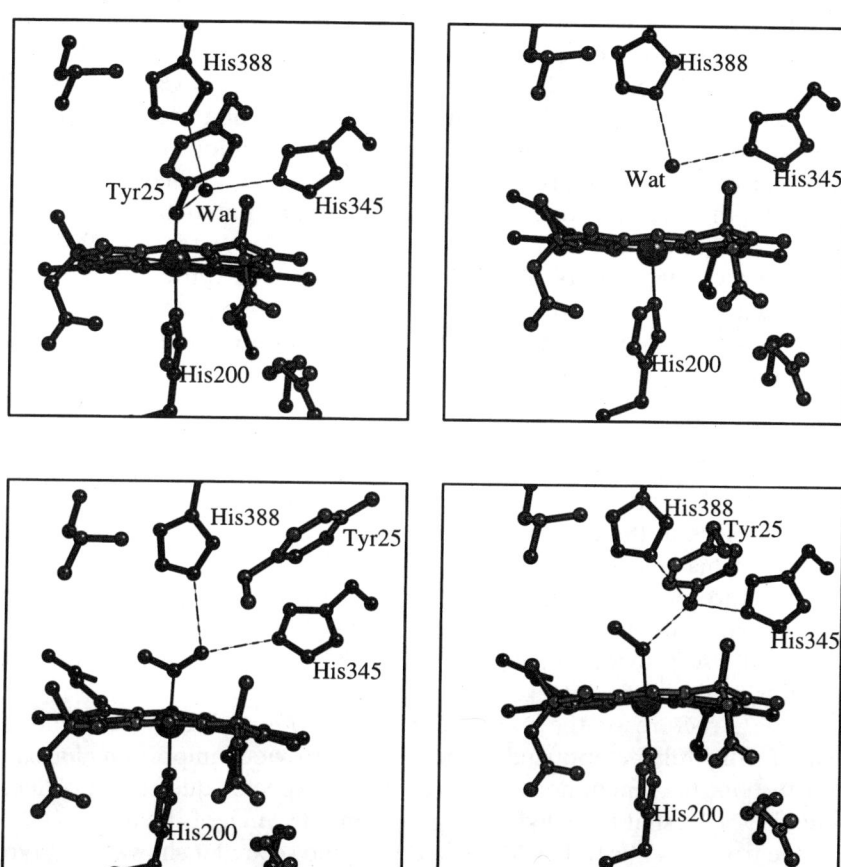

**FIGURE 4.** The environment of the $d_1$ heme in various states of the *P. pantotrophus* cytochrome $cd_1$. a. The oxidised enzyme. b. The reduced enzyme. c. The reduced enzyme to which nitrite has been added. d. The reduced enzyme that has converted bound nitrite to nitric oxide. Tyr25 provides an axial oxygen ligand to the $d_1$ heme iron in the oxidised enzyme (4a). This Tyr25 is displaced in the reduced enzyme (4b) but can be seen returning to the $d_1$ heme in the structures with nitrite (4c) or nitric oxide (4d) bound. His345 and His388 are proposed proton donors to the substrate. In (4b) the $d_1$ heme iron is 5-coordinate. The water molecule (Wat) that is seen above the heme ring is not coordinated to the iron. Dashed lines represent potential hydrogen bonds. (Drawn from PDB entries 1qks, 1aof, 1aom and 1aoq)

As mentioned earlier, the electron donor proteins to $cd_1$ are thought to include $c$-type cytochromes and cupredoxins. In the case of *P. pantotrophus* the two obvious molecules in this category are cytochrome $c_{550}$ and pseudoazurin. Whilst both these molecules will act as *in vitro* electron donors to the cytochrome $cd_1$, there is not yet any direct proof that they so act *in vivo*. It is the case that whole cells of *P. denitrificans* show no significant attenuation of the rate of nitrite reduction when the expression of cytochrome $c_{550}$ is not possible as a result of disruption of its gene (van Spanning *et al.*, 1990). Thus either cytochrome $c_{550}$ is not *in vivo* an electron donor to the nitrite reductase or it can be substituted by an alternate protein. The latter might be pseudoazurin (Moir and Ferguson, 1994). The properties of mutants specifically deleted in both of cytochrome $c_{550}$ and pseudoazurin are awaited with interest. However, the *in vitro* experiments alone raise the interesting question as to how proteins of such different structures as cytochrome $c_{550}$ and pseudoazurin could interact with cytochrome $cd_1$. It has been recognised that each of these putative donor proteins has a surface hydrophobic patch that is studded with positively charged residues and that these make a pseudospecific docking interaction with a complementary hydrophobic patch, studded with negative residues, on the surface of the cytochrome $c$ domain of cytochrome $cd_1$ (Williams *et al.*, 1995). This proposal remains to be tested by experiment. To summarise, the structure of the oxidised enzyme does not show an obvious catalytic site, and suggests that electron transfer between the $c$ and $d_1$ hemes could be fast. It also provides a proposal for where the donor electron transport proteins might bind or dock.

The structure of the fully reduced *P. pantotrophus* cytochrome $cd_1$ could fortunately be obtained (Figure 5) and provided important clues into how the enzyme functions as well as raising unexpected questions (Williams *et al.*, 1997). First, it revealed an unprecedented switch of the ligands at the $c$ heme from His/His to His/Met (Figure 6), and second it showed that tyrosine 25 had vacated the $d_1$ heme iron coordination sphere thus generating a five coordinate site which could be anticipated to provide the nitrite binding site (Figure 4b). The latter proposition was quickly supported by the finding that if nitrite was diffused into the reduced crystals, and followed by rapid freezing, then several different structures could be obtained with either nitrite or nitric oxide bound (Figure 4c and 4d). In one of these structures one of the oxygens in nitrite was clearly within hydrogen bonding distance of both histidines already mentioned above as putative proton donor residues (Figure 4c). The position of the tyrosine 25 residue varied depending upon whether the ligand bound was nitrite or nitric oxide. With the latter ligand present the tyrosine was in a position as if poised to enter the active site and displace the bound nitric oxide (Figure 4d). The X-ray structures naturally do not tell us if the heme centers are in the oxidised or reduced

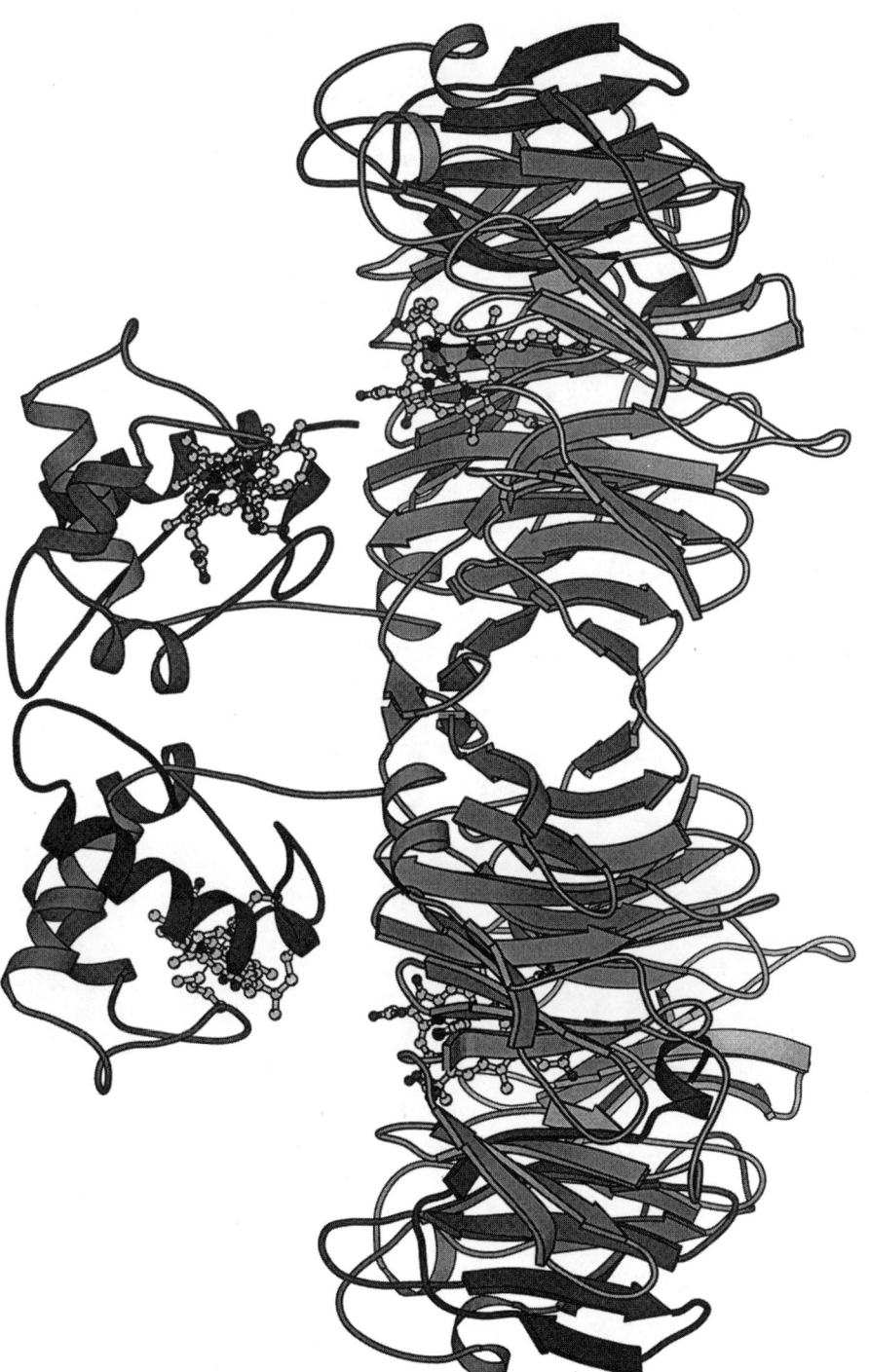

**FIGURE 5.** The X-ray crystal structure of the reduced state of cytochrome $cd_1$ nitrite reductase from *P. pantotrophus*. (Drawn from PDB entry 1aof)

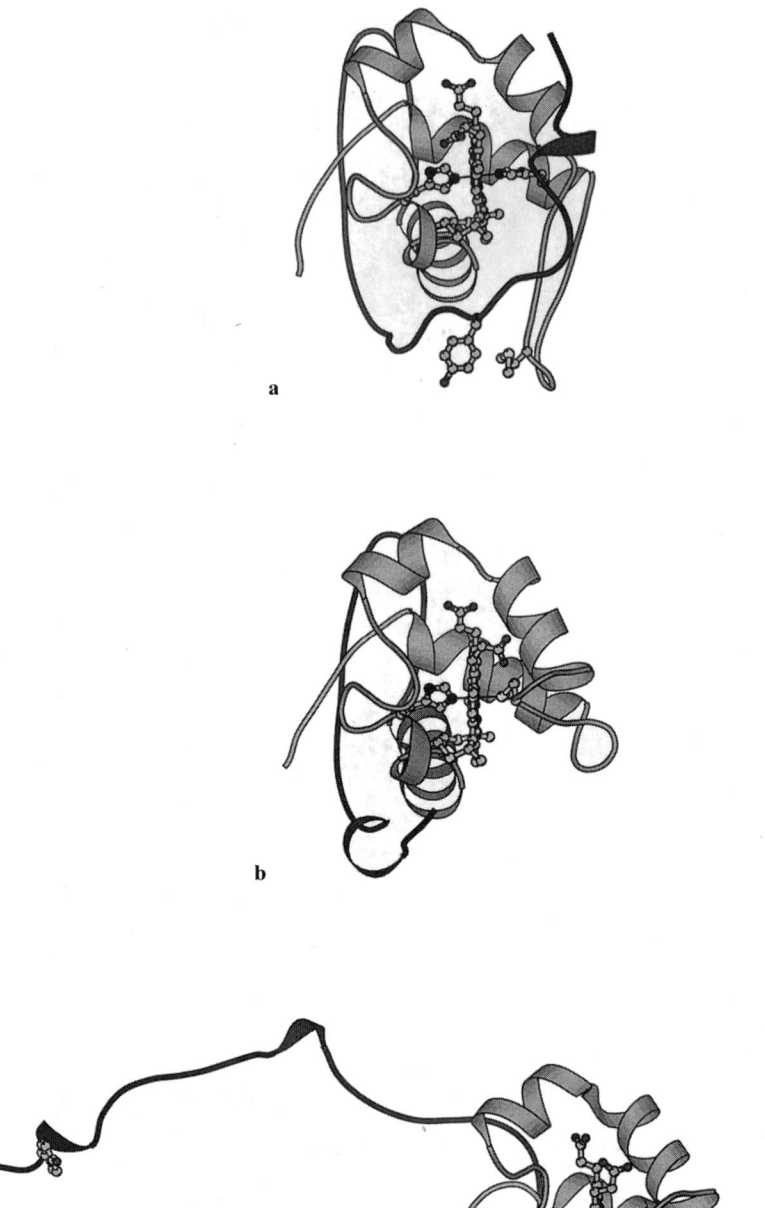

a

b

c

states but we may evaluate some possibilities. In the structure with nitrite bound to the $d_1$ heme iron it seems very likely that the $d_1$ heme iron must be in the reduced state. This is because there is no evidence that nitrite will readily bind to the oxidised form of the enzyme. If the $d_1$ heme is in the reduced state then the next question to be addressed is that of the oxidation state of the $c$ heme center on the same polypeptide chain. This could be either oxidised or reduced because the $d_1$ heme might already have carried out one reductive turnover of a nitrite ion to nitric oxide, received an electron from the $c$ heme and bound a second nitrite, in which case the $c$ heme would be oxidised. Alternatively the $c$ heme might still be reduced if the structure is of an enzyme that has bound nitrite but not effected any catalysis at this point. When the cytochrome $c$ domain is inspected for this structure, which has nitrite bound, it is found that the ligation is His/His, suggesting that it is in the oxidised state. This observation is important because it shows that the His/His liganded state does not need tyrosine anchored to the Fe of $d_1$ heme. The same issue, of course, pertains to the $c$ heme domain when nitric oxide is bound to the $d_1$ heme iron. Again all polypeptides observed with nitric oxide bound also had His/His ligation of the $c$-type heme center. It is also not possible to deduce the oxidation sate of the $d_1$ heme from the structure although ruffling of the heme varies depending upon which ligand is bound.

The question of the oxidation state of the heme iron with NO bound might in principle thought to be accessible by consideration of the Fe—N distance and bond angles. However, with respect to distance it is important to realise that X-ray structures do not in themselves determine distances such as that from Fe to N to sub ångstrom resolution. The procedure is to fit electron density using restraints frequently obtained from small molecule studies. Thus unless the resolution of a protein crystal structure is remarkably high, for example in the range of 1 Å, it is not possible to state definitively that the Fe—N distance is 1.8 or 2.0 Å because the electron density is subject to ripples around the metal center due to limited resolution effects. Partial occupancy of the bound ligand and mixed oxidation or spin states of the metal can further complicate deduction of metal to ligand

◄─────────────────────────────────────────

**FIGURE 6.** The $c$-heme domain in cytochromes $cd_1$ from *P. pantotrophus* and *P. aeruginosa*. a. Oxidised form of *P. pantotrophus* cytochrome $cd_1$ showing bis-histidine ligation of the heme iron. Tyr25 that is a ligand to the $d_1$ heme can be seen at the bottom of the figure adjacent to Met106. b. Reduced form of *P. pantotrophus* cytochrome $cd_1$ showing the movement of Met106 which replaces His17 as a ligand. c. Oxidised form of *P. aeruginosa* cytochrome $cd_1$ showing His/Met ligation coordination of the heme iron. To the left of the figure can be seen Tyr10 which is adjacent to the $d_1$ heme in the other subunit of the dimer. (Drawn from PDB entries 1qks, 1aof and 1nir)

bond lengths. Thus the temptation to deduce too much about the metal ligand distances needs to be resisted. Nevertheless, the data did fit better to 2.0 Å for the Fe—N distance and this might be taken as tentative evidence for the (III) oxidation state. The bond angle for Fe—N—O cannot automatically be used as diagnostic of the oxidation state. Whereas such bond angles can be diagnostic for small molecules, other factors such as the presence of amino acid side chains can seriously affect the geometry of a heme ligand inside a protein. Thus whilst small molecule studies on heme-NO complexes show that the Fe(III) derivatives have an essentially linear Fe—N—O group and a very short bond Fe—N of 1.63 to 1.65 Å in either five or six coordinate species (Scheidt and Ellison, 1999), the deviation from this pattern seen in nitrite reductase cannot be taken as definite evidence against the Fe—NO species being in the (III) state. Small molecule Fe(II)—NO complexes have an Fe—N—O bond angle of 140–150° and a short Fe—N bond distance of 1.72 to 1.74 Å (Scheidt and Ellison, 1999). A further feature of the Fe(II)—NO state is that the bond trans to the NO is very long in the six coordinate iron (II) species; this bond lengthening is not present in the structures of cytochrome $cd_1$ with NO bound. In fact notwithstanding the caveats above about metal-ligand bond distances, the Fe-histidine bond distance in this complex shortens slightly to 1.96 Å, compared with 1.98–2.00 Å in other ligand states, but everything is restrained to 2.00 Å. Thus although the Fe—N—O angle suggests that the NO-bound form is a ferrous state (formally at least) not all the evidence supports this view.

The observations made by X-ray crystallography would clearly be consistent with a reaction mechanism for cytochrome $cd_1$ nitrite reductase in which heme ligand switching occurred at the $c$ heme and a the $d_1$ heme each time a nitric oxide molecule was produced. It is notable that microspectrophotometry of crystals has shown that, following reduction and exposure to nitrite, the crystals regain the characteristic spectrum of the fully oxidised enzyme (Williams *et al.*, 1997). This means that in the crystalline state the nitrite reductase does not get trapped in a dead end state but rather performs rounds of catalysis before regaining the initial structrual state. The observation of the tyrosine 25 residue poised to return to the $d_1$ heme iron is also consistent with such a conclusion. This tyrosine residue could play a key role in ensuring that the nitric oxide is displaced from a Fe(III)-$d_1$ heme before the arrival of one electron would generate the Fe(II) oxidation state.

Such an involvement of amino acid side chain ligand switching within each catalytic cycle was a novel proposal and as such needs to be scrutinised by a variety of experimental procedures as well as analysis in the context of information known for cytochrome $cd_1$ nitrite reductase from another source (see below). However, it is interesting to note that something similar has recently been proposed for the protocatechuate 3,4-

dioxygenase enzyme from *P. putida* (Frazee *et al.*, 1998). On the other hand bacterial cytochrome *c* peroxidase offers an example where ligand switching seemingly relates only to an activation phenomenon. It is interesting that recently it has been proposed that the peroxidase from *P. denitrificans* may undergo a Met/His ligand switching at center (II) (N-terminal domain) (Lopes *et al.*, 1998). This however remains to be tested experimentally.

## 3. KINETIC STUDIES ON *P. PANTOTROPHUS* CYTOCHROME $cd_1$

Information about the kinetics of electron transfer between the hemes and of chemical events at the hemes is sparse for the enzyme from *P. pantrotrophus*, or even for that from the better known related organism *P. denitrificans* which has been studied for many years. The only reported kinetic study on the *P. pantotrophus* enzyme is that of Kobayashi *et al.* (1997) in which pulse radiolysis methodology was used. In this work exceedingly rapid reduction of the *c*-type cytochrome center in the enzyme was followed by electron transfer to the $d_1$ heme on the millisecond timescale. This method involves one electron processes only, and so the *c* heme, once reoxidised by the $d_1$ heme, remained oxidised. Interpretation of this millisecond rate rate of electron transfer is not straightforward because we do not know the driving force, i.e. the redox potentials of the *c* and $d_1$-type hemes. However, under the conditions of the pulse radiolysis experiment the electron transfer from the *c*-type center to the $d_1$ heme occurs essentially to completion. This implies that, at under these conditions at least, the redox potential difference is of the order of at least 100 mV. A difference of 100 mV and an edge to edge heme distance of 11 Å would suggest that electron transfer might be faster than is observed, as judged by current theories (Page *et al.*, 1999). However, these theories suppose that no chemical bond rearrangements accompany the electron transfer event. In the case of cytochrome $cd_1$ from *P. pantotrophus*, at least two chemical bond rearrangements might accompany the oxidation/reduction processes. These are the ligand switching at the *c* heme and the dissociation of tyrosine 25 from the $d_1$ heme iron. For the following reasons it is likely in the pulse radiolysis experiment that the ligands did not change at the *c*-type center but did so at the $d_1$ heme:

(i)     the spectrum in the Soret region, indicative of the *c*-type center of the enzyme immediately following the reduction is not identical to that obtained when the enzyme is fully reduced under steady state/equilibrium conditions. This suggests that the reduced *c*-type cytochrome center formed under the pulse radiolysis exerimental conditions retained the His/His coordination.

(ii)  The speed of the reduction by the solution radical generated in the pulse radiolysis experiment is also consistent with this proposal; conformational rearrangement is unlikely on the microsecond timescale.

(iii) the redox potential of a his/his coordinated heme is likely to be less than 100 mV and the $d_1$ heme, given that it is catalysing a reaction with a mid point potential of approx. 350 mV, can be expected to have a potential considerably more positive than 100 mV, thus accounting for the stoichiometric transfer of electrons from the $c$ to the $d_1$ under the conditions of the pulse radiolysis experiment.

(iv)  the $d_1$ heme probably loses its tyrosine 25 ligand under these conditions, because if nitrite is present during the pulse radiolysis experiment then, although the rate of electron transfer between the hemes is essentially unaltered, there is evidence for a chemical process taking place at the $d_1$ heme center (Kobayashi et al., 1997). This is suggestive that, at least in the presence of nitrite, the arrival of an electron at the $d_1$ heme triggers the dissociation of the tyrosine 25 ligand.

All these observations prompt the question of what values are obtained for the redox potentials of the $c$ and $d_1$ hemes under equilibrium conditions. This issue is currently under study and all that can be said here is that the enzyme does not give a straightforward redox titration. In particular, and in contrast to the observations made under pulse radiolysis conditions, the $c$ and $d_1$ hemes titrate together, suggesting a cooperativity of behaviour between them (A. Koppenhöfer and K. Turner et al. unpublished data). It remains for future work to elucidate the molecular basis of this effect.

## 4.  SOLUTION SPECTROSCOPY OF *P. PANTOTROPHUS* CYTOCHROME $cd_1$

It is clear that a good deal of information needs to be obtained in order to interpret the structural data; in particular, the question has to be faced as to whether the oxidised crystal structure is that of a resting state or whether the heme iron ligand switching occurs on each catalytic cycle. We do, however, know that the oxidised protein in solution has the same His/His coordination of the $c$-type heme as in the oxidised state of the crystalline enzyme. This was determined from MCD spectroscopy (Cheesman et al., 1997). Other solution spectroscocpic measurements have shown that the ligation of the $d_1$ heme is very likely the same as in the crystalline state (Cheesman et al., 1997). These studies also showed that the $d_1$ heme iron appeared to be in an unusual room temperature high/low spin equilibrium.

An important question about cytochrome $cd_1$ is why the $d_1$ heme has been uniquely recruited to this enzyme. Spectroscopic analysis has suggested that the relative energy levels of some of the d orbitals are inverted, relative to normal heme, for the Fe atom in this type of modified porphyrin ring. This would lead to NO binding being weaker than to standard $b$-type heme and thus may be a factor promoting dissociation of the product nitric oxide. Other important aspects of the $d_1$ heme could be the electron withdrawing effect of the carbonyl groups which thus could increase the affinity of the ferrous state of the $d_1$ heme iron for nitrite. It should be remembered that nitric oxide has a much higher affinity for ferrous heme than for ferric but even the latter state can offer sufficiently strong binding for the rate of nitric oxide dissociation to be slow. Thus it must be criticial feature of cytochrome $cd_1$ that nitric oxide release is facilitated; it is a good proposition that is achieved in part by ensuring departure of this product before the $d_1$ heme with nitric oxide bound can be reduced.

In concluding this section we note that cytochrome $cd_1$ nitrite reductase also has an oxidase activity (Fülöp $et$ $al.$, 1995). This four electron reaction, which contrasts with the one electron reduction of nitrite to nitric oxide, is outside the scope of this article.

## 5. THE CYTOCHROME $cd_1$ FROM PSEUDOMONAS AERUGINOSA

Over the years the cytochrome $cd_1$ that has received most attention is that from $Pseudomonas$ $aeruginosa$. Despite the fact that cytochrome $cd_1$ is a specialised enzyme, it is remarkable tnat the structure of the $P.$ $aeruginosa$ enzyme is not identical in several critical respects to the enzyme from $P.$ $pantotrophus$, most notably in terms of the coordination of the $c$ heme and $d_1$ heme iron centers. In the oxidised structure of the $P.$ $aeruginosa$ enzyme one sees that the overall structure of the dimer is the same (Figure 7) in the sense that the $c$-type cytochrome domain is helical and separated from the 8-bladed β-propeller domain that binds the $d_1$ heme center (Nurizzo $et$ $al.$, 1997). However, there are intriguing differences. First, the fold of the cytochrome $c$ domain of the oxidised $P.$ $aeruginosa$ enzyme is essentially the same as in the reduced $P.$ $pantotrophus$ enzyme and the heme iron coordination is His/Met. The iron of the $d_1$ heme does not have a tyrosine ligand; in contrast to the enzyme from $P.$ $pantotrophus$ the sixth ligand is hydroxide, but this is in turn is coordinated to tyrosine 10. However, the latter residue is in no sense equivalent to the tyrosine 25 of the $P.$ $pantotrophus$ enzyme. The tyrosine 10, which is not an essential residue (Cutruzzola $et$ $al.$, 1997), is provided by the other monomer to that in which

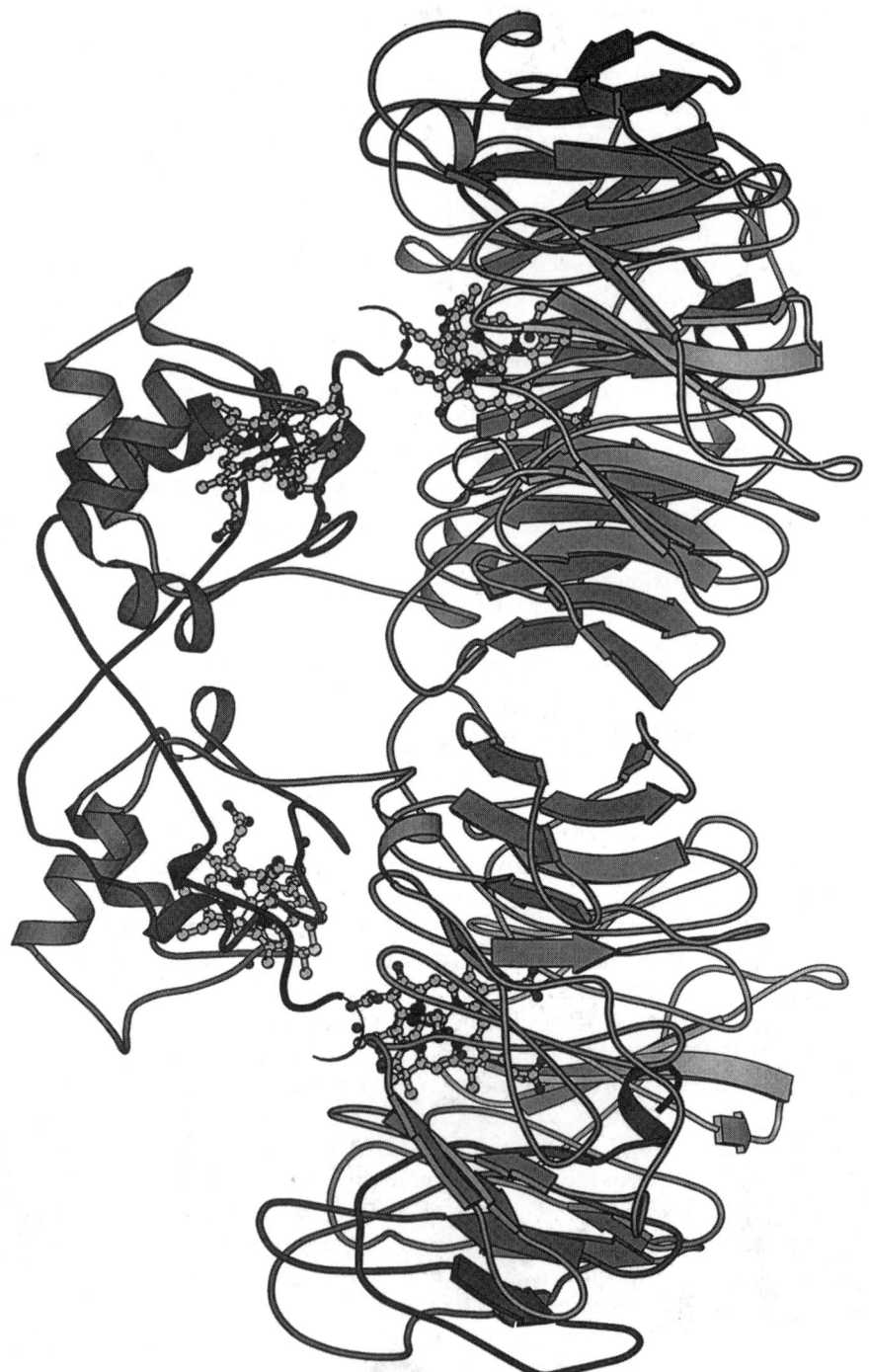

**FIGURE 7.** The structure of the oxidised cytochrome $cd_1$ from *P. aeruginosa*. (Drawn from PDB entry 1nir)

it is positioned close to the $d_1$ heme iron. In other words there is a crossing over of the domains. A reduced state structure of the *P. aeruginosa* enzyme has only been obtained with nitric oxide bound to the $d_1$ heme iron (Nurizzo *et al.*, 1998). As expected, the heme $c$ domain is unaltered by the reduction but the tyrosine 10 has moved away from the heme $d_1$ iron and clearly the hydroxide ligand to the $d_1$ heme has dissociated so as to allow the binding of the nitric oxide (Figure 8). This form of the enzyme was prepared by first reducing with ascorbate and then adding nitrite.

Although the conformational changes in the *P. aeruginosa* enzyme are clearly less pronounced than those in the *P. pantotrophus* enzyme, the driving forces for the changes still require to be understood. It had been postulated (Nurizzo *et al.*, 1998) that reduction of the $c$-type cytochrome domain might lead to conformational changes leading to the release of the hydroxide from the $d_1$ heme Fe. However, a recent study contradicts this view. It has been possible to obtain crystals of the *P. aeruginosa* enzyme in which the $c$ heme is reduced but $d_1$ heme is oxidised and this condition persists for sufficient time for the structure to be obtained. This shows that hydroxide is still bound to the $d_1$ heme and that therefore reduction of the $d_1$ heme, either alone or in combination of reduction of the $c$ type heme, is responsible for the conformational change at the $d_1$ heme iron (Nurizzo *et al.*, 1999). Reduction of the $d_1$ heme iron alone seems likely to be the essential factor, not least because no change in the $c$ heme domain occurs upon reduction and therefore it is difficult to see how there could be any effect of reduction of the $c$ heme center relayed to the $d_1$ heme. Recently pulse radiolysis work with the *P. aeruginosa* enzyme has shown that the electon transfer from the $c$ heme to the $d_1$ heme is very slow (order of seconds) (K. Kobayashi *et al.*, unpublished observations). This would be consistent with

**FIGURE 8.** Structural changes at the $d_1$ heme of *P. aeruginosa* cytochrome $cd_1$ upon reduction and binding of nitric oxide. There are no significant changes in the $c$ heme domain. The residue positions shown by thin lines are those of the oxidised protein with hydroxide ion, whilst those in bold represent the reduced enzyme with nitric oxide bound to the $d_1$ heme iron. Changes in the $d_1$ conformation can also be seen. The figure shows the $d_1$ heme from the A subunit; note that Tyr10 is provided from the B subunit. Dashed lines represent potential hydrogen bonds. (Drawn from PDB entries 1nir and 1nno)

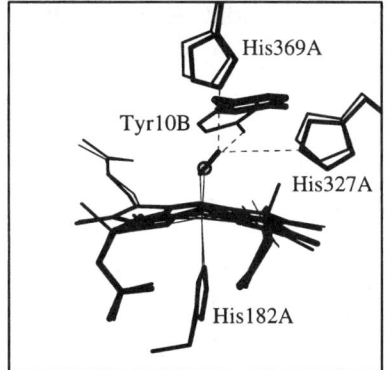

the idea that it is arrival of an electron at the $d_1$ heme that triggers the conformational change. This change could easily limit the electron transfer rate. There is clearly a parallel here with the *P. pantotrophus* enzyme where the reduction of the $d_1$ heme has been argued to trigger the conformational change (see above).

It has long been known that, under some conditions at least, electron transfer between the $c$ and $d_1$ hemes of the *P. aeruginosa* enzyme is slow, in the order of seconds (Cutruzzola, 1999). What does this mean? It is not necessarily related to the loss of the hydroxide ligand from the $d_1$ heme iron because under some experimental conditions used the enzyme was reduced at the outset, with azurin present, and it is the movement of an electron from the $c$ to the $d_1$ heme that is slow despite the presence of a ligand, nitric oxide, on the $d_1$ heme. This raises a problem because the nitric oxide gets trapped and thus what seems to be a dead end ferrous $d_1$ heme nitric oxide complex is formed. An expected gating mechanism that might be anticipated to prevent the formation of an Fe(II)-$d_1$ heme nitric oxide complex appears not to operate. However, it is clear that under some conditions the rate of interheme electron transfer can be much faster. A recent example is provided by the work of Wilson *et al.* (1999). The latter workers triggered electron transfer by photodissociation of carbon monoxide from a mixed valence form of the enzyme. Electrons moved from the $d_1$ to the $c$ heme at a rate of thousands per second. The dissociation of carbon monoxide and study of the kinetics of its rebinding also gave insight into the dynamics of the $d_1$ heme pocket. Again the meaning of these observations for the functioning of the enzyme is not clear.

It has long been assumed that azurin is an *in vivo* electron donor to cytochrome $cd_1$ of *P. aeruginosa*. However, construction of mutants of *P. aeruginosa* in which one or both of the genes for azurin and cytochrome $c_{551}$ have been deleted has led to the conclusion that *in vivo* cytochrome $c_{551}$ is essential, but that azurin is ineffective, for the donation of electrons to the nitrite reductase (Vijgenboom *et al.*, 1997). The discrepancy between *in vivo* and *in vitro* observations could be reconciled if it is the failure of azurin to except electrons from the cytochrome $bc_1$ complex, or other donor, that is responsible for its ineffectiveness *in vivo*.

## 6.   COPPER NITRITE REDUCTASE

As mentioned earlier, the copper containing nitrite reductase is a trimer of identical subunits. In each subunit there is a type 1 copper which acts analogously to the $c$-type heme in cytochrome $cd_1$ and thus is the point of entry of electron into the enzyme. The three catalytic sites have type 2 copper and are located on interfaces between two subunits. The type 2 copper ligands are provided by amino acid side chains; there is no organic cofactor. Current mechanisms for the copper-type nitrite reductase envisage that nitrite binds in bidentate fashion, via its two oxygen atoms, to copper(II) with concommittant displacement of a hydroxide ion. Subsequently it is proposed that an electron is received from the type 1 copper center. The reduction of the type 2 copper, together with protonation and loss of hydroxide, generates nitric oxide bound via its oxygen to copper (Cutruzzola, 1999; Watmough et al., 1999). There is debate (Dodd et al., 1997 and 1999; Murphy et al., 1997) as to whether there is rearrangement of bound nitric oxide so as to give a copper-nitrogen bond before product release. In any event, it is clear that the mechanism of nitrite reduction catalysed by the copper enzyme is distinct that from the cytochrome $cd_1$.

## 7.   CONCLUSIONS AND OUTSTANDING ISSUES

a. It is remarkable that the cytochromes $cd_1$ from P. pantotrophus and P. aeruginosa have different structures. It is not clear at present whether one of these structures is superior for catalysing nitrite reduction. This may not be the end of the story concerning structural variation. The sequence of cytochrome $cd_1$ from Pseudomonas stutzeri shows no counterpart of either Tyr10 or Tyr25 and thus the c heme domain may be distinct from those observed to date.

b. In the P. pantotrophus enzyme the ligand switching at both ligand centres upon changing oxidation state is novel. The X-ray structures of enzyme with bound substrate and product would suggest that these swithces occur during catalysis, but this is not definitely demonstrated. If these switches do not relate to catalysis then their role is not clear. Arguably, the oxidised structure of the enzyme may represent a resting or protected state. If so, why is a similar state seemingly not needed for the P. aeruginosa enzyme?

c. In the cytochromes $cd_1$ from both sources it is not clear how the electron transfer from the c to the $d_1$ haem is regulated so as to avoid formation of a dead-end Fe(II)-$d_1$ heme nitric oxide complex.

d. Why in both cytochrome $cd_1$ and copper nitrite reductase is there the requirement for a metal center to act as a receiver from an electron donor protein? As a one electron reaction is catalysed, a single metal center might have been expected to be sufficient.

e. Is there any significance to cytochrome $cd_1$ being a dimer? In the case of P. aeruginosa the structure shows that the two monomers are not independent. But in the case of the *P. pantotrophus* enzyme no obvious functional significance can be assigned to the dimeric organisation.

f.  The chemical rational for the adoption of the $d_1$ haem ring has not been rigorously determined. Nitrite reduction is one of many examples where biology can use copper or iron. It is striking that no organic cofactor is needed for the copper enzyme.

ACKNOWLEDGEMENTS. The authors' work was supported by the BBSRC (grant B05860) and the European Union (BIO4-CT96-0281). V.F. is a Royal Society University Research Fellow. We thank János Hajdu, James W. B. Moir and Pamela A. Williams for their important contributions to the structural work on *P. pantotropha* cytochrome $cd_1$. Figures were drawn with MolScript (Kraulis, 1991; Esnouf, 1997).

# 8.   REFERENCES

Averill, B. A., 1996, Dissimilatory nitrite and nitric oxide reductases, *Chemical Reviews* **96**:2951–2964.

Baker, S. C., Saunders, N. F. W., Willis, A. C., Ferguson, S. J., Hajdu, J., and Fülöp, V., 1997, Cytochrome $cd_1$ structure: unusual haem environments in a nitrite reductase and analysis of factors contibuting to β-propeller folds, *J. Mol. Biol.* **269**:440–455.

Berks, B. C., Ferguson, S. J., Moir, J. W. B., and Richardson, D. J., 1995, Enzymes and associated electron transport systems that catalyse the respiratory reduction of nitrogen oxides and oxyanions, *Biochim. Biophys. Acta* **1231**:97–173.

Cheesman, M., Ferguson, S. J., Moir, J. W. B., Richardson, D. J., Zumft, W. G., and Thomson, A. J., 1997, Two enzymes with a common function but different heme ligands. The optical and magnetic properties of the heme groups in the oxidised forms of nitrite reductase, cytochrome $cd_1$, from *Pseudomonas stutzeri* and *Thiosphaera pantotropha*, *Biochemistry* **36**:16267–16276.

Cole, J. A., 1988, Assimilatory and dissimilatory reduction of nitrate to ammonia, *Symp. Soc. Gen. Microbiol.* **42**:281–329.

Cutruzzola, F., Arese, M., Grasso, S., Bellelli, A., and Brunori, M., 1997, Tyrosine 10 in the *c*-haem domain is not involved in the catalytic mechanism of nitrite reductase from *Pseudomonas aeruginosa*, *FEBS Lett.* **412**:365–369.

Cutruzzola, F., 1999, Bacterial nitric oxide synthesis, *Biochim. Biophys. Acta* **1411**:231–249.

Dodd, F. E., Hasnain, S. S., Abraham, Z. H. L., Eady, R. R., and Smith, B. E., 1997, Structures of a blue copper nitite reductase and its substrate-bound complex, *Acta Cryst.* **D53**:406–418.

Dodd, F. E., van Beeumen, J., Eady, R. R., and Hasnain, S. S., 1998, X-ray structure of a blue copper nitrite reductase in two crystal forms. The nature of the copper sites, mode of substrate binding and recognition by redox partner, *J. Mol. Biol.* **282**:369–382.

Einsle, O., Messerschmidt, A., Stach, P., Bourenkov, G. P., Bartunik, H. D., Huber, R., and Kroneck, P. M. H., 1999, Structure of cytochrome *c* nitrite reductase, *Nature* **400**:476–480.

Esnouf, R. M., 1997, An extensively modified version of MolScript that includes greatly enhanced colouring capabilities, *J. Mol. Graphics* **15**:133–138.

Frazee, R. W., Orville, A. M., Dolbeare, K. B., Yu, H., Ohlendorf, D. H., and Lipscomb J. D., 1998, The axial tyrosinate $Fe^{3+}$ ligand in protcatechurate 3,4-dioxygenase influences substrate binding and product release: evidence for new reaction cycle intermediates, *Biochemistry* **37**:2131–2144.

Fülöp, V., Moir, J. W. B., Ferguson, S. J., and Hajdu, J., 1995, The anatomy of a bifunctional enzyme: structural basis for reduction of oxygen to water and synthesis of nitric oxide by cytochrome $cd_1$, *Cell* **81**:369–377.

Kobayashi, K., Koppenhöfer, A., Ferguson, S. J., and Tagawa, S., 1997, Pulse radiolysis studies on cytochrome $cd_1$ nitrite reductase from *Thiosphaera pantotropha*: Evidence for a fast intramolecular electron transfer from $c$ heme to $d_1$ heme, *Biochemistry* **36**:13611–13616.

Kraulis, P. J., 1991, MolScript: a program to produce both detailed and schematic plots of protein structures, *J. Appl. Cryst.* **24**:946–950.

Lopes, H., Pettigrew, G. W., Moura, I., and Moura, J. J. G., 1998, Electrochemical study on cytochrome $c$ peroxidase from *Paracoccus denitrificans*: a shifting pattern of structural and thermodynamic properties as the enzyme is activated, *J. Biol. Inorg. Chem.* **3**:632–642.

Moir, J. W. B., and Ferguson, S. J., 1994, Properties of a *Paracoccus denitrificans* mutant deleted in cytochrome $c_{550}$ indicate that a copper protein can substitute for this cytochrome in electron transport to nitrite, nitric oxide and nitrous oxide, *Microbiology* **140**:398–397.

Murphy, M. E. P., Turley, S., and Adman, E. T., 1997, Structure of nitrite bound to copper-containing nitrite reductase from *Alcaligenes faecalis*, *J. Biol. Chem.* **272**:28455–28460.

Nurizzo, D., Silvestrini, M.-C., Mathieu, M., Cutruzzola, F., Bourgeois, D., Fülöp, V., Hajdu, J., Brunori, M., Tegoni, M., and Cambillau, C., 1997, N-terminal arm exchange is observed in the 2.15 Å crystal structure of oxidised nitrite reductase from *Pseudomonas aeruginosa*, *Structure* **5**:1157–1171.

Nurizzo, D., Cutruzzola, F., Arese, M., Bourgeois, D. J, Brunori, M., Cambillau, C., and Tegoni, M., 1998, Conformational changes occurring upon reduction and NO binding in nitrite reductase from *Pseudomonas aeruginosa*, *Biochemistry* **37**:13987–13996.

Nurizzo, D., Cutruzzola, F., Arese, M., Bourgeois, D. J., Brunori, M., Cambillau, C., and Tegoni, M., 1999, Does the reduction of $c$ heme trigger the conformational change of crystalline nitrite reductase? *J. Biol. Chem.* **274**:14999–15004.

Page, C. C., Moser, C. C., Chen, X., and Dutton, P. L., 1999, Natural engineering principles in electron tunelling in biological oxidation-reduction, *Nature* **402**:47–52.

Rainey, F. A., Kelly, D. P., Stackebrandt, E., Burghardt, J., Hiraishi, A., Katayama, Y., and Wood, A. P., 1999, A re-evaluation of the taxonomy of *Paracoccus denitrificans* and a proposal for the combination *Paracoccus pantotrophus* comb. nov. *Int J. Sys. Bacteriol.* **49**:645–651.

Richardson, D. J., and Watmough, N. J., 1999, Inorganic nitrogen metabolism in bacteria, *Curr. Opin. Chem. Biol.* **3**:207–219.

Scheidt, W. R., and Ellsion, M. K., 1999, The synthetic and structural chemistry of heem derivatives with nitric oxide ligands, *Acc. Chem. Res.* **32**:350–359.

van Spanning, R. J. M., Wansell, W. N. M., Harms, N., Ras, J., Oltmann, L. F., and Stouthamer, A. H., 1990, Mutagenesis of the gene encoding cytochrome $c_{550}$ of *Paracoccus denitrificans* and analysis of the resultant physiological effects, *J. Bacteriol.* **172**:986–996.

Vijgenboom, E., Busch, J. E., and Canters, G. W., 1997, *In vivo* studies disprove an obligatory role of azurin in denitrification in *Pseudomonas aeruginosa* and show that azurin expression is under control of RpoS and ANR, *Microbiol.* **143**:2853–2863.

Watmough, N. J., Butland, G., Cheesman, M. R., Moir, J. W. B., Richardson, D. J., and Spiro, S., 1999, Nitric oxide in bacteria: synthesis and consumption, *Biochim. Biophys. Acta* **1411**:456–474.

Williams, P. A., Fülöp, V., Leung, Y.-C., Chan, C., Moir, J. W. B., Howlett, G., Ferguson, S. J., Radford, S. E., and Hajdu, J., 1995, Pseudospecific docking surfaces on electron transfer proteins as illustrated by pseudoazurin, cytochrome $c_{550}$ and cytochrome $cd_1$ nitrite reductase, *Nat. Struct. Biol.* **2**:975–982.

Williams, P. A., Fülöp, V., Garman, E. F., Saunders, N. F. W., Ferguson, S. J., and Hajdu, J., 1997, Haem ligand switching during catalysis by a nitrogen cycle enzyme *Nature* **389**:406–412.

Wilson, E. K., Bellelli, A., Liberti, S., Arese, M., Grasso, S., Cutruzzola, F., Brunori, M., and Brezezinski, P., 1999, Internal electron transfer and structural dynamics of $cd_1$ nitrite reductase revealed by laser CO photodissociation, *Biochemistry* **38**:7556–7564.

Zumft, W. G., 1997, Cell biology and molecular basis of denitrification, *Microbiol. Mol. Biol. Rev.* **61**:533–616.

*Chapter 16*

# Mitochondrial Cytochrome $bc_1$ Complex

ZhaoLei Zhang*[†], Edward A. Berry*,
Li-Shar Huang*[‡], and Sung-Hou Kim*[†‡]

## 1. INTRODUCTION

In aerobic organism, the majority of the ATP in cells comes from the process called oxidative phosphorylation, in which ATP is synthesized as electrons are sequentially transferred from low potential molecules, mainly NADH, to the high potential oxygen through the respiratory chain. The respiratory chain consists of several large protein complexes (complex I, II, III, IV and sometimes V), embedded in the inner membrane of mitochondria (Figure 1). As electrons are passed through these complexes, free energy is released, and protons are pumped out from the matrix side (negative or N side) to the inter-membrane space (positive or P side). Thus a proton gradient is created across the membrane. The electro-chemical energy stored in this proton gradient is subsequently utilized by another protein complex in the membrane, the ATP synthase, to synthesize ATP from ADP and phosphate. This protonmotive mechanism was first proposed by Peter Mitchell (1961) in his chemiosmotic hypothesis and has since become widely accepted in the field of bioenergetics.

ZHAOLEI ZHANG, EDWARD A. BERRY, LI-SHAR HUANG and SUNG-HOU KIM
*E. O. Lawrence Berkeley National Laboratory, [†]Graduate Group of Biophysics, and [‡]Department of Chemistry, University of California, Berkeley, CA 94720, USA.

*Subcellular Biochemistry, Volume 35: Enzyme-Catalyzed Electron and Radical Transfer*, edited by Holzenburg and Scrutton. Kluwer Academic / Plenum Publishers, New York, 2000.

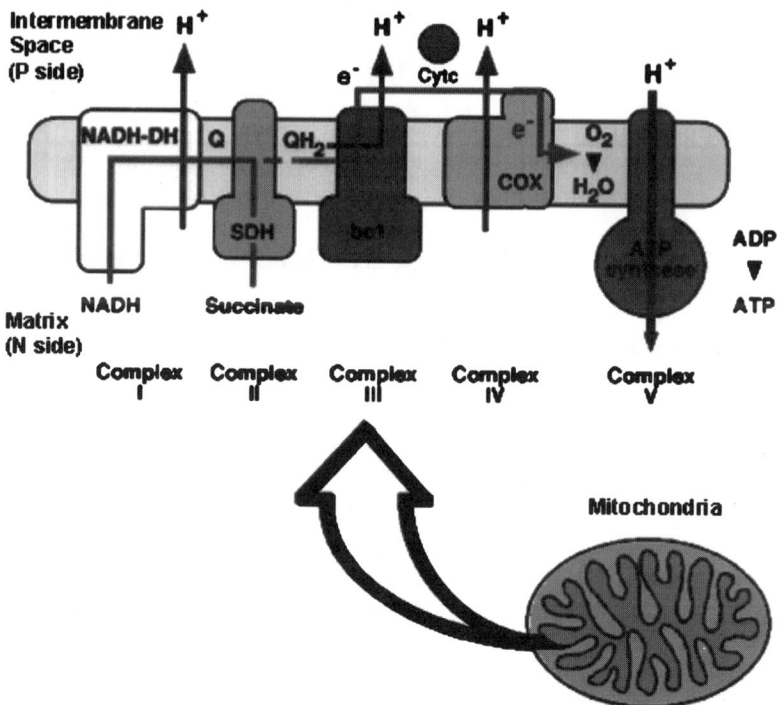

**FIGURE 1.** Respiratory chain: the enzymes of the mitochondrial inner membrane involved in oxidative phosphorylation. From complex I to V, they are NADH-dehydrogenase, succinate dehydrogenase, cytochrome $bc_1$ complex, and cytochrome $c$ oxidase. Protons are translocated across the membrane while electrons are transferred to $O_2$ through the chain. The proton gradient is used by ATP synthase (complex V) to make ATP. (Reprinted with permission from Saraste, 1999, American Association for the Advancement of Science.)

The third protein complex in this electron-transfer chain (complex III) is ubiquinol: cytochrome $c$ oxidoreductase (E.C. 1.10.2.2), or commonly known as cytochrome $bc_1$ complex named after the its $b$-type and $c$-type cytochrome subunits. Probably the best-understood one among the complexes, $bc_1$ complex catalyses electron transfers between two mobile electron carriers: the hydrophobic molecule ubiquinone (Q) and the small soluble haem-containing protein cytochrome $c$. Two protons are translocated across the membrane per quinol oxidized (Hinkel, 1991; Crofts, 1985; Mitchell, 1976).

Mitochondrial cytochrome $bc_1$ complexes have been isolated from animals, plants, yeast, and fungus by the use of detergents. These complexes all have the common three redox-active subunits that contain prosthetic

**Table 1**
**Subunit Composition of Bovine Heart Cytochrome**
**$bc_1$ Complex**

| Subunit | Residues | M.W. (Da) |
|---|---|---|
| 1 Core 1 | 446 | 49,132 |
| 2 Core 2 | 439 | 46,471 |
| 3 Cytochrome b | 379 | 42,592 |
| 4 Cytochrome $c_1$ | 241 | 27,288 |
| 5 Rieske Fe-S | 196 | 21,611 |
| 6 13.4 kDa | 110 | 13,347 |
| 7 "Q-binding" | 81 | 9,590 |
| 8 $c_1$ hinge | 78 | 9,170 |
| 9 Fe-S preseq. | 78 | 7,956 |
| 10 $c_1$-assoicate | 62 | 7,198 |
| 11 6.4 kDa | 56 | 6,363 |
| Apo-$bc_1$ | 2,166 | 240,718 |
| $Fe_2S_2$ | | 76 |
| Heme $c_1$ | | 616 |
| Heme $b_H$ | | 616 |
| Heme $b_L$ | | 616 |
| Prosthetic groups | | 2,024 |
| Holo-$bc_1$ complex | | 242,742 |

From Zhang *et al.*, 1998.

groups: cytochrome $b$, cytochrome $c_1$, and the Rieske iron-sulfur protein in addition to some other subunits that don't have prosthetic groups. The $bc_1$ complex purified from bovine heart consists of 11 different subunits with total molecular weight of about 250 kDa (Schägger, 1986), as listed in Table 1. There are four redox centers in the complex, two haem groups, high potential haem $b_H$ and low potential haem $b_L$ of cytochrome $b$, one haem group in cytochrome $c_1$ and one $Fe_2S_2$ cluster in the Rieske protein. Only the cytochrome $b$ is coded by the mitochondrial genome, all other subunits are coded by nuclear genes.

Homologues of mitochondrial $bc_1$ complex are found in photosynthetic bacteria and other prokaryotes. Some bacterial $bc_1$ complexes contain only the three redox-active subunits, cytochrome $b$, cytochrome $c_1$, and the Rieske protein. Another $bc_1$ homologue, $b_6f$ complex in chloroplasts and cyanobacteria, has a shorter cytochrome $b_6$, cytochrome $f$, Rieske protein and a subunit 4. The absence of the extra subunits in the bacterial complexes seems to indicate that the non-redox subunits present in the mitochondrial $bc_1$ complex are generally not directly involved in electron transfer and proton translocation per se.

The overall reaction catalyzed by the mitochondrial $bc_1$ complex is:

$$QH_2 + 2 \text{ ferri-cytochrome } c^{3+} + 2 H^+_N \rightarrow Q + 2 \text{ ferro-cytochrome } c^{2+} + 4 H^+_P$$

The abbreviations used above are: $QH_2$, fully reduced ubiquinol; Q, fully oxidized ubiquinone. Subscripts N and P refer to the negative (matrix) side and positive (inter-membrane) side. Extensive biochemical and bio-physical studies have been carried out on $bc_1$ complex in the last several decades, which has greatly benefited from use of various respiratory inhibitors (von Jagow and Link, 1986). It is now generally accepted that the complex works through a modified "protonmotive Q-cycle" mechanism (Crofts, 1985; Mitchell, 1976). This mechanism requires the $bc_1$ complex to have two separate quinone binding sites: a quinol oxidation site ($Q_o$ site) from which the oxidized quinone goes out of the complex, and a quinone reduction site ($Q_i$ site) to which the reduced quinol enters the complex. The $Q_o$ site was proposed to be close to the inter-membrane side (P side) and the $Q_i$ site to be close to the matrix side (N side). Suppose at the beginning there is one ubiquinone molecule at the $Q_i$ site, and one ubiquinol mole-cule at the $Q_o$ site. The quinol oxidation reaction at $Q_o$ site is rate-limiting and the reaction is a bifurcated, two-step reaction because the two electrons of the quinol molecule are transferred through two separate paths (Figure 2). The first electron is passed along a high potential chain to the $Fe_2S_2$ cluster of Rieske protein, then to cytochrome $c_1$ subunit and from there to the substrate cytochrome c. The newly generated semiquinone molecule in the $Q_o$ site then reduces the nearby low-potential haem $b_L$ and the result-ing fully oxidized quinone molecule diffuses away into the bulk membrane. The reduced haem $b_L$ rapidly passes the electron to the high-potential haem $b_H$, which is located close to the ubiquinone molecule bound at the $Q_i$ site. This ubiquinone is then reduced by the haem $b_H$ to become semiquinone but remains bound at the $Q_i$ site. It takes another ubiquinol molecule to be fully oxidized at $Q_o$ site, and another electron coming to the $Q_i$ site to have this semiquinone molecule completely reduced to ubiquinol. Overall, one complete Q-cycle requires two ubiquinol molecules to be oxidized at the $Q_o$ site and two successive electrons transferred to the $Q_i$ site. Two protons are taken up from the matrix space after one turnover at the $Q_i$ site and four protons are released into the inter-membrane space after two turnovers at the $Q_o$ site. So this Q-cycle mechanism predicts that for each electron transferred to cytochrome c, two protons are released outside with only one proton taken from the inside (Brand, 1977).

Although the Q-cycle mechanism successfully explained most of the experimental data, still our understanding of the mechanism of $bc_1$ complex's function was hindered by lack of the detailed structural infor-

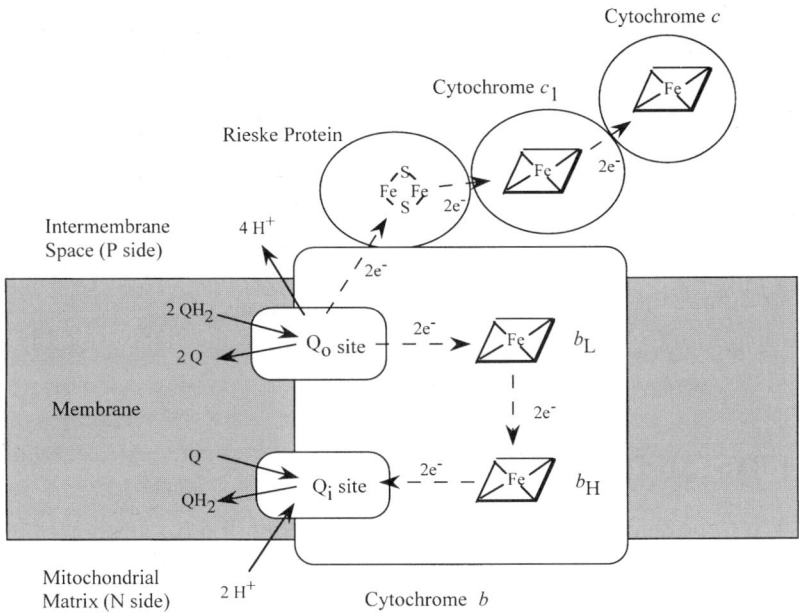

**FIGURE 2.** Q-cycle schematic diagram. Two ubiquinol molecules need to be oxidized at the $Q_o$ site to fully reduce one ubiquinone molecule at $Q_i$ site. The quinone oxidation reaction at Qo site is bifurcated with electron passed through two separate chains: the high potential chain of Rieske protein and cytochrome $c_1$, and the low potential chain of haem $b_L$ and haem $b_H$. The abbreviations are defined in the text.

mation at atomic level, primarily due to the difficulty of obtaining high quality three-dimensional crystals of the complex suitable for x-ray diffraction studies. Large unit cell and huge molecular weight made it further a daunting task for data collection and structure determination. Until three years ago, only a low-resolution structure for the cytochrome $bc_1$ complex from *Neurospora crassa* was available from electron microscopy of two-dimensional crystals (Weiss and Leonard, 1987). Progress has been made during the 90s as three-dimensional crystals of different space groups and from different sources were obtained by several laboratories (Berry *et al.*, 1992; Yue *et al.*, 1991; Kubota *et al.*, 1991). Breakthrough came along since 1997 when three research groups independently reported high-resolution crystal structures of the $bc_1$ complex from bovine and chicken sources (Iwata *et al.*, 1998; Zhang *et al.*, 1998; Xia *et al.*, 1997).

The crystal structures confirmed many hypotheses and predictions from previous studies, and also offered some major surprises. The most fascinating and unexpected is the suggestion of a "domain shuttle" mechanism of electron transfer between the $Q_o$ site and cytochrome $c_1$, which was first observed by comparing the different locations of head domain of the Rieske protein in different crystals (Zhang *et al.*, 1998; Kim *et al.*, 1998; Iwata *et al.*, 1998). Further observation of ubiquinone molecule and quinone analogues at the two quinone-binding sites also provided great insight and directions for the studies of Q cycle mechanism (Berry *et al.*, 1999a; Berry *et al.*, 1999b).

In this chapter, we will divide our discussion into two parts. First, we will describe the structural features of the whole complex and of each individual subunit. Then we will discuss the implications of the new structural information on our understanding of the reaction mechanism of the $bc_1$ complex.

## 2. CRYSTAL STRUCTURES OF $bc_1$ COMPLEX

Up to now, five different versions of mitochondria $bc_1$ structures from two different sources have been made public by three research groups. The PDB entry codes and related information of these structures are listed in the Table 2 below. Deisenhofer and Yu's group obtained the first structure of $bc_1$ complex from bovine heart mitochondria in 1997 (Xia *et al.*, 1997). More than 80% of the total residues were assigned in their structure including the two largest core 1 and core 2 subunits, cytochrome $b$, and part of the cytochrome $c_1$ and Rieske protein; but most of the protein expected to

### Table 2
### Summary of Crystal Structures of Cytochrome $bc_1$ Complex

| PDB code | Protein source | Space-group | Cell Parameters (Å) | | | Resolution (Å) | Quinone site occupants |
|---|---|---|---|---|---|---|---|
| | | | a | b | c | | |
| 1QCR[a] | Bovine | I4₁22 | 153.50 | 153.50 | 597.70 | 2.70 | None |
| 1BCC[b] | Chicken | P2₁2₁2₁ | 169.59 | 182.52 | 240.57 | 3.16 | Ubiquinone |
| 2BCC[b] | Chicken | P2₁2₁2₁ | 173.46 | 182.45 | 241.33 | 3.50 | Ubiquinone & Stigmatellin |
| 3BCC[b] | Chicken | P2₁2₁2₁ | 173.18 | 179.73 | 238.22 | 3.70 | Antimycin & Stigmatellin |
| 1BE3[c] | Bovine | P6₅22 | 211.20 | 211.20 | 339.28 | 3.00 | None |
| 1BGY[c] | Bovine | P6₅ | 130.11 | 130.11 | 720.94 | 3.00 | None |

[a] Xia *et al.*, 1997.
[b] Zhang *et al.*, 1998.
[c] Iwata *et al.*, 1998.

be in the inter-membrane region, especially the extrinsic catalytic domains of the cytochrome $c_1$ and Rieske protein, were missing in the model. Berry and Kim's group independently determined the crystal structure of $bc_1$ complex from chicken heart mitochondria, which included all three redox subunits and four redox centers (Zhang *et al.*, 1998). The extrinsic domain of the Rieske protein was found to be in two different positions among crystals grown from native protein and crystals grown from protein treated with inhibitor stigmatellin, the "domain shuttle mechanism" was proposed based on this observation. Iwata and Jap's group reported the most complete bovine $bc_1$ complex structure, which includes all 11 bovine subunits (Iwata *et al.*, 1998). The conformational change of the Rieske protein head domain was also observed by these authors.

### 2.1.   $bc_1$ Complex is a Homodimer as a Functional Unit

The bovine and chicken $bc_1$ structures have similar overall shape. Figure 3A shows the structure model of 11-subunit bovine $bc_1$ complex based from the PDB entry 1BE3. The model reveals $bc_1$ complex as two identical monomers which are related by a two-fold axis running vertically in the plane of the paper. The molecular mass of bovine $bc_1$ complex in solution state has been determined to be about 500 kDa by analytical ultra-centrifugation, neutron and light scattering after correcting for bound detergent and phospholipids (Karlsson *et al.*, 1983; Perkins and Weiss, 1983; Weiss and Kolb, 1979). This is about twice of the total molecular mass of the subunits, which suggested the complex exists in a dimeric state. Isolation of monomeric $bc_1$ complex has been reported (Musatov and Robinson, 1994). As will be discussed later, the crystal structures not only confirmed the dimeric state of the complex as a physical presence, but also a requirement for the function of electron transfer. Simplified schematic structures of individual subunit and a monomer complex are drawn in Figure 3B.

The overall dimension of the $bc_1$ complex is 150 Å long with maximal diameter of 130 Å. The general topological relationship between the subunits has been studied by labeling and cross-linking experiments (Gonzalez-Halphen *et al.*, 1988). Cytochrome $b$ forms the central part of the complex as well as most of the contact sites between two monomers. Other subunits are arranged peripherally, both within the membrane and in the aqueous phase. The structure of $bc_1$ complex can be divided into three parts as in three distinct regions: the matrix region, the transmembrane region, and the inter-membrane region. In each monomer, four subunits are found to be in the matrix region which accounts for more than half of the total protein mass: subunit 1 (core 1), subunit 2 (core 2), subunit 6, and a

**FIGURE 3.** (**A**) The structure of the bovine dimeric $bc_1$ complex viewed perpendicular to the twofold axis and parallel to the membrane. The helices in cytochrome $b$, cytochrome $c_1$ and Rieske protein are drawn as cylinders, other subunits as ribbons, haems and the Rieske $Fe_2S_2$ cluster as ball-and-stick models. The top part of the complex is in the mitochondrial inter-membrane space, the middle part spans the membrane and the bottom part in the matrix; approximate dimensions are given for each part. The labels are color-coded corresponding to the color of the subunit used in the figure. This figure and the following figures 5, 6, 7, 10, 11, 13, and 15 are produced using program MOLSCRIPT (Kraulis, 1991). Based on structure 1BE3 (Iwata *et al.*, 1998). (**B**) Schematic structure and location of each subunit shown both individ-ually and in the assembled monomer complex. Transmembrane helices are represented as cylinders.

⸻⸻⸻⸻⸻⸻⸻⸻⸻⸻⸻⸻⸻⸻⸻⸻⸻⸻⸻⸻➤

small peptide subunit 9 which is the pre-sequence of the Rieske protein. The transmembrane part of the dimeric complex is predominantly α helical, which falls into two clearly separately packed bundles. The separation of the α helices can be used as a criterion to divide the dimer complex into two monomers. Figure 4 shows the arrangement of transmembrane helices at the level of haem $b_H$ and $b_L$: the ones belonging to cytochrome $b$ are labeled from A to H and rest of the helices are labeled numerically. Each

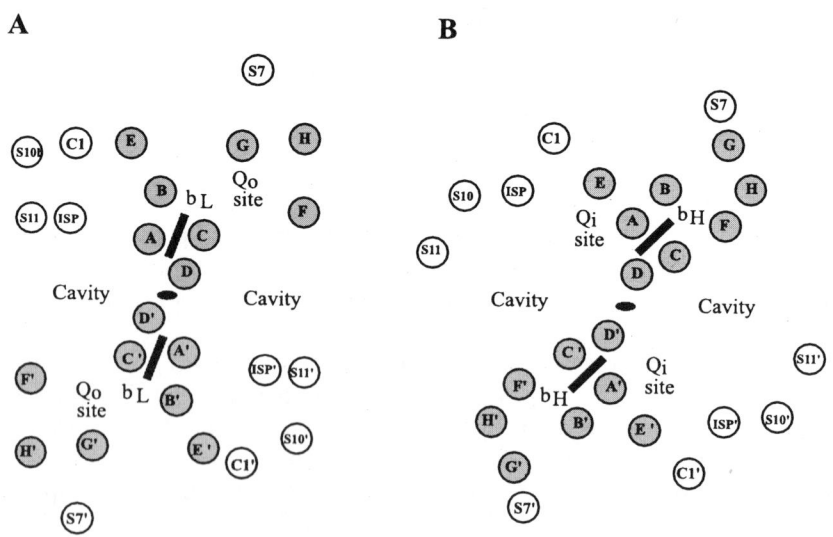

**FIGURE 4.** (**A**) Arrangement of the transmembrane helices of the $bc_1$ dimer at the level of haem $b_L$, close to the inter-membrane (P) side. (**B**) Arrangement of the transmembrane helices of the $bc_1$ dimer at the level of haem $b_H$, close to the matrix (N) side. The cavities formed by the helices are also indicated. $C_1$: cytochrome $c_1$; ISP: Rieske iron-sulfur protein; $b_L$: low poten-tial haem; $b_H$: high potential haem.

**Intermembrane space
(P side)**

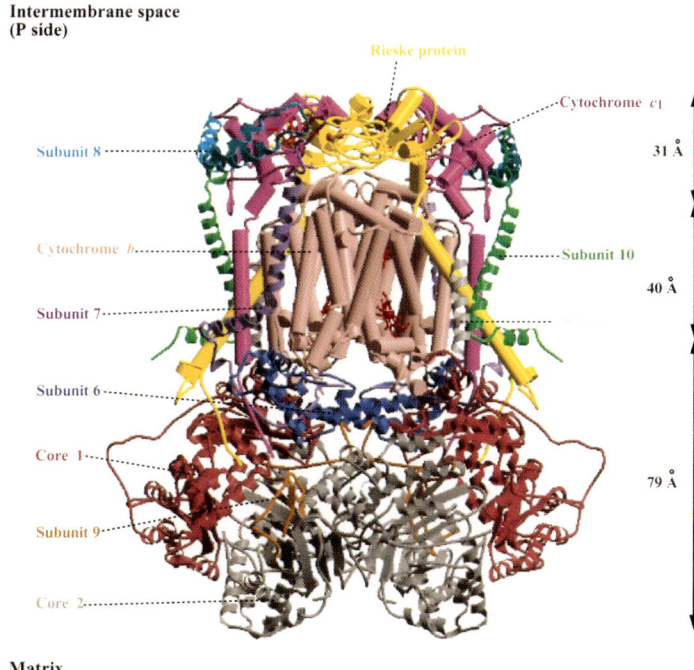

Rieske protein

Cytochrome $c_1$

Subunit 8

31 Å

Cytochrome $b$

Subunit 10

Subunit 7

40 Å

Subunit 6

Core 1

79 Å

Subunit 9

Core 2

**Matrix
(N side)**

130 Å

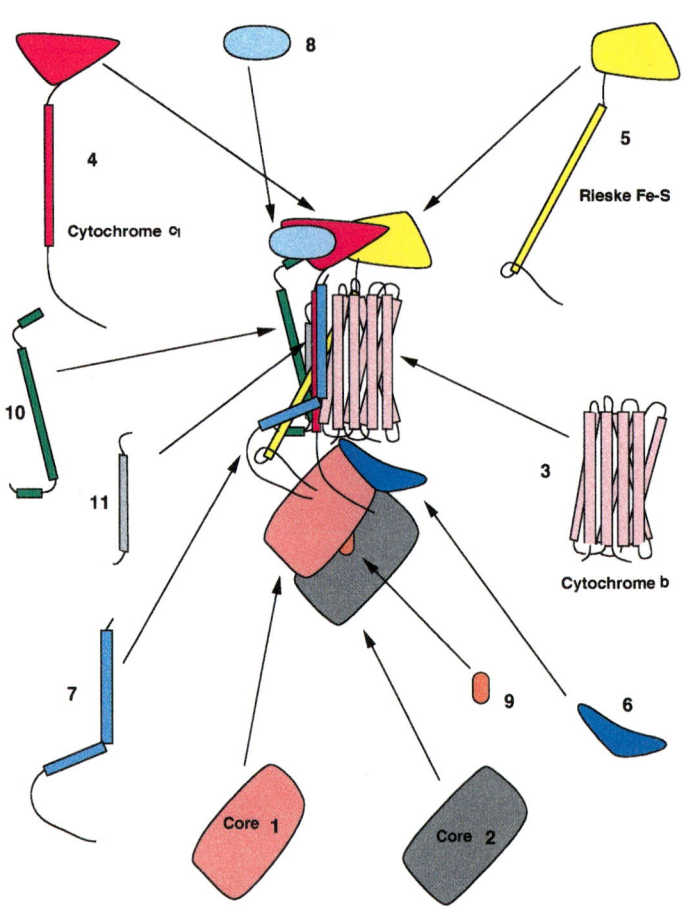

8

5

4

Cytochrome c₁

Rieske Fe-S

10

11

3

Cytochrome b

7

9

6

Core 1

Core 2

monomer of the $bc_1$ complex has 13 transmembrane helices, eight from cytochrome $b$, one from cytochrome $c_1$, one from Rieske protein and one from each of the three small single helix subunits: subunit 7, 10 and 11. The two haems of the cytochrome $b$ are identified in figure 4, labeled as $b_L$ and $b_H$. As seen in the figure, $bc_1$ dimer has two large cavities on both sides of the complex, the cavities are related by two-fold symmetry and open to the bulk lipid phase through side entrances. Two supposed quinone-binding sites, $Q_i$ and $Q_o$, are located in the cavities and close to the haems.

The inter-membrane region has the external domains of the cytochrome $c_1$ and the Rieske protein as well as the so-called "hinge protein"(subunit 8) which associates closely with cytochrome $c_1$. Only one conformation of the head domain of Rieske protein is drawn in Figure 3, more will be discussed on this matter later in this chapter.

## 2.2. Cytochrome $b$ and Two Haems

Figure 5 shows the structure of the dimer of cytochrome $b$ in the $bc_1$ complex. Both the N-terminus and the C-terminus of the subunit are located in the matrix side. The eight transmembrane helices are approximately at the same places as predicted from hydropathy analysis and mutagenesis studies (Link $et$ $al.$, 1994; Crofts $et$ $al.$, 1992). The helices are connected by four long linkers (AB, CD, DE, and EF) and three short ones (BC, FG, and GH). The AB and EF linker each contains one $\alpha$ helix, labeled as ab and ef in Figure 5. The CD linker has two short helices cd1 and cd2, which form a hairpin structure. The DE linker has no regular secondary structure. In the chicken complex, helix F was bent by over 45° in the middle around residue 296. A short amphipathic helix a, preceding all the transmembrane helices, runs parallel to the membrane.

The first four helices of the cytochrome $b$, helices A to D, form a right-handed anti-parallel four-helical bundle motif, which is also found in many other proteins. Within the four-helix bundle, the two b-type haems are located on top of each other with an iron-iron distance of approximately 21 Å and the closet edge-to-edge distance of 12 Å. As predicted, four absolutely conserved histidine residues on the helix B and helix D provide the axial ligands for the haems (Yun $et$ $al.$, 1991). In the chicken $bc_1$ structure, they are His84 and His183 for haem $b_L$, and His98 and His197 for haem $b_H$. Other widely conserved residues as well as inhibitor-resistant mutation sites are also found to be close to the haems in the structure (Brasseur $et$ $al.$, 1996; Esposti $et$ $al.$, 1993). Two pairs of well conserved glycine residues, Gly35 and Gly49 on helix A and Gly117 and Gly131 on helix C (chicken $bc_1$ numbering), are critical to accommodate the two haems into the tightly constrained four-helix bundle. An alanine residue can be tolerated at

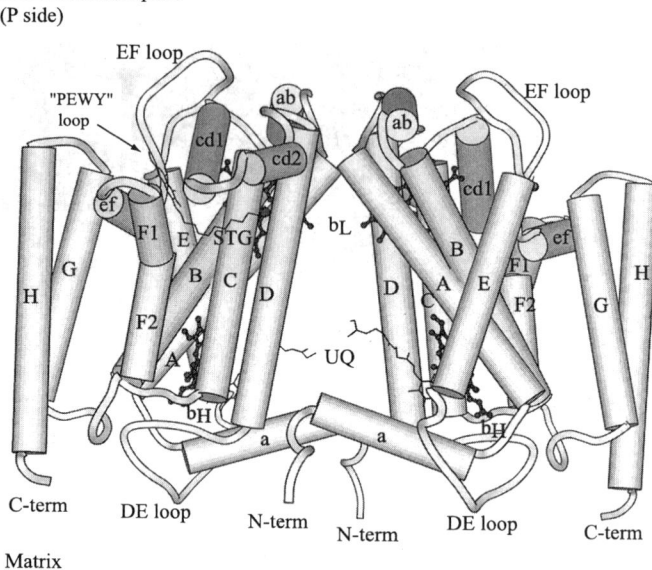

**FIGURE 5.** Structure of the dimer of cytochrome *b* in *bc₁* complex, oriented with the inter-membrane side at top and matrix side at bottom. The eight transmembrane helices are labeled sequentially, A through H. Helix F is drawn as two short helices, F1 and F2. Loops are denoted according to the two helices they connect, only loops DE and EF are labeled. Peripheral helices are labeled as helix a, ab (in AB loop), cd1, cd2 (both in CD loop) and ef (in EF loop). The haems $b_L$ and $b_H$, are shown in ball-and-stick models. Two quinone site occupants: substrate ubiquinone and inhibitor stigmatellin are labeled as UQ and STG respectively, both are drawn in thick wire models. The conserved "PEWY" loop is also shown in the figure. Based on chicken *bc₁* structure 2BCC (Zhang *et al.*, 1998).

position 49 (Yun *et al.*, 1992; Tron *et al.*, 1991). Located between helix E and the peripheral helix ef, the highly conserved "PEWY" loop (residues 271 to 274 in chicken and yeast sequences) are also present in the plant homologue $b_6f$ complex.

## 2.3. Cytochrome $c_1$

Cytochrome $c_1$ contains a c-type haem as prosthetic group in its wedge-shaped N-terminal domain located in the inter-membrane space. This extrinsic domain is anchored to membrane by a transmembrane helix at the C-terminal end (residues 204 to 222 in bovine $bc_1$). This helix runs alongside cytochrome *b* and can be removed by mild protease treatment or gene truncation to produce a protein fragment (Hase *et al.*, 1987; Li *et al.*, 1981).

This water-soluble fragment cannot be assembled into the $bc_1$ complex, but displays essentially the same spectroscopic properties as the native protein, and is able to transfer electrons to cytochrome $c$ (Bechmann *et al.*, 1992). Cytochrome $c_1$ is classified as c-type cytochrome like others containing the "fingerprint" haem-binding penta-peptide —CXXCH- (residues 37 to 41 in bovine), however there is little sequence homology between $c_1$ and rest of the c-type cytochromes outside this region. The $c_1$ haem is covalently attached to the two cysteines in this "fingerprint" region (Cys37 and Cys40 in bovine) through thioether bonds. The two axial iron ligands of the haem are His41 and Met160 as predicted by Gray *et al.* (1992).

The three-dimensional structure of cytochrome $c_1$ is similar to other c-type cytochromes despite lack of sequence homology. Figure 6 compares the

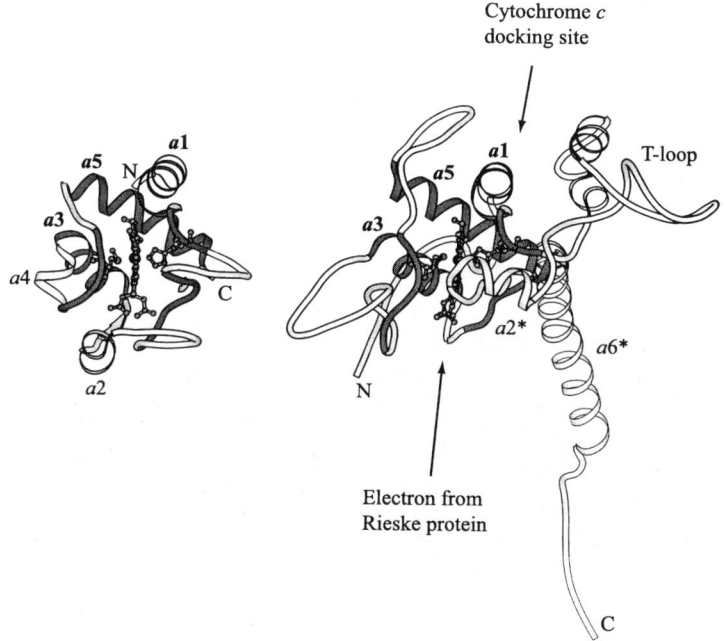

**FIGURE 6.** Structural comparison between mitochondrial cytochrome $c$ and cytochrome $c_1$. Left, the ribbon diagram of mitochondrial cytochrome $c$ with the open corner of the pyrrole of the haem group facing the viwer, and the haem propionates directed downwards. Right, structure of cytochrome $c_1$ (based on chicken $bc_1$ structure 1BCC), rotated to put the common features between the two cytochromes in the same orientations. The conserved secondary structures including the helices $\alpha1$, $\alpha3$, and $\alpha5$, which are present in both cytochromes, are painted in dark gray. Helices $\alpha2^*$ and $\alpha6^*$ in cytochrome $c_1$ have no counterpart in cytochrome $c$. The hypothetical docking sites of cytochrome $c$ and Rieske protein are also indicated.

backbone folding patterns of cytochrome $c_1$ and mitochondrial cytochrome *c*, the prototype of Ambler's (1991) class I cytochromes. Three $\alpha$ helices ($\alpha 1$, $\alpha 3$, and $\alpha 5$) which are conserved in class I cytochromes in general, are present in cytochrome $c_1$ and occupy similar positions relative to each other and to the haem, although only helix $\alpha 1$ has sufficient sequence identity for correct alignment based on sequence alone. Conserved residues involved in interaction between helices $\alpha 1$ and $\alpha 5$ (Gly6, Phe10 and Tyr97 in mitochondrial cytochrome) are also present in $c_1$ as Gly 29, Tyr33 and Phe189. The residues -PDL-, starting at Pro111 in cytochrome $c_1$, is also present as a tripeptide -PXL- in some other class I cytochromes. The proline carbonyl accepts a hydrogen bond from $N^\delta$ of the histidine haem axial ligand and the leucine provides a hydrophobic environment for the haem ring.

Major differences between cytochrome *c* and $c_1$ are the insertions or deletions at regions between the conserved helices. Between the conserved proline (see above) and helix $\alpha 3$, cytochrome *c* has a loop, which covers the propionate edge of the haem. Lacking this loop and the helix $\alpha 2$, cytochrome $c_1$ only has a short helix $\alpha 2^*$ (residues 98 to 103 in bovine) covering the front face of the haem as shown in Figure 6. This results in the exposure of the propionate edge of the haem at the bottom, allowing electron transferred from the Rieske protein (see Figure 7). In the chicken $bc_1$

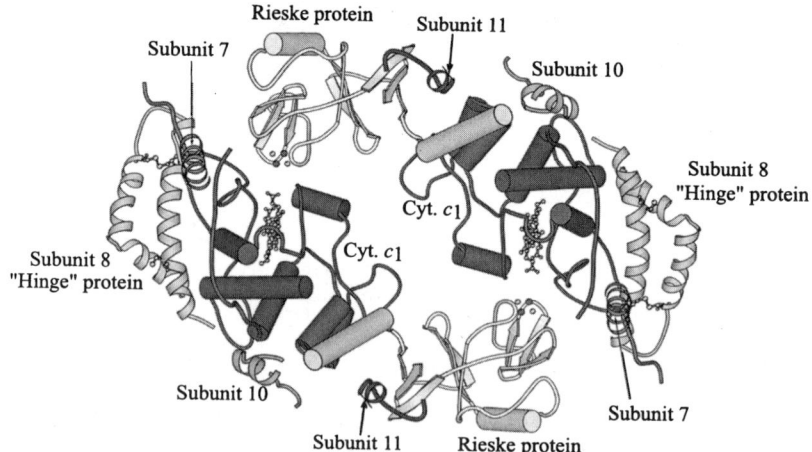

**FIGURE 7.** Structure of the inter-membrane (external surface) domains of the $bc_1$ complex viewed from within the membrane, with the transmembrane helices truncated at roughly the membrane surface. Cytochrome $c_1$ and Rieske protein are drawn as cylinders, subunit 7, 8, 10, and 11 as ribbons. The $c_1$ haem, Rieske $Fe_2S_2$ cluster and the two disulfide cysteines of subunit 8 are drawn as ball-and-stick models. Cytochrome $c_1$ is painted in dark gray, the Rieske protein in light gray.

structures, cytochrome $c_1$ has a T-shaped insertion loop of 58 residues (50 to 107, labeled as T-loop in Figure 6) between helix $\alpha 1$ and helix $\alpha 2^*$, corresponding to between residue 26 and 27 in cytochrome $c$. The first branch of the T-loop, residues 48 to 67 and mainly $\alpha$ helical, folds back towards the globular domain of cytochrome $c_1$. The second branch, residues 65 to 87, reaches out toward the other monomer and can be considered as a dimerization domain: the carbonyl oxygen of Gly78 at the tip of the loop forms a hydrogen bond with Gln97 of cytochrome $c_1$ from the other monomer. Both residues are highly conserved. Acidic residues on this loop: Glu66, Glu67, Glu76 and Asp77 have been suggested being involved in binding of cytochrome $c$ (Stonehuerner *el al.*, 1985). There is no structure in cytochrome $c_1$ corresponding to helix $\alpha 4$ of cytochrome $c$, long inserting loops take up the place instead. Acidic residues 167 to 174, near the end of the loop and before helix $\alpha 5$, have been implicated in cytochrome $c$ binding (Broger *et al.*, 1983). Thus the crevice formed by these two insertions loops on top of the haem becomes the most likely docking site of cytochrome $c$.

### 2.4. Alternative Conformations of Rieske Protein and Cross-transfer of Electrons

This subunit was first isolated by Rieske *et al.* (1964) and later identified as the "oxidation factor" of the $bc_1$ complex (Trumpower and Edwards, 1979). Rieske protein contains a $Fe_2S_2$ cluster, which shows a distinct EPR spectrum and unusual high redox potential (+300 mV). As predicted, Rieske protein is bound to the membrane via a single N-terminal membrane-spanning helix (Van Doren *et al.*, 1993; Link *et al.*, 1987). In the crystal structures, residues 1 to 24 are on the matrix side, interacting with subunit 1, residues 25 to 62 form a slightly curved and highly slanted transmembrane helix passing through the membrane at an angle of about 32° relative to the two-fold axis. Residues 60 to 66 are in close contact with both two cytochrome $b$ subunits in the dimer, whereas residues 67 to 73 provide a flexible linker connecting the extrinsic catalytic domain to its transmembrane helix.

As seen in Figure 7, the transmembrane helix of the Rieske protein associates with those of other subunits from the same monomer. However, the extrinsic domain of the protein extends out so that the $Fe_2S_2$ cluster comes close to the haem group of cytochrome $c_1$ from the other monomer, providing the pathway for electron transfer between these two subunits. This feature provides the structural basis for understanding that the dimeric state of the $bc_1$ complex is necessary for the electron transfer function since electrons are cross-transferred between the two $bc_1$ monomers.

Link *et al.* (1996) have isolated and crystallized a water-soluble fragment of Rieske protein from bovine source (residues 68 to 196), crystal structure was determined to 1.5 Å (Iwata *et al.*, 1996). This flat spherical domain contains three layers of anti-parallel β sheets and a long α helix inserted in-between (Figure 7). An extensive network of salt bridges and hydrogen bonds help stabilize the protein, e.g. five arginine residues (99, 101, 126, 170 and 172) and four acidic residues (Glu105, Glu109, Asp123 and Asp166) are involved in the network. Mutations in one or more these residues have resulted in partial or complete loss of wild type activity (Graham *et al.*, 1992; Gatti *et al.*, 1989). The Rieske domain can be further divided into two separate folds or sub-domains: the cluster-binding fold (residues 137 to181) and the base fold (residues 74 to136 and 182 to 196). Figure 8 shows the structure of cluster-binding fold; $Fe_2S_2$ cluster is coordinated between two short loops at the tip of the domain, with each loop providing one cysteine and one histidine ligand (Cys139, His141; and Cys158, His161). Two adjacent cysteine residues, Cys144 and Cys160, form a disulfide bond which stabilizes the fold around the cluster (Davidson *et al.*, 1992; Graham and Trumpower, 1991). The "Pro loop", containing the fully conserved sequence -GPAP- (174 to 177) is also critical to the stability of the cluster as it insulates the cluster from solvent on the other side (Graham *et al.*, 1992; Gatti *et al.*, 1989; Beckmann *et al.*, 1989). Inside the $Fe_2S_2$ cluster, one iron atom (Fe-1) is coordinated by the $S^\gamma$ atoms of the two cysteine residues, the other iron (Fe-2) by the $N^\delta$ atoms of the two histidine imidazoles as predicted by the ENDOR and ESEEM spectroscopy (Britt *et al.*, 1991; Gurbiel *et al.*, 1991; Gurbiel *et al.*, 1991). Except the $S^\gamma$ of Cys158, all other sulfur atoms are tightly constrained by multiple hydrogen bonds. This contributes to the stability of the cluster as well as the high redox potential as demonstrated by Schroter *et al.* (1998) and Denke *et al.* (1998). The electro-negativity of the histidine ligands and the solvent exposure of the cluster are also factors contributing to the high redox potential.

It was unexpected that the extrinsic domain of Rieske protein was found at different positions in different crystal forms. In bovine structure 1QCR (see Table 2), the location of the Rieske extrinsic domain was identified to be close to cytochrome *b* (Xia *et al.*, 1997). This domain was suggested to have high mobility based on further inhibitor binding experiments and anomalous scattering studies (Kim *et al.*, 1998). In the native chicken $bc_1$ structure (1BCC), the Rieske domain was found to be far away from haem $b_L$ and close to cytochrome $c_1$. However, in the chicken $bc_1$ complex co-crystallized with $Q_o$ site inhibitor stigmatellin (structure 2BCC and 3BCC) the Rieske domain was close to haem $b_L$ and far away from cytochrome $c_1$, which is approximately the position found in the structure

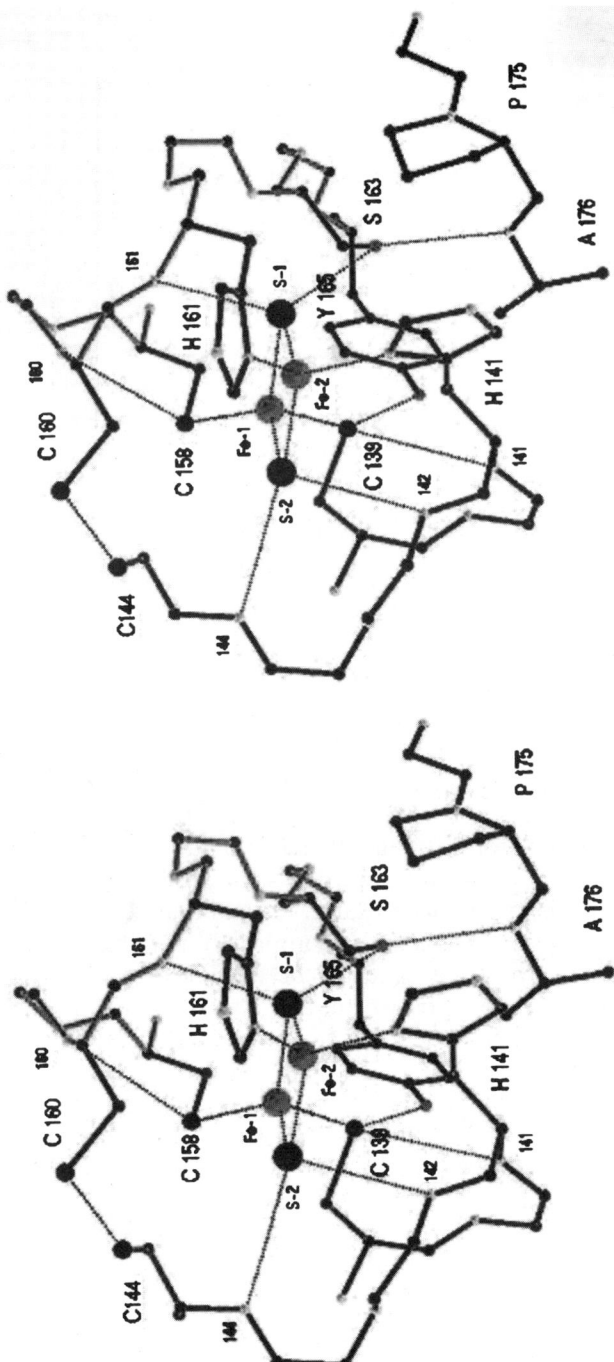

**FIGURE 8.** Stereoview of the structure of the cluster-binding fold of Rieske protein: the Fe₂S₂ cluster is shown as four spheres in the center, coordinated by ligands Cys158, Cys139, His141 and His161. The conserved Pro175 in the "Pro loop" is shown in the lower right. (Reprinted from Iwata *et al.*, 1996 with permission from Elsevier Science.)

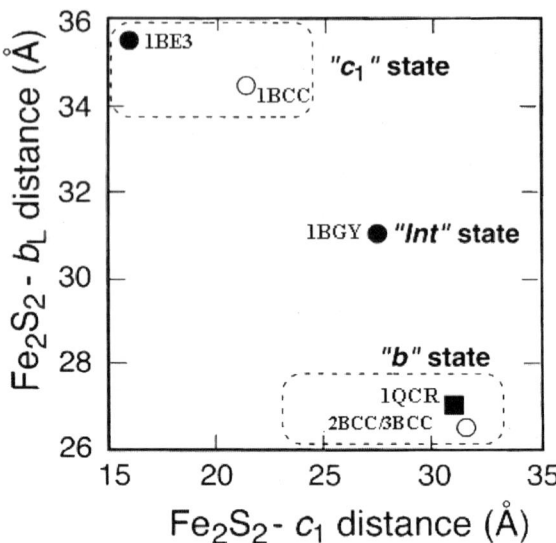

**FIGURE 9.** Plot of $Fe_2S_2$-haem $c_1$ and $Fe_2S_2$-haem $b_L$ distances for different structures of $bc_1$ complex. (Adapted from Iwata *et al.*, 1998)

1QCR. Iwata *et al.*' (1998) also observed the Rieske domain in different positions in their two crystal forms, a position similar to the $c_1$ state in structure 1BE3, and an "intermediate" position between haem $b_L$ and cytochrome $c_1$ in structure 1BGY. Figure 9 plots the $Fe_2S_2$-haem $c_1$ and $Fe_2S_2$-haem $b_L$ distances for different crystal structures, as the positions are classified as "$c_1$" state, "b" state and "Int" state following Iwata *et al.*'s nomenclature. As will be discussed later, the alternative positions of the Rieske domain suggest a dramatic conformational change, which is required for the electron transfer in the $bc_1$ complex. In structure 1BE3, a hydrogen bond between His161 of the Rieske protein, one of the ligands of the $Fe_2S_2$ cluster, and one of the propionates of the haem $c_1$ was observed.

Conformational changes within the Rieske extrinsic domain are found between structures 1BE3 ("$c_1$" state) and 1BGY ("Int" state). Clear positional displacement between the cluster-binding fold and the base fold is observed between the two structures. The structure of the Rieske domain in 1BE3 is the same as that of the water soluble fragment, while in 1BGY the cluster-binding domain appears to be detached from the base fold and rotated by about 6° into an "open" conformation (Iwata *et al.*, 1998). The r.m.s. (root-mean-square) difference of the Cα positions of the cluster-binding fold is 1.6 Å if the two domains are superimposed using the base fold residues. Some of the salt bridges and hydrogen bonds between the cluster-binding fold and the base fold are broken in structure 1BGY. In addition, residue Pro175, which is modeled as *trans* form in structure 1BE3 and the water-soluble fragment is best fitted as *cis* form in structure1BGY. This *cis*-Pro175 is also found in the structure of the water-soluble "Rieske" fragment from the $b_6f$ complex, which is the counterpart of the $bc_1$ complex in chloroplasts (Carrell *et al.*, 1997).

### 2.5. The Two Core Proteins and Subunit 9

Core 1 and core 2 proteins, the two largest subunits of the vertebrate $bc_1$ complex, contribute about 40% of the total protein mass. They are not involved in electron transfer, but deletions of their encoding genes affect the assembly of the $bc_1$ complex (Gatti and Tzagoloff, 1990; Oudshoorn, *et al.*, 1987); in some species (e.g. *Neurospora crassa*) they can be released from the isolated $bc_1$ complexes by salt treatment (Hovmöller *et al.*, 1981). The precursor peptides of the core proteins are synthesized in cytosol with additional N-terminal pre-sequence as mitochondrial targeting signals, which are cleaved off in the mature proteins (Gencic *et al.*, 1991). The two core proteins exhibit moderate sequence identities to each other (21% in bovine). They also show sequence identities to the two subunits of the general mitochondrial processing peptidase (MPP), whose function is to recognize and cleave off the targeting sequence of nucleus-encoded mitochondrial proteins upon their import into the organelle (Braun and Schmitz, 1995). Both the core proteins and MPPs belong to a family of $Zn^{2+}$ dependent metallo-endoproteases, which includes insulin-degrading enzymes from mammals and *E. coli.* pitrilysin (Rawlings and Barrett, 1993). Mammalian MPPs are soluble heterodimers located in the mitochondrial matrix, with an $\alpha$ and $\beta$ subunit of approximately 50 kDa each. The two sub-units are homologous to each other and to the core proteins: $\alpha$-MPP is more similar to the core 2 while the $\beta$-MPP resembles the core 1. Both MPP sub-units are required for protease activity. The $\beta$-subunit has been suggested to be the catalytically active subunit since it contains an "inverted zinc binding motif" (-HXXEH-$X_{76}$-E-) while the $\alpha$-subunit only has an incomplete such site. In mammalian $bc_1$, the core proteins show no protease activity as neither core 1 nor core 2 contains a complete zinc-binding site. The core1 protein and the $\beta$-subunit of MPP are identical in *Neurospora crassa* as the $bc_1$ complex exhibits protease activity if combined with $\alpha$-MPP (Schulte *et al.*, 1989). In plant mitochondria, the MPP activity is membrane-bound and the MPP are integral part of the $bc_1$ complex (Braun *et al.*, 1995; Braun *et al.*, 1992). It has been reported that the core proteins from bovine heart $bc_1$ complex show peptidase activity after mild detergent treatment (Deng *et al.*, 1998). Braun and Schmitz (1995) have proposed a model suggesting that the core proteins originated from an ancient proteolytic enzyme that was integrated into the $bc_1$ complex during early stages of endosymbiosis.

The two core proteins also have similar three-dimensional structures. Each core protein consists of two structural domains of about equal size and almost identical folding topology, which are related by approximately twofold rotation symmetry (Xia *et al.*, 1997). Both domains are folded into one mixed β sheet of six or five β strands, flanked by three α helices from the other domain on the other side. Rotation of the C-terminal domain onto the N-terminal domain superimposes 134 α carbon from each domain of core 1 with r.m.s. deviation of 2.0 Å, and 124 α carbons from each domain of core 2 with an r.m.s. deviation of 2.1 Å. The overall shape of each core protein resembles a bowl, while the two core proteins enclose a big cavity that was also observed by electron microscopy (Akiba *et al.*, 1996; Leonard *et al.*, 1981). The amino acid residues lining the wall of the cavity are mostly hydrophobic. A small peptide, subunit 9 is found inside the cavity, mainly bound to core 2 protein (Figure 10). The bovine subunit 9 is an extended

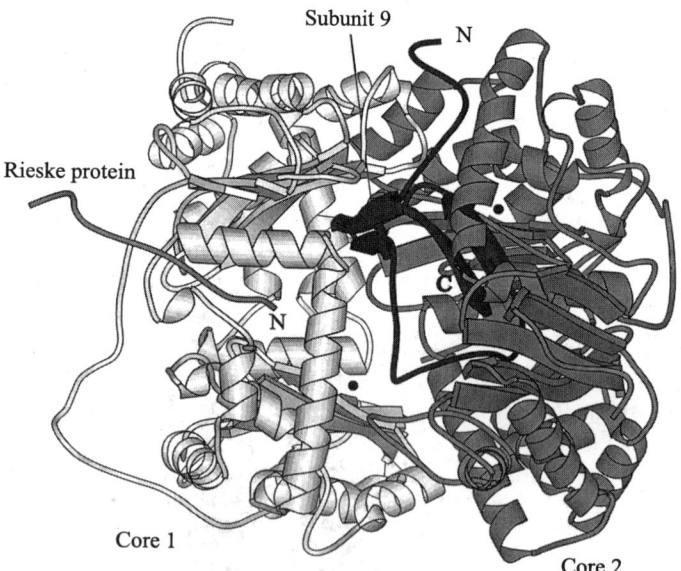

**FIGURE 10.** Interaction of subunit 9 with the two core protein subunits, viewed from the mitochondrial matrix side of the complex. Subunit 9 is shown in black, core 1 in light gray and core 2 in dark gray. The N-terminal end of the Rieske protein is shown in the left, close to core 1. Two possible $Zn^{2+}$-binding sites are marked by two black spheres.

peptide consisting of two β-sheets, one at the C-terminal end and the other in the middle region. The C-terminal β-sheet pairs up together with two β-strands of the N-terminal domain of the core 2 and form an extended β-sheet. Subunit 9 has 78 amino acids and is the pre-sequence of the Rieske protein subunit (Brandt et al., 1993). The nuclear-coded Rieske protein requires the pre-sequence for mitochondrial targeting: the protein is processed and the pre-sequence is cleaved off in a single proteolytic step after it is inserted into the $bc_1$ complex (Glick et al., 1992). This is the first instance in which a cleaved pre-sequence has been shown to be retained as part of a mature complex. No metal ion is observed at the supposed zinc binding site though the residues making up the binding motif (Tyr57, Glu60, His61, Glu137 in core 1) are found in close proximity and suitable for a metal binding site. The formation of a shared β-sheet between subunit 9 with core 2 protein is suggested to be essential for the proper recognition of the pre-sequence by the core 2 subunit. However, there is no direct evidence showing subunit 9 is processed by the core proteins during the assembly of the $bc_1$ complex.

## 2.6. Other Subunits without Prosthetic Groups

Subunit 6 of the bovine $bc_1$ complex consists of four helices and connecting loops. Located in the matrix, it contacts the exposed part of helix F, G, and H of cytochrome $b$ of one monomer and core 1 and core 2 from the other monomer. It has been suggested this subunit is involved in quinone binding (Yu and Yu, 1982) and proton translocation (Cocco et al., 1991) but no supporting evidences has been observed in the crystal structures. Subunit 7 of the complex is anchored in the matrix side as its N-terminal end associates with the core 1 protein as part of a β-sheet; its C-terminal 50 residues form a long, bent transmembrane helix (see Figure 3).

Subunit 8 and subunit 10 of the bovine $bc_1$ complex have been known to be tightly associated with cytochrome $c_1$ to form the so-called "$c_1$-subcomplex", though the term mostly only includes $c_1$ and subunit 8 (see Figure 7). Subunit 8 is also called "hinge protein" as it was thought to be essential for proper complex formation between cytochrome $c_1$ and its electron transfer partner cytochrome $c$ (Wakabayashi et al., 1982). But complex formation between these two cytochromes was also observed in the absence of the "hinge protein"(Konig et al., 1981). This subunit is extremely acidic as it contains 24 acidic residues out of 78 total, there are also 8 consecutive glutamic acid residues near the N-terminal end. As seen in Figure 7, the subunit has a hairpin-like structure with two long helices (Pro16 to Arg47, and Cys54 to Leu77) connected by two disulfide bonds

(Cys54-Cys68, and Cys40-Cys54), as expected from biochemical studies (Mukai *et al.*, 1985). The helices are kinked at Leu27 and Lys72; the first helix is bent almost 90° forming two separate helices. The first 14 residues in the sequence, including the glutamic acid stretch, are not observed in the electron density, suggesting this part of the protein may be highly mobile. Together with the $\alpha 1$ helix and the two insertion loops of cytochrome $c_1$, "hinge protein" participates in forming a large negatively charged docking surface, probably to accommodate the highly basic cytochrome *c*. Effects of the hinge protein on X-ray absorption of cytochrome $c_1$ and electron transfer between cytochrome $c_1$ and *c* have been observed (Kim *et al.*, 1989; Kim *et al.*, 1987).

Subunit 10 and 11 both consist of single transmembrane helix with the N-termini in the matrix side. These subunits have contacts with cytochrome $c_1$ and Rieske protein (Schagger *et al.*, 1986), and have been implicated to play a role in the proper assembly of the $bc_1$ complex (Phillips *et al.*, 1993). Subunit 10 contains 62 residues and forms three helices: two short helices found at the N- and C-terminal ends (residues 4 to 15, and 51 to 56) in addition to a longer transmembrane helix in the middle (residues 17 to 48). The N-terminal helix and middle helix interact with both the transmembrane helices of cytochrome $c_1$ and Rieske protein, and the C-terminal helix makes contacts with two helices on the extrinsic domain of cytochrome $c_1$. These extensive interactions may explain the tight association observed between subunit 10 and cytochrome $c_1$ in the dissociation experiments (Schagger *et al.*, 1986). Subunit 11 is peripherally located on the complex and is easily removed by detergent without loss of activity (Schagger *et al.*, 1990), despite interactions with both the Rieske protein and subunit 10. When the yeast counterpart of subunit 11 was deleted, the activity of the $bc_1$ complex was decreased by 40% and the Rieske protein was lost during purification (Brandt *et al.*, 1994). The C-terminal parts of the subunit 10 and 11 form a polar invagination, located near the transmembrane helices of the Rieske protein and cytochrome $c_1$, as well as subunit 8 in the inter-membrane side. This concave surface is large enough to accommodate the cluster-binding fold of the Rieske protein, and was suggested to be related to the insertion of the $Fe_2S_2$ cluster into the Rieske extrinsic domain (Iwata *et al.*, 1998).

It is estimated about 9 molecules of cardiolipin are bound to each $bc_1$ monomer (Gomez and Robinson, 1999) in addition to about 100 molecules of neutral phospholipids per dimer (Schagger *et al.*, 1990), which are required for full electron transfer activity. Two phospholipid molecules have been identified in chicken $bc_1$, crystals.

## 3.  QUINONE REACTIONS AND ELECTRON TRANSFER BY *bc*₁ COMPLEX

The crystal structures provide new understanding to a large amount of information previously gathered from mutation and inhibitor binding studies (summarized by Brasseur *et al.*, 1996). A quinone molecule at the $Q_i$ site ins chicken *bc*₁ complex has been identified by Zhang *et al.* (1998) in structure 1BCC and 2BCC; Qi site inhibitor antimycin and $Q_o$ site inhibitors stigmatellin, myxothiazol *et al.*, have also been observed in both the chicken and bovine structures (Iwata *et al.*, 1998; Zhang *et al.*, 1998; Xia *et al.*, 1997). These observations provides direct snapshots of the reactions carried out in the *bc*₁ complex, and models of the reaction mechanism at the $Q_i$ site and $Q_o$ site have been proposed based on them.

### 3.1.  Quinone Reduction at $Q_i$ Site

Despite lack of sequence homology, the function of the quinone reduction site ($Q_i$ site) is similar to that of the secondary quinone-binding site ($Q_B$ site) of bacterial reaction centers. Both sites have a conserved histidine residue as quinone ligand and both quinone molecules are reduced to hydroquinone in two consecutive one-electron transfer steps. The midpoint potential for the first step is pH-independent at near neutrality, whereas that for the second reduction varies by 120 mV per pH unit (Robertson *et al.*, 1984). This suggests a reaction pathway $Q \rightarrow Q^{\bullet-} \rightarrow QH_2$, with both protons added concomitantly with the second electron. A stable semiquinone anion intermediate can be detected by EPR spectroscopy of samples frozen during turnover (Yu *et al.*, 1980; de Vries *et al.*, 1980) or with the redox potential adjusted near the midpoint of ubiquinone (Robertson *et al.*, 1984; Ohnishi and Trumpower, 1980). The semiquinone signal is not observed in the presence of antimycin, which is consistent with the proposal that antimycin inhibits the reaction at the $Q_i$ site (Mitchell, 1976; Mitchell, 1975).

Indirect evidence for the location of the quinone at $Q_i$ site was first deduced from the decreases in electron density in bovine *bc*₁ crystals containing antimycin (Kim *et al.*, 1998; Xia *et al.*, 1997). Electron density at the $Q_i$ site in chicken *bc*₁ crystals is good enough so that a model for the bound quinone molecule can be built into the density. Figure 11 shows a stereo diagram of $Q_i$ site structure based on structure 1BCC. As expected, the $Q_i$ site is close to haem b$_H$ as required by the observed fast electron transfer rate and also by the "cytochrome b-150" phenomenon (Crofts *et al.*, 1995; Rich *et al.*, 1990; Salerno *et al.*, 1989; Rosa and Palmer, 1983; Dutton and

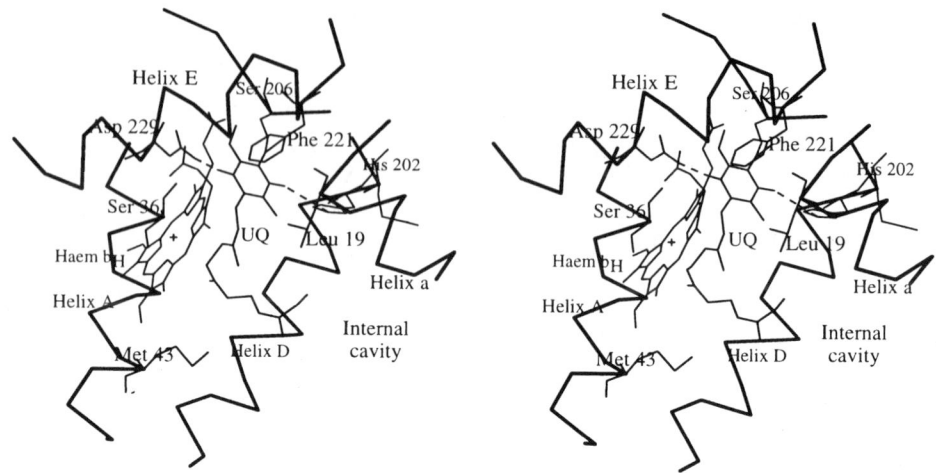

**FIGURE 11.** Stereo diagram of the $Q_i$ site structure in chicken $bc_1$ complex (structure 1BCC), viewed parallel to the membrane with the matrix side on the top. The amphipathic helix a, transmembrane helices A, D, and E are shown in bold lines connecting $\alpha$ carbon atoms, while haem $b_H$, ubiquinone (UQ) and selected residues are shown as thin wire models. The hydrogen bonds between ubiquinone and residues His202, Ser206 and Asp229 are drawn as dashed lines.

Jackson, 1972). The quinone molecule is located in a pocket surrounded by the amphipathic helix a and the transmembrane helices A, D, and E. The significant interactions between quinone and the surrounding residues are the hydrogen bonds between the 3-methoxy oxygen and Ser206, between the $C_4$ carbonyl oxygen and His202 and between the $C_1$ carbonyl and Asp229. The last one can also be modeled as a bifurcated H-bond in which a proton donated by Asp229 being shared by both Ser36 and quinone. Such multiple hydrogen bonds on quinone molecule have been previously suggested by electron nuclear double resonance (ENDOR) experiment (Salerno *et al.*, 1990). The His202 and Asp229 are well conserved in mitochondria and bacterial cytochrome $b$. Ser206 becomes either Asn or Gln in bacteria but still the hydrogen-bonding donor capability is conserved at that position. Mutations in the residues corresponding to His202, Asp229 and Ser206 result in perturbations of the $Q_i$ site function or changes in the sensitivity to the $Q_i$ site inhibitors (von Jagow and Link, 1986). Mutation of His202 to the more basic Arg increases the stability of the semiquinone, whereas the mutation to the acidic Asp or to hydrophobic Leu de-stabilize the semiquinone and inactivate the $bc_1$ complex (Gray *et al.*, 1994). This is consistent with the observation in the crystal structure that His202 is an essential quinone ligand. Residue Ser36 is not well conserved, so its hydro-gen-bonding sidechain may not be essential. The residue Phe221 is also highly conserved, which makes hydrophobic contact with the quinone ring. The ubiquinone tail runs in the hydrophobic pocket between the residues Ile15, Leu19 in helix a, Ala39, Met43 in the helix A and Ile195, Leu198 in helix D; the tail further extends into the internal cavity between the two $bc_1$ monomers as described in section 2.1 (also see Figure 4). The electron density of the tail ends after three isoprenoid units, which may result from disorder of the quinone tail due to lack of interactions with protein.

For the protonmotive Q cycle mechanism to account for the proton translocation, it is necessary that protons consumed in quinone reduction come from the matrix phase. The $Q_i$ sites in chicken structure is nearly entirely covered by the amphipathic helix a; however partial surface of quinone ligand His202 is exposed to the matrix side, and the quinone itself is exposed by a narrow channel bounded by residues His202, Gly205, and Pro23. Calculation of the accessible surface area by the program SURFACE of the CCP4 suite (Collaborative Computational Project Number 4, 1994), using a test water molecule of 1.4 Å radius, results in expo-sure values of between 0.0 and 0.1 Å$^2$ for quinone atoms facing this channel. This suggests those water molecules from the matrix aqueous phase, but nothing larger, could access the $Q_i$ pocket. This provides a path for proton access to the site. Figure 12 describes one possible scheme for the mecha-nism of quinone reduction proposed based on the crystal structure: water enters the empty $Q_i$ site pocket and protonates the residues Asp229 and His202, which are involved in quinone binding. Then, quinone comes into the pocket and form hydrogen bonds with these two residues. While the quinone is reduced by sequential one-electron transfers, the Asp229 and His202 lose their protons to quinone through the shift of hydrogen bond. The fully reduced hydroquinone molecule then diffuses away as Asp229 and His202 become protonated again (Berry *et al.*, 1999b).

Once quinone is reduced at the $Q_i$ site, it must leave the site and be oxidized at the $Q_o$ site in order to release the protons taken from the matrix side into the inter-membrane space to complete the proton translocation. Both the $Q_o$ and $Q_i$ sites are open to the bulk membrane phase around the dimer. As described by Xia *et al.* (1997), the peripheral transmembrane helices of the complex divide the surrounding membrane phase into two semi-enclosed cavities between the monomers (see Figure 4). Figure 13 shows that the $Q_i$ site of one monomer and the $Q_o$ site of the other monomer are segregated into the same cavity. The regions labeled "inter-nal cavity" in the lower right corners of Figure 11 is within this cavity, into which the disordered tail group of ubiquinone extends. The quinol mole-cule fits rather loosely into the binding sites, and should have no difficulty

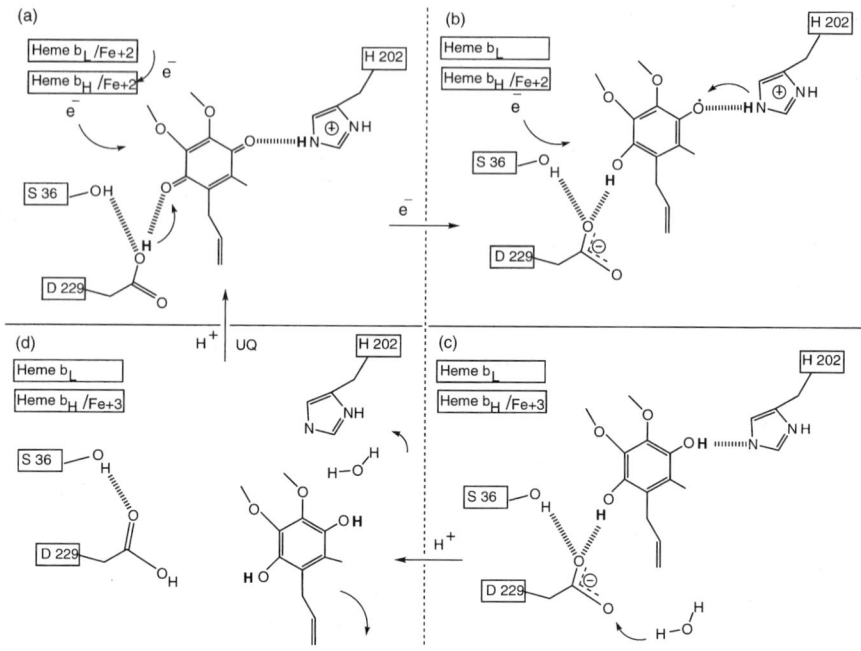

**FIGURE 12.** A model for H$^+$ and e$^-$ transfer reaction mechanism at the Q$_i$ site. The protons consumed in the reduction of the quinone come from the hydrogen-bonding residues D229 and H202, which are re-protonated after dissociation of the quinol.

diffusing out of the pocket and into the cavity. It is interesting to ask that whether this newly reduced quinol molecule in the Q$_i$ site is the same quinol molecule that would be oxidized at the Q$_o$ site in the same cavity. Such pathway would suggest a cross-transfer of protons as protons are shuttled between the Q$_i$ site of one monomer and the Q$_o$ site of the other monomer through the reduction and oxidation of quinone substrate. Ubiquinol is a non-polar molecule, easily soluble in the lipid phase of the membrane (Chazotte *et al.*, 1991), and is able to "flip-flop" across the membrane bilayer (Kingsley and Feigenson, 1981). Whether such internal-exchange is a preferred pathway is not clear, but it is unlikely to be a required pathway. Exchanging of the bound quinone molecules with the "Q-Pool" in the bulk lipid phase of the bilayer is still expected.

The two *bc*$_1$ monomers are represented as two large rectangle boxes from left to right; two cavities are shown as two separated smaller rectangles in the front and back. Two quinone binding sites, Q$_i$ and Q$_o$ are shown as two L-shaped tubes opening to both the internal cavity and the outside

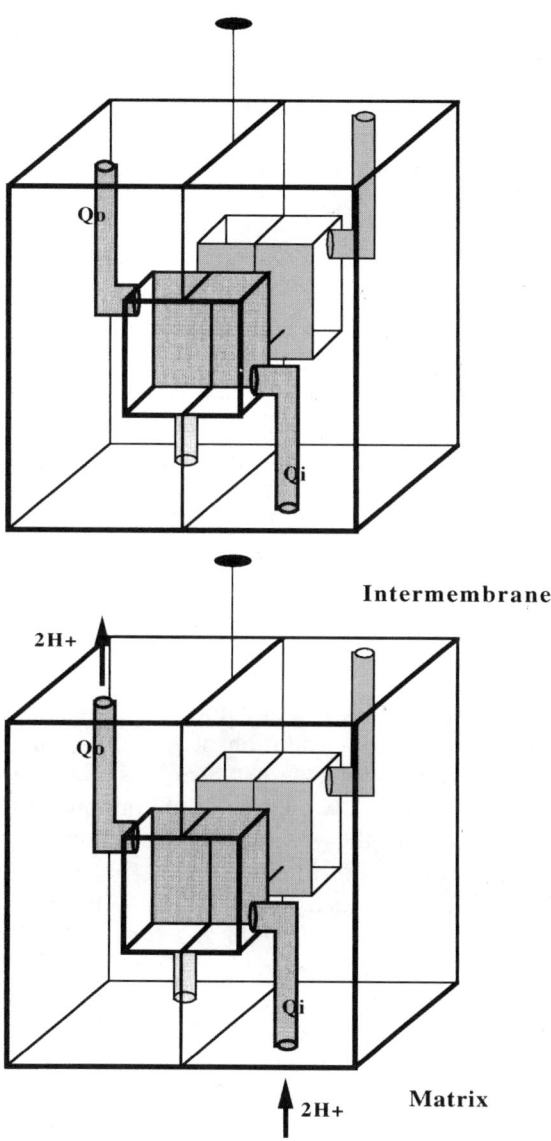

**FIGURE 13.** Schematic diagram showing the internal cavities between the two monomers.

aqueous phase. Protons are taken from the matrix at Qi site and release to the inter-membrane space at $Q_o$ site.

The antimycin binding site has been located from increased electron density in the vicinity of the haem $b_H$ in crystals of $bc_1$ complex with the inhibitor bound (Zhang *et al.*, 1998; Xia *et al.*, 1997). The aromatic ring of antimycin binds next to the haem with the polar —OH and formamido groups in a position to form hydrogen bonds with polar groups in Helix E, while the non-polar faces of the antimycin ring interact with Phe221 and the elbow of the haem propionate. This close proximity to the haem explains the effects of antimycin on the spectrum of the haem and the quenching of antimycin fluorescence upon specific binding to the $bc_1$ complex (Slater, 1973). The aromatic ring of antimycin is sufficient for inhibitory activity when esterified with a sufficiently hydrophobic amine (Miyoshi *et al.*, 1995; Neft and Farley, 1973), as suggested to mimic the aromatic ring of ubiquinone. There is overlap between the ubiquinone and antimycin binding sites, with some major differences. Residues Asp229 and Ser36 as well as the hydrophobic contacts with Phe221, Leu19, and Met43 are shared between the two $Q_i$ site occupants. Lys228 is involved in binding antimycin but not ubiquinone; Ser206 and His202 are involved in quinone binding but not antimycin binding. Mutations of Lys228 and Asp229 lead to antimycin resistance (Crofts *et al.*, 1995; Hacker *et al.*, 1993).

## 3.2.    Quinone Oxidation at $Q_o$ Site

One of the central reactions of the modified Q-cycle is the bifurcation of the electron pathway upon oxidation of hydroquinone at the $Q_o$ site. Unlike $Q_i$ site, no quinone molecule has been observed at $Q_o$ site in crystals, which is probably due to low content of ubiquinone in the preparation used for crystallization and the greater affinity of quinone at $Q_i$ site. The binding sites of several $Q_o$ site inhibitors have been identified from co-crystals of $bc_1$ complex and each individual inhibitor (Crofts and Berry, 1998; Zhang *et al.*, 1998; Iwata *et al.*, 1998; Xia *et al.*, 1997). These inhibitors could be classified into two classes based on their different effects on $Fe_2S_2$ center and electron flow (von Jagow and Link, 1986). Stigmatellin belongs to one such class that also includes UHDBT (5-undecyl-6hydroxy-4, 7-dioxobenzothiazole). This class of inhibitors interacts with the $Fe_2S_2$ center and prevents electron transfer to cytochrome $c_1$. myxothiazol, strobilurin, MOA ($\beta$-methoxyacrylate)-stilbene and other MOA type inhibitors belong to the other class, which don't perturb the $Fe_2S_2$ center or prevent electron flow to cytochrome $c_1$, but instead block turnover at the $Q_o$ site. Both classes of inhibitors block hydroquinone oxidation at $Q_o$ site and are presumed to bind as analogues of hydroquinone or its intermediates.

The inhibitor binding sites define the $Q_o$ site as sandwiched between four strands of peptides forming two 45° elbows (see Figure 14, also Figure 5): the transmembrane helix C with the peripheral helix cd1, and the PEWY loop on the EF linker with the peripheral helix ef (Berry *et al.*, 1999b; Zhang *et al.*, 1998). $Q_o$ site can be further devided into distal and proximal position based on distance to the haem $b_L$ (Crofts and Berry, 1998). Binding modes of myxothiazol and stigmatellin have been proposed to be different, though the binding of one inhibitor exclude the other (Link *et al.*, 1993; Meinhardt and Crofts, 1982). The side-chain methoxy group of myxothiazol and the ring of stigmatellin all make hydrophobic contact with the ring of Pro271 of the "PEWY" loop. The first thiazo ring of myxothiazol stack with Phe275, which the hydrophobic tail of stigmatellin also touches. However the rest of the protein contacts are quite different between these two inhibitors. The carbonyl end of the pharmacophore of the MOA inhibitors reaches toward the haem $b_L$ through the crack between PEWY loop and the peripheral helix cd1, probably forming hydrogen bonds with the backbone N of Glu272 and/or the phenolic —OH of Tyr274. The ring of stigmatellin reaches in the opposite direction toward the His161 of Rieske protein, one of the ligands of $Fe_2S_2$ cluster (Zhang *et al.*, 1998). This results in stigmatellin being within hydrogen bond distance of the Rieske protein, where as the MOA inhibitors reach much closer (5 ~ 8Å) to the haem $b_L$. Such proximal position adopted by the MOA inhibitors has been suggested as the site where electron transfer from the semiquinone to the cytochrome $b_L$ haem occurs (Zhang *et al.*, 1998; Yu *et al.*, 1998).

Electron density map shows a strong density connection between stigmatellin and His161 of Rieske protein, representing a hydrogen bond between the two. As mentioned in previous section, the extrinsic domain of Rieske protein comes closer towards cytochrome *b* in the stigmatellin-bound *bc*₁ complex (structure 2BCC and 3BCC) thus provides the ligand for the inhibitor. Formation of such hydrogen bond between stigmatellin or ubiquinol at the $Q_o$ site and a Rieske histidine residue has been previously suggested (Link, 1997; Robertson *et al.*, 1990). There is another hydrogen bond from the —OH group on the other side of stigmatellin ring to the carboxylates of residue Glu272 of cytochrome *b* (Figure 14). This Glu residue seems to adopt a different configuration in the presence of stigmatellin: in the absence of this inhibitor (and even in the presence of another $Q_o$ site inhibitor, myxothiazol), the carboxylates of Glu272 turns the opposite direction and reach into the hydrophilic region around the haem propionates (structure 1BCC). If we consider stigmatellin as a quinone analogue, then we can speculate Glu272 of cytochrome *b* and His161 of Rieske protein are the ligands for the hydroxyl oxygen of hydroquinone; after quinone oxidation, the protons are carried away by the Rieske domain and

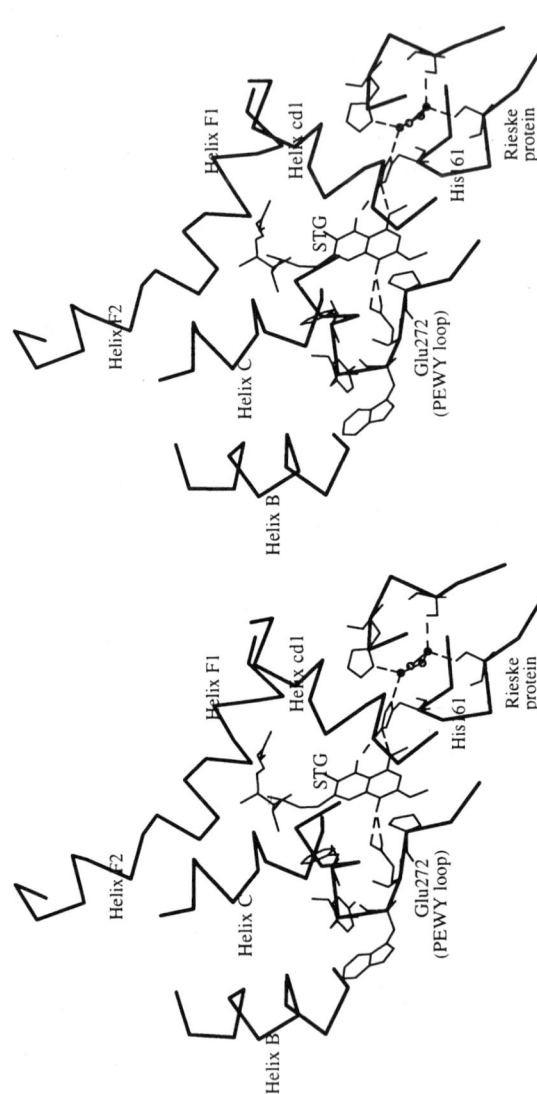

**FIGURE 14.** Stereo diagram of the $Q_o$ site structure of chicken $bc_1$ complex with bound inhibitor stigmatellin (structure 3BCC), viewed parallel to the membrane with the inter-membrane side at the bottom. The backbones of cytochrome $b$ and Rieske protein, including the amphipathic helix cd1, transmembrane helices B, C, F1 and F2 and PEWY loop are shown in bold lines connecting the $\alpha$ carbon atoms. Selected residues and inhibitor stigmatellin (labeled as STG) are shown as thin wire model, and $Fe_2S_2$ cluster as small spheres. The hydrogen bonds between stigmatellin and Glu272 and His161 are drawn as dashed lines.

by the movement of Glu272. Assuming such mechanism, the movement of Rieske domain would be required for the oxidation of hydroquinone at $Q_o$ site. Figure 15 shows a proposed reaction path of such mechanism.

Based on the analysis of the size and redox of the quinone pool on the EPR spectrum of Rieske $Fe_2S_2$ cluster, a "double-occupancy model" was proposed that the $Q_o$ site can accommodate two ubiquinone molecules: a strong bound species and a weakly bound species (Ding et al., 1995; Ding et al., 1992). It was further suggested these two ubiquinone molecules bind at the proximal and distal positions in the $Q_o$ pocket, which coincide with stigmatellin and myxothiazol binding sites respectively. The crystal structures don't show enough space in the Qo site to accommodate two ubiquinone molecules at the same time, crystallographic observation of quinol at $Q_o$ site is still to be made.

### 3.3. The "Domain Shuttle" Mechanism

Based on the two conformations of the Rieske protein observed in chicken $bc_1$ crystals, it was first proposed that the electron transfer during quinol oxidation must require movement of the extrinsic domain of the Rieske protein. In structures 1QCR and 3BCC, the Rieske $Fe_2S_2$ cluster is near the $Q_o$ site; in structures 1BCC and 1BE3, the cluster is near the haem of cytochrome $c_1$ (see Figure 9). In either position, the $Fe_2S_2$ cluster is too far from one of its redox partners for electron transfer to occur at the observed rates. Thus the Rieske extrinsic domain must move back and forth during the catalytic cycle, ferrying electrons from ubiquinol at the Qo site to cytochrome $c_1$. The movement in chicken $bc_1$ complex has been characterized as a rotational displacement of about 57° (65° in bovine structures), while the N-terminal membrane anchor is fixed and the action occurs in a short linker region between the membrane anchor and the extrinsic domain (Zhang et al., 1998; Crofts and Berry, 1998) (Figure 16). The $Fe_2S_2$ cluster moves about 16 Å in chicken and 21 Å in bovine complex, between a catalytic interface on cytochrome $b$ and an interface on cytochrome $c_1$. Such "domain shuttle" mechanism has been supported by mutation studies in *Rhodobacter sphaeroide* (Tian et al., 1999; Tian et al., 1998), when decreasing of $bc_1$ activity were observed after increasing the rigidity of the short linker region of Rieske protein described above. A conserved, concave surface on cytochrome $b$ has been identified, into which the cluster-binding domain of Rieske protein binds tightly (Crofts and Berry, 1998). A third "intermediate" position was identified in one type of bovine crystal (structure 1BGY), the two state "domain shuttle" mechanism was extended to a "three states" model by Iwata et al. (1998). The driving force for the movement is still under investigation (Berry et al., 1999b).

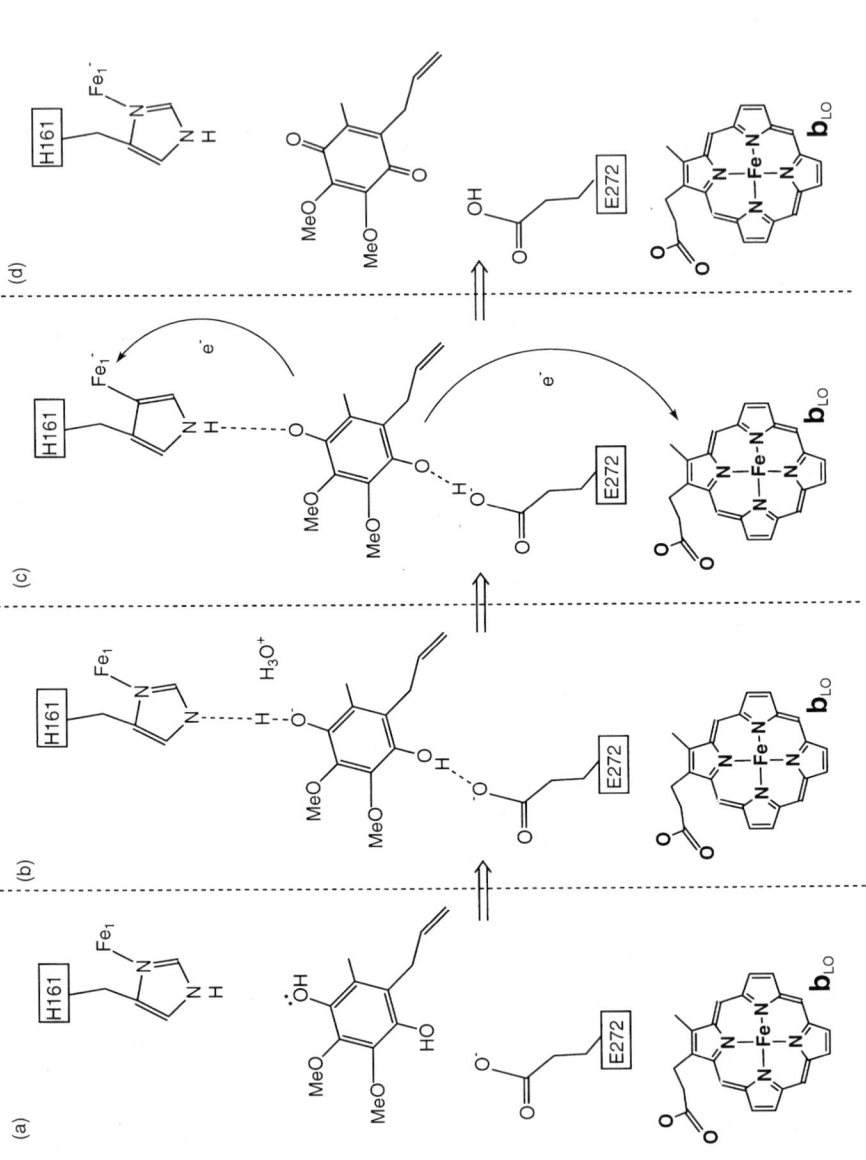

**FIGURE 15.** A model for H$^+$ and e$^-$ transfer reaction mechanism at the Q$_o$ site. Hydroquinone binds to deprotonated Glu272 of cytochrome *b* and His161 of Rieske protein, one of the ligands of Fe$_2$S$_2$ cluster. After electron transfer, the oxidized quinone unbinds, leaving Glu272 and His161 protonated.

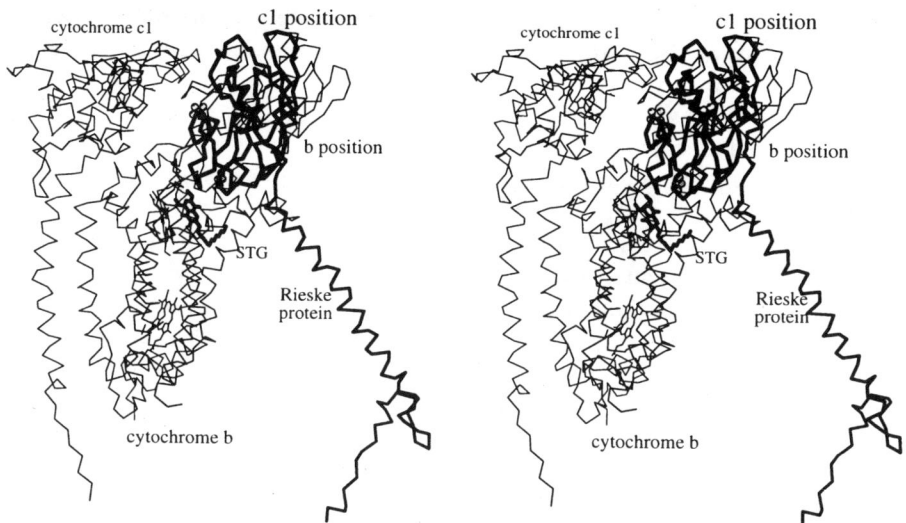

**FIGURE 16.** Stereo diagram showing the movement of Rieske extrinsic domain based on structure 1BCC and 3BCC (Zhang *et al.*, 1998). The three catalytic subunits of the native structure 1BCC, cytochrome *b*, cytochrome $c_1$ and the Rieske protein including the haems, are shown in thin wire models. Only the Rieske protein and the inhibitor stigmatellin from the stigmatellin-bound structure (3BCC) are shown in bold, superimposed onto the native structure. The two conformations of the Rieske extrinsic domain are labeled accordingly. The $Fe_2S_2$ clusters are shown as small spheres.

## 3.4. Bifurcated Reaction at $Q_o$ Site

One of the most important features and a continuing paradox of Q cycle theory is the nature of bifurcation at $Q_o$ site, by which the two electrons from one hydroquinone molecule are funneled into two separate paths. A good summary of possible mechanisms has been made by Berry *et al.* (1999b). Jünemann *et al.* (1998) has proposed a "thermodynamic explanation", in which it was argued that the instability of semiquinone at $Q_o$ site, together with the fact that the Rieske protein can only accept one electron at one time, will prevent both electrons from going down the high-potential chain. The domain movement of the Rieske protein makes this argument stronger, as the movement would leave the haem $b_L$ as the only electron acceptor nearby the semiquinone. Another model (Brandt, 1991; Brandt and von Jagow, 1991) requires a "catalytic switch" mechanism to prevent inappropriate reduction of the Rieske protein by the newly generated semiquinone at $Q_o$ site. A conformational change was suggested to serve as a gate opening and closing to the Rieske protein alternatively. No significant conformation change on cytochrome *b* has been observed, though the movement of Rieske extrinsic domain could obviously be involved in the gating (Brandt, 1998).

The third mechanism is based on the inference of the different binding modes for quinone species at $Q_o$ site from the different binding positions of the two classes of $Q_o$ site inhibitors as discussed in previous section. It was proposed by Crofts and Berry (1998) that after transfer of the first electron to Rieske protein at the stigmatellin binding site, the hydrogen bond breaks and the semiquinone flips into the position of the MOA inhibitors before passing the second electron to haem $b_L$. This model predicts high occupancy of a semiquinone under conditions of oxidant-induced reduction, and the failure to observe the radical by EPR spectroscopy under these conditions needs to be explained (Junemann *et al.*, 1998).

Based on the reaction center stigmatellin model (Lancaster and Michel, 1997), Link (1997) proposed a proton-gated affinity-change mechanism. In this mechanism, the semiquinone, formed upon the transfer of the first electron to Rieske protein, binds tightly to the Rieske protein and raises its midpoint potential above that of cytochrome $c_1$. The Rieske protein is thus blocked, both physically and thermodynamically, from disposing of the first electron until after deprotonation and transfer of the second electron converts the semiquinone to the weakly bound ubiquinone. This model also predicts accumulation of the semiquinone under the condition of oxidant-induced reduction, but the failure to observe the semiquinone can be explained by assuming that the radical is spin-coupled with the unpaired electron of reduced Rieske $Fe_2S_2$ cluster. This model requires the second electron be transferred from the semiquinone in the stigmatellin position.

## 4. SUMMARY AND PERSPECTIVE

The $bc_1$ complex is one of the best studied membrane proteins. Research in previous decades using biochemical and molecular biology techniques has allowed the construction of sophisticated models of the complex, such as the protonmotive Q cycle model (Crofts, 1985; Mitchell, 1976). Recent crystallographic studies provided a huge body of new information, which confirmed, in most part, these models, and also offered a few interesting surprises. The novel "domain shuttle" mechanism (Zhang *et al.*, 1998; Kim *et al.*, 1998; Iwata *et al.*, 1998) has been proposed based on the crystal structures and is supported by biochemical studies. Some basic questions remain to be answered, such as the structural details and the nature of the bifurcation at $Q_o$ site, the "cytochrome b-150" phenomenon at the $Q_i$ site, and the binding mode between $bc_1$ complex and its redox partner cytochrome $c$. Future research should include obtaining crystals of substantial higher resolution, identifying quinone substrate at $Q_o$ site, and co- crystallizing $bc_1$ with cytochrome $c$.

ACKNOWLEDGEMENTS. The authors' research is supported by grants from the Office of Biosciences and Environmental Research, U.S. Department of Energy (to SHK; DE AC03-76SF00098) and from The National Institute of Health (to EAB; R01DK44842). We are grateful to the staffs at Stanford Synchrotron Radiation Laboratory at Stanford, CA, National Synchontron Light Source at Brookhaven National Laboratory, NY and Advanced Light Source, Lawrence Berkeley National Laboratory at Berkeley, CA.

## 5.  REFERENCES

Akiba, T., Toyoshima, C., Matsunaga, T., Kawamoto, M., Kubota, T., Fukuyama, K., Namba, K., and Matsubara, H., 1996, Three-dimensional structure of bovine cytochrome bc1 complex by electron cryomicroscopy and helical image reconstruction, *Nature Struct. Biol.* 3:553–561.

Ambler, R. P., 1991, Sequence variability in bacterial cytochromes c, *Biochim. Biophys. Acta* **1058**:42–47.

Bechmann, G., Schulte, U., and Weiss, H., 1992, Mitochondrial ubiquinol-cytochrome c oxidoreductase, in: *Molecular Mechanisms in Bioenergetics*, (L. Ernster, ed.) Elsevier, Amsterdam, pp. 199–216.

Beckmann, J. D., Ljungdahl, P. O., and Trumpower, B. L., 1989, Mutational analysis of the mitochondrial Rieske iron-sulfur protein of *Saccharomyces cerevisiae*. I. Construction of a RIP1 deletion strain and isolation of temperature-sensitive mutants, *J. Biol. Chem.* **264**:3717–3722.

Berry, E. A., Huang, L.-S., Earnest, T. N., and Jap, B. K., 1992, X-ray Diffraction by Crystals of Beef Heart Ubiquinol:Cytochrome c Oxidoreductase, *J. Mol. Biol.* **224**:1161–1166.

Berry, E. A., Huang, L.-S., Zhang, Z., and Kim S.-H., 1999a, The structure of the Avian Mitochondrial cytochrome $bc_1$ complex, *J. Bioenerg. Biomembr.* **31**(in press).

Berry, E. A., Zhang, Z., Huang, L.-S., and Kim, S.-H., 1999b, Structures of quinone-binding sites in *bc* complexes: functional implications, *Biochem. Soc. Trans.* **27**:565–572.

Brand, M. D., 1977, The stoichiometric relationships between electron transport, proton translocation and adenosine triphosphate synthesis and hydrolysis in mitochondria, *Biochem. Soc. Trans.* **5**:1615–1620.

Brandt, U., 1998, The chemistry and mechanics of ubihydroquinone oxidation at center P (Qo) of the cytochrome bc1 complex, *Biochim. Biophys. Acta* **1365**:261–208.

Brandt, U., Haase, U., Schagger, H., and von Jagow, G., 1991, Significance of the "Rieske" iron-sulfur protein for formation and function of the ubiquinol-oxidation pocket of mitochondrial cytochrome c reductase (bc1 complex), *J. Biol. Chem.* **266**:19958–19964.

Brandt, U., and von Jagow, G., 1991, Analysis of inhibitor binding to the mitochondrial cytochrome c reductase by fluorescence quench titration. Evidence for a "catalytic switch" at the Qo center, *Eur. J. Biochem.* **195**:163–170.

Brandt, U., Uribe, S., Schagger, H., and Trumpower, B. L., 1994, Isolation and characterization of QCR10, the nuclear gene encoding the 85-kDa subunit 10 of the *Saccharomyces cerevisiae* cytochrome $bc_1$ complex, *J. Biol. Chem.* **269**:12947–12953.

Brandt, U., Yu, L., Yu, C. A., and Trumpower, B. L., 1993, The mitochondrial targeting presequence of the Rieske iron-sulfur protein is processed in a single step after insertion into the cytochrome bc1 complex in mammals and retained as a subunit in the complex, *J. Biol. Chem.* **268**:8387–8390.

Brasseur, G., Saribas, A. S., and Daldal F., 1996, A Compilation of mutations located in the Cytochrome b Subunit of the Bacterial and Mitochondrial bc1 Complex, *Biochim. Biophys. Acta* **1275**:61–69.

Braun, H. P., and Schmitz, U. K., 1995, Are the "core" proteins of the mitochondrial bc1 complex evolutionary relics of a processing protease? *Trends Biochem.* **20**:171–175.

Braun, H. P., Emmermann, M., Kruft, V., and Schmitz, U. K., 1992, The general mitochondrial processing peptidase from potato is an integral part of cytochrome c reductase of the respiratory chain, *EMBO J.* **11**:3219–3227.

Braun, H. P., Emmermann, M., Kruft, V., Bodicker, M., and Schmitz, U. K., 1995, The general mitochondrial processing peptidase from wheat is integrated into the cytochrome bc1-complex of the respiratory chain, *Planta* **195**:396–402.

Britt, R. D., Sauer, K., Klein, M. P., Knaff, D. B., Kriauciunas, A., Yu, C. A., Yu, L., and Malkin, R., 1991, Electron spin echo envelope modulation spectroscopy supports the suggested coordination of two histidine ligands to the Rieske Fe-S centers of the cytochrome b6f complex of spinach and the cytochrome bc1 complexes of *Rhodospirillum rubrum*, *Rhodobacter sphaeroides*, and bovine heart mitochondria, *Biochemistry* **30**:1892–1901.

Broger, C., Salardi, S., and Azzi, A., 1983, Interaction between isolated cytochrome c1 and cytochrome c, *Eur. J. Biochem.* **131**:349–352.

Carrell, C. J., Zhang, H., Cramer, W. A., and Smith, J. L., 1997, Biological identity and diversity in photosynthesis and respiration: structure of the lumen-side domain of the chloroplast Rieske protein, *Structure* **5**:1613–1625.

Chazotte, B., Wu, E. S., and Hackenbrock, C. R., 1991, The mobility of a fluorescent ubiquinone in model lipid membranes. Relevance to mitochondrial electron transport, *Biochim. Biophys. Acta* **1058**:400–409.

Cocco, T., Lorusso, M., Sardanelli, A. M., Minuto, M., Ronchi, S., Tedeschi, G., and Papa, S., 1991, Structural and functional characteristics of polypeptide subunits of the bovine heart ubiquinol–cytochrome-c reductase complex, *Eur. J. Biochem.* **195**:731–734.

Collaborative Computational Project Number 4, 1994, The CCP4 suite: programs for protein crystallography, *Acta Cryst.* D **50**:760–763.

Crofts, A. R., 1985, The mechanism of ubiquinol:cytochrome c oxidoreductases of mitochondria and of *Rhodopseudomonas sphaeroides*, in: *The Enzymes of Biological Membranes*, Volume 4 (A. N. Martonosi, ed.), Plenum Press, New York, pp. 347–382.

Crofts, A. R., Barquera, B., Bechmann, G., Guergova, M., Salcedo-Hernandez, R., Hacker, B., Hong, S., and Gennis, R. B., 1995, Structure and function in the bc1-complex of Rb. Sphaeroides, in: *Photosynthesis: from light to biosphere*. Vol. II (P. Mathis, ed.) Kluwer Academic Publishers, Dordrecht, pp. 493–500.

Crofts, A. R., and Berry, E. A., 1998, Structure and function of cytochrome $bc_1$ complex of mitochondria and photosynthetic bacteria, *Curr. Opin. in Struct. Biol.* **8**:501–509.

Crofts, A. R., Hacker, B., Barquera, B., Yun, C.-H., and Gennis, R. B., 1992, Structure and function of the bc-complex of *Rhodobacter sphaeroides*, in: *Research in Photosynthesis*, Vol. II (N. Murata, ed.), Kluwer Academic Publishers, Dordrecht, pp. 463–470.

Davidson, E., Ohinshi, T., Atta-Asafo-Adjei, E., and Daldal, F., 1992, Potential ligands of the [2Fe-2S] Rieske cluster of the cytochrome $bc_1$ complex of *Rhodobacter capsulatus* probed by site-directed mutagenesis, *Biochemistry* **31**:3342–3351.

Deng, K., Zhang, L., Kachurin, A. M., Yu, L., Xia, D., Kim, H., Deisenhofer, J., and Yu, C. A., 1998, Activation of a matrix processing peptidase from the crystalline cytochrome bc1 complex of bovine heart mitochondria, *J. Biol. Chem.* **273**:20752–20757.

Denke, E., Merbitz-Zahradnik, T., Hatzfeld, O. M., Snyder, C. H., Link, T. A., and Trumpower, B. L., 1998, Alteration of the midpoint potential and catalytic activity of the rieske iron-sulfur protein by changes of amino acids forming hydrogen bonds to the iron-sulfur cluster, *J. Biol. Chem.* **273**:9085–9093.

Ding, H., Robertson, D. E., Daldal, F., and Dutton, P. L., 1992, Cytochrome bc1 complex [2Fe-2S] cluster and its interaction with ubiquinone and ubihydroquinone at the Qo site: a double-occupancy Qo site model, *Biochemistry* **31**:3144–3158.

Ding, H., Moser, C. C., Robertson, D. E., Tokito, M. K., Daldal, F., and Dutton, P. L., 1995, Ubiquinone pair in the Qo site central to the primary energy conversion reactions of cytochrome bc1 complex, *Biochemistry* **34**:15979–15996.

Dutton, P. L., and Jackson, J. B., 1972, Thermodynamic and kinetic characterization of electron transfer components *in situ* in *Rhodopseudomonas spheroides* and *Rhodospirillum rubrum*, *Eur. J. Biochem.* **30**:495–510.

de la Rosa, F. F., and Palmer, G., 1983, Reductive titration of CoQ-depleted Complex III from Baker's yeast. Evidence for an exchange-coupled complex between QH and low-spin ferricytochrome b, *FEBS Lett.* **163**:140–143.

de Vries, S., Berden, J. A., and Slater, E. C., 1980, Properties of a semiquinone anion located in the QH2:cytochrome c oxidoreductase segment of the mitochondrial respiratory chain, *FEBS Lett.* **122**:143–148.

Esposti, M. D., De Vries, S., Crimi, M., Ghelli, A., Patarnello, T., and Meyer, A., 1993, Mitochondrial cytochrome b: evolution and structure of the protein, *Biochim. Biophys. Acta* **1143**:243–271.

Gatti, D. L., Meinhardt, S. W., Ohnishi, T., and Tzagoloff, A., 1989, Structure and function of the mitochondrial $bc_1$ complex. A mutational analysis of the yeast Rieske iron-sulfur protein, *J. Mol. Biol.* **205**:421–435.

Gatti, D. L., and Tzagoloff, A., 1990, Structure and function of the mitochondrial $bc_1$ complex. Properties of the complex in temperature-sensitive cor1 mutants, *J. Biol. Chem.* **265**:21468–21475.

Gencic, S., Schagger, H., and von Jagow, G., 1991, Core I protein of bovine ubiquinol-cytochrome-c reductase; an additional member of the mitochondrial-protein-processing family. Cloning of bovine core I and core II cDNAs and primary structure of the proteins, *Eur. J. Biochem.* **199**:123–131.

Glick, B. S., Beasley, E. M., and Schatz, G., 1992, Protein sorting in mitochondria, *Trends Biochem. Sci.* **17**:453–459.

Gomez, B., Jr, and Robinson, N. C., 1999, Quantitative determination of cardiolipin in mitochondrial electron transferring complexes by silicic acid high-performance liquid Chromatography, *Anal Biochem.* **267**:212–216.

Gonzalez-Halphen, D., Lindorfer, M. A., and Capaldi, R. A., 1988, Subunit arrangement in beef heart complex III, *Biochemistry* **27**:7021–7031.

Graham, L. A., Brandt, U., Sargent, J. S., and Trumpower, B. L., 1992, Mutational analysis of assembly and function of the iron-sulfur protein of the cytochrome $bc_1$ complex in *saccharomyces cerevisiae, J. Bioenerg. Biomembr.* **25**:245–257.

Graham, L. A., and Trumpower, B. L., 1991, Mutational analysis of the mitochondrial Rieske iron-sulfur protein of *saccharomyces cerevisiae*. III. Import, protease processing, and assembly into the cytochrome $bc_1$ complex of iron-sulfur protein lacking the iron-sulfur cluster, *J. Biol. Chem.* **266**:22485–22492.

Gray, K. A., Davidson, E., and Daldal, F., 1992, Mutagenesis of methionine—183 drastically affects the physicochemical properties of cytochrome c1 of the bc1 complex of *Rhodobacter capsulatus, Biochemistry* **31**:11864–11873.

Gurbiel, R. J., Batie, C. J., Sivaraja, M., True, A. E., Fee, J. A., Hoffman, B. M., and Ballou, D. P., 1989, Electron-nuclear double resonance spectroscopy of 15N-enriched phthalate dioxygenase from *Pseudomonas cepacia* proves that two histidines are coordinated to the [2Fe-2S] Rieske-type clusters, *Biochemistry* **28**:4861–4871.

Gurbiel, R. J., Ohnishi, T., Robertson, D. E., Daldal, F., and Hoffman, B. M., 1991, Q-band ENDOR spectra of the Rieske protein from *Rhodobactor capsulatus* ubiquinol-cytochrome c oxidoreductase show two histidines coordinated to the [2Fe-2S] cluster, *Biochemistry* **30**:11579–11584.

Hacker, B., Barquera, B., Crofts, A. R., and Gennis, R. B., 1993, Characterization of mutations in the cytochrome b subunit of the bc1 complex of *Rhodobacter sphaeroides* that affect the quinone reductase site (Qc), *Biochemistry* **32**:4403–4410.

Hase, T., Harabayashi, M., Kawai, K., and Matsubara, H., 1987, A carboxyl–terminal hydrophobic region of yeast cytochrome $c_1$ is necessary for functional assembly into complex III of the respiratory chain, *J. Biochem.* (Tokyo) **102**:411–419.

Hinkle, P. C., Kumar, M. A., Resetar, A., and Harris, D. L., 1991, Mechanistic stoichiometry of mitochondrial oxidative phosphorylation, *Biochemistry* **30**:3576–3582.

Hovmöller, S., Leonard, K., and Weiss, H., 1981, Membrane crystals of a subunit complex of mitochondrial cytochrome reductase containing the cytochromes b and c1, *FEBS Lett.* **123**:118–122.

Iwata, S., Lee, J. W., Okada, K., Lee, J. K., Iwata, M., Rasmussen, B., Link, T. A., Ramaswamy, S., and Jap, B. K., 1998, Complete structure of the 11-subunit bovine mitochondrial cytochrome $bc_1$ complex, *Science* **281**:64–71.

Iwata, S., Saynovits, M., Link, T. A., and Michel, H., 1996, Structure of a water soluble fragment of the "Rieske" iron-sulfur protein of the bovine heart mitochondrial cytochrome bc1 complex determined by MAD phasing at 1.5 Å resolution, *Structure* **4**:567–579.

Junemann, S., Heathcote, P., and Rich, P. R., 1998, On the mechanism of quinol oxidation in the bc1 complex, *J. Biol. Chem.* **273**:21603–21607.

Karlsson, B., Hovmoller, S., Weiss, H., and Leonard, K., 1983, Structural studies of cytochrome reductase. Subunit topography determined by electron microscopy of membrane crystals of a subcomplex, *J. Mol. Biol.* **165**:287–302.

Kim, C. H., Balny, C., and King, T. E., 1987, Role of the hinge protein in the electron transfer between cardiac cytochrome c1 and c. Equilibrium constants and kinetic probes, *J. Biol. Chem.* **262**:8103–8108.

Kim, H., Xia, D., Yu, C.-A., Xia, J.-Z., Kachurin, A. M., Zhang, L., Yu, L., and Deisenhofer, J., 1998, Inhibitor binding changes domain mobility in the iron-sulfur protein of the mitochondrial bc1 complex from bovine heart, *Proc. Natl. Acad. Sci.* (U.S.) **95**:8026–8033.

Kim, C. H., Yencha, A. J., Bunker, G., Zhang, G., Chance, B., and King, T. E., 1989, Effect of the hinge protein on the heme iron site of cytochrome c1, *Biochemistry* **28**:1439–1441.

Kingsley, P. B., and Feigenson, G. W., 1981, 1H-NMR study of the location and motion of ubiquinones in perdeuterated phosphatidylcholine bilayers, *Biochim. Biophys. Acta* **635**:602–618.

Konig, B. W., Wilms, J., and Van Gelder, B. F., 1981, The reaction between cytochrome c1 and cytochrome c, *Biochim. Biophys. Acta* **636**:9–16.

Kraulis, P. J., 1991, MOLSCRIPT: a program to produce both detailed and schematic plots of protein structures, *J. Appl. Cryst.* **24**:946–950.

Kubota, T., Kawamoto, M., Fukuyama, K., Shinzawa-Itoh, K., Yoshikawa, S., and Matsubara, H., 1991, Crystallization and preliminary X-ray crystallographic studies of bovine heart mitochondrial cytochrome $bc_1$ complex, *J. Mol. Biol.* **221**:379–382.

Lancaster, C. R., and Michel, H., 1997, The coupling of light-induced electron transfer and proton uptake as derived from crystal structures of reaction centres from Rhodopseudomonas viridis modified at the binding site of the secondary quinone, $Q_B$, *Structure* **5**:1339–1359.

Leonard, K., Wingfield, P., Arad, T., and Weiss, H., 1981, Three-dimensional structure of ubiquinone:cytochrome c reductase from *Neurospora* mitochondria determined by electron microscopy of membrane crystals, *J. Mol. Biol.* **149**:259–274.

Li, Y., Leonard, K., and Weiss, H., 1981, Membrane–bound and water-soluble cytochrome $c_1$ from *Neurospora* mitochondria, *Eur. J. Biochem.* **116**:199–205.

Link, T. A., 1997, The role of the "Rieske": iron sulfur protein in the hydroquinone oxidation ($Q_p$-) site of the cytochrome $bc_1$ complex: the "proton-gated affinity change" mechanism, *FEBS Lett.* **412**:257–264.

Link, T. A., Haase, U., Brandt, U., and von Jagow, G., 1993, What information do inhibitors provide about the structure of the hydroquinone oxidation site of ubihydroquinone: cytochrome c oxidoreductase? *J. Bioenerg. Biomembr.* **25**:221–232.

Link, T. A., Saynovits, M., Assmann, C., Iwata, S., Ohnishi, T., and von Jagow, G., 1996. Isolation, characterization, and crystallization of a water-soluble fragment of the Rieske iron-sulfur protein of bovine heart mitochondrial bc1 complex, *Eur. J. Biochem.* **237**: 71–75.

Link, T. A., Schägger, H., and Von Jagow, G., 1987, Structural analysis of the $bc_1$ complex from beef heart mitochondria by the sidedness hydropathy plot and by comparison with other bc complexes. in: *Cytochrome Systems: Molecular Biology and Bioenergetics*, (S. Papa, B. Chance, and L. Ernster, eds.) Plenum Press, New York, pp. 289–301.

Link, T. A., Wallmeier, H., and von Jagow, G., 1994, Modeling the three dimensional structure of cytochrome b, *Biochem. Soc. Trans.* **22**:197–203.

Meinhardt, S. W., and Crofts, A. R., 1982, The site and mechanism of action of myxothiazol as an inhibitor of electron transfer in *Rhodopseudomonas sphaeroides*, *FEBS Lett.* **149**:217–222.

Mitchell, P., 1961, Coupling of phosphorylation to electron and proton transfer by a chemiosmotic type of mechanism, *Nature* **191**:144–148.

Mitchell, P., 1975, Protonmotive redox mechanism of the cytochrome b-c1 complex in the respiratory chain: protonmotive ubiquinone cycle, *FEBS Lett.* **56**:1–6.

Mitchell, P., 1976, Possible molecular mechanisms of the protonmotive function of cytochrome systems, *J. Theor. Biol.* **62**:327–367.

Miyoshi, H., Tokutake, N., Imaeda, Y., Akagi, T., and Iwamura, H., 1995, A model of antimycin A binding based on structure-activity studies of synthetic antimycin A analogues, *Biochim. Biophys. Acta* **1230**:149–154.

Mukai, K., Miyazaki, T., Wakabayashi, S., Kuramitsu, S., and Matsubara, H., 1985, Dissociation of bovine cytochrome c1 subcomplex and the status of cysteine residues in the subunits, *J. Biochem.* (Tokyo) **98**:1417–1425.

Musatov, A., and Robinson, N. C., 1994, Detergent-solubilized monomeric and dimeric cytochrome $bc_1$ isolated from bovine heart, *Biochemistry* **33**:13005–13012.

Neft, N., and Farley, T. M., 1971, Inhibition of electron transport by substituted salicyl-N-(n-octadecyl)amides, *J. Med. Chem.* **14**:1169–1170.

Ohnishi, T., and Trumpower, B. L., 1980, Differential effects of antimycin on ubisemiquinone bound in different environments in isolated succinate: cytochrome c reductase complex, *J. Biol. Chem.* **255**:3278–3284.

Oudshoorn, P., Van Steeg, H., Swinkels, B. W., Schoppink, P., and Grivell, L. A., 1987, Subunit II of yeast QH2: cytochrome-c oxidoreductase. Nucleotide sequence of the gene and features of the protein, *Eur. J. Biochem.* **163**:97–103.

Perkins, J. S., and Weiss, H., 1983, Low-resolution structural studies of mitochondrial ubiquinol:cytochrome c reductase in detergent solutions by neutron scattering, *J. Mol. Biol.* **168**:847–866.

Phillips, J. D., Graham, L. A., and Trumpower, B. L., 1993, Subunit 9 of the *Saccharomyces cerevisiae* cytochrome $bc_1$ complex is required for insertion of EPR-detectable iron-sulfur cluster into the Rieske iron-sulfur protein, *J. Biol. Chem.* **268**:11727–11736.

Rich, P. R., Jeal, A. E., Madgwick, S. A., and Moody, A. J., 1990, Inhibitor effects on redox-linked protonations of the b haems of the mitochondrial bc1 complex, *Biochim. Biophys. Acta* **1018**:29–40.

Rieske, J. S., Maclennan, D. H., and Coleman, R., 1964, Isolation and properties if an iron-protein from the (reduced coenzyme Q)-cytochrome c reductase complex of the respiratory chain, *Biochem. Biophys. Res. Commun.* **15**:338–344.

Rawlings, N. D., and Barrett, A. J., 1993, Evolutionary families of peptidases, *Biochem. J.* **290**:205–218.

Robertson, D. E., Daldal, F., and Dutton, P. L., 1990, Mutants of ubiquinol-cytochrome c2 oxidoreductase resistant to Qo site inhibitors: consequences for ubiquinone and ubiquinol affinity and catalysis, *Biochemistry* **29**:11249–11260.

Robertson, D. E., Prince, R. C., Bowyer, J. R., Matsuura, K., Dutton, P. L., and Ohnishi, T., 1984, Thermodynamic properties of the semiquinone and its binding site in the ubiquinol-cytochrome c (c2) oxidoreductase of respiratory and photosynthetic systems, *J. Biol. Chem.* **259**:1758–1763.

Salerno, J. C., Osgood, M., Liu, Y. J., Taylor, H., and Scholes, C. P., 1990, Electron nuclear double resonance (ENDOR) of the Qc.-ubisemiquinone radical in the mitochondrial electron transport chain, *Biochemistry* **29**:6987–6993.

Salerno, J. C., Xu, Y., Osgood, M. P., Kim, C. H., and King, T. E., 1989, Thermodynamic and spectroscopic characteristics of the cytochrome bc1 complex. Role of quinone in the behavior of cytochrome b562, *J. Biol. Chem.* **264**:15398–153403.

Saraste, M., 1999, Oxidative Phosphorylation at the *fin de siècle*, *Science* **283**:1488–1493.

Schagger, H., Hagen, T., Roth, B., Brand, U., Link, T. A., and von Jagow, G., 1990, Phospholipid specificity of bovine heart bc1 complex, *Eur. J. Biochem.* **190**:123–130.

Schägger, H., Link, T. A., Engel, W. D., and von Jagow, G., 1986, Isolation of the eleven protein subunits of the bc₁ complex from beef heart, *Methods in Enzymology* **126**:224–237.

Schroter, T., Hatzfeld, O. M., Gemeinhardt, S., Korn, M., Friedrich, T., Ludwig, B., and Link, T. A., 1998, Mutational analysis of residues forming hydrogen bonds in the Rieske [2Fe-2S] cluster of the cytochrome bc1 complex in *Paracoccus denitrificans*, *Eur. J. Biochem.* **255**:100–106.

Schulte, U., Arretz, M., Schneider, H., Tropschug, M., Wachter, E., Neupert, W., and Weiss, H., 1989, A family of mitochondrial proteins involved in bioenergetics and biogenesis, *Nature* **339**:147–149.

Slater, E. C., 1973, The mechanism of action of the respiratory inhibitor, antimycin, *Biochim. Biophys. Acta* **301**:130–154.

Stonehuerner, J., O'Brien, P., Geren, L., Millett, F., Steidl, J., Yu, L., and Yu, C. A., 1985, Identification of the binding site on cytochrome c1 for cytochrome c, *J. Biol. Chem.* **260**:5392–5398.

Tian, H., White, S., Yu, L., and Yu, C. A., 1999, Evidence for the head domain movement of the rieske iron-sulfur protein in electron transfer reaction of the cytochrome bc1 complex, *J. Biol. Chem.* **274**:7146–7152.

Tian, H., Yu, L., Mather, M. W., and Yu, C. A., 1999, Flexibility of the neck region of the rieske iron-sulfur protein is functionally important in the cytochrome bc1 complex, *J. Biol. Chem.* **273**:27953–27959.

Tron, T., Crimi, M., Colson, A. M., and Degli Esposti, M., 1991, Structure/function relationships in mitochondrial cytochrome b revealed by the kinetic and circular dichroic properties of two yeast inhibitor-resistant mutants, *Eur. J. Biochem.* **199**:753–760.

Trumpower, B. L., and Edwards, C. A., 1979, Identification of oxidation factor as a reconstitutively active form of the iron-sulfur protein of the cytochrome $b$-$c_1$ segment of the respiratory chain, *FEBS Lett.* **100**:13–16.

Van Doren, S. R., Yun, C.-H., Crofts, A. R., and Gennis, R., 1993, Assembly of the Rieske iron-sulfur subunit of the cytochrome bc1 complex in Escherichia·coli and Rhodobacter sphaeroides membranes independent of the cytochrome b and c1 subunits. *Biochemistry* **32**:628–636.

von Jagow, G., and Link, T. A., 1986, Use of specific inhibitors on the mitochondrial $bc_1$ complex, *Methods in Enzymology* **126**:253:271.

Wakabayashi, S., Takeda, H., Matsubara, H., Kim, C. H., and King, T. E., 1982, Identity of the heme-not-containing protein in bovine heart cytochrome $c_1$ preparation with the protein mediating c1-c complex formation—a protein with high glutamic acid content, *J. Biochem.* (Tokyo) **91**:2077–2085.

Weiss, H., and Kolb, H. J., 1979, Isolation of mitochondrial succinate: ubiquinone reductase, cytochrome c reductase and cytochrome c oxidase from Neurospora crassa using nonionic detergent, *Eur. J. Biochem.* **99**:139–149.

Weiss, H., and Leonard, K., 1987, Structure and function of mitochondrial ubiquinol: cytochrome c reductase and NADH:ubiquinone reductase, *Chemica Scripta* **27B**:73–81.

Xia, D., Yu, C.-A., Kim, H., Xia, J.-Z., Kachurin, A. M., Zhang, L., Yu, L., and Deisenhofer, J., 1997, Crystal structure of the cytochrome bc1 complex from bovine heart mitochondria, *Science* **277**:60–66.

Yu, C. A., Nagaoka, S., Yu, L., and King, T. E., 1980, Evidence of ubisemiquinone radicals in electron transfer at the cytochromes $b$ and $c_1$ region of the cardiac respiratory chain, *Arch. Biochem. Biophys.* **204**:59–70.

Yu, C. A., Xia, D., Kim, H., Deisenhofer, J., Zhang, Li., Kachurin, A. M., and Yu, L., 1998, Structural basis of functions of the mitochondrial cytochrome bc1 complex, *Biochim. Biophys. Acta* **1101**:162–165.

Yu, C. A., Xia, J. Z., Kachurin, A. M., Yu, L., Xia, D., Kim, H., and Deisenhofer, J., 1996, Crystallization and preliminary structure of beef heart mitochondrial cytochrome-$bc_1$ complex, *Biochim. Biophys. Acta* **1275**:47–53.

Yu, L., and Yu, C. A., 1982, The interaction of arylazido ubiquinone derivative with mitochondrial ubiquinol-cytochrome c reductase, *J. Biol. Chem.* **257**:10215–10221.

Yue, W. H., Zou, Y. P., Yu, L., and Yu, C. A., 1991, Crystallization of mitochondrial ubiquinol-cytochrome c reductase, *Biochemistry* **30**:2303–2306.

Yun, C. H., Crofts, A. R., and Gennis, R. B., 1991, Assignment of the histidine axial ligands to the cytochrome $b_H$ and cytochrome $b_L$ components of the $bc_1$ complex from *Rhodobacter sphaeroides* by site-directed mutagenesis, *Biochemistry* **30**:6747–6754.

Yun, C. H., Wang, Z., Crofts, A. R., and Gennis, R. B., 1992, Examination of the functional roles of 5 highly conserved residues in the cytochrome b subunit of the bc1 complex of *Rhodobacter sphaeroides*, *J. Biol. Chem.* **267**:5901–5909.

Zhang, Z., Huang, L.-S., Shulmeister, V.-M., Chi, Y.-I., Kim, K. K., Hung, L.-W., Crofts, A. R., Berry, E. A., and Kim, S.-H., 1998, Electron transfer by domain movement in cytochrome $bc_1$, *Nature* **392**:677–684.

*Chapter 17*

# Bovine Heart Cytochrome c Oxidase

Shinya Yoshikawa

## 1. INTRODUCTION

Mitochondrial cytochrome c oxidase reduces molecular oxygen ($O_2$, hereafter) to water, and this process is coupled to the pumping of protons through the mitochondrial inner membrane from matrix space to inter-membrane space (Ferguson-Miller and Babcock, 1996). The elucidation of the reaction mechanism of this enzyme continues to be one of the most important and intriguing subjects in biological science. Elucidation of the reaction mechanism of an enzyme means complete description of changes in the three-dimensional structure of the enzyme during the enzymic turnover triggered by its substrates. Thus, solving of the three dimensional structures in various intermediate states during enzymic turnover is crucial for a complete understanding of the reaction mechanism. A powerful method for determination of the three-dimensional structure of a protein is X-ray crystallography. However, using this method for the identification of the structures of short-lived transient intermediates at the active site can be quite difficult. Furthermore, the resolution of the X-rays scattered from crystals of enzymes is, in most cases, not high enough for a complete

**SHINYA YOSHIKAWA**      Department of Life Science, Himeji Institute of Technology, and CREST, Japan Science and Technology Corporation (JST) Kamigohri, Akoh, Hyogo 678-1297 Japan.

*Subcellular Biochemistry, Volume 35: Enzyme-Catalyzed Electron and Radical Transfer*, edited by Holzenburg and Scrutton. Kluwer Academic / Plenum Publishers, New York, 2000.

evaluation of the chemical reactivity at the catalytic site. Thus, implementations of additional spectroscopic methods are indispensable for a complete description of chemical events that take place the at catalytic site of an enzyme. However, any spectroscopic method is sensitive only to the chromophores and it provides essentially one-dimensional information. Thus, if an X-ray structure for a particular enzyme is not available, chemical events at the catalytic site cannot be completely elucidated. Thus crystallographic and spectroscopic methods are quite complementary and both types of methods are required for a complete examination of any enzyme reaction mechanism.

Several requirements need to be considered in order to obtain the X-ray structure of an enzyme at high resolution. First of all, the enzyme must be purified from cells for crystallization, and the chemical composition of the isolated enzyme needs to be determined. These initial steps in elucidation of the reaction mechanisms of large multicomponent membrane proteins such as cytochrome c oxidase are usually quite time-consuming. In this article, the first and fundamental step required for structure-function investigations of cytochrome c oxidase, determination of the composition of the purified enzyme and its crystallization, will be reviewed. Then, mechanisms of $O_2$ reduction and proton pumping will be discussed with emphasis on the importance of the X-ray crystallographic findings for increasing our understanding of the reaction mechanism.

## 2.  COMPOSITION OF BOVINE HEART CYTOCHROME C OXIDASE

### 2.1.  Purification

Usually it is very difficult to isolate membrane proteins since typical membrane proteins have large hydrophobic surfaces surrounding a central core region. These hydrophobic surfaces interact with the phospholipid bilayer of the biological membrane. On the other hand, the two surfaces at the outer ends of membrane proteins are exposed to the aqueous phases on both sides of the biological membrane. This common arrangement of hydrophobic and hydrophilic surfaces on membrane proteins indicates that in their isolated form membrane proteins are unstable in an isolated aqueous medium or in an isolated hydrophobic medium. Actually, an example of a membrane protein that has been isolated in an organic solvent cannot be found. Thus, the best method for isolation of membrane proteins is to extract the protein from the phospholipid bilayer by exchanging the membrane phospholipids bound to the hydrophobic surface of the membrane protein with amphipathic detergent molecules. The addition of detergent stabilizes the membrane protein in aqueous solutions. All the previously-reported membrane protein preparations which have yielded crystals diffracting X-rays to high resolution are detergent solubilized preparations (Garavito and Picot, 1990).

Bovine heart cytochrome c oxidase was first isolated by Okunuki and Yakushiji (Okunuki and Yakushiji, 1941; Yakushiji and Okunuki, 1941) using the detergent sodium cholate about 60 years ago. The isolated preparation did not exhibit $O_2$ consumption activity because of the inhibitory effect of sodium cholate. However, by exchanging this anionic detergent with non-ionic synthetic detergents (Yonetani, 1959), the enzymic activity was fully recovered. After these initial reports, many additional purification procedures have been reported (Caughey et al., 1976; Steffens and Buse, 1976; Yoshikawa et al., 1977). Unfortunately, the characteristics reported for these preparations differ, and each group reporting a purification procedure believed their preparation to be the best. However, in reality, the fundamental properties associated with each of these preparations are for the most part essentially the same.

The most uniform preparation of an enzyme with regard to three dimensional structure produces the best crystals, since the quality of a protein crystal depends primarily on structural homogeneity at the molecular level. Structural homogeneity is likely to be achieved when all the protein molecules in a sample are in the thermodynamically most stable conformation which is identical to, at least, one of the physiologically relevant conformations. Interestingly, cholate, which is a strong naturally-occurring detergent found in the small intestine, seems to be the best detergent for solubilization of bovine heart cytochrome c oxidase, and it is still used in many procedures for the isolation of this enzyme.

Our purification procedure (Yoshikawa et al., 1977) is essentially identical to the Okunuki method (Okunuki et al., 1958). Steps in the Okunki method include solubilization of the enzyme from the mitochondrial inner membrane, followed by fractionation of the solubilized sample with ammonium sulfate in the presence of cholate, and then followed by additional fractionations with ammonium in the presence of the non-ionic detergent, decyl maltoside. The nonionic detergents are not as effective as the ionic detergents for the solubilization of this enzyme. Namely, the most effective detergent for solubilization may not necessarily be the best detergent for stabilization of the isolated protein in an aqueous solution.

Reproducibility of the purification procedure for a protein is one of the most critical factors not only for the determination of the chemical composition of the protein but also for the optimization of the crystallization conditions. Based on our experience, the most reproducible procedure for purification produces the most integral (or native) preparation. Alternate procedures with lower reproducibility produce preparations containing denatured proteins in varying amounts. The two most critical requirements for attaining this reproducibility during bovine heart cytochrome c oxidase preparation have been the freshness of the source material, bovine heart muscle, and the purity of the detergents, especially that of cholic acid.

As discussed below, crystallization of the enzyme is also an effective method for removing contaminating and denatured proteins. Crystallization has the potential to produce a preparation not only of high purity but also of extreme reproducibility in both composition and enzyme activity. An important property of crystallization is its inherent capability to select for protein molecules that possess the same three dimensional structure. This is in contrast to other purification steps which are likely to induce some degree of denaturation.

## 2.2. Metal Content

The purity of cytochrome c oxidase preparations is significantly dependent on the purification procedure. Most of the preparations which are not purified by crystallization contain contaminating proteins containing varying amounts of iron. Furthermore, bovine heart cytochrome c oxidase chelates copper ions. Thus, the ratio of copper to iron reported so far for the enzyme ranges between 4.5 and 1.0 (Okuniki *et al.*, 1958; Yonetani, 1961; Caughey *et al.*, 1976; Yoshikawa *et al.*, 1977; Yoshikawa *et al.*, 1988; Steffens *et al.*, 1987; Mochizuki *et al.*, 1999). From knowledge of the iron content, the molecular extinction coefficient of this enzyme has been determined. (Okuniki *et al.*, 1958; Caughey *et al.*, 1976; Mochizuki *et al.*, 1999). Contaminating iron atoms in non-crystalline preparations have yielded significantly lower values for the molecular extinction coefficients than that of the crystalline enzyme. Bovine heart cytochrome c oxidase stabilized in the presence of an alkylpolyoxyethylene-type detergent (BL8SY, $CH_3(CH_2)_{11}O$ $(CH_2CH_2O)_8H$), shows a value for the molecular extinction coefficient of $32.2 \, mM^{-1} cm^{-1}$ prior to crystallization of the enzyme. This molecular extinction coefficient value corresponds to the difference in absorbance at 604 nm versus 630 nm for the $\alpha$-band of fully reduced form. This value increases to $42.2 \, mM^{-1} cm^{-1}$ for a sample following crystallization, $45.6 \, mM^{-1} cm^{-1}$ for sample crystallized twice, and $46.6 \, mM^{-1} cm^{-1}$ for a sample crystallized three times. Standard deviations for these determinations are about $1.0 \, mM^{-1} m^{-1}$ (Mochizuki *et al.*, 1999). Within the error of the determination, there is no difference in the value of the extinction coefficient between the sample crystallized two times versus the sample crystallized three times.

This indicates that the sample crystallized twice is free from contaminating iron atoms, which were removed by crystallization. Similarly, the content of other metals also reach their lowest values following two crystallization steps, which produces a metal composition of $Fe:Cu:Mg:Zn = 2:3:1:1$ (Mochizuki *et al.*, 1999). The presence of Mg and Zn in bovine heart cytochrome c oxidase was discovered in 1985 (Einarsdottir and Caughey, 1985) and has been confirmed by other laboratories (Steffens *et al.*, 1987; Yoshikawa *et al.*, 1988). However, the physiological roles of these two metals are not known. The possibility that these metals are exogenous and co-purified with the enzyme during preparation cannot be excluded until the physiological function of these metals is understood. However, the continuous presence of these metals in the preparation even after repeated crystallization strongly indicates that these metals are intrinsic components. These observations show that crystallization can be a powerful tool for the determination of the composition of proteins, especially, of large multi-component proteins like cytochrome c oxidase.

Recently, a sodium ion has been detected in the X-ray structures of the fully oxidized and reduced forms of bovine heart cytochrome c oxidase at 2.3 Å and 2.35 Å resolutions, respectively (Yoshikawa *et al.*, 1998). In the bacterial enzyme, a site corresponding to the Na site in the bovine enzyme is occupied by a $Ca^{2+}$ ion (Ostermeier *et al.*, 1997). A comparison of the coordination geometry for the bacterial versus the mammalian species shows slight differences, and this observation is consistent with differences in coordination chemistry associated with each of the two metals. Prior to the determination of the X-ray structures, the existence of these metal sites had never been proposed. The presence of these metals in X-ray structure strongly indicates that the metals are intrinsic components of the enzyme. However, as in the case of the Mg and Zn ions, the physiological functions of Na and Ca are not understood.

## 2.3. Structures and Spectral Properties of the Redox-active Metal Sites

Iron atoms found in the structure of cytochrome c oxidase are located on the two heme A prosthetic groups. The structure of heme A is presented in Figure 1. Warburg (1924) discovered that this respiratory enzyme is a hemoprotein from its photochemical action spectrum. However, the structure of the porphyrin portion of this heme is much more complex and unstable than protoheme. The structure shown in Figure 1, was not determined until as late as 1975 by Caughey *et al.* (1975). However, at that time, the absolute configuration of a chiral carbon in the hydroxyfarnesyl ethyl group located at position 2 of the porphyrin ring was not known, since isolated heme A had never been crystallized for a small molecule X-ray structure

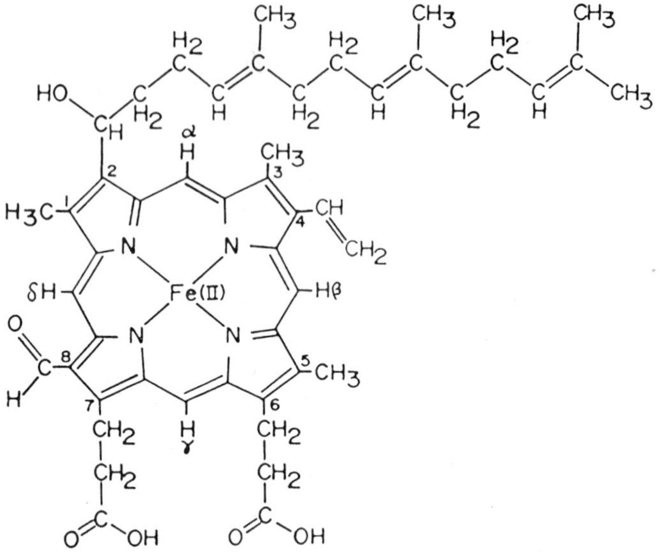

**FIGURE 1.** Structure of Heme A.

determination. The structure of the hydroxyfarnesylethyl group has been a fascinating subject for many investigators. Several possible roles for this moiety have been proposed (Caughey *et al.*, 1976; Woodruff *et al.*, 1991). In one of these proposals, it was postulated that the hydroxyfarnesylethyl group folds into a conformation to form a conjugated π-electron system with one of the pyrrole rings of heme. The conjugated π-electron system could promote a facile pathway for electron transfer (Caughey *et al.*, 1976). Cytochrome c oxidase enzyme contains two heme A molecules within the functional unit of the enzyme. One of these heme A molecules, designated heme a, coordinates two hisitine imidazole ligands and the other, designated heme a3,-coordinates only one imidazole side chain. Thus, heme a3 contains a vacant coodination site and can bind exogenous ligands such as $O_2$, CO, $CN^-$ an $N_3^-$.

Heme prosthetic groups have strong absorption in visible-Soret spectral regions. This heme absorption is extremely sensitive to the structure of porphyrin as well as to the redox and ligand binding states of the iron. Thus, hemes a and a3 must exhibit differences in their absorption spectra even if their respective iron atoms are in the same oxidation state, since the number of ligands bound to heme a is different from the number of ligands bound to heme a3. Careful measurements of the absorption spectra and the accurate assignment of spectra of the two hemes are essential for a determination of the function of the two hemes during the turnover of the enzyme.

Perhaps the most comprehensive spectroscopic investigations on the two hemes were performed by Vanneste as early as 1966. Most of the kinetic investigations (Ferguson-Miller and Babcock, 1996) were based on the extinction coefficients determined in these 1966 investigations (Vanneste, 1966). The redox difference absorption spectrum of heme a was calculated from the difference between the spectrum of the cyanide-bound fully oxidized enzyme treated with a slight excess amount of dithionite and that of cyanide-bound fully oxidized enzyme in the absence of reducing agents. It has been well established that bound cyanide strongly stabilizes the oxidized state of the heme $a_3$ iron (Caughey et al., 1976). Thus, treatment of the cyanide-bound fully oxidized enzyme with dithionite gives a mixed-valence state with a ferrous heme a and a cyanide derivative of a ferric heme $a_3$. Then, subtraction of the spectrum of the cyanide-bound fully oxidized enzyme, which contains a ferric heme a and a cyanide derivative of a ferric heme $a_3$, from the spectrum of the mixed-valence enzyme gives the redox difference spectrum of heme a. The redox difference spectrum of heme $a_3$ was calculated by subtracting the redox difference spectrum of heme a from the redox difference spectrum of the whole enzyme. The calculation for these difference spectra is based on an assumption that both the oxidation state and the ligand binding state of one of the hemes do not influence the spectrum of the other heme. This assumption should be examined experimentally, since the two hemes are in such close proximity based on the X-ray structure (presented below) that the two hemes could interact spectroscopically, and the absorption of one heme could be perturbed by the other heme (Tsukihara et al., 1995).

Absolute spectra of both hemes, each in both the ferrous and ferric oxidation states are also reported in the 1966 paper. Vanneste claimed that these spectra were calculated by using the photochemical action spectrum of the enzyme (Vanneste, 1966). However, no information is written about the methods used for calculation of the extinction coefficient of ferrous heme $a_3$-CO complex from this photochemical action spectrum (Vanneste, 1966). Even though this important information is missing, these absolute spectra have been cited by many investigators. These absolute spectra of hemes a and $a_3$ can be experimentally determined by monitoring other physical parameters such as the spin state of the heme irons. For example, accurate determination of efficiencies of resonance Raman excitation for the high/low spin components as a function of wavelength could provide a non-empirical method for resolving the whole spectrum into those of the components, hemes a and $a_3$.

One of the copper sites, $Cu_A$ shows an EPR signal in the g = 2 region but exhibits no significant absorption in the visible region (Beinert *et al.*, 1962). Metal analysis of an enzyme preparation in which recrystallization was not performed showed a Fe/Cu ratio close to unity (Caughey *et al.*, 1976). This result is consistent with the metal composition of two heme iron sites and two mononuclear copper sites, although presence of "extra Cu atoms" was suggested. Thus, for a long time, $Cu_A$ was considered to be a mononuclear type 2 copper site. However, several years before the X-ray structure of this enzyme appeared, the existence of a Cu—Cu interaction was proposed. This proposal was based on a comparison of the multifrequency EPR spectrum of the $Cu_A$ site in bovine heart cytochrome c oxidase in the fully oxidized state with the EPR spectrum of a bacterial nitrous oxide reductase (Kroneck *et al.*, 1990). The results of this comparison suggest that the $Cu_A$ site in the oxidized form is in a mixed valence state in which an electron is delocalized between the two $Cu^{2+}$ ions, that is, $Cu^{1.5+}$—$Cu^{1.5+}$. This site is a one-electron reduction site, the same as a mononuclear copper site, since on reduction of this site both coppers are in the cuprous state ($Cu^{1+}$).

In the near infrared region, a broad and weak absorption band attributed to $Cu_A$ in the oxidized state was discovered (Wharton and Tzagoloff, 1964) and this band has been an effective probe for monitoring the oxidation state of $Cu_A$ (Hill, 1994; Ferguson-Miller and Babcock, 1996). However, contributions from heme absorption in this region cannot be ignored in the analysis of this near infrared the $Cu_A$ band (Caughey *et al.*, 1976). Recently a genetically engineered peptide fragment of the $Cu_A$ site was prepared and its X-ray structure has been solved (Wilmanns *et al.*, 1995). The fragment shows an absorbance spectrum in the visible-Soret region which is much weaker than heme absorption (Lappalainen *et al.*, 1995).

It has been well accepted since an early EPR investigation in 1969 by Van Gelder and Beinert that the other copper site, $Cu_B$, is antiferromagnetically coupled to the heme $a_3$ iron in the fully oxidized state. EXAFS data suggest that the $Fe_{a3}$—$Cu_B$ distance is about 3.7 Å in the fully oxidized state (Scott, 1989). The EPR signal of $Cu_B$ shows strong rhombic character and is different from both the signals of Type 1 and Type 2 $Cu^{2+}$ ions. These signals have been detected only under specific conditions in which heme $a_3$ is in the reduced state and $Cu_B$ is in the oxidized state (Reinhammar *et al.*, 1980). No additional spectroscopic signal assigned to $Cu_B$ has been reported. The oxidizing equivalents in the fully oxidized enzyme indicated that $Cu_B$ in the fully oxidized state is a one electron reduction site (Yoshikawa *et al.*, 1995; Mochizuki *et al.*, 1999). The proximity of $Cu_B$ to heme $a_3$, shown by EPR and EXAFS, strongly suggests an important role for $Cu_B$ in the $O_2$ reduction mechanism, which will be described below.

However, compared with our knowledge of structure-function relationships at the other metal sites, many questions still remain about the $Cu_B$ site since this site shows essentially no spectral signals upon changes in its oxidation and ligand binding states. Mutagenesis work has suggested that $Cu_B$ has three histidines and one tyrosine as the ligands (Hosler *et al.*, 1993).

## 2.4. Subunit Composition and Amino Acid Sequences

Kadenbach and his colleagues proposed that the bovine heart cytochrome c oxidase is composed of 13 different protein subunits based on SDS-PAGE results. This conclusion was reached long before the crystal structure of bovine heart cytochrome c oxidase appeared (Kadenbach *et al.*, 1983). They did not use a crystalline preparation of the enzyme for SDS-PAGE. Thus, it is conceivable that most of the subunits with unknown physiological function, which include all subunits except for subunits I and II containing the redox active sites, are copurified contaminants. On the other hand, experimental evidence has not been presented that suggests that the number of subunits that compose the enzyme is different from 13. Kadenbach's 13 subunit proposal is based on the reproducible results of many repeated experiments, and experimental verification for their proposal has now been provided by the X-ray structure of this enzyme, which will be described below.

The amino acid sequences of the subunits of bovine heart enzyme have been determined primarily by Buse and his colleagues (Buse *et al.*, 1986). Many of the sequences have been determined not only by DNA sequencing but also by peptide analysis. Peptide analysis is an indispensable method for the detection of post translational modifications, and an example of such a modification has been recently shown in this enzyme, and will be described below. Hydropathy plots for these amino acid sequences have provided an astonishingly successful predictions of which regions contain $\alpha$-helices. This is quite remarkable considering the limited accuracy of this method for structural identification (Hosler *et al.*, 1993; Tsukihara *et al.*, 1996).

Amino acid sequence analysis and determination of subunit composition are painstaking but these steps are usually necessary before further structural investigations are undertaken. It should not be forgotten that chemical composition and amino acid sequencing provided a foundation for recent structure-function findings in the cytochrome oxidase field. The complete amino acid sequence and a successful prediction of the number of $\alpha$-helices greatly contributed to the successful and rapid crystallographic analysis of bovine heart cytochrome c oxidase at 2.8 Å resolution, four years ago (Tsukihara *et al.*, 1995; Tsukihara *et al.*, 1996).

## 3. FUNCTION OF BOVINE HEART CYTOCHROME C OXIDASE

Long before the crystal structure of cytochrome c oxidase was solved, the function of this enzyme had already been investigated by using the transition metal sites as spectroscopic probes (Caughey *et al.*, 1976; Malmström, 1990; Ferguson-Miller and Babcock, 1996). The information obtained from these spectroscopic investigations initiated prior to the crystal structure determination will be summarized.

### 3.1. Enzymic Activity

The enzymatic activity of cytochrome c oxidase under turnover conditions can be evaluated by measuring the rates $O_2$ consumption or ferrocytochrome c oxidation. For measurements of $O_2$-consumption, ascorbate is used as a reducing agent in the presence of a catalytic amount of cytochrome c because of the sensitivity required for the measurement. However, ascorbate is a relatively slow electron donor to cytochrome c. Thus, the rate-limiting step in this system under turnover conditions often is the transfer of electrons from ascorbate to cytochrome c. The other system, in which the oxidation of ferrocytochrome c is monitored, is less complicated and has been used for a detailed kinetic analysis of this enzyme. The first comprehensive kinetic analysis of this enzyme was reported by Minnaert, who demonstrated that the kinetics of ferrocytochrome c oxidation by this enzyme are first-order with respect to the concentration of ferrocytochrome c. A Michaelis-Menten relation was also found to exist between the initial velocity and ferrocytochrome c concentration (Minnaert, 1961). The two findings indicate that ferricytochrome c has the same affinity for the enzyme as that of ferrocytochrome c, and this competition for binding results in product inhibition. In a later investigation, an extremely strong binding of ferrocytochrome c to the enzyme was discovered in addition to the binding associated with the enzymatic activity that was observed by Minnaert. This result suggests the existence of two binding sites for cytochrome c on the enzyme (Ferguson-Miller *et al.*, 1976). However, it has also been shown that a single site model explains the kinetic results equally well (Speck *et al.*, 1984). Thus, identification of possible cytochrome c binding sites can be achieved by a detailed examination of the molecular surface from the X-ray structure and by co-crystallization of this enzyme with cytochrome c for X-ray structure determination.

It is important to recognize that a complete initial steady state kinetic analysis of the cytochrome c oxidase reaction has yet to be undertaken. The complete set of enzyme reactants includes four ferrocytochrome c molecules and one $O_2$ molecule as substrates, and because of the extremely low

Km of the enzyme for $O_2$, This system is not easily amenable to steady state kinetic analysis. As has been widely demonstrated, initial steady state kinetic analysis provides a variety of unique information about the function of enzymes impossible to obtain from other methods (Cleland, 1970).

The reaction of cytochrome c oxidase under single turnover conditions was initiated by the pioneering work of Gibson and Greenwood in 1963. The enzyme reduces $O_2$ to waters on a time scale as rapid as 0.1 msec at physiological temperature. Thus, these investigations were limited by the dead time of the stopped flow-apparatus 30 years ago, which was about 3 msec. On the other hand, the binding of CO at heme $a_3$ is not strong enough, compared with the binding of $O_2$, to prevent heme $a_3$ from reacting with $O_2$. Thus, the flash photolysis method, which is useful for kinetic analysis of hemoglobins and myoglobins with $O_2$, does not work for this enzyme. The method developed by Gibson and Greenwood (1963) is designated the flow-flash method in which a fully reduced enzyme solution saturated with CO is mixed with an $O_2$ containing buffer in a stopped flow apparatus. When the mixing is finished, the mixed solution injected into a thin glass cell is irradiated with a Xenon lamp flash for initiating the reaction of the fully reduced enzyme with $O_2$. The Xenon flash induces the photolysis of CO from the enzyme. Changes in the absorption of the Soret band, $\alpha$-band and a band in the near infrared region can then be carefully monitored. This method has been extensively implemented for the past 20 years, to determine the pathway of electron transfer between the redox active metal sites. This pathway has now been established, and electrons are transferred first from cytochrome c, to $Cu_A$, to heme a and finally to the $O_2$ reduction site (heme $a_3$ and $Cu_B$) (Hill, 1994). Gibson and his coworkers attempted to detect the $O_2$-bound form of the enzyme by analyzing the appearance of a photosensitivity that was detected with much stronger light intensities than used for photolyzing CO (Blackmore et al., 1991). However, the discovery of this photosensitivity does not offer conclusive evidence for the existence of $Fe^{2+}$—$O_2$ complex since photosensitivity does not provide detailed information on the chemical structure of the photosensitive species.

The transfer of electrons to heme $a_3$ from ferrocytochrome c in the fully oxidized enzyme under anaerobic conditions in an atmosphere of CO was monitored by following development of photosensitivity due to formation of the CO complex (Gibson et al., 1965). Surprisingly, the photosensitivity appeared within a timeframe of one second, which is $10^4$ times slower than the electron transfer to heme $a_3$ from heme a in the reaction of the fully reduced enzyme with $O_2$ (Hill, 1994). On the other hand, Brunori and his coworkers discovered that the enzyme molecules in the preparation isolated by usual purification methods show much weaker reactivity toward ferrocytochrome c and cyanide than the form of the enzyme that exists during turnover. This indicates that the former does not exist during enzymic turnover, and thus, it is designated the resting oxidized form (Brunori et al., 1979). The slow development of photosensitivity for preparation as isolated, which, therefore, is in the resting oxidized state is consistent with the findings of Brunori. However, mechanisms for controlling the rate of electron transfer from heme a to heme $a_3$ in the resting oxidized form have not been proposed. Furthermore, evidence for structural differences between the resting oxidized form and the oxidized form functioning during enzymatic turnover have not been observed, although several structural models have been proposed (Scott, 1989; Landrum et al., 1978; Seiter and Angelos, 1980; Franceschi et al., 1996).

## 3.2. Reduction of $O_2$

It is well established that a one electron reduction of $O_2$ is energetically unfavorable, but a two electron reduction is extremely favorable. Thus, for the autooxidation of transition metal compounds in aqueous solution, the metal in a 1:1 metal-$O_2$ compound is not readily oxidized. However, two metals in a 2:1 metal-$O_2$ complex are rapidly oxidized, since $O_2$ can receive two electrons, with one electron from each of the two metals, to form a peroxide compound, $M^+$—O—O—$M^+$, where M+ denotes a one electron-oxidized metal ion. Thus, the rate of oxidation of the metal in such systems is rate-limited by the formation of the 2:1 metal-$O_2$ compound (Caughey, 1976). In addition, ferrous heme, not ferric heme, reacts with $O_2$. The stability of the oxygenated heme ($O_2$-bound hemes) on hemoglobins and myoglobins is mainly due to this two-electron reduction requirement of $O_2$. In these globins, each oxygenated heme is so completely buried inside the protein that two hemes could never form a 2:1 heme-$O_2$ compound. On the other hand, in cytochrome c oxidase, $Cu_B$ is near the $O_2$ binding site on heme $a_3$, as discussed above. Thus, the $Cu_B$ could be a second electron donor to $O_2$ bound at heme $a_3$ $Fe^{2+}$. The importance of $Cu_B$ was recognized by Caughey (Caughey et al., 1976; Yoshikawa et al., 1977) and summarized in a proposed mechanism, the bridging peroxide mechanism as shown in Figure 2.

Although this mechanism has been widely accepted, an important point suggested by this mechanism is not well understood. The formation of the μ-peroxo intermediate ($Fe^{3+}$—O—O—$Cu^{2+}$) is likely to be extremely rapid, since this reaction step is limited by the rate of electron transfer from $Cu_B$ to one of the oxygen atoms of the bound $O_2$. Electron transfer rates for such a short distance (about 2 Å) could be as fast as a pico second. On

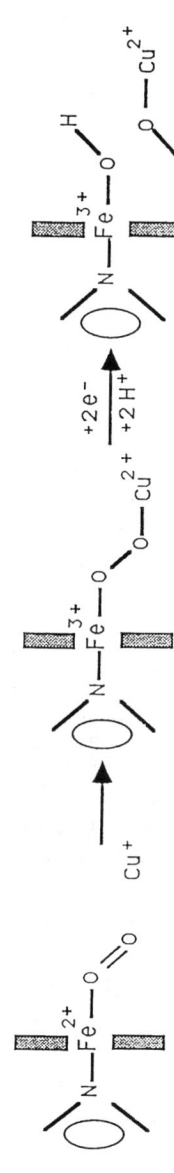

**FIGURE 2.** The bridging peroxide mechanism. Shaded rectangles denote porphyrin planes. Imidazole nitrogen, $Fe_{a3}$ and $Cu_B$ are shown by N, Fe and Cu respectively.

the other hand, formation of the $Fe^{2+}$—$O_2$ complex is rate-limited by transfer of $\dot{O}_2$ to the heme $a_3$ $Fe^{2+}$ site inside the protein through a specific pathway that leads to the enzyme surface. In hemoglobins and myoglobins, as well as in cytochrome c oxidase, the transfer of $O_2$ through such a pathway is likely to be controlled by protein dynamics. In other words, the pathway or channel is not likely to be wide enough to permit the free diffusion of $O_2$ to the binding site without conformational changes in the protein. In this case, $O_2$-transfer is likely to be much slower than the electron transfer from $Cu_B$ to $O_2$ at heme $a_3$. Thus, detection of the oxygenated intermediate, $(Fe^{2+}$—$O_2)$ is likely to be impossible. Thus, the first intermediate species that will most likely be detected in the reaction between $O_2$ and the enzyme is unlikely to be $Fe^{2+}$—$O_2$.

The majority of the kinetic investigations on this enzyme in the presence of $O_2$ have been undertaken by monitoring the absorption changes, which provides little information on the chemical structure of the external ligand at heme $a_3$. Therefore, reliable structural information is absolutely necessary for the identification of the initial intermediate species that the enzyme forms. Chance and his colleagues proposed that the initial intermediate they found at low temperature is the oxygenated form, since it had an absorption spectrum similar to that of the CO bound form. This intermediate was found not to be sensitive to the flash used during photolysis of the CO-bound reactant (Chance et al., 1975). These results do not provide conclusive evidence to support the proposal that the initial intermediate is the oxygenated form. However, 15 years later, it was shown by using resonance Raman techniques that the initial intermediate is indeed the oxygenated form, $Fe^{2+}$—$O_2$ (Kitagawa and Ogura, 1997). Thus, implementation of resonance Raman spectroscopy was the key to the discovery of the oxygenated form, and this point will be further elaborated upon.

Time-resolved resonance Raman spectroscopy provides a powerful method for identifying the intermediate species, which is an external ligand to heme $a_3$, during the enzymic reaction. As in the case of infrared spectroscopy, isotopic substitution increases the sensitivity of weak signals in Raman spectroscopy. Figure 3(a) shows an isotopic shift effect on the resonance Raman spectra during $O_2$-reduction. The spectra presented are the difference between the spectra at 0.1 msec after initiating the reaction of the enzyme with the naturally abundant $^{16}O_2$ and the spectra obtained under the same conditions except for the replacement of $^{16}O_2$ with $^{18}O_2$ (Ogura et al., 1993). The peak-trough pair corresponds to a signal due to $O_2$. These peaks and troughs are extremely difficult to identify in the absolute spectra where strong resonance Raman signals due to porphyrin modes overlap with weak signals due to $O_2$. These porphyrin signals are insensitive to $^{16}O/^{18}O$ isotopic substitution. The $571\,cm^{-1}$ band in $^{16}O$

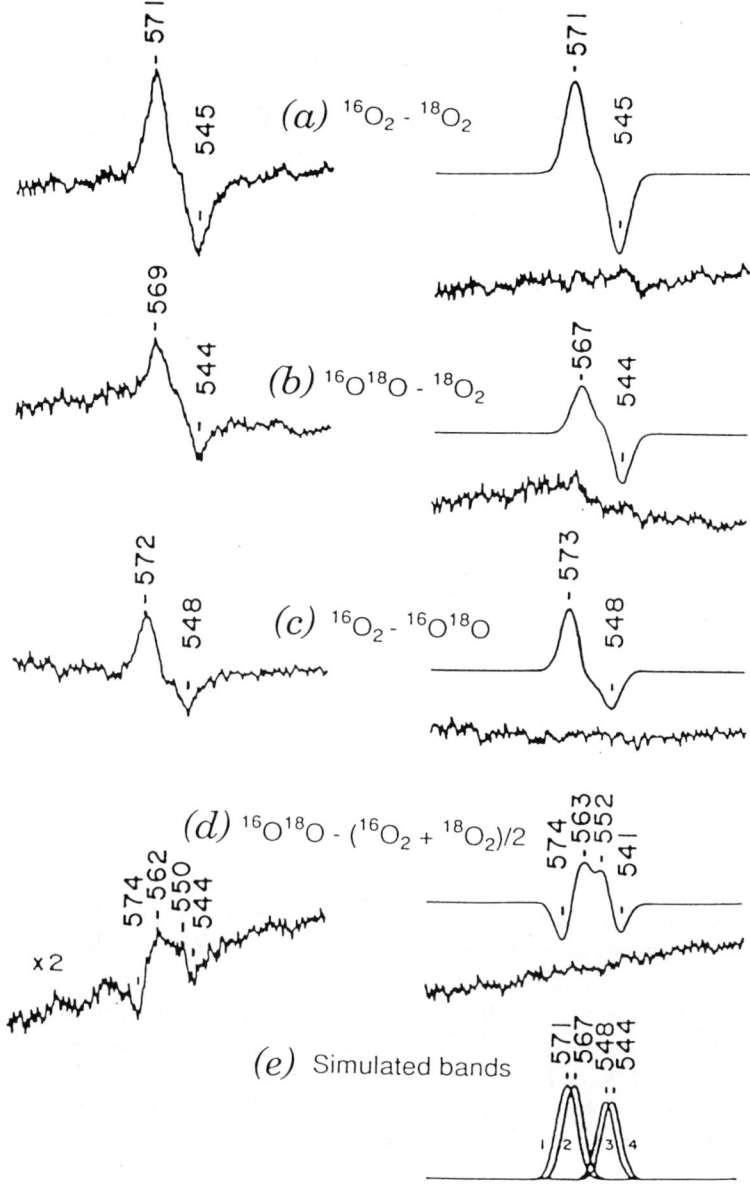

**RAMAN SHIFT (cm⁻¹)**

**FIGURE 3.** Time-resolved resonance Raman difference spectra in $Fe_{a3}^{2+}$—$O_2$ stretching frequency region of cytochrome c oxidase 0.1 msec after the initiation of the reaction of fully reduced enzyme with $O_2$. Observed spectra and calculated spectra are given in left side and right side, respectively. (a)$^{16}O_2$—$^{18}O_2$; (b)$^{16}O^{18}O$—$^{18}O_2$; (c)$^{16}O_2$—$^{16}O^{18}O$: (d)$^{16}O^{18}O$—($^{16}O_2$ + $^{18}O_2$)/2. (e)Fe—$^{16}O_2$(1), Fe—$^{16}O^{18}O$(2), Fe—$^{18}O^{16}O$(3), and Fe—$^{18}O_2$(4) stretching Raman bands assumed in the simulation. In the calculation for the $^{16}O^{18}O$ spectrum, (Spectrum(2) + Spectrum(3))/2 was used. The difference between the observed and calculated spectra are depicted along the same ordinate scale as that of the observed spectra under each individual calculated spectra.

spectrum has been assigned to the $Fe^{2+}$—$O_2$ vibrational mode. Oxyhemoglobin and oxymyoglobin also show similar vibrational bands within the same wavenumber region as the vibrational bands shown by cytochrome c oxidase. Figures 3(b) and (c) show isotopic shift effect by $^{16}O$=$^{18}O$. These results indicate that the $O_2$ derivative is bound at heme $a_3$ neither in a side-on fashion nor in a linear Fe=O mode, but in a bent end-on fashion closely similar to the binding geometry of $O_2$ in hemoglobins and myoglobins (Ogura et al., 1993). The half life measured for this oxygenated species was about 0.4 msec, about $10^8$ times longer than predicted. Furthermore, the bent end-on binding, similar to the binding in globins, indicates that significant interactions between the bound $O_2$ and $Cu_B$ do not exist. These results were completely unexpected since the bridging peroxide mechanism, sated above, had already been widely accepted. This is one of the most important findings in the elucidation of the reaction mechanism of this enzyme. Why is the oxygenated form so stable? What are the roles of $Cu_B$ in the $O_2$ reduction process? It is impossible to answer these intriguing questions without the three dimensional structure of the $O_2$ reduction site.

## 4. CRYSTALLIZATION OF BOVINE HEART CYTOCHROME C OXIDASE

As discussed previously, crystallization of multicomponent proteins is indispensable not only for determining the structure by X-ray crystallography but also for obtaining stable preparations at the high level of purity required for chemical composition and functional analyses. However, crystallization conditions are influenced by a multitude of factors, some of which will be described below.

There are some fundamental principles which should be followed to crystallize proteins. Crystallization conditions are primarily influenced by the three dimensional structure of the protein. A correlation between the size and shape of a protein and its conditions for crystallization is not straightforward. Thus, knowledge of the crystallization conditions for a single protein does not necessarily contribute effectively to predicting the crystallization conditions for a related protein. Reliable and systematic methods for the screening of crystallization conditions that will lead to success do not yet exist. If one is lucky, superior crystals of an enzyme could appear tomorrow or may not appear even after 10 or more years of effort. Thus, usually, crystallization is the rate-limiting step in an X-ray structural determination of a protein. Here, the general strategies used for the crystallization of membrane proteins will be described including those strategies used for the crystallization of cytochrome c oxidase.

## 4.1.  Crystallization of Membrane Proteins

As previously described, the best method to isolate membrane proteins is to solubilize with detergents. The hydrophobic surface in the transmembrane region of a solubilized membrane protein molecule suspended in aqueous solution is surrounded by detergent molecules. Most of these detergent molecules on the membrane protein surface are likely to bind loosely and non-specifically to the surface of the protein since these bound detergents are replacements for the phospholipids of the biological membrane. Thus, these loosely-bound detergent molecules are unlikely to participate in specific stabilizing interactions that occur between membrane protein molecules in the crystal lattice (Garavito and Picot, 1990). Specific interactions between hydrophilic regions of the protein surface are believed to constitute the primary stabilizing factor for membrane protein crystals. This indicates that smaller-sized detergents should be more effective for crystallization since these smaller detergents should promote the formation of strong interactions between the hydrophilic surfaces of adjacent molecules in the crystal lattice. For large membrane protein complexes, these strong interactions are expected between the extramembranous hydrophilic domains of the molecules. These ideas have been confirmed by the finding which relates the size of the detergents to their effectiveness for growing crystals (Garavito and Picot, 1990). Further evidence is provided from the work of Michel and his coworkers who crystallized a bacterial cytochrome c oxidase with an Fv fragment of a monoclonal antibody which enlarges the hydrophilic surface of the enzyme (Ostermeier et al., 1995). Indeed the packing of the protein molecules in the crystal lattice of the bacterial enzyme shows that protein-protein lattice contacts only occur between the hydrophilic Fv fragments. This method is truly ingenious and it can be applied to crystallization of many other membrane proteins. It should also be emphasized that as for the crystallization of water-soluble proteins, membrane protein crystallization requires conformational integrity of the isolated protein preparation, and this is a critical issue for the success of the crystallization. Even if an Fv fragment and small detergent are available, inhomogeneous protein preparations will probably not produce high quality crystals.

## 4.2.  Crystallization of Cytochrome c Oxidase

In the currently accepted theories on membrane protein crystallization, discussed above, the only role of the detergent is to stabilize the membrane protein in aqueous solution. However, the effect of detergent structure on the crystallization of bovine heart cytochrome c oxidase, presented here, suggests an additional role for detergent molecules on the transmembrane hydrophobic surface. These detergent molecules also contribute to the stability of membrane protein crystals.

Bovine heart cytochrome c oxidase stabilized with alkyl polyoxyethylene type detergent produced three types of crystals depending on the size of polyoxyethylene portion of the added detergent (Shinzawa-Itoh et al., 1995). Crystals with the shape of tetragonal plates were obtained from enzyme preparations stabilized with the detergent molecules containing with 6–23 oxyethylene units. However, this crystal form was not obtained from preparations stabilized with similar detergents containing only 5 oxyethylene units. At the other extreme in detergent size, hexagonal bipyramidal crystals were obtained only from preparations stabilized with a detergent molecule containing 23 oxyethylene units, which is the biggest detergent used. Tetragonal rod-shaped crystals were obtained when detergents containing 7 or 8 oxyethylene units were used. Only in the case of the tetragonal rods did an upper limit for the size of detergent molecule suitable for crystallization exist. On the other hand, all the three crystal forms have lower limits for the size of the detergent molecule added to the crystallization. All the detergents with 5–23 oxyethylene units and with 10–12 carbon atoms in saturated hydrocarbon chain are equally effective in stabilizing the enzyme in aqueous solution. These results indicate that detergent molecules bound to the transmembrane hydrophobic surface also contribute to the stability of crystals by forming bridging interactions between adjacent molecules in the crystal, but non-specifically.

The tetragonal plate crystals have the lowest specificity for detergent structure among the three types of crystals. Only the detergent with 5 oxyethylene units did not yield crystals. On the other hand, the tetragonal rod-shaped crystals and the hexagonal bipyramidal crystals have much higher dependence on the structure of the detergent compared with the tetragonal plate crystals. The highest resolution of X-ray diffraction from the tetragonal plates, hexagonal bipyramids, and tetragonal rod-shaped crystals are 15 Å, 8 Å and 6 Å respectively. Thus, the diffraction quality of the crystals shows the same dependence on the specificity of the detergent structure. This correlation is expected and it also provides further evidence for the influence of these detergents on the stabilization of crystals.

Orthorhombic crystals obtained from the preparations stabilized with the detergent decyl maltoside diffracted X-rays up to a maximum resolution of 2.3 Å resolution from non-frozen crystals (Yoshikawa et al., 1998). No commercially available detergent other than decyl maltoside has been found that will produce orthorhombic crystals. Even dodecyl maltoside, only two alkyl chain units longer than decyl maltoside, did not produce crystals even though the detergent stabilizes the enzyme effectively. It should be noted that for crystallization of this large membrane protein with a molecular weight of 410 kDa for the asymmetric unit, only two extra carbon atoms in the detergent structure critically influences the crystallization. This result strongly suggests some highly specific interactions involving the detergent promote a unique positioning of the protein molecules in the crystal. However, detergent molecules which interact with adjacent protein molecules in the crystal lattice were not observed in the X-ray structure (Yoshikawa et al., 1998). Thus, these detergent molecules are not located at a specific position on the molecular surface, but they must be positioned to promote the formation of specific interactions, perhaps, between the hydrophilic surfaces. The non-specific binding of the detergent suggests that a detergent other than decylmaltoside could stabilize the crystals more effectively. In other words, if decyl maltoside molecules were tightly bridged between the two adjacent proteins then these detergent molecules would have been observed in the crystal structure, and no other detergent could be substituted in this crystal form.

## 5. X-RAY STRUCTURE OF BOVINE HEART CYTOCHROME C OXIDASE

In 1995, two X-ray structures of cytochrome c oxidase, one structure from bovine heart and the other structure from bacterial cells were reported in the same week (Tsukihara et al., 1995; Iwata et al., 1995). The bacterial enzyme is the smaller and contains four protein subunits while bovine enzyme is much larger and has 13 different subunits. However the core portions of the two enzymes, composed of the three largest subunits, show an amazingly similar three dimensional structure, especially in the region near the redox active metal sites. Furthermore, these X-ray structures display a remarkable agreement with predictions for the locations of $\alpha$-helical regions from the amino acid sequence. The structure is also in amazing agreement with mutagenesis investigations on the ligand structure at the metal binding sites. (Tsukihara et al., 1996; Hosler et al., 1993). However, the level of structural knowledge based on these empirical methods can be quite limited. This level is far less than required for a complete understanding of the catalytic mechanism of this enzyme.

### 5.1. Three Dimensional Structure of the Protein Portion

Figure 4 shows a $C_\alpha$ backbone trace of the asymmetric unit of the orthorhombic crystal form of bovine heart cytochrome c oxidase (Tsukihara et al., 1996). The structure revealed that cytochrome c oxidase is a

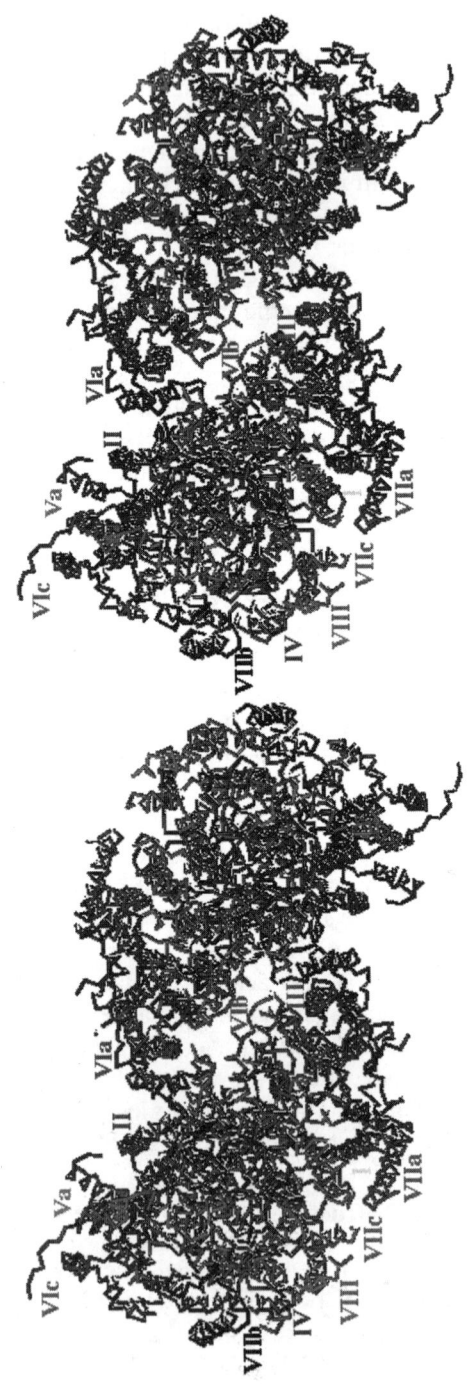

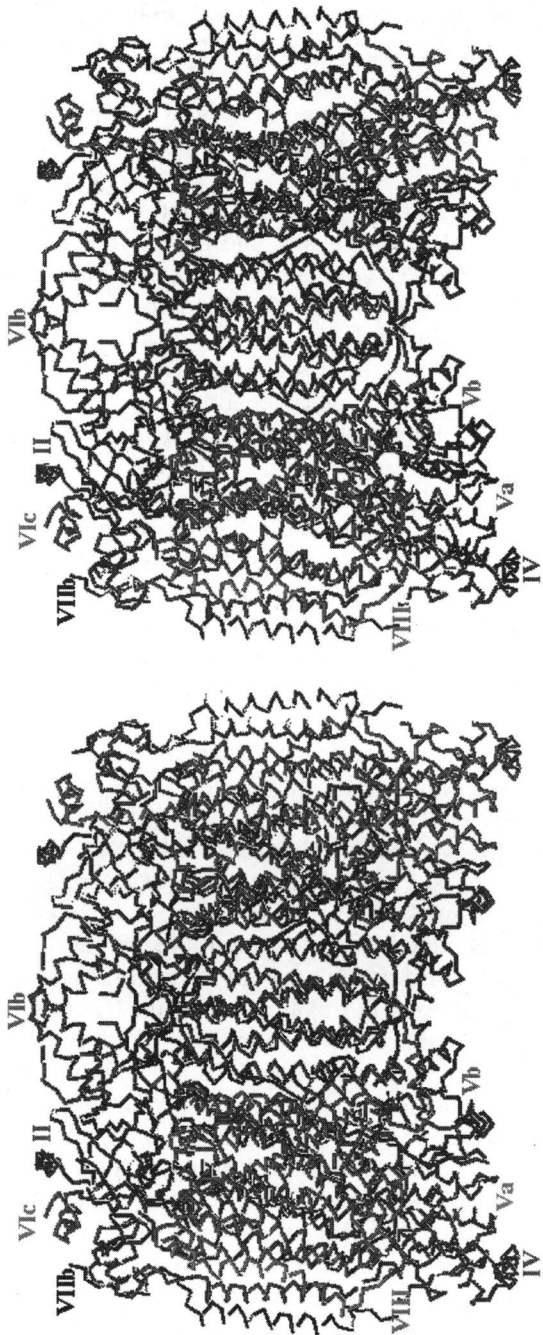

**FIGURE 4.** The $C_\alpha$ back bone trace of a dimer of bovine heart cytochrome c oxidase. (left) A view to the transmembrane surface and (right) a view from the cytosolic side.

dimer, with a molecular mass in the asymmetric unit of approximately 420 kDa. The central core of the structure seen in Figure 4 (left) consists essentially of segments of polypeptide chain that are folded into $\alpha$-helices. This portion lies within the transmembrane moiety of the mitochondrial inner membrane. One of the regions protruding into the bulk aqueous phase contains the $Cu_A$ site identifying this region of the protein to be protruding into the intermembrane space. A fairly large region of the structure between the monomers is devoid of electron density, and this region is easily seen in a top view of the dimer (Figure 4 (right)). This space will most likely be filled with phospholipids to prevent protons from leaking across the space. The largest three subunits, subunits I, II and III, which are encoded by mitochondrial genes, form the core of this enzyme The other ten smaller subunits, which are synthesized by nuclear encoded genes, surround this core moiety composed of the largest three subunits. Three subunits out of the ten nuclear-coded subunits do not contain transmembrane helices and are positioned at the hydrophilic surfaces of the core. Subunits Va and Vb are on the matrix side and subunit VIb is on the intermembrane side. Each of the other nuclear encoded subunits contains a transmembrane helix. These helices are not distributed evenly around the core. For example, the transmembrane surface surrounding the intermonomer space and the surface of subunit III opposite to the inter-monomer space are scarcely distributed by transmembrane $\alpha$-helices of the nuclear coded subunits and are for the most part absent from this region.

As described above, eleven of the thirteen subunits of the bovine enzyme do not contain redox-active metal sites. The X-ray structures show very tight and specific interactions between these eleven subunits and the remaining two cofactor-containing subunits. This indicates that these eleven subunits are not copurified contaminating peptides, although physiological roles of these subunits are unknown. These nuclear encoded subunits and subunit III, which is encoded by a mitochondrial gene but does not contain a redox active metal, possess many peculiar structural features which suggests some possible functions. For example, subunit III is essentially composed of only $\alpha$-helices, and does not contain large areas of extramembrane hydrophilic surface. The seven helices of this subunit form two helix bundles, each of which include 5 or 2 helices. The packing of the two helix bundles produces a V-shaped crevice which is exposed to the intermembrane side. The crevice holds three phospholipids, with the head groups of two phospholipids positioned at the surface of the mitochondrial inner membrane on the intermembrane side and the head group of the remaining phospholipid at the surface of the membrane on the matrix side. The fatty acid tails are located inside the crevice to form a region of extreme hydrophobicity. This hydrophobic space is very close to one of the possible pathways for $O_2$ transfer which leads to the $O_2$ reduction site within subunit I (Tsukihara *et al.*, 1996). Thus, a possible role of subunit III is its involvement in forming a substrate pool of $O_2$ molecules for subunit I. Subunits VIa and VIb are likely to contribute to the stabilization of the dimer. The amino terminal segment of subunit VIa is in an extended conformation and makes contact with helices V and VII of the other monomer. The segment between Cys39 and Cys53 of subunit VIb surrounding subunits II and III at the intermembrane side makes contact with the same region on the other monomer near the two fold symmetry axis that relates the two monomers in the dimer. Subunit Vb is located at the surfaces of subunits I and III on the matrix side, and contains a zinc atom which is coordinated to four cysteine sulfur atoms. Subunit Va contains five short α-helices, each with four to five turns. These α-helices are arranged in a right-handed superhelix.

A concave surface which is created by subunits II, VIa and VIb on the intermembrane side of subunit I, is about 25 Å in diameter, suggesting that it is the binding site for cytochrome c. The transmembrane α-helices of the nuclear encoded subunits form helix-helix interactions with any one pair of these helices crossing at one of three inclination angles of approximately O°, 20° and 50°, which provide the most stable mode of interaction. Thus, the nuclear-encoded subunits along with their transmembrane helices contribute to the overall stability of the enzyme. This contribution is an example of one of the possible physiological roles provided by these subunits.

## 5.2. Structures and Locations of the Metal Sites

Figure 5, shows the locations and structures of metal sites in bovine heart cytochrome c oxidase, in the X-ray structure determined at 2.8 Å resolution (Tsukihara *et al.*, 1996). During the structure determination the magnesium site was assigned a location based on previous site-directed mutagenesis work, which implicated which amino acid residues interact with this metal (Hosler *et al.*, 1995). The two hemes are oriented essentially perpendicular to the plane of the membrane surface with the two propionates pointing toward the intermembrane surface. Thus, for both heme a and heme $a_3$, positions 6 and 7 (Figure 1) of the porphyrin ring are facing toward the intermembrane side (upper side in Figure 5) and positions 2 and 3 are facing toward the matrix side. It is important to note that determination of the arrangement of the heme planes with this accuracy would be possible only by the X-ray crystallographic method.

The long alkyl side chain of heme a is in an almost extended conformation whereas that of heme $a_3$ is in a U-shaped conformation (Tsukihara

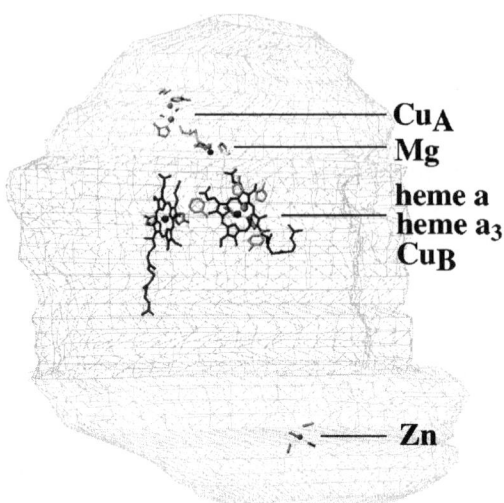

**FIGURE 5.** A schematic representation of the metal site locations in bovine heart cytochrome c oxidase. The molecular surface is determined from the electron density map at 5 Å resolution and is shown by the cage.

*et al.*, 1995). The conformations of both of these side chains do not indicate that formation of a conjugated $\pi$-electron system, as has been suggested. The discovery of heme O (a derivative of heme A in which the formyl group at position 8 is substituted with methyl) in bacterial cytochrome $bo_3$ (Puusti-nen and Wikstrom, 1991) suggests that the formyl group of the alkyl side chain plays a critical role in the proton pumping function (Woodruff *et al.*, 1991). However, the current X-ray structure does not support Woodruff's proposal.

The dinuclear copper site, $Cu_A$, displayed a spherical cloud of electron density at 2.8 Å resolution. However, six ligands (two histidine imidazoles, one methionine sulfur atom, two cysteine sulfur atoms and a peptide car-bonyl group) were found to be located at this site. The value of electron density at the site was significantly higher than the magnitude of electron density expected for a single copper atom. Furthermore, metal analysis of preparations used for crystallization, which have been purified by crystal-lization showed an iron/copper ratio of 2/3 (Mochizuki *et al.*, 1999). These results indicate that the $Cu_A$ site contains two copper atoms that are bridged by two sulfur atoms of cysteines in the form of a rhombic arrangement. Each copper atom has two additional ligands, giving each copper atom a tetrahedral coordination geometry. Thus, the $Cu_A$ site is similar to an 2Fe/2S type iron-sulfur cluster in which Fe and inorganic sulfur are replaced with Cu and cysteine sulfur. This symmetrical model for Cu-coordination is consistent with the electron density observed from the 2.8 Å resolution maps. The dinuclear copper structure for $Cu_A$ has been suggested by the EPR spectrum, indicating a delocalized electron between the two cupric ions in the oxidized form as described above. Possible structural asymmetry at the $Cu_A$ site could be detected at higher resolution. Such asymmetry has been suggested by an EPR investigation (Mims et al., 1980).

Figure 6 shows the X-ray structure of the $O_2$ reduction site in the fully oxidized state at 2.3 Å resolution. Two unexpected features of structure have been observed; a direct covalent bond between His 240 and Tyr244 and a peroxide bridged between $Fe_{a3}$ and $Cu_B$. The electron density between the two metals points to the presence of two oxygen atoms. However, two hydroxide ions or two water molecules, with one molecule coordinated to $Fe_{a3}$ and the other to $Cu_B$, do not model well into this electron density. In other words, the electron density indicates the existence of a covalent bond between the two oxygen atoms. The accuracy of the electron density at this resolution is not high enough to identify the bond order of this covalent bond. Thus, it is possible to model an $O_2$ into this electron density. However, $O_2$ does not bind to ferric heme $a_3$, but peroxide, $O_2^{2-}$ does. It has been well known that bovine heart cytochrome c oxidase isolated with a usual method is in the oxidized form (the resting oxidized form) different from the one under turnover. It had never been thought that $O_2^{2-}$ would bridge the two metals to form the resting oxidized form, which is very stable and inert. Furthermore, $O_2^{2-}$ contains two oxidation equivalents. Thus, six electron equivalents are required for complete reduction of the resting oxidized form, which is in disagreement with the reported results. A magnetic coupling between high spin $Fe_{a3}^{3+}$ and $Cu_B^{2+}$ in the resting oxidized form has been proposed (Van Gelder and Beinert, 1969). However, binding of $O_2^{2-}$ is likely to induce the formation of a low spin form of $Fe^{3+}$, not the high spin state. The X-ray structure at the current resolution (Yoshikawa et al., 1998) indicates a significantly longer distance between one of the bridging oxygen atoms and $Fe_{a3}$ than that of a typical coordination bond. This observation is consistent with a high-spin state assigned to the Fe atom of heme $a_3$.

A direct covalent linkage between Tyr244 and His240 would have never been conceived, before the structure appeared (Yoshikawa et al., 1998). This post-translational modification was not discovered in the initial sequence analysis, since the region of the protein containing these two amino acids was determined by DNA sequencing. However, the sequence in this region has been reexamined using Edman degradation to verify chemical modification of these amino acids. No ordinary amino acid was detectable at amino acid positions at 240 and 244 in the Edman analysis (Buse et al., 1999). When and how the bond is formed is unknown. However, the enzyme containing this bond is completely functional suggesting that

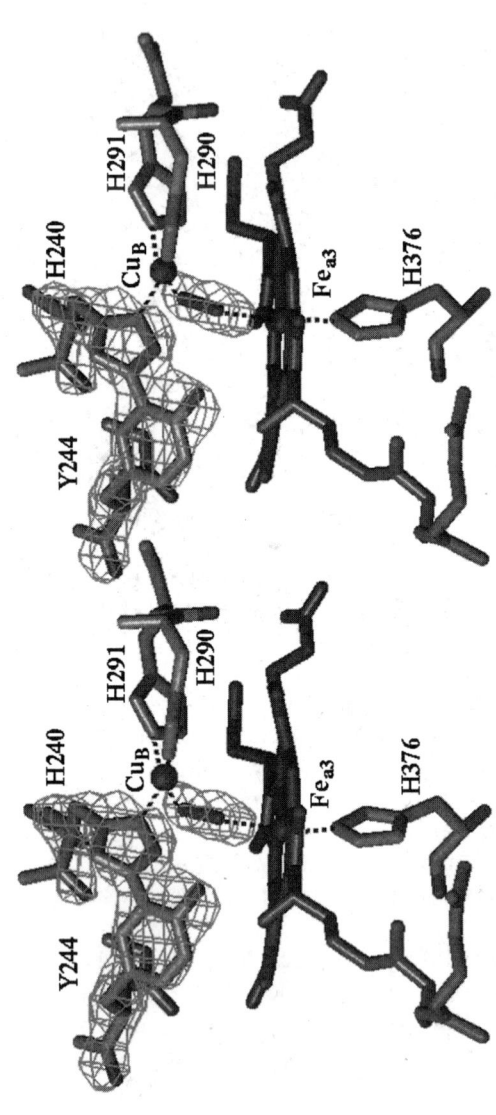

**FIGURE 6.** X-ray structure of Fe$_{a3}$—Cu$_B$ site of the fully oxidized form at 2.3 Å resolution. The (*Fo-Fc*) difference Fourier map of the oxidized form calculated by omitting His240 and Tyr244, and any ligand between Fe$_{a3}$ and Cu$_B$ from the *Fc* calculation. Contours are drawn at 7 σ level (1σ = 0.0456e⁻/Å³).

this modification is not caused by the result of a non-physiological oxidation event that occurs outside the cell.

## 6. A POSSIBLE $O_2$ REDUCTION MECHANISM

The X-ray structure of the redox active metal sites suggests the following $O_2$ reduction mechanism. The direct covalent linkage between His240 and Tyr244 places the hydroxyl group of Tyr244 in close enough proximity to the heme $a_3$ iron to form a hydrogen bond with bound $O_2$, which is coordinated to the iron atom of heme $a_3$. A modeling of $O_2$ into the active site of the fully reduced form at 2.35 Å resolution (not shown) indicated that the formation of this hydrogen bond is possible. The X-ray structure seen in Figure 6 also shows that $Cu_B$ has only three histidine imidazole ligands, which form a trigonal planar cuprous copper compound in which the $Cu_B^{1+}$ is placed at the center of triangular plane with the three nitrogen atoms of the imidazoles at the vertices of the triangle. In general, trigonal planar cuprous copper compounds are very stable, thus, $Cu_B$ is likely to be a poor electron donor as well as a poor ligand acceptor. Thus, the $O_2$ bound at heme $a_3$ could form a hydrogen bond to Tyr244, rather than coordinate to $Cu_B$. The stability of $Cu_B^{1+}$ suggests that the second electron for reducing $O_2$ is from heme a. The proximity of heme a and heme $a_3$, as revealed by X-ray structure, indicates that heme a is an effective electron donor to heme $a_3$. One of the ligands of heme a is a histidine which is separated from a nearby histidine residue by a single amino acid residue in the sequence. This second histidine is also a ligand of heme $a_3$. This single amino acid insertion provides a facile electron transfer pathway through chemical bonds from the heme a iron to the heme $a_3$ iron. The nearest distance between the two porphyrin side groups on each of the two hemes is about 4.5 Å, which suggests an effective electron transfer through space. However, these electron transfers could be much slower than the one through a coordination bond between $Cu_B$ and $O_2$ as suggested in the bridging peroxide mechanism (Caughey, 1976).

Thus, major features of the X-ray structure at the dioxygen reduction site, contribute to the stability of the oxygenated form of this enzyme, and this stability has been confirmed by the resonance Raman investigations described above (Kitagawa and Ogura, 1997). The major features are: trigonal planar coordination of $Cu_B$, and the close proximity of both Tyr244 and heme a to heme $a_3$.

Tyr244 is connected to the matrix space by a hydrogen bond network which is likely to be a facile proton transfer path, and this is described below. Thus Tyr244 could be an effective proton donor to the $O_2$ bound to heme $a_3$ and result in the formation of a hydroperoxo intermediate, $(Fe_{a3}^{3+}—O—O—H)$, instead of the μ-peroxo intermediate $(Fe_{a3}^{3+}—O—O—Cu_B^{2+})$ as had been previously suggested in the bridging peroxide mechanism (Caughey, 1976). The hydroperoxo intermediate is the initial compound produced in the reaction between peroxidase and its substrate, $H_2O_2$. By analogy to the peroxidase reaction, the peroxide in the hydroperoxo intermediate is likely to be cleaved to produce an abnormally high oxidation state intermediate of the heme $a_3$ iron with the concomitant release of a water molecule. Even with the implementation of time-resolved resonance Raman methods, the hydroperoxo intermediate has not been detected. The intermediate form detected following the reaction of the oxygenated form was an iron oxo compound with the same over all oxidation state as that of $Fe_{a3}^{3+}—O—OH$ (Kitagawa and Ogura, 1997). The following three structures of this intermediate are possible; $Fe_{a3}^{5+}=O$, $Fe_{a3}^{4+}=O$ with a π-cation radical state of the heme $a_3$ porphyrin, and $Fe_{a3}^{4+}=O$ with a Tyr244-O radical. However, a tyrosine radical species has not been detected in the EPR spectrum of the enzyme. Also, an observable decrease in the Soret band intensity of heme $a_3$, which indicates formation of π-cation radical, has not been observed. On the other hand, no spectroscopic method for identification of $Fe^{5+}$ has been established. All these possible structures are likely to have extremely strong affinity for electrons. Thus, the stable $Cu_B^{1+}$ could donate an electron to the oxo compound with the same overall oxidation state as that of $Fe_{a3}^{3+}—O—OH$ to form $Fe_{a3}^{4+}=O$ without a radical species, as described above. The resulting $Cu_B^{2+}$ would coordinate the water molecule produced from cleavage of $O_2$ as a fourth ligand. At least two protons and an electron are required for producing the second water molecule and $Fe_{a3}^{3+}$ from $Fe_{a3}^{4+}=O$. The water molecule at $Cu_B^{2+}$ could be an effective proton transfer site for the oxygen atom bound to $Fe_{a3}$. However, it is located too far to interact with the Tyr244 OH group. The extreme instability of the $Fe_{a3}—O—O—H$ species can be attributed to the high propensity of $Fe_{a3}$ to donate electrons, and this contributes to the extremely facile reduction of $O_2$ to the water level. On the other hand, $Fe_{a3}$ in a high oxidation states is reduced slowly in order for the enzyme to provide effective energy transformation.

## 7.  PROTON TRANSFER IN BOVINE HEART CYTOCHROME C OXIDASE

### 7.1.  Possible Proton Transfer Pathways in Membrane Proteins

One of the most important findings of investigations on the reaction mechanism of cytochrome c oxidase is the discovery of its proton pumping function by Wikstrom (1977). There are two primary reasons why the enzyme must transfer protons through the protein. First, protons are required for making waters from $O_2$. Second, protons are pumped across the inner mitochondrial membrane. The most effective way of transferring protons from an extremely hydrophobic environment, which exists inside of protein, is through a chain of hydrogen bonds formed between two nitrogen or oxygen atoms, each of which has a hydrogen atom, such as —OH and —NH. For example, in a hydrogen bond between the two —OH groups, when one of the OH groups is protonated, the positive charge is transferred to the oxygen atom of the protonated OH group to provide the following two equally possible interconvertible states.

$$
\begin{array}{c}
\underset{\displaystyle -\overset{+}{O}-H\text{----------}O-}{\overset{\displaystyle \overset{H}{\diagup}\qquad\overset{H}{\diagdown}}{}}
\quad\rightleftharpoons\quad
\underset{\displaystyle -O\text{----------}H-\overset{+}{O}-}{\overset{\displaystyle \overset{H}{\diagup}\qquad\overset{H}{\diagdown}}{}}
\end{array}
\qquad \text{scheme 1}
$$

Then, a deprotonation at the point different from the protonation point results in a proton transfer through the hydrogen bond. Similarly, the hydrogen bond network composed of three or more nitrogen or oxygen atoms, each with a hydrogen atom, can be an effective proton pathway. It is well known that the strength of a hydrogen bond is sensitively influenced by geometry. Thus, small changes in protein conformation can easily influence the effectiveness of proton transfer through a hydrogen bond network. The X-ray structures of bovine heart cytochrome c oxidase determined at 2.8 Å resolution or above show the existence of cavities large enough to contain water molecules. However, electron density is not detected inside these cavities, indicating that the water molecules within these cavities are mobile. Then, these cavities could be an effective route for proton transfer. Furthermore, amino acid side chain dynamics could play a role in proton transfer. For example, two protonable amino acid side chains that are not initially hydrogen bonded can be placed in close enough proximity for the formation of a new hydrogen bond, even without any movement of the main chain. Such side chain movements can be induced by changes in the oxidation and ligand binding states at the metal sites. Thus, we designate a pair of such amino acid side chains a possible hydrogen bonded structure (Tsukihara et al., 1996).

In the X-ray structure of bovine heart cytochrome c oxidase in the fully oxidized state at 2.8 Å resolution, two hydrogen bond networks connecting the molecular surface on the matrix side to that on the intermembrane surface were identified (Tsukihara et al., 1996). Both networks contain possible hydrogen bond structures, as defined by the criteria stated above, but do not include elements of structure comprising the $O_2$ reduction site. However, small conformational changes induced by changes in the redox and ligand binding states could promote pK changes for protonable groups along these two pathways. Thus, both networks could be components of the apparatus used for proton pumping, which is driven by $O_2$ reduction at the heme $a_3$-$Cu_B$ site. A third hydrogen bond network connects Tyr244 with the matrix surface of the enzyme. As previously stated above, Tyr244 is likely to form a hydrogen bond with $O_2$ bound at heme $a_3$. Thus, this hydrogen bond network is utilized for transferring protons to the $O_2$ reduction site for water production. It should be noted that this network also has a possible hydrogen bond network. Thus, the transfer of protons for making waters is also controlled by changes in the redox or ligand binding states at the metal sites. In other words, proton transfer for making water is likely to be tightly coupled with electron transfers in the enzyme.

The couplings between electron and proton transfers in the enzyme have been shown by analysis of site-directed mutants. A possible hydrogen bond structure exists between Lys319 and Thr316 along the hydrogen bond network placed from the matrix surface of the enzyme to Tyr244, which is the possible proton donor to $O_2$ at heme $a_3$, as described above (Tsukihara et al., 1996). Mutations of the corresponding amino acids in the enzyme of a bacteria, *Rhodobacter spheroides*, to methionine and alanine, respectively, result in a drastic reduction in the overall turnover rate. The mutant enzymes showed very slow reduction of heme $a_3$ in the presence of ascorbate and cytochrome c under anaerobic conditions. However, the reaction of the fully reduced enzyme with $O_2$ was essentially the same as that of the wild type enzyme (Adelroth et al., 1998). This result indicates that electron transfer to heme $a_3$ is coupled with the proton transfer through these amino acids, whereas electron transfer from the enzyme to $O_2$ is not controlled by the proton transfer. On the other hand, mutation of Glu286 in the *Rb. sphaeroides* enzyme corresponding to Glu242 in the bovine enzyme impaired reduction of the two electron-reduced form of the oxygenated enzyme. Thus, the mutated enzyme does not reduce $O_2$ completely to water molecules (Adelroth et al., 1997; Konstantinov et al., 1997). The simplest explanation of this result is that Glu242 is the proton donor for producing the second water molecule from the peroxide intermediate which is likely to be $Fe^{5+}=O$, as described above. Glu242 is hydrogen-bonded to Met71 in the X-ray structures of all oxidation and ligand binding states examined. A hydrogen bond network leading to the heme $a_3$ iron is not observed in the X-ray structure (Yoshikawa et al., 1998). Thus, conformational changes, in addition to those that take place at Glu 242, may be required for proton transfer from this sidechain to the $O_2$ reduction site. The most important conclusion from the mutagenesis result is that the first two protons and the last two protons for producing two water molecules from an $O_2$ are transferred through different pathways. This is consistent to the finding by

Wikström and his coworkers that proton pumping is coupled only to the transfer of the latter two of the four electrons that are required for complete reduction of an $O_2$ to two water molecules (Wikström and Morgan, 1992). The finding was reexamined by another group in detail (Michel, 1999) to yield somewhat different interpretation for the experimental results.

## 7.2. A Redox Coupled Conformational Change in Cytochrome c Oxidase

The structure of the enzyme in the fully oxidized state at 2.8 Å resolution has provided some unique information on how protons are transferred in this system. However, an experimentally observed redox-coupled conformational change has tremendously improved our understanding of the mechanism of proton pumping (Yoshikawa et al., 1998). In the search for hydrogen bond networks in the X-ray structure of the fully oxidized state at 2.8 Å resolution, a possible hydrogen bond structure is defined as a pair of amino acid side chains which could form a hydrogen bond upon side chain conformational changes without any additional movements of the main chain (Tsukihara et al., 1996). This assumption is reasonable because large conformational changes involving main chains are unlikely to occur during enzymic turnover. However, quite unexpectedly, a redox coupled conformational change including a portion of the mainchain was discovered in a segment of the protein from Gly49 to Asn55. This segment is located in the loop region between helices II and III of subunit I (Yoshikawa et al., 1998), and this segment is very close to the intermembrane space.

A second most readily prepared oxidation state of the enzyme, after the resting oxidized state, is the fully reduced state. Most fortunately, crystals of fully oxidized enzyme molecules can be reduced with ascorbate and a catalytic amount of cytochrome c without deterioration of the quality of the crystals. However, the fully reduced form in crystals is stable only in the presence of mother liquor containing the reducing system, because of the extremely strong reactivity of the enzyme to $O_2$. Thus, X-ray diffraction data for the fully reduced enzyme crystals were collected from crystals mounted inside quartz capillaries tubes containing Sephadex G-25 gel equilibrated with the mother liquor containing the reducing system. Complete reduction of the crystals inside the capillaries filled with the mother liquor containing the reducing system, but without the Sephadex gel, was confirmed by the absorption spectrum taken with a microspectrophotometer.

Out of the six amino acid residues in the loop segment, which undergo conformation upon reduction, the only residue whose solvent accessibility to the intermembrane space changes is Asp 51. In the fully oxidized state,

Asp51 is completely buried inside the protein. However, upon complete reduction, Asp51 moves toward the molecular surface at the intermembrane space. One of the oxygen atoms of the carboxyl group then becomes exposed to the bulk water phase (Figure 7). In the fully oxidized state, Asp51 is connected to the matrix surface through a proton transfer network which includes hydrogen bonds, a cavity containing disordered water molecules, a peptide bond and a water path, as given in Figure 8. The water path is a tunnel through which water molecules are reversibly transferred from the matrix side to the cavity. Thus, Asp51 can take up protons from the matrix space. In this sense, Asp51 has accessibility to the matrix space. The affinity of Asp51 for protons in the fully oxidized state, in which it is buried inside the protein, should be much higher than when it is exposed to the bulk water phase because of the low dielectric constant of the protein interior. Upon reduction, Asp51 moves away from the hydrogen bond network and loses its accessibility to the matrix side, but it gains the accessibility to the aqueous environment at the intermembrane space. Con-

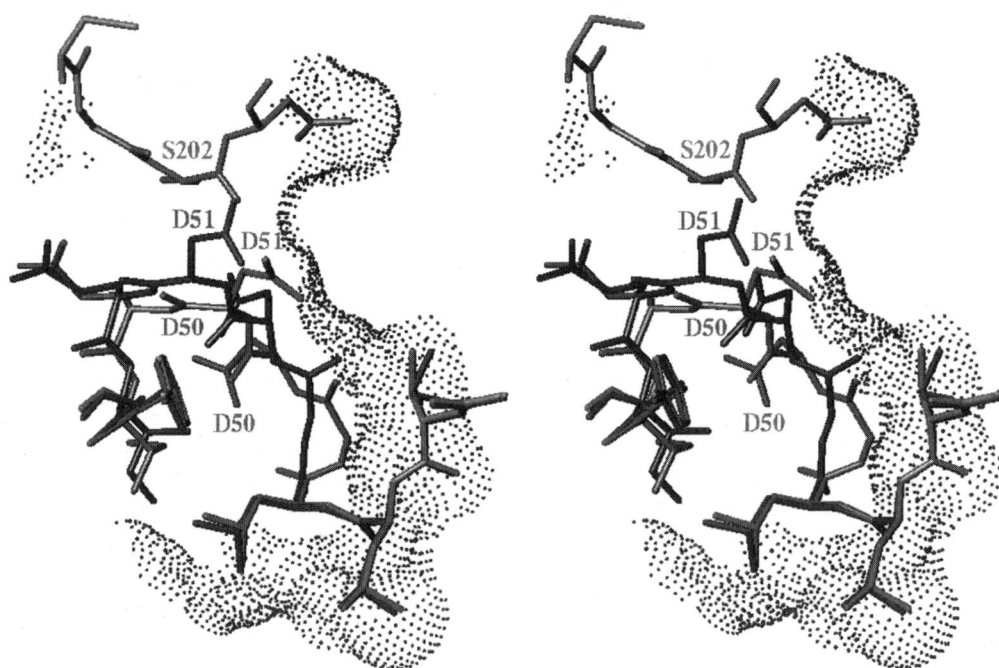

**FIGURE 7.** Redox coupled conformational change in the segment from Gly49 to Asn55. The accessible surface on the intermembrane side for the fully oxidized state is indicated by dots.

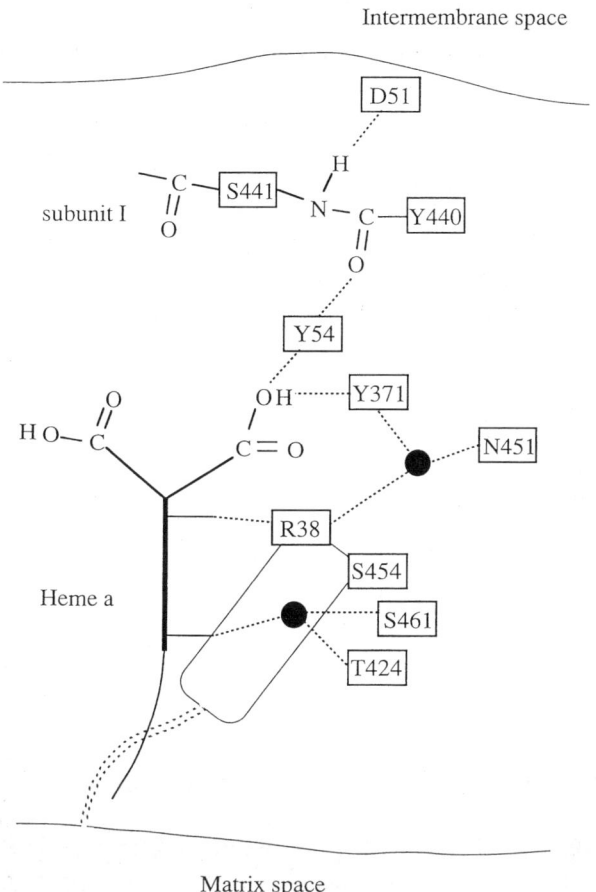

**FIGURE 8.** A schematic representation of the hydrogen bond network from Asp51 to the matrix surface. Dotted lines denote hydrogen bonds. The small dark circles indicate fixed waters. A thick stick denotes a side view of the porphyrin plane of heme a and thin sticks (from top to bottom) are the side chains, propionates, formyl and hydroxyl fornesylethyl. The hydroxyl fornesylethyl group is in the water path with its hydroxyl group hydrogen bonded to Ser461 and Thr424 via a water molecule positioned at the boundary of the cavity.

comitantly, the pK value of the carboxyl side chain of Asp 51 should be lower in the fully reduced state. Thus, this conformational change strongly suggests that Asp51 is the site of proton pumping.

The hydrogen bond network leading to Asp51 in the structure of the fully oxidized form at 2.8 Å was identified but believed to be a dead-end network, since there was no evidence to suggest movements of the main chain of the protein. Prediction of conformational change, including those associated with the main chain is practically impossible given the multitude of possibilities. Thus, the redox-coupled conformational change described here is not easy to predict theoretically.

A peptide bond is located within this hydrogen bond network and provides for the unidirectional proton transfer through this network. In the fully reduced state, carboxyl group of Asp51 is likely to be in the deprotonated state. Upon oxidation of the enzyme, this carboxyl group moves away from the membrane surface into its position in the fully oxidized state. In this position away from the surface, the carboxyl group is located near the N—H group of the peptide bond between Tyr 440 and Ser 441 on Subunit I. On the other hand, the C=O group of the peptide is hydrogen-bonded to Tyr54. When Tyr 54 protonates the C=O group, a protonated peptide, —C(OH)=N(H$^+$)—, will result. The proton on the peptide nitrogen would be quite easily removed by the nearby carboxyl side chain of Asp51 in the deprotonated state. On reduction, the protonated carboxyl side chain of Asp 51 moves away from the peptide, resulting in the enol tautomer, — C(OH)=N—. The deprotonated Tyr54 will be readily re-protonated through a hydrogen bond network leading to Arg35 at the edge of a large cavity. Then, the enol form of the peptide should tautomerize back to the keto form, since the keto form is much more stable than the enol form. The stability of keto versus enol tautomer determines the uni-directional flow of protons through this system.

This proton pumping system does not contain the $O_2$ reduction site. The absence of the $O_2$ binding site serves a very important advantage for this proton pumping system structure in that this arrangement allows for differentiation between protons that are pumped from those that are need for water formation. However, heme a interacts to this pumping pathway at several places. Thus, changes in the redox state at heme a may affect the proton pumping. At the present time, X-ray structures in only two oxidation states are available. Thus, it is not clear which metal site (or sites) controls the conformation of Asp51.

It should be noted that the hydrogen bond network in this system is between Arg38 and Asp51 in the fully oxidized state. Thus, mutagenesis of amino acids that form walls of the cavity and do not alter the water pathway would most likely not have any effect of the function of this enzyme, since protons are transferred by mobile water molecules inside the cavity. The hydrogen-bond network comprises only about 1/3 part of the total length of the proton pumping system. The short hydrogen bond network contributes to rapid and effective proton transfer through the system.

Asp51 is conserved only in animals. Plant and bacterial enzymes do not contain this amino acid. This observation does not provide supporting evidence for the essential role of Asp51 as the proton pumping site if the system responsible for proton pumping must be conserved in all the biological species. A reasonable explanation for the evolutionary divergence is as follows. The process of $O_2$ reduction accomplished by the enzyme to form water without the release of an activated oxygen intermediate is chemically a very complex process. Nature engineered the Fe—Cu system as being a very effective system. No better system has yet been discovered. On the other hand, proton pumping is chemically a quite simple process, since it is essentially the dissociation and association of protons from amino acids. Thus, many amino acids could be replaced and proton transfer retained. Therefore, the $O_2$ reduction site is almost completely conserved while the proton-pumping site is not as well conserved.

On the other hand, the site of the proton pumping site was proposed to be at the dioxygen reduction site prior to the determination of the X-ray structure of this enzyme. This proposal is based on the assumption that both the proton pumping and the dioxygen reduction mechanisms are well conserved in all heme/Cu terminal oxidases. In this proposal, one of the conserved histidine imidazoles coordinated to $Cu_B$ is reversibly released depending on the oxidation state of $Cu_B$. This movement serves to transfer protons away from the $O_2$ reduction site (Wikstrom *et al.*, 1994). Electron density for one of the three imidazole ligands to $Cu_B$ was found to be missing in the X-ray structure of the bacterial enzyme in the fully oxidized-azide bound form (Iwata *et al.*, 1995). The lack of electron density suggests mobility of the imidazole side chain. However, the X-ray structure of the bovine heart enzyme clearly showed electron density for the three imidazoles in the fully oxidized, the fully reduced, the fully oxidized-azide bound and the fully reduced-CO bound forms. Furthermore, it has been shown that the bacterial enzyme in the fully oxidized state in the absence of azide has three imidazoles in the X-ray structure (Ostermeier *et al.*, 1997). The discrepancy in the X-ray structure between azide derivatives of the bovine and bacterial enzymes suggests differences in the stability of the imidazole-$Cu_B$ ligation in the azide bound form between the mammalian and bacterial enzymes. However, under physiological conditions without azide, both enzymes are likely to possess three imidazoles as the stable ligands to $Cu_B$ without the release of imidazoles from $Cu_B$ coordination. Furthermore, another difficulty with Wickstrom's proposal is that a mechanism for differentiating between pumping protons from protons for making waters is not given. All the intermediate species which appear during the $O_2$ reduction are extremely reactive toward protons. Thus, proton pumping and the $O_2$ reduction at the same site would most likely be hopelessly entangled. Thus, little experimental evidence other than conservation of key amino acids involved in $O_2$ reduction supports the proposal for the proton pumping site to be at the dioxygen reduction site. On the other hand, proton pumping at Asp 51 must be proven experimentally through such investigative methods as vibrational spectroscopy since the X-ray structures only suggest the involvement of Asp 51 in the proton pumping mechanism.

The author thanks Dr. Herbert L. Axelrod for his valuable discussion and his critical reading of this manuscript.

## 8. REFERENCES

Ädelroth, P., Ek, M. S., Mitchell, D. M., Gennis, R. B., and Brzezinski, P., 1997, Glutamate 286 in Cytochrome aa₃ from *Rhodobacter spheroides* is involved in proton uptake during the reaction of the fully-reduced enzyme with dioxygen, *Biochemistry* **36**:13824–13829.

Ädelroth, P., Gennis, R. B., and Brzezinski, P., 1998, Role of the pathway through K (I-362) in proton transfer in cytochrome c oxidase from *R.spheroides*, *Biochemistry* **37**:2470–2476.

Beinert, H., Griffiths, D. E., Wharton, D. C., and Sands, R. H., 1962, Properties of the Copper associated with cytochrome oxidase as studied by paramagnetic resonance spectroscopy, *J. Biol. Chem.* **237**:2337–2346.

Blackmore, R. S., Greenwood, C., and Gibson, Q. H., 1991, Studies of the primary oxygen intermediate in the reaction of fully reduced cytochrome oxidase, *J. Biol. Chem.* **266**:19245–19249.

Brunori, M., Colosimo, A., Rainoni, G., Wilson, M. T., and Antonini, E., 1979, Functional intermediates of cytochrome oxidase, *J. Biol. Chem.* **254**:10769–10775.

Buse, G., Soulimane, T., Dewor, M., Meyer, H. E., and Blüggel, M., 1999, Evidence for a copper-coordinated histidine-tyrosine cross-link in the active site of cytochrome oxidase, *Prot. Sci.* **8**:985–990.

Buse, G., Steffens, G. J., Steffens, G. C. M., Meinecke, L., Hensel, S., and Reumkens, J., 1986, Seguence analysis of complex membrane proteins (cytochrome c oxidase), in: *Advanced Methods in Protein Microsequence Analysis* (B. Wittman-Liebold, ed.), Springer-Verlag Berlin, pp. 340–351.

Caughey, W. S., Smythe, G., O'Keefe, D. H., Maskasky, J., and Smith, M. L., 1975, Heme A of cytochrome c oxidase, *J. Biol. Chem.* **250**:7602–7622.

Caughey, W. S., Wallace, W. J., Volpe, J. A., and Yoshikawa, S., 1976, Cytochrome c oxidase, in: *The Enzymes*, 3rd ed., Volume 13 (P.D. Boyer, ed.), Academic Press, New York, pp. 299–344.

Chance, B., Saronio, C., and Leigh Jr., J. S., 1975, Functional intermediotes in the reaction of membrane-bound cytochrome oxidase with oxygen, *J. Biol. Chem.* **250**:9226–9237.

Cleland, W. W., 1977, Determining the chemical mechanism of enzyme-catalyzed reactions by kinetic studies, *Adv. Enzymol.* **45**:273–387.

Einarsdottir, O., and Caughey, W. S., 1985, Bovine heart cytochrome c oxidase preparations contain high affinity binding sites for magnesium as well as for zinc, copper and heme iron, *Biochem. Biophys. Res. Comm.* **129**:840–847.

Ferguson-Miller, S., and Babcock, G. T., 1996, Heme/Copper terminal oxidases, *Chem. Rev.* **96**:2889–2907.

Ferguson-Miller, S., Brautigan, D. L., and Margoliash, E., 1976, Correlation of the Kinetics of electron transfer activity of various eukariotic cytochrome c with binding to mitochondrial cytochrome c oxidase, *J. Biol. Chem.* **251**:1104–1115.

Franceschi, F., Gullotti, M., Monzani, E., Casella, L., and Papaetthymiou, V., 1996, Cytochrome c oxidase models, *Chem. Commun.* 1645–1646.

Garavito, R. M., and Picot, D., 1990, The art of crystallizing membrane proteins, *Methods*, **1**:57–69.

Gibson, Q. H., and Greenwood, C., 1963, Reactions of cytochrome oxidase with oxygen and carbon monoxide, *Biochem. J.* **86**:541–554.

Gibson, Q. H., Greenwood, C., Wharton, D. C., and Palmer, G., 1965, The reaction of cytochrome oxidase with cytochrome c, *J. Biol. Chem.* **240**:888–894.

Hill, B. C., 1994, Modeling the sequence of electron transfer reactions in the single turnover of reduced, mammalian cytochrome c oxidase with oxygen, *J. Biol. Chem.* **269**:2419–2425.

Hosler, J. P., Espe, M. P., Zhen, Y., Babcock, G. T., and Ferguson-Miller, S., 1995, Analysis of site-directed mutants locates a non-redox-active metal near the active site of cytochrome c oxidse of *Rhodobacter Spheroides*, *Biochemistry* **34**:7586–7592.

Hosler, J. P., Ferguson-Miller, S., Calhoun, M. W., Thomas, J. W., Hill, J., Lemieux, L., Ma, J., Georgiou, C., Fetter, J., Shapleigh, J., Tecklenburg, M. M. J., Babcock, G. T., and Gennis, R. B., 1993, Insight into the active site structure and function of cytochrome oxidase by analysis of site-directed mutants of bacterial cytochrome aa$_3$ and cytochrome bo, *J. Bioener. Biomembr.* **25**:121–136.

Iwata, S., Ostermeier, C., Ludwig, B., and Michel, H., 1995, Structure at 2.8 Å resolution of cytochrome c oxidase from *Paracoccus denitrificans*, *Nature* **376**:660–669.

Kadenbach, B., Ungibauer, M., Jarausch, J., Büge, U., and Kuhn-Nentwig, 1983, The complexity of respiratory complexes, *TIBS* **8**:398–400.

Kitagawa, T., and Ogura, T., 1997, Oxygen activation mechanism at the binuclear site of heme-copper oxidase superfamily as revealed by time-resolved resonance Raman spectroscopy, in: *Progress in Inorganic Chemistry*, volume 45(K. D. Karlin ed.), John Wiley of Son, Inc, New York pp. 431–479.

Konstantinov, A. A., Siletsky, S., Mitchel, D., Kaulen, A., and Gennis, R., 1997, The roles of the two proton input channels in cytochrome c oxidase from *Rhodobactor spheroides* probed by the effects of site-directed mutations on time-resolved electrogenic intraprotein proton transfer, *Proc. Natl. Acad. Sci. USA* **94**:9085–9090.

Kroneck, P. M. H., Antholine, W. E., Kastrau, D. H. W., Buse, G., Steffens, G. C. M., and Zumft, W. G., 1990, Multifrequency EPR evidence for a bimetallic center at the Cu$_A$ site in Cytochrome c oxidase, *FEBS Lett.* **268**:274–276.

Landrum, J. T., Reed, C. A., Hatano, K., and Scheidt, W. R., 1978, Imidazolate anion bridged metalloporphyrins of relevance to a model for cytochrome oxidase, *J. Am. Chem. Soc.* **100**:3232–3234.

Lappalainen, P., Watmough, N. J., Greenwood, C., and Saraste, M., 1995, Electron transfer between cytochrome c and the isolated Cu$_A$ domain: identification of substrate-binding residues in cytochrome c oxidase, *Biochemistry*, **34**:5824–5830.

Malmström, B. G., 1990, Cytochrome c oxidase as a redox-linked proton pump, *Chem. Rev.* **90**:1247–1260.

Michel, H., 1999, The mechanism of proton pumping by cytochrome c oxidase, *Proc. Natl. Acad. Sci. USA* **95**:12819–12824.

Mims, W. B., Peisach, J., Shaw, R. W., and Beinert, H., 1980, Electron spin echo studies of cytochrome c oxidase, *J. Biol. Chem.* **255**:6843–6846.

Minnaert, K., 1961, The kinetics of cytochrome c oxidase I. The system: cytochrome c-cytochrome oxidase-oxygen, *Biochim. Biophys. Acta.* **50**:23–34.

Mochizuki, M., Aoyama, H., Shinzawa-Itoh, K., Usui, T., Tsukihara, T., and Yoshikawa, S., 1999, Quantitative reevaluation of the redox active sites of crystalline bovine heart cytochrome c oxidase, *J. Biol. Chem.* in press.

Ogura, T., Takahashi, S., Hirota, S., Shinzawa-Itoh, K., Yoshikawa, S., Appelman, E. H., and Kitagawa, T., 1993, Time-resolved resonance Raman elucidation of the pathway for dioxygen reduction by cytochrome c oxidase, *J. Am. Chem. Soc.* **115**:8527–8536.

Ogura, T., Takahashi, T., Shinzawa-Itoh, K., Yoshikawa, S., and Kitagawa, T., 1991, Time resolved resonance Raman investigation of cytochrome oxidase catalysis: observation of a new oxygen-isotope sensitive Raman band, *Bull. Chem. Soc. Jpn.*, **64**:2901–2907.

Okunuki, K., and Yakushiji, E., 1941, *Proc. Imp. Acad.*, **17**:263.

Okunuki, K., Sekuzu, I., Yonetani, T., and Takemori, S., 1958, Studies on cytochrome A, I: Extraction, purification and some properties of cytochrome A, *J. Biochem.* **45**:847–854.

Ostermeier, C., Harrenga, A., Ermer, U., and Michel, H., 1997, Structure at 2.7 Å resolution of the *Paracoccus denitrificans* two-subunit cytochrome c oxidase complexed with an antibody Fv fragment., *Proc. Natl. Acad. USA* **94**:10547–10553.

Ostermeier, C., Iwata, S., Lubwig, B., and Michel, H., 1995, Fv fragment-mediated crystallization of the membrane protein bacterial cytochrome c oxidase, *Nature Structural Biology*, **2**:842–846.

Puustinen, A., and Wikström, M., 1991, The heme groups of cytochrome o from Eschericia coli, *Proc. Natl. Acad. Sci. USA* **88**:6122–6126.

Reinhammar, B., Malkin, R., Jensen, P., Karlson, B., Andréasson, L., Aasa, R., Vanngard, T., and Malmström, B. G., 1980, A new copper (II) electron paramagnetic resonance signal in two laccases and in cytochrome c oxidase, *J. Biol. Chem.* **255**:5000–5003.

Scott, R. A., 1989, X-ray absorption spectroscopic investigations of cytochrome c oxidase structure and function, *Ann. Rev. Biophys. Biophys. Chem.*, **18**:137–158.

Seiter, C. H. A., and Angelos, S. G., 1980, Cytochrome oxidase: An alternative model, *Proc. Natl. Acad. Sci. USA* **77**:1806–1808.

Shinzawa-Itoh, K., Ueda, H., Yoshikawa, S., Aoyama, H., Yamashita, E., and Tsukihara, T., 1995, Effects of Ethylene glycol chain length of dodecyl polyethyleneglycol monoether on the crystallization of bovine heart cytochrome c oxidase, *J. Mol. Biol.* **246**:572–575.

Speck, S. H., Dye, D., and Margoliash, E., 1984, Single catalytic site model for the oxidation of ferrocytochrome c by mitochondrial cytochrome c oxidase, *Proc. Natl. Acad. Sci. USA* **81**:347–351.

Steffens, G., and Buse, G., 1976, Studies on Cytochrome c oxidase, I: Purification and characterization of the enzyme from bovine heart and identification of peptide chains in the complex, *Hoppe-Seyler's Z. Physiol. Chem.* **357**:1125–1137.

Steffens, G. C. M., Biewald, R., and Buse, G., 1987, Cytochrome c oxidase is a three-copper, two-heme A protein, *Eur. J. Biochem.* **64**:295–300.

Tsukihara, T., Aoyama, H., Yamashita, E., Tomizaki, T., Yamaguchi, H., Shinzawa-Itoh, K., Nakashima, R., Yaono, R., and Yoshikawa, S., 1995, Structures of metal sites of oxidized bovine heart cytochrome c oxidase at 2.8 Å, *Science* **269**:1069–1074.

Tsukihara, T., Aoyama, H., Yamashita, E., Tomizaki, T., Yamaguchi, H., Shinzawa-Itoh, K., Nakashima, R., Yaono, R., and Yoshikawa, S., 1996, The whole structure of the 13-subunit oxidized cytochrome c oxidase at 2.8 Å, *Science* **272**:1136–1144.

Van Gelder, B. F., and Beinert, H., 1969, Studies on the heme components of cytochrome c oxidase by EPR spectroscopy, *Biochim. Biophys. Acta*, **189**:1–24.

Vanneste, W. H., 1966, The stoichiometry and absorption spectra of components a and $a_3$ in cytochrome c oxidase, *Biochemistry* **5**:838–848.

Warburg, O., 1924, Über eisen, den sauerstoffübertragenden bestandteil des atmungsfermentes, *Biochem. Z.* **152**:479–494.

Wharton, D. C., and Tzagoloff, A., 1964, Studies on the electron transfer system LVII. The near infrared absorption band of cytochrome oxidase, *J. Biol. Chem.* **239**:2036–2041.

Wikström, M., and Morgan, J. E., 1992, The dioxygen cycle, *J. Biol. Chem.* **267**:10266–10273.

Wikström, M. K. F., 1977, Proton pump coupled to cytochrome c oxidase in mitochondria, *Nature* **266**:271–273.

Wilmanns, M., Lappalainen, P., Kelly, M., Sauer-Eriksson, E., and Saraste, M., 1995, Crystal structure of the membrane-exposed domain from a respiratory quinol oxidase complex with an engineered dinuclear copper center, *Proc. Natl. Acad. Sci. USA* **92**:11955–11959.

Woodruff, W. H., Einersdottir, O., Dyer, R. B., Bagley, K. A., Palmer, G., Atherton, S. J., Goldbeck, R. A., Dawes, T. D., and KLiger, D. S., 1991, Nature and functional implications of the cytochrome $a_3$ transients after photodissociation of CO-cytochrome oxidase, *Proc. Natl. Acad. Sci. USA* **88**:2588–2592.

Yakushiji, E., and Okunuki, K., 1941, *Proc. Imp. Acad.*, **17**:38.

Yonetani, T., 1959, Studies on cytochrome A, III: Effect of synthetic detergents upon the activity of cytochrome A, *J. Biochem. (Tokyo)* **46**:917–924.

Yonetani, T., 1961, Studies on cytochrome oxidase, III: Improved preparation and some properties, *J. Biol. Chem.* **266**:1680–1688.

Yoshikawa, S., Choc, M. G., O'Toole, M. C., and Caughey, W. S., 1977, An infrared study of CO binding to heart cytochrome c oxidase and hemoglobin A, *J. Biol. Chem.* **252**:5498–5508.

Yoshikawa, S., Mochizuki, M., Zhao, X. J., and Caughey, W. S., 1995, Effects of overall oxidation state on infrared spectra of heme $a_3$ cyanide in bovine heart cytochrome c oxidase, *J. Biol. Chem.* **270**:4270–4279.

Yoshikawa, S., Shinzawa-Itoh, K., Nakashima, R., Yaono, R., Yamashita, E., Inoue, N., Yao, M., Fei, M. J., Peters Libeu, C., Mizushima, T., Yamaguchi, H., Tomizaki, T., and Tsukihara, T., 1998, Redox-coupled crystal structural changes in bovine heart cytochrome c oxidase, *Science* **280**:1723–1729.

Yoshikawa, S., Tera, T., Takahashi, Y., Tsukihara, T., and Caughey, W. S., 1988, Crystalline cytochrome c oxidase of bovine heart mitochondrial membrane: composition and X-ray diffraction studies, *Proc. Natl. Acad. Sci. USA* **85**:1354–1358.

*Chapter 18*

# Reaction Centres of Purple Bacteria

## Marion E. van Brederode* and Michael R. Jones

## 1. INTRODUCTION

The reaction centre found in many purple non-sulphur bacteria is a simple example of a group of proteins that are nature's solar batteries. The reaction centre uses the energy of sunlight to generate positive and negative charges on opposite sides of the bacterial cytoplasmic membrane. This potential difference drives a circuit of electron transfer reactions that are linked to proton translocation across this membrane.

A feature that makes the reaction centre different from most other redox proteins is that light is used as a substrate by the complex, triggering the trans-membrane electron transfer reaction. As a result, the reaction centre has been at the forefront of research on biological electron transfer, and in particular has made a major contribution to the study of ultrafast, picosecond time-scale electron transfer.

**MARION E. VAN BREDERODE**   Faculty of Sciences, Division of Physics and Astronomy, Department of Biophysics and Physics of Complex Systems, Free University of Amsterdam, de Boelelaan 1081, 1081 HV Amsterdam, The Netherlands.   *Present address: Interfaculty Reactor Institute, Department of Radiation Chemistry, University of Technology TU Delft, The Netherlands.   **MICHAEL R. JONES**   Department of Biochemistry, School of Medical Sciences, University of Bristol, University Walk, Bristol, BS8 1TD, United Kingdom.

*Subcellular Biochemistry, Volume 35: Enzyme-Catalyzed Electron and Radical Transfer*, edited by Holzenburg and Scrutton. Kluwer Academic / Plenum Publishers, New York, 2000.

This article does not seek to provide a comprehensive review of studies on the bacterial reaction centre, as this is a massive topic that has been covered in recent years in a number of excellent books and reviews (Hoff and Deisenhofer, 1997; Parson, 1996; Blankenship *et al.*, 1995; Deisenhofer and Norris, 1993; Levanon, 1992). Rather, after an introduction to the structure and mechanism of the bacterial reaction centre, the present article looks at research concerned with real-time monitoring of biological electron transfer on a picosecond time-scale, an area where the bacterial reaction centre has made a unique contribution. This time window is accessible in this system because pulses of laser light of less than a picosecond in duration can be used both as a trigger to drive electron transfer, and as a probe of the rate of electron transfer reactions through the monitoring of absorbance changes that accompany changes in the redox state of the cofactors involved.

## 2. STRUCTURE OF THE BACTERIAL REACTION CENTRE

### 2.1. The Structures of the *Rhodopseudomonas Viridis* and *Rhodobacter Sphaeroides* Reaction Centres

The larger part of research on light-driven electron transfer in purple photosynthetic bacteria has involved three species, *Rhodopseudomonas* (*Rps.*) *viridis*, *Rhodobacter* (*Rb.*) *sphaeroides* and *Rb. capsulatus*. The bulk of this article is written in reference to the *Rb. sphaeroides* reaction centre, the subject of the majority of spectroscopic and mutagenesis work carried out to date. However, much of the research described below has involved the reaction centre from *Rb. capsulatus* or *Rps. viridis*, or reaction centres from other species of purple bacteria.

The reaction centre from *Rps. viridis* was the first integral membrane protein for which a high resolution X-ray crystal structure was determined (Deisenhofer and Michel, 1989a,b; Michel *et al.*, 1986; Deisenhofer *et al.*, 1985; Deisenhofer *et al.*, 1984). This Nobel prize-winning work was followed by the determination of the three dimensional structure of the reaction centre from *Rb. sphaeroides* (Chang *et al.*, 1991; El Kabbani *et al.*, 1991; Allen *et al.*, 1988; Komiya *et al.*, 1988; Yeates *et al.*, 1988, 1987; Allen *et al.*, 1987a,b, 1986; Chang *et al.*, 1986). Following on from the first series of reports, in recent years structures have been published for the *Rb. sphaeroides* reaction centre at approximately 2.6 Å resolution (Ermler *et al.*, 1994a,b), and for the *Rps. viridis* reaction centre at 2.3 Å resolution (Deisenhofer *et al.*, 1995). The X-ray crystal structure of the *Rb. capsulatus* reaction centre has not been reported, although a detailed structural model based on the structure of the closely-related *Rb. sphaeroides* reaction centre has been described (Foloppe *et al.*, 1995).

The three-dimensional structure of the reaction centre from *Rb. sphaeroides* is depicted in Figure 1. The complex consists of three polypeptides that are denoted L, M and H. The L and M polypeptides each have five trans-membrane α-helices that related by an axis of pseudo two-fold symmetry that runs perpendicular to the plane of the membrane (Figure 1C). The H polypeptide has a single trans-membrane α-helix and an extra-

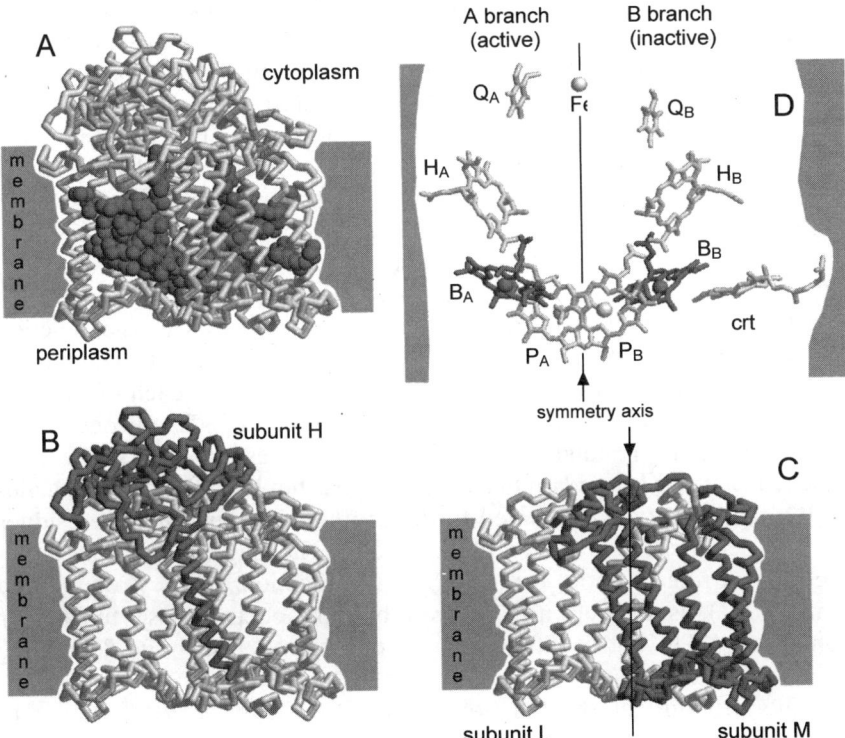

**FIGURE 1.** Structure of the reaction centre from *Rhodobacter sphaeroides*. (A) The reaction centre is an integral membrane protein consisting of ten cofactors (dark grey spheres) encased in a scaffold provided by three polypeptide subunits (backbones shown in light grey), (B) The H subunit (highlighted in dark grey) has a cytoplasmic domain and a single trans-membrane α-helix, (C) The L (light grey) and M (dark grey) subunits each have 5 trans-membrane α-helices and arranged around an axis of two-fold symmetry (vertical line), (D) enlarged view of the ten reaction centre cofactors (see text for an explanation of the nomenclature); the BChl, BPhe and $UQ_{10}$ cofactors are arranged around the symmetry axis (vertical line) in two membrane-spanning branches.

membrane domain that caps the cytoplasmic faces of the L and M polypeptides (Figure 1B). The reaction centre has an amphipathic structure; the extra-membrane surfaces of the reaction centre are coated with mainly polar amino acids, whilst the intra-membrane surfaces of the reaction centre polypeptides are coated with mainly hydrophobic amino acids.

The L and M polypeptides encase the ten reaction centre cofactors (Figures 1A and D). These are 4 molecules of bacteriochlorophyll $a$ (BChl), 2 molecules of bacteriopheophytin $a$ (BPhe), two molecules of ubiquinone-10 ($UQ_{10}$), a single carotenoid and a non non-heme iron atom. Two of the BChls, denoted $P_A$ and $P_B$ (Figure 1D) are located near the periplasmic face of the protein, and are arranged sufficiently close to together that their $\pi$ electronic orbitals mix. This "special pair" of BChls straddles the axis of two-fold symmetry (Figure 1D), and is the primary donor of electrons (denoted P) in light-driven trans-membrane electron transfer.

The remaining monomeric BChl ($B_A$, $B_B$), BPhe ($H_A$, $H_B$) and $UQ_{10}$ ($Q_A$, $Q_B$) cofactors are arranged in a pseudo-symmetrical manner around the two-fold symmetry axis, in two membrane-spanning branches termed A and B (Figure 1D). As discussed in Section 5.2, spectroscopic studies have shown that light-driven trans-membrane electron transfer proceeds essentially exclusively along only one of these branches, identified as the A-branch. Binding of the BChl, BPhe and $UQ_{10}$ cofactors to the reaction centre occurs via one or more non-covalent interactions with residues of the surrounding protein. The central magnesium atom of each of the four BChl molecules has a fifth, axial ligand donated by the NE2 nitrogen of an adjacent histidine residue (BPhe lacks this magnesium atom). The non-heme iron atom is located on the symmetry axis between the two molecules of $UQ_{10}$ and is hexa-coordinated to the protein via four histidine residues and a bi-dentate ligand from a glutamic acid. The single carotenoid molecule plays a role in photo-protection of the reaction centre (Cogdell and Frank, 1987), is located in a curved binding pocket close to the $B_B$ monomeric BChl, and is either spheroidene or spheroidenone depending on growth conditions.

The structure of the reaction centre from *Rps. viridis* is similar to that of *Rb. sphaeroides*. The principal differences are that the *Rps. viridis* reaction centre contains BChl $b$ and BPhe $b$ in place of BChl $a$ and BPhe $a$, menaquinone-9 in place of $UQ_{10}$ at the $Q_A$ site, and the carotenoid hydroneurosporene in place of spheroidene/spheroidenone. The other major difference is the presence of a fourth extra-membrane subunit in the *Rps. viridis* reaction centre that consists of a tetra-heme cytochrome that is attached to the periplasmic faces of the L and M subunits (Deisenhofer *et al.*, 1995; Deisenhofer and Michel, 1989a,b; Michel *et al.*, 1986; Deisenhofer *et al.*, 1985; Deisenhofer *et al.*, 1984).

## 2.2. Recent Structural Information Relating to Function

A number of recent studies have used X-ray crystallography in an attempt to obtain information on structural changes that accompany the reduction of $UQ_{10}$ in the reaction centre. The $Q_A$ and $Q_B$ $UQ_{10}$ cofactors are located near the cytoplasmic side of the membrane, in binding pockets which are different in character. The head group of the $Q_A$ $UQ_{10}$ is locked into the interior of the protein, in a binding pocket which is largely hydrophobic. This $UQ_{10}$ operates as a one electron carrier, receiving an electron from the $H_A$ BPhe and passing it on to the $Q_B$ $UQ_{10}$. In keeping with its role in the electron transport chain, the $Q_A$ $UQ_{10}$ is not able to dissociate from the reaction centre. In contrast, the $Q_B$ $UQ_{10}$ operates as a two electron/two proton redox centre and is able to exchange with $UQ_{10}$ in the intra-membrane pool. It is located in a binding site that contains a number of polar residues that are thought to play a role in the coupling of reduction of the $Q_B$ $UQ_{10}$ to its protonation (Okamura and Feher, 1995, 1992).

The inclusion of modelled water molecules in more recent, higher resolution structures for the *Rb. sphaeroides* and *Rps. viridis* reaction centres has revealed chains of connected water molecules and polar residues linking the $Q_B$ binding pocket with the cytoplasmic surface of the protein (Lancaster and Michel, 1999; Abresch *et al.*, 1998; Fritzsch *et al.*, 1998; Lancaster and Michel, 1997; Stowell *et al.*, 1997; Lancaster and Michel, 1996; Deisenhofer *et al.*, 1995; Ermler *et al.*, 1994a,b). It has been proposed that these chains deliver protons to acidic amino acids that have been implicated in protonation of the reduced $Q_B$ $UQ_{10}$ (Abresch *et al.*, 1998; Lancaster and Michel, 1997; Stowell *et al.*, 1997; Lancaster and Michel, 1996; Ermler *et al.*, 1994a; Beroza *et al.*, 1992; Okamura and Feher, 1992).

Two recent studies have led to proposals that the $Q_B$ $UQ_{10}$ undergoes changes in binding conformation at various stages in the process of reduction and protonation. One study involved a comparison of the X-ray structure of *Rb. sphaeroides* reaction centres frozen to cryogenic temperatures in the light and frozen in the dark, revealing a difference in the binding position of the $Q_B$ $UQ_{10}$ under the two sets of conditions (Stowell *et al.*, 1997). The second study looked at the structure of the $Q_B$ binding site in *Rps. viridis* reaction centres that had either been depleted of $UQ_{10}$, reconstituted with a short chain analogue of $UQ_{10}$ ($UQ_2$), or treated with an inhibitor of the $Q_B$ site (Lancaster and Michel, 1997). Taken together, these studies have shown that the $Q_B$ $UQ_{10}$ can bind in at least two distinct positions in the $Q_B$ pocket, and that a transition between the two involves not only a translation of the head group of the $UQ_{10}$, but also an approximately 180° rotation of the head group. This observation is in line with the idea that the rate limiting step in the electron transfer reaction from $Q_A^-$ to $Q_B$ is a protein conformational change or movement of the $Q_B UQ_{10}$, which was proposed on the basis of the observation that this reaction is blocked in reaction centres that have been frozen to 77 K in the dark, but not in reaction centres that have been frozen to 77 K under continuous illumination (Kleinfeld *et al.*, 1984). Recently, this proposal of conformational gating has received additional support from the observation that the rate of electron transfer from $Q_A^-$ to $Q_B$ is not affected as the free energy of the reaction is varied (Graige *et al.*, 1998).

## 2.3.  Structures of Mutant Complexes

The X-ray crystal structure of the bacterial reaction centre has guided a major research effort aimed at understanding, at a molecular level, the electron transfer reactions that take place in the complex. A large part of this research effort has involved the study of reaction centres altered by site-directed mutagenesis. A number of spectroscopic studies have shown that the replacement of one or more amino acids in the vicinity of the reaction centre cofactors can have large effects on the characteristics of light-driven trans-membrane electron transfer (see Section 5). A limited number of reports have described the structure of a reaction centre containing a single site mutation (Ridge *et al.*, 1999; McAuley-Hecht *et al.*, 1998; Fyfe *et al.*, 1998; Chirino *et al.*, 1994; Sinning, 1992; Sinning *et al.*, 1990). The general finding of these studies is that the structural changes that accompany mutation tend to be localised, with no major changes in the structure of the protein cofactor system, even in mutants where there are significant effects on the rate of trans-membrane electron transfer. This indicates that the fine details of the interactions between the cofactors and the surrounding protein are of great importance for optimisation of the energetics of primary electron transfer, and that this reaction can be affected to a significant extent without any major change in the positions of the cofactors. Two points to note, however, are that only a minority of the structures obtained thus far have been at a high resolution (i.e. better than 3.0 Å), and the range of mutant reaction centres examined has necessarily been limited to those complexes that are not adversely affected in terms of yield or stability by the mutation(s) carried.

Very recently, it has been shown that it is possible to exclude cofactors such as the $Q_A$ ubiquinone or the carotenoid by mutating a small residue such as Ala or Gly in the cofactor binding pocket to a bulkier residue such as Trp or Leu (Ridge *et al.*, 1999). X-ray crystallography has shown that in such cases there are no changes in structure of the protein outside the immediate vicinity of the cofactor binding pocket (Ridge *et al.*,

1999). In cases where a large residue has been changed to a smaller residue, or a structural change has occurred that creates a cavity in the interior of the protein, electron density has been observed in the cavity that can be modelled as one or more new water molecules (McAuley-Hecht *et al.*, 1998; Fyfe *et al.*, 1998a,b). This raises the possibility that some effects of certain mutations may be attributable to a change in the number and position of bound waters in the vicinity of the mutation site (McAuley-Hecht *et al.*, 1998), an effect which is difficult to model during the design of a mutation.

## 2.4. Electronic Structure: The Reaction Centre Absorbance Spectrum

The absorbance properties of photosynthetic reaction centres are of key importance to our understanding of the mechanism of light energy transduction by these complexes. One of the reasons that the reaction centre from purple bacteria has been used so extensively in this research effort is that the cofactors (or groups of cofactors) have absorbance bands which are clearly distinguishable (see Figure 2). This is in contrast to the reaction centres of higher plants, where all of the chlorophyll-type cofactors absorb in a relatively narrow wavelength region. The visible and near-infrared absorption spectrum of the bacterial reaction centre is contributed to by the BChl, BPhe and carotenoid cofactors. The spectrum provides important information on the design of the complex, provides a means of monitoring individual cofactors or groups of cofactors, and provides a means of initiating processes such as electron transfer in a specific manner by monochromatic excitation of a selected cofactor.

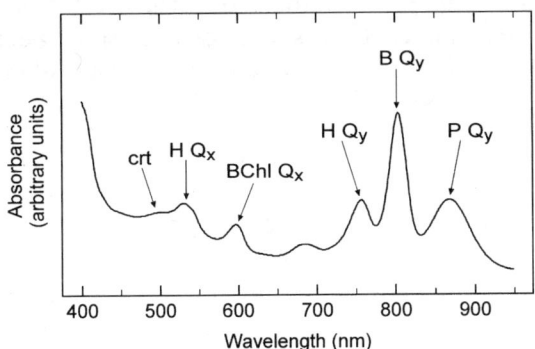

**FIGURE 2.** Room temperature absorbance spectrum of purified *Rb. sphaeroides* reaction centres.

The absorption spectra of the BChl and BPhe cofactors in the visible and near-infrared regions contains two types of electronic ($\pi\pi^*$) transitions. These are termed $Q_x$ or $Q_y$ depending upon whether they are polarised along the molecular X or Y axis, respectively (the $\pi$ electron system of BChl has an elongated shape, and so the mutually perpendicular X and Y axes are not equivalent). The absorbance spectrum of the purified wild-type *Rb. sphaeroides* reaction centre is shown in Figure 2. Although an exact description of the origins of all of the features in the absorption spectrum is a complex issue, a simplified account provides an adequate description for many purposes. The spectrum contains two main bands in the $Q_x$ region between 500 and 650 nm. The band at 540 nm is attributable to the $Q_x$ transitions of the reaction centre BPhes (H $Q_x$ band); at cryogenic temperatures this band splits into a component at 532 nm attributable to the $H_B$ BPhe and a component at 545 nm attributable to the $H_A$ BPhe. The second band at 590 nm contains contributions from the $Q_x$ transitions of all four reaction centre BChls (BChl $Q_x$ band).

In the $Q_y$ region, the band at 756 nm is attributable to the $Q_y$ transitions of both reaction centre BPhes (H $Q_y$ band), whilst the band at 804 nm mainly consists of contributions from the $Q_y$ transitions of the monomeric BChls $B_A$ and $B_B$ (B $Q_y$ band). The two BChl molecules that constitute P are excitonically-coupled, which results in a splitting of their $Q_y$ optical transitions into a lower exciton component (denoted $P_-$) and a higher exciton component (denoted $P_+$). The $P_-$ band carries most of the intensity and gives rise to the absorbance band at 867 nm (P $Q_y$ band), whilst the much smaller $P_+$ band underlies the B $Q_y$ band of the monomeric BChls. In linear and circular dichroism experiments this upper excitonic component has been estimated to be at approximately 814 nm, with an intensity that is approximately 10% of the total intensity of the absorbance band centred at 804 nm (Breton *et al.*, 1989; Breton, 1988). The carotenoid present in the *Rb. sphaeroides* reaction centre is spheroidene or spheroidenone, depending on growth conditions, and shows absorbance bands between 450 and 550 nm.

## 3. THE MECHANISM OF ENERGY STORAGE BY THE BACTERIAL REACTION CENTRE

As depicted in Figure 3, the bacterial reaction centre can be thought of in a number of ways. In one view, the reaction centre can be thought of as a solar battery (Figure 3A), using light to separate charge across a charge-impermeable membrane, creating the so-called $P^+Q_A^-$ state. The electrical potential difference established across the membrane then drives an exter-

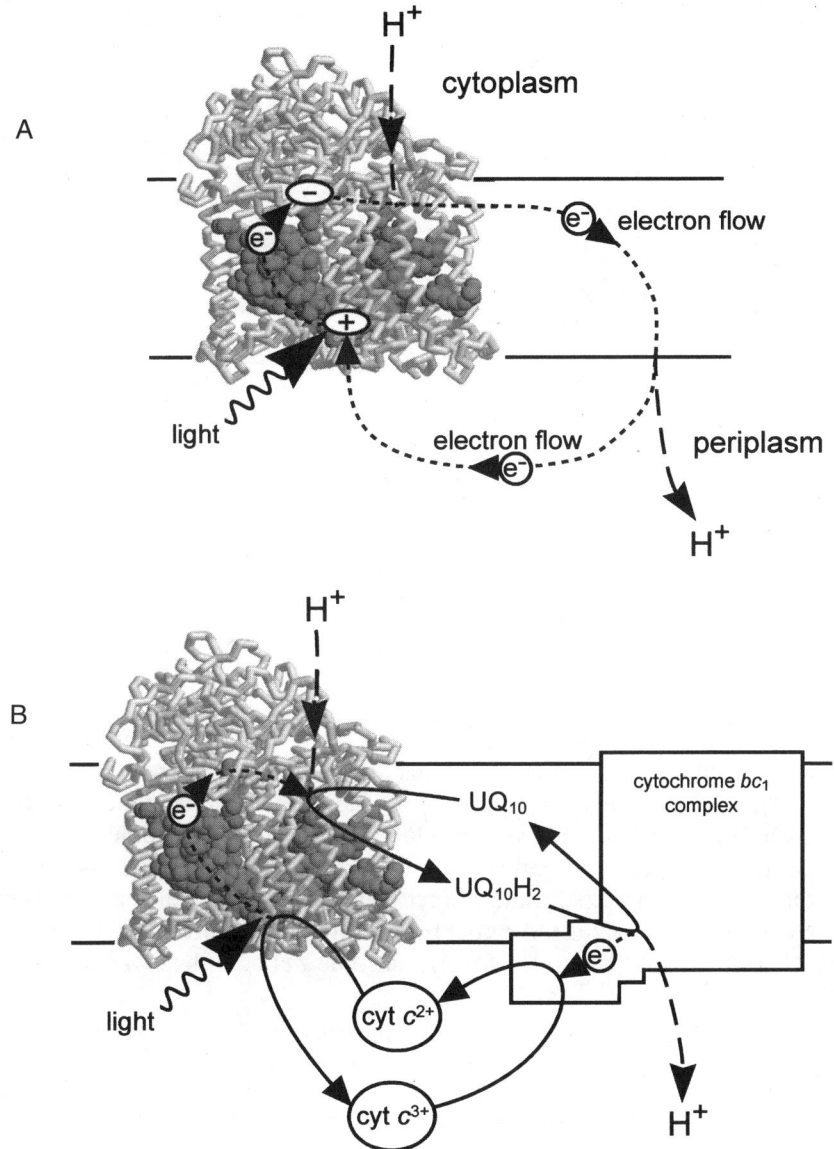

**FIGURE 3.** Reaction centre function. (A) The reaction centre operates as a solar battery, driving an external circuit that is linked to proton translocation across the membrane. (B) The reaction centre can also be viewed as an enzyme, catalysing the reduction of $UQ_{10}$ by cyt $c^{2+}$ in a light-dependent reaction. (C) When operating in conjunction with the cytochrome $bc_1$ complex, the reaction centre operates as a light-driven proton pump.

**Marion E. van Brederode and Michael R. Jones**

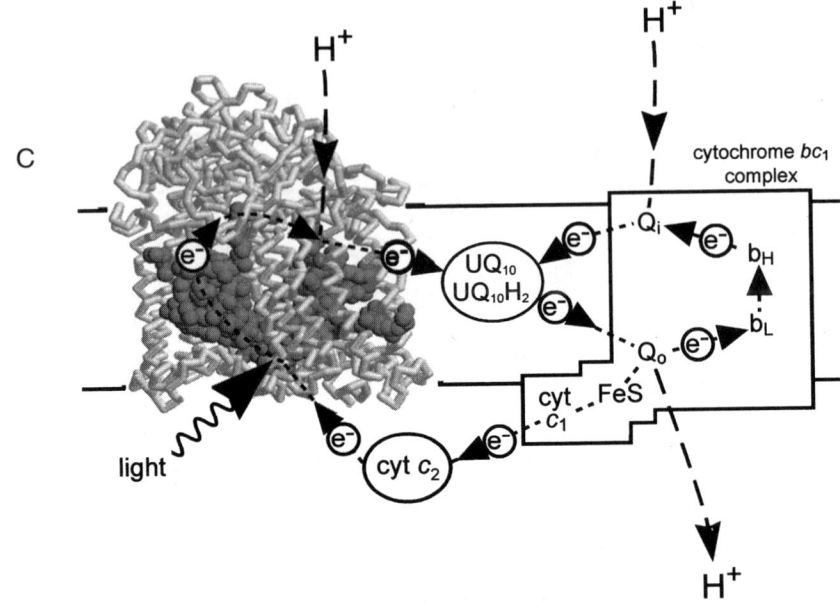

**FIGURE 3.** (*Continued*)

nal circuit that returns the electron from $Q_A^-$ to $P^+$ in a manner which is linked to the translocation of protons across the membrane.

The reaction centre can also be viewed as an enzyme, namely a cytochrome $c_2$:ubiquinone oxidoreductase (Figure 3B). This enzyme uses light to power an energetically-unfavourable reaction, the reduction of $UQ_{10}$ by reduced cytochrome $c_2$ (cyt $c^{2+}$). The products of reaction centre enzymatic activity, oxidised cytochrome $c_2$ (cyt $c^{3+}$) and reduced ubiquinone ($UQ_{10}H_2$) are the substrates for cytochrome $bc_1$ complex, which catalyses the reduction of cyt $c^{3+}$ by $UQ_{10}H_2$.

Finally, when operating in conjunction with the cytochrome $bc_1$ complex, cytochrome $c_2$ and the intra-membrane $UQ_{10}/UQ_{10}H_2$ pool, the reaction centre uses light energy to drive a proton pump (Figure 3C). Light energy is used to pump protons from the cytoplasm of the bacterium to the periplasm, establishing an electrochemical gradient of protons across the bacterial inner membrane. In fact, the reaction centre itself operates as a light-driven electron pump, in which light energy is used to pump electrons across the membrane against the potential established by the proton electrochemical gradient.

## 3.1.  Primary Photochemistry

Electron transfer in the reaction centre is triggered by the promotion of the P BChls from their ground electronic state to the first singlet excited state P* (Figure 4). This changes the reduction potential of P by approximately 1.4 eV, converting P into a powerful reductant that is capable of reducing the adjacent reaction centre cofactors. The P $\rightarrow$ P* transition can either occur through the direct absorption of a photon by the P BChls, or by femtosecond time-scale resonant energy transfer from the reaction centre monomeric BChl, BPhe and carotenoid molecules, or by picosecond time-scale resonant energy transfer from the BChl molecules of the attendant light harvesting complexes. Irrespective of how the P* state is generated, it triggers the formation of the charge separated state $P^+Q_A^-$ on a picosecond time-scale and with a quantum yield that is close to 100% (Wraight and Clayton, 1973).

The changes in electronic state of the cofactors that occur during trans-membrane electron transfer are accompanied by changes in their absorption spectra. In addition, the movement of charge establishes electric fields which shift the absorbance spectra of adjacent cofactors. The trans-membrane reaction can therefore be studied by monitoring changes in the colour of the reaction centre that occur as the electron moves through the complex. The early application of laser spectroscopy to trans-membrane electron transfer in the reaction centre, using lasers capable of generating pulses of a few tens of picoseconds in duration, resolved the first easily detectable intermediate in charge separation, $P^+H_A^-$, and determined the rate of electron transfer from $H_A^-$ to $Q_A$ (Kaufmann et al., 1976; Kaufmann et al., 1975; Rockley et al., 1975). In subsequent years, steady improvement was made in the time-resolution and sensitivity of instrumentation used to monitor the primary electron transfer in the reaction centre. A number of recent reviews have provided excellent insights into the development of our understanding of light-driven electron transfer in the reaction centre (Bixon and Jortner, 1999; Hoff and Deisenhofer, 1997; Parson, 1996; Woodbury and Allen, 1995; Fleming and Van Grondelle, 1994; Deisenhofer and Norris, 1993; DiMagno and Norris, 1993; Kirmaier and Holten, 1993; Shuvalov, 1993; Zinth and Kaiser, 1993).

At room temperature, the P* state decays by electron transfer from P to the $H_A$ BPhe, located approximately half-way across the membrane, forming the radical pair state $P^+H_A^-$. Why electron transfer in the reaction centre proceeds exclusively along the A-branch of cofactors, despite the approximately symmetrical arrangement of the cofactors in the A- and B-branches, is a matter of ongoing debate (see Section 5.2), as is the role

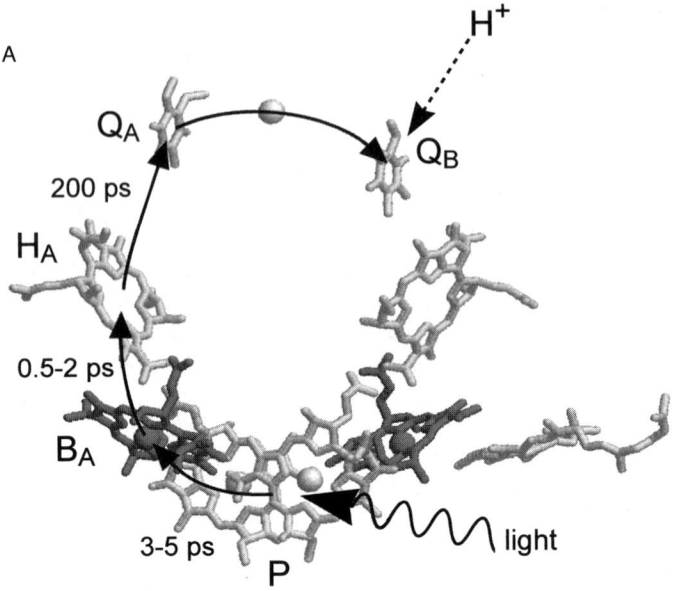

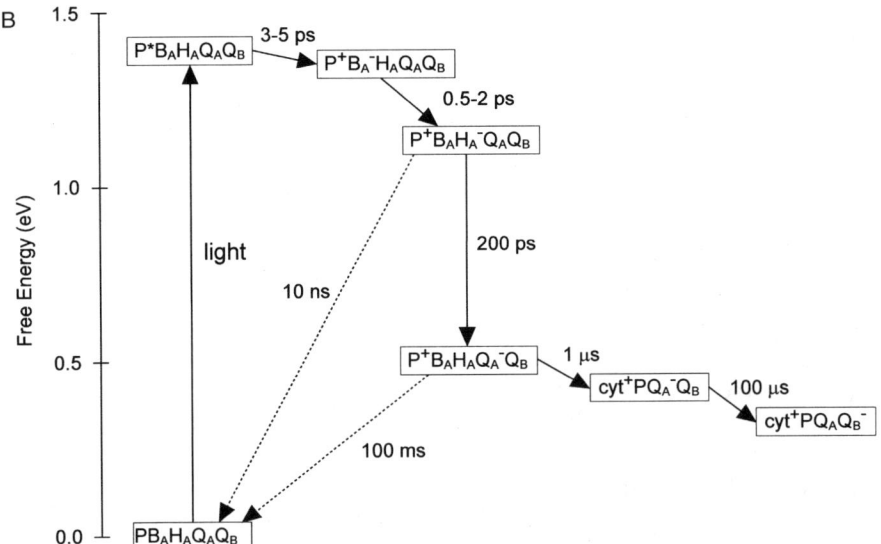

**FIGURE 4.** Reaction centre mechanism. (A) Room temperature light-driven trans-membrane electron transfer in the *Rb. sphaeroides* reaction centre. (B) The energetics of electron transfer in the *Rb. sphaeroides* reaction centre at room temperature, showing approximate free energies of intermediate states in forward electron transfer. Solid arrows show forward electron transfer, and dotted arrows show charge recombination to the ground state when forward electron transfer is blocked.

played by the monomeric $B_A$ BChl (see Section 5.1). The P* state has a lifetime of 3–5 ps at room temperature, the precise value varying with sample identity and measuring conditions. If forward electron transfer from $H_A^-$ is blocked, the $P^+H_A^-$ radical pair has a lifetime of approximately 10 nanoseconds, the main route of decay being direct recombination to the ground state (Figure 4B) with some decay to the ground state after a hyperfine-induced singlet-triplet mixing via the triplet excited state of P ($P^T$) (Volk et al., 1995).

Electron transfer from $H_A^-$ to $Q_A$ occurs with a lifetime of 200 ps at room temperature, completing trans-membrane electron transfer (Figure 4). As with primary electron transfer, if forward electron transfer from $Q_A^-$ is blocked then recombination of $P^+Q_A^-$ to the ground state occurs, but at a rate ($\tau$ = 100 ms) that is several orders of magnitude slower than $P^+Q_A^-$ formation (Figure 4B).

The time constants for forward electron transfer and charge recombination reactions in the reaction centre are summarised in Figure 4. The high overall quantum yield of close to 100% is a consequence of the fact that at each stage in trans-membrane electron transfer the forward charge separation reaction is more than an order of magnitude faster than competing charge recombination to the ground state. One reason for this is that the component reactions of charge separation, $P^* \rightarrow P^+B_A^-$, $P^+B_A^- \rightarrow P^+H_A^-$ and $P^+H_A^- \rightarrow P^+Q_A^-$ involve electron transfer between redox centres that are close together, and so occur with fast (picosecond time-scale) rates. In contrast, competing back reactions to the ground state such as recombination of $P^+H_A^-$ and $P^+Q_A^-$ involve electron transfer over relatively long distances, and so are slow. To take an example, productive forward electron transfer from $P^+H_A^-$ to $P^+Q_A^-$ (200 ps) is a much faster reaction than recombination of $P^+H_A^-$ to the ground state (10 ns). In the case of primary electron transfer, the formation of $P^+H_A^-$ (3 ps) is faster than decay of P* to the ground state by internal conversion and fluorescence emission, which is a reaction that occurs over several hundred picoseconds. It is also possible that differences in the energetics of charge separation and charge recombination may contribute to the effective irreversibility and high yield of the forward process, as will be considered in more detail below.

## 3.2. Ubiquinol Formation and Re-reduction of $P^+$

With ubiquinone ($UQ_{10}$) located at the $Q_B$ site, electron transfer occurs from $Q_A^-$ to $Q_B$ with a lifetime of approximately 100 microseconds ($\mu$s) (Okamura and Feher, 1992). Also on a $\mu$s time-scale, the $P^+$ cation is re-reduced by an electron donated by the heme of a cytochrome (Mathis, 1994; Wang et al., 1994; Dutton et al., 1975; Prince et al., 1974). In the case of the

*Rb. sphaeroides* reaction centre, the electron is provided by a cytochrome $c_2$ that docks to the periplasmic face of the reaction centre. In the absence of this reduction of $P^+$, the $P^+Q_B^-$ radical pair state recombines to the ground state in approximately 1 second. Re-reduction of $P^+$ by cytochrome $c_2$ primes the reaction centre for a second light-driven trans-membrane electron transfer from P to $Q_A$, followed by double reduction of $Q_B$ with a time constant of approximately 1 ms (Graige *et al.*, 1998; Okamura and Feher, 1992). Two protons are taken up from the cytoplasm, and doubly reduced/doubly protonated ubiquinone (ubiquinol, $UQ_{10}H_2$) is released into the intra-membrane space.

### 3.3. Light-Driven Cyclic Electron Transfer Coupled to Proton Translocation

The sequence of light-activated reactions described above generates the substrates for the cytochrome $bc_1$ complex, namely cyt $c^{3+}$ and $UQ_{10}H_2$. Oxidation of $UQ_{10}H_2$ takes place at the $Q_o$ site at the periplasmic side of the membrane and is a bifurcated reaction (Crofts *et al.*, 1983) (Figure 3C). One electron is used to reduce cyt $c^{3+}$, one electron is used to reduce a low potential heme (termed $b_L$) and two protons are released into the periplasm. The $b_L$ heme then reduces a high potential heme ($b_H$) in a transmembrane, electrogenic reaction, and $b_H$ in turn reduces $UQ_{10}$ at a ubiquinone reductase ($Q_i$) site near the cytoplasmic side of the membrane (Figure 3C). A second $UQ_{10}H_2$ oxidation by the cytochrome $bc_1$ complex will lead to double reduction of $UQ_{10}$ at the cytoplasmic side of the membrane, followed by proton uptake and release of $UQ_{10}H_2$ into the intra-membrane pool. Light-driven cyclic electron transfer is therefore actually a double cycle, involving electrogenic steps in both the reaction centre and cytochrome $bc_1$ complex. $UQ_{10}$ reduction at the cytoplasmic side of the membrane in both the reaction centre and cytochrome $bc_1$ complex, and $UQ_{10}H_2$ oxidation at the periplasmic side of the membrane in the cytochrome $bc_1$ complex, results in proton translocation across the membrane, with a stoichiometry of two protons translocated per electron pumped across the membrane by the reaction centre. The structural basis for the activity of the cyt $bc_1$ complex is now becoming clear, following the publication of the structure of the mitochondrial enzyme (Crofts and Berry, 1998; Iwata *et al.*, 1998; Zhang *et al.*, 1998; Xia *et al.*, 1997).

## 4. BIOLOGICAL ELECTRON TRANSFER

Light excitation of the bacterial reaction centre triggers a series of electron transfer reactions that, due to variations in distance between the cofactors, their chemical identity and differences in their environments, occur over a range of time spanning 12 orders of magnitude. As a result, studies of the bacterial reaction centre have been at the forefront of research on biological electron transfer (reviewed in Bixon and Jortner, 1999; Marcus, 1996; Moser and Dutton, 1996; Parson, 1996). In this section we discuss some of the general principles of electron transfer theory as applied to biological systems. In Section 5, we follow this with a discussion of experiments which illustrate the application and limitations of this theory for the description of the picosecond time-scale electron transfer reactions that take place in the bacterial reaction centre.

### 4.1. Non-adiabatic Electron Transfer

Electron transfer reactions in biological systems usually occur between redox cofactors that are surrounded by an electrically-insulating protein medium. For electron transfer to occur, the electron must tunnel from the electron donor to the electron acceptor through the barrier presented by this insulating medium. As this medium is not homogeneous, consisting of a complex structure of different types of chemical bonds and through-space contacts between atoms, each element in this structure will be characterised by different electron tunnelling parameters. However, it seems likely that for many biological electron transfer reactions the protein can be treated as an effectively homogenous medium that presents a quasi-uniform, average barrier to electron tunnelling (Moser et al., 1992). With this simplification of being able to ignore detailed structure, the probability of the quantum mechanical electron tunnelling process falls off exponentially with the distance that separates the donor and acceptor molecules, according to:

$$V_{DA}^2 = V_0^2 \exp(-\beta R) \tag{1}$$

where $V_{DA}^2$ represents the square of the electronic coupling of the donor and acceptor states, $V_0^2$ is the maximum electronic coupling at direct contact between donor and acceptor, $R$ is the average edge-to-edge distance between the donor and acceptor, and $\beta$ is an exponential coefficient that is characteristic for a certain medium. From a comparison of energetically-optimised rates for various electron transfer reactions that occur in the reaction centre, $\beta$ was estimated as $1.4 \, \text{Å}^{-1}$ for electron transfer in proteins (Moser et al., 1992).

As in any other chemical reaction, the equilibrium configuration of the nuclei before electron transfer is generally not the same as that after electron transfer. Consequently, upon electron transfer the system must convert from a nuclear configuration that is in equilibrium with the reactant electronic state to a nuclear configuration that is in equilibrium with the product electronic state. In electron transfer theory (see for reviews Marcus, 1993; Marcus and Sutin, 1985; DeVault, 1980; Marcus, 1956) the nuclei of the reactant and its immediate environment are represented by a simple harmonic potential surface, as depicted in Figure 5. The potential energy surface of the product state is assumed to have the same shape as that of the reactant state, but it may be shifted in energy by an amount $\Delta G$ that reflects the difference in free energy between the two states, and it may be shifted in the horizontal direction to reflect a different equilibrium nuclear configuration (Figure 5). This shift in nuclear configuration determines the reorganisation energy ($\lambda$), which is defined as the amount of energy that must be added to the system to change from the equilibrium nuclear configuration of the reactant state to that of the product state without transferring the electron, i.e. while remaining on the potential surface of the reactant state (Figure 5).

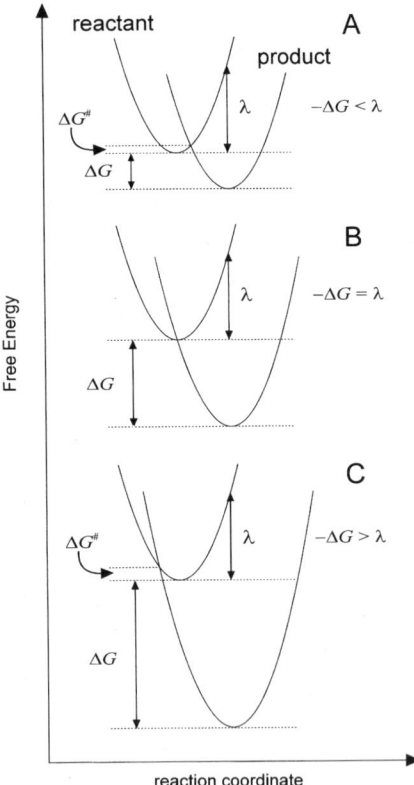

FIGURE 5. The Marcus description of non-adiabatic electron transfer. The reactant and product states are depicted as simple harmonic oscillator potential wells, where the equilibrium nuclear geometry corresponds to the bottom of the well. The product potential well is shifted to a more negative free energy by an amount $\Delta G$, and is shifted in equilibrium nuclear geometry along the reaction coordinate. The reorganisation energy, $\lambda$, is the amount of energy required to move the system from the equilibrium nuclear configuration of the reactant state to that of the product state without transferring the electron. In (A), (B) and (C) $\lambda$ is constant whilst $-\Delta G$ is varied such that its absolute value is smaller, equal to and larger than $\lambda$, respectively. In (A) and (C) the reaction has an activation energy, $\Delta G^{\#}$, whilst in (B) the potential wells intersect at the equilibrium nuclear geometry of the reactant state, and there is no activation energy.

As the electron that is transferred is a very light particle, the nuclei will (according to the Franck-Condon principle) not have enough time to change their position during the actual transfer of the electron. Consequently, the nuclear configurations of the reactant and product states must be the same when the electron transfer occurs. In combination with a restriction for the conservation of energy, this implies that the electron transfer reaction can only occur in the transition configuration where the two potential surfaces intersect (Figure 5).

In the non-adiabatic view of electron transfer, the coupling between the reactant state and product state is assumed to be weak. As a consequence the system is envisaged to move many times back and forth on the potential energy surface of the reactant state, thereby crossing the transition region many times before the weak electronic coupling between the donor and acceptor permits the electron transfer reaction to take place. The free energy difference between the equilibrium configuration of the reactant state and the transition configuration is the activation free energy, $\Delta G^{\#}$ (Figure 5), and for parabolic potentials $\Delta G^{\#}$ is given by $(\Delta G + \lambda)^2/4\lambda$. When $\lambda$ equals $-\Delta G$ (Figure 5B), the barrier $\Delta G^{\#}$ is zero and no thermal energy is required to reach the transition configuration, and the reaction displays no activation energy. When $-\Delta G$ is either smaller or larger than $\lambda$ (Figs. 5A and C, respectively), $\Delta G^{\#}$ is not equal to zero and the reaction will be thermally-activated.

Applying Fermi's golden rule, the rate of the electron transfer reaction is determined by the product of the probability of the nuclear transition occurring (the Franck-Condon term, $(FC)$) and the probability of the electron tunnelling occurring:

$$k_{ET} = \frac{2\pi}{\hbar} V_{DA}^2 FC \tag{2}$$

In the classical limit when the thermal energy $K_B T$ is much higher than the energy $\hbar\omega$ of the vibrational frequencies that are coupled to the electron transfer reaction, the Franck-Condon factor can be expressed in terms of $\Delta G$ and $\lambda$ and equation 2 converts to the classical Marcus formula for the electron transfer rate:

$$k_{ET} = \frac{2\pi}{\hbar} V_{DA}^2 \frac{1}{\sqrt{4\pi\lambda k_B T}} \exp{-\frac{(\Delta G + \lambda)^2}{4\lambda k_B T}} \tag{3}$$

According to equation 3 the dependence of $k_{ET}$ on $\Delta G$ is a parabolic function, and $k_{ET}$ is maximal when $\lambda$ is equal to $-\Delta G$ (Figure 6). The region

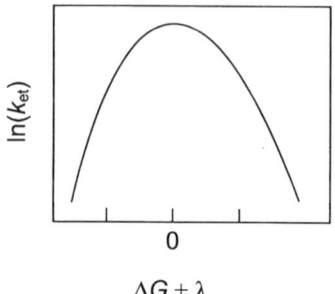

**FIGURE 6.** Relationship between the rate of electron transfer, $\Delta G$ and $\lambda$. The rate is maximal when $-\Delta G = \lambda$.

in which the electron transfer rate becomes slower upon further increase of the driving force $\Delta G$ is called the inverted region (Figure 6). The presence of this inverted region was a specific prediction of the Marcus theory for electron transfer.

The temperature dependence of the electron transfer rate is dominated by the exponential factor in equation 3, and for most values of $\Delta G$ it will show an Arrhenius-like behaviour with an activation energy $\Delta G^{\#}$ of ($\Delta G$ + $\lambda)^2/4\lambda$. However, when the exponential factor approximates to zero as the magnitude of $\lambda$ approaches that of $-\Delta G$, the electron transfer rate will show a reverse temperature dependence determined by the pre-exponential factor in equation 3. This reverse temperature dependence can be understood by considering that raising the temperature will decrease the probability that the system is in the lowest nuclear configuration that is most favourable for the electron transfer reaction.

In certain cases, the classical Marcus formula is not sufficient to explain the observed dependence of the electron transfer rate on temperature or $\Delta G$, which could indicate that it is necessary to use a Franck-Condon term in which the contribution of the nuclei is treated in quantum mechanical terms. In this treatment, the Franck-Condon term equals the thermally-weighted sum of the contributions from all possible vibrational states of the reactants, each multiplied by their Franck-Condon factor i.e. the square of the overlap integral of a nuclear wave function of the reactant with the nuclear wave function of the product state that has the same total energy.

In the quantum mechanical description, the nuclear transition is allowed at temperatures at which the vibrational level near the crossing point is not populated. This is because the overlap of the nuclear functions is not zero for the vibrational levels below the crossing point. This phenomenon has been called nuclear tunnelling, because the nuclear configuration of the reactant state tunnels to the product state through the barrier presented by $\Delta G^{\#}$. At temperatures for which $k_B T \ll \hbar\omega$, only the lowest vibrational level is occupied and the Franck-Condon factor is temperature independent, giving rise to a temperature-independent electron transfer rate in which both the electron and the nuclei tunnel from the reactant to the product configuration. The observation that many of the electron transfer reactions that take place in the reaction centre can occur at temperatures down to 4 K, with a temperature-independent electron transfer rate for temperatures below 100 K, has been taken to reveal the importance of nuclear tunnelling for biological electron transfer reactions (DeVault, 1980).

## 4.2. Adiabatic Electron Transfer

Non-adiabatic electron transfer occurs when the electronic coupling between the donor and acceptor states is sufficiently weak that the nuclei reach the crossing point of the two potential curves many times before the actual electron transfer reaction occurs (Figure 7A). Thus the time-scale of the motion of the nuclei does not determine the electron transfer rate, but rather this is determined by the relative probability of being at the crossing point. In contrast, if the electronic coupling between reactant and product is sufficiently strong, or the nuclear motion sufficiently slow, the electron transfer reaction takes place the first time the transition nuclear configuration is achieved, and the reaction is described as adiabatic (Figure 7B). The strong electronic interaction results in a quantum mechanical mixing between the reactant and product electronic states that at the crossing point have the same energy, resulting in a splitting of these states into

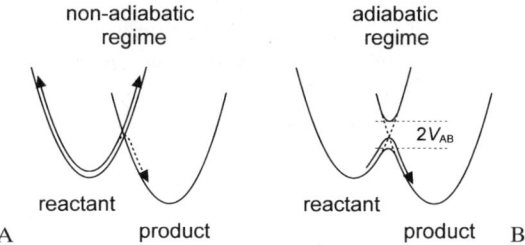

**FIGURE 7.** Non-adiabatic and adiabatic electron transfer. (A) in non-adiabatic electron transfer the system moves back and forth along the potential surface (solid arrow), passing through the crossing point many times before the weak electronic coupling permits electron transfer to occur (dotted arrow). (B) in adiabatic electron transfer, coupling between reactant and product is sufficiently strong that motion through this region leads directly to the product state (solid arrow).

a higher and a lower level with an energy difference $2V_{DA}$ as depicted in Figure 7B. In this way two new potential surfaces are created, and the system moves on the lowest-energy potential surface. The conversion from the reactant to product state is now determined by the time-scale of the movement of the nuclei on this new lowest-energy electronic surface. Whether the electron transfer reaction occurs in the adiabatic or the non-adiabatic regime is determined by the Landau-Zener parameter given in equation 4 (see Zinth *et al.*, 1998; Bixon *et al.*, 1991; Marcus and Sutin, 1985).

$$\gamma_{LZ} = \frac{2\pi V_{DA}^2}{\hbar\omega_0 \sqrt{2\lambda k_B T}} \qquad (4)$$

in which $\hbar\omega_0$ is the energy of the effective frequency coupled to the nuclear transition. When $\gamma_{LZ} \ll 1$ the reaction is non-adiabatic, but when $\gamma_{LZ}$ approaches values of 0.5–1 the reaction reaches the adiabatic limit (Bixon *et al.*, 1991). According to the Landau-Zener parameter, strong electronic coupling, low frequency nuclear motions, a small reorganisation energy or a combination of these can place the reaction in the adiabatic regime.

The relevance of adiabatic electron transfer to the primary charge separation reaction has been the subject of considerable discussion, mainly due to the observation of undamped low-frequency nuclear motions associated with the P* state (see Section 5.5). More recently, sub-picosecond time-scale electron transfer has been observed at cryogenic temperatures, driven either by the P* state in certain mutant reaction centres (see Section 5.6) or by the monomeric BChls in both wild-type and mutant reaction centres (see Section 5.7). These observations have led to the proposal that such ultra-fast electron transfer reactions require strong electronic coupling between the co-factors and occur on a time-scale in which vibrational relaxation is not complete, which would place these reactions in the adiabatic regime. Finally, as discussed in Section 2.2, evidence has been obtained that electron transfer from $Q_A^-$ to $Q_B$ is limited by nuclear rearrangement, rather than by the driving force for the reaction.

## 5. STUDIES OF ULTRAFAST ELECTRON TRANSFER IN A LIGHT-ACTIVATED PROTEIN

The application of time-resolved spectroscopy has revealed many of the details and complexities of the picosecond time-scale electron transfer reactions that are catalysed by the bacterial reaction centre. Most studies have been carried out on purified reaction centres in solution or in films, but some have been carried out on reaction centres embedded in natural membranes that are deficient of antenna complexes (Van Stokkum et al., 1997; Beekman et al., 1995; Vos et al., 1994c; Schmidt et al., 1993). Picosecond time-scale electron transfer has also been studied under a wide range of experimental conditions, and in a large number of mutant reaction centres. In the remainder of this article, we discuss some of the main themes of this research, and how this work relates to the theory of biological electron transfer.

## 5.1. The Role of the $B_A$ Monomeric BChl

For many years the failure to obtain clear evidence for a transient intermediate prior to the formation of $P^+H_A^-$ fuelled an intensive discussion on the mechanism of electron transfer from $P^*$ to $H_A$, and in particular the role played by the intervening $B_A$ monomeric BChl. Given the distance between the P BChls and the $H_A$ BPhe, direct electronic coupling between the two is insufficient to explain the formation of $P^+H_A^-$ in 3–5 ps, and therefore the $B_A$ BChl is expected to play some role in this reaction (Figure 1D).

The most obvious role for the $B_A$ BChl is that it forms part of a $P^+B_A^-$ radical pair which acts as an intermediate in a two-step primary electron transfer reaction; i.e. $P^* \rightarrow P^+B_A^- \rightarrow P^+H_A^-$. According to this mechanism, the free energy of the $P^+B_A^-$ state must be located below or just above (within $k_BT$ of) the free energy of the $P^*$ state for electron transfer to occur. However, initial calculations based upon the crystal structure placed the energy level of $P^+B_A^-$ significantly above that of $P^*$. Furthermore, for many years there was a lack of any clear spectroscopic evidence from time-resolved absorption measurements for a transient population of $P^+B_A^-$. In the light of this, it was proposed that the $P^+B_A^-$ state functions as virtual intermediate that enhances the electronic coupling between $P^*$ and $P^+H_A^-$ via a super-exchange mechanism (Bixon and Jortner, 1999; Bixon et al., 1991, 1989, 1988; Marcus, 1988; Michel-Beyerle et al., 1988; Warshel et al., 1988; Marcus, 1987; Woodbury et al., 1985).

The super-exchange electronic coupling term describes the coupling of the $P^+B_A^-$ state to the $P^*$ and $P^+H_A^-$ states at the position of the intersection between the potential surfaces of the $P^*$ and $P^+H_A^-$ states. When the $P^*$ and $P^+H_A^-$ potential surfaces intersect at the minimum of the $P^*$ potential surface (i.e. assuming that electron transfer from $P^*$ to $P^+H_A^-$ is activationless), the super-exchange coupling ($V_{super}$) is given by:

$$V_{super} = \frac{V_{PB}V_{BH}}{\delta E} \tag{5}$$

where $V_{PB}$ is the electronic coupling between P and $B_A$, $V_{BH}$ the electronic coupling between $B_A$ and $H_A$, and $\delta E$ is the energy difference between P* and $P^+B_A^-$ at the intersection of the curves of P* and $P^+H_A^-$ and is equal to $\lambda + \Delta G$. It has been suggested that the super-exchange mechanism is particularly advantageous relative to the two-step mechanism when the energy level of the $P^+B_A^-$ state lies significantly above that of P*.

In recent years, increasing evidence has been gathered that the $P^+B_A^-$ state does form as a discrete, but short-lived intermediate during electron transfer from P* to $H_A$, supporting the view that primary electron transfer is a two-step reaction (Kennis *et al.*, 1997; van Stokkum *et al.*, 1997; Arlt *et al.*, 1996; Holzwarth and Müller, 1996; Schmidt *et al.*, 1994; Arlt *et al.*, 1993; Holzapfel *et al.*, 1990, 1989). Particularly persuasive evidence has come from transient absorption experiments recorded over the whole of the $Q_y$ spectral region, including around 1020 nm where the $BChl^-$ anion is expected to show a distinct absorbance band. The consensus view from these studies is that $P^+B_A^-$ is formed from P* with a lifetime of 3–5 ps, and that $P^+B_A^-$ decays to $P^+H_A^-$ with a lifetime of 0.5–2 ps, the precise values depending on the nature of the sample and on experimental conditions (Figure 4). Since the rate of decay of the $P^+B_A^-$ intermediate is faster than the rate of its formation, the transient population of $P^+B_A^-$ is small and is therefore difficult to detect. Resolution of the $P^+B_A^-$ intermediate in wild-type reaction centres requires a careful analysis of transient absorbance measurements made over an appropriate range of wavelengths.

An enhanced population of the $P^+B_A^-$ intermediate has been achieved in experiments with reaction centres in which the $H_A$ BPhe has been removed and replaced with pheophytin *a* (Phe) from higher plants. This modification significantly raises the energy level of the $P^+H_A^-$ (now termed $P^+Phe_A^-$) intermediate and leads to a dramatic decrease in the rate of electron transfer from $P^+B_A^-$ to $P^+Phe_A^-$. Since the rate of electron transfer from P* to $P^+B_A^-$ appears to be unaffected in these reaction centres, the $P^+B_A^-$ state now appears as a relatively long-lived and highly populated intermediate (Kennis *et al.*, 1997; Schmidt *et al.*, 1994). The presence of a long-lived $P^+B_A^-$ intermediate also allowed estimation of the free energy of the $P^+B_A^-$ state, which was calculated to be approximately 55 meV below that of P* (Schmidt *et al.*, 1994). An alternative method of increasing the transient population of the $P^+B_A^-$ intermediate is by increasing the rate of electron transfer from P* to $P^+B_A^-$, which has been achieved in a mutant of the *Rps. viridis* reaction centre in which His L168 is replaced by Phe (Arlt *et al.*, 1996).

In closing, it should be commented that although evidence in support of $P^+B_A^-$ being formed as a true intermediate in a two-step primary electron transfer reaction has grown in recent years, the mechanism of this ultra-fast electron transfer reaction remains the subject of considerable discussion and ongoing investigation. In particular, it is possible that the super-exchange mechanism becomes important under certain experimental conditions, or may play an increased role in reaction centres in which the energetics of primary electron transfer are strongly affected by treatments such as mutagenesis (see Section 5.6.2).

## 5.2. The Asymmetry of Primary Electron Transfer

One of the most intriguing features of the bacterial reaction centre is the asymmetry of trans-membrane electron transfer. Although the BChl, BPhe and $UQ_{10}$ cofactors are arranged in two approximately symmetrical trans-membrane branches, only the so-called A-branch is used for trans-membrane electron transfer. The factors determining this functional asymmetry continue to be the subject of great interest, as the reaction centre presents a chain of cofactors that catalyses electron transfer with great efficiency, and a similar chain of cofactors that is much less effective. This functional asymmetry is due to small but crucial differences in the structure of the protein:cofactor system along the two branches.

### 5.2.1. Evidence for the Asymmetry of Primary Electron Transfer

The evidence for essentially exclusive use of the A-branch for primary electron transfer came primarily from monitoring the absorption changes due to reduction of the $H_A$ and $H_B$ BPhes at cryogenic temperatures, where $H_A$ and $H_B$ show distinct absorbance bands in the $Q_x$ region with maxima at 542 nm ($H_A$) and 532 nm ($H_B$). The first indications that only one of the two BPhes present in the reaction centre was an intermediate electron carrier in the electron transfer pathway to $Q_A$ came from so-called photo-trapping experiments, and were obtained several years before the elucidation of the crystal structure that revealed the structural arrangement of the reaction centre cofactors (Tiede *et al.*, 1976; Shuvalov and Klimov, 1976; van Grondelle *et al.*, 1976). In these experiments, $Q_A$-reduced reaction centres were subjected to continuous illumination in the presence of cyt $c^{2+}$. These conditions result in trapping of the reaction centre in the state $PH_A^-Q_A^-$, since there is some probability that cyt $c^{2+}$ can donate an electron to $P^+$ before the $P^+H_A^-$ state that is formed by illumination recombines to the ground state. Although, the quantum yield of this process is very low, repeated excitation will eventually convert a large part of the reaction centre population into the $PH_A^-Q_A^-$ state. The absorbance spectrum of such reaction centres at cryogenic temperature clearly demonstrated that only one of the two BPhe cofactors in the reaction centre was reduced as a result of this treatment, with bleaching of the 542 nm band only (Tiede *et al.*, 1976; Shuvalov and Klimov, 1976; van Grondelle *et al.*, 1976). The near-exclusive reduction of only one of the two reaction centre BPhes was also shown directly in transient absorption measurements in the $Q_x$ region at cryogenic temperature (Kirmaier *et al.*, 1985b). The interpretation of possible small absorbance changes in the region of the $Q_x$ band of the $H_B$ BPhe is difficult, due to limitations arising from signal-to-noise ratio and the superposition of electrochromic band-shifts at the position where bleaching of the $Q_x$ band would occur. It is therefore difficult to accurately determine the amount of electron transfer along the B-branch in the wild-type reaction centre from such experiments, other than to say that it is at most a few percent of the total electron transfer that occurs.

Although it is a minor process, there is good evidence that reduction of $H_B$ by P* does occur. In photo-trapping experiments, prolonged illumination induces accumulation of some $H_B^-$ in addition to the accumulation of $H_A^-$, indicating that electron transfer along the B-branch is possible in principle, and probably occurs in $H_A^-$ reduced reaction centres (Kellogg *et al.*, 1989; Mar and Gingras, 1990). Robert and co-workers have succeeded in performing photo-trapping experiments under conditions where $H_B^-$ accumulates but $H_A^-$ does not (Robert *et al.*, 1985). The key feature of this experiment was that methyl viologen was used as the electron donor to P$^+$ rather than cyt $c^{2+}$, and it was proposed that methyl viologen was acting as both the electron donor to P$^+$ and an electron acceptor from $H_A^-$ (but not $H_B^-$, which gradually accumulated after prolonged illumination). Significant reduction of $H_B$ has also been observed in transient absorption experiments employing high intensity excitation conditions, leading to the suggestion that two-photon excitation generates the state $P*B_B*$ which then drives the formation of $P^+B_B^-$ and, subsequently, $P^+H_B^-$ (Lin *et al.*, 1999).

### 5.2.2.  Origins of the Asymmetry of Primary Electron Transfer

Although the two-fold symmetry displayed by the reaction centre is striking, it is only a pseudo-symmetry, because differences in the amino acid sequences of the L and M subunits result in small differences in the positions and relative orientations of equivalent cofactors on the two branches, and in differences of the protein environment of equivalent cofactors. The root cause of the functional asymmetry that is observed when electron transfer is monitored is therefore asymmetry in the detailed structure of the cofactor:protein system on the two branches. Assuming that the transmembrane electron transfer process can basically be described as a non-adiabatic electron transfer reaction according to the Marcus equation, this subtle structural asymmetry must result in a difference in one or more of the rate-governing parameters, $\lambda$, $\Delta G$ and $V_{DA}$, for equivalent reactions on the two branches. The question that many workers have sought to address, and which remains to be solved, is which of the small but significant differences in the structure of the cofactor:protein system induces this functional asymmetry, and which parameter in the Marcus equation is affected by these differences.

Several groups have approached this problem from both experimental and theoretical viewpoints, and have proposed a number of factors which may result in asymmetric electron transfer, including asymmetry in electronic coupling between cofactors on the A and B-branches (Ivashin *et al.*, 1998; Bixon *et al.*, 1989; Michel-Beyerle *et al.*, 1988; Plato *et al.*, 1988), asymmetry in the dielectric environment on the two branches (Steffen *et al.*, 1994) and asymmetry in static intra-protein electric fields (Gunner *et al.*, 1996). A much-discussed idea, summarised in Figure 8, is that asymmetric electron transfer is largely attributable to a difference in energy between the $P^+B_A^-$ state and the analogous $P^+B_B^-$ state, with the latter having a significantly higher free energy than either $P^+B_A^-$ or $P^*$ (Alden *et al.*, 1995; Warshel *et al.*, 1994; Parson *et al.*, 1990). Findings with mutant reaction centres, discussed in Section 5.2.3, provide some evidence in support of this.

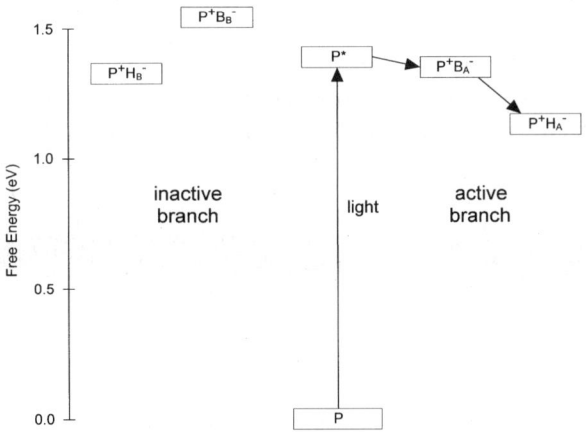

**FIGURE 8.** Schematic of the possible energetics of electron transfer along the A- and B-branches of cofactors in the *Rb. sphaeroides* reaction centre. The free energies of the B-branch radical pairs are higher than that of their A-branch counterparts. Mutations that affect the redox potential of $B_A$, $B_B$, $H_A$ or $H_B$ will affect electron transfer on one branch only, whereas changes in the redox potential of P will affect electron transfer on both branches.

A possible explanation for selective electron transfer along the A branch that has attracted much attention is that the excited state that drives the primary charge separation, P*, is asymmetrically-mixed with the internal charge transfer (CT) states $P_A{}^+P_B{}^-$ and $P_A{}^-P_B{}^+$. The internal CT state $P_A{}^-P_B{}^+$ involves the movement of electron density in the direction of $B_A$, and could favour electron transfer along the A branch, whereas the state $P_A{}^+P_B{}^-$ could favour electron transfer along the B branch. Stark spectroscopy has shown that the lowest energy excited state of P is more polarisable, and has more dipolar character, than the lowest energy excited states of the monomeric pigments in the reation centre, which suggests that an internal CT state is mixed with the P* exciton state (Middendorf et al., 1993). However, the relevance of the internal CT states to asymmetric primary electron transfer was challenged by findings with two BChl:BPhe heterodimer mutants, in which one of the two P BChls is replaced by a BPhe, following mutation of the axial His ligand to Leu. Due to the difference in redox-potential of BPhe and BChl, the CT state that can be coupled to the dimer excited state has opposite directions in the two mutants i.e. $BChl_A{}^+BPhe_B{}^-$ in one and $BPhe_A{}^-BChl_B{}^+$ in the other. Despite this difference in the orientation of the coupled CT state, electron transfer was found to proceed exclusively down the A-branch in both heterodimer reaction centres (McDowell et al., 1991), which suggests that the CT states associated with P do not determine the asymmetry of primary electron transfer. In addition, it is worth noting that in the photosystem II reaction centre from higher plants no large Stark effect is observed (Van Grondelle, R.; personal communication), although this reaction centre almost certainly performs asymmetric electron transfer along one of two symmetrically-arranged cofactor branches. This further supports the view that asymmetric electron transfer is not attributable to a primary donor charge transfer state.

### 5.2.3.  Re-routing Primary Electron Transfer

A number of groups have used site-directed mutagenesis to investigate whether particular amino acids or groups of amino acids play a significant role in determining asymmetric electron transfer. Two sets of experiments are of particular note. The first address the role of residue Tyr M210, which is located in a "pocket" created by the P BChls, the $B_A$ BChl and the $H_A$ BPhe and is conserved amongst the available sequences for the reaction centre M subunit from different purple bacteria (Figure 9A). The symmetry-related residue, Phe L181, occupies a similar position close to the $B_B$ BChl and $H_B$ BPhe of the inactive B-branch of cofactors and also shows sequence conservation (Figure 9). A number of roles have been proposed

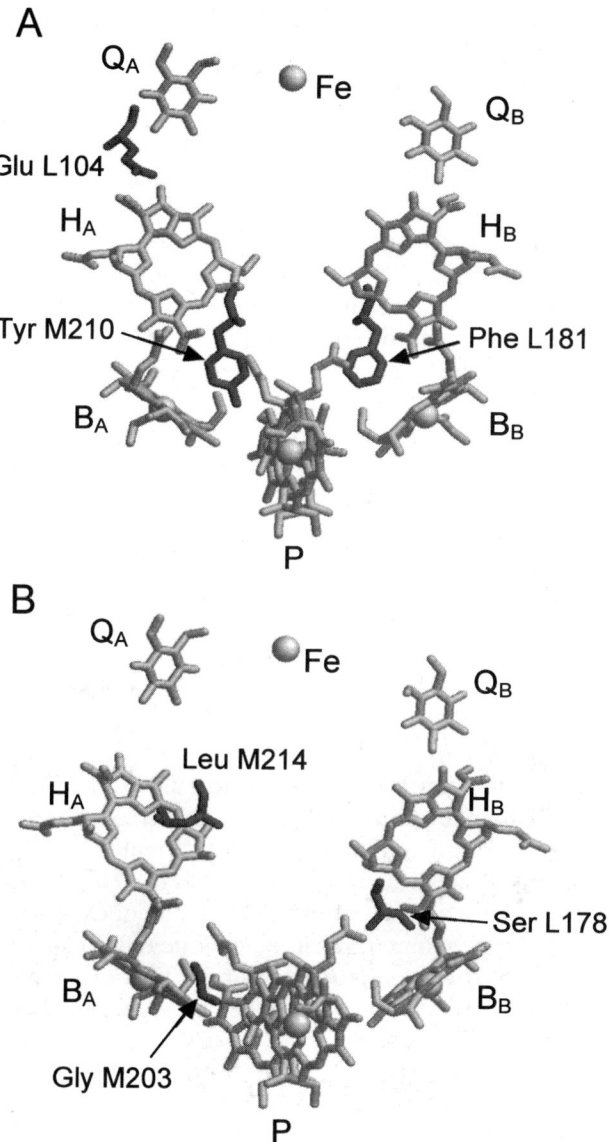

**FIGURE 9.** Amino acid residues that have been the subject of mutation studies. (A) The symmetry-related pair Tyr M210 and Phe L181, and Glu L104 which forms a hydrogen bond with the $H_A$ BPhe. (B) Mutation of Leu M214, Gly M203 and Ser L178 affects the balance of electron flow along the A- and B-branches (see text).

for Tyr M210, including stabilisation of the $P^+B_A^-$ radical pair through a favourable interaction of $B_A^-$ with the OH group of the Tyr side-chain (Parson, 1996; Alden et al., 1996; Gunner et al., 1996; Parson et al., 1990). The homologous Phe L181 residue, which lacks the OH group, cannot have a similar stabilising effect on the $P^+B_B^-$ radical pair which, as discussed above, probably lies higher in free energy than both P* and $P^+B_A^-$ (Figure 8). Mutation of Tyr M210 to a non-polar residue also brings about an increase in the mid-point redox potential of the $P/P^+$ redox couple ($E_m$ $P/P^+$), which affects the driving force of the first step in primary charge separation (Visschers et al., 1999; Beekman et al., 1996; Jia et al., 1993; Nagarajan et al., 1993). Estimations of electronic coupling elements have also indicated a possible important contribution of the Tyr M210 residue to enhancement of the electronic coupling between P* and $P^+B_A^-$ (Ivanshin et al., 1998).

Given these roles, one might think that an exchange of residues Tyr M210 and Phe L181 would cause complete or partial re-routing of primary electron transfer down the inactive B-branch of cofactors, but this has proven not to be the case. When this residue pair is made identical, either by mutation of Tyr M210 to Phe or mutation of Phe L181 to Tyr, the pronounced asymmetry of primary electron transfer is not affected although, as might also be expected, mutation of Tyr M210 in particular has a strong effect on the rate of primary electron transfer (see Section 5.6.1, below). Reversing these residues, through the double mutation Tyr M210 → Phe/PheL181 → Tyr (Jia et al., 1993; Gray et al., 1992), also does not affect the asymmetry of primary electron transfer.

A second residue which is worthy of note is Glu L104, which appears to be responsible for the splitting of the $Q_x$ absorbance bands of the reaction centre BPhes, described above. Glu L104 donates a hydrogen bond to the 9-keto carbonyl substituent group of the $H_A$ BPhe, modulating the redox properties of this cofactor, but the $H_B$ BPhe does not form an equivalent interaction as the symmetry-related residue to Glu L104 is a Leu. One of the first mutations to be constructed after publication of the X-ray crystal structure of the reaction centre was Glu L104 → Leu (Bylina et al., 1988). Although this mutation removed the energetic inequivalence between the $Q_x$ absorbance bands of $H_A$ and $H_B$, and slowed the rate of primary electron transfer from P* to $H_A$, it did not affect the asymmetry of this reaction.

A mutant for which it is claimed that electron transfer proceeds along the B-branch with a quantum yield of 15% is a derivative of the so-called β-mutant, which has a BChl (denoted β) in place of the $H_A$ BPhe as a result of mutation of Leu M214 to His (Kirmaier et al., 1991). This derivative has a second mutation, Gly M203 → Asp, in the vicinity of the $B_A$ BChl (Heller et al., 1995). In the β-mutant, electron transfer proceeds with time constant of 6 ps to a state that is interpreted as a mixture of $P^+B_A^-$ and $P^+\beta^-$ ($P^+\beta^-$ lies higher in free energy than the equivalent $P^+H_A^-$ state in the wild type reaction centre because BChl is harder to reduce than BPhe). In the double Leu M214 → His/Gly M203 → Asp mutant, electron transfer along the A-branch is slowed to 20 ps, probably due to an increase of the free energy of $P^+B_A^-$. In addition, a small, long-lived bleach of the $H_B$ absorbance is apparent, which is interpreted as approximately 15% electron transfer along the B branch with a time constant of approximately 100 ps. Recently, it has been reported that a third mutation, Ser L178 → Lys, in the vicinity of $B_B$, increases the yield of electron transfer to $H_B$ to 23%, and decreases the time constant of the reaction to 65 ps (Kirmaier et al., 1999). Findings with these mutants support the idea that an inequivalence in the free energy of the $P^+B_A^-$ and $P^+B_B^-$ radical pairs is at least partially responsible for asymmetric electron transfer. The mutations that retard electron transfer along the A-branch affect mainly $B_A$ and $H_A$ rather than P (mutations that affect P have consequences for the activity of both branches), whilst the Ser L178 → Lys mutation will mainly affect the redox properties of $B_B$.

Another approach to study the origin of the asymmetry of primary electron transfer has been the construction of large-scale symmetry mutants of the Rb. capsulatus reaction centre (Taguchi et al., 1996; Lin et al., 1996). In these mutants, segments of the M subunit (which interact with both the A- and B-branch pigments) have been replaced by the homologous region of the L subunit. The ten mutants that were created in this way symmetrised in total about 80% of the amino acids that come in close contact with the reaction centre cofactors. Most of the mutants that were made in this way were capable of performing trans-membrane electron transfer and growing under photosynthetic conditions. Only the large scale mutations around the Fe atom and the ubiquinone binding sites resulted in strains with unstable reaction centres and loss of photosynthetic capacity. Time-resolved spectroscopy at 20 K carried out on these large-scale symmetry mutants showed that some had yields and lifetimes for primary electron transfer that were similar to those of the wild-type reaction centre. One mutant of particular note however, termed sym 2-1, exhibited a rate of primary electron transfer that was decreased by nearly a factor of 100, and had a yield of $P^+$ of only 35%. The changes in this mutant included the mutation Tyr M210 → Phe, described above, and replacement of six adjacent amino acids. Inspection of the absorbance difference spectrum of the sym 2-1 mutant at 650 ps in the BPhe $Q_x$ region showed a small absorbance decrease that was attributed to 10–20% electron transfer along the B-branch.

## 5.3.  Temperature Dependence—Activationless Reactions

A remarkable feature of primary electron transfer is that the reaction speeds up with decreasing temperature (Breton *et al.*, 1988; Fleming *et al.*, 1988; Woodbury *et al.*, 1985). In *Rb. sphaeroides* reaction centres, for example, the time constant for decay of the P* state changes from approximately 3 ps at room temperature to 1.2 ps at 10 K (Fleming *et al.*, 1988). In the simplest view, this suggests that the activation free energy, $\Delta G^{\#}$, is zero, which in turn implies that $\lambda$ for this reaction equals $-\Delta G$, and that in this respect the reaction is optimised. An attractive idea is that the acceleration in rate that is observed on decreasing the temperature is due to a contraction of the protein, which would increase the electronic coupling between donor and acceptor, and which could also explain the increase in the rate of $P^+Q_A^-$ recombination with decreasing temperature (Fleming *et al.*, 1988). In an interesting parallel to this, the rate of P* decay also accelerates with increasing pressure (Timpmann *et al.*, 1997), where again the distance between donor and acceptor might be reduced as the pressure is increased. However, it should be noted that there is also strong evidence that the primary electron transfer reaction occurs near the top of the Marcus parabola, since through modulation of the driving force of the reaction by site-directed mutagenesis it is possible to move the reaction out of the activationless region, resulting in a reaction which is thermally-activated. It is therefore probable that the temperature dependence of the primary electron transfer reaction arises from both an increase of the electronic coupling with decreasing temperature, and energetic optimisation of the reaction.

Secondary electron transfer from $H_A^-$ to $Q_A$ also speeds up with decreasing temperature in *Rb. sphaeroides* reaction centres, the measured time constant decreasing from approximately 230 ps at room temperature to 100 ps at 80 K (Kirmaier *et al.*, 1985a). Below 80 K, the rate of the reaction is independent of temperature. In contrast, the rate of this reaction in the *Rps. viridis* reaction centre is independent of temperature between 295 and 5 K (Kirmaier and Holten, 1988). Possible explanations for the temperature dependence of this reaction have been discussed, and it has been pointed out that the acceleration of this reaction seen in *Rb. sphaeroides* reaction centres between 295 and 80 K cannot be taken as evidence that $\lambda$ equals $-\Delta G$ for this reaction (Parson, 1996; Warshel *et al.*, 1989; Gunner and Dutton, 1989; Kirmaier *et al.*, 1988; Kirmaier *et al.*, 1985a). In particular, it has been shown that the temperature dependence of the rate of electron transfer from $H_A$ to $Q_A$ is relatively insensitive to changes in $-\Delta G$ for the reaction (Gunner and Dutton, 1989), in contrast to the expected behaviour if $-\Delta G$ equals $\lambda$ for this reaction in the native *Rb. sphaeroides* reaction centre at room temperature.

## 5.4. Dispersive Kinetics: Heterogeneity or Protein Dynamics?

The temporal resolution of measurements of electron transfer in the reaction centre has greatly increased in recent years. Recent time-resolved absorption and fluorescence measurements on wild-type reaction centres have revealed that the decay of the P* state is multi-exponential (Van Stokkum *et al.*, 1997; Holzwarth and Müller, 1996; Lin *et al.*, 1996; Vos *et al.*, 1996; Beekman *et al.*, 1995; Woodbury *et al.*, 1994; Arlt *et al.*, 1993; Hamm *et al.*, 1993; Jia *et al.*, 1993; Du *et al.*, 1992; Müller *et al.*, 1991; Vos *et al.*, 1992, 1991). Measurements carried out at room temperature on purified reaction centres typically yield a main component (>65%) in the 2.5–4.5 ps range and a minor component (<35%) in the 7–15 ps range (Lin *et al.*, 1996; Beekman *et al.*, 1995; Arlt *et al.*, 1993; Hamm *et al.*, 1993; Jia *et al.*, 1993; Du *et al.*, 1992; Müller *et al.*, 1991). If forward electron transfer beyond $H_A$ is blocked by reduction or removal of $Q_A$ then additional components of P* decay are observed up to the nanosecond timescale, although the two fastest lifetimes are dominant, accounting for >97% of the decay (Ogrodnik *et al.*, 1994; Peloquin *et al.*, 1994; Woodbury and Parson, 1984).

Two principal explanations have been put forward to account for this multi-exponential behaviour, which have turned out to be very difficult to distinguish experimentally (and anyway are not mutually exclusive). However this topic has attracted considerable interest because regardless of which explanation is correct, both originate from fundamental properties of protein systems which are difficult to access through experiment. A number of the papers cited above have discussions on the possible origins of multi-exponential P* decay, as do recent reviews and articles (Parson, 1996; Bixon *et al.*, 1995; Woodbury and Allen, 1995; Gehlen *et al.*, 1994; Gudowska-Nowak *et al.*, 1994; Kolaczkowski *et al.*, 1994; Small *et al.*, 1992).

The first explanation offered for the phenomenon of dispersive kinetics is that it is caused by a distribution of rates of primary electron transfer, and that the "slow" P* lifetimes originate from a minority of reaction centres from the "slow" tail of this distribution. The energetic basis for this distribution could be an inhomogeneous distribution of a rate-determining parameter such as $\lambda$, $\Delta G$ or $V_{DA}$ (or any combination of these) (Figure 10A). The principal alternative explanation is that the multiple lifetimes represent a time-dependent energetic relaxation of the $P^+H_A^-$ intermediate due

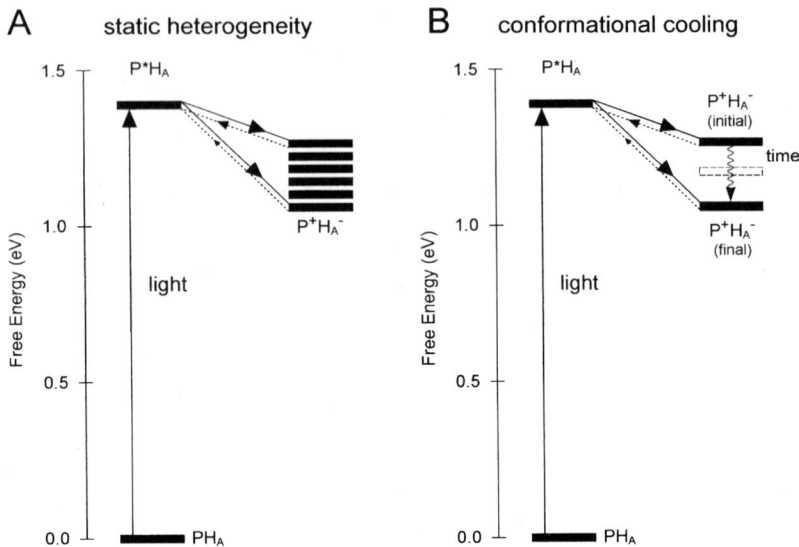

**FIGURE 10.** Possible origin of dispersive kinetics of primary electron transfer. Forward electron transfer is indicated by the solid arrows, thermal repopulation of the P* state (a minor process) by the dotted arrows. (A) Static heterogeneity—electron transfer takes place from the P* state to $P^+H_A^-$ states with distribution of free energies. The reaction therefore occurs with a distribution of driving forces, and hence a distribution of rates. Most thermal repopulation of the P* state occurs from the $P^+H_A^-$ states that are highest in energy. A similar model can be constructed based upon a P* $\rightarrow$ $P^+H_A^-$ reaction with two or more values for the reorganisation energy or electronic coupling. (B) Conformational cooling—the $P^+H_A^-$ state formed initially relaxes, possibly through one or more intermediate states (dashed lines) to a final $P^+H_A^-$ state. Thermal repopulation of the P* state is more likely from $P^+H_A^-$ states that lie highest in energy, and so the amount of repopulation decreases with time as $P^+H_A^-$ relaxes.

to solvation of the radical pair by the protein environment, which causes a time-dependent increase in the free energy gap between P* and $P^+H_A^-$ (Figure 10B). This energetic relaxation of $P^+H_A^-$ is probed in the concentration profile of P* due to some amount of thermal repopulation of P* from $P^+H_A^-$ (i.e. a reversal of the primary electron transfer reaction).

Addressing the possibility of static heterogeneity, the magnitude of the possible inhomogeneity in the free energy of the $P^+H_A^-$ radical pair has been estimated by comparing the magnetic field-dependency of the kinetics of $P^+H_A^-$ charge recombination in absorption measurements (which probe the bulk of the reaction centre population) and in measurements of delayed recombination fluorescence from P* (which probes only those reaction centres residing in the high energy tail of the $P^+H_A^-$ population) (Ogrodnik *et al.*, 1994). A simultaneous fit of both experiments resulted in a width of 100 meV for a Gaussian inhomogeneous distribution function, and an average free energy gap between P* and $P^+H_A^-$ of 250 meV. A similar amount of energetic inhomogeneity has been proposed for the $P^+B_A^-$ radical pair, in order to explain dispersive kinetics of P* decay in pigment-modified reaction centres in which the $H_A$ BPhe was replaced by plant pheophytin (Phe) (Huber *et al.*, 1998). In these reaction centres the $P^+B_A^-$ intermediate is relatively long-lived and reaches a much higher population, because electron transfer from $B_A^-$ to $Phe_A$ is slow and reversible due to the states $P^+B_A^-$ and $P^+Phe_A^-$ being approximately equal in free energy. Finally, it has been pointed out that the kinetics of P* decay can also be adequately fit with a Gaussian distribution of lifetimes, rather than two or more discrete lifetimes as described above (Van Stokkum *et al.*, 1997; Beekman *et al.*, 1996). Experiments with mutant complexes have suggested that it is unlikely that such a distribution of lifetimes is solely attributable to a distribution in $\Delta G$ (Beekman *et al.*, 1996). An obvious possibility is that population heterogeneity of the protein:cofactor system gives rise to distributions in $\Delta G$, $\lambda$ and $V_{DA}$, rather than in just one of these.

The model in which energetic relaxation of the $P^+H_A^-$ radical pair gives rise to the multi-exponential decay of P* has been developed on the basis of experiments with wild-type reaction centres and reaction centres with a greatly elevated mid-point redox potential for the $P/P^+$ redox couple (i.e. in which the free energy of $P^+H_A^-$ and $P^+B_A^-$ is increased considerably) (Peloquin *et al.*, 1994; Woodbury *et al.*, 1994; Woodbury and Parson, 1984). In this so-called conformational cooling model, $P^+$ and/or $H_A^-$ and their protein environments are able to adopt two or more distinct conformations, and it is assumed that there is significant thermal repopulation of the P* state from the $P^+H_A^-$ state (Figure 10B). The transition (relaxation) through these conformational states increases the $\Delta G$ between the P* state and the $P^+H_A^-$ state, and so lessens the likelihood of thermal repopulation of P* (i.e. the relaxation shifts the equilibrium of the reversible electron transfer reaction towards $P^+H_A^-$). This in turn leads to a time-dependent decrease in repopulation of the P* state.

Indirect evidence in support of both explanations for the dispersive kinetics of P* decay comes from analysis of $H_A^-$ in reaction centres in the phototrapped state $PH_A^-Q_A^-$, involving optical and magnetic resonance measurements at different temperatures and after different illumination times (Müh *et al.*, 1999, 1998; Tiede *et al.*, 1987). These experiments have shown that the $H_A$ BPhe can adopt more than one conformation, and in particular that the 2-acetyl carbonyl substituent group adopts different conformations relative to the plane of the BPhe macrocycle depending on experimental conditions (Müh *et al.*, 1999, 1998). Recent results from FTIR spectroscopy support the proposal that $H_A$ exhibits conformational heterogeneity (Breton et al., 1999).

A recent report based upon a comparison of the kinetics of fluorescence decay in $Q_A$-containing and $Q_A$-depleted reaction centres has argued against the conformational cooling model (Hartwich et al., 1998). The amplitudes of fitted lifetimes for P* decay on the 40–600 ps time-scale were found to be similar in the two types of reaction centre, despite the fact that the lifetime of the $P^+H_A^-$ state is changed from approximately 100 ps in $Q_A$-containing reaction centres to approximately 10 ns in $Q_A$-depleted reaction centres. As a result it was concluded that relaxation of $P^+H_A^-$ is not the cause of the slower components in the decay of P*. Instead, it was proposed that P* fluorescence up to 600 ps is dominated by prompt fluorescence from P*, with the slowest components arising from a small fraction of reaction centres which have an energetically high-lying $P^+B_A^-$ state, and which carry out electron transfer via a super-exchange mechanism.

In conclusion, experimental evidence has been gathered in favour of both principal models put forward to explain multi-exponential P* decay, and a clear distinction between the two may be difficult to achieve. In illustration, it was concluded from a recent set of experiments in which $P^+B_A^-$ was detected as a transient intermediate that energetic relaxation of the $P^+H_A^-$ state could account for the characteristics of P* decay on the 10–20 ps time-scale (Van Stokkum et al., 1997; Holzwarth and Müller, 1996). However, an almost equally good and realistic description of the data could be obtained by assuming an inhomogeneous distribution of P* decay times (Van Stokkum et al., 1997). Both protein:cofactor heterogeneity and protein:cofactor relaxation are plausible root causes of the effects observed, and it is entirely possible that the two occur in tandem. Clearly, however, these experiments address fundamental properties of protein:cofactor systems, and this will remain an intriguing and active area of research for some time to come.

### 5.5.    Femtosecond Biology: Coherent Nuclear Dynamics Studied in Populations of Proteins

A number of experiments measuring the decay of the P* state have revealed complicating factors in addition to the multi-exponential decay of the state described above. In measurements monitoring the decay of P* using short (50–100 fs) laser pulses conducted by Vos and co-workers, oscillations were observed superimposed on the resulting kinetic traces (Vos et al., 1994a,b,c; 1993, 1991). These oscillations have been attributed to coherent nuclear motion associated with the P* state (Vos and Martin, 1999;

Stanley and Boxer, 1995; Vos *et al.*, 1993). Under conditions where the excitation pulse is long relative to the period of this nuclear motion, this phenomenon is not observed.

Excitation of P in a population of reaction centres will, according to the Franck-Condon principle, coherently populate a group of vibrational modes (a wavepacket) on the P* excited state that most closely resemble the configuration of the ground state, after which rapid vibrational relaxation results in a thermal equilibrium over all possible vibrational sub-states of the P* excited state. The wavepacket formed by excitation is not stationary on the excited state potential surface but moves in a regular manner from one edge of the surface to the other and back again, modulating the energy of the P* state, until the wavepacket is broadened across the whole potential surface due to vibrational dephasing and relaxation. Since primary electron transfer is a fast reaction, it is possible that some nuclear motions that occur on a time-scale similar to the electron transfer reaction ($<100\,cm^{-1}$, $>330\,fs$) have not reached thermal equilibrium on the time-scale that the electron transfer reaction takes place. If now the same vibrations that are coupled to the optical transition are also coupled to the electron transfer reaction, then the electron transfer reaction will be modulated by the movement of the nuclei of the cofactors or the protein environment. The semi-classical description of non-adiabatic electron transfer (Marcus, 1993; Marcus and Sutin, 1985; Marcus, 1956) assumes that electron transfer takes place from a vibrationally-relaxed reactant state. The observation of coherent nuclear motion associated with the P* state on time-scales similar to that of primary electron transfer indicates that this assumption may not be valid for the first step of light energy transduction in the reaction centre.

If site directed mutagenesis is used to markedly slow down the rate of primary electron transfer, it is possible to examine the time-scale of coherent nuclear motion in the absence of decay of the non-oscillatory part of the signal due to depopulation of the P* state. As shown in Figure 9, residue Tyr M210 is located within ~4 Å of $P_A$, $P_B$, $B_A$ and $H_A$, and as discussed in Section 5.6.1, mutation of this residue to Phe, Leu or Trp brings about a dramatic decrease in the rate of primary electron transfer. Although mutations at the M210 position have a strong effect on the rate of primary electron transfer, they do not affect the frequency of the vibrational modes that are coupled to the P* state (Vos *et al.*, 1996). Their main effect is to allow the oscillations in P* emission to be more clearly resolved, as no significant depopulation of the P* state through electron transfer occurs. Under such conditions the length of time for which oscillations can be resolved is determined by vibrational dephasing. At 10 K, experiments with the M210 mutant reaction centres showed that the lifetime for most modes was approximately 2 ps, longer than the 1 ps time constant for electron transfer in the wild-type reaction centre at 10 K (Vos et al., 1996).

Mutagenesis has also been used in an attempt to examine the molecular origin of the vibrational modes that are coupled to the P* state. Mutations at the M210 position do not affect the frequencies of these modes, despite having a strong effect on the rate of the primary reaction. However, mutations which change the pattern of hydrogen bond interactions between the P BChls and the surrounding protein do alter the frequency spectrum of these coupled modes (Vos and Martin, 1999; Rischel et al., 1998). This effect is most striking for mutations made near the interface region of the P BChls, that change hydrogen bonding to the 2-acetyl carbonyl substituent groups (Rischel et al., 1998). Mutations near the edge of the P dimer have weaker effects.

The experiments of Vos and co-workers raise the question of whether the coherent nuclear motion associated with the P* state that persists on the time-scale of electron transfer is coupled to the primary electron transfer reaction. In particular, do any of the nuclear vibrations coupled to the P* state facilitate the transfer of electrons from P* to $B_A$? The observation of coherent nuclear motion that persists on the time-scale of primary electron transfer raises the possibility that this nuclear motion may be an important parameter that governs the characteristics of this reaction, which would place this process in a near-adiabatic regime. Of obvious importance is the question of whether it is possible to observe coherence in the formation of a product state such as $P^+B_A^-$. A number of recent studies have addressed this difficult problem with conflicting conclusions (Spörlein et al., 1998; Streltsov et al., 1998; Vos et al., 1998; Streltsov et al., 1996) and, as discussed in recent review (Vos and Martin, 1999) at present this question remains to be answered.

### 5.6.    Modulation of the Time Constant for Primary Electron Transfer between 200 fs and 500 ps through Site-Directed Mutagenesis

Site-directed mutagenesis has proven to be a very powerful tool for the study of electron transfer in the bacterial reaction centre. It has been used to investigate the role played by specific amino acid residues during electron transfer, and as a means of modifying the energetics of electron transfer. A number of reviews of the application of mutagenesis to study of the bacterial reaction centre have been published (Fyfe et al., 1998; Parson, 1996; Woodbury and Allen, 1995; Takahashi and Wraight, 1994; Kirmaier and Holten, 1993; Coleman and Youvan, 1990). In this section, we concentrate on two heavily-studied classes of reaction centre mutant that have given considerable insights into the mechanism and energetics of primary electron transfer.

### 5.6.1. Tyrosine M210

In experiments with mutant reaction centres, perhaps the most heavily studied residue to date is Tyr M210 and its symmetry-related counterpart, Phe L181. As described in Section 5.2.3, above, Tyr M210 is a conserved residue which is intimately associated with the P BChls, the $B_A$ BChl and the $H_A$ BPhe (Figure 9A).

Although, as described above, the Tyr M210/Phe L181 pair do not dictate the asymmetry of primary electron transfer, Tyr M210 in particular has a strong effect on the rate of this reaction. In a Tyr M210 → Phe mutant, the time constant for primary electron transfer was slowed from 5 ps to 28 ps at room temperature in membrane-bound reaction centres, and similar effects have been seen in purified complexes (Beekman *et al.*, 1996; Hamm *et al.*, 1993; Jia *et al.*, 1993; Nagarajan *et al.*, 1993; Chan *et al.*, 1991; Finkele *et al.*, 1990). Mutations of Tyr M210 to other non-polar residues such as Leu, Ile or Trp also slow the rate of this reaction and reverse the temperature dependence of the rate of the reaction (Van Brederode *et al.*, 1997b; Beekman *et al.*, 1996; Vos *et al.*, 1996; Hamm *et al.*, 1993; Nagarajan *et al.*, 1993; Finkele *et al.*, 1990; Nagarajan *et al.*, 1990). The most dramatic result in this respect has been observed for the primary electron transfer in the Tyr M210 to Trp mutant, in which this reaction is slowed by 100 to 400-fold at cryogenic temperature (Van Brederode, 1997b; Vos *et al.*, 1996; Nagarajan *et al.*, 1993). These effects are not due to any gross structural perturbations, as a combination of spectroscopic and crystallographic studies have shown that these mutations do not have any strong effects on the structure of the reaction centre (Fyfe *et al.*, 1998; Chirino *et al.*, 1994; Jones *et al.*, 1994; Shochat *et al.*, 1994; Gray *et al.*, 1990). The effects of mutating the M210 residue can largely be rationalised in terms of a significant increase in the free energy of the $P^+B_A^-$ radical pair, which would slow the rate of primary electron transfer and make the reaction thermally-activated, changing its temperature dependence. This increase in the free energy of $P^+B_A^-$ is brought about by an (measureable) increase in $E_m$ $P/P^+$ (Visschers *et al.*, 1999; Beekman *et al.*, 1996; Jia *et al.*, 1993; Nagarajan *et al.*, 1993) and a (proposed) destabilisation of $B_A^-$ (Parson, 1996; Alden *et al.*, 1996; Gunner *et al.*, 1996; Parson *et al.*, 1990). Mutation of Phe L181 to Tyr produces a slight acceleration in the rate of electron transfer along the A-branch (Hamm *et al.*, 1993; Jia *et al.*, 1993), which appears to be due to a (measured) decrease in $E_m$ $P/P^+$ which would lower the free energy of the $P^+B_A^-$ state.

Studies of primary electron transfer in a series of mutant reaction centres with different combinations of residues at the M210 and L181 positions have provided support for the idea that the principal effect of altering these residues is to modulate the free energy of the $P^+B_A^-$ state (Beekman *et al.*, 1996; Jia *et al.*, 1993). An analysis with the classical Marcus function of the relationship between the measured $\ln(k_{et})$ and a $\Delta G$ for the reaction that was inferred from a measured value for $E_m$ $P/P^+$, yielded a reasonable fit and a rather small reorganisation energy of ~30 meV for the primary reaction.

### 5.6.2. Hydrogen Bond Mutants

Another class of mutant reaction centres in which primary electron transfer is slowed and is energetically up-hill are the so-called hydrogen bond mutants, which have been most extensively studied in *Rb. sphaeroides*. In these, the number and pattern of hydrogen bond interactions between the protein and the 2-acetyl and 9-keto carbonyl groups of the $P_A$ and $P_B$ BChls is altered by the introduction of a hydrogen-bonding His residue in place of residues Phe M197, Leu L131 or Leu M160, or removal of the naturally-occurring His L168 (Figure 11). The pattern of hydrogen bond interactions can be determined from FTIR and FT-Raman spectroscopy (Mattioli *et al.*, 1995, 1994; Nabedryk *et al.*, 1993) and it has been shown that the introduction of a hydrogen bond increases $E_m$ $P/P^+$ by between 60 and 125 mV, whilst the removal of the hydrogen bond donated by His L168 lowers $E_m$ $P/P^+$ by 95 mV (Lin *et al.*, 1994; Murchison *et al.*, 1993; Williams *et al.*, 1992). Using all possible combinations of single and multiple mutations, it is therefore possible to modulate $E_m$ $P/P^+$ between +410 and +765 mV, the wild-type reaction centre having an $E_m$ $P/P^+$ of approximately +505 mV (Lin *et al.*, 1994).

If it is assumed that the difference in free energy between $P^*$ and $P^+B_A^-$ is of the order of 55 meV in the wild type reaction centre (Schmidt *et al.*, 1993), then the larger increases in $E_m$ $P/P^+$ seen in some of the hydrogen bond mutants would be expected to raise the free energy of the $P^+B_A^-$ state significantly above that of $P^*$ (by up to 200 meV in a mutant with three additional hydrogen bonds). In the more extreme cases this would place the first step of a two-step sequential mechanism for primary electron transfer in a region where the reaction would be expected to be thermally-activated. In a simple view, one would therefore expect that this mechanism cannot operate at cryogenic temperatures, since even for nuclear tunnelling or strong electronic coupling it is necessary that there is a positive driving force for the reaction. Despite this, the formation of $P^+H_A^-$ still occurs at cryogenic temperature even in the most extreme case of the triple hydrogen

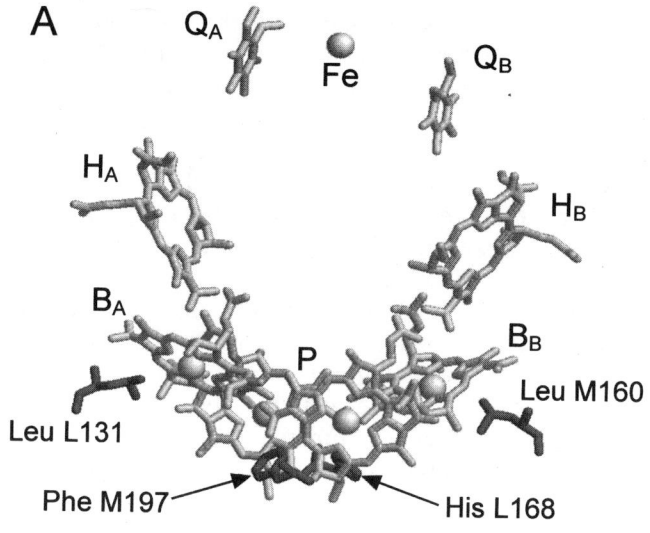

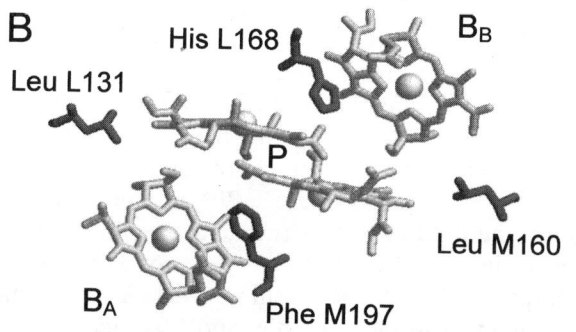

**FIGURE 11.** Amino acid residues at key locations near the P BChls. (B) is a view of P and the monomeric BChls from the periplasmic side of the membrane (bottom in (A)). Mutation of Leu M160, Leu L131 or Phe M197 to His introduces a new hydrogen bond between the protein and the carbonyl groups of the P BChls. Mutation of His L168 to Phe breaks the naturally-occurring hydrogen bond donated to the 2-acetyl carbonyl group of $P_A$.

bond mutant, although the reaction occurs at a greatly reduced rate and with a reduced quantum yield. These observations could be taken as evidence that super-exchange coupling via $P^+B_A^-$ mediates the electron transfer process in these mutant reaction centres, particularly at cryogenic temperature (Bixon *et al.*, 1995). Alternatively, a dynamic relaxation mechanism and adiabatic electron transfer, as described in Section 5.4, has been proposed to explain the primary charge separation process in these high potential mutants (Woodbury *et al.*, 1995). Another explanation could be that the $P^+B_A^-$ state is very inhomogeneous in terms of free energy, as discussed in Section 5.3, and that a fraction of the $P^+B_A^-$ states lie below or equal to the energy of the $P^*$ state. Finally, another mechanism that has been explored on theoretical grounds, and which could also account for charge separation under the circumstances that $P^+B_A^-$ is not thermally-accessible from $P^*$, involves the state $B_A^+H_A^-$ as an intermediate in the reaction scheme $P^* \rightarrow B_A^+H_A^- \rightarrow P^+H_A^-$ (Scherer and Fischer, 1989; Creighton *et al.*, 1988; Warshel *et al.*, 1988; Fischer and Scherer, 1987). For this pathway to operate, the $P^*$ state would have to contain contributions from the excited states of H and/or B, in addition to the excited states of the P BChls. This idea has gained further support with the observation that, under some circumstances, the $B_A^+H_A^-$ state can function as an intermediate in an alternative primary electron transfer route that is driven by $B_A^*$ (see Section 5.7, below).

For those hydrogen bond mutants in which $E_m$ $P/P^+$ is decreased relative to the wild-type reaction centre, the rate of primary electron transfer is similar to that observed in the wild-type reaction centre. This is consistent with the idea that primary electron transfer in the wild-type reaction centre is positioned near the top of the Marcus parabola, where the rate of electron transfer is relatively insensitive to changes in driving force (and where coupling with high frequency modes could also flatten the Marcus parabola in the inverted region (Allen and Williams, 1995)). An exception to this is the His L168 $\rightarrow$ Phe reaction centre of *Rps. viridis*, in which, as with its counterpart in *Rb. sphaeroides*, $E_m$ $P/P^+$ is decreased by 80 mV due to the breakage of one naturally-occurring hydrogen bond. In *Rps. viridis*, this mutation results in an acceleration of the rate of electron transfer from $P^*$ to $B_A$ from 3.5 ps to 1.1 ps at room temperature. At cryogenic temperature this increase is even more dramatic, and this electron transfer step occurs with a time constant of 250 fs (Zinth *et al.*, 1998). Inspection of this electron transfer reaction with the Landau-Zener parameter indicates that, at cryogenic temperatures, this reaction probably occurs in the adiabatic regime and therefore is limited by nuclear motions and not by energetic or electronic parameters (Zinth *et al.*, 1998).

Although the structures of the *Rb. sphaeroides* and *Rps. viridis* reaction centres are highly homologous, and the kinetics of primary electron transfer show strong similarities, the different effects of the His L168 $\rightarrow$ Phe mutation suggests a difference in the energetics of the reaction in the two species. One possibility is that the energetics of primary electron transfer are optimised in the *Rb. sphaeroides* reaction centre but not in the *Rps. viridis* reaction centre, hence the acceleration observed in the latter on altering $\Delta G$ for the reaction. This would imply that primary electron transfer in the wild-type *Rb. sphaeroides* reaction centre takes place at the top of the Marcus parabola, whereas it is in the normal region in the wild-type *Rps. viridis* reaction centre. For the rate of electron transfer to be coincidental in wild-type reaction centres from the two species, one would have to postulate that another parameter, such as the electronic coupling, is better optimised in *Rps. viridis* than in *Rb. sphaeroides*. However, an alternative possibility is that the acceleration of primary electron transfer that occurs in the *Rps. viridis* His L168 → Phe reaction centre is mainly due to an additional change in another parameter, such as the electronic coupling, which does not occur in the *Rb. sphaeroides* reaction centre. As yet there is no crystallographic information on the structural consequences of this mutation in either species that could perhaps address this point further.

### 5.7. Parallel Pathways for Primary Electron Transfer

Until recently, it was generally assumed that excitation of the BPhe and monomeric BChl cofactors of the reaction centre was followed by femtosecond time-scale resonant energy transfer to the P BChls, forming the P* state that drives trans-membrane electron transfer. However, it has now been demonstrated that formation of the $B_A$* excited state leads to the formation of the $P^+H_A^-$ state in a process that does not involve P* (reviewed in Van Brederode and Van Grondelle, 1999). These experiments have shown that the process of light energy transduction in the bacterial reaction centre is more complex than first thought, with the possibility of more than one pathway by which excited state energy can be converted into a trans-membrane electrical potential.

The first indications that excitation of the monomeric BChls and BPhes of the reaction centre triggers processes other than energetically-downhill energy transfer to P came from measurements of the fluorescence excitation spectrum of P* at cryogenic temperatures in a reaction centre with the mutation Tyr M210 → Trp (van Brederode *et al.*, 1997a). In the wild-type reaction centre, measurement of the fluorescence excitation spectrum is complicated by the fact that the amount of "prompt" fluorescence from the initially-created P* state is rather low, due to the short lifetime of this state (Figure 12A). Much of the P* fluorescence detected in experiments with the wild-type reaction centre is attributable to "long-lived" P* that is formed by thermal repopulation from the $P^+H_A^-$ state (Figure 12B). In contrast, in the Tyr M210 → Trp reaction centre the rate of primary electron transfer is dramatically slowed down, and so the P* state created by excitation has a long lifetime (approximately 400 ps at 77 K). Most of the fluo-

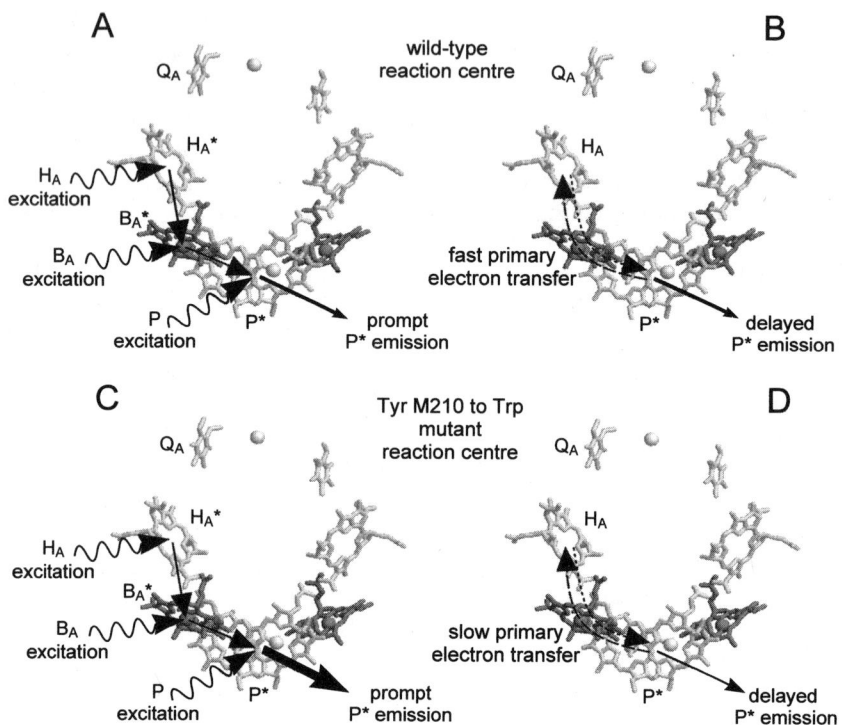

**FIGURE 12.** Fluorescence emission from P* in the bacterial reaction centre. (A,B) In the wild-type reaction centre, P* emission is a mixture of prompt and delayed emission. In (A), prompt emission occurs from the P* state that is created by direct excitation of P or by energy transfer from $B_A^*$ or $H_A^*$ (or $B_B^*$ and $H_B^*$ (not shown)), and is weak because of the very fast decay of P* via primary electron transfer. In (B), delayed emission occurs from the P* state that is formed following fast primary electron transfer to $H_A$ (dashed arrow) and thermal repopulation of P* from the $P^+H_A^-$ state (dotted arrow), and is relatively strong because of the long lifetime of this recombination fluorescence component. (C,D) In the Tyr M210 → Trp mutant, primary electron transfer is slow. P* emission is mainly attributable to prompt emission from the relatively long-lived P* state formed following excitation (C), whilst delayed emission makes a much smaller contribution (D). Note that the recombination fluorescence from $P^+H_A^-$ is not a good probe for the determination of the efficiency of energy transfer.

rescence emanating from P is therefore prompt fluorescence, that is attributable to P* that is created immediately on direct excitation of P, or to P* that is created by femtosecond time-scale energy transfer following excitation of the monomeric bacteriochlorins (Figure 12C,D). This makes the Tyr M210 → Trp reaction centre an excellent vehicle for studying the efficiency of energy transfer from the accessory bacteriochlorins to P using the steady state fluorescence excitation spectrum (Van Brederode et al., 1997b). If the energy transfer from the BPhes and monomeric BChls to P occurs with 100% efficiency, then the fluorescence excitation spectrum of the reaction centre in the $Q_y$ wavelength region should be identical to the reaction centre absorbance spectrum.

As can be seen in Figure 13, following normalisation using the P $Q_y$ band, the fluorescence excitation spectrum of the Tyr M210 → Trp reaction centre does not match the absorbance spectrum in either the B or H $Q_y$ band. As argued in detail elsewhere, it appears that energy absorbed by $B_A$ and $H_A$ is not passed to P. However, the action spectrum for formation of the $P^+Q_A^-$ state shows clearly that excitation of these cofactors does lead to efficient formation of the product state of trans-membrane electron transfer (Figure 13). These observations lead to the unavoidable conclusion that, following excitation of $H_A$ and $B_A$, an electron transfer process driven by $H_A^*$ and/or $B_A^*$ competes efficiently with resonant energy transfer to P (and subsequent electron transfer driven by P*). Similar results were obtained with a number of other mutant reaction centres in which the rate of P* decay is slowed (Van Brederode et al., 1999a, 1999b).

Subsequent to this work, picosecond time-scale transient absorbance measurements have demonstrated the existence of at least two new pathways of trans-membrane electron transfer that can operate in parallel to the well-established route that is driven by P* (Figure 13). These new pathways are driven by the state $B_A^*$, and involve the primary reactions $B_A^* \rightarrow B_A^+H_A^- \rightarrow P^+H_A^-$ and $B_A^* \rightarrow P^+B_A^- \rightarrow P^+H_A^-$ (Van Brederode et al., 1999a,b, 1997b). The relative contributions of these pathways, and the conventional pathway driven by the P* state, to overall light-driven charge separation in the reaction centre has been found to vary in different reaction centres, as discussed in detail in a recent review (Van Brederode and Van Grondelle, 1999). Much of this variation can be rationalised in terms of the effect of particular mutations on the free energy of different radical pair states in the reaction centre (Van Brederode and Van Grondelle, 1999). The existence of these pathways could also explain the excitation wavelength dependence of spectral evolution during primary electron transfer in the wild-type reaction centre, that has been observed by a number of groups (Van Brederode et al., 1999a; Vos et al., 1997; Lin et al., 1996). However,

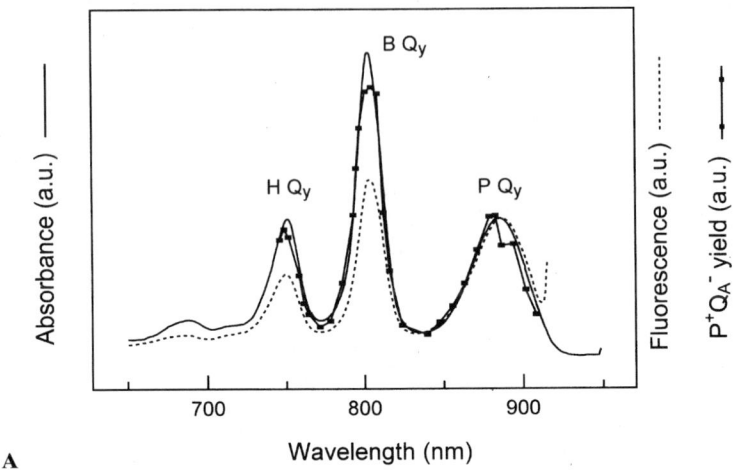

**A**

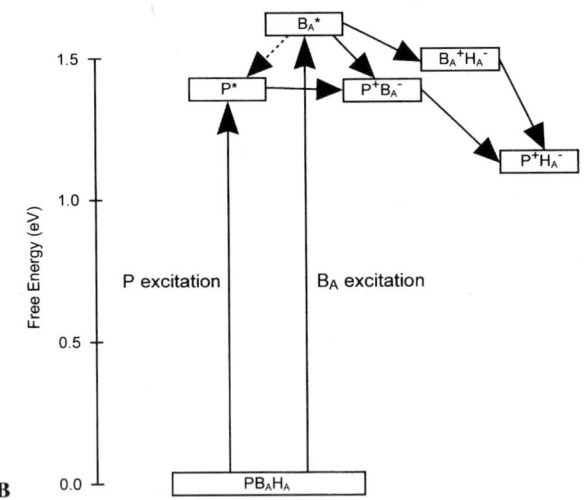

**B**

**FIGURE 13.** Energy and electron transfer in the Tyr M210 → Trp mutant reaction centre. (A) The fluorescence excitation spectrum of this reaction centre (dotted) does not match the absorbance spectrum (solid) in the H and B $Q_y$ bands, indicating that some energy delivered to the BPhes and monomeric BChls is not passed to P. Despite this, excitation of the BPhes and monomeric BChls does lead to efficient formation of $P^+Q_A^-$ (filled boxes). (B) Schematic of routes of energy and electron transfer in the bacterial reaction centre. Excitation of $B_A$ either results in energy transfer to P or in direct formation of $P^+H_A^-$ via the intermediates $P^+B_A^-$ and/or $B_A^+H_A^-$.

other explanations such as spectral heterogeneity have also been put forward to explain this phenomenon (Lin *et al.*, 1996).

Finally, an intriguing point that come from this work is that these relatively long-distance electron transfer reactions compete with femtosecond time-scale energy transfer between $B_A^*$ and P, and thus must also take place on sub-picosecond time-scale. Therefore, it is most likely that these electron transfer reactions occur in the adiabatic regime, and are driven by a vibrationally-unrelaxed $B_A^*$ state.

## 6. SUMMARY

The bacterial reaction centre is undoubtedly one of the most heavily studied electron transfer proteins and, as this article has tried to describe, it has made some unique contributions to our understanding of biological electron transfer and coupled protonation reactions, and has provided fascinating information in areas that concern basic properties such as protein heterogeneity and protein dynamics. Despite intensive study, much remains to be learned about how this protein catalyses the conversion of solar energy into a form that can be used by the cell. In particular, the dynamic roles played by the protein are still poorly understood. The wide range of time-scales over which the reaction centre catalyses electron transfer, and the relative ease with which electron transfer can be triggered and monitored, will ensure that the reaction centre will continue to be used as a laboratory for testing ideas about the nature of biological electron transfer for many years to come.

ACKNOWLEDGEMENTS. The authors wish to thank Professor Rienk van Grondelle for critical reading of this article and helpful suggestions, and Dr. Paul Fyfe for helpful comments.

## 7. REFERENCES

Abresch, E. C., Paddock, M. L., Stowell, M. H. B., McPhillips, T. M., Axelrod, H. L., Soltis, S. M., Rees, D. C., Okamura, M. Y., and Feher, G., 1998, Identification of proton transfer pathways in the X-ray crystal structure of the bacterial reaction center from *Rhodobacter sphaeroides. Photosynth. Res.*, **55**:119–125.

Alden, R. G., Parson, W. W., Chu, Z. T., and Warshel, A., 1995, Calculations of electrostatic energies in photosynthetic reaction centers. *J. Am. Chem. Soc.*, **117**:12284–12298.

Alden, R. G., Parson, W. W., Chu, Z. T., and Warshel, A., 1996, Orientation of the OH dipole of tyrosine (M)210 and its effect on electrostatic energies in photosynthetic bacterial reaction centers. *J. Phys. Chem.*, **100**:16761–16770.

Allen, J. P., and Williams, J. C., 1995, Relationship between the oxidation potential of the bacteriochlorophyll dimer and electron transfer in photosynthetic reaction centers. *J. Bioenerg. Biomemb.*, **27**:275–283.

Allen, J. P., Feher, G., Yeates, T. O., Rees, D. C., Deisenhofer, J., Michel, H., and Huber, R., 1986, Structural homology of reaction centers from *Rhodopseudomonas sphaeroides* and *Rhodopseudomonas viridis* as determined by X-ray diffraction. *Proc. Natl. Acad. Sci. USA*, **83**:8589–8593.

Allen, J. P., Feher, G., Yeates, T. O., Komiya, H., and Rees, D. C., 1987a, Structure of the reaction center from *Rhodobacter sphaeroides* R-26: The cofactors. *Proc. Natl. Acad. Sci. USA*, **84**:5730–5734.

Allen, J. P., Feher, G., Yeates, T. O., Komiya, H., and Rees, D. C., 1987b, Structure of the reaction center from *Rhodobacter sphaeroides* R-26: The protein subunits. *Proc. Natl. Acad. Sci. USA*, **84**:6162–6166.

Allen, J. P., Feher, G., Yeates, T. O., Komiya, H., and Rees, D. C., 1988, Structure of the reaction center from *Rhodobacter sphaeroides* R-26: Protein-cofactor (quinones and $Fe^{2+}$) interactions. *Proc. Natl. Acad. Sci. USA*, **85**:8497–8491.

Arlt, T., Schmidt, S., Kaiser, W., Lauterwasser, C., Meyer, M., Scheer, H., and Zinth, W., 1993, The accessory bacteriochlorophyll: A real electron carrier in primary photosynthesis. *Proc. Natl. Acad. Sci. USA*, **90**:11757–11761.

Arlt, T., Dohse, B., Schmidt, S., Wachtveitl, J., Laussermair, E., Zinth, W., and Oesterhelt, D., 1996, Electron transfer dynamics of *Rhodopseudomonas viridis* reaction centers with a modified binding site for the accessory bacteriochlorophyll. *Biochemistry*, **35**:9235–9244.

Beekman, L. M. P., Visschers, R. W., Monshouwer, R., Heer-Dawson, M., Mattioli, T. A., McGlynn, P., Hunter, C. N., Robert, B., Van Stokkum, I. H. M., Van Grondelle, R., and Jones, M. R., 1995, Time-resolved and steady-state spectroscopic analysis of membrane-bound reaction centers from *Rhodobacter sphaeroides*: Comparisons with detergent-solubilized complexes. *Biochemistry*, **34**:14712–14721.

Beekman, L. M. P., Van Stokkum, I. H. M., Monshouwer, R., Rijnders, A. J., McGlynn, P., Visschers, R. W., Jones, M. R., and Van Grondelle, R., 1996, Primary electron transfer in membrane-bound reaction centers with mutations at the M210 position. *J. Phys. Chem.*, **100**:7256–7268.

Beroza, P., Fredkin, D. R., Okamura, M. Y., and Feher, G., 1992, Proton transfer pathways in the reaction center of *Rhodobacter sphaeroides*: a computational study. In *The Photosynthetic Bacterial Reaction Center II* (J. Breton and A. Verméglio, eds.) pp. 363–374. Plenum Press, New York.

Bixon, M., and Jortner, J., 1999, Electron transfer—from isolated molecules to biomolecules. *Adv. Chem. Phys.*, **106**:35–202.

Bixon, M., Michel Beyerle, M. E., and Jortner, J., 1988, Formation dynamics, decay kinetics and singlet-triplet splitting of the (bacteriochlorophyll dimer)-positive (bacteriopheophytin)-negative radical pair in bacterial photosynthesis. *Isr. J. Chem.*, **28**:155–168.

Bixon, M., Jortner, J., Michel-Beyerle, M. E., and Ogrodnik, A., 1989, A superexchange mechanism for the primary charge separation in photosynthetic reaction centers. *Biochim. Biophys. Acta*, **977**:273–286.

Bixon, M., Jortner, J., and Michel Beyerle, M. E., 1991, On the mechanism of the primary charge separation in bacterial photosynthesis. *Biochim. Biophys. Acta*, **1056**:301–316.

Bixon, M., Jortner, J., and Michel-Beyerle, M. E., 1995, A kinetic-analysis of the primary charge separation in bacterial photosynthesis—energy gaps and static heterogeneity. *Chem. Phys.*, **197**:389–404.

Blankenship, R. E., Madigan, M. T., and Bauer, C. E. (eds.), 1995, *Anoxygenic Photosynthetic Bacteria*, Kluwer Academic Publishers, Dordrecht, The Netherlands.

Breton, J., 1988, Low temperature linear dichroism study of the orientation of the pigments in reduced and oxidised reaction centers of *Rps. viridis* and *Rb. sphaeroides*. *NATO ASI Ser., Ser. A*, **149**:59–69.

Breton, J., Martin, J. L., Fleming, G. R., and Lambry, J. C., 1988, Low-temperature femtosecond spectroscopy of the initial step of electron transfer in reaction centers from photosynthetic purple bacteria. *Biochemistry*, **27**:8276–8284.

Breton, J., Bylina, E. J., and Youvan, D. C., 1989, Pigment organization in genetically modified reaction centers of *Rhodobacter capsulatus*. *Biochemistry*, **28**:6423–6429.

Breton, J., Bibikova, M., Oesterhelt, D., and Nabedryk, E., 1999, Conformational heterogeneity of the bacteriopheophytin electron acceptor $H_A$ in reaction centers from *Rhodopseudomonas viridis* revealed by Fourier transform infrared spectroscopy and site-directed mutagenesis. *Biochemistry*, **38**:11541–11552.

Bylina, E. J., Kirmaier, C., McDowell, L., Holten, D., and Youvan, D. C., 1988, Influence of an amino-acid residue on the optical-properties and electron-transfer dynamics of a photosynthetic reaction center complex. *Nature*, **336**:182–184.

Chan, C. K., Chen, L. X. Q., Di Magno, T. J., Hanson, D. K., Nance, S. L., Schiffer, M., Norris, J. R., and Fleming, G. R., 1991, Initial electron transfer in photosynthetic reaction centers of *Rhodobacter capsulatus* mutants. *Chem. Phys. Letts.*, **176**:366–372.

Chang, C. H., Tiede, D., Tang, J., Smith, U., Norris, J., and Schiffer, M., 1986, Structure of *Rhodopseudomonas sphaeroides* R-26 reaction center. *FEBS Letts.*, **205**:82–86.

Chang, C. H., El Kabbani, O., Tiede, D., Norris, J., and Schiffer, M., 1991, Structure of the membrane-bound protein photosynthetic reaction center from *Rhodobacter sphaeroides*. *Biochemistry*, **30**:5352–5360.

Chirino, A. J., Lous, E. J., Huber, M., Allen, J. P., Schenck, C. C., Paddock, M. L., Feher, G., and Rees, D. C., 1994, Crystallographic analyses of site-directed mutants of the photosynthetic reaction center from *Rhodobacter sphaeroides*. *Biochemistry*, **33**:4584–4593.

Cogdell, R. J., and Frank, H. A., 1987, How carotenoids function in photosynthetic bacteria. *Biochim. Biophys. Acta*, **895**:63–79.

Coleman, W. J., and Youvan, D. C., 1990, Spectroscopic analysis of genetically modified photosynthetic reaction centers. *Ann. Rev. Biophys. Biophys. Chem.*, **19**:333–367.

Creighton, S., Hwang, J. K., Warshel, A., Parson, W. W., and Norris, J., 1988, Simulating the dynamics of the primary charge separation process in bacterial photosynthesis. *Biochemistry*, **27**:774–781.

Crofts, A. R., and Berry, E. A., 1998, Structure and function of the cytochrome bc(1) complex of mitochondria and photosynthetic bacteria. *Curr. Opin. Str. Biol.*, **8**:501–509.

Crofts, A. R., Meinhardt, S. W., Jones, K. R., and Snozzi, M., 1983, The role of the quinone pool in the cyclic electron transfer chain of *Rhodopseudomonas sphaeroides*. A modified Q-cycle mechanism. *Biochim. Biophys. Acta*, **723**:202–218.

Deisenhofer, J., and Michel, H., 1989a, The photosynthetic reaction center from the purple bacterium *Rhodopseudomonas viridis*. *EMBO J.*, **8**:2149–2170.

Deisenhofer, J., and Michel, H., 1989b, The photosynthetic reaction center from the purple bacterium *Rhodopseudomonas viridis*. *Science*, **245**:1463–1473.

Deisenhofer, J., and Norris, J. R. (eds.), 1993, *The Photosynthetic Reaction Center*, Volumes 1 and 2, Academic Press, San Diego, USA.

Deisenhofer, J., Epp, O., Miki, K., Huber, R., and Michel, H., 1984, X-ray structure-analysis of a membrane-protein complex-electron-density map at 3 Å resolution and a model of the chromophores of the photosynthetic reaction center from *Rhodopseudomonas viridis*. *J. Mol. Biol.*, **180**:385–398.

Deisenhofer, J., Epp, O., Miki, K., Huber, R., and Michel, H., 1985, Structure of the protein subunits in the photosynthetic reaction center of *Rhodopseudomonas viridis* at 3 Å resolution. *Nature*, **318**:618–624.

Deisenhofer, J., Epp, O., Sinning, I., and Michel, H., 1995, Crystallographic refinement at 2.3 Å resolution and refined model of the photosynthetic reaction centre from *Rhodopseudomonas viridis*. *J. Mol. Biol.*, **246**:429–457.

DeVault, D., 1980, Quantum mechanical tunnelling in biological systems. *Q. Rev. Biophys.*, **13**:387–564.

DiMagno, T. J., and Norris, J. R., 1993, Initial electron transfer events in photosynthetic bacteria. In *The Photosynthetic Reaction Center*, (J. Deisenhofer and J. R. Norris, eds.) Volume 2, pp. 105–132, Academic Press, San Diego, USA.

Du, M., Rosenthal, S. J., Xie, X., Di Magno, T. J., Schmidt, M., Hanson, D. K., Schiffer, M., Norris, J. R., and Fleming, G. R., 1992, Femtosecond spontaneous-emission studies of reaction centers from photosynthetic bacteria. *Proc. Natl. Acad. Sci. USA*, **89**:8517–8521.

Dutton, P. L., Petty, K. M., Bonner, H. S., and Morse, S. D., 1975, Cytochrome $c_2$ and reaction center of *Rhodopseudomonas sphaeroides* Ga membranes. Extinction coefficients, content, half-reduction potentials, kinetics and electric field alterations. *Biochim. Biophys. Acta*, **387**:536–556.

El Kabbani, O., Chang, C. H., Tiede, D., Norris, J., and Schiffer, M., 1991, Comparison of reaction centers from *Rhodobacter sphaeroides* and *Rhodopseudomonas viridis*: Overall architecture and protein-pigment interactions. *Biochemistry*, **30**:5361–5369.

Ermler, U., Fritzsch, G., Buchanan, S. K., and Michel, H., 1994a, Structure of the photosynthetic reaction centre from *Rhodobacter sphaeroides* at 2.65 Å resolution: Cofactors and protein-cofactor interactions. *Structure*, **2**:925–936.

Ermler, U., Michel, H., and Schiffer, M., 1994b, Structure and function of the photosynthetic reaction center from *Rhodobacter sphaeroides*. *J. Bioenerg. Biomemb.*, **26**:5–15.

Finkele, U., Lauterwasser, C., Zinth, W., Gray, K. A., and Oesterhelt, D., 1990, Role of tyrosine M210 in the initial charge separation of reaction centers of *Rhodobacter sphaeroides*. *Biochemistry*, **29**:8517–8521.

Fischer, S. F., and Scherer, P. O. J., 1987, On the early charge separation and recombination processes in bacterial reaction centers. *Chem. Phys.*, **115**:151–158.

Fleming, G. R., and Van Grondelle, R., 1994, The primary steps of photosynthesis. *Physics Today*, 48–55.

Fleming, G. R., Martin, J. L., and Breton, J., 1988, Rates of primary electron transfer in photosynthetic reaction centres and their mechanistic implications. *Nature*, **333**:190–192.

Foloppe, N., Ferrand, M., Breton, J., and Smith, J. C., 1995, Structural model of the photosynthetic reaction center of *Rhodobacter capsulatus*. *Prot. Str. Fun. Gen.*, **22**:226–244.

Fritzsch, G., Kampmann, L., Kapaun, G., and Michel, H., 1998, Water clusters in the reaction centre of *Rhodobacter sphaeroides*. *Photosynth. Res.*, **55**:127–132.

Fyfe, P. K., McAuley-Hecht, K. E., Jones, M. R., and Cogdell, R. J., 1998a, Purple bacterial photosynthetic reaction centres. In *Biomembrane Structures*, (P. I. Haris and D. Chapman, eds.) 64–87, IOS Press, Amsterdam, The Netherlands.

Fyfe, P. K., McAuley-Hecht, K. E., Ridge, J. P., Prince, S. M., Fritzsch, G., Isaacs, N. W., Cogdell, R. J., and Jones, M. R., 1998b, Crystallographic studies of mutant reaction centres from *Rhodobacter sphaeroides*. *Photosynth. Res.*, **55**:133–140.

Gehlen, J. N., Marchi, M., and Chandler, D., 1994, Dynamics affecting the primary charge transfer in photosynthesis. *Science*, **263**:499–502.

Graige, M. S., Feher, G., and Okamura, M. Y., 1998, Conformational gating of the electron transfer reaction $Q_A^{-\bullet}Q_B \rightarrow Q_AQ_B^{-\bullet}$ in bacterial reaction centers of *Rhodobacter sphaeroides* determined by a driving force assay. *Proc. Natl. Acad. Sci. USA*, **95**:11679–11684.

Gray, K. A., Farchaus, J. W., Wachtveitl, J., Breton, J., and Oesterhelt, D., 1990, Initial characterization of site-directed mutants of tyrosine M210 in the reaction center of *Rhodobacter sphaeroides*. *Embo J.*, **9**:2061–2070.

Gray, K. A., Wachtveitl, J., and Oesterhelt, D., 1992, Photochemical trapping of a bacteriopheophytin anion in site- specific reaction-center mutants from the photosynthetic bacterium *Rhodobacter sphaeroides*. *Eur. J. Biochem.*, **207**:723–731.

Gudowska-Nowak, E., 1994, Effects of heterogeneity on relaxation dynamics and electron transfer rates in photosynthetic reaction centers. *J. Phys. Chem.*, **98**:5257–5264.

Gunner, M. R., and Dutton, P. L., 1989, Temperature and $-\Delta G°$ dependence of the electron transfer from BPh•$^-$ to $Q_A$ in reaction center protein from *Rhodobacter sphaeroides* with different quinones as $Q_A$. *J. Am. Chem. Soc.*, **111**:3400–3412.

Gunner, M. R., Nicholls, A., and Honig, B., 1996, Electrostatic potentials in *Rhodopseudomonas viridis* reaction centers: implications for the driving force and directionality of electron transfer. *J. Phys. Chem.*, **100**:4277–4291.

Hamm, P., Gray, K. A., Oesterhelt, D., Feick, R., Scheer, H., and Zinth, W., 1993, Subpicosecond emission studies of bacterial reaction centers. *Biochim. Biophys. Acta*, **1142**:99–105.

Hartwich, G., Lossau, H., Michel-Beyerle, M. E., and Ogrodnik, A., 1998, Nonexponential fluorescence decay in reaction centers of *Rhodobacter sphaeroides* reflecting dispersive charge separation up to 1 ns. *J. Phys. Chem.*, **102**:3815–3820.

Heller, B. A., Holten, D., and Kirmaier, C., 1995, Control of electron transfer between the L- and M-sides of photosynthetic reaction centers. *Science*, **269**:940–945.

Hoff, A. J., and Deisenhofer, J., 1997, Photophysics of photosynthesis. Structure and spectroscopy of reaction centers of purple bacteria, *Physics Reports-Review Section Of Physics Letters*, **287**:2–247.

Holzapfel, W., Finkele, U., Kaiser, W., Oesterhelt, D., Scheer, H., Stilz, H. U., and Zinth, W., 1989, Observation of a bacteriochlorophyll anion radical during the primary charge separation in a reaction center. *Chem. Phys. Lett.*, **160**:1–7.

Holzapfel, W., Finkele, U., Kaiser, W., Oesterhelt, D., Scheer, H., Stilz, H. U., and Zinth, W., 1990, Initial electron transfer in the reaction center from *Rhodobacter sphaeroides*. *Proc. Natl. Acad. Sci USA*, **87**:5168–5172.

Holzwarth, A. R., and Müller, M. G., 1996, Energetics and kinetics of radical pairs in reaction centers from *Rhodobacter sphaeroides*. A femtosecond transient absorption study. *Biochemistry*, **35**:11820–11831.

Huber, H., Meyer, M., Scheer, H., Zinth, W., and Wachtveitl, J., 1998, Temperature dependence of the primary electron transfer reaction in pigment-modified bacterial reaction centers. *Photosynth. Res.*, **55**:153–162.

Ivashin, N., Kallenbring, B., Larsson, S., and Hansson, O., 1998, Charge separation in photosynthetic reaction centers. *J. Phys. Chem.*, **102**:5017–5022.

Iwata, S., Lee, J. W., Okada, K., Lee, J. K., Iwata, M., Rasmussen, B., Link, T. A., Ramaswamy, S., and Jap, B. K., 1998, Complete structure of the 11-subunit bovine mitochondrial cytochrome $bc_1$ complex. *Science*, **281**:64–71.

Jia, Y. W., DiMagno, T. J., Chan, C. K., Wang, Z. Y., Du, M., Hanson, D. K., Schiffer, M., Norris, J. R., Fleming, G. R., and Popov, M. S., 1993, Primary charge separation in mutant reaction centers of *Rhodobacter sphaeroides*. *J. Phys. Chem.*, **97**:13180–13191.

Jones, M. R., Heer Dawson, M., Mattioli, T. A., Hunter, C. N., and Robert, B., 1994, Site-specific mutagenesis of the reaction centre from *Rhodobacter sphaeroides* studied by Fourier transform Raman spectroscopy: Mutations at tyrosine M210 do not affect the electronic structure of the primary donor. *FEBS Letts.*, **339**:18–24.

Kaufmann, K. J., Dutton, P. L., Netzel, T. L., Leigh, J. S., and Rentzepis, P. M., 1975, Picosecond kinetics of events leading to reaction centers bacteriochlorophyll oxidation. *Science*, **188**:1301–1304.

Kaufmann, K. J., Petty, K. M., Dutton, P. L., and Rentzepis, P. M., 1976, Picosecond kinetics in reaction centers of *Rps. sphaeroides* and the effects of ubiquinone extraction and reconstitution. *Biochem. Biophys. Res. Commun.*, **70**:839–845.

Kellogg, E. C., Kolaczkowski, S., Wasielewski, M. R., and Tiede, D. M., 1989, Measurement of the extent of electron transfer to the bacteriopheophytin in the M-subunit in reaction centers of *Rhodopseudomonas viridis*. *Photosynth. Res.*, **22**:47–60.

Kennis, J. T. M., Shkuropatov, A. Y., Van Stokkum, I. H. M., Gast, P., Hoff, A. J., Shuvalov, V. A., and Aartsma, T. J., 1997, Formation of a long-lived $P^+B_A^-$ state in plant pheophytin-exchanged reaction centers of *Rhodobacter sphaeroides* R26 at low temperature. *Biochemistry*, **36**:16231–16238.

Kirmaier, C., and Holten, D., 1988, Temperature effects on the ground state absorption spectra and electron transfer kinetics of bacterial reaction centers. In *NATO ASI Ser., Ser. A*, **149**:219–228.

Kirmaier, C., and Holten, D., 1993, Electron transfer and charge recombination reactions in wild-type and mutant bacterial reaction centers. In *The Photosynthetic Reaction Center*, (J. Deisenhofer and J. R. Norris, eds.) Volume 2, 49–70, Academic Press, San Diego, USA.

Kirmaier, C., Holten, D., and Parson, W. W., 1985a, Temperature and detection-wavelength dependence of the picosecond electron transfer kinetics measured in *Rhodopseudomonas sphaeroides* reaction centers: Resolution of new spectral and kinetic components in the primary charge-separation process. *Biochim. Biophys. Acta*, **810**:33–48.

Kirmaier, C., Holten, D., and Parson, W. W., 1985b, Picosecond-photodichroism studies of the transient states in *Rhodopseudomonas sphaeroides* reaction centers at 5 kelvin: Effects of electron transfer on the six bacteriochlorin pigments. *Biochim. Biophys. Acta*, **810**:49–61.

Kirmaier, C., Gaul, D., Debey, R., Holten, D., and Schenck, C. C., 1991, Charge separation in a reaction center incorporating bacteriochlorophyll for photoactive bacteriopheophytin. *Science*, **251**:922–927.

Kirmaier, C., Weems, D., and Holten, D., 1999, M-side electron transfer in reaction center mutants with a lysine near the nonphotoactive bacteriochlorophyll. *Biochemistry*, **38**:11516–11530.

Kleinfeld, D., Okamura, M. Y., and Feher, G., 1984, Electron transfer kinetics in photosynthetic reaction centres cooled to cryogenic temperatures in the charge-separated state: evidence for light-induced structural changes. *Biochemistry*, **23**:5780–5786.

Kolaczkowski, S. V., Hayes, J. M., and Small, G. J., 1994, A theory of dispersive kinetics in the energy transfer of antenna complexes. *J. Phys. Chem.*, **98**:13418–13425.

Komiya, H., Yeates, T. O., Rees, D. C., Allen, J. P., and Feher, G., 1988, Structure of the reaction center from *Rhodobacter sphaeroides* R-26 and 2.4.1: Symmetry relations and sequence comparisons between different species. *Proc. Natl. Acad. Sci. USA*, **85**:9012–9016.

Lancaster, C. R. D., and Michel, H., 1996, Three-dimensional structures of photosynthetic reaction centers. *Photosynth. Res.*, **48**:65–74.

Lancaster, C. R. D., and Michel, H., 1997, The coupling of light-induced electron transfer and proton uptake as derived from crystal structures of reaction centres from *Rhodopseudomonas viridis* modified at the binding site of the secondary quinone, Q(B). *Structure*, **5**:1339–1359.

Lancaster, C. R. D., and Michel, H., 1999, Refined crystal structures of reaction centres from *Rhodopseudomonas viridis* in complexes with the herbicide atrazine and two chiral atrazine derivatives also lead to a new model of the bound carotenoid. *J. Mol. Biol.*, **286**:883–898.

Levanon, H. (ed.), 1992, Primary events in photosynthesis: problems, speculations, controversies, and future trends. *Isr. J. Biochem.*, **32**:369–518.

Lin, X., Murchison, H. A., Nagarajan, V., Parson, W. W., Allen, J. P., and Williams, J. C., 1994, Specific alteration of the oxidation potential of the electron donor in reaction centers from *Rhodobacter sphaeroides. Proc. Natl. Acad. Sci. USA*, **91**:10265–10269.

Lin, S., Taguchi, A. K. W., and Woodbury, N. W., 1996, Excitation wavelength dependence of energy-transfer and charge separation in reaction centers from *Rhodobacter-sphaeroides* -evidence for adiabatic electron-transfer. *J. Phys. Chem.*, **100**:17067–17078.

Lin, S., Jackson, J. A., Taguchi, A. K. W., and Woodbury, N. W., 1999, B-side electron transfer promoted by absorbance of multiple photons in *Rhodobacter sphaeroides* R-26 reaction centers. *J. Phys. Chem.*, **103**:4757–4763.

Mar, T., and Gingras, G, 1990, *Biochim. Biophys. Acta*, **1017**:112–117.

Marcus, R. A., 1956, On the theory of oxidation reduction reactions involving electron transfer. *J. Chem. Phys.*, **24**:966–978.

Marcus, R. A., 1987, Superexchange versus an intermediate BChl⁻ mechanism in reaction centres of photosynthetic bacteria. *Chem. Phys. Lett.*, **133**:471–477.

Marcus, R. A., 1988, Mechanisms of the early steps in bacterial photosynthesis and their implications for experiment. *Isr. J. Chem.*, **28**:205–213.

Marcus, R. A., 1993, Electron-transfer reactions in chemistry—theory and experiment (Nobel lecture). *Angewandte Chemie-International Edition in English*, **32**:1111–1121.

Marcus, R. A., 1996, Electron transfer reactions in chemistry. Theory and experiment. In *Protein Electron Transfer* (D. S. Bendall, ed.), BIOS Scientific Publishers, Oxford, pp. 249–283.

Marcus, R. A., and Sutin, N., 1985, Electron transfers in chemistry and biology. *Biochim. Biophys. Acta*, **811**:265–322.

Mathis, P., 1994, Electron transfer between cytochrome $c_2$ and the isolated reaction center of the purple bacterium *Rhodobacter sphaeroides. Biochim. Biophys. Acta*, **1187**:177–180.

Mattioli, T. A., Williams, J. C., Allen, J. P., and Robert, B., 1994, Changes in primary donor hydrogen-bonding interactions in mutant reaction centers from *Rhodobacter sphaeroides*: Identification of the vibrational frequencies of all the conjugated carbonyl groups. *Biochemistry*, **33**:1636–1643.

Mattioli, T. A., Lin, X., Allen, J. P., and Williams, J. C., 1995, Correlation between multiple hydrogen bonding and alteration of the oxidation potential of the bacteriochlorophyll dimer of reaction centers from *Rhodobacter sphaeroides. Biochemistry*, **34**:6142–6152.

McAuley-Hecht, K. E., Fyfe, P. K., Ridge, J. P., Prince, S. M., Hunter, C. N., Isaacs, N. W., Cogdell, R. J., and Jones, M. R., 1998, Structural studies of wild-type and mutant reaction centers from an antenna-deficient strain of *Rhodobacter sphaeroides*: Monitoring the optical properties of the complex from bacterial cell to crystal. *Biochemistry*, **37**:4740–4750.

McDowell, L. M., Gaul, D., Kirmaier, C., Holten, D., and Schenck, C. C., 1991, Investigation into the source of electron transfer asymmetry in bacterial reaction centers. *Biochemistry*, **30**:8315–8322.

Michel, H., Epp, O., and Deisenhofer, J., 1986, Pigment-protein interactions in the photosynthetic reaction center from *Rhodopseudomonas viridis. EMBO J.*, **5**:2445–2452.

Michel-Beyerle, M. E., Plato, M., Deisenhofer, J., Michel, H., Bixon, M., and Jortner, J., 1988, Unidirectionally of charge separation in reaction centers of photosynthetic bacteria. *Biochim. Biophys. Acta*, **932**:52–70.

Middendorf, T. R., Mazzola, L. T., Lao, K., Steffen, M. A., and Boxer, S. G., 1993, Stark effect (electroabsorption) spectroscopy of photosynthetic reaction centers at 1.5 K: Evidence that the special pair has a large excited-state polarizability. *Biochim. Biophys. Acta*, **1143**:223–234.

Moser, C. C., and Dutton, P. L., 1992, Engineering protein-structure for electron-transfer function in photosynthetic reaction centers. *Biochim. Biophys. Acta*, **1101**:171–176.

Moser, C. C., and Dutton, P. L., 1996, Outline of theory of protein electron transfer. In *Protein Electron Transfer* (D. S. Bendall, ed.) pp. 1–22, BIOS Scientific Publishers Ltd. Oxford, U.K.

Moser, C. C., Keske, J. M., Warncke, K., Farid, R., and Dutton, P. L., 1992, Nature of biological electron transfer. *Nature*, **355**:796–802.

Moser, C. C., Page, C. C., Farid, R., and Dutton, P. L., 1995, Biological electron transfer. *J. Bioenerg. Biomemb.*, **27**:263–274.

Müh, F., Williams, J. C., Allen, J. P., and Lubitz, W., 1998, A conformational change of the photoactive bacteriopheophytin in reaction centers from *Rhodobacter sphaeroides*. *Biochemistry*, **37**:13066–13074.

Müh, F., Jones, M. R., and Lubitz, W. L., 1999, Reorientation of the acetyl group of the photoactive bacteriopheophytin in reaction centers of *Rhodobacter sphaeroides*: An ENDOR/TRIPLE resonance study. *Biospectroscopy*, **5**:35–46.

Müller, M. G., Griebenow, K., and Holzwarth, A. R., 1991, Primary processes in isolated bacterial reaction centers from *Rhodobacter sphaeroides* studied by picosecond fluorescence kinetics. *Chem. Phys. Letts.*, **199**:465–469.

Murchison, H. A., Alden, R. A., Allen, J. P., Peloquin, J. M., Taguchi, A. K. W., Woodbury, N. W., and Williams, J. C., 1993, Mutations designed to modify the environment of the primary electron donor of the reaction center from *Rhodobacter sphaeroides*: Phenylalanine to leucine at L167 and histidine to phenylalanine at L168. *Biochemistry*, **32**:3498–3505.

Nabedryk, E., Allen, J. P., Taguchi, A. K. W., Williams, J. C., Woodbury, N. W., and Breton, J., 1993, Fourier transform infrared study of the primary electron donor in chromatophores of *Rhodobacter sphaeroides* with reaction centers genetically modified at residues M160 and L131. *Biochemistry*, **32**:13879–13885.

Nagarajan, V., Parson, W. W., Gaul, D., and Schenck, C., 1990, Effect of specific mutations of tyrosine-(M)210 on the primary photosynthetic electron-transfer process in *Rhodobacter sphaeroides*. *Proc. Natl. Acad. Sci. USA*, **87**:7888–7892.

Nagarajan, V., Parson, W. W., Davis, D., and Schenck, C. C., 1993, Kinetic and free energy gaps of electron-transfer reactions in *Rhodobacter sphaeroides* reaction centers. *Biochemistry*, **32**:12324–12336.

Ogrodnik, A., Keupp, W., Volk, M., Auermeier, G., and Michel-Beyerle, M. E., 1994, Inhomogeneity of radical pair energies in photosynthetic reaction centers revelaed by differences in recombination dynamics of $P^+H_A^-$ when detected in delayed emission and in absorption. *J. Phys. Chem.*, **98**:3432–3439.

Okamura, M., and Feher, G., 1992, Proton transfer in reaction centers from photosynthetic bacteria. *Ann. Rev. Biochem.*, **61**:861–896.

Okamura, M. Y., and Feher, G., 1995, Proton-coupled electron transfer reactions of $Q_B$ in reaction centers from purple bacteria. In *Anoxygenic Photosynthetic Bacteria*, (R. E. Blankenship, M. T. Madigan, and C. E. Bauer, eds.) pp. 577–594, Kluwer Academic Publishers, Dordrecht, The Netherlands.

Parson, W. W., 1996, Photosynthetic bacterial reaction centers. In *Protein Electron Transfer* (D. S. Bendall, ed.) pp. 125–160, BIOS Scientific Publishers Ltd. Oxford, U.K.

Parson, W. W., Chu, Z. T., and Warshel, A., 1990, Electrostatic control of charge separation in bacterial photosynthesis. *Biochim. Biophys. Acta*, **1017**:251–272.

Peloquin, J. M., Williams, J. C., Lin, X., Alden, R. G., Taguchi, A. K. W., Allen, J. P., and Woodbury, N. W., 1994, Time-dependent thermodynamics during early electron transfer in reaction centers from *Rhodobacter sphaeroides*. *Biochemistry*, **33**:8089–8100.

Plato, M., Möbius, K., Michel Beyerle, M. E., Bixon, M., and Jortner, J., 1988, Intermolecular electronic interactions in the primary charge separation in bacterial photosynthesis. *J. Am. Chem. Soc.*, **110**:7279–7285.

Prince, R. C., Cogdell, R. J., and Crofts, A. R., 1974, The photo-oxidation of horse heart cytochrome *c* and native cytochrome $c_2$ by reaction centers from *Rhodopseudomonas spheroides* $R_{26}$. *Biochim. Biophys. Acta*, **347**:1–13.

Ridge, J. P., van Brederode, M. E., Goodwin, M. G., van Grondelle, R., and Jones, M. R., 1999, Mutations that modify or exclude binding of the $Q_A$ ubiquinone and carotenoid in the reaction center from *Rhodobacter sphaeroides*. *Photosynth. Res.*, **59**:9–26.

Rischel, C., Spiedel, D., Ridge, J. P., Jones, M. R., Breton, J., Lambry, J. C., Martin, J. L., and Vos, M. H., 1998, Low-frequency vibrational modes in proteins: large frequency-shifts induced by point-mutations in the protein-cofactor matrix of bacterial reaction centers. *Proc. Natl. Acad. Sci. USA*, **95**:12306–12311.

Robert, B., Lutz, M., and Tiede, D. M., 1985, Selective photochemical reductin of either of the two bacteriopheophytins in reaction centres of *Rps. sphaeroides* R-26. *FEBS Letts.*, **183**:326–300.

Rockley, M. G., Windsor, M. W., Cogdell, R. J., and Parson, W. W., 1975, Picosecond detection of an intermediate in the photochemical reaction of bacterial photosynthesis. *Proc. Natl. Acad. Sci. USA*, **72**:2251–2255.

Scherer, P. O. J., and Fischer, S. F., 1989, Long-range electron-transfer within the hexamer of the photosynthetic reaction center *Rhodopseudomonas viridis*. *J. Phys. Chem.*, **93**:1633–1637.

Schmidt, S., Arlt, T., Hamm, P., Lauterwasser, C., Finkele, U., Drews, G., and Zinth, W., 1993, Time-resolved spectroscopy of the primary photosynthetic processes of membrane-bound reaction centers from an antenna-deficient mutant of *Rhodobacter capsulatus*. *Biochim. Biophys. Acta*, **1144**:385–390.

Schmidt, S., Arlt, T., Hamm, P., Huber, H., Nagele, T., Wachtveitl, J., Meyer, M., Scheer, H., and Zinth, W., 1994, Energetics of the primary electron-transfer reaction revealed by ultrafast spectroscopy on modified bacterial reaction centers. *Chem. Phys. Lett.*, **223**:116–120.

Shochat, S., Arlt, T., Francke, C., Gast, P., Van Noort, P. I., Otte, S. C. M., Schelvis, H. P. M., Schmidt, S., Vijgenboom, E., Vrieze, J., Zinth, W., and Hoff, A. J., 1994, Spectroscopic characterization of reaction centers of the (M)Y210W mutant of the photosynthetic bacterium *Rhodobacter sphaeroides*. *Photosynth. Res.*, **40**:55–66.

Shuvalov, V. A., 1993, Time and frequency domain study of different electron transfer processes in bacterial reaction centers. In *The Photosynthetic Reaction Center*, (J. Deisenhofer and J. R. Norris, eds.) Volume 2, 89–103, Academic Press, San Diego, USA.

Shuvalov, V. A., and Klimov, V. V., 1976, The primary photoreactions in the complex cytochrome P-890. P-760 (bacteriopheophytin$_{760}$) of *Chromatium minutissimum* at low redox potentials. *Biochim. Biophys. Acta*, **440**:587–599.

Sinning, I., 1992, Herbicide binding in the bacterial photosynthetic reaction center. *TIBS*, **17**:150–154.

Sinning, I., Koepke, J., Schiller, B., and Michel, H., 1990, 1st glance on the 3-dimensional structure of the photosynthetic reaction center from a herbicide-resistant *Rhodopseudomonas viridis* mutant. *Zeitschrift fur Naturforschung C*, **45**:455–458.

Small, G. J., Hayes, J. M. amd Silbey, R. J., 1992, The question of dispersive kinetics for the initial phase of charge separation in bacterial reaction centers. *J. Phys. Chem.*, **96**:7499–7501.

Spörlein, S., Zinth, W., and Wachtveitl, J., 1998, Vibrational coherence in photosynthetic reaction centers observed in the bacteriochlorophyll anion band. *J. Phys. Chem.*, **102**:7492–7496.

Stanley, R. J., and Boxer, S. G., 1995, Oscillations in the spontaneous fluorescence from photosynthetic reaction centers. *J. Phys. Chem.*, **99**:859–863.

Steffen, M. A., Lao, K., and Boxer, S. G., 1994, Dielectric asymmetry in the photosynthetic reaction center. *Science*, **264**:810–816.

Stowell, M. H. B., McPhillips, T. M., Rees, D. C., Soltis, S. M., Abresch, E., and Feher, G., 1997, Light-induced structural changes in photosynthetic reaction center: Implications for mechanism of electron-proton transfer. *Science*, **276**:812–816.

Streltsov, A. M., Yakovlev, A. G., Shkuropatov, A. Y., and Shuvalov, V. A., 1996, Femtosecond kinetics of electron transfer in the bacteriochlorophyll-M-modified reaction centers from *Rhodobacter sphaeroides* (R-26). *FEBS Letts.*, **383**:129–132.

Streltsov, A. M., Vulto, S. I. E., Shkuropatov, A. Y., Hoff, A. J., Aartsma, T. J., and Shuvalov, V. A., 1998, $B_A$ and $B_B$ absorbance perturbations induced by coherent nuclear motions in reaction centers from *Rhodobacter sphaeroides* upon 30-fs excitation of the primary donor. *J. Phys. Chem. B*, **102**:7293–7298.

Taguchi, A. K. W., Eastman, J. E., Gallo, D. M.Jr., Sheagley, E., Xiao, W., and Woodbury, N. W., 1996, Asymmetry requirements in the photosynthetic reaction center of *Rhodobacter capsulatus*. *Biochemistry*, **35**:3175–3186.

Takahashi, E., and Wraight, C. A., 1994, Molecular genetic manipulation and characterization of mutant photosynthetic reaction centers from purple non-sulphur bacteria. In *Advances in Molecular and Cell Biology: Molecular Processes in Photosynthesis*, (J. Barber, ed.) 197–251, JAI Press, Greenwich.

Tiede, D. M., Kellogg, E., and Breton, J., 1987, Conformational changes following reduction of the bacteriopheophytin electron acceptor in reaction centers of *Rhodopseudomonas viridis*. *Biochim. Biophys. Acta*, **892**:294–302.

Tiede, D. M., Prince, R. C., and Dutton, P. L., 1976, EPR and optical spectroscopic properties of the electron carrier intermediate between the reaction center bacteriochlorophylls and the primary acceptor in *Chromatium vinosum*. *Biochim. Biophys. Acta*, **449**:447–467.

Timpmann, K., Ellervee, E., Laisaar, A., Jones, M. R., and Freiberg, A., 1997, High pressure-induced acceleration of primary photochemistry in membrane-bound wild type and mutant bacterial reaction centres in *Ultrafast Processes in Spectroscopy* (R. Karli, P. Freiberg, and P. Saari, eds.) pp. 236–247.

Van Brederode, M. E., and Van Grondelle, R., 1999, New and unexpected routes for ultrafast electron transfer in photosynthetic reaction centers. *FEBS Letts.*, **455**:1–7.

Van Brederode, M. E., Jones, M. R., and Van Grondelle, R., 1997a, Fluorescence excitation spectra of membrane bound photosynthetic reaction centers of *Rhodobacter sphaeroides* in which tyrosine M210 is replaced by tryptophan: evidence for a new pathway of charge separation. *Chem. Phys. Letts.*, **268**:143–149.

Van Brederode, M. E., Jones, M. R., Van Mourik, F., Van Stokkum, I. H. M., and Van Grondelle, R., 1997b, A new pathway for transmembrane electron transfer in photosynthetic reaction centers of *Rhodobacter sphaeroides* not involving the excited special pair. *Biochemistry*, **36**:6855–6861.

Van Brederode, M. E., Van Mourik, F., Van Stokkum, I. H. M., Jones, M. R., and Van Grondelle, R., 1999a, Multiple pathways for ultrafast transduction of light energy in the photosynthetic reaction center of *Rhodobacter sphaeroides*. *Proc. Natl. Acad. Sci. USA*, **96**:2054–2059.

Van Brederode, M. E., Van Stokkum, I. H. M., Katilius, E., Van Mourik, F., Jones, M. R., and Van Grondelle, R., 1999b, Primary charge separation routes in the Bchl:Bphe heterodimer reaction centres of *Rhodobacter sphaeroides*. *Biochemistry*, **38**:7545–7555.

Van Grondelle, R., Romjin, J. C., and Holmes, N. G., 1976, Photoreduction of the long wavelength bacteriopheophytin in reaction centers and chromatophores of the photosynthetic bacterium *Chromatium vinosum* strain D. *FEBS Letts.*, **72**:187–190.

Van Stokkum, I. H. M., Beekman, L. M. P., Jones, M. R., Van Brederode, M. E., and Van Grondelle, R., 1997, Primary electron transfer kinetics in membrane-bound *Rhodobacter sphaeroides* reaction centers: A global and target analysis. *Biochemistry*, **36**:11360–11368.

Visschers, R. W., Vulto, S. I. E., Beekman, L. M. P., Jones, M. R., van Grondelle, R., and Kraayenhof, R., 1999, Functional LH1 antenna complexes influence electron transfer in bacterial photosynthetic reaction centers. *Photosynth. Res.*, **59**:95–104.

Volk, M., Ogrodnik, A., and Michel-Beyerle, M. E., 1995, The recombination dynamics of the radical pair $P^+H^-$ in external magnetic and electric fields. In *Anoxygenic Photosynthetic Bacteria*, (R. E. Blankenship, M. T. Madigan, and C. E. Bauer, eds.) pp. 595–626, Kluwer Academic Publishers, Dordrecht, The Netherlands.

Vos, M. H., and Martin, J. L., 1999, Femtosecond processes in proteins. *Biochim. Biophys. Acta*, (in press).

Vos, M. H., Lambry, J. C., Robles, S. J., Youvan, D. C., Breton, J., and Martin, J. L., 1991, Direct observation of vibrational coherence in bacterial reaction centers using femtosecond absorption spectroscopy. *Proc. Natl. Acad. Sci. USA*, **88**:8885–8889.

Vos, M. H., Lambry, J. C., Robles, S. J., Youvan, D. C., Breton, J., and Martin, J. L., 1992, Femtosecond spectral evolution of the excited state of bacterial reaction centers at 10 K. *Proc. Natl. Acad. Sci. USA*, **89**:613–617.

Vos, M. H., Rappaport, F., Lambry, J. C., Breton, J., and Martin, J. L., 1993, Visualization of coherent nuclear motion in a membrane protein by femtosecond spectroscopy. *Nature*, **363**:320–325.

Vos, M. H., Jones, M. R., Hunter, C. N., Breton, J., Lambry, J. C., and Martin, J. L., 1994a, Coherent dynamics during the primary electron-transfer reaction in membrane-bound reaction centers of *Rhodobacter sphaeroides*. *Biochemistry*, **33**:6750–6757.

Vos, M. H., Jones, M. R., Hunter, C. N., Breton, J., and Martin, J. L., 1994b, Coherent nuclear dynamics at room temperature in bacterial reaction centers. *Proc. Natl. Acad. Sci. USA*, **91**:12701–12705.

Vos, M. H., Jones, M. R., McGlynn, P., Hunter, C. N., Breton, J., and Martin, J. L., 1994c, Influence of the membrane environment on vibrational motions in reaction centres of *Rhodobacter sphaeroides*. *Biochim. Biophys. Acta*, **1186**:117–122.

Vos, M. H., Jones, M. R., Breton, J., Lambry, J. C., and Martin, J. L., 1996, Vibrational dephasing of long- and short-lived primary donor excited states in mutant reaction centers of *Rhodobacter sphaeroides*. *Biochemistry*, **35**:2687–2692.

Vos, M. H., Breton, J., and Martin, J. L., 1997, Electronic energy transfer within the hexamer cofactor system of bacterial reaction centers. *J. Phys. Chem.*, **101**:9820–9832.

Vos, M. H., Jones, M. R., and Martin, J. L., 1998, Vibrational coherence in bacterial reaction centers: spectroscopic characterisation of motions active during primary electron transfer. *Chem. Phys.*, **233**:179–190.

Wang, S., Li, X., Williams, J. C., Allen, J. P., and Mathis, P., 1994, Interaction between cytochrome $c_2$ and reaction centers from purple bacteria. *Biochemistry*, **33**:8306–8312.

Warshel, A., Creighton, S., and Parson, W. W., 1988, Electron-transfer pathways in the primary event of bacterial photosynthesis. *J. Phys. Chem.*, **92**:2696–2701.

Warshel, A., Chu, Z. T., and Parson, W. W., 1989, Dispersed polaron simulations of electron transfer in photosynthetic reaction centers. *Science*, **246**:112–116.

Warshel, A., Chu, Z. T., and Parson, W. W., 1994, On the energetics of the primary electron-transfer process in bacterial reaction centers. *J. Photochem. Photobiol.*, **82**:123–128.

Williams, J. C., Alden, R. G., Murchison, H. A., Peloquin, J. M., Woodbury, N. W., and Allen, J. P., 1992, Effects of mutations near the bacteriochlorophylls in reaction centers from *Rhodobacter sphaeroides*. *Biochemistry*, **31**:11029–11037.

Woodbury, N. W. T., and Parson, W. W., 1984, Nanosecond fluorescence from isolated photosynthetic reaction centers of *Rhodopseudomonas sphaeroides*. *Biochim. Biophys. Acta*, **767**:345–361.

Woodbury, N. W., and Allen, J. P., 1995, The pathway, kinetics and thermodynamics of electron transfer in wild type and mutant bacterial reaction centers of purple nonsulfur bacteria. In *Anoxygenic Photosynthetic Bacteria*, (R. E. Blankenship, M. T. Madigan, and C. E. Bauer, eds.) pp. 527–557, Kluwer Academic Publishers, Dordrecht, The Netherlands.

Woodbury, N. W., Becker, M., Middendorf, D., and Parson, W. W., 1985, Picosecond kinetics of the initial photochemical electron-transfer reaction in bacterial photosynthetic reaction centers. *Biochemistry*, **24**:7516–7521.

Woodbury, N. W., Peloquin, J. M., Alden, R. G., Lin, X., Lin, S., Taguchi, A. K. W., Williams, J. C., and Allen, J. P., 1994, Relationship between thermodynamics and mechanism during photoinduced charge separation in reaction centers from *Rhodobacter sphaeroides*. *Biochemistry*, **33**:8101–8112.

Woodbury, N. W., Lin, S., LIN, X. M., Peloquin, J. M., Taguchi, A. K. W., Williams, J. C., and Allen, J. P., 1995, The role of reaction-center excited-state evolution during charge separation in a *Rhodobacter sphaeroides* mutant with an initial electron-donor midpoint potential 260 mV above wild-type. *Chem. Phys.*, **197**:405–421.

Wraight, C. A., and Clayton, R. K., 1973, The absolute quantum efficiency of bacteriochlorophyll photooxidation in reaction centres of *Rhodopseudomonas spheroides*. *Biochim. Biophys. Acta*, **333**:246–260.

Xia, D., Yu, C.-A., Kim, H., Xia, J.-Z., Kachurin, A. M., Zhang, L., Yu, L., and Deisenhofer, J., 1997, Crystal structure of the cytochrom $bc_1$ complex from bovine heart mitochondria. *Science*, **277**:60–66.

Yeates, T. O., Komiya, H., Rees, D. C., Allen, J. P., and Feher, G., 1987, Structure of the reaction center from *Rhodobacter sphaeroides* R-26: Membrane-protein interactions. *Proc. Natl. Acad. Sci. USA*, **84**:6438–6442.

Yeates, T. O., Komiya, H., Chirino, A., Rees, D. C., Allen, J. P., and Feher, G., 1988, Structure of the reaction center from *Rhodobacter sphaeroides* R-26 and 2.4.1: Protein-cofactor (bacteriochlorophyll, bacteriopheophytin and carotenoid) interactions. *Proc. Natl. Acad. Sci. USA*, **85**:7993–7997.

Zhang, Z., Huang, L., Shulmeister, V. M., Chi, Y.-I., Kim, K. K., Hung, L.-W. , Crofts, A. R., Berry, E. A., and Kim, S.-H., 1998, Electron transfer by domain movement in cytochrom $bc_1$. *Nature*, **392**:677–684.

Zinth, W., and Kaiser, W., 1993, Time-resolved spectroscopy of the primary electron transfer in reaction centers of *Rhodobacter sphaeroides* and *Rhodopseudomonas viridis*. In *The Photosynthetic Reaction Center*, (J. Deisenhofer and J. R. Norris, eds.) Volume 2, 71–88, Academic Press, San Diego, USA.

Zinth, W., Huppmann, P., Arlt, T., and Wachtveitl, J., 1998, Ultrafast spectroscopy of the electron transfer in photosynthetic reaction centres: towards a better understanding of electron transfer in biological systems *Phil. Trans. Roy. Soc. Lond. A*, **356**:465–476.

# Index

*Acetobacterium dehalogenans*, 362, 363
Acetyl-CoA, 496
Acetyl-CoA synthase, 487–488, 489, 497–499;
    *see also* CO dehydrogenases
Acetylene hydratase, 451
Acyl-CoA mutases, 389; *see also #specific*
    *mutases*
Adenosylcobalamin-dependent enzymes, 351–
    397
    carbon-centered radicals in enzyme
        reactions, 359–360
    enzymes, 355–357
    longstanding questions, 360–361
    mechanistic aspects of catalysis, 375–394
        EPR studies, 375–377
        homolysis of adenosylcobalamin and
            formation of substrate radicals, 375–
            386
        magnetic field effects on, 381–382, 383
        resonance Raman experiments, 382, 384,
            385
        role of axial ligand in, 384, 385
        stopped-flow studies, 377–381
    methylcobalamin-dependent enzymes, 354–355
    outline mechanism of rearrangements, 357–
        358
    perspective, 394–397
    rearrangements of substrate radicals, 386–390
        mechanistic studies in methylmalonyl
            CoA reductase, 390–391, 392
        mechanistic studies on glutamate mutase,
            391–394

Adenosylcobalamin-dependent enzymes (*cont.*)
    ribonucleotide reductase, 358–359
    structural features, 361–374
        cobalamin binding by enzymes containing
            D-x-H-x-x-G motif, 362–369
        diol dehydrase, 371, 373, 374
        substrate binding and initiation of
            catalysis, 369–371, 372
        structure and reactivity of cobalamins, 352–354
*S*-Adenosylmethionine, 360, 407
Adhesion protein, human vascular, 221
Adiabatic reactions, 132, 639–640
ADP, *see* Nucleoside cofactors
Alcohols, PQQ adducts, 86, 87
Aldehyde oxidoreductase, 448, 453–455
Algae, ascorbate peroxidase, 320–321, 323
Amicyanin, 120, 125–127, 148
    MADH-amicyanin-cytochrome *c*-551i
        complex
        structure, 128; 128–131
        amicyanin-cytochrome *c*-511i interactions,
            128–129
        pathways analysis of electron transfer
            protein complex, 129–131
    MADH interactions, 126–127
    physical properties, 125–126
Amiloride-binding protein, 220
Amine dehydrogenases, 74
Amine oxidases, 147, 197–221; *see also*
    Galactose oxidase
    biogenesis of TPQ and related cofactors,
        217–218

Amine oxidases (*cont.*)
  biological roles, 219–221
    mammals, 220–221
    microorganisms, 219
    plants, 219–220
  catalytic mechanism, oxidative half cycle,
    214–217, 213–214
  catalytic mechanism, reductive half cycle,
    208–216
    product Schiff-base, 211–213
    reduced forms of enzyme, 213–214
    substrate binding/substrate Schiff-base,
      209–211
  comparisons with galactose oxidase, 221–
    222
  model studies, 219
  quinoprotein prosthetic groups, 74
  structure, 199–208
    active site structure, 205–207
    copper site, 205
    general properties, 199
    protein, 199–205
    substrate access channels to active site,
      207
    TPQ site, 205–207
Amines
  dehydrogenases, *see* Methylamine
    dehydrogenase; Trimethylamine
    dehydrogenase
  and methanol dehydrogenase, 78–80, 86, 87,
    94
Amino acids
  methyl group oxidation, 147, 148
  radical electron transfer, 8, 17; *see also*
    Tyrosyl radical
Amino acid sequences
  amine oxidases, 203
  cobalamin-binding domains, 363
  copper oxidases, 200–202
  methanol dehydrogenase, 98–99
Amino-mutases, 355, 356, 387–388; *see also*
  #*specific enzymes*
Aminoquinol form of TPQ, structure, 198
Aminoquinone, 134
Aminotransferase mechanism, amine oxidase,
  208
Ammonia
  methanol dehydrogenase
    activation, 73, 78–80
    reductive half-reaction, 94

Ammonia (*cont.*)
  methylamine dehydrogenase and, *see*
    Methylamine dehydrogenase
  PQQ adducts, 86, 87
Anaerobes, *see* CO dehydrogenases
Anthraquinones, 8
Antimycin, 561, 566
#*Arabidopsis*, 186, 197, 449
  amine oxidase, 220
  nitrate reductase, 467, 468
Arachidonate, P450 BM3, 308
#*Arthrobacter*, 148
#*Arthrobacter* amine oxidases, 200–202, 219
#*Arthrobacter globiformis* amine oxidase, 200–
  202, 203, 206, 208, 217
Ascorbate peroxidase, 318–342
  catalytic mechanisms, 329–332
    pre-steady-state kinetics, 331–332
    steady-state kinetics, 329–331
  cDNA sequences and bacterial expression of
    recombinant APXs, 319–320
  general properties, 321
  inactivation, 339–341
    ascorbate-depleted media, 340–341
    cyanide, 339
    hydrogen peroxide, 341
    salicylate, 341
    sulfhydryl reagents, 339–340
  isolation and characterization of, 320–321
  radical chemistry, 332–335
    fate of monodehydroascorbate radical,
      332–333
    nature of intermediates, 333–335
  redox properties, 337–339
  spectroscopic and spin-state considerations,
    328–332
  structural studies, 321–328
  substrate recognition, 335–337
#*Aspergillus*, 193
Azurin, 31, 138

#*Bacillus megaterium*, 38–42, 303–304
#*Bacillus subtilis* P450, 299
Bacteria, *see also* Flavin proteins and
    cofactors; Purple bacteria reaction
    centers; #*specific organisms*
  adenosylcobalamin-dependent enzymes, 355,
    356, 357
  amine oxidases, 200–202
  cytochrome *c* oxidase, 610

Bacteria (*cont.*)
    cytochrome P450 systems, 301
    flavocytochome P450 BM3 model systems,
        302–304
    molybdenum enzymes, 448
    nitrite reductase, 522–536
    oxotransferases, 472–477
    P450s, 299–300
    quinoprotein prosthetic groups, 74
Bacteriochlorophylls, 14–17, 624; *see also*
        Purple bacteria reaction centers
Bacteriophage T4 ribonucleotide reductase,
        409, 432–433, 434–435
Barium, 86, 88, 89, 93
Benzyl viologen, 495
β (factor), MADH complexes, 131–132
Blue copper protein, 120; *see also* Amicyanin
BMP, *see* Cytochrome P450 domain of P450
        BM3; Flavocytochrome P450 BM3
Boltzmann constant, 55, 131
Bovine aorta amine oxidase, 212
Bovine eye ascorbate peroxidase, 321, 323
Bovine heart cytochrome $bc_1$ complex, *see*
        Cytochrome $bc_1$ complex
Bovine heart cytochrome *c* oxidase, *see*
        Cytochrome *c* oxidase, bovine heart
Bovine milk xanthine oxidase, 463
Bovine serum amine oxidase, 200–202, 211,
        212, 214–215

#*Caenorhabditis elegans*, 363
Calcium, methanol dehydrogenase, 82–83, 84, 86,
        88, 89, 90, 92, 93, 107, 108–109, 110
Calvin-Benson-Bassham cycle, 491
Camphor hydroxylase, 302–303
Cancer, lysyl oxidase roles, 221
#*Candida boidinii*, 208
Carbanion mechanism, flavocytochrome $b_2$,
        282–285
Carbon-hydrogen bonds
    methane monooxygenase, 236, 261–270
    TMADH, 176
    vibrationally enhanced ground state
        tunneling, 164
Carbonium ion, 505–506
Carbon monoxide dehydrogenase, *see* CO
        dehydrogenases
Carbon source, carbon monoxide as, 488
Cardiac cytochrome *c* oxidase, *see* Cytochrome
        *c* oxidase, bovine heart

cDNA sequences, recombinant ascorbate
        peroxidases, 319–320
Chains, robust electron transfer protein design,
        14–17
Channels, molecular, 207, 488
Chick pea seedling AO, 212
*Chlorella vulgaris*, 321, 323
Chlorophylls, 8
Chloroplasts, *see* Nitrite reductase
*Chromatium vinosum* hydrogenase, 511
Cinnamyl alcohol, 109
Clock, radical, 264–266
*Clostridium* cobalamin-binding domains, 363,
        366–368
*Clostridium pasteurianum* hydrogenase, 501,
        504–505, 506, 508–509
*Clostridium subterminale*, 360
*Clostridium thermoaceticum*, 446
*Clostridium thermoaceticum* CO
        dehydrogenase, 488, 489, 493, 494,
        497
Clotimazole, 300
Cobalamins, 386–390, 407; *see also*
        Adenosylcobalamin-dependent
        enzymes
Cobalt-carbon bond, 396
Cobinamides, 384
CO dehydrogenases (acetyl-CoA synthase),
        445, 453, 477
    CO oxidation and $CO_2$ reduction by, 491,
        492, 491, 492, 493–499
    acetyl CoA synthase, catalytic Ni-Fe-S
        cluster, 497–499
    carbon entry and exit, 496
    catalytic redox machine, Ni-Fe-S cluster,
        493–495
    coupling $CO_2$ reduction and acetyl CoA
        synthesis, 496–497
    intermolecular wire, electron entry and
        exit, 495–496
    intramolecular wire, 495
    corrinoid/FeS protein, 362
    importance of CO and $H_2$ metabolism for
        anerobes, 487–491
        CO, 487–489
        $H_2$, 489–491
    mechanistic aspects, 458–459, 464
    molybdenum enzymes, 451
    structure, 456–457
Coenzyme M, 362

Collagen, 221
Copper enzymes
  amine oxidases, 183, 205, 212; see also
    Amine oxidases
  cytochrome c oxidase, 584, 585
  galactose oxidases, 183; see also Galactose
    oxidase
  MADH electron transfer
    to heme, 138
    from TTQW to, 134–138
  nitrite reductase, 520, 536–537
Copper protein, blue, 120; see also Amicyanin
Corrin ring, 352, 353
Corrins, 496; see also Adenosylcobalamin-
    dependent enzymes
Corynebacterium ammoniagenes ribonucleotide
    reductase, 432
CPR, see Cytochrome P450 reductase
p-Cresol, 31, 330–331
p-Cresol methylhydroxylase (PCMH), 330–331
  background and structural properties, 45–47
  coupling and transfer rates, 58
  domain interactions, 67
  donors, acceptors, flavin domains, 31
  electronic coupling, 61–63
  electron transfer rates, 68
  features of electron transfer, 67
  flavin protein donors and acceptors, 32
  structural and catalytic properties, 57
  structure, 35
Crystal structures, 8–9
Cubylmethanols, 265
Cyancobalamin, see Adenosylcobalamin-
    dependent enzymes
Cyanide
  ascorbate peroxidase inhibition, 339
  methanol dehydrogenase, 77, 81
  molybdenum enzymes, 448
  PQQ adducts, 86, 87
Cyanobacteria ascorbate peroxidase, 321
Cyclic electron transfer, photosynthetic
    reaction centers in purple bacteria, 634
Cyclopropanol, 88
Cysteine, 17
6-S-Cysteinyl FMN, TMADH, 150–151, 154–
    155, 165, 173–175
Cytochrome a and $a_3$, 280
Cytochrome b
  cytochrome $bc_1$ complex, 549–550
  DMSO reductase and, 452

Cytochrome $b_2$, 31; see also Flavocytochrome
    $b_2$; Flavocytochome P450 BM3
Cytochrome $b_5$, 31, 32
Cytochrome b-150 phenomenon, 561, 573
Cytochrome $bc_1$ complex, 541–573
  crystal structure, 543, 546–560
    alternative conformations of Rieske
      protein and cross-transfer of
      electrons, 553–557
    core proteins and subunit 9, 557–559
    cytochrome b and two hemes, 549–550
    cytochrome $c_1$, 550–553
    homodimer, 547–549
    subunits without prosthetic groups, 559–
      560
  escapement mechanism, 22
  homologues, 543
  overall reaction, 544
  purple bacteria reaction centers, 630
  quinone reactions and electron transfer, 541,
    542, 561–572
    bifurcated reaction at $Q_o$ site, 571–572
    domain shuttle mechanism, 545–546, 569,
      571
    reduction at $Q_i$ site, 561–566
    reduction at $Q_o$ site, 566–569, 570
  subunit composition, 543
Cytochrome c, see also Flavocytochrome c
  flavin protein donors and acceptors, 31, 32
  flavocytochrome $b_2$ and, 280, 281, 282, 287–
    290
  MADH-amicyanin-cytochrome c, 120
Cytochrome c2, redox cofactor chains, 14–17
Cytochrome c4, 31, 47
Cytochrome c-550, 526
Cytochrome c-551i, MADH-amicyanin
    complex, 128–131
Cytochrome c-552, 31
Cytochrome $cd_1$ nitrite reductase, see Nitrite
    reductase
Cytochrome $c_L$
  methanol dehydrogenase, 77–78, 90, 94,
    95–97
    kinetics, in vivo rate, 80
    processing and assembly, 111
    reaction mechanism, 95–97
Cytochrome c oxidase, bovine heart, 581–616
  composition, 582–589
    metal content, 584–585
    purification, 582–584

Cytochrome *c* oxidase, bovine heart (*cont.*)
  composition (*cont.*)
    structures and spectral properties of
      redox-active metal site, 585–589
    subunit composition and amino acid
      sequences, 589
  crystallization, 596–599
    enzyme, 597–599
    membrane proteins, 597
  function, 590–596
    enzymic activity, 590–592
    reduction of $O_2$, 592–596
  $O_2$ reduction mechanism, 607–608
  proton transfer in, 608–616
    possible pathways in membrane proteins,
      608–611
    redox coupled conformational change,
      611–616
  x-ray structure, 599–607
    structures and locations of metal sites,
      603–607
    three-dimensional, 599–603
Cytochrome oxidase, CO and, 488
Cytochrome P450, 23; *see also*
      Flavocytochrome P450 BM3
  domain interactions, 67
  flavin protein donors and acceptors, 31, 32
  structure, 36, 37
Cytochrome P450 domain of P450 BM3/FMN
  coupling and transfer rates, 58, 68
  domain interactions, 66, 67
  donors, acceptors, flavin domains, 31
  FMN electronic coupling, 59–61
  escapement mechanism, 23
  structural and catalytic properties, 57
Cytochrome P450 reductase (CPR), 31, 32
  donors, acceptors, flavin domains, 31
  electron transfer rates, 56
  expression systems, 300–301
  as flavin proteins, 32–38
  structural and catalytic properties, 56, 57
  structure, 36–38

*Dactylium dendoides*, 196
DBU (diazabicyclo-undecene), 92–93
*Dectylium dendroides*, 185, 186
Dehydroascorbate reductase, 332–333
Density functional theory, methane
      monooxygenase, 268–269
*Desulfomicrobium baculatum* hydrogenase, 500

*Desulfovibrio desulfuricans*
  hydrogenase, 504–505, 506, 509, 510
  molybdenum enzymes, 448, 472–473
*Desulfovibrio fructosovorans* hydrogenase,
      509–510
*Desulfovibrio gigas*, 446, 448, 453, 454, 458–
      459, 509
*Desulfovibrio vulgaris*, 500
Deuterium isotope effect, methane
      monooxygenase, 259–260, 263–264
Diamine oxidase, human kidney, 220
Diamond core structure, methane
      monooxygenase, 258–259
Diethylmethylamine, 158–159
Diferrous/ferric iron, *see* Iron centers,
      binuclear
Diflavin domain, P450 BM3, 304, 309
Diflavin reductase, 301
3,4-Dihydroxyphenylalanine, ribonucleotide
      reductase and, 431
Diiron center, *see* Iron centers, binuclear
Dimethylamine, 48
Dimethylamine dehydrogenase, 147, 150, 153
Dimethylamine monooxygenase, 146, 147
Dimethylbenzene, 67–68
Dimethylbenzimidazole ligand, 395
Dimethylbutylamine, 171
Dimethylsulfite, 470
Dimethylsulfoxide reductases, 447–448
Dinitrobenzenes, substitution for ubiquinones, 8
Diodehydrase, 376
Diol dehydrase, 355, 356, 362, 371, 373, 374,
      386, 387, 388, 395
Disulfide ring, methanol dehydrogenase active
      site, 105–107
Dithionite, 151
DMSO reductases, 447–448, 449, 452, 472–475
dNTPs (deoxyribonucleoside phosphates), 406
Domain shuttle mechanism, cytochrome $bc_1$
      complex, 545–546, 569, 571
*Drosophila*, 450, 457–458
D-x-H-x-x-G motif, 362–369, 395
Dye assays, methanol dehydrogenase, 75, 77–
      79

Elastin, lysyl oxidase roles, 221
Electronic coupling, 2
Electron nuclear double resonance (ENDOR)
      studies
  CO dehydrogenase, 493, 494

Electron nuclear double resonance studies
    (*cont.*)
  cytochrome $bc_1$ complex, 562
  methane monooxygenase, 242–243
  molybdenum enzymes, 454
  xanthine oxidase, 461, 462
Electron paramagnetic resonance (EPR)
    spectroscopy
  adenosylcobalamin-dependent enzymes,
    375–377, 395
  ascorbate peroxidase, 328–329
  cytochrome $c$ oxidase, 588
  flavocytochrome $b_2$, 285
  glutamate mutase, 391–394
  hydrogenase, 505
  methane monooxygenase, 242, 245, 250,
    252
  molybdenum enzymes, 454, 459, 460, 461,
    462
  ribonucleotide reductase, tyrosyl radical
    conformers, 431–433
  TMADH, 165
  xanthine oxidase, 459, 460, 461, 462
Electron Spin Echo Envelope Modulation
    (ESEEM), 208
Electron transfer theory, 1–24
  caution and hope, 23
  chains and robust electron transfer protein
    design, 14–17
  electron transfer clusters, 17–23
    control of electron transfer through
      escapement mechanisms, 21–23
    examples, 19–21
  electron tunneling theory, 2–9
    classical free energy and temperature
      dependence, 3–6
    generic protein tunneling rate expression,
      8–9
    quantum free energy and temperature
      dependence, 7–8
  protein structural heterogeneity, 9–14
    monitoring via packing density, 10–13
    natural tunneling distances, 13–14
  purple bacteria reaction centers, 635
Electron tunneling, *see* Tunneling
*Escherichia coli*
  amine oxidase, 199, 203–207, 213, 215, 216,
    217, 219
  cobalamin-binding domains, 363, 364
  flavocytochrome $b_2$ expression in, 283–284

*Escherichia coli* (*cont.*)
  fumarate reductase, 52
  molybdenum enzymes, 448, 449, 452
    cofactor biosynthesis, 450
    oxotransferases, 472
    nitrite reductase, 520
    P450 lacking in, 299–300
  ribonucleotide reductase, 360, 409, 411, 412,
    413–414, 415, 425
    metal site interactions, 428–429, 430
    radical transfer pathway, 415, 416, 417,
      418–420, 421, 422
    tyrosyl radical, 432, 433, 434
*Escherichia coli* expression systems
  DMSO reductase, 476
  sulfite oxidase, 470–471
ESEEM (Electron Spin Echo Envelope
    Modulation), 208
ETF
  flavin protein donors and acceptors, 31
  TMADH, 167–170, 176
Ethane, 266, 268
Ethanolamine ammonia-lyase (deaminase), 355,
    356, 361, 375, 377, 381–382, 383
*p*-Ethylphenoxyl radical, 432
*Euglena gracilis*, 321, 323

FAD, *see* Flavin proteins and cofactors;
    Nucleoside cofactors
Fatty acid hydroxylase P450, 302, 303–304
Fatty acid P450 BM3, 305–307; *see also*
    Flavocytochrome P450 BM3
$FCB_2$, *see* Flavocytochrome $b_2$
FCSD, *see* Flavocytochrome $c$ sulfide
    dehydrogenase
Fe, *see* Iron centers
Fermi's Golden Rule, 2, 3
Ferredoxin, 31, 32, 495, 519
Ferredoxin oxidoreductase, pyruvate, 495, 496
Ferredoxin reductase, 32, 34, 301, 489
Ferricenium ion, 168, 170
Ferrocytochrome $c$, 590
Ferryl-oxo species, methane monooxygenase,
    257, 258
Flavin proteins and cofactors, 8, 29–68; *see*
    *also specific enzymes*
  background and structural properties, 32–54,
    55
  *p*-cresol methylhydroxylase (PCMH), 45–
    47

Flavin proteins and cofactors (*cont.*)
  background and structural properties (*cont.*)
    flavocytochrome $b_2$ (FCB$_2$), 42–45
    flavocytochrome $c$ sulfide dehydrogenase
      (FCSD), 47–48
    flavocytochrome P450BM3, 38–42
    fumarate reductase (FUM), 52–54, 55
    NADPH-cytochrome P450 reductase
      (CPR), 32–38
    phthalate dioxygenase reductase (PDR),
      50–52
    trimethylamine dehydrogenase (TMADH),
      48–50
  cluster examples, 19–20
  conclusions, 65–68
    domain interactions, 65–67
    features of electron transfer, 67–68
  electron transfer, 55–65, 66
    BMP/FMN electronic coupling, 59–61
    FCB$_2$ electronic coupling, 61
    FCSD electronic coupling, 63–65
    FUM electronic coupling, 65, 66
    general aspects, 55–59
    PCMH electronic coupling, 61–63
  escapement mechanism, 23
Flavinylation of TMADH, 173, 174, 175
Flavocytochrome $b_2$ (FCB$_2$), 49, 279–292
  background and structural properties, 42–45
  coupling and transfer rates, 58
  cytochrome $c$ interaction, 286–290
  domain interactions, 67
  donors, acceptors, flavin domains, 31
  electronic coupling, 61
  engineering substrate specificity in, 290–292
  features of electron transfer, 67
  flavin reduction and substrate oxidation,
    282–295
  flavin to heme electron transfer, 285–286
Flavocytochrome $c$, 45; *see also* Cytochrome $c$
Flavocytochrome $c$ sulfide dehydrogenase
    (FCSD)
  background and structural properties, 47–48
  coupling and transfer rates, 58, 68
  domain interactions, 67
  donors, acceptors, flavin domains, 31
  electronic coupling, 63–65
  features of electron transfer, 67
  structural and catalytic properties, 57
Flavocytochrome P450 BM3, 297–312, 304
  background and structural properties, 38–42

Flavocytochrome P450 BM3 (*cont.*)
  bacterial model systems, 302–304
  electron transfer and its control, 307–309
  flavin protein donors and acceptors, 31, 32
  introduction, 297–302
  site-directed mutagenesis in substrate
      selectivity and electron transfer, 309–
      312
  structure and mechanisms, 304–307
Flavodoxin (FDX), 32, 33, 495
Fluorenones, 8
Fluorescence decay, 654
FMN, *see* Flavin proteins and cofactors
Formaldehyde, 76; *see also* Methylamine
    dehydrogenase
Formate, 407, 488, 492
Formate dehydrogenase, 447, 448, 449, 452,
    472, 473, 477
Formate reductase, 453
Formylmethanofuran dehydrogenase, 450, 453
14-3-3 protein, 468
Franck Condon (FC) principle, 655
Franck Condon (FC) term, 3, 7–8, 637, 638,
    639
Freeze-quench techniques, methane
    monooxygenase, 252
Fumarate reductase, 511
  background and structural properties, 52–54,
    55
  coupling and transfer rates, 58
  donors, acceptors, flavin domains, 31
  electronic coupling, 65, 66
  electron transfer rates, 56
  features of electron transfer, 67
  flavin protein donors and acceptors, 32
  structural and catalytic properties, 57
*Fusarium*, 186, 196

Galactose oxidase, 183–197; *see also* Amine
    oxidases
  biogenesis of thio-ether bond and other
      processing events, 192–193
  biological role, 196–197
  catalytic mechanism, 189–192
  model studies, 193–196
  structure, 185–189
    active site, 186–189
    general properties, 185
    primary, 185–186
    secondary and tertiary, 186

*Galdieria partita*, 320, 323
Gas constant, 131
Gated electron transfer, 137–138, 146, 250
Glucose oxidase, 221–223
Glutamate mutase, 355, 356, 363–364, 366–
    367, 368, 369, 371, 372, 376, 395
  axial ligand, 384, 386
  rearrangements of substrate radicals, 387,
    389–390, 391–394
Glutaredoxin, 407, 410, 411
Glutathione reductase, 31, 32, 35, 49
Glycerol dehydrase, 355, 356
Glycolate oxidase, 289
Glycyl radical, 407, 435–436
GREENPATH, 56, 59–65, 126
Guaiacol, 336–337

*Hansenula anomala*, 279, 280
*Hansenula anomala* flavocytochrome $b_2$, 286
*Hansenula polymorpha* amine oxidase, 200–
    202, 203, 213
Harmonic oscillator expression, quantum
    mechanical, 7
Heart cytochrome *c* oxidase, *see* Cytochrome *c*
    oxidase, bovine heart
Heme groups, *see also specific enzymes*
  ascorbate peroxidase, 318
  CO and, 488
  cytochrome $cd_1$ and, 520, 521, 522
  flavin proteins, *see* Flavin proteins and
    cofactors; *specific enzymes*
  flavocytochrome $b_2$, 43, 45, 280–281, 282,
    285–286
  MADH electron transfer from copper to
    heme, 138
  monooxygenases, *see* Methane
    monooxygenase
  P450 BM3 reduction potential comparisons,
    309
  photosynthetic reaction center cofactor
    chains, 14–17
  *Rp. viridis*, 14–17
Hemiketal adducts, PQQ, 86, 87, 89, 90, 92
Hemoglobin, CO and, 488
Hemopexin, 186
Herpes simplex virus ribonucleotide reductase,
    422
Hopfield expression, 7
Human kidney amine oxidase, 200–202, 220
Humans, CO as neurotransmitter, 488

Human vascular adhesion protein, 221
Hydride transfer, 19–20
  flavocytochrome $b_2$, 284–285
  methanol dehydrogenase, 89, 91
Hydrogen abstraction, methane
    monooxygenase and, 262
Hydrogenases, 491–492, 499–511
  catalytic redox machine, Ni-Fe-S and Fe-
    FeS clusters, 499–506
  electron tunneling, 17
  Fe-only, 504–505
  importance of CO and $H_2$ metabolism for
    anerobes, 487–491
    CO, 487–489
    $H_2$, 489–491
  intermolecular wires, electron entry and exit,
    510
  intramolecular wire, 508–510
  metal-free enzyme, 505–506
  models of, 505
  Ni-Fe, 499–504
  proton transfer pathway, 506–508
  tunnel diodes and catalytic bias, 510–511
Hydrogen atom transfer theory, ribonucleotide
    reductase, 423
Hydrogen cycling, 492
Hydrogen kinetic isotope effects, 19–20
Hydrogen peroxide, ascorbate peroxidase
    inhibition, 341
Hydrogen sensors, 491
Hydrogen sulfide, flavin proteins, 31
Hydropathy analysis, cytochrome $bc_1$ complex,
    549
Hydroquinone, trimethylamine dehydrogenase,
    49
4-Hydroxybutyrate, methanol dehydrogenase
    and, 76
Hydroxylases, 448, 449, 453–464; *see also*
    *specific enzymes*
  mechanistic aspects, 458–464
  structural studies, 453–458
6-Hydroxyphenylalanine, 74
Hydroxyurea, 433–434
*Hyphomicrobium X*, 146, 147, 153

Imidazoles, P450 inhibition, 300
Iminosemiquinone, 134
Insects, ascorbate peroxidase, 321
Intermediate radical state, 18
Interspecies $H_2$ transfer, 490–491

Intramolecular electron transfer, TMADH, 164–166
Iron centers
  binuclear
    methane monooxygenase, 238–241; *see also* Methane monooxygenase
    in radical electron transfer, 17
    ribonucleotide reductase, 407, 418–412
  cytochrome $c$ oxidase, 584, 585
  FeMoCo clusters, nitrogenase, 20–21
  Fe-only hydrogenase, 504–505
  hydrogenase, 491, 499–506
  methane monooxygenase, 238–241
  in radical electron transfer, 17
  ribonucleotide reductase, 418–420, 424–425
Iron-sulfur centers
  CO dehydrogenase, 493
  cytochrome $bc_1$, 22
  flavin protein donors and acceptors, 31
  fumarate reductase, 52–54, 55
  methane monooxygenase and, 248–249
  molybdenum enzymes, 450, 452, 458–459
    hydroxylases, 453, 455
    oxotransferase structural studies, 472–473
  nitrogenase, 21
  phthalate dioxygenase reductase, 50-52
  ribonucleotide reductases, 407, 435–436
  TMADH, 48–50, 66, 155–156, 157–158, 165, 166, 172; *see also* Trimethylamine dehydrogenases
Isoalloxazine ring, 29, 30
Isobutyryl-CoA mutase, 355, 356, 388–389
Isotope effects
  methane monooxygenase, 262–264
  ribonucleotide reductase, 380–381

Ketoconazoles, 300
Kidney amine oxidases, 200–202
*Klebsiella aerogenes* amine oxidase, 200–202
Krebs cycle, 279, 280; *see also* Flavocytochrome $b_2$

Lactate, 31, 279; *see also* Flavocytochrome $b_2$
Lactate dehydrogenase, 279
Lactate oxidase, 159
*Lactobacillus leichmannii* ribonucleotide reductase, 358–359, 408, 434
Lambda, *see* Reorganization energy
Landau-Zener parameter, 640
Lentil seedling amine oxidase, 200–202

L-Leucine 2,3-aminomutase, 355, 356
Light-activated reactions, *see* Purple bacteria reaction centers
Lipid metabolism, sulfite oxidase genetic defects, 466–467
Liver
  sulfite oxidase, 449, 465–467
  xanthine dehydrogenase, 457–458
  xanthine oxidase, 457–458, 463
Lupanine, 74
Lysine 2,3-aminomutase, 360, 387–388
Lysine-6-oxidase, 198
D-α-Lysine 5,6-aminomutase, 355, 356
Lysyl oxidase, 183–184, 198, 212
  quinoprotein prosthetic groups, 74, 75
  roles of, 221
Lysyl tyrosylquinone (LTQ), 74–75

MADH, *see* Methylamine dehydrogenase
Magnesium, 86, 88, 585
Magnetic interactions
  adenosylcobalamin-dependent enzymes, 381–382, 383
  hydroxylases, 455
  ribonucleotide reductase, 434
Malignant transformation, lysyl oxidase roles, 221
Mammals
  adenosylcobalamin-dependent enzymes, 357
  amine oxidases, 220–221
  ascorbate peroxidase, 321, 323
Mandelate dehydrogenase, 279
Marcus theory, 3–5, 6, 138–139, 644, 645
  MADH complexes, 133, 136, 137–138, 139
  methanol dehydrogenase, 78
  purple bacteria reaction centers, 635, 637, 638
MCD spectroscopy, *cytochrome $cd_1$*, 532
*M. crystallinum*, 320, 327, 337
*Megasphaera elsdenii*, 511
Menaquinone, 31, 52, 53
Metal cations, 8
  methanol dehydrogenase, 89, 92–93
  monooxygenases, *see* Methane monooxygenase
  PQQ and, 86, 88
Methane monooxygenase, 233–270
  carbon-hydrogen bond cleavage and oxygen insertion, 261–270
  concerted oxygen insertion mechanism, 267–268

Methane monooxygenase (*cont.*)
  carbon-hydrogen bond cleavage and oxygen
    insertion (*cont.*)
    isotope effects, 262–264
    mechanistic theory based on calculations,
      268–270
    radical clock substrates, 264–266
    radical rebound mechanism, 262
    radical rebound mechanism, modified,
      266–267
  component complexes, 245–246
  components, 237–245
    MMOB, 243–244
    MMOH, 237–243
    MMOR, 244–245
    spectroscopy, 241–243
    X-ray crystallography, 237–241
  electron transfer kinetics, 248–250
  oxidation-reduction potentials, 246–248
  reaction cycle intermediates, 252–261
    oxyten cleavage mechanism, 259–261
    structures, 256–259
    transient intermediates of MMOH, 252–
      256
  turnover systems, 250–252
*Methanobacterium thermoautotrophicum*, 449,
  451, 453
*Methanobacterium wolfei*, 450
Methanofuran, 453
Methanogens, *see* CO dehydrogenases
Methanol, coenzyme M methyltransferase, 362,
  363
Methanol dehydrogenase, 73–112, 186
  absorption spectra, 80–88, 84
  general enzymology, 75–77
  kinetics, 77–80
  processing and assembly of, 110–112
  pyrrolo-quinoline quinone (PQQ) prosthetic
    group, 84–88
  reaction mechanism, 88–97
    interaction with cytochrome $c_L$, 95–97
    methanol oxidase electron transport chain,
      94–95
    oxidation of reduced PQQ, 97
    oxidative half-reaction, 94–97
    reductive half-reaction, 88–93
    role of ammonia in reductive half-
      reaction, 94
  structure, 97–110
    active site, 105–110

Methanol dehydrogenase (*cont.*)
  structure (*cont.*)
    α subunit, 102–104
    β subunit, 105
*Methanosarcina barkeri*, 362, 363
Methanotrophs, monooxygenases, *see* Methane
  monooxygenase
Methionine synthase, 362, 363, 364, 366–367,
  395
Methyl alanine dehydrogenase, 147
Methylamine dehydrogenase, 95, 119–140, 147,
  380
  amicyanin, 125–127
  catalytic reaction mechanism, 121–123
  electron transfer from copper to heme, 138
  electron transfer reactions, 131–138
    coupled electron transfer, 133
    gated electron transfer, 137–138
    kinetic complexity of reactions, 132–134
    theory, 131–132
    true electron transfer, 132–133
  electron transfer from TTQ to copper, 134–
    138
    gated, 137–138
    mutation of amicyanin alters $H_{ab}$ for
      electron transfer from MADH, 124–
      137
    nonadiabatic, 134
  enzyme-amicyanin-cytochrome $c$-551i
    complex, 128–131
    amicyanin-cytochrom c-511i interactions,
      128–129
    pathways analysis of electron transfer
      protein complex, 129–131
    structure, 128
  Marcus theory, application to other protein
    transfer reactions, 138–139
  spectral and redox properties, 124–125
  structure and function, 120
  TTQ prosthetic group, 121
Methylamine oxidase, 208
Methylcobalamin, 352; *see also*
  Adenosylcobalamin-dependent
  enzymes
Methylene blue, 495
Methyleneglutarate mutase, 355, 356
2-Methyleneglutarate mutase, 362, 363, 376,
  388–389
2-Methyleneglutaryl radical, 394
N-Methyl glutamate dehydrogenase, 147

N-Methyl glutamate synthetase, 147
Methyl group oxidation pathways in
    methylotrophs, 147
N{5}-Methyl-H4-methanopterin, coenzyme M
    methyltransferase, 362
Methylmalonyl-CoA mutase, 357, 362, 364,
    365, 366, 367, 369–371, 376, 384,
    385, 388, 390–392, 395
Methylmalonyl-CoA reductase, 380, 390–391,
    392
*Methylobacterium extorquens*, 73, 76, 83, 97,
    108, 109, 120; *see also* Methylamine
    dehydrogenase
*Methylococcus capsulatus*, 234, 238–241
*Methylophaga marina*, 83, 84, 96
*Methylophilus methylotrophus*, 163; *see also*
    Trimethylamine dehydrogenase
*Methylophilus methylotrophus* W3A1, 48–50,
    97
*Methylophilus* W3A1, 108, 109
*Methylosinus trichosporium*, 234, 238, 239,
    245, 246
Methylotrophs, trimethylamine dehydrogenase,
    *see* Trimethylamine dehydrogenase
Methyl-tetrahydrofolate reductase, 363
Methyltransferases, 362, 363
Methyl viologen, 495, 644
Methymalonyl-CoA mutase, 355, 356, 363
Microsomal CPR structure, 36–38
Midpoint potentials, 3, 8–9, 55–56
Mitochondria, *see also* Cytochrome $bc_1$
        complex; Cytochrome $c$ oxidase,
        bovine heart
    cytochrome P450 systems, 301
    escapement mechanisms, 23
    flavocytochrome $b_2$, 42–45
MMOH Bath, 238–241
MOA ($\beta$-methoxyacrylate)-stilbene, 562, 572
Molecular channels
    acetyl-CoA production, 488
    amine oxidases, 207
Molecular graphics, 10
Molecular orbital calculations, PQQ, 86, 88
Molybdenum enzymes, 445–478
    hydrolylases, 453–464
        mechanistic aspects, 458–464
        structural studies, 453–458
    oxotransferases, eukaryotic, 465–472
        mechanistic aspects, 468–472
        structural studies, 465–468

Molybdenum enzymes (*cont.*)
    oxotransferases, prokaryotic, 472–477
        mechanistic aspects, 475–477
        structural studies, 472–475
    reactions catalyzed by, 445–446
    sequence homologies, 446–450
    structural and catalytic variations in families,
        451–453
    sulfite oxidase, 13
Monoamine oxidase, 176
Monodehydroascorbate radical, 332–333
Monooxygenases, *see also* Methane
        monooxygenase
    dimethylamine monooxygenase, 146, 147
    P450, 234
Monsanto process, 497
Mossbauer spectroscopy, 241, 245, 257, 505
Mouse ribonucleotide reductase, 417, 420, 421,
        422, 425, 430, 433
M-protein, 76–77
*Mycobacterium tuberculosis*, 300, 422, 432
Myxothiazol, 520, 521, 561, 566, 567

NADH dehydrogenase, 520, 521
NADH/NADPH cofactors, 30, 34, *see also*
        Flavin proteins and cofactors; *specific*
        *enzymes*
    catalytic turnover, 36, 37
    cluster examples, 19–20
    cytochrome *cd*1 and, 521
    cytochrome P450 systems, 67, 301
    flavin-binding domains, 34
    methane monooxygenase and, 248–249, 250
NAD/NAD{+} cluster examples, 19
Naphthoquinones, 8
*Neurospora crassa*, 545
Neurotransmitter, CO as, 488
Nickel enzymes, *see* CO dehydrogenases
Nickel-iron class of hydrogenases, 17, 491,
        499–504
Nickel-iron-selenium class of hydrogenases,
        491
Nickel-iron-sulfur clusters
    acetyl CoA synthase, 497–499
    hydrogenases, 499–506
*Nicotiana tabacum*, 450
Nitrate reductase, 447, 448, 449, 467–468
    mechanistic aspects, 468–472
    structural studies, 472–473
    sulfite oxidase and, 451–452

Nitrite reductase (cytochrome $cd_1$), 101–102,
　　519–538
　copper nitrite reductase, 536–537
　*Paracoccus pantotrophus*, 522–533
　　kinetic studies, 531–532
　　solution spectroscopy, 532–533
　　structure, 522–531
　*Pseudomonas aeruginosa*, 533–536
Nitrogenase, 445
　clusters, 20–21
　molybdenum enzymes, 451
　oxygen-insensitive, 477
Non-adiabatic electron transfer reactions, 2, 3,
　　134, 136, 138–139, 655
NrdH redoxin, 407, 408
Nuclear magnetic resonance studies, 244, 285
Nucleoside cofactors, *see also* Flavin proteins
　　and cofactors
　ADP-ribityl group, flavin-binding domains,
　　35
　ribonucleotide reductases, 407
　molybdenum enzymes, 450
　P450 BM3 reduction potential comparisons,
　　304, 309
　TMADH prosthetic groups, 149, 153

Old yellow enzyme, 49
*Oligotropha carboxidovorans*, 446, 451, 458–
　459
d-Ornithine 4,5-aminomutase, 355, 356
Oxotransferases, 448, 449, 450
　eukaryotic, 451–452, 465–472
　　mechanistic aspects, 468–472
　　structural studies, 465–468
　prokaryotic, 451, 472–477
　　mechanistic aspects, 475–477
　　structural studies, 472–475
Oxygen
　flavin protein donors and acceptors, 31
　methane monooxygenase
　　cleavage, 259–261
　　insertion, 267–268
　monooxygenases, *see* Methane
　　monooxygenase
　ribonucleotide reductases, 408, 424
Oxygen activation theory, methane
　　monooxygenase, 269

P450, *see* Cytochrome P450 domain of P450
　　BM3; Flavocytochrome P450 BM3

Packing density, protein structural
　　heterogeneity, 10–13
Palmitoleate-bound P450 BM3, 305–307
*Paracoccus denitrificans*, 47–48, 95, 120, 121,
　　126, 526, 531; *see also* Methylamine
　　dehydrogenase
*Paracoccus pantotrophus* cytochrome $cd1$
　　nitrite reductase, 522–533
PCMH, *see* p-Cresol methylhydroxylase
PDR, *see* Phthalate dioxygenase reductase
　　(PDR)
Pea seedling amine oxidase, 200–202, 206, 212
*Pelobacter acidigallici*, 451
*Penicillium crysogenum* amine oxidase, 219
*Penicillium simplicissimum*, 456
Pentanol, 109
Perturbation theory, 470
PES (phenazine ethosulfate), 90, 97
PFOR (pyruvate ferredoxin oxidoreductase),
　　495, 496
·pH
　amicyanin redox properties, 125–126
　ascorbate peroxidase spin state, 328
　methane monooxygenase oxygen cleavage
　　mechanism, 259–260
　methanol dehydrogenase, 78–79
　methylamine dehydrogenase complexes, 134,
　　137–138
　sulfite oxidase half-reaction, 471–472
　TMADH, 164–166, 176
　　with DEMA, 158–159
　　with TMA, native and mutant enzyme
　　　studies, 159–163
Phage T4 ribonucleotide reductase, 409, 432–
　　433, 434–435
*Phanerochetes chrysosporium*, 196
Phenazine ethosulfate (PES), 90, 97
Phenylhydrazine, 167
Phloroclucinol, 451
Phosphine, 489
Photochemistry
　nitrate reductase, 468
　PCMH, 47–48
Photosynthetic reaction centers, *see also* Purple
　　bacteria reaction centers
　redox cofactor chains, 14–17
　reorganization energies in, 6
Photosystem II
　tunneling, 3
　tyrosyl radical, 432

Phthalate dioxygenase reductase (PDR)
background and structural properties, 50–52
donors, acceptors, flavin domains, 31
electron transfer rates, 68
features of electron transfer, 67
flavin protein donors and acceptors, 31, 32
structural and catalytic properties, 56, 57
Pichia pastoris expression system, 467
Pi electrons, 449, 450, 604
Pig kidney AO, 212
Pig serum amine oxidase, 208–209, 212
Ping-pong mechanism, 88, 190–191, 208
Piracy, redox, 170
Placenta, amine oxidases, 221
Planck's constant, 55, 131
Plants, see also Nitrite reductase
amine oxidases, 200–202, 212, 219–220
ascorbate peroxidase, see Ascorbate
peroxidase
cytochrome c oxidase, 610
galactose oxidase, 186
molybdenum enzymes, 448, 450
quinoprotein prosthetic groups, 74, 75
Plasma amine oxidases, 212
Polyamine oxidases, 219, 220
Polyprous circinatus, 184
Polysulfide reductase, 452–453, 477
Porphyrins, 8
Potentiometry, MADH-amicyanin interactions, 126
PQQ, see Pyrrolo-quinoline quinone
Prokaryotes, see Bacteria; Flavin proteins and
cofactors; Oxotransferases,
prokaryotic; Purple bacteria reaction
centers; specific organisms
1,2-Propane diol, 76
Propionibacterium shermanii, 363, 366–367
Protein Data Bank (PDB), 11, 12, 13
Protium/deuterium/tritium kinetic isotope
effects, 19–20
Protocatechate 3,4-dioxygenase, 530–531
Proton abstraction, methanol dehydrogenase,
89, 90, 91
Proton translocation, photosynthetic reaction
centers in purple bacteria, 634
Prototropic control, TMADH, 164–166
Pseudoazurin, 526
Pseudomonads, p-Cresol methylhydroxylase,
45–47
Pseudomonas aeruginosa cytochrome $cd_1$
nitrite reductase, 533–536

Pseudomonas cepacia, 50-52
Pseudomonas putida, 531
Pseudomonas putida P450 cam camphor
hydroxylase, 302–303
Pseudomonas stutzeri, 47, 532
Pterins, 8
molybdenum enzymes, 448, 449, 450, 472–
473, 474–475
monooxygenases, see Methane
monooxygenase
sulfite oxidase, 466
Purple bacteria reaction centers, 621–665
biological electron transfer, 634–640
adiabatic, 639–640
nonadiabatic, 635–639
energy storage mechanism by reaction
center, 628–634
light-driven cyclic electron transfer
coupled to proton translocation, 634
primary photochemistry, 631–633
ubiquinol formation and re-reduction of
P{+}, 633–634
structure, 622–628
electronic, reaction center absorbance
spectrum, 627–628
mutant complexes, 626–627
recent information relating to function,
625–626
Rhodopseudomonas viridis and
Rhodobacter sphaeroides reaction
centers, 622–624
studies of ultrafast electron transfer in light-
activated protein, 640–650
dispersive kinetics, heterogeneity of protein
dynamics, 651–654
femtosecond biology, coherent nuclear
dynamics studied in populations of
proteins, 654–656
modulation of time constant for primary
electron transfer between 200 fs and
500 ps, 656–661
parallel pathways for primary electron
transfer, 661–665
primary electron transfer asymmetry, 643–
650
primary electron transfer asymmetry,
evidence, 643–644
primary electron transfer asymmetry, origins,
644–646
primary electron transfer rerouting, 646–650

Purple bacteria reaction centers (*cont.*)
  role of BA monomeric BChl, 641–643
  site-directed mutagenesis, hydrogen bond
    mutants, 658–661
  site-directed mutagenesis, tyrosine M210,
    657–658
  temperature dependence—activationless
    reactions, 650–651
Pyridine nucleotides, *see* Flavin proteins and
  cofactors; Nucleoside cofactors
Pyridoxal phosphate, amino mutase
  requirements, 387–388
*Pyrococcus furiosus*, 450
Pyrogallol transhydrolase, 451
Pyrrolo-quinoline quinone (PQQ)
  methanol dehydrogenase, 84–88; *see also*
    Methanol dehydrogenase
  active site, bonding, 107–109
  oxidation of, 97
  methanol dehydrogenase processing and
    assembly, 110, 111, 112
  quinoprotein prosthetic groups, 73
Pyruvate
  CO dehydrogenase and, 487–488, 489
  flavocytochrome $b_2$ and, 279; *see also*
    Flavocytochrome $b_2$
Pyruvate ferredoxin oxidoreductase, 495,
  496
Pyruvate ferredoxin reductase, 489
Pyruvate formate-lyase, 360

Q-cycle mechanism, cytochrome $bc_1$ complex,
  544, 546, 569
Quantum free energy, tunneling theory, 7–8
Quantum mechanical description, purple
  bacteria reaction centers, 638, 639
Quantum tunneling, *see* Tunneling
Quinones, 8, 49; *see also specific quinones*
  cytochrome $bc_1$ complex, 22, 541, 542, 544,
    546, 561–572
    bifurcated reaction at $Q_o$ site, 571–572
    domain shuttle mechanism, 545–546, 569,
      571
    reduction at $Q_i$ site, 561–566
    reduction at $Q_o$ site, 566–569, 570
  methanol dehydrogenase, *see* Methanol
    dehydrogenase
  prosthetic groups, 74, 75
  tryptophan tryptophylquinone, *see*
    Tryptophan tryptophylquinone

Radical chemistry, ascorbate peroxidase, 332–
  335
Radical clock substrates, 264–266
Radical electron transfer, 17
Radical intermediate states, 18
Radical rebound mechanism, methane
  monooxygenase, 237, 262, 266–267
Radical scavengers, 433–434, 458
Rat kidney amine oxidase, 200–202
Reaction centers of purple bacteria, *see* Purple
  bacteria reaction centers
Recombinant ascorbate peroxidases, 319–320
Redox piracy by oxygen, 170
Redox properties
  amicyanin, 125–126
  MADH, 124–126
  TMADH, 155–156
Reorganization energy ($\lambda$), 9, 55–56
  cofactors, 6
  defined, 3–4
  hydride transfer considerations, 20
  MADH complexes, 133
  methanol dehydrogenase, 78
  temperature dependence of reaction and, 7–8
Resonance Raman spectra
  adenosylcobalamin-dependent enzymes, 382,
    384, 385
  ascorbate peroxidase, 328, 329, 326
  cytochrome *c* oxidase, 587, 594–596, 607
  methane monooxygenase, 241, 257
  methylmalonyl-CoA mutase, 367
  PQQ, 85
Respiratory chain, *see* Cytochrome $bc_1$
  complex
Reverse TCA cycle, 491
*Rhodobacter capsulatus* molybdenum enzymes,
  448, 450, 472, 473–474
*Rhodobacter sphaeroides*, 569
  cytochrome *c* oxidase, 610
  molybdenum enzymes, 448, 449, 472, 473–
    474
  packing density estimation, 11, 12
  reaction centers, 622–624; *see also* Purple
    bacteria reaction centers
*Rhodopseudomonas viridis* reaction centers,
  622–624; *see also* Purple bacteria
    reaction centers
*Rhodospira viridis*
  c heme chain, 14–17
  packing density estimation, 11, 12

*Rhodospirillum rubrum* CO dehydrogenase, 493
*Rhodotorula graminis*, 279, 291
Ribonucleotide reductase, 351, 358–359, 360, 361, 362, 375–376, 377, 395, 405–436
  classes, 406–410
  class I, radical transfer pathway, 415–424
    protein R1, 418
    protein R2, 416, 418–422
    theoretical considerations, 422
  generation of stable tyrosyl radical in protein R2, 424–431
  interactions between metal sites, 428–429
  non-native radicals and secondary radical transfer pathways observed in mutant R2 proteins, 428–429
  radical generation via radical transfer pathway, 424–428
  unexpected hydroxylation reactions, 431
  radical transfer reactions in class II and III enzymes, 434–436
  rate constants and kinetic isotope effects, 380–381
  reaction mechanisms, 410–415
    active site mutant studies, 413–415
    radical chemistry at work, 410–412
    substrate analogues, 412–413
  tunneling, 3
  tyrosyl radical generation in protein R2, 424–431
    interactions between metal sites, 428–429
    radical transfer pathway, 424–428
    unexpected hydroxylation reactions, 431
  tyrosyl radical stability, 431–434
    during catalysis, 433–434
    conformers, 431–433
    environment, 433
Rieske protein, 544, 545, 546, 547, 548, 567, 568, 569–573
Rubredoxin, 495

*Saccharomyces cerevisiae*, 42–45, 279; *see also* Flavocytochrome $b_2$
Salicylate, 341
*Salmonella typhimurium*, 299, 361, 422, 430, 432
Saperconazole, 300
Scavengers, radical, 433–434, 458
Schiff bases, 86, 87, 209–213

Selanylcysteine, 456
*Selenastrum capricornutum*, 321
Selenium
  CO dehydrogenase, 456, 464
  hydrogenases, 491, 505
Semicarbazide-sensitive amine oxidases, 212
Semiquinones
  cytochrome $bc_1$ complex, 544
  flavins, 29
  flavocytochrome P450BM3, 39
  methanol dehydrogenase, 90
  phthalate dioxygenase reductase, 50-52
Serum amine oxidases, 212
*Shewaniella massilia*, 448, 472, 474
Sodium ion, cytochrome *c* oxidase, 585
Soluble MMO, 234; *see also* Methane monooxygenase
Solvation, packing density estimation, 10
Soybean seedling AO, 212
Spectroscopy, *see also specific techniques*
  adenosylcobalamin-dependent enzymes, 375–385
  ascorbate peroxidases, 328–332
  methane monooxygenase, 241–243, 257–258
  molybdenum enzymes
    mechanistic aspects, 475–477
    pterin cofactors, 474–475
Spinach glycolate oxidase, 289
Static titrations, TMADH, 154–155
Stigmatellin, 561, 566, 567, 572
Stopped-flow studies
  adenosylcobalamin-dependent enzymes, 377–381
  methanol dehydrogenase, 77–78
  TMADH
    ETF complex assembly, 167
    pH-jump, 164–166
    single-turnover, 156–164
*Streptomyces thermoautotrophicus*, 449, 451, 453
Strobilurin, 562
Strontium, 86, 88, 89, 93
Succinate, 52, 492
Succinate dehydrogenase, 511
Suicide reaction, ribonucleotide reductase mutants, 414, 415
Sulfhydryl reagents
  ascorbate peroxidase inhibition, 339–340
  PQQ reduction, 97
Sulfite oxidase, 13

Sulfite oxidase (*cont.*)
    mechanistic aspects, 468–472
    molybdenum enzymes, 449
    structural and catalytic variations, 451–452
    structural studies, 465–467
Supercellular biochemistry, $H_2$ transfer, 490–491
Sybyl molecular graphics program, 10

TCA cycle, reverse, 491
Tegoni model, 287
Temperature
    ascorbate peroxidase spin state, 328
    electron transfer kinetics, 1
    methane monooxygenase studies, 238–239
    tunneling theory, 3–6, 7–8
    ultrafast electron transfer in light-activated proteins, 650–651
Temperature jump experiments, flavocytochrome $b_2$, 285
1,2,3,5-Tetrahydroxybenzene, 451
Tetramethylamine, 158
Tetramethylammonium chloride, 151, 152, 153, 155
T4 ribonucleotide reductase, 409, 432–433, 434–435
Thermodynamic control, 286, 569
*Thiobacillus versutus*, 120; *see also* Methylamine dehydrogenase
Thiocyanate, molybdenum enzymes, 448
Thiolates, molybdenum enzymes, 489
Thioredoxin, 407, 410, 411
*Thiosphaera pantotropha, see Paracoccus pantotrophus*
Thiyl radical, ribonucleotide reductase, 409, 435
TMADH, *see* Trimethylamine dehydrogenase
Topaquinone, 74
TPQ, *see* Trihydroxyphenylalanine quinone
Trihydroxyphenylalanine quinone (TPQ), 183–184, 198, 199, 205–207, 217–218
Trimethylamine, 31, 151, 158
Trimethylamine dehydrogenase (TMADH), 145–177, 380
    background and structural properties, 48–50
    cofactor assembly and role of 6-*S*-cysteinyl FMN, 173–175
    control of intramolecular electron transfer domain interactions, 66
    donors, acceptors, flavin domains, 31

Trimethylamine dehydrogenase (*cont.*)
    electron flow, 154–156
        reduction potential and selective inactivation studies, 155–156
        static titrations, 154–155
    electron transfer proteins in methylotrophic bacteria, 145–148
    electron transfer rates, 68
    enzyme over-reduction and substrate inhibition, multiple turnover studies, 170–173
    oxidative half reaction and TMADH-ETF complex assembly, 167–170
        model for electron transfer complex, 168–170
        stopped-flow studies, 167
    pH-jump-stopped-flow studies, 164–166
    prosthetic groups and structure, 148–154
        identification of prosthetic groups, 148–149
        structure, 149–154
    single turnover stopped-flow studies, 156–164
        early stopped-flow investigations, 156–158
        pH-dependence of reductive half-reaction with DEMA, 158–159
        pH-dependence of reductive half-reaction with TMA, native and mutant enzyme studies, 159–163
        quantum tunneling of hydrogen, 163–164
    structural and catalytic properties, 56, 57
    structure, 149–154
        domain and quaternary structure, 150
        evolution of crystal structure, 149–150
        large domain and active site, 150–153
        medium and small domains, 153–154
    summary and future prospects, 176–177
Trimethylamine-N-oxide, 448, 474
Trimethylamine-N-oxide reductase, 472
Trimethylphosphine, 489
Triose phosphate isomerase, 31, 32
Triquat, 496
Tritium partitioning technique, 390–391
*Trypanosoma cruzi* ascorbate peroxidase, 321
Tryptophan, 17, 429
Tryptophan tryptophylquinone (TTQ), 74, 120, 135; *see also* Methylamine dehydrogenase
    MADH electron transfer from TTQ to copper, 134–138

Tryptophan tryptophylquinone (*cont.*)
MADH electron transfer from TTQ to
copper (*cont.*)
gated, 137–138
mutation of amicyanin alters $H_{ab}$ for
electron transfer from MADH, 124–
137
nonadiabatic, 134
MADH prosthetic group, 121
methylamine dehydrogenase, 148
spectral and redox properties, 124–125
Tungsten enzymes, 450
Tunnel diodes, 510–511
Tunneling, 1–2
flavocytochrome $b_2$, 289
methylmalonyl-CoA reductase, 380
purple bacteria reaction centers, 635, 638,
639
vibrationally enhanced ground state
(VEGST), 164
Tunneling theory, 2–9
classical free energy and temperature
dependence, 3–6
generic protein tunneling rate expression, 8–
9
quantum free energy and temperature
dependence, 7–8
Two-electron bond rearrangements, 18
Tyrosine, posttranslational processing, 221
Tyrosyl radical, 17, 407, 418–420, 424–429,
430
generation of, 424–431
generation via radical transfer pathway,
424–428
H-bonded to radical transfer pathway, 422
interactions between metal sites, 428–429
non-native radicals and secondary radical
transfer pathways observed in mutant
R2 proteins, 428–429
stability of, 431–434
unexpected hydroxylation reactions, 431

Ubiquinol, photosynthetic reaction centers,
633–634
Ubiquinones
cytochrome $bc_1$ complex, 544, 546
photochemistry, *see* Purple bacteria reaction
centers
substituting, 8
UHDBT (5-Undecyl-6-hydroxy-4,7-
dioxobenzothiazole), 562
Ultrafast electron transfer in light-activated
protein, 640–650; *see also* Purple
bacteria reaction centers
United atom approximation, 10

Vacuum, electron transfer in, 2–3
Vanillate specific O-demethylase corrinoid
protein, 363
Vanillyl alcohol oxidase, 456
Vascular adhesion protein, 221
Vibrationally enhanced ground state tunneling
theory (VEGST), 164
Vitamin $B_{12}$, *see* Adenosylcobalamin-dependent
enzymes
Voltammetry, hydrogenase, 510–511

Water molecules
amine oxidase active site, 221
galactose oxidase, 189
packing density estimation, 10
purple bacteria reaction centers, 625
*Wolinella succinogenes*, 453
Wood-Ljungdahl pathway, 491
Wurster's blue, 75, 81

Xanthine dehydrogenase, 457–458
Xanthine oxidase, 448
mechanistic aspects, 458–464
structure, 454–455, 457–458

Zinc, 585